T0182138

Handbook of Statistical Methods for Randomized Controlled Trials

Chapman & Hall/CRC
Handbooks of Modern Statistical Methods

Series Editor

Garrett Fitzmaurice, *Department of Biostatistics, Harvard School of Public Health, Boston, MA, U.S.A.*

The objective of the series is to provide high-quality volumes covering the state-of-the-art in the theory and applications of statistical methodology. The books in the series are thoroughly edited and present comprehensive, coherent, and unified summaries of specific methodological topics from statistics. The chapters are written by the leading researchers in the field and present a good balance of theory and application through a synthesis of the key methodological developments and examples and case studies using real data.

Published Titles

Handbook of Neuroimaging Data Analysis
Hernando Ombao, Martin Lindquist, Wesley Thompson, and John Aston

Handbook of Statistical Methods and Analyses in Sports
Jim Albert, Mark E. Glickman, Tim B. Swartz, and Ruud H. Koning

Handbook of Methods for Designing, Monitoring, and Analyzing Dose-Finding Trials
John O'Quigley, Alexia Iasonos, and Björn Bornkamp

Handbook of Quantile Regression
Roger Koenker, Victor Chernozhukov, Xuming He, and Limin Peng

Handbook of Statistical Methods for Case-Control Studies
Ørnulf Borgan, Norman Breslow, Nilanjan Chatterjee, Mitchell H. Gail, Alastair Scott, and Chris J. Wild

Handbook of Environmental and Ecological Statistics
Alan E. Gelfand, Montserrat Fuentes, Jennifer A. Hoeting, and Richard L. Smith

Handbook of Approximate Bayesian Computation
Scott A. Sisson, Yanan Fan, and Mark Beaumont

Handbook of Graphical Models
Marloes Maathuis, Mathias Drton, Steffen Lauritzen, and Martin Wainwright

Handbook of Mixture Analysis
Sylvia Frühwirth-Schnatter, Gilles Celeux, and Christian P. Robert

Handbook of Infectious Disease Data Analysis
Leonhard Held, Niel Hens, Philip O'Neill, and Jacco Walllinga

Handbook of Forensic Statistics
David L. Banks, Karen Kafadar, David H. Kaye, and Maria Tackett

Handbook of Meta-Analysis
Christopher H. Schmid, Theo Stijnen, and Ian White

Handbook of Statistical Methods for Randomized Controlled Trials
KyungMann Kim, Frank Bretz, Ying Kuen K. Cheung, and Lisa Hampson

For more information about this series, please visit: https://www.crcpress.com/Chapman--HallCRC-Handbooks-of-Modern-Statistical-Methods/book-series/CHHANMODSTA

Handbook of Statistical Methods for Randomized Controlled Trials

KyungMann Kim
Frank Bretz
Ying Kuen K. Cheung
Lisa V. Hampson

CRC Press
Taylor & Francis Group
Boca Raton London New York

CRC Press is an imprint of the
Taylor & Francis Group, an **informa** business

A CHAPMAN & HALL BOOK

First edition published 2021
by CRC Press
6000 Broken Sound Parkway NW, Suite 300, Boca Raton, FL 33487-2742

and by CRC Press
2 Park Square, Milton Park, Abingdon, Oxon, OX14 4RN

© 2021 Taylor & Francis Group, LLC

Library of Congress Cataloging-in-Publication Data

Names: Kim, KyungMann, editor.
Title: Handbook of statistical methods for randomized controlled trials /
edited by KyungMann Kim, Frank Bretz, Ying Kuen K. Cheung, and Lisa
Hampson.
Description: First edition. | Boca Raton : CRC Press, 2021. | Series:
Chapman & Hall/CRC handbooks of modern statistical methods | Includes
bibliographical references and index.
Identifiers: LCCN 2020055560 (print) | LCCN 2020055561 (ebook) | ISBN
9781498714624 (hardback) | ISBN 9781315119694 (ebook)
Subjects: LCSH: Clinical trials--Statistical methods--Handbooks, manuals,
etc.
Classification: LCC R853.C55 H365 2021 (print) | LCC R853.C55 (ebook) |
DDC 610.72/4--dc23
LC record available at https://lccn.loc.gov/2020055560
LC ebook record available at https://lccn.loc.gov/2020055561

ISBN: 9781498714624 (hbk)
ISBN: 9781032009100 (pbk)
ISBN: 9781315119694 (ebk)

DOI: 10.1201/9781315119694

Typeset in CMR10 font
by KnowledgeWorks Global Ltd.

*To our families
for their love and support*

Contents

III   Design of Randomized Controlled Trials   **137**

VI   Miscellaneous Topics in Randomized Controlled Trials   485

25 Incorporating Historical Data into Randomized Controlled Trials   545
Heinz Schmidli, Sandro Gsteiger and Beat Neuenschwander

26 Evaluation of Surrogate Endpoints   567
Geert Molenberghs, Ziv Shkedy, Tomasz Burzykowski, Marc Buyse,
Ariel Alonso Abad and Wim Van der Elst

Preface

Perhaps the single most important contribution to advances in clinical medicine is the advent of randomized controlled trials in the 1940s in the United Kingdom and in the 1960s in the USA. Since then there have been many breakthroughs in clinical medicine through the screening, diagnosis, prevention, and treatment of conditions affecting health and wellbeing. It is fair to say that randomized controlled trials have been indispensable in advancing nearly all branches of clinical medicine.

There are many books on clinical trials and clinical trials methodology, some with emphasis on statistical methods, and others with emphasis on diseases or conditions. Most books that emphasize statistical methods present a collection of topics related to clinical trials methodology, but oftentimes without a logical development of statistical concepts and the associated statistical methods involved in planning, monitoring, and analyzing clinical trials.

This handbook will provide an in-depth treatment of up-to-date and evolving statistical methods relevant to the planning, monitoring, and analysis of clinical trials, focusing exclusively on randomized controlled trials and is laid out to follow the logical development of statistical concepts from beginning to end.

The statistical concepts addressed in this handbook should be familiar to statisticians who are trained at the master's level, but are not familiar with randomized controlled trials. This handbook is intended to serve as a reference text for statistical methods for randomized controlled trials for those who are involved in design, monitoring, and analysis, and are interested in learning more about the statistical concepts involved. It can also be used a textbook for a graduate course in statistical methods for randomized controlled trials.

In the face of the coronavirus disease 2019 (COVID-19) pandemic and the many uncertainties surrounding its pathogenic virus, severe acute respiratory syndrome coronavirus 2 (SARS-Cov-2), the importance of randomized controlled trials and their critical and even central role in answering urgent clinical questions are becoming widely recognized and emphasized now more than ever as indicated in the website RECOVERY (Randomized Evaluation of COVID-19 Therapy) (https://www.recoverytrial.net). In this context it is worth noting how statistical methods for design and analysis of randomized controlled trials have evolved significantly since their inception almost 80 years ago and have broadened the application of randomized controlled trials with their flexibility check this word for intended meaning, robustness and efficiency. As a result randomized controlled trials are well placed to be able to find answers to future public health crises.

We would like to express our gratitude to the contributors to the chapters in this handbook. We are also grateful to John Kimmel for his patience and unwavering support for this project.

List of Figures

List of Tables

Contributors

Ariel Alonso Abad
KU Leuven
Leuven, Belgium

Keaven Anderson
Merck & Co.
Kenilworth, New Jersey

Oliver Bautista
Merck & Co.
Kenilworth, New Jersey

Jesse A. Berlin
Johnson & Johnson
New Brunswick, New Jersey

John Martin Bland
University of York (retired)
York, United Kingdom

Tomasz Burzykowski
Hasselt University
Hasselt, Belgium

Emily Butler
GlaxoSmithKline
Collegeville, Pennsylvania

Marc Buyse
International Drug Development Institute
Louvain-la-Neuve, Belgium

Michael J. Campbell
University of Sheffield (retired)
Sheffield, United Kingdom

Bibhas Chakraborty
Duke-National University of Singapore
    Medical School
Singapore

Richard J. Cook
University of Waterloo
Waterloo, Ontario, Canada

Cindy Cooper
University of Sheffield
Sheffield, United Kingdom

Brenda J. Crowe
Eli Lilly and Company
Indianapolis, Indiana

Wim Van der Elst
Jassen Pharmaceuticals
Beerse, Belgium

Loïc Ferrer
Sophia Genetics
Bordeaux, France

Garrett M. Fitzmaurice
Harvard T. H. Chan School of Public
    Health
Boston, Massachusetts

Tim Friede
University Medical Center Göttingen
Göttingen, Germany

Palash Ghosh
Indian Institute of Technology
Guwahati, Assam, India

Sandro Gsteiger
Roche
Basel, Switzerland

Xin He
University of Maryland School of Public
    Health
College Park, Maryland

Jason Hsu
The Ohio State University and JCH
Statistical Decision Sciences, LLC
    Columbus, Ohio

Hélène Jacqmin-Gadda
University of Bordeaux
Bordeaux, France

Steven A. Julious
University of Sheffield
Sheffield, United Kingdom

Sin-Ho Jung
Duke University
Durham, North Carolina

Mona Kanaan
University of York
York, United Kingdom

Haesook Kim
Dana-Farber Cancer Institute
Boston, Massachusetts

KyungMann Kim
University of Wisconsin-Madison
Wisconsin

Soeun Kim
Azusa Pacific University
Azusa, California

Franz Koenig
Medical University of Vienna
Vienna, Austria

Michael Kosorok
University of North Carolina at Chapel Hill
Chapel Hill, North Carolina

Stuart R. Lipsitz
Harvard Medical School and Brigham and
    Women's Hospital
Boston, Massachusetts

Yi Liu
Nektar Therapeutics
San Francisco, California

Geert Molenberghs
Hasselt University
Hasselt, Belgium

Tobias Mütze
Novartis Pharma AG
Basel, Switzerland

Beat Neuenschwander
Novartis Pharma AG
Basel, Switzerland

Fang-Shu Ou
Mayo Clinic
Rochester, Minnesota

Myunghee Cho Paik
Seoul National University
Seoul, Korea

Martin Posch
Medical University of Vienna
Vienna, Austria

Michael Proschan
National Heart, Lung, and Blood Institute
Bethesda, Maryland

Cécile Proust-Lima
University of Bordeaux
Bordeaux, France

Joanne C. Rothwell
Parexel
Sheffield United Kingdom

Yevgen Ryeznik
AstraZeneca
Göteborg Sweden

Daniel Scharfstein
University of Utah
Salt Lake City, Utah

Heinz Schmidli
Novartis Pharma AG
Basel, Switzerland

Stephen Senn
Luxembourg Institute of Health (retired)
Edinburgh, United Kingdom

Ziv Shkedy
Hasselt University
Hasselt, Belgium

(Tony) Jianguo Sun
University of Missouri
Columbia, Missouri

Oleksandr Sverdlov
Novartis Pharmaceuticals Corporation
East Hanover, New Jersey

Szu-Yu Tang
Roche Tissue Diagnostics
Oro Valley, Arizona

Anastasios A. Tsiatis
North Carolina State University (retired)
Raleigh, North Carolina

Rui Wang
Harvard Pilgrim Health Care
Institute and Harvard Medical School
    Boston, Massachusetts

Gernot Wassmer
University of Cologne
Cologne, Germany

Brian Wiens
Acelyrin, Inc Westlake Village
California

H. Amy Xia
Amgen
Thousand Oaks, California

Yujie Zhong
Shanghai University of Finance and
    Economics
Shanghai, China

Yuxin Zhu
Johns Hopkins Bloomberg School of Public
    Health
Baltimore, Maryland

Part I

Introduction to Randomized Controlled Trials

1

Introduction

KyungMann Kim

CONTENTS

1.1   Historical Background

Randomized controlled trials, the topic of this handbook, owe their existence and development to the pioneering work of Ronald A. Fisher in the design of planned experiments in agricultural setting [5]. The first modern randomized controlled trial was the British Medical Research Council's Streptomycin Treatment of Pulmonary Tuberculosis trial [3]. As suggested in the report, the trial became a model for many therapeutic trials. Richard Doll described the trial as the watershed event for randomized controlled trials in his 50th anniversary review article about the trial [4]. In the same issue, Alan Yoshioka describes that "[r]andomised allocation of patients is rarely mentioned in the Medical Research Council's documents on streptomycin clinical trials" and "[r]andomization relieved the MRC's clinicians of responsibility for deciding who would be treated" with streptomycin which was in very limited supply at the time [13]. Austin Bradford Hill expounded that randomized controlled trials are ethical [10], and later Peter Armitage noted that "[r]andomization is entirely compatible with medical ethics in circumstances when the treatment of choice is not clearly identified" [1].

In the United States, modern randomized controlled trials got their start in the early 1960s, supported by the National Institutes of Health. This was also partly driven by the 1962 Kefauver-Harris amendments to the United States Food, Drug and Cosmetic Act of 1938 which required "adequate and well-controlled investigations" as the nature of scientific evidence required for a drug to be approved for human use. Landmark randomized controlled trials include the University Group Diabetes Program [9] initiated in 1961, the Coronary Drug Project [7] started in 1965 by the then National Heart Institute, and the Diabetic Retinopathy Study [8] started in 1973 by the National Eye Institute, to name a few. These trials and their conduct and monitoring had a profound and lasting impact in all randomized controlled trials supported by the National Institutes of Health and by biopharmaceutical industry thereafter.

Even after three quarters of a century since the publication of the Streptomycin Treatment of Pulmonary Tuberculosis trial and numerous other randomized controlled trials that have changed the standard of care for many illnesses, there continues to be detractors,

3

especially under the mantra of "big data" from electronic health records and large observational registries in recent times. Collins et al. point out biases inherent in non-randomized observational analyses of large electronic patient databases despite statistical adjustments for all the known differences between treatment groups primarily due to residual confounding from incompletely measured or unmeasured covariates [2]. They go on to underline the "magic" of randomization in that it guarantees the balance in treatment groups subject to random error with respect both to measured and unmeasured confounders.

1.2   Statistical Concepts

Following the first National Conference on Clinical Trials Methodology held at the US National Institutes of Health campus in October 1977, Robert S. Gordon Jr. published an article "Clinical Trials" in which he stated that "[w]ith the advent of sophisticated biostatistical methods, clinical trials have acquired scientific respectability during the past two decades" [6]. In the introduction to the textbook, *Clinical Trials*, by Daniel Schwartz, Robert Flamant and Joseph Lellouch, originally published in French in 1970 and published by the Academic Press in 1980 [12], the authors stated the following: "The development of a truly scientific methodology for clinical trials, capable of providing them with a degree of rigour hitherto characteristics of laboratory experiments, is a development mainly of the years since the Second World War. The essential core of this methodology has been its statistical component." Also the translator, M. J. R. Healy in the Translator's Preface stated that "[a] clinical trial is a major undertaking and any doctor contemplating such a trial is liable to stand in need of considerable guidance on the statistics side." Judging from these, it is evidently clear that statistical concepts are fundamental to clinical trials. On the side, Gina Kolata, reporting on the National Conference on Clinical Trials Methodology for the journal *Science* in December 1977, stated that "[t]his is the age of clinical trial" [11]. More than forty years later and especially during the coronavirus disease 2019 (COVID-19) pandemic, her statement rings truer than ever.

In the planning, monitoring and analysis of randomized controlled trials, there are a number of statistical concepts involved. Following the clinical questions to be addressed in terms of the clinical hypotheses and the objectives of the randomized controlled trials, these statistical concepts need to be developed in a logical and sequential manner. The clinical questions, the hypotheses, and the objectives lead naturally to appropriate trial design. This includes the outcome measures or endpoints, statistical distributions suited for them, statistical hypotheses (both null and alternative) in terms of the unique parameters that characterize the statistical distribution and capture the clinical effects of treatment or intervention under investigation, and the statistical methods for analysis and the statistical test of the null hypothesis. These are the concepts that will be covered in Part II.

The clinical questions and the hypotheses will lead to experimental designs. Another important requirement in planning is to estimate the required sample size based on the statistical test of the null hypothesis at a desired significance level α to detect the "treatment effect" with the desired power $1 - \beta$ of the statistical test. Experimental design, sample size estimation, and power analysis for various types of outcome measures or endpoints commonly employed in randomized controlled trials are the statistical concepts to be covered in Part III.

Invariably, randomized controlled trials are monitored for the safety of participants and efficacy of treatments or interventions under investigation. This is accomplished by interim analysis based on several different approaches including group sequential methods. Interim analyses also offer the opportunity to evaluate the assumptions made during the planning for possible sample re-estimation and to introduce adaptive designs if desired. These concepts will be covered in Part IV.

Beyond the standard methods for statistical analysis as described in Part II, there are other practical and important issues in statistical analysis. They include multiple testing, subgroup analyses, competing risks, and joint models for longitudinal markers and clinical outcomes and will be covered in Part V.

Other miscellaneous topics in design and analysis will be covered in Part VI. These include sequential multiple assignment randomization trials, analysis of safety outcomes, non-inferiority trials, incorporating historical data into randomized controlled trials, and validation of surrogate outcomes.

1.3 Organization of the Handbook

The rest of the handbook is divided into Parts II–VI.

Part II (Chapters 2–7) describes typical outcomes used to address clinical questions and hypotheses in randomized controlled trials and their statistical distributions suited for the outcome measures of interest. This leads naturally to the development of statistical hypotheses, statistical analysis methods used to test the statistical hypotheses, and statistical models that may be employed in the statistical data analysis. Chapter 2 will describe statistical methods suitable for dichotomous (qualitative) and ordinal data; Chapter 3 will describe statistical methods suitable for continuous (quantitative) outcomes; Chapter 4 will describe statistical methods suitable for time to event outcomes subject to right censoring, widely used in randomized controlled trials in chronic diseases; Chapter 5 will describe statistical methods suitable for count data; Chapter 6 will describe statistical methods suitable for longitudinal data; and Chapter 7 will describe statistical methods suitable for repeated events data. These are the outcomes typically used in randomized controlled trials in various diseases and health conditions. Part II is preparatory for describing the sample size estimation and power analysis to be discussed in Part III.

Part III (Chapters 8–14) covers planning of randomized controlled trials, such as trial design, sample size estimation, and power analysis based on the primary endpoints chosen, the types of which were described in Part II. Chapter 8 will describe cross-over design; Chapter 9 will describe factorial design; Chapter 10 will describe cluster randomized design, used often in health services research; Chapter 11 will describe methods of treatment allocation such as randomization, stratification, and outcome-adaptive allocation; Chapter 12 will describe sample size estimation and power analysis for dichotomous, ordinal, continuous, and count data; Chapter 13 will describe sample size estimation for time to event data subject to right censoring; and Chapter 14 will describe sample size estimation and power analysis for longitudinal data.

Part IV (Chapters 15–17) describes monitoring of randomized controlled trials including issues in data and safety monitoring, interim analysis, methods for early stopping, sample size re-estimation, and adaptive designs. Chapter 15 will describe methods for interim analysis such as group sequential methods, triangular methods, and stochastic curtailment tests; Chapter 16 will address sample re-estimation in interim analyses; and Chapter 17 will describe adaptive designs.

Part V (Chapter 18–21) covers practical and important issues in data analysis beyond what is covered as standard statistical methods in Part II. Chapter 18 will address multiple testing due to multiplicity of outcomes; Chapter 19 will address how to properly conduct subgroup analyses; Chapter 20 will address a challenging problem of competing risks in the analysis of time to event data; and Chapter 21 will address joint models for longitudinal markers and clinical outcomes typically as time to event data subject to right censoring.

Part VI (Chapters 22–26) covers miscellaneous topics in design and analysis of randomized controlled trials. More specifically, Chapter 22 will describe sequential multiple assignment randomization trials for dynamic treatment allocation in certain types of randomized controlled trials in which treatment consists of multiple interventions given in sequence, e.g. in psychiatric conditions; Chapter 23 will address how to conduct statistical analysis of safety data such as adverse events; Chapter 24 will address design and analysis of non-inferiority trials in which the objective of the study is to establish or demonstrate that the new treatment is not inferior to the standard treatment; Chapter 25 will describe how to incorporate historical data into the design and analysis of randomized controlled trials; and finally Chapter 26 will address validation of outcomes as surrogate for clinical outcomes.

Bibliography

[1] Peter Armitage. The role of randomization in clinical trials. *Statistics in Medicine*, 1(4):245–352, 1982.

[2] Rory Collins, Louise Bowman, Martin Landray, and Richard Peto. The magic of randomization versus the myth of real-world evidence. *New England Journal of Medicine*, 382(7):674–678, 2020.

[3] Medical Research Council. Streptomycin treatment of pulmonary tuberculosis. *British Medical Journal*, 2(4582):769–782, 1948.

[4] Richard Doll. Controlled trials: the 1948 watershed. *British Medical Journal*, 317(7167):1217–1220, 1998.

[5] Ronald Aylmer Fisher. The arrangement of field experiments. *Journal of the Ministry of Agriculture*, 33:503–513, 1926.

[6] Robert S. Gordon, Jr. Clinical trials. *New England Journal of Medicine*, 298(7):400–401, 1978.

[7] The Coronary Drug Project Research Group. The coronary drug project: design, methods, and baseline results. *Circulation*, 47(suppl 1):I–1–I–50, 1973.

[8] The Diabetic Retinopathy Study Research Group. Preliminary report on effects of photocoagulation therapy. *American Journal of Ophthalmology*, 81(4):383–396, 1976.

[9] The University Group Diabetes Program Research Group. A study of the effects of hypoglycemic agents on vascular complications in patients with adult-onset diabetes: I. design, methods and baseline characteristics. *Diabetes*, 19(suppl 2):747–783, 1970.

[10] Austin Bradford Hill. Medical ethics and controlled trials. *British Medical Journal*, 1(5337):1043–1049, 1963.

[11] Gina Bari Kolata. Clinical trials: methods and ethics are debated. *Science*, 198(4322):1127–1131, 1977.

[12] Daniel Schwartz, Robert Flamant, and Joseph Lellouch. *Clinical Trials.* Academic Press, London, UK, 1980.

[13] Alan Yoshioka. Use of randomization in the medical research council's clinical trial of streptomycin in pulmonary tuberculosis in the 1940s. *British Medical Journal*, 317(7167):1220–1223, 1998.

Part II

Analytic Methods for Randomized Controlled Trials

2

Binary and Ordinal Outcomes

Garrett M. Fitzmaurice and Stuart R. Lipsitz

CONTENTS

2.1 Introduction

The objective of most randomized clinical trials is to compare treatment groups on the basis of some well-defined outcome. In many clinical trials, the outcome variable of interest is categorical rather than continuous. For example, the outcome may be binary, perhaps denoting the shrinkage of a tumor, and the objective is to estimate and compare the risk or probability of tumor shrinkage. Alternatively, it may be a categorical variable whose categories can be naturally ordered, e.g., "cancer stage" is a four-level ordinal variable that describes how far cancer has spread anatomically and can be used to categorize subjects with similar prognosis. In this chapter we present an overview of many of the widely used statistical methods for analyzing categorical "outcome" data from clinical trials; we are primarily concerned with the analysis of confirmatory, typically phase III, randomized clinical trials, although the methods we discuss also have applications in earlier phase trials. For the most part, we focus on the typical randomized clinical trial with two "treatment arms", one consisting of subjects randomized to "treatment" and the other of subjects randomized to "control"; we note that the control group may be composed of subjects receiving no treatment, a different treatment, or the same treatment given at a different dose. We begin with a description of relatively simple methods for making treatment comparisons when the outcome is binary or

TABLE 2.1
Binary outcome data from a randomized clinical trial of subjects with a respiratory disorder.

| | Response | | |
Treatment	Poor	Good	Total
Placebo	29	28	57
Active	17	37	54
Total	46	65	111

ordinal. In typical randomized trials where baseline variables (e.g., demographical variables such as age or gender, disease characteristics such as duration or severity, or other prognostic variables) are balanced across treatment arms due to randomization, these simple methods may form the basis of the main analysis of the trials. However, to facilitate proper adjustment for baseline factors or variables (e.g., blocking factors, baseline characteristics of subjects, baseline values of the outcome), we also consider regression models for binary and ordinal outcomes. Specifically, we discuss the application of *generalized linear models* for analyzing categorical data.

To fix ideas, consider the data in Table 2.1 which are from a multi-center randomized clinical trial of subjects with a respiratory disorder (Davis, 1991; Stokes et al., 2000). In this trial, 111 subjects from two different clinics were randomly assigned to active treatment or placebo and followed over time. The trial recruited 54 subjects in the active treatment group and 57 in the placebo group. The outcome of interest is a binary variable indicating respiratory status (categorized as 1 = good, 0 = poor) after one month of treatment. In addition, data were collected on baseline age and baseline respiratory status of the subjects. The goal of the analysis of this trial data is to determine whether the active treatment increases the likelihood of a more favorable response.

Table 2.1 is commonly referred to as a 2×2 contingency table. Much of the statistical theory underlying the analysis of categorical data is more easily formulated for 2×2 contingency tables. Indeed, methods for the analysis of 2×2 contingency tables provide the cornerstone for many of the advanced statistical methods required for more complicated problems. These include extensions for analyzing outcomes with more than 2 levels (e.g., self-assessment of a subjects's current arthritis, measured on a five-level ordinal scale) which may be ordered. In addition, there can be more than 2 levels of the treatment variable (e.g., placebo, low, medium, and high dose) and other factors or covariates (e.g., age, health status before treatment) that influence the outcome variable. For example, consider the data in Table 2.2 which are from a clinical trial of subjects with rheumatoid arthritis (Bombardier et al., 1986). In this six-month, randomized, double-blind trial, 303 subjects with classic or definite rheumatoid arthritis were randomized to either auranofin therapy (3 mg of oral gold, twice daily) or placebo and followed over time. The outcome variable of interest is a global impression scale (Arthritis Categorical Scale) at month 6. This is a self-assessment of a subject's current arthritis, measured on a five-level ordinal scale: (1) very good, (2) good, (3) fair, (4) poor, and (5) very poor. Baseline data on this outcome variable are available for 303 subjects who participated in this trial; follow-up data at 6 months are available for 294 of the subjects. The data in Table 2.2 display the number of subjects with each level of response at 6 months (from very good to very poor) in each of the two treatment groups; this is an example of a 2×5 contingency table, where the columns have more than 2 levels and are ordered. The goal of the analysis of the data from this trial is to determine whether treatment with auranofin therapy increases the likelihood of a more favorable response.

TABLE 2.2
Ordinal outcome data from a randomized clinical trial of subjects with rheumatoid arthritis.

Treatment	Very Good	Good	Fair	Poor	Very Poor	Total
			Response			
Placebo	10	48	52	29	8	147
Auranofin	28	45	51	21	2	147
Total	38	93	103	50	10	294

Some of the most widely used probability distributions for discrete outcomes include the Bernoulli, binomial, hypergeometric, and multinomial distributions. Throughout this chapter we assume the reader has some familiarity with these probability distributions. The chapter is organized as follows. We begin with a description of methods for analyzing 2×2 contingency tables. We then consider the extensions to $R \times C$ contingency tables (i.e., contingency tables with R rows and C columns), in particular where the rows represent 2 or more levels of the treatment factor and the columns represent an outcome measured on a C-level ordinal scale, with $C \geq 3$. We also discuss the analysis of sets of 2×2 tables, as might often arise when there is blocking of subjects within more homogeneous groups (e.g., blocking of subjects by clinic in a multicenter clinical trial); specifically, we describe the Cochran-Mantel-Haenszel test for sets of 2×2 tables. Finally, we present an overview of regression models for categorical data, including logistic regression models for binary outcomes and multinomial logistic regression models for ordinal outcomes. We conclude the chapter by mentioning how baseline values of the outcome can be incorporated in the analysis and the potential benefits of adjustment for baseline response.

2.2 Analysis of 2×2 Contingency Tables

In this section we discuss inference based on data from a 2×2 contingency table. In order to motivate the methods, consider the example data from the respiratory disorder clinical trial in Table 2.1. In this example, note that the number of subjects assigned to each treatment is fixed by design. The question of scientific interest is: "Does treatment affect the outcome?" Suppose we let X denote the row variable and Y denote the column variable, where both X and Y are binary (taking values 0 or 1). Then, the data in a general 2×2 contingency table can be represented as in Table 2.3.

In Table 2.3 n_{jk} is the count of the number of subjects with $X = j$ and $Y = k$; n_{jk} is referred to as a cell count. For example, n_{11} is the number of subjects with $X = 1$ (say, active treatment) and $Y = 1$ (say, positive response). Also, in Table 2.3 the marginal row counts are $n_{j+} = n_{j0} + n_{j1}$ (the number of subjects with $X = j$), and the marginal column counts are $n_{+k} = n_{0k} + n_{1k}$ (number of subjects with $Y = k$). Note that for a prospective clinical trial, n_{0+} and n_{1+} are usually considered to be fixed by design and the counts in the two rows have independent binomial distributions; n denotes the total sample size. With n_{0+} and n_{1+} fixed, we can write the probabilities for the 2×2 table as in Table 2.4. For a clinical trial, only the two conditional row probabilities $\pi(X = 0)$ and $\pi(X = 1)$ are estimable. Furthermore, it can be shown that the maximum likelihood estimates of $\pi(X = 0)$ and $\pi(X = 1)$ are simply the proportion of successes in each of the treatment groups, $\widehat{\pi}(X = j) = \frac{n_{j1}}{n_{j0} + n_{j1}} = \frac{n_{j1}}{n_{j+}}$.

TABLE 2.3
General representation of counts in a 2×2 contingency table.

		Y		
		0	**1**	
X	0	n_{00}	n_{01}	n_{0+}
	1	n_{10}	n_{11}	n_{1+}
		n_{+0}	n_{+1}	n

TABLE 2.4
Probabilities in a 2×2 contingency table, with n_{0+} and n_{1+} fixed by design.

		Y		
		0	**1**	
X	0	$1 - \pi(X = 0)$	$\pi(X = 0)$	1
	1	$1 - \pi(X = 1)$	$\pi(X = 1)$	1

To determine whether X and Y are associated ("Does treatment affect the outcome?"), it becomes necessary to formulate measures of association that quantify any departure from independence of X and Y. Because it is generally more natural to compare values of π in *relative* rather than *absolute* terms, this suggests a measure of association know as the *relative risk* (RR). The relative risk is defined as the ratio of the probability of success when $X = 1$, $\pi(X = 1)$, to the probability of success when $X = 0$, $\pi(X = 0)$,

$$\text{RR} = \frac{\pi(X = 1)}{\pi(X = 0)},$$

with estimate of the relative risk given by

$$\hat{\text{RR}} = \frac{n_{01}/n_{0+}}{n_{11}/n_{1+}}.$$

Note that RR $= 1$ if there is no association between X and Y and hence no effect of treatment on the binary outcome. Relative risks are often considered the preferred measures of association, especially when the binary outcome of interest is common. In particular, many empirical researchers prefer relative risks because they find them to be simple and intuitively interpretable. However, if adjustment for baseline variables is required due to imbalance across treatment groups, extending this measure of association to the regression setting with additional covariates (beyond treatment group) is problematic. Specifically, in models for the log of the probability of success (see Section 2.5), the regression parameters can be interpreted as log relative risks; these log-probability models are often referred to as "relative risk regression models". Although relative risk regression models constrain the predicted probabilities to be greater than 0, they do not constrain them to be less than

1. Thus, unless constraints are placed on the regression coefficients or on the covariates, relative risk regression models can yield predicted probabilities greater than 1.

To circumvent the aforementioned difficulty, the most commonly used measure of association is expressed in terms of a relative comparison of the *odds*. Recall that the odds of success is defined as the ratio of the probability of success to the probability of failure, $\frac{\pi}{1-\pi}$. For example, when the probability of success is 0.8 then the odds (of success) is $0.8/0.2 = 4$, indicating that a success is four times as likely as a failure. The relative comparison of the odds produces a measure of association known as the *odds ratio* (OR) or *cross-product ratio*, defined as

$$\text{OR} = \frac{\pi(X=1)}{1-\pi(X=1)} \Big/ \frac{\pi(X=0)}{1-\pi(X=0)} = \frac{\pi(X=1)[1-\pi(X=0)]}{[1-\pi(X=1)]\pi(X=0)};$$

the latter name refers to the fact that OR equals the ratio of the product of probabilities from diagonally opposite cells of the 2×2 table. Note that $\text{OR} = 1$ if there is no association between X and Y. Quite often, the log of the odds ratio is used as a measure of association, since $\log(\text{OR}) = 0$ under the assumption of no association between X and Y and, moreover, the sampling distribution of the estimator of $\log(\text{OR})$ converges far more quickly to a normal distribution.

One appealing feature of the (log) odds ratio is that it extends in a very natural way to the regression setting with additional covariates. Specifically, logistic regression models have regression parameters that can be directly interpreted in terms of (log) odds ratios; moreover, the logistic regression model does not share the structural issues associated with the relative risk regression model. That is, logistic regression constrains the probabilities to lie within the range from 0 to 1. In some clinical trial settings, the odds ratio can be interpreted as an approximation to the relative risk. Specifically, when the probability of a positive response ($Y = 1$) is low, and $\pi(X = j)$ is reasonably close to 0 in both of the treatment groups (often known as the "rare disease" assumption when Y denotes the presence or absence of a disease or disorder), the odds ratio provides a close approximation to the relative risk. However, extra care is necessary when interpreting the odds ratio as an approximation to the relative risk in clinical trials. In many trials the binary event is relatively common (say greater than 10%) and the "rare disease" assumption no longer holds; in these settings, the odds ratio can be a very poor and unreliable approximation to the relative risk and should not be given such an interpretation.

Before discussing inference for the odds ratio, we note that one other measure of association that is sometimes reported in randomized clinical trials is the *absolute* or *risk difference*, defined as the difference in the probabilities of success between the two treatment groups

$$\text{RD} = \pi(X=1) - \pi(X=0).$$

Note that when there is no treatment effect, then the probability of success is the same on both arms, i.e., $\text{RD} = 0$; often, the null hypothesis of no treatment effect is expressed in this form. However, similar to the relative risk, extending this measure of association to the regression setting with additional covariates is problematic. The risk difference can be obtained from a linear model for the probability of success, but, unless constraints are placed on the regression coefficients or on the covariates, linear regression models for probabilities can yield predicted probabilities outside of the range from 0 to 1. The risk difference is often given an interpretation in terms of the additional number of subjects expected to have successful outcomes on, say, the intervention arm versus the control arm. For example, if $\pi(X = 1) = 0.6$ and $\pi(X = 0) = 0.4$, then we would expect 20 more successes in 100 subjects assigned to the intervention arm than in 100 subjects assigned to the control arm. The inverse of the risk difference is also sometimes reported as the *number needed to treat*

(NNT). Specifically, NNT is defined as the number of subjects needed to treat in the two groups for there to be one more success in the intervention group than in the control group. In this simple example, the NNT=$1/(0.6-0.4) = 5$. With 5 subjects in each group, we would expect 60% or 3 subjects in the intervention group and 40% or 2 subjects in the control group to have successes, i.e., with 5 subjects in each group, we expect 1 more success in the intervention group. The larger the NNT, the less effective the treatment. Formally, if we let n^* be the number of subjects needed to treat (assumed to be the same in both groups), then n^* satisfies $n^*(\pi(X = 1) - \pi(X = 0)) = 1$, or equivalently, $n^* = 1/(\pi(X = 1) - \pi(X = 0))$.

Returning to the odds ratio, it is usually of interest to obtain a point estimate and confidence interval (CI) for the odds ratio, or to test the null hypothesis that the odds ratio equals 1. The maximum likelihood estimate of the odds ratio is given by the cross-product ratio,

$$\hat{\text{OR}} = \frac{n_{00}n_{11}}{n_{10}n_{01}}. \tag{2.1}$$

Because the sampling distribution of $\hat{\text{OR}}$ is skewed unless the total sample size, n, is very large, it is preferable to obtain a confidence interval for $\log(\text{OR})$ because the distribution of the sample $\log(\text{OR})$ converges far more quickly to normality as n increases. Given a confidence interval for $\log(\text{OR})$, a 95% confidence interval for OR is obtained by exponentiating the endpoints, i.e.,

$$\exp\left\{\log(\hat{\text{OR}}) \pm 1.96\sqrt{\hat{\text{Var}}[\log(\hat{\text{OR}})]}\right\},$$

where, from the so-called delta method, $\hat{\text{Var}}[\log(\hat{\text{OR}})] = \frac{1}{n_{00}} + \frac{1}{n_{10}} + \frac{1}{n_{01}} + \frac{1}{n_{11}}$.

Suppose it is of interest to construct a test for no association (independence) between X and Y. The Wald statistic used to test the null hypothesis, $H_0: \log(\text{OR}) = 0$, is given by

$$Z = \frac{\log(\hat{\text{OR}})}{\sqrt{\hat{\text{Var}}[\log(\hat{\text{OR}})]}}, \tag{2.2}$$

which, in large samples, has an approximate standard normal distribution, denoted by $N(0,1)$, under the null hypothesis of no association; equivalently, Z^2 has an approximate chi-squared distribution with 1 degree of freedom (df). Alternatively, the likelihood ratio statistic can be used to test the null hypothesis and is simply twice the difference in the log-likelihood under the alternative (association) and null (independence) hypotheses. The likelihood ratio statistic reduces to

$$G^2 = 2\sum_{j=0}^{1}\sum_{k=0}^{1} O_{jk} \log\left(\frac{O_{jk}}{E_{jk}}\right), \tag{2.3}$$

where $O_{jk} = n_{jk}$ is the "observed" count in the 2×2 table and $E_{jk} = \hat{E}(n_{jk}|H_0) = n_{j+}n_{+k}/n$ is the "estimated expected" count (under the assumption of independence). Finally, a third alternative is the score test statistic which reduces to

$$X^2 = \sum_{j=0}^{1}\sum_{k=0}^{1} \frac{(O_{jk} - E_{jk})^2}{E_{jk}}, \tag{2.4}$$

the latter is also known as the Pearson chi-squared test for a 2×2 table. In large samples, both the likelihood ratio and the Pearson chi-squared statistics have an approximate chi-squared distribution with 1 df; in general, these two statistics have better properties in finite samples than the Wald test in the sense of having actual error rates closer to the nominal level (see, for example, Agresti, 2013, p. 12).

If the sample size, n, is relatively small, these asymptotic approximations can no longer be relied upon and may result in incorrect Type I and II error rates. In particular, a rule-of-thumb in statistical folklore is that the asymptotic approximations cannot be relied upon if one (or 25%) of the cells in the 2×2 table have estimated expected counts (E_{jk}) less than 5. When at least one E_{jk} is less than 5, and it is of interest to make inferences about the OR, a common technique is to fix both margins of the 2×2 table and use "exact" tests and confidence intervals. That is, for a clinical trial where the row margins are fixed by design, we further condition on the column margins. In doing so, it can be shown that the data in the table have a non-central hypergeometric distribution. Under the null hypothesis H_0: $\log(OR) = 0$ (or, equivalently, H_0: OR=1), the non-central hypergeometric becomes a central hypergeometric distribution (see, for example, Fisher, 1934), which forms the basis of Fisher's exact test of no association in a 2×2 contingency table. This test is appropriate in small samples; the non-central hypergeometric can also be used to obtain an estimate of the odds ratio, and a confidence interval, that has better small sample properties than the usual cross-product estimate of the odds ratio given by (2.1). However, a potential drawback with Fisher's exact test is that it can be *conservative* in the sense that the true significance level is often far smaller than the nominal level (e.g., 0.05), thereby making it more difficult to reject the null hypothesis when in fact it is not true.

To illustrate some of the key ideas discussed so far, we return to the example data in Table 2.1. From Table 2.1 the percentage of subjects with a good response is 49.1% and 68.5% in the placebo and active treatment groups, respectively. The relative comparison of the proportion of good responses between the two treatment groups, comparing active to placebo, yields a relative risk, RR = 1.39 (95% CI: 1.01, 1.92); this interval is obtained by exponentiating the endpoints of the corresponding 95% CI for $\log(RR)$, with $\hat{Var}[\log(\hat{RR})] = \frac{1}{n_{01}} + \frac{1}{n_{11}} - \frac{1}{n_{0+}} - \frac{1}{n_{1+}}$. Subjects in the active treatment group have a 39% higher likelihood of a good response when compared to subjects in the placebo group. Because the 95% confidence interval does not include the null value for RR, this difference in response is significant at the 0.05 level. Thus, there is evidence of a statistically significant treatment effect. When the comparison is made in terms of odds, this yields an OR = 2.25 (95% CI: 1.04, 4.89) and this difference in response is also significant at the 0.05 level. Note that the estimated odds ratio is discernibly larger than the estimated relative risk; in this instance, the good response $(Y = 1)$ is relatively common (with rates of good response > 45% in both treatment groups) and the odds ratio is a poor and unreliable approximation to the relative risk and should not be given such an interpretation. For this example, the Wald, Pearson, and likelihood ratio chi-squared statistics are very similar (approximately 4.3), yielding p-values of 0.040, 0.038, and 0.037, respectively; however, this degree of similarity cannot be expected in general. Although the sample size for this example is adequate for relying on asymptotic approximations, for purely illustrative purposes we calculate Fisher's exact test. Fisher's exact test yields an exact 2-sided p-value of 0.054; the exact OR = 2.24 (exact 95% CI: 0.97, 5.28). As might be expected, exact methods yielded a p-value for the test of treatment effect that was slightly larger, and a confidence interval that was slightly wider, than based on asymptotic approximations.

2.3 Analysis of $R \times C$ Contingency Tables

The outcome data from a clinical trial can often be summarized in the form of a $R \times C$ contingency table, where the number of rows R and/or the number of columns C is greater than 2. For example, in Table 2.2 there are 2 treatment groups (placebo and active treatment) and the outcome is expressed in terms of an ordinal scale with five categorized levels of self-assessment of a subject's current arthritis: (1) very good, (2) good, (3) fair, (4) poor, and (5) very poor. Suppose we again let X denote the row variable (treatment group) and Y denote the column variable (outcome). The notation for an $R \times C$ table is a straightforward extension of the 2×2 table. In particular, let n_{jk} be the number of subjects with $X = j$ and $Y = k$.

As before, we assume the rows of the $R \times C$ contingency table are fixed by the design of the clinical trial, so that each row now follows a multinomial distribution, (a generalization of the binomial distribution). The marginal row counts are $n_{j+} = n_{j1} + n_{j2} + \cdots + n_{jC}$ (the number of subjects with $X = j$). With the marginal row counts fixed, only the conditional row probabilities $\pi_k(X = j) = \Pr[(Y = k)|(X = j)]$, for $j = 1, ..., R$ and $k = 1, ..., C$, are estimable from the data and $\sum_{k=1}^{C} \pi_k(X = j) = 1$ for each j. Furthermore, it can be shown that the maximum likelihood estimate of $\pi_k(X = j)$ is simply the proportion in each of the treatment groups with $Y = k$, $\hat{\pi}_k(X = j) = \frac{n_{jk}}{n_{j+}}$.

In general, for any $R \times C$ table, the question of scientific interest is whether treatment has an effect on outcome and it can also be framed as: "Are X and Y associated or are they independent?" To study departures from independence, it is possible to define sets of $(R - 1) \times (C - 1)$ odds ratios for the $R \times C$ table. For example, one such set of odds ratios are those that are relative to the last row and the last column of the table, i.e.,

$$\mathrm{OR}(j, R : k, C) = \frac{\pi_k(X = j)\pi_C(X = R)}{\pi_C(X = j)\pi_k(X = R)} , \tag{2.5}$$

for $j = 1, ..., R - 1$, and $k = 1, ..., C - 1$. The cross-product ratio of probabilities in (2.5) is the odds ratio, conditional on rows j and R and columns k and C of the table. It is often referred to as a *local* odds ratio in the sense that it describes the magnitude of association in a particular localized region of the $R \times C$ table. If the rows and columns are independent, then all of these odds ratios equal 1 (or, equivalently, the log odds ratios equal 0). Thus, the usual global null hypothesis is H_0: $\mathrm{OR}(j, R : k, C) = 1$ for $j = 1, ..., R - 1$, and $k = 1, ..., C - 1$. The maximum likelihood (ML) estimate of the local odds ratio is

$$\hat{\mathrm{OR}}(j, R : k, C) = \frac{n_{jk}n_{RC}}{n_{jC}n_{Rk}},$$

and confidence intervals or hypothesis tests can be formed based on the large sample normality of the estimator of the log odds ratio. Inference for the local odds ratio, $\mathrm{OR}(j, R : k, C)$, is exactly the same as for an odds ratio from a 2×2 table as described earlier in Section 2.2. Alternatively, for testing the global null hypothesis of independence, the likelihood ratio or the Pearson chi-squared tests can be used and their form is identical to (2.3) and (2.4), except that the sum is now over $j = 1, ..., R$, and $k = 1, ..., C$. In large samples, both the likelihood ratio and the Pearson chi-squared statistics have an approximate chi-squared distribution with $(R - 1) \times (C - 1)$ degrees of freedom.

If the sample size is relatively small (and at least 25% of the cells in the $R \times C$ table have estimated expected counts E_{jk} less than 5), then the asymptotic chi-squared approximations for these test statistics cannot be relied upon. In that case, by conditioning on both the rows and columns, the data in the table have a non-central hypergeometric distribution. Under the global null hypothesis of no association between treatment group and the outcome variable, the non-central hypergeometric becomes a central hypergeometric distribution, which forms the basis for Fisher's exact test for no association in an $R \times C$ table. Note that the likelihood ratio, Pearson chi-squared, and Fisher's exact test consider the most general alternative to the null hypothesis of independence, i.e., *any possible* dependence among the cell probabilities in the $R \times C$ table. However, these tests may have relatively low power for detecting monotone trend alternatives when Y is ordinal rather than nominal.

For the case where the outcome Y is ordinal and X denotes treatment group, it is reasonable to expect that the largest odds ratios will arise from a comparison of pairs of treatment groups in terms of the odds of the highest versus lowest categories of the outcome Y, $\mathrm{OR}(j, R : 1, C)$. That is, local odds ratios formed by the most extreme categories of Y ($Y = 1$ and $Y = C$) are expected to be stronger, in the sense of being further from the null, than odds ratios formed from adjacent categories of Y. Moreover, when Y is ordinal, it is more natural to describe the association with X in terms of an odds ratio that uses all of the categories of Y instead of only a localized region of the $R \times C$ table. This can be achieved by constructing so-called *cumulative odds ratios* defined as,

$$
\begin{aligned}
\mathrm{OR}^c(j, R : k) &= \frac{\Pr(Y \leq k | X = j)/\Pr(Y > k | X = j)}{\Pr(Y \leq k | X = R)/\Pr(Y > k | X = R)} \\
&= \frac{F_k(X = j)/[1 - F_k(X = j)]}{F_k(X = R)/[1 - F_k(X = R)]} ,
\end{aligned}
\tag{2.6}
$$

where $F_k(X = j) = \Pr(Y \leq k | X = j) = \pi_1(X = j) + \pi_2(X = j) + \cdots + \pi_k(X = j)$ is the cumulative distribution function (CDF) of Y given X. For any pair of rows of X, there are $C - 1$ cumulative odds ratios that can be formed by collapsing the column classification into dichotomies. For example, the 5-level ordinal scale assessing a subject's current arthritis in Table 2.2 can be collapsed, grouping contiguous categories, to form the following four dichotomies: (i) very good versus very poor–good, (ii) good–very good versus very poor–fair, (iii) fair–very good versus very poor–poor, and (iv) poor–very good versus very poor. If the rows and columns are independent, then all of these cumulative odds ratios equal 1 (or the log cumulative odds ratios equal 0). Thus, the usual null hypothesis is H_0: $\mathrm{OR}^c(j, R : k) = 1$ for $j = 1, ..., R - 1$, and $k = 1, ..., C - 1$. The estimate of this odds ratio is

$$
\widehat{\mathrm{OR}}^c(j, R : k) = \frac{\widehat{F}_k(X = j)/[1 - \widehat{F}_k(X = j)]}{\widehat{F}_k(X = R)/[1 - \widehat{F}_k(X = R)]} ,
$$

where $\widehat{F}_k(X = j) = \widehat{\pi}_1(X = j) + \cdots + \widehat{\pi}_k(X = j) = (n_{j1} + n_{j2} + \cdots + n_{jk})/n_{j+}$. As before, inference for the cumulative odds ratio, $\mathrm{OR}^c(j, R : k)$, is based on the large sample normality of the estimator of the log cumulative odds ratio and is exactly the same as for an odds ratio from a 2×2 table (here with columns formed by a collapsing of the ordinal response categories to the binary response categories $Y \leq k$ and $Y > k$) as described earlier in Section 2.2. However, when Y is ordinal a test that is more sensitive to a monotone trend in Y can be constructed that imposes restrictions on the ordinality of Y by assuming a common cumulative odds ratio $\mathrm{OR}^c(j, R : 1) = \mathrm{OR}^c(j, R : 2) = \cdots = \mathrm{OR}^c(j, R : C - 1) = \mathrm{OR}^c(j, R)$, for $j = 1, ..., R - 1$. That is, for any pair of rows of X, it is assumed that the effect of X on Y is the same for the $C - 1$ different cumulative probabilities and can be described with a single parameter, $\mathrm{OR}^c(j, R)$. This assumption, known as the *proportional odds* assumption, implies that the cumulative odds ratios for any pair of rows of X are the same for all $C - 1$ possible collapsings of the ordinal response to the binary response $Y \leq k$ and $Y > k$. As we will see later in Section 2.6, the *proportional odds* assumption also forms the basis of one of the most widely used multinomial logistic regression models for ordinal outcomes; as a result, we defer discussing inference for $\mathrm{OR}^c(j, R)$ to Section 2.6.

Next, we mention two commonly used tests for comparing two treatment groups on an ordinal response. The first is the *Wilcoxon rank-sum test* (or *Mann-Whitney test*), with correction for ties, for testing equality of the two distributions (Wilcoxon, 1945; Mann and Whitney, 1947). Interestingly, it can be shown that the score test for the treatment group effect in the proportional odds model described in Section 2.6 is equivalent to the Wilcoxon rank-sum test (McCullagh, 1980); moreover, the proportional odds model has the advantage of also providing a simple to interpret treatment effect estimate. Instead of a test based on ranks, a second test that is often used is one that employs fixed scores (e.g., equally-spaced integer scores) for the ordinal response categories. Ordinarily, the scores are chosen *a priori* such that the test of trend is locally most powerful for detecting particular types of associations (e.g., monotone trend alternatives); in general, the test based on fixed scores is relatively insensitive to different choices of monotone scores. A commonly used version of this test is the *Cochran-Armitage trend test* (Cochran, 1954; Armitage, 1955). This test also relates to the logistic regression model discussed in Section 2.5; specifically, by reversing their roles and regarding treatment group as the outcome and the scores for the ordinal categories as the covariate, it can be shown that the Cochran-Armitage trend test is simply the score test for association in the logistic regression relating the outcome to the covariate (see, for example, Agresti, 2010). Because so many of the commonly used tests for comparing two treatment groups on an ordinal response are simply special cases of the regression models considered in Sections 2.5 and 2.6, we do not discuss any more detail of them here.

Finally, we note that if both the rows and columns are ordinal, which might occur if the rows correspond to the same treatment given at three or more different doses, then a 1 degree of freedom chi-square test that assigns scores to both the rows and columns can be used. In particular, a 1 degree of freedom chi-square test for association equals $(n-1)r^2$, where n is the total sample size and r equals the Pearson correlation coefficient between assigned row and column scores; if row and column ranks are used, then r equals the Spearman correlation coefficient.

2.4   Analysis of Stratified 2×2 Contingency Tables

We complete our discussion of contingency table methods by considering the case where there are sets of 2×2 contingency tables, specifically a set of J, 2×2 tables that arise from splitting or stratifying the data by two or more levels of an additional categorical variable. In the clinical trials setting, this might arise when there is blocking of subjects within more homogeneous groups, e.g., blocking of subjects by clinic in a multicenter clinical trial. In the latter setting, we may be interested in inference about the effect of treatment (X) on a binary outcome (Y) after adjusting or controlling for clinic effects (say W, a categorical variable with J levels).

In general, when there are three categorical variables, we can form a "multidimensional contingency table". Depending on the study design, some margins of the table are considered to be fixed by design. One particular type of multidimensional contingency table results from a set of J, 2×2 tables. For example, consider the data in Table 2.5 where subjects from the clinical trial of a respiratory disorder are now cross-classified by clinic (W), treatment group (X), and outcome (Y).

The 2×2 tables of (X, Y) at each level of W are referred to as *partial* or *conditional* tables; for example, they express the relationship between treatment and outcome controlling for clinic. The (X, Y) table formed by combining or collapsing over the partial tables is referred to as the *marginal* table; for example, Table 2.1 is the marginal table formed by collapsing the data in the partial tables in Table 2.5 over clinic.

Similarly, the odds ratios in the partial tables are called partial odds ratios, and the odds ratio in the marginal table is called the marginal odds ratio. When the odds ratios in the partial tables differ, there is said to be *interaction* between W and (X, Y). On the other hand, when the partial odds ratios are the same, but the common partial odds ratio is

TABLE 2.5
Illustration of two 2×2 contingency tables from a clinical trial of subjects with a respiratory disorder.

		Outcome (Y)	
Clinic (W)	Treatment (X)	Poor	Good
1	Placebo	17	12
	Active	13	14
2	Placebo	12	16
	Active	4	23

TABLE 2.6
General representation of counts in the jth 2×2 contingency table, $j = 1, ..., J$.

		Y		
		1	2	
X	1	n_{j11}	n_{j12}	n_{j1+}
	2	n_{j21}	n_{j22}	n_{j2+}
		n_{j+1}	n_{j+2}	n_{j++}

different from the marginal odds ratio, there is said to be *confounding*. Confounding occurs when two variables are associated with a third in a way that obscures their relationship. In particular, W (clinic) can potentially confound the relationship between X (treatment) and Y (outcome) when W is related to both X and Y.

In Table 2.5 there are two 2×2 tables. Suppose, in general, that there are J, 2×2 tables, with notation as in Table 2.6. Let $n_{jk\ell}$ denote the number of subjects with ($W = j, X = k, Y = \ell$). Suppose it is of interest to test for no partial association, i.e., H$_0$: no association between Y and X given $W = j$, $j = 1, ..., J$.

For the stratified 2×2 contingency tables, Cochran-Mantel-Haenszel (Mantel and Haenszel, 1959) proposed a test statistic based on conditioning upon both margins. Conditional on both margins of the jth table, the data follow a (central) hypergeometric distribution under the null hypothesis of no association, leading to the Cochran-Mantel-Haenszel test for H$_0$: no association between Y and X, given W,

$$Z = \frac{\sum_{j=1}^{J}[n_{j11} - E_j]}{\sqrt{\sum_{j=1}^{J} V_j}}, \tag{2.7}$$

where $E_j = \frac{n_{j1+}n_{j+1}}{n_{j++}}$ is the hypergeometric mean and $V_j = \frac{n_{j1+}n_{j2+}n_{j+1}n_{j+2}}{n_{j++}^2(n_{j++}-1)}$ is the hypergeometric variance. The Cochran-Mantel-Haenszel statistic has an approximate $N(0,1)$ under the null hypothesis provided that the number of tables is large, say $J > 30$, and/or the sample size in each table, n_{j++}, is large. We note that the Cochran-Mantel-Haenszel statistic is also equivalent to a score test from a logistic regression model that includes the effect of treatment (X) in addition to the effects of the levels of W.

Returning to the example data from the clinical trial of subjects with respiratory illness in Table 2.5, we consider an analysis of the effect of treatment that adjusts for the effect of center. Specifically, the Cochran-Mantel-Haenszel test for no treatment effect on outcome, conditional on clinic, yields a chi-squared statistic of 4.40, with 1 df, $p = 0.0359$. Thus, when adjusted for center, there is evidence of a significant treatment effect at the 0.05 level. The estimate of the adjusted odds ratio is 2.35 (95% CI: 1.03, 5.34), slightly larger than the unadjusted $\widehat{\text{OR}} = 2.25$, and indicating that within centers subjects in the active treatment group have more than twice the odds of a good response when compared to subjects in the placebo group.

Next, we consider the more general setting where it is of interest to estimate the effect of treatment on outcome while adjusting for multiple covariates, both categorical and quantitative; the effect of treatment can be estimated using regression models for binary and ordinal outcomes. Regression models have important advantages over stratified contingency table methods when the number of categorical covariates is larger or when adjustment for quantitative covariates is required.

2.5   Regression Models for Binary Outcomes

Among the many binary response models that have been proposed in the statistical literature, logistic regression has become the standard model for analyzing the effects of a set of predictor variables. Logistic regression is ubiquitous not only in the clinical trials setting but also in most areas of application in the social, behavioral and medical sciences. In common with standard linear regression, the primary objective of logistic regression is to model the mean of the binary response variable, or the probability of success, given a set of predictor variables. However, what distinguishes logistic regression from linear regression is that a transformation of the mean responses, the logit or log odds of success, is modelled as a linear function of the predictors.

In this section we introduce some general notation and consider regression models for a binary response. Let Y_i denote a binary response variable for the ith subject ($i = 1, ..., n$); for convenience, the two response categories are often referred to as "success" or "failure". Denoting these two possible outcomes for Y_i by 1 and 0 respectively, the probability distribution of the response variable is the Bernoulli distribution, with $\Pr(Y_i = 1) = \pi_i$ (and, correspondingly, $\Pr(Y_i = 0) = 1 - \pi_i$). We assume that the primary goal is to describe the effects of a set of p predictor variables, $X_{i1}, X_{i2}, ..., X_{ip}$, on π_i. In the clinical trial setting, a key predictor is the treatment group; other covariates may include blocking factors, baseline characteristics of subjects (e.g., prognostic variables not balanced by the randomization), and baseline values of the outcome. Note, for much of the remainder of this section we suppress the subscript i when the meaning is clear from the context.

Given that the main analytic goal is to investigate the relationship between π and a set of p predictor variables (alternatively, they may represent q distinct predictor variables, where $q < p$, in addition to a subset of interactions among the predictors), there are a number of alternative binary response models that can be considered. Because linear models play such an important and dominant role in applied statistics, it may at first seem natural to assume a linear model relating the mean of Y to the X's,

$$
\begin{aligned}
E(Y|X_1, X_2, ..., X_p) &= \pi = \Pr(Y = 1|X_1, X_2, ..., X_p) \\
&= \beta_0 + \beta_1 X_1 + \beta_2 X_2 + \cdots + \beta_p X_p.
\end{aligned}
$$

(2.8)

However, this linear model for the probabilities has at least one very obvious difficulty. Expressing π as a linear function violates the restriction that probabilities must lie within the range from 0 to 1, unless constraints are placed on $X_1, X_2, ..., X_p$ or on $\beta_0, ..., \beta_p$. That is, for sufficiently large or small values of the X's (when the β's $\neq 0$), the linear model given by (2.8) will yield predicted probabilities outside of the range from 0 to 1. A further difficulty with the linear model for the probabilities is that we often expect there to be a nonlinear relationship between π and the X's. For example, a 0.2 unit increase in π might be considered more "extreme" when $\pi = 0.1$ than when $\pi = 0.5$. In terms of ratios, the change from $\pi = 0.1$ to $\pi = 0.3$ represents a three-fold $\left(\frac{0.3}{0.1}\right)$ or 200% increase, whereas the change from $\pi = 0.5$ to $\pi = 0.7$ represents only a 40% (or $\frac{0.7}{0.5}$) increase. In a sense, the units of measurement for a probability or proportion are often not considered to be constant over the range from 0 to 1. The linear probability model given by (2.8) simply does not take this into consideration when relating π to the X's.

Because it is more natural to compare values of π in *relative* rather than *absolute* terms, this suggests fitting a linear model relating the *logarithm* of π to the X's

$$\log(\pi) = \beta_0 + \beta_1 X_1 + \beta_2 X_2 + \cdots + \beta_p X_p. \tag{2.9}$$

The model given by (2.9) is often referred to as "relative risk" regression. For example, suppose X_1 is a binary indicator of treatment group, say active treatment versus placebo; it can be easily shown that β_1 has interpretation as the adjusted log *relative risk* (comparing active treatment to placebo) when controlling for $X_2, ..., X_p$. That is, adjusting for or holding constant $X_2, ..., X_p$, β_1 has interpretation as the *difference* between the log of the probability of success ($Y = 1$) when assigned to the active treatment ($X_1 = 1$) and the log of the probability of success when assigned to placebo ($X_1 = 0$),

$$\begin{aligned} \beta_1 &= \log\{\pi(X_1 = 1; x_2, ..., x_p)\} - \log\{\pi(X_1 = 0; x_2, ..., x_p)\} \\ &= \log\left\{\frac{\pi(X_1=1;x_2,...,x_p)}{\pi(X_1=0;x_2,...,x_p)}\right\}, \end{aligned}$$

where $\pi(X_1 = 1; x_2, ..., x_p)$ and $\pi(X_1 = 0; x_2, ..., x_p)$ denote the respective conditional success probabilities for fixed values of $X_2, ..., X_p$. Therefore, e^{β_1} has interpretation as the adjusted *ratio* of the probability of success when assigned to the active treatment ($X_1 = 1$) to the probability of success when assigned to placebo ($X_1 = 0$), $e^{\beta_1} = \frac{\pi(X_1=1;x_2,...,x_p)}{\pi(X_1=0;x_2,...,x_p)}$. Although relative risks are often considered the preferred measures of association, due to their intuitive interpretation, the relative risk regression model given by (2.9) shares some of the structural issues associated with the linear probability model. Although the predicted probabilities given by $\pi = e^{\beta_0 + \beta_1 X_1 + \beta_2 X_2 + \cdots + \beta_p X_p}$ cannot be negative, they can still exceed one for sufficiently large or small values of the X's (when the β's $\neq 0$). As a result, relative risk regression is not widely used in clinical trial applications.

To circumvent the aforementioned difficulty, alternative nonlinear transformations can be applied to π where the constraints on the predicted probabilities are automatically satisfied. In particular, a transformation of π, say $g(\pi)$, can be chosen so that it maps the range of π from (0,1) to $(-\infty, \infty)$. Since there are many possible transformations, $g(\pi)$, that achieve this goal, this leads to an extensive choice of *generalized linear models* (McCullagh and Nelder, 1989) that are all of the form

$$g(\pi) = \beta_0 + \beta_1 X_1 + \beta_2 X_2 + \cdots + \beta_p X_p, \tag{2.10}$$

and where the transformed probabilities are related linearly to the X's. However, the most commonly used in practice are

(i) logit or logistic function: $g(\pi) = \log[\pi/(1-\pi)]$

(ii) probit or inverse normal function: $g(\pi) = \Phi^{-1}(\pi)$, where Φ is the standardized Gaussian cumulative distribution function

(iii) complementary log-log function: $g(\pi) = \log[-\log(1-\pi)]$.

We note that all of these transformations are very closely related when $0.2 < \pi < 0.8$, and in a sense only differ in the degree of "tail-stretching" outside of this range. Indeed, for most practical purposes it is not possible to discriminate between a data analysis that is based on, for example, the logit and probit functions. To discriminate empirically between probit and logistic regression would, in general, require very large numbers of observations. However, the logit function does have a number of distinct advantages over the probit and complementary log-log functions which probably account for its more widespread use in practice. Later in this chapter we will consider some of the advantages of the logit function.

2.5.1 Logistic regression

The analytic goal is to investigate the relationship between π and a set of p predictor variables. When the transformed probabilities are related linearly to the X's by the logit function,

$$\log[\pi/(1-\pi)] = \beta_0 + \beta_1 X_1 + \beta_2 X_2 + \cdots + \beta_p X_p, \qquad (2.11)$$

this is known as the *logistic regression* model (see, for example, Hosmer and Lemeshow, 2000). Thus, the logistic regression model assumes a linear relationship between the log odds (of success) and the X's.

Next we consider the interpretation of the logistic regression coefficients. The logistic regression intercept, β_0, has interpretation as the log odds (of success) when all predictor variables assume the value zero, i.e., when $X_1 = X_2 = \cdots = X_p = 0$. Alternatively, $\frac{\exp(\beta_0)}{1+\exp(\beta_0)}$ is the probability of success when $X_1 = X_2 = \cdots = X_p = 0$. Each of the logistic regression coefficients, β_k (for $k = 1, ..., p$), has interpretation as the change in the log odds (of success) for a unit change in X_k given that all of the other predictor variables remain constant. We note that this is completely analogous to the interpretation of the regression coefficients in multiple linear regression. Thus, holding the remaining predictors at some fixed set of values, a single unit increase in X_k is predicted to increase or decrease the log odds of success by an amount β_k. Equivalently, a single unit increase in X_k increases or decreases the odds of success *multiplicatively* by a factor of $\exp(\beta_k)$, i.e., $\exp(\beta_k)$ has interpretation as the odds ratio of the response for a single unit increase in X_k.

We remark that the multiple logistic regression model given by (2.11) can also be expressed equivalently in terms of π,

$$\pi = \frac{\exp(\beta_0 + \beta_1 X_1 + \beta_2 X_2 + \cdots + \beta_p X_p)}{1 + \exp(\beta_0 + \beta_1 X_1 + \beta_2 X_2 + \cdots + \beta_p X_p)}. \qquad (2.12)$$

Expression (2.11) describes how the log odds, $\log(\frac{\pi}{1-\pi})$, has a linear relationship with the X's. Expression (2.12), on the other hand, describes how π has an S-shaped relationship with increasing values of $\beta_k X_k$ (for $k = 1, ..., p$); although, in general, this relationship is approximately linear within the range $0.2 < \pi < 0.8$ (see Figure 2.1 for a plot of π versus a single predictor X when $\beta_0 = 0.5$ and $\beta_1 = 0.9$). Observe that the expression on the right of (2.12) cannot yield a value that is either negative or greater than 1. That is, the logistic transformation ensures that the predicted probabilities are restricted to the range from 0 and 1.

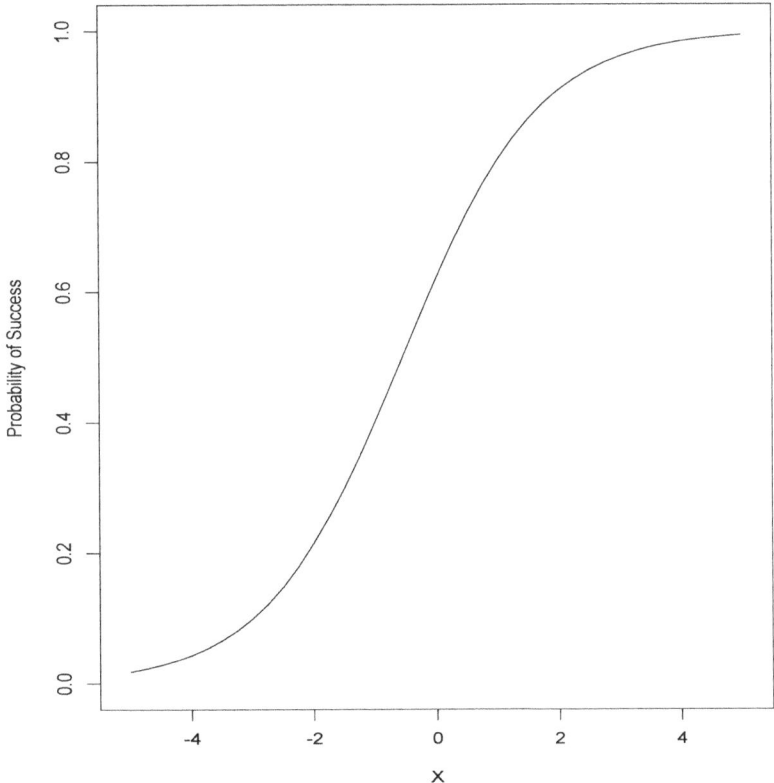

FIGURE 2.1
Plot of Logistic Regression Function for Logistic Model: $\log(\frac{\pi}{1-\pi}) = 0.5 + 0.9\,X$.

Finally, the odds ratio has many appealing properties that probably account for the widespread adoption of the logit function (and, hence, logistic regression) over other functions such as the probit or complementary log-log functions. Because the odds ratio does not change when rows and columns of a 2×2 table are interchanged, this implies that it is not necessary to distinguish which variable is the response and predictor variable in order to estimate the odds ratio. An appealing feature of logistic regression is that the odds ratio, $\exp(\beta_k)$, is equally valid whether the study design is prospective (as is the case in randomized clinical trials), cross-sectional, or retrospective. That is, logistic regression provides an estimate of the same association between Y and X_k regardless of whether we randomly sample individuals having $X_k = 0$ and $X_k = 1$ and subsequently classify them according to whether $Y = 1$ or $Y = 0$ (prospective design), or randomly sample individuals having $Y = 0$ and $Y = 1$ and subsequently classify them according to whether $X_k = 0$ and $X_k = 1$ (retrospective or case-control design). The odds ratio is the only measure of association that has this property. Thus, in a retrospective study $\beta_1, ..., \beta_p$ can be estimated by analyzing the data as if they had been obtained from a prospective study; a formal proof of this result can be found in Farewell (1979) and Prentice and Pyke (1979), also see Breslow and Day (1980). The only parameter that cannot be estimated with data from a retrospective (case-control) study is the intercept, β_0, since it is determined by the proportions of "successes" ($Y = 1$) and "failures" ($Y = 0$) selected into the study. In many studies, however, the intercept is not of direct scientific interest. Occasionally, the odds ratio from a retrospective study may be used to determine the expected effect size when designing a randomized clinical trial.

2.5.2   Estimation and inference for logistic regression

In this section, we very briefly discuss estimation of the logistic regression coefficients. Estimates of the β's are obtained using the method of maximum likelihood. Construction of the likelihood function requires an assumption about the probability distribution of Y. Here, the n binary responses, Y_i ($i = 1, ..., n$), are assumed to be independent Bernoulli random variables, with $\pi_i = E(Y_i | X_{i1}, X_{i2}, ..., X_{ip}) = \Pr(Y_i = 1 | X_{i1}, X_{i2}, ..., X_{ip})$. By independence of the n observations, the likelihood function can be expressed as the product,

$$\prod_{i=1}^{n} \pi_i^{y_i} (1 - \pi_i)^{1-y_i},$$

and is a function of the unknown logistic regression coefficients, since

$$\pi_i = \frac{\exp(\beta_0 + \beta_1 X_{i1} + \beta_2 X_{i2} + \cdots + \beta_p X_{ip})}{1 + \exp(\beta_0 + \beta_1 X_{i1} + \beta_2 X_{i2} + \cdots + \beta_p X_{ip})}.$$

Maximization of the likelihood, or the log-likelihood, requires an iterative procedure and this has been implemented in almost every statistical software package (e.g., R, SAS, STATA, and SPSS). A description of the iterative procedure can be found in, for example, McCullagh and Nelder (1989; pp. 40–43). Finally, estimates of the standard errors (SEs) of the estimated logistic regression coefficients are readily obtained using the inverse of the information matrix.

For inference about a particular β_k, confidence intervals or hypothesis tests can be formed based on the large sample normality of the estimator of β_k. For example, a 95% confidence interval for the odds ratio, $\exp(\beta_k)$, is obtained by exponentiating the endpoints of the corresponding confidence interval for β_k,

$$\exp\left\{ \widehat{\beta}_k \pm 1.96 \sqrt{\widehat{\mathrm{Var}}(\widehat{\beta}_k)} \right\}.$$

A test of no association, $H_0\colon \beta_k = 0$, can be constructed based on the *Wald test* statistic,

$$Z = \frac{\widehat{\beta}_k}{\sqrt{\widehat{\mathrm{Var}}(\widehat{\beta}_k)}},$$

which, in large samples, has an approximate standard normal distribution under the null hypothesis of no association; equivalently, the squared Wald test statistic, Z^2, has an approximate chi-square distribution with 1 degree of freedom. Alternatively, the *likelihood ratio test* (LRT) or *score test* statistics can be used. The likelihood ratio test of $H_0\colon \beta_k = 0$ is

$$G^2 = -2(l_0 - l_1),$$

where l_0 and l_1 denote the maximized log-likelihood values under the null ($H_0\colon \beta_k = 0$) and alternative hypothesis ($H_1\colon \beta_k \neq 0$), respectively. In addition, multivariate versions of these test statistics can be used to test hypotheses concerning various subsets of the β's. In large samples, the Wald, likelihood ratio, and score chi-square statistics have an approximate chi-square distribution with 1 degree of freedom. However, for logistic regression, the Wald test is known to exhibit unreliable and aberrant behavior that leads to very low power in certain practical circumstances; as a result, the likelihood ratio or score test is preferred. Although most statistical software packages routinely report Wald tests, many provide options for producing likelihood ratio tests (and likelihood-based confidence intervals) and score tests. If the sample size, n, is relatively small, these asymptotic approximations cannot be relied upon; in the next section, we discuss alternative methods when data are sparse.

2.5.3 Exact logistic regression

In general, logistic regression analysis with many predictors relies on large sample theory for the validity of its results. If the sample size, n, is relatively small, these asymptotic approximations cannot be relied upon. In particular, maximum likelihood estimation of the logistic regression parameters can perform poorly when the sample size is small, the probability of success is near one or zero, or when there is an insufficient number of successes or failures for certain combinations of the predictor variables. Recall that in earlier sections describing the analysis of 2×2 contingenciy tables, we discussed the use of Fisher's exact test when data are sparse. We conclude our discussion of logistic regression by briefly outlining an extension known as exact logistic regression.

Exact logistic regression (see, for example, Mehta and Patel, 1995) is an alternative method for fitting logistic regression models that produces valid estimates, test statistics, and confidence intervals when data are sparse. The relationship of exact logistic regression to maximum likelihood logistic regression is similar in certain ways to the relationship of Fisher's exact test to large-sample methods for 2×2 contingency tables. However, whereas Fisher's exact test conditions on the row and column totals in order to derive the distribution of the test statistic, exact logistic regression conditions on the so-called *sufficient statistics* for the remaining parameters in the model when estimating each logistic regression parameter. The sufficient statistics for the parameters are determined by the number of successes for different values of the corresponding predictors. Inference is based on the exact distributions of sufficient statistics for parameters of interest in the logistic regression model, conditional on those for the remaining parameters. In general, enumerating the exact distributions can be computationally infeasible; however, efficient algorithm for generating the distributions have been developed that greatly reduce the computational effort required to produce hypothesis tests and parameter estimates.

Similar to Fisher's exact test and other exact methods, exact logistic regression guarantees that tests conducted at the α significance level have a type I error rate less than or equal to α, and that 95% confidence intervals have at least 95% coverage even for small sample sizes and sparse data. Parameter estimates and confidence intervals have interpretations identical to those for standard maximum likelihood logistic regression.

In summary, exact logistic regression overcomes problems due to sparse data. In principle, exact logistic regression can be applied in settings with multiple predictors; however, greater care is required when attempting to fit complex models. Although the theory underlying exact logistic regression has been well established since the 1970's, the computations required of the method were so demanding that its implementation was considered impractical. More recently, however, with the development of efficient algorithms coupled with increasingly fast computers, exact logistic regression routines are implemented in many statistical software packages. The method has some potential disadvantages. In particular, it may be overly conservative in settings when maximum likelihood logistic regression performs adequately. In addition, it can be computationally intensive, especially when quantitative predictors or a large number of categorical predictors are included in the logistic regression model. When feasible, exact logistic regression is an attractive alternative to maximum likelihood logistic regression in small sample and sparse data settings.

2.5.4   Example

To illustrate the application of logistic regression we return to the example data from the clinical trial of patients with respiratory illness in Table 2.5. Consider the following logistic regression relating the log odds of a good response to both treatment group and center,

$$\log[\pi/(1-\pi)] = \beta_0 + \beta_1 X_1 + \beta_2 X_2,$$

where $X_1 = 1$ if assigned to active treatment and 0 otherwise, and $X_2 = 1$ if subject is treated in center 2 and 0 otherwise. In this model, the parameter of primary interest is β_1 and $\exp(\beta_1)$ has interpretation as the adjusted odds ratio for treatment, controlling for center. The ML estimate of the adjusted odds ratio is 2.36 (95% CI: 1.06, 5.28), and the likelihood ratio test of H_0: $\beta_1 = 0$ yields $\chi_1^2 = 4.53$, $p = 0.033$ (the Wald test yields $Z = 2.10$ or $\chi_1^2 = 4.39$). As might be expected, this is very similar to the results obtained using the Cochran-Mantel-Haenszel estimator and test for stratified 2×2 tables in Section 2.4. The results of the logistic regression analysis indicate that, within centers, subjects in the active treatment group have approximately 2.4 times the odds of a good response when compared to subjects in the placebo group. Of less substantive interest is the estimate of the center effect; here the ML estimate of $\exp(\beta_2)$ is 2.92 (95% CI: 1.31, 6.52), indicating that subjects in center 2 have almost 3 times the odds of a good response when compared to those in center 1. Although the data in Table 2.5 are not particularly sparse, we use these data to illustrate the application of exact logistic regression. When the logistic regression model given above is estimated using exact methods, this yields an estimate of the adjusted odds ratio for treatment of 2.32 (exact 95% CI: 0.98, 5.67), and an exact test of H_0: $\beta_1 = 0$ with $p = 0.056$. The estimate of the adjusted odds ratio for treatment is very similar to that obtained from standard logistic regression. However, in this instance, inference based on exact methods would lead to a different conclusion about the effect of treatment on outcome (at the 0.05 significance level). The small increase in the p-value for the test of H_0: $\beta_1 = 0$ is completely consistent with the well known conservativeness of exact methods.

Finally, for illustration purposes, we consider an additional logistic regression analysis that also adjusts for baseline age of the subjects,

$$\log[\pi/(1-\pi)] = \beta_0 + \beta_1 X_1 + \beta_2 X_2 + \beta_3 X_3,$$

where X_3 is a subject's age in years. Note that the inclusion of baseline age changes the interpretation of the treatment effect. That is, $\exp(\beta_1)$ has interpretation as the adjusted odds ratio for treatment in the subpopulation within each center of a given age (say $X_3 = x_3$). The ML estimate of the center and age adjusted odds ratio for treatment is 2.35 (95% CI: 1.05, 5.26), and the likelihood ratio test of H_0: $\beta_1 = 0$ yields $\chi_1^2 = 4.48$, $p = 0.034$ (Wald $\chi_1^2 = 4.35$, $p = 0.037$), results that are remarkably similar to those obtained with adjustment for center alone. This is to be expected because the age distributions were very similar (due to randomization) in the two treatment groups (mean age 33.6 and 32.9 years in the placebo and active groups respectively; standard deviation 13.4 and 13.9) and age was also not strongly related to outcome.

2.6   Regression Models for Ordinal Outcomes

In the previous section we discussed regression models for a binary response; in this section we consider regression models for an ordinal response with three or more levels. In particular, we focus on a regression model for the ordinal response that is a natural extension of the familiar logistic regression model for a binary response. However, with an ordinal response the logit transformation is applied to the *cumulative* response probabilities (McCullagh, 1980; Agresti, 2010). This leads to a regression model known as the *proportional odds model*. The proportional odds model is probably the most widely used model for the analysis of an ordinal response; we also briefly discusss some alternative regression models for ordinal outcomes.

Throughout the remainder of this section we assume that we have n independent observations of an ordinal response. We let Y_i $(i = 1, ...n)$ denote an ordinal response with K ordinal categories $(1, ..., K)$ for the ith subject; note, this is a slight departure from the notation used in Section 2.3 where the ordinal response had C categories. The actual integer values, $1, ..., K$, are not particularly relevant except that larger values are assumed to correspond to "higher" outcomes and smaller values to "lower" outcomes. The distribution of Y_i is multinomial, with K multinomial probabilities, $\Pr(Y_i = k)$ for $k = 1, ..., K$, for the distinct ordinal categories; note, there are only $K - 1$ non-redundant multinomial probabilities because the K probabilities are constrained to sum to 1. Associated with each response, Y_i, is a set of p covariates $X_{i1}, X_{i2}, ..., X_{ip}$. As before, a key covariate in the clinical trial setting is the treatment group, say X_{i1}; other covariates may include blocking factors, baseline characteristics of subjects, and baseline values of the outcome.

2.6.1 Proportional odds model

To develop the proportional odds regression model, suppose that we dichotomize the ordinal outcome at 1 versus greater than 1, creating the binary response

$$U_{i1} = \begin{cases} 1 & \text{if } Y_i = 1, \\ 0 & \text{if } Y_i > 1. \end{cases}$$

Letting $F_{i1} = \Pr(U_{i1} = 1 | X_{i1}, X_{i2}, ..., X_{ip})$, it is natural to formulate a logistic regression model for the binary response U_{i1} by relating the logit transformation of F_{i1} to the covariates,

$$\text{logit}(F_{i1}) = \log \left(\frac{F_{i1}}{1 - F_{i1}} \right) = \alpha_1 + \beta_1 X_{i1} + \beta_2 X_{i2} + \cdots + \beta_p X_{ip};$$

although, in principle, any suitable link function (e.g., probit) could be used. Next, we can dichotomize the ordinal outcome at less than or equal to 2 versus greater than 2, creating a second binary response

$$U_{i2} = \begin{cases} 1 & \text{if } Y_i \leq 2, \\ 0 & \text{if } Y_i > 2, \end{cases}$$

with $F_{i2} = \Pr(U_{i2} = 1 | X_{i1}, X_{i2}, ..., X_{ip})$. Because U_{i2} is also binary, we can formulate a logistic regression model relating the logit of F_{i2} to the covariates,

$$\text{logit}(F_{i2}) = \log \left(\frac{F_{i2}}{1 - F_{i2}} \right) = \alpha_2 + \beta_1 X_{i1} + \beta_2 X_{i2} + \cdots + \beta_p X_{ip}.$$

Note that we have allowed the intercepts for $\text{logit}(F_{i1})$ and $\text{logit}(F_{i2})$ to be different, but have assumed the β's for the covariates to be the same; later we discuss the implications of this assumption. If we continue up the ordinal scale, dichotomizing the ordinal outcome above and below the remaining categories, we can generate a series of additional binary variables

$$U_{ik} = \begin{cases} 1 & \text{if } Y_i \leq k, \\ 0 & \text{if } Y_i > k, \end{cases}$$

and formulate a logistic regression model relating the logit of F_{ik} to the covariates (for $k = 1, ..., K - 1$),

$$\text{logit}(F_{ik}) = \log\left(\frac{F_{ik}}{1 - F_{ik}}\right) = \alpha_k + \beta_1 X_{i1} + \beta_2 X_{i2} + \cdots + \beta_p X_{ip}, \qquad (2.13)$$

where $F_{ik} = \Pr(U_{ik} = 1 | X_{i1}, ..., X_{ip}) = \Pr(Y_i \leq k | X_{i1}, ..., X_{ip})$ is referred to as a "cumulative probability" of response and $\text{logit}(F_{ik})$ is referred to as a "cumulative log odds" (or "cumulative logit"). The model given by (2.13) is commonly called the *proportional odds model* (McCullagh, 1980), and it applies simultaneously to all $K - 1$ cumulative probabilities (or cumulative logits).

Thus the basic idea underlying the proportional odds model is a cumulative dichotomization of the ordinal variable going up (or down) the ordinal scale. A logistic regression model is assumed to hold simultaneously for each of these $K - 1$ binary variables, in which the $K - 1$ intercepts (α_k's) are allowed to differ, but the covariate effects (β's) are assumed to be the same. Therefore the proportional odds model can be thought of as a logistic regression model for the *cumulative probabilities* of response; specifically, it relates the cumulative log odds of response, $\text{logit}(F_{ik})$, to the covariates and assumes that $\text{logit}(F_{ik})$ and $\text{logit}(F_{ik'})$ (for $k \neq k'$) have the same slopes (β's), only the intercepts (α_k and $\alpha_{k'}$) differ.

Next we consider the interpretation of the regression parameters in the proportional odds model. For ease of exposition, suppose that we have a proportional odds model with two covariates, X_{i1} (say, treatment) and X_{i2} (say, some baseline subject characteristic),

$$\text{logit}(F_{ik}) = \alpha_k + \beta_1 X_{i1} + \beta_2 X_{i2}.$$

We can interpret β_1 in a manner similar to how we interpret standard logistic regression parameters, but recognizing that we are modeling the cumulative log odds (or cumulative logit). Thus β_1 has interpretation as the change in the cumulative log odds for each one unit increase in X_{i1}, while holding X_{i2} constant. That is, holding X_{i2} fixed, β_1 is the cumulative log odds ratio,

$$\log\left[\frac{F_{ik}(X_{i1} = c + 1; x_2)/\{1 - F_{ik}(X_{i1} = c + 1; x_2)\}}{F_{ik}(X_{i1} = c; x_2)/\{1 - F_{ik}(X_{i1} = c; x_2)\}}\right].$$

When X_{i1} is binary, say an indicator of treatment group, then $\exp(\beta_1)$ is the cumulative odds ratio for treatment, adjusting for X_{i2}.

The sign of the coefficient for β_1 sometimes causes confusion about the direction of the relationship between the ordinal response and the covariate. Recall that (2.13) models the cumulative log odds of being in *lower-numbered* categories. Therefore larger values of $\beta_1 X_{i1}$ are associated with an *increased* probability of being in the *lower-numbered* categories or, equivalently, a *decreased* probability of being in the *higher-numbered* categories. For example, when β_1 is positive this implies an inverse or negative relationship between X_{i1} and Y_i, with increases in X_{i1} associated with lower values of the ordinal scale. We caution the reader that some textbooks, statistical software, and alternative derivations of the model, use the following convention for the proportional odds model,

$$\begin{aligned}
\text{logit}(F_{ik}) &= \alpha_k - (\beta_1 X_{i1} + \beta_2 X_{i2} + \cdots + \beta_p X_{ip}) \\
&= \alpha_k - \beta_1 X_{i1} - \beta_2 X_{i2} - \cdots - \beta_p X_{ip},
\end{aligned}$$

placing a negative sign in front of the β's so that larger values of $\beta_1 X_{i1}$ are associated with an *increased* probability of being in the *higher-numbered* categories.

Note that our interpretation of β_1 is not specific about which of the $K-1$ cumulative log odds it refers to. The reason for this is that β_1 has the same interpretation for all $K-1$ cumulative log odds, i.e., the cumulative log odds ratio is independent of the response level k. In the proportional odds model, the log odds ratio for a one unit increase in a covariate (while holding the other covariates constant) is the same for any of the cumulative probabilities,

$$\beta_1 = \log\left[\frac{F_{i1}(X_{i1}=c+1;x_2,...,x_p)/\{1-F_{i1}(X_{i1}=c+1;x_2,...,x_p)\}}{F_{i1}(X_{i1}=c;x_2,...,x_p)/\{1-F_{i1}(X_{i1}=c;x_2,...,x_p)\}}\right]$$

$$= \log\left[\frac{F_{i2}(X_{i1}=c+1;x_2,...,x_p)/\{1-F_{i2}(X_{i1}=c+1;x_2,...,x_p)\}}{F_{i2}(X_{i1}=c;x_2,...,x_p)/\{1-F_{i2}(X_{i1}=c;x_2,...,x_p)\}}\right]$$

$$= \log\left[\frac{F_{i3}(X_{i1}=c+1;x_2,...,x_p)/\{1-F_{i3}(X_{i1}=c+1;x_2,...,x_p)\}}{F_{i3}(X_{i1}=c;x_2,...,x_p)/\{1-F_{i3}(X_{i1}=c;x_2,...,x_p)\}}\right].$$

What this means is that if a unit increase in a covariate triples the odds of being in response level 1 (versus level 2 or higher), it also triples the odds of being in response level 2 or below (versus level 3 or higher), or in level 3 or below (versus level 4 or higher), and so on. This is the "proportionality assumption" that gives the model its name. This property of the proportional odds model also implies that if you were to dichotomize an ordinal response (above and below a given level k) and use standard logistic regression as the method of analysis, the resulting odds ratios are invariant to where you dichotomize the ordinal scale; only the intercept would depend on where you choose to dichotomize the scale.

One appealing property of the proportional odds model is that its regression parameters are invariant to collapsing of adjacent response categories. That is, we would not expect the results of an analysis to change much if we were to combine two adjacent categories. This feature of the model can be helpful when it is of interest to compare regression estimates from studies using different ordinal scales (e.g., one trial using a five-level version of a quality-of-life scale, another using a three-level version of the same scale). Also this property of the model can be used to justify combining adjacent categories prior to analyses when data are sparse for certain response categories.

As additional motivation for the proportional odds assumption, we note that the proportional odds model can also be developed from a linear regression model for a latent continuous variable (Anderson and Philips, 1981). Suppose that L_i is a latent (i.e., unobserved) continuous variable, such that values of the ordinal response are observed only when L_i falls within one of K intervals determined by a set of "cut-points", α_k. In this latent variable formulation, the observed ordinal response can be thought of as a $K-$level categorization of the unobserved latent variable, with

$$Y_i = \begin{cases} 1 & \text{if } -\infty < L_i \leq \alpha_1, \\ 2 & \text{if } \alpha_1 < L_i \leq \alpha_2, \\ 3 & \text{if } \alpha_2 < L_i \leq \alpha_3, \\ \vdots & \\ K & \text{if } \alpha_{K-1} < L_i < \infty. \end{cases}$$

Next suppose that the following linear regression model holds for L_i:

$$L_i = \beta_0 + \beta_1 X_{i1} + \beta_2 X_{i2} + \cdots + \beta_p X_{ip} + e_i,$$

where e_i has a standard logistic distribution with mean zero and variance $\pi^2/3$; the corresponding cumulative distribution function is $G(z) = \frac{\exp(z)}{1+\exp(z)}$. Here we regard the "cut-points" (α_k's) of L_i as fixed and the mean of the distribution of L_i as changing with the X's.

It can be shown that this linear model for the latent variable (with logistic errors) implies a proportional odds model for the observed ordinal response with the same covariate effects (β's),

$$\Pr(Y_i \leq k) = \Pr(L_i \leq \alpha_k) = \frac{\exp[\alpha_k - (\beta_1 X_{i1} + \beta_2 X_{i2} + \cdots + \beta_p X_{ip})]}{1 + \exp[\alpha_k - (\beta_1 X_{i1} + \beta_2 X_{i2} + \cdots + \beta_p X_{ip})]},$$

or put in a linear form,

$$\text{logit}[\Pr(Y_i \leq k)] = \text{logit}(F_{ik}) = \alpha_k - (\beta_1 X_{i1} + \beta_2 X_{i2} + \cdots + \beta_p X_{ip});$$

note, this yields the version of the proportional odds model with a negative sign in front of the β's. Therefore the latent variable formulation provides at least some motivation for the assumption of common covariate effects across the different cumulative logits in the proportional odds model.

As was mentioned earlier, the proportional odds model makes a strong assumption that the covariate effects (β's) are invariant to where you dichotomize the ordinal scale. This makes interpretation of the effects of covariates relatively straightforward when only a single parameter is required for each covariate. It is possible to relax the proportionality assumption and consider a "non-proportional" odds model, in which the covariate effects depend on the response level k,

$$\text{logit}(F_{ik}) = \alpha_k + \beta_{k1} X_{i1} + \beta_{k2} X_{i2} + \cdots + \beta_{kp} X_{ip}; \qquad (2.14)$$

this model allows separate covariate effects for each cumulative logit. In the model given by (2.14) the log odds ratio now depends on k,

$$\beta_{k1} = \log \left[\frac{F_{ik}(X_{i1} = c+1; x_2, ..., x_p)/\{1 - F_{ik}(X_{i1} = c+1; x_2, ..., x_p)\}}{F_{ik}(X_{i1} = c; x_2, ..., x_p)/\{1 - F_{ik}(X_{i1} = c; x_2, ..., x_p)\}} \right].$$

Model (2.14) can be used to test the proportionality assumption based on a test of the null hypothesis,

$$H_0 \colon \beta_{1j} = \beta_{2j} = \cdots = \beta_{K-1,j} = \beta_j \quad \text{for all } p \text{ covariates } (j = 1, ..., p).$$

Under the null hypothesis that the proportional odds model holds, there are p distinct β's for the covariate effects. Under the alternative hypothesis there are $(K-1) \times p$ distinct β's. So the test of the proportionality assumption has $df = (K-1) \times p - p = (K-2) \times p$. Furthermore the proportionality assumption can be relaxed for only a subset of the covariates; this leads to a *partial* proportional odds model where separate effects for each cumulative logit are fit for some but not all of the covariates (Peterson and Harrell, 1990). For example, the following *partial* proportional odds model,

$$\text{logit}(F_{ik}) = \alpha_k + \beta_{k1} X_{i1} + \beta_2 X_{i2} + \beta_3 X_{i3} + \cdots + \beta_p X_{ip},$$

allows for separate (or "non-proportional") effects of X_{i1} but makes the proportionality assumption for the remaining covariates, $X_{i2}, ..., X_{ip}$. One word of caution about model (2.14) and *partial* proportional odds models: By relaxing the proportionality assumption, the model no longer constrains the cumulative probabilities. As a result the fitting of model (2.14) can potentially lead to incoherent results where, for example, the estimate of $F_{i3} = \Pr(Y_i \leq 3 | X_{i1}, ..., X_{ip})$ is less than the estimate of $F_{i2} = \Pr(Y_i \leq 2 | X_{i1}, ..., X_{ip})$ for some values of the covariates. This violates the proper order of cumulative probabilities and implies that $\Pr(Y_i = 3 | X_{i1}, ..., X_{ip}) = F_{i3} - F_{i2}$ must be negative. Because of the potential for incoherent results from model (2.14), a test of the proportionality assumption based on the score statistic is preferred because it does not require fitting the non-proportional odds model.

Finally, the regression parameters of the proportional odds model can be estimated by ML. This requires maximizing the multinomial likelihood for the ordinal response, with the response probabilities viewed as functions of the α's and β's. When regarded as a generalized linear model, the proportional odds model has certain non-standard features. In the proportional odds model we do not relate the mean of Y_i to the covariates via a logit link function (or any other suitable link function). Instead, we *jointly* relate the means of the $K - 1$ cumulative random variables, $(U_{i1}, ..., U_{i,K-1})$, to the covariates. Put another way, it is the cumulative probabilities, and not the mean of the ordinal response, that are simultaneously related to the covariates (via a logit link function). Fitting the proportional odds model requires computer algorithms that can handle the fact that it is a *multivariate*, rather than a *univariate*, generalized linear model for the $K - 1$ cumulative dichotomizations of the ordinal response. Procedures for fitting the model have been implemented in most of the widely-available statistical software packages.

2.6.2 Some alternative models for ordinal outcomes

Although the proportional odds model is probably the most widely used model for the analysis of ordinal responses, there are several alternative regression models. Here we briefly consider two models, the *adjacent-category* and *continuation-ratio* models, that are both based on a type of logistic regression model for the ordinal response. The basic idea underlying the adjacent-category logistic regression model is to compare each category of the response to the next largest level. So, with a K-level ordinal response we compare level 1 versus level 2, level 2 versus level 3, level 3 versus level 4, and so on. When these comparisons are made on the log odds scale, the following adjacent-category model is obtained,

$$\log \left\{ \frac{\Pr(Y_i = k | Y_i = k \text{ or } k + 1)}{\Pr(Y_i = k + 1 | Y_i = k \text{ or } k + 1)} \right\} = \alpha_k + \beta_1 X_{i1} + \beta_2 X_{i2} + \cdots + \beta_p X_{ip}, \qquad (2.15)$$

where the left-hand side of (2.15) is the conditional log odds of response at level k, given that the response is at either level k or level $k + 1$. This model assumes the effects of the covariates do not depend on the particular pair of adjacent categories being compared.

Instead of comparing each category of the response to the next largest level, the continuation-ratio model compares each category to all higher response levels. So with a K-level ordinal response we compare level 1 versus level 2 through K, level 2 versus level 3 through K, level 3 versus level 4 through K, and so on. When these comparisons are made on the log odds scale, the following continuation-ratio model is obtained,

$$\begin{aligned} \text{logit}\{\Pr(Y_i = k | Y_i \geq k)\} &= \log \left\{ \frac{\Pr(Y_i = k)}{\Pr(Y_i > k)} \right\} \\ &= \alpha_k + \beta_{k1} X_{i1} + \beta_{k2} X_{i2} + \cdots + \beta_{kp} X_{ip}. \end{aligned}$$

$$(2.16)$$

This model ordinarily assumes separate effects of the covariates on each of the $K - 1$ logits; when it seems plausible, it is possible to constrain covariate effects to be the same for each of the logits. The continuation-ratio model can be appealing when the categories of the ordinal response represent a natural sequence of stages in some progression (e.g., cancer stage). One less appealing feature of the model is that the results are not invariant to whether the categories have been ordered from low to high or from high to low.

Finally, we note that there is an alternative class of models that are referred to as "mean response" or "mean score" models. These models treat the ordinal categorical outcome in a quantitative way by assigning monotone scores to the outcome categories, such as equally spaced scores as described in Section 2.3. The models then assume a linear relationship between the mean of the response scores and the covariates. Weighted least squares (Grizzle et al, 1969) or ordinary least squares (OLS) can be used to unbiasedly estimate the regression parameters. However, with OLS estimation of the regression parameters the standard errors must properly account for the fact that the variance is not constant but depends on the mean (Lipsitz, 1992). The mean score model is more appealing when the ordinal response arises from a crude categorizing of an underlying quantitative variable, making the choice of monotone scores for the outcome categories more apparent. In general, these models are not widely used for analyses of ordinal outcomes in randomized clinical trials.

2.6.3 Example

We illustrate the application of the proportional odds model to the data in Table 2.2 from the clinical trial comparing auranofin therapy and placebo for the treatment of rheumatoid arthritis. Recall that the outcome variable of interest is a global impression scale (Arthritis Categorical Scale) assessed at month 6; the outcome is measured on a five-level ordinal scale: (1) very good, (2) good, (3) fair, (4) poor, and (5) very poor. Data on this outcome variable are available for 293 of the patients who participated in this trial. The goal of the analysis is to determine whether treatment with auranofin therapy increases the odds of a more favorable response, after controlling for the baseline age of the patients.

Consider the following proportional odds model:

$$\text{logit}(F_{ik}) = \log\left\{\frac{\Pr(Y_i \leq k)}{\Pr(Y_i > k)}\right\} = \alpha_k + \beta_1 X_{i1} + \beta_2 X_{i2},$$

where the two model covariates are treatment group ($X_{i1} = 1$ if randomized to auranofin, $X_{i1} = 0$ if randomized to placebo) and baseline age in units of 10 years (X_{i2}). Maximum likelihood estimates of the model parameters are presented in Table 2.7. The results in Table 2.7 indicate that there is a significant treatment effect (Wald $Z = 2.84$ or $\chi_1^2 = 8.05$, $p < 0.005$; likelihood ratio $\chi_1^2 = 8.14$, $p < 0.005$), with patients in the auranofin therapy group having an increased odds of a lower or more favorable response. Specifically, when adjusted for baseline age, patients in the auranofin therapy group have approximately twice (or $e^{0.608} = 1.84$) the odds of a self-assessment of arthritis at response level k or lower (corresponding to a more favorable response) relative to patients in the placebo group. Although of far less interest in this clinical trial, the estimated effect of age indicates that older patients in both treatment groups tend to report less favorable response. For example, a 10-year difference in baseline age decreases the odds of a more favorable response by a factor of 0.82 (or $e^{-0.205}$).

The validity of the above analysis requires that the proportional odds assumption holds. We can assess the assumption of proportionality by considering a more complex model that allows for separate effects of treatment and age on the four dichotomizations of the ordinal response. The resulting test of proportionality yields a chi-squared statistic of 8.55, with 6 degrees of freedom ($p > 0.20$). This test has 6 df because it allows for 6 additional regression parameters, 3 additional parameters for the treatment effect, and 3 additional parameters for the age effect. Because the more complex non-proportional odds model does not fit significantly better ($p > 0.20$), the proportional odds assumption appears to hold for these data.

As was mentioned earlier, a property of the proportional odds model is that if you dichotomize the ordinal response (above and below a given level k) and use standard logistic regression as the method of analysis, the resulting odds ratios are invariant to where you

TABLE 2.7
ML estimates and standard errors from the proportional odds model for the arthritis clinical trial data.

Variable	Estimate	SE	Z
α_1	−1.212	0.532	−2.28
α_2	0.525	0.525	1.00
α_3	2.149	0.538	3.99
α_4	4.137	0.612	6.76
Treatment	0.608	0.214	2.84
Age	−0.205	0.098	−2.08

TABLE 2.8
Binary outcome formed by dichotomizing the ordinal outcome (categories 1, 2 versus 3, 4, 5) for the arthritis clinical trial data.

	Response		
Treatment	Poor	Good	Total
Placebo	89	58	147
Auranofin	74	73	147
Total	163	131	294

dichotomize the ordinal scale. To illustrate this property of the model, in Table 2.8 we consider a collapsing of the five-level ordinal scale to a discrete outcome with two levels only, "good" (categories 1 or 2) or "poor" (categories 3, 4, or 5); later, we mention some of the consequences of dichotomizing an ordinal outcome. In Table 2.8, the percentage of subjects with a "good" response is 39.5% and 49.7% in the placebo and auranofin therapy groups respectively. The relative comparison of the odds of good responses between the two treatment groups, comparing auranofin to placebo, yields an OR = 1.51 (95% CI: 0.95, 2.40); for the data in Table 2.8, the Wald, Pearson, and likelihood ratio chi-squared statistics are very similar (approximately 3.1), yielding p-values of 0.079, 0.078, and 0.078 respectively. Thus, at the 0.05 significance level, the difference between treatment groups in the dichotomized response can be explained by sampling variability. Accounting for baseline age, we can fit the following standard logistic regression model to the "dichotomized" outcome, relating the log odds of a "good" response to both treatment group and baseline age,

$$\log[\pi_i/(1-\pi_i)] = \beta_0 + \beta_1 X_{i1} + \beta_2 X_{i2},$$

where $X_{i1} = 1$ if assigned to auranofin and 0 otherwise, and X_{i2} is baseline age in years. In this model, the parameter of primary interest is β_1 and the resulting odds ratio $\exp(\beta_1)$

has similar interpretation to the adjusted odds ratio for treatment in the proportional odds model. The ML estimate of the adjusted odds ratio for treatment is 1.52 (95% CI: 0.96, 2.42), and the test of H_0: $\beta_1 = 0$ yields likelihood ratio $\chi_1^2 = 3.14$, $p = 0.076$ (Wald $\chi_1^2 = 3.12$, $p = 0.077$). Note that the adjusted odds ratio based on the "dichotomized" outcome is qualitatively similar to the adjusted odds ratio for treatment based on the proportional odds model; the latter is $e^{0.608} = 1.84$. However, an analysis of the dichotomized outcome is discernibly less efficient and has reduced statistical power. In general, there is additional statistical power gained from a regression analysis of an ordinal outcome over a binary outcome. Collapsing an ordinal scale to a binary scale will almost always sacrifice efficiency due to the fact that it discards valuable information about how individuals differ. For example, after dichotomizing an ordinal scale individuals who fall close to, but on different sides of, the binary cut-off are assumed to be different in the analysis, despite the fact that they are actually quite similar; by the same token, individuals on one side of the binary cut-off are assumed to be similar in the analysis, although they may differ in important ways. Thus, analysis of the data in Table 2.8 rather than Table 2.2 discards much information and yields a test of the treatment effect with lower statistical power. When the K categories of an ordinal scale have equal marginal probabilities (when averaged over treatment groups), dichotomizing the ordinal scale can be shown to reduce efficiency by a factor of $\frac{3}{4}/(1 - \frac{1}{K^2})$ (Whitehead, 1993). Thus, dichotomizing a five-level ordinal scale, as in the case of the arthritis global impression scale, reduces efficiency by approximately 20% (assuming ordinal response categories are approximately equally probable).

Finally, we note that in addition to baseline age, information on baseline levels of the outcome are also available. In general, adjustment for baseline levels of the response will lead to a more powerful test of treatment effect; this topic is explored in far greater detail in the next section. To illustrate the potential benefits of adjustment for baseline response, we consider the following proportional odds model:

$$\text{logit}(F_{ik}) = \log\left\{\frac{\Pr(Y_i \leq k)}{\Pr(Y_i > k)}\right\} = \alpha_k + \beta_1 X_{i1} + \beta_2 X_{i2} + \beta_3 Y_{i0},$$

where the covariates are treatment group ($X_{i1} = 1$ if randomized to auranofin, $X_{i1} = 0$ if randomized to placebo), baseline age in units of 10 years (X_{i2}), and baseline levels of the ordinal outcome, Y_{i0}. Maximum likelihood estimates of the model parameters are presented in Table 2.9 and the results indicate that there is a significant treatment effect ($p < 0.005$). Specifically, when adjusted for baseline age and baseline response, patients in the auranofin therapy group have over twice (or $e^{0.741} = 2.10$) the odds of a self-assessment of arthritis at response level k or lower (corresponding to a more favorable response) relative to patients in the placebo group. Of note, the test of the treatment effect yields a Wald $Z = 3.38$ and $p = 0.0007$ (likelihood ratio $\chi_1^2 = 11.63$, $p = 0.0006$), whereas the analysis that ignores the baseline response produces a Wald $Z = 2.84$ and $p = 0.0045$ (likelihood ratio $\chi_1^2 = 8.14$, $p = 0.0043$); this highlights the potential increase in power from an analysis that adjusts for baseline response. As we discuss in the next section, the potential increase in statistical power is determined by the strength of the association between the baseline and follow-up measure of the outcome. Here, the estimated effect of baseline response indicates that there is strong positive association between the baseline and 6 month follow-up assessment of the outcome, i.e., patients with more favorable response at baseline also tend to report more favorable response at 6 months.

TABLE 2.9
ML estimates and standard errors from the proportional odds model, with adjustment for baseline response, for the arthritis clinical trial data.

Variable	Estimate	SE	Z
α_1	1.251	0.619	2.02
α_2	3.256	0.641	5.08
α_3	5.164	0.679	7.60
α_4	7.345	0.762	9.64
Treatment	0.741	0.219	3.38
Age	−0.138	0.099	−1.39
Baseline response, Y_0	−1.015	0.130	−7.79

2.7 Adjustment for Baseline Response

A characteristic feature of many randomized clinical trials is the presence of baseline variables. A baseline variable is one that is measured prior to the start of treatment and usually at or prior to randomization. Common baseline variables include demographical variables such as age or gender, disease characteristics such as duration or severity, prognostic variables, and other factors such as trial center or site. In the clinical trial setting, the main concern is with baseline variables that are thought likely to influence the outcome variable of interest. In many trials, baseline values of the outcome are available and tend to be strongly positively correlated with the follow-up assessment of the outcome. For example, in a clinical trial of a novel analgesic the main objective may be to compare treatment groups in terms of the change in a self-assessed 7-level ordinal pain scale from baseline to end of follow-up. The baseline measurement is an outcome variable like those measured subsequently, although sometimes the baseline outcome is range restricted, as when only subjects with values greater than or less than a threshold are included in the trial, e.g., subjects with back pain rated moderate to severe on a 7-point scale. However, the baseline outcome is unique in that, being pre-randomization, it can safely be assumed not to depend on treatment group. The focus of the remainder of this section is on baseline values of the outcome; nevertheless, many of the key ideas concerning baseline response apply equally to other types of baseline variables. A more detailed discussion of some of the issues concerning adjustment for baseline response can be found in the textbook by Fitzmaurice, Laird and Ware (2011; Sections 5.6–5.7).

A question that naturally arises is how to handle the baseline measurement in the assessment of the effect of treatment. The issue is important since it will affect how we construct hypothesis tests, and also how they should be interpreted. In addition, how we handle the baseline response in the analysis will have an impact on efficiency and the power of tests of hypotheses. In the following we begin by considering the case where the outcome is assumed to be continuous rather than categorical; we do so because it is more straightforward to demonstrate some of the potential benefits of alternative approaches to analyses with this type of outcome. Later we will see that many, but not all, of the key ideas presented here are the same for binary (and ordinal) outcomes.

To fix ideas, suppose that in a randomized trial comparing two treatment groups, say active treatment versus control, a baseline and an end of follow-up assessment of the outcome variable are available, denoted by Y_0 and Y_1 respectively (note, to simplify notation, we suppress the subscript i for subjects); methods for the analysis of two or more follow-up measures of the outcome are discussed in Fitzmaurice, Laird, and Ware (2011; Chapters 12–16). Conditional on an indicator of treatment group, X, we assume that Y_0 and Y_1 have a bivariate normal distribution, with mean vector $(\mu_0, \mu_1 + \delta X)$, and with a "compound symmetry" covariance matrix determined by the two parameters, $\sigma^2 = \mathrm{Var}(Y_0) = \mathrm{Var}(Y_1)$ and $\rho = \mathrm{Corr}(Y_0, Y_1)$. Here, δ is the treatment effect when comparing the mean of Y_1 in the active treatment group to the control. In this setting, there are two obvious ways to proceed with the analysis. The first is to simply ignore the baseline response, Y_0, and conduct inference about δ based only on the follow-up assessment of the outcome Y_1. This analysis can be expressed in terms of the following regression model for Y_1 given X,

$$E(Y_1|X) = \beta_0 + \beta_1 X. \tag{2.17}$$

Here, the treatment effect β_1 is defined *unconditionally*. The ordinary least squares (OLS) estimate of β_1 provides an unbiased estimate of the treatment effect, δ, where "unbiased" in this context means over repeated replications of the randomized experiment. Testing the null hypothesis, H_0: $\delta = 0$ is equivalent to testing the null hypothesis H_0: $\beta_1 = 0$; this is simply the standard t-test for two independent groups. A second approach to the analysis is to define the treatment effect *conditionally* by adjusting the treatment comparison of Y_1 for the baseline response. This leads to the following regression model for Y_1 given both X and Y_0,

$$E(Y_1|X, Y_0) = \alpha_0 + \alpha_1 X + \alpha_2 Y_0; \tag{2.18}$$

this linear regression model is often referred to as an analysis of covariance (ANCOVA). Because of randomization, X and Y_0 are independent and the OLS estimate of α_1 also provides an unbiased estimate of the treatment effect, δ. Testing the null hypothesis, H_0: $\delta = 0$ is equivalent to testing the null hypothesis H_0: $\alpha_1 = 0$. There is an alternative method of adjusting for baseline response that leads to a third approach to the analysis. This third approach is based on the simple change or difference score, $D = Y_1 - Y_0$. This leads to the following regression model for D given X,

$$E(D|X) = E(Y_1 - Y_0|X) = \gamma_0 + \gamma_1 X. \tag{2.19}$$

This model can be derived as follows,

$$E(Y_1 - Y_0|X) = E(Y_1|X) - E(Y_0|X) = \beta_0 + \beta_1 X - \mu_0 \, ,$$

where $E(Y_0|X) = \mu_0$, independent of X because of randomization, so that $\gamma_0 = \beta_0 - \mu_0$, and $\gamma_1 = \beta_1$. Thus, the OLS estimate of γ_1 provides an unbiased estimate of the treatment effect, δ. Testing the null hypothesis, H_0: $\delta = 0$ is equivalent to testing the null hypothesis H_0: $\gamma_1 = 0$; once again, this is simply the standard two independent groups t-test of the difference score, D. Finally, an approach sometimes used in practice is to apply an ANCOVA model to the difference score, D, adjusting the treatment comparison of D for the baseline response Y_0. However, it can be shown (e.g., see Fitzmaurice, Laird and Ware, 2011; Section 5.7) that this approach is equivalent to the second approach. That is, ANCOVA of Y_1 and

D yield identical estimates of the treatment effect; the estimated slope for Y_0 from the ANCOVA of Y_1 is simply one unit larger than the slope from the ANCOVA of D,

$$E(D|X, Y_0) = \alpha_0 + \alpha_1 X + (\alpha_2 - 1)Y_0.$$

Note that all three approaches to analyses yield unbiased estimates of the treatment effect δ and valid tests of the null hypothesis, H_0: $\delta = 0$. That is, the three approaches represent alternative tests of the same null hypothesis but differ in terms of efficiency and the power of the test of H_0: $\delta = 0$. Specifically, the test based on the ANCOVA approach in (2.18) will always be more powerful. The ANCOVA approach yields an estimate of the treatment effect, $\widehat{\alpha}_1$, with smaller standard errors than those obtained from an analysis of Y_1 alone ($\widehat{\beta}_1$) or of the difference score, $D = Y_1 - Y_0$ ($\widehat{\gamma}_1$). For example, the greater efficiency of ANCOVA can be highlighted by examining the asymptotic relative efficiency (ARE) of (2.17) and (2.19) to (2.18). The ARE is defined as the ratio of the variance of the estimator based on (2.18) to the variance of the estimator based on either (2.17) or (2.19); a relative efficiency less than one implies that the ANCOVA estimator based on (2.18) has smaller variance. Under the assumption of a compound symmetry covariance, the ARE of $\widehat{\beta}_1$ to $\widehat{\alpha}_1$ is given by

$$\mathrm{ARE}(\widehat{\beta}_1 \text{ to } \widehat{\alpha}_1) = \frac{\mathrm{Var}(Y_1|X, Y_0)}{\mathrm{Var}(Y_1|X)} = \frac{\sigma^2(1 - \rho^2)}{\sigma^2} = \left(1 - \rho^2\right) \leq 1; \tag{2.20}$$

where $\rho = \mathrm{Corr}(Y_0, Y_1)$. Recall that the proportion of variability accounted for by linear regression on the baseline response is equal to the square of the correlation between Y_1 and Y_0. This simple expression indicates that the two methods of analysis are equally efficient only when $\rho = 0$. When $\rho > 0$, the greater efficiency of ANCOVA depends on the strength of the correlation between Y_0 and Y_1. For example, when $\rho = 0.7$, ANCOVA is approximately twice as efficient as an analysis of Y_1 without any adjustment for baseline response. Also, we note that in practice, a correlation as high as 0.7 is quite plausible for the same outcome variable measured at baseline and at end of follow-up.

Comparing the efficiency of the two alternative approaches for adjusting for baseline response, the ARE of $\widehat{\gamma}_1$ to $\widehat{\alpha}_1$ is given by

$$\mathrm{ARE}(\widehat{\gamma}_1 \text{ to } \widehat{\alpha}_1) = \frac{\mathrm{Var}(Y_1|X, Y_0)}{\mathrm{Var}(D|X)} = \frac{\sigma^2(1 - \rho^2)}{2\,\sigma^2(1 - \rho)} = \frac{1}{2}\left(1 + \rho\right). \tag{2.21}$$

This simple expression indicates that the two methods of adjustment for baseline response are equally efficient only when $\rho = 1$. When $\rho = 0$, ANCOVA is approximately twice as efficient as subtracting the baseline response. When $\rho = 0.5$, the analysis based on the difference score is only $\frac{3}{4}$ as efficient as ANCOVA.

Thus, for randomized trials with a continuous outcome, ANCOVA is the preferred method of analysis yielding the more efficient estimate of the treatment effect and the more powerful test of H_0: $\delta = 0$. Although the derivations above were for a continuous rather than a categorical outcomes, and under the simple assumption of compound symmetry covariance (see Yang and Tsiatis (2001) for efficiency results under more general conditions for the covariance), similar conclusions can be drawn about the benefits of adjusting for baseline response as an additional covariate in regression analyses of categorical outcomes. Before discussing adjustment for baseline response in logistic regression of a binary outcome, we mention that our discussion above has important implications for the design of randomized clinical trials. Because the ANCOVA approach can potentially lead to a substantial reduction in the variability of the follow-up assessment of the outcome, i.e., $\mathrm{Var}(Y_1|X, Y_0) << \mathrm{Var}(Y_1|X)$, and consequently a more powerful test of treatment group differences, a clinical trial requiring far fewer subjects can be conducted when a baseline assessment of the outcome is included in the design of the trial. In particular, for the baseline adjusted analysis to achieve the same statistical power as an unadjusted analysis, the required sample size is reduced proportionately by $(1 - \rho^2)$. Thus, with a heterogeneous trial population and relatively modest within-subject variability in the outcome, correlations of 0.7 or higher are quite common and the required number of trial subjects can be approximately halved.

So far, we have seen the desirable consequences of adjustment for baseline response in linear regression for a continuous outcome. However, for logistic regression with a binary outcome (and logistic regression models for ordinal outcomes) the statistical properties of adjustment for baseline response are somewhat different. In particular, the baseline outcome-adjusted estimate of the treatment effect has *larger*, not smaller, standard errors than those obtained from an unadjusted analysis. In addition, the comparison of approaches is complicated by the fact that the baseline outcome-adjusted estimate of the treatment effect odds ratio differs from the unadjusted odds ratio. To better understand these differences, consider the same clinical trial scenario as described above except let Y_0 and Y_1 denote baseline and end of follow-up assessments of a *binary* outcome variable. Their joint distribution is multinomial with variances determined by the means, $\text{Var}(Y_0) = \mu_0(1 - \mu_0)$ and $\text{Var}(Y_1) = (\mu_1 + \delta X)(1 - \mu_1 - \delta X)$, and $\rho = \text{Corr}(Y_0, Y_1)$ (where $\mu_0 = \Pr(Y_0 = 1 | X = 0) = \Pr(Y_0 = 1 | X = 1)$ and $\mu_1 = \Pr(Y_1 = 1 | X = 0)$). As before, we consider the following *unconditional* logistic regression analysis,

$$\text{logit}[\Pr(Y_1 = 1 | X)] = \beta_0 + \beta_1 X. \tag{2.22}$$

Here, the treatment effect β_1 is defined *unconditionally*. The ML estimate of β_1 provides an unbiased estimate of the treatment effect, δ, and testing the null hypothesis, $H_0: \delta = 0$ is equivalent to testing the null hypothesis $H_0: \beta_1 = 0$. A second approach is to consider a *conditional* analysis based on the following logistic regression,

$$\text{logit}[\Pr(Y_1 = 1 | X, Y_0)] = \alpha_0 + \alpha_1 X + \alpha_2 Y_0. \tag{2.23}$$

In general, the conditional log odds ratio for treatment, α_1, differs from the unconditional log odds ratio for treatment, β_1. This is due to the "non-collapsibility" of the odds ratio (Gail et al., 1984; Neuhaus and Jewell, 1993). Indeed, the independence of X and Y_0 guaranteed by randomization is not sufficient to ensure that $\beta_1 = \alpha_1$, as was the case in linear regression; instead it would require the conditional independence of X and Y_0 given Y_1. Because $\beta_1 \neq \alpha_1$ somewhat greater care is required in the interpretation of the estimated treatment effect. Specifically, $\exp(\widehat{\beta}_1)$ is an estimate of the *unconditional* odds ratio for treatment which is the treatment effect averaged across the population. In contrast, $\exp(\widehat{\alpha}_1)$ is an estimate of the *conditional* odds ratio for treatment which is a relative comparison of the odds conditional on the subset of the population with $Y_0 = y_0$. In the special case of linear models for continuous outcomes, $\beta_1 = \alpha_1 = \delta$ when the ANCOVA model holds; moreover, $\widehat{\alpha}_1$ is a consistent estimator of δ even when the functional relationship between Y_1 and Y_0 has been misspecified. As a result, this distinction between conditional and unconditional treatment effects is not important in linear models. However, for logistic regression, due to the non-linear link function, $\beta_1 \neq \alpha_1$ and the targets of inference (conditional versus unconditional) differ; as an aside, we note that Tsiatis et al (2008) have described an interesting semiparametric approach for incorporating baseline response to improve the efficiency of estimators of the unconditional treatment effect β_1. Although in general $\beta_1 \neq \alpha_1$, for logisitic regression it can be shown that $|\beta_1| \leq |\alpha_1|$. Also, when $\alpha_1 = 0$, it can be shown that $\beta_1 = 0$, and thus the null hypothesis of no treatment effect can also be expressed as $H_0: \alpha_1 = \beta_1 = \delta = 0$. That is, the conditional and unconditional logistic regression analyses represent alternative tests of the same null hypothesis of no treatment effect.

Next we compare the conditional and unconditional logistic regression analyses in terms of efficiency and the power of the test of $H_0\colon \delta = 0$. Unlike in linear regression, the asymptotic relative efficiency of $\widehat{\beta}_1$ to $\widehat{\alpha}_1$ is *greater* than, not less than, 1. That is, for logistic regression,

$$\text{ARE}(\widehat{\beta}_1 \text{ to } \widehat{\alpha}_1) \geq 1; \tag{2.24}$$

consequently, adjustment for baseline response in logistic regression yields *less*, not more, precise estimates of the treatment effect. Given that $\text{ARE}(\widehat{\beta}_1 \text{ to } \widehat{\alpha}_1) \geq 1$, it is tempting to draw the conclusion that the test of $H_0\colon \alpha_1 = \beta_1 = 0$ based on $\widehat{\beta}_1$ is more powerful. However, this ignores that the baseline outcome-adjusted estimate of the treatment effect odds ratio is further from the null, since $|\beta_1| \leq |\alpha_1|$. That is, although the standard error of $\widehat{\alpha}_1$ may be larger, the point estimate $\widehat{\alpha}_1$ will also tend to be larger than $\widehat{\beta}_1$. To determine which of these two tests, the one based on $\widehat{\alpha}_1$ or the one based on $\widehat{\beta}_1$, is more powerful, Robinson and Jewell (1991) compared the asymptotic relative efficiency of the two tests and found that

$$\text{ARE}(\widehat{\beta}_1 \text{ to } \widehat{\alpha}_1, \text{ at } \alpha_1 = 0) \leq 1. \tag{2.25}$$

As a result, when testing the null hypothesis of no treatment effect it is always as or more efficient to adjust for baseline response in the logistic regression analysis (and also in logistic regression models for ordinal outcomes; Neuhaus, 1998). In this regard, the logistic and linear regression models have similar statistical properties.

In summary, efficiency results for logistic regression differ in some important ways when compared to linear regression. Nonetheless, as with ANCOVA for a quantitative outcome, logistic regression with the inclusion of baseline outcome as a covariate increases the power of the test of treatment effect. This has important implications for the design of randomized clinical trials since discernibly fewer subjects are required for the baseline adjusted analysis to achieve the same statistical power as an unadjusted analysis. Another potential benefit of adjustment for baseline response it that it corrects for any baseline imbalances in the outcome between treatment groups due to chance. When the randomization process has worked correctly, a discernible baseline imbalance is not expected *a priori*; any observed imbalance is simply a chance phenomenon, but one whose likelihood is higher in trials with a relatively small sample size. As such, randomisation guarantees balance in expectation, but not balance in any given clinical trial. Adjustment for baseline response will correct for chance baseline imbalance in the outcome. However, it is important to remember that inclusion of baseline response increases power of the test of the treatment effect even when there is perfect balance.

2.8 Concluding Remarks

This chapter presents an overview of statistical methods commonly used for analysis of categorical outcomes, primarily binary and ordinal outcomes. In clinical trials, the main focus is on estimation and testing of treatment effects. For certain clinical trials, contingency table methods are adequate and provide simple and intuitive ways of examining the effect of treatment on a categorical outcome. Logistic regression models for binary and ordinal outcomes offer many of the same advantages of contingency table methods but also provides an efficient method for incorporating an arbitrary number of categorical and quantitative covariates in the analysis. These model-based methods are especially appealing when there are important features of the trial design (e.g., blocking factors, trial sites) and baseline characteristics of subjects that need to be incorporated in the analysis. Regression adjustment for baseline variables can be used to correct the treatment comparison for chance baseline imbalance. When baseline values of the outcome are available, regression adjustment for baseline response will generally yield a discernibly more powerful test of treatment effect.

Bibliography

Agresti, A. (2013). *Categorical Data Analysis, 3rd Edition*. New Jersey: Wiley.

Agresti, A. (2010). *Analysis of Ordinal Categorical Data, 2nd Edition*. New Jersey: Wiley.

Anderson, J.A. and Philips, P.R. (1981). Regression, discrimination and measurement models for ordered categorical variables. *Applied Statistics*, **30**, 22-31.

Armitage, P. (1955). Tests for linear trends in proportions and frequencies. *Biometrics*, **11**, 375–386.

Bombardier, C., Ware, J., Russell, I.J., Larson, M., Chalmers, A. and Read, J.L. (1986). Auranofin therapy and quality of life in patients with rheumatoid arthritis. Results of a multicenter trial. *American Journal of Medicine*, **81**, 565–578.

Breslow, N.E. and Day, N.E. (1980). *Statistical Methods in Cancer Research (Vol. 1): The Analysis of Case-Control Studies*. Lyon: World Health Organization.

Cochran, W.G. (1954). Some methods of strengthening the common χ^2 tests. *Biometrics*, **10**, 417–451.

Davis, C.S. (1991). Semi-parametric and non-parametric methods for the analysis of repeated measurements with applications to clinical trials. *Statistics in Medicine*, **10**, 1959–1980.

Farewell, V.T. (1979). Some results on the estimation of logistic models based on retrospective data. *Biometrika*, **66**, 27–32.

Fisher, R.A. (1934). *Statistical Methods for Research Workers*. Edinburgh: Oliver and Boyd.

Fitzmaurice, G.M., Laird, N.M. and Ware, J.H. (2011). *Applied Longitudinal Analysis, 2nd Edition*. New Jersey: Wiley.

Gail, M.H., Wieand, S. and Piantadosi, S. (1984). Biased estimates of treatment effect in randomized experiments with nonlinear regressions and omitted covariates. *Biometrika*, **71**, 431–444.

Grizzle, J.E., Starmer, C.F., and Koch, G.G. (1969). Analysis of Categorical Data by Linear Models. *Biometrics*, **25**, 489–504.

Hosmer, D.W. and Lemeshow, S. (2000). *Applied Logistic Regression, Second Edition*. New York: Wiley.

Lipsitz, S.R. (1992). Methods for estimating the parameters of a linear model for ordered categorical data. *Biometrics*, **48**, 271–281.

Mann, H.B. and Whitney, D.R. (1947). On a test of whether one of two random variables is stochastically larger than the other. *Annals of Mathematical Statistics*, **1**, 50–60.

Mantel, N. and Haenszel, W. (1959). Statistical aspects of the analysis of data from retrospective studies of disease. *J. Natl. Cancer Inst.*, **22**, 719–748.

McCullagh, P. (1980). Regression models for ordinal data. *Journal of the Royal Statistical Society B*, **43**, 109–142.

McCullagh, P. and Nelder, J. (1989). *Generalized Linear Models*. 2nd edition. London: Chapman & Hall.

Mehta, C.R. and Patel, N.R. (1995). Exact logistic regression: theory and examples. *Statistics in Medicine*, **14**, 2143-2160.

Neuhaus, J.M. (1998). Estimation efficiency with omitted covariates in generalized linear models. *Journal of the American Statistical Association*, **93**, 1124–1129.

Neuhaus, J.M. and Jewell, N.P. (1993). A geometric approach to assess bias due to omitted covariates in generalized linear models. *Biometrika*, **80**, 807–815.

Peterson, B. and Harrell, F.E. (1990). Partial proportional odds models for ordinal response variables. *Applied Statistics*, **39**, 205–217.

Prentice, R.L. and Pyke, R. (1979). Logistic disease incidence models and case-control studies. *Biometrika*, **66**, 403–412.

Robinson, L.D. and Jewell, N.P. (1991). Some surprising results about covariate adjustment in logistic regression models. *International Statistical Review*, **59**, 227–240.

Stokes, M.E., Davis, C.S., and Koch, G.G. (2000). *Categorical Data Using the SAS System, Second Edition*. Cary, NC: SAS Institute Inc.

Tsiatis, A., Davidian, M., Zhang, M. and Lu, X. (2008). Covariate adjustment for two-sample treatment comparisons in randomized clinical trials: a principled yet flexible approach. *Statistics in Medicine*, **27**, 4658–4677.

Whitehead, J. (1993). Sample size calculations for ordered categorical data. *Statistics in Medicine*, **12**, 2257–2271.

Wilcoxon, F. (1945). Individual comparisons by ranking methods. *Biometrics Bulletin*, **1**, 80–83.

Yang, L. and Tsiatis, A. (2001). Efficiency study of estimators for a treatment effect in a pretest-posttest trial. *The American Statistician*, **55**, 314–321.

3

Continuous Outcomes

Fang-Shu Ou

CONTENTS

3.1 Introduction

This chapter is devoted to analysis methods when the response variable is continuous. Continuous response variables are common in medical studies, for example, body mass index in a weight loss study, cholesterol levels in a heart disease study, or pain scale in a eye surgery study. These methods which can be used to analyze continuous response variables can be categorized based on the modeling assumptions one makes. In a randomized control trial, the analysis of primary endpoint is typically a comparison across different treatment arms. For secondary endpoint or other exploratory analysis, an analysis done to better understand or compare features of individual arms is also possible. To choose the appropriate analysis method, one must determine whether a sample is from a single population, or should be divided into two or more separate groups based on experimental arms. If any covariates are available, one may like to determine if and how they affect the response. To better understand these methods, it is helpful to understand the underlying hypothesis test one is attempting to conduct and the assumptions that are required for each of the tests.

Many of these methods rely on observing normally distributed data with an unknown variance, and use a t-statistic for constructing p-values and confidence intervals. If one observes data that takes the basic shape of a normal distribution, then these methods can be used with little cause for concern even with small sample sizes. If the data does not appear to be normal, the central limit theorem holds for larger samples and the test becomes a z-test. Trusting the inferential results for small samples would not be prudent when the underlying data does not appear to be normal. In fact, when the data is highly skewed, which is a common trait for certain laboratory test results, alternative methods which do not rely on normal assumption may be more appropriate even for fairly large studies.

3.2 The t-Test (One Population)

As mentioned before, in a randomized control trial, comparisons are commonly performed across different treatment arms. The t-test for one-population tests will be helpful for secondary analysis, such as determining whether a control group in the current trial agrees with previously established results, and aid in understanding the t-test in a two-sample setting. One of the most common goals of statistical analysis for continuously distributed response variables is to determine whether the mean of a population is equal to a certain value, μ_0, or not. As an example, a study may be conducted to determine whether a population's average weight has changed compared to a historical figure. This two-sided hypothesis test is

$$H_0 : \mu_Y = \mu_0 \text{ vs } H_a : \mu_Y \neq \mu_0. \tag{3.1}$$

If only one direction of change is of interest, for example, we would like to test the hypothesis that the current mean body weight is more than the historical figure, a one-sided hypothesis may be used,

$$H_0 : \mu_Y \leq \mu_0 \text{ vs } H_a : \mu_Y > \mu_0,$$

where $\leq$ and $>$ may be replaced by $\geq$ and $<$ if one is interested in a decrease.

If the underlying distribution for the population of the random variable Y is normally distributed[1], then the sample mean, $\bar{Y}$, of a simple random sample $y_1, \ldots, y_n$ would be distributed as

$$\bar{Y} = \frac{1}{n}\sum_i y_i \sim N(\mu_Y, \sigma_Y^2/n),$$

where σ_Y is the standard deviation of the underlying random variable Y. Since it is unlikely that the actual standard deviation σ_Y is known it is necessary to estimate it through

$$\hat{\sigma}_Y = \sqrt{\frac{1}{n-1}\sum_i (y_i - \bar{Y})^2}. \tag{3.2}$$

The t-statistic from a sample of size n for Hypothesis (3.1) would be

$$t_{n-1} = \frac{\sqrt{n}(\bar{Y} - \mu_0)}{\hat{\sigma}_Y},$$

where t_k follows a t distribution with k degrees of freedom (df). As the df increase, the t distribution converges to a standard normal distribution as would be expected from the central limit theorem for the behavior of sample means. The t-statistic can be used to find a p-value, p, to be compared against a test of size α, for the selected hypothesis test through

$$p = 2P(t > |t_{n-1}|), \quad p = P(t > t_{n-1}) \quad \text{or} \quad p = P(t < t_{n-1}), \tag{3.3}$$

---

[1] As noted in the chapter introduction, it is not necessary to assume that the data is drawn from a normal distribution. If the sample size is large enough, then the t-test is, in effect, a Z-test and the central limit theorem for means validates the test.

depending on whether H_a is $\mu_Y \neq \mu_0$, $\mu_Y > \mu_0$ or $\mu_Y < \mu_0$ respectively. In addition to a hypothesis test, two-sided confidence intervals can be constructed as

$$\bar{Y} \pm t_{n-1,\alpha/2} \frac{\hat{\sigma}_Y}{\sqrt{n}},$$

where $t_{k,\alpha/2}$ is selected as the value which satisfies $P(t > t_{k,\alpha/2}) = \alpha/2$. This interval may be interpreted as the set of μ_0 which could not be rejected through a two-sided hypothesis test at the significance level α.

A cholesterol example

Suppose we want to test the two-sided hypothesis $H_0 : \mu = 205$ for the cholesterol levels of a high risk population and we observe the values in Treatment A of Table 3.1. The point estimator for the mean is $\bar{Y} = 211.7$ and the standard deviation is 10.62. The t-statistic would be

$$t_{11} = \frac{211.7 - 205}{10.62/\sqrt{12}} = 2.17.$$

The two-sided p-value for $t_{11} = 2.17$ with 11 *df* is 0.0528, so at the $\alpha = .05$ significance level, we cannot conclude that the population cholesterol level is significantly different from 205. On the other hand, if our alternative hypothesis had been that the cholesterol levels are greater than 205 (i.e., a one-sided test with alternative $H_A : \mu > 205$), the p-value would have been .0264, and we could conclude that the cholesterol levels of the group are significantly greater than 205. The $\alpha = .05$ confidence interval for the mean cholesterol level would be $211.7 \pm 2.200985 * 10.62/\sqrt{12} = (204.95, 218.45)$, which covers the value 205 as expected by the two sided p-value .0528.

3.3 The t-Test (Two Populations)

In a randomized clinical trial, the hypothesis testings are commonly done across different treatment arms. Suppose one collects a sample of subjects with a given criteria then

TABLE 3.1
Cholesterol Levels (mg/dL)

Treatment A	Treatment B
215	218
212	207
200	192
218	217
215	212
206	196
191	193
215	203
224	223
226	220
219	214
199	194

randomly assigns each subject to either Treatment A $(x_1, \ldots, x_{n_1})$ or Treatment B $(y_1, \ldots, y_{n_2})$, usually designed to have $n_1 = n_2$ (i.e., a one-to-one (1:1) randomization). Often through natural erosion, $n_1 \neq n_2$ even if the design is intended to be balanced because, among other issues, treatments fail to be applied to specification, subjects drop out, or even imputation errors occur. We can consider patients randomized to Treatment A as being from population X and patients randomized to Treatment B as being from population Y. A fundamental statistical analysis is to determine whether the mean from X, μ_X, is the same as the mean from Y, μ_Y. The two-sided hypothesis would be

$$H_0 : \mu_Y - \mu_X = 0 \text{ vs } H_a : \mu_Y - \mu_X \neq 0. \tag{3.4}$$

This hypothesis is used to capture either a positive effect or a negative effect of a treatment. Unlike the one-sample t-test, the exact value of μ_X or μ_Y is not of direct interest, only the difference between the population means. As with the single population scenario, one may choose a one-sided test if only one direction is of interest.

If each population is normally distributed, then the t-statistic would be

$$t_{df} = \frac{\bar{Y} - \bar{X}}{s},$$

where s is an estimate for the standard deviation of $\bar{Y} - \bar{X}$. The hypothesis test p-values are calculated for t_{df} identically to Equation (3.3), and size α confidence intervals are constructed as

$$(\bar{Y} - \bar{X}) \pm t_{df, \alpha/2} \cdot s.$$

The main complication is deciding how s and df should be estimated. If one supposes that under H_0, the two samples should show identical results, hence have identical standard deviations because there is no real difference between the groups, one may pool the standard deviation estimators through

$$s_p^2 = \frac{(n_x - 1)s_x^2 + (n_y - 1)s_y^2}{n_x + n_y - 2}, \tag{3.5}$$

where s_x and s_y are the estimated sample standard deviations for X or Y, respectively, defined as $s_x = \hat{\sigma}_X / \sqrt{n}$ and $s_y = \hat{\sigma}_Y / \sqrt{n}$. n_x and n_y are the number of observations for X or Y, respectively. The resulting t-statistic would have df equal to $n_x + n_y - 2$. On the other hand, if one does not expect that the two populations have the same standard deviation, an estimate for the standard deviation can be calculated using the Welch method (7)

$$s_W^2 = \frac{s_x^2}{n_x} + \frac{s_y^2}{n_y}. \tag{3.6}$$

Unfortunately, the calculation for df is much more complicated than that for the pooled estimator and is calculated as

$$df_W = \frac{\left[\frac{s_x^2}{n_x} + \frac{s_y^2}{n_y}\right]^2}{\left(\frac{s_x^2}{n_x}\right)^2 \frac{1}{n_x - 1} + \left(\frac{s_y^2}{n_y}\right)^2 \frac{1}{n_y - 1}}.$$

When the sample sizes in the groups is large, then the t-statistic is essentially the same as a normal random variable, so the difference in df will be of little consequence. In a randomized control trial, since patients are randomly assigned to the different treatment arms, it is typically reasonable to assume that the sample standard deviations are the same for different treatment arms (under H_0); therefore, the pooled standard deviation, Equation (3.5) can be used.

A cholesterol example continued

Going back to the example comparing cholesterol levels, the two treatment groups in Table 3.1 may be compared using either method of calculation for the standard deviation. The t-statistic using either s_p or s_W is .94223. The difference between the degrees of freedom is 22 for the pooled method and 21.873 for the Welch method resulting in two-sided p-values of .3564 and .3563 for the pooled and Welch method, respectively. Neither two-sample t-test method shows any evidence for a mean difference in cholesterol levels between Treatment A and Treatment B.

3.4 Mann-Whitney U-Test

The Mann-Whitney U-test (12) is a nonparametric analysis method meant to compare samples from two populations and relies on the hypothesis test,

$$H_0 : P(Y < X) = P(X < Y) \text{ vs } H_a : P(Y < X) \neq P(X < Y). \tag{3.7}$$

Hypothesis (3.7) determines whether it is equally likely that a randomly selected value from one sample, say X, will be less than or greater than a randomly selected value from a second sample, say Y^2. This hypothesis test is very different than the test for differences between means, and the two are not generally comparable. Consider the following scenario:

- Population 1 (X) has a point mass of .51 at 0 and .49 at 1 making the mean .49.

- Population 2 (Y) has a single point mass at δ.

If $\delta = .48$, then $\mu_X > \mu_Y$, yet $.51 = P(Y > X) > P(X < Y) = .49$. On the other hand when $\delta = .5$, then $\mu_Y > \mu_X$, however $.51 = P(Y > X) > P(X < Y) = .49$ is the same. Under stronger assumptions on the underlying population distributions, Hypothesis (3.7) may be identical to Hypothesis (3.4). For instance, if we suppose that the distribution of X is the same as $Y - \delta$ (one distribution is a shift of the other), then the two hypotheses coincide, $H_0 : \delta = 0$.

Deciding which hypothesis is better for trial design requires careful consideration. The hypothesis concerning means is of importance when one is concerned about the behavior of a large group of individuals; whereas the second hypothesis is concerned with whether a single subject is likely to do better in X or Y. While Hypothesis (3.4) would be of greater importance for most tests, Hypothesis (3.7) does not require a parametric assumption and allows for tests of any totally ordered response types where the mean may not be meaningful.

The Mann-Whitney U-test may be carried out using the following algorithm:

1. Combine all observed data from X and Y, and order them in ascending rank.

2. Assign a numerical value for each rank starting with 1. When there are ties, assign the midpoint value for each of the ties. Examples:

---

[2]The Mann-Whitney test is one of many rank-sum tests. A good review of rank-sum tests detailing the subtle differences between the underlying hypotheses can be found in (3).

(a) The ranks for (1,4,7,8,9,12) would be (1,2,3,4,5,6).

(b) The ranks for (1,4,4,4,7,7,8,9) would be (1,3,3,3,5.5,5.5,7,8).

3. Sum the ranks associated with samples from X and refer to this as R_X.

4. The observed sample statistic is $U = R_X - \frac{n_X(n_X - 1)}{2}$.

5. Compute p-value using a normal distribution with

$$Z = \frac{U - \mu_U}{\sigma_U}$$

where

$$\mu_U = \frac{n_1 n_2}{2} \quad \text{and} \quad \sigma_U = \sqrt{\frac{n_1 n_2}{12}\left(n + 1 - \sum_i \frac{t_i^3 - t_i}{n(n-1)}\right)},$$

where $n = n_1 + n_2$, and t_i is the number of observations at i-th rank.

Unlike the t-tests, this version of the Mann-Whitney U-test uses the central limit theorem to establish the approximate normality of the test statistic U. This means that it is unsuitable for very small sample sizes. For small sample sizes, the distribution of the U statistic has been tabulated within many software platforms.

There are several advantages for using the Mann-Whitney U-test over the t-test. The Mann-Whitney test is much more robust to outliers than the t-test. A single outlier can change the value of $\bar{X} - \bar{Y}$ drastically, whereas the rank of an observation cannot drop below 1, or exceed $n_1 + n_2$. The Mann-Whitney test is also more versatile since it only needs the order of observations. If one is comparing the results of a survey and individuals rate a feature, say pain level, on a scale of 1–10, it is difficult to determine whether the distance between 1 and 2 is the same as the distance between 7 and 8. If the differences between values is not consistent, then performing tests based on means (t-tests) may not provide as good of a summary as a comparison using $P(Y < X)$ (Mann-Whitney U-test). The robustness of the Mann–Whitney U-test does come with a price. Since the Mann-Whitney U-test does not require parametric assumptions, it is slightly less powerful than the t-test and tends to wider confidence intervals. The efficiency loss compared to the t-test is not drastic, it is in the range of 5% when the underlying distribution is normally distributed (4). However, if the underlying distribution is far from normal, the Mann-Whitney test may actually become more efficient than the t-test. It is of the utmost importance to consider the type of endpoint a trial will be using and the nature of the underlying distribution before choosing the best test in the design stage(11).

A cholesterol example continued
The Mann-Whitney U-test for the cholesterol results of Treatment A and Treatment B in Table 3.1 gives a test statistic of $U = 57$, $\mu_U = 78$, and $\sigma_U = 17.3$ which leads to a p-value of .386. This is only slightly larger than the p-value obtained through the t-test.

3.5   Paired Tests

Returning to the example of the two treatments in Table 3.1, assume now that we only enrolled identical twins for the study and randomized the twins to either Treatment A or

Treatment B. Under this scenario the rows of Table 3.1 have a natural pairing associated with them, as it is reasonable to assume that identical twins are similar to each other. The twin pair should be looked at as a single experimental unit rather than 2 separate units. This data structure is termed paired data, and specialized analyses methods have been derived to accommodate pairing.

3.5.1 Paired t-test

The paired t-test is performed by taking the difference between pairs and performing a one-sample t-test on the differences. The paired t-test allows one to remove the variability based on the wide difference between pairs in the population so that one may magnify the change in response due solely to a treatment effect. In other words, one is able to mitigate the effect of confounding variables which are possessed by each pair. When one designs a study, they may manufacture paired data by matching subjects based on observable traits then assigning treatments randomly to each set of pairs.

A cholesterol example continued
For the cholesterol example, the differences (Treatment A – Treatment B) in each twin pair are $(-3, 5, 8, 1, 3, 10, -2, 12, 1, 6, 5, 5)$. These values have a mean of 4.25, and a standard deviation of 4.53 creating a t statistic with 11 df equal to 3.25. The two-sided p-value for this is .0077 which indicates a very strong likelihood that there has been a decrease in cholesterol levels due to Treatment B compared to Treatment A. This result is quite different compared to the 2-sample t-test which does not take into account the pairing between subjects in Treatment A and B.

3.5.2 Wilcoxon signed rank test

A nonparametric alternative to the paired t-test, which takes into account the pairing of data, is the Wilcoxon signed rank test. The Wilcoxon signed rank test is similar to the Mann-Whitney test in that it also uses hypothesis (3.7). As this test relies on the central limit theorem, it is not recommended when one possesses less than 10 pairs of data without tied values. The method is implemented through the following algorithm:

1. For each pair of data $(i = 1, \ldots, n)$, set $y_i = x_{1,i} - x_{2,i}$. Remove any observation where $y_i = 0$ to create a reduced set of variables $\tilde{y}_i$ where $i = 1, \ldots, N_r$.

2. Order $|y_i|$ then assign each a rank R_i as performed for the Mann-Whitney test.

3. The test statistic will be
$$W = \sum_{i=1}^{N_r} \text{sign}(y_i) R_i.$$

4. For values of $N_r \geq 10$, W approximately follows a normal distribution with mean 0 and standard deviation σ_W (i.e., $N(0, \sigma_W)$), where
$$\sigma_W = \sqrt{\frac{N_r(N_r + 1)(2N_r + 1)}{6}}$$

Generally speaking, results from nonparametric tests tend to be more conservative than parametric tests. The major benefit of nonparametric methods are a reduced amount of assumptions on population distributions and ability to provide inference when the distribution of a test statistic cannot be determined.

A cholesterol example continued

The results for the cholesterol test indicate that $W = 63$, and $\sigma_W = 25.5$. These values indicate that the two sided p-value would be 0.013 meaning that, at the $\alpha = .05$ significance level, we may conclude that application of the Treatment B is likely to decrease an individual's cholesterol level compared to Treatment A. This result is considerably more statistically significant than the result from the Mann-Whitney U-test; however, it is not quite as strong as the t-test p-value.

3.6   Multiple Comparisons

In some clinical trials, there are more than 2 treatment arms being tested, say k arms. When looking at a comparison of means, a hypothesis test of interest could be

$$H_0 : \mu_1 = \mu_2 = \cdots \mu_k \text{ vs } H_a : \mu_i \neq \mu_j \text{ for at least one set } i, j \in \{1, \ldots, k\}. \qquad (3.8)$$

Under these circumstances, one can perform $k(k-1)/2$ simple hypothesis tests of the form in Equation (3.4). If one were to perform each individual test with a significance level α, the chance that the null hypothesis would be falsely rejected would increase. If each of the individual hypothesis tests were independent, and the size of the individual test is α, the Type I error rate would be

$$1 - (1 - \alpha)^{k(k-1)/2} \geq \alpha. \qquad (3.9)$$

As an example, if $k = 4$, and $\alpha = .05$, the actual Type 1 error rate of independent tests would be 26.5% which exceeds α.

There are a number of methods to correct the nominal α used for each individual test to control the overall study error rate, also termed the Family Wise Error Rate (FWER). The most conservative means to correct the nominal α to the FWER is the Bonferroni method[3],

$$\alpha_{B,m} = \frac{\alpha}{m},$$

where m is the number of hypothesis tests being performed. (If all comparisons were tested, $m = k(k-1)/2$.) Regardless of dependence among the individual tests, the FWER would be

$$FWER \leq \sum_{i=1}^{m} \frac{\alpha}{m} = \alpha.$$

This is a very conservative correction, hence, a rejection of H_0 would be a very good indicator that an effect has been observed. This severe penalty is required when the individual tests are correlated in such a way that no two rejection regions overlap.

There are many more approaches available to control FWER. A good overview of these can be found in (2).

---

[3]The Bonferroni correction assumes that each comparison is of equal importance. If some tests are of greater importance, one could partition α as seems fitting by using any desired nominal significance levels $(\alpha_1, \ldots, \alpha_m)$ such that $\alpha_i \geq 0$, and $\sum_i \alpha_i = \alpha$.

3.7 Regression

Suppose we have the weight (lbs), height (inches), and gender data of Table 3.2 for 12 toddlers aged between 2 and 5 years old and wish to understand the relationship between toddler's weight and height while taking into account gender differences. This is an instance where a regression method would be useful.

The continuous univariate response Y is a vector of weights with the i-th entry representing the weight of the i-th subject as presented in the first column of Table 3.2. The collection of covariates can be represented as a matrix X where the columns represent the height and gender. The first column vector of X (height) is continuous, while the second is a 2 level categorical variable for the gender of a subject. A single "dummy variable",

$$I(\text{Gender} = \text{Male}) = \begin{cases} 1 & \text{if Gender is Male} \\ 0 & \text{otherwise} \end{cases},$$

is used in place of the original gender variable. The i-th row of X, is a p-vector ($p = 2$) of covariates for a specific subject, and x_i will refer to the transpose (column vector).

The most common and basic model is to assume that weight relates to height and gender through a linear relationship. Using the notation for individual subjects and understanding that the response is unlikely to fall directly onto a line, the linear model may be written as

$$y_i = \beta_0 + \beta^t x_i + \epsilon_i, \tag{3.10}$$

where $\beta = [\beta_1; \beta_2]$ is a real valued $p = 2$ column vector of slopes with respect to the covariate values $x_i = [\text{Height}_i; \text{Male}_i]$, and ϵ_i are independent and identically distributed random variables with a mean of 0 and finite variance $\sigma_\epsilon^2 < \infty$. There is no need to include a dummy variable for females, because β_2 is interpreted as a shift in the linear model from the female baseline.

Figure 3.1 displays the data Y plotted against the first column of X from Table 3.2 with "M" and "F" for the gender term. The lines depicts a possible fit for Model (3.10). Notice that there is a separate line for each level of gender because the linear coefficient for the Male indicator acts as an offset for the intercept of the Male cohort.

TABLE 3.2
Toddler's weight and height data for linear regression

Weight (lbs.)	Height (inches)	Gender
24.72	31.61	M
20.88	30.44	F
25.34	34.71	F
35.03	38.43	M
26.41	31.95	F
28.32	38.37	F
24.20	29.23	M
22.60	32.93	F
28.43	32.68	M
28.48	33.43	M
35.66	39.82	M
25.84	32.22	F

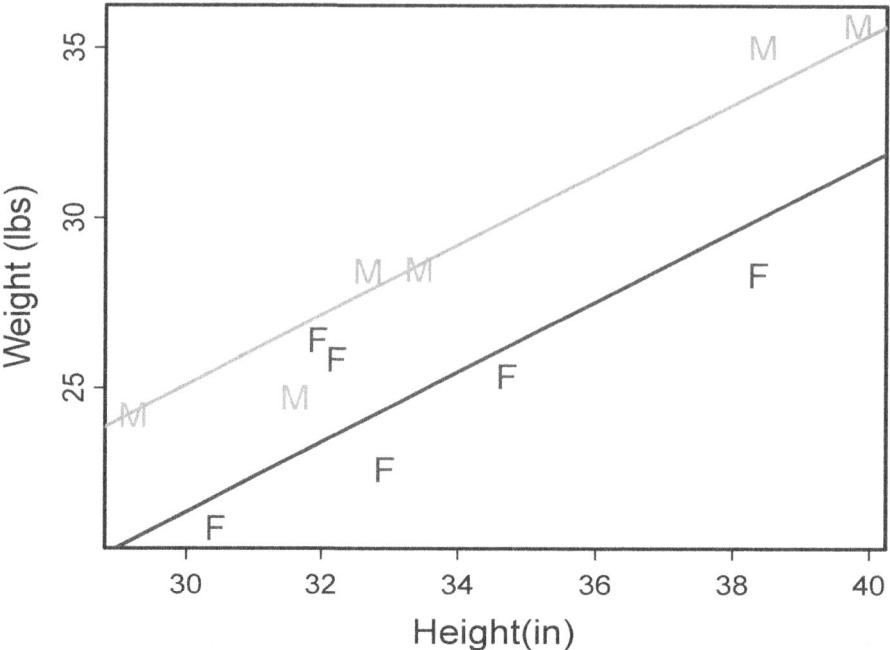

FIGURE 3.1
Relationship between toddler's weight (response variable) and covariates (height and gender)

The primary question for linear regression is how to find the "best" estimator for the linear coefficients β. The most commonly used estimator can be found by minimizing

$$\hat{\beta} = \arg\min_{\beta} \sum_{i} (y_i - \beta_0 + \beta^t x_i)^2, \qquad (3.11)$$

and the estimate $\hat{\beta}$ is termed the Least Squares Estimator (LSE). Luckily, the LSE, $\hat{\beta}$, can be estimated quite easily using matrix algebra. Let the vector β include the intercept β_0 as the first element in the $p+1-$dimensional column vector and append a column of 1's to X creating the $n \times (p+1)$ covariate matrix,

$$\tilde{X} = \begin{bmatrix} 1 & x_{1,1} & \cdots & x_{1,p} \\ \vdots & \vdots & \vdots & \vdots \\ 1 & x_{n,1} & \cdots & x_{n,p} \end{bmatrix}.$$

The solution of Equation (3.11) may be succinctly written as $\hat{\beta} = (\tilde{X}^t \tilde{X})^{-1} \tilde{X}^t Y$. While the derivation is beyond the scope of this introduction, an interested reader can find a more thorough presentation in (1).

3.7.1 Residuals

Once $\hat{\beta}$ is estimated from observed data, the expected value of the response y for an observation with covariate values x can be estimated as $\hat{y} = \hat{\beta}^t x$ since the expected value of the error term ϵ of Equation (3.10) is 0. Hence, the errors ϵ_i associated with subject i can be estimated as $\hat{\epsilon}_i = y_i - \hat{y}_i = y_i - \hat{\beta}^t x_i$. The underlying standard deviation, σ_ϵ, of data about

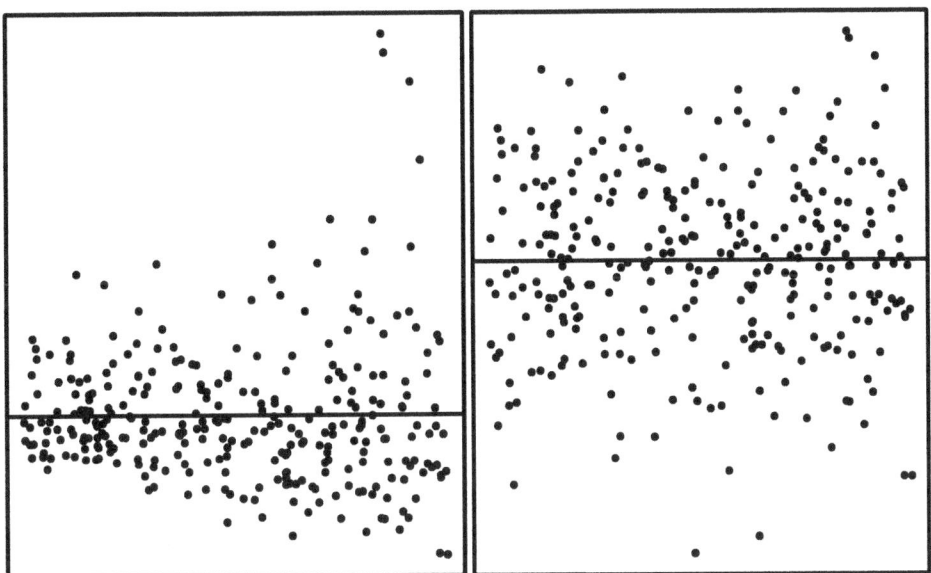

FIGURE 3.2
Two figures for the residuals of the same data. The lefthand side shows the residuals with the original scale of Y. The second with the response $\log(Y)$.

the line is typically estimated as

$$s_\epsilon = \sqrt{\frac{1}{n-p-1} \sum_i y_i - \hat{y}_i{}^2}. \tag{3.12}$$

(Note: If there are no covariates, this is identical to Equation (3.2) for the one population t-test.)

Model (3.10) supposes that ϵ_i are independent and identically distributed random variables, so they should be homoskedastic, i.e., the variance should not depend on the values of x_i. As an informal test of this assumption, it is helpful to look at a scatter plot of $\hat{\epsilon}_i$ plotted against the values of x_i (for each covariate) as depicted in Figure 3.2. One should look for evidence of unequal spread over a predictor x_i, also termed heteroskedasticity. If there is obvious heteroskedasticity, one should question the underlying model assumptions, particularly those for the error term. When there are a large number of predictors p, it may be tedious to visually inspect the residuals for a model, but inspection is a good means to find issues in a model that mathematical techniques would fail to find. Plotting against single variables $\mathbf{X}_i$ may not help when the residual variance changes in a more complicated pattern involving multiple predictor variables.

The left hand plot of Figure 3.2 displays the residuals of a linear model of the form in Equation (3.10) with a single covariate. The residuals towards the right appear to have a much larger variance than the residuals on the left. The general shape of the residuals may be referred to as a trumpeting shape. With such a shape, the inference techniques in the proceeding exposition would have questionable validity. The right hand plot of Figure 3.2 shows the same data fit with the model

$$\log y_i = \beta^t x_i + \epsilon_i. \tag{3.13}$$

The residuals from this model appear to have a more uniform variance as a function of

X. In this case, the log transform provides variance stabilization. If a variance stabilizing transformation is not possible, one should not use the inferential methods of linear regression without skepticism. The trumpeting pattern is fairly common in observed data, so this simple log transformation can be very useful. The log transform is actually a special case of the Box-Cox transformation (10), which suggests the transform

$$y_\lambda = \begin{cases} \frac{y^\lambda - 1}{\lambda} & \lambda \neq 0 \\ \log(y) & \lambda = 0 \end{cases}.$$

Selecting the value λ for the Box-Cox transform is usually done through maximum likelihood estimation methods while assuming that the residuals follow a normal distribution.

A major criticism of variance stabilization methods is that they do not actually model the relationship between the original Y and X as linear. In the case of Equation (3.14), the model for the unadulterated response would be

$$y_i = e^{\beta^t x_i + \epsilon_i}. \tag{3.14}$$

Furthermore, the least squares regression is based on minimizing $\sum_i (\log(y_i) - \widehat{\log(y_i)})^2$, rather than $\sum_i (y_i - \hat{y}_i)^2$ which leads to a subtle bias in the original domain. If one desires a more exacting approach to modeling, generalized least squares modeling can provide a more direct platform to model fitting and inference.

3.7.2 Inference for linear regression

Simply having the estimator $\hat{\beta}$ is of little consequence unless one can determine whether the linear model is legitimate rather than a chance observance. The hypothesis test generally used to test for model validity is

$$H_0 : \beta_1 = \cdots = \beta_p = 0 \text{ vs } H_a : \exists i \in \{1, \ldots, p\} : \beta_i \neq 0. \tag{3.15}$$

It should be noted that the intercept β_0 is not included in this hypothesis test. This allows for the underlying response values Y to have a non-zero mean in the absence of the linear relationship. The hypothesis is tested by considering how much of the underlying variability is accounted for by the model. The total amount of variability the model is attempting to explain may be characterized by the Sum of Squares (SS), $SS = \sum_i (y_i - \bar{Y})^2$, where $\bar{Y}$ is the sample mean for the observed response variable. Using some arithmetic and properties of how $\hat{y}_i$ was derived (10) one can see

$$\sum_i (y_i - \bar{Y})^2 = \sum_i (y_i - \hat{y}_i)^2 + \sum_i (\hat{y}_i - \bar{Y})^2. \tag{3.16}$$

The first term of the right hand side of Equation (3.16) is the residual sum of squares from the estimated model, $\hat{\epsilon}_i$, (RSS). The second term is the sum of squares accounted for by the model (SSM). The test statistic

$$f_m = \frac{SSM/p}{RSS/(n-p-1)},$$

(asymptotically) follows an F distribution with parameters p and (n-p-1). If f_m is large, it means that the variability which can be explained by the linear model components is high relative to the variability not accounted for by the full model. If f_m is small, the linear components do not help describe much of the variability seen in the response variable and it is more likely that the linear components are a product of noise. Hence, the p-values for Hypothesis (3.15) are found by calculating $P(F \geq f_M)$, when F follows an F distribution with parameters p and (n-p-1).

Another commonly used criterion to assess the usefulness of one's linear model is through the R^2 statistics, which is

$$R^2 = 1 - \frac{SSR}{SS}$$

The R^2 statistic tells one the proportion of the variability in the Y variable which may be accounted for by the model. A model may be very strongly significant according to the preceding F test, yet have a low R^2 value. This tells you that the model does not provide much additional prediction value compared to simply guessing the average value of Y. The R^2 value for our toddler weight linear regression model is .87 which accounts for an 87% drop in squared losses when the model is used compared to simply guessing the average weight. Some rules of thumb for the R^2 exist in the literature, for example, $R^2 > 0.7$ indicates that a model has strong ability to explain the variability exhibited in the data (6). Even when there is a small R^2, a model may be significant with a large enough data set. For the linear model, a small R^2 value means that the random noise, σ_ϵ, is large compared to the variation of $\beta^t X$.

Testing the significance of individual slopes β_i is equivalent to the hypothesis test

$$H_0 : \beta_i = 0 | \beta_j = \hat{\beta}_j \text{ for } j \neq i \text{ vs } H_a : \beta_i \neq 0 | \beta_j = \hat{\beta}_j \text{ for } j \neq i, \qquad (3.17)$$

where | stands for "given". If the observed data are independent realizations of Model (3.10), then the LSE, $\hat{\beta}$, converges in distribution to a multivariate normal distribution with mean β and variance $\sigma_\epsilon^2(\tilde{X}^t\tilde{X})^{-1}$. In lieu of the true value σ_ϵ^2, s^2 from Equation (3.12) can be used as an approximation. Let S_i^2 be the i-th diagonal element of $s^2(\tilde{X}^t\tilde{X})^{-1}$. Then the test statistic $t_i = \hat{\beta}_i/S_i$, approximately follows a t distribution with $n - p - 1$ degrees of freedom.

Toddler Weight vs Height and Gender

Table 3.3 displays the results of fitting the linear model to the data in Table 3.2. The observed p-values show that each of the covariate terms in the model have a significant effect on the response variable except for the intercept. Depending on the terms included in the model, the p-value for a parameter could change drastically, particularly if two of the variables are highly correlated.

TABLE 3.3
Results of linear regression. The overall model has an F-statistic of 30.94 based on 2 and 9 *df* with a p-value < 0.0001.

Parameter	Estimate	St. Dev.	t-statistic	p-value
β_0 (Intercept)	−9.56	5.36	−1.78	0.1083
β_1 (Height)	1.03	0.16	6.49	0.0001
β_2 (Gender)	3.74	1.02	3.65	0.0053

---

[3]Notice that the variance does not contain an explicit division by the sample size, n, yet the terms of X^tX grow linearly with respect to n. The asymptotic variance may also be rewritten as $\frac{\sigma_\epsilon^2}{n} E[(x_i^t x_i)^{-1}]$

3.7.3 ANCOVA models

In the toddler weight example we had a categorical term, gender, with two levels (male or female). In clinical trials it is often the case that categories have more than two levels. For instance, one may wish to compare the responses of several different drugs or dosing schemes to a control. Categorical variables divide subjects into one and only one of m different values. Then we can construct a vector $z_i = [I(\text{category} = 2), \ldots, I(\text{category} = m)]^t$ is a $q = (m-1)$-vector of dummy variables. As in the example with the female category of gender, the category $= 1$ term (perhaps the control group) is set as the baseline to avoid redundancies in subsequent modeling.

ANalysis of COVAriance (ANCOVA) models for linear regression separate the categorical variable[4] from the other variables as

$$y_i = \beta_0 + \beta^t x_i + \gamma^t z_i + \epsilon_i, \tag{3.18}$$

where x_i is a p-vector, and z_i is a q-vector. The hypothesis test of greatest importance for ANCOVA is whether the overall category variable is significant or not. In other words, one should test

$$H_0 : \gamma_j = 0 \text{ for } j \in \{1, \ldots, m\} \text{ vs } H_a : \gamma_j \neq 0 \text{ for any } j \in \{1, \ldots, m\}. \tag{3.19}$$

Hypothesis (3.19) can be evaluated by using an F-test. Let RSS_c be the residual sums of squares estimated after fitting Model (3.18). Then let RSS_b be the residual sums of squares estimated after fitting the base Model (3.10) which does not include the categorical term z_i. Then the most marginal variability that $\gamma^t z_i$ may be used to account for is in RSS_b. An F-test can be conducted using the F-statistic

$$f_c = \frac{(RSS_b - RSS_c)/q}{(RSS_c)/(n - p - q - 1)}$$

where f_c follows the F distribution with parameters $q, n - p - q - 1$, and p-values are calculated by finding $P(F \geq f_c)$.

If γ is found to be statistically significant in the preceding, it is usually of interest to compare individual categories. One can perform t-tests to see if the individual parameters of γ are significant, but this only tells you if each category is significantly different from the baseline category 1. If category 1 is the control, then this might be enough for one's purposes. It is usually also important to determine whether there are differences between individual treatments which would involve $\binom{m}{2}$ 2-sided comparisons. Previously, this was stated as a multiple comparisons problem which means that the nominal α should be adjusted prior to individual t-tests, however, we have already performed an F-test to determine that there is at least one significant difference between categories.

Under Model (3.18), we assumed that the variance of ϵ_i was identical for all observations, so we can use $\hat{\sigma}_\epsilon$ from Equation (3.12) as a pooled estimate for the variance. The t-test for treatment i vs j would then use

$$t_{n-p-q} = \frac{\hat{\gamma}_i - \hat{\gamma}_j}{\hat{\sigma}_\epsilon \sqrt{\frac{1}{n_i} + \frac{1}{n_j}}},$$

where n_k is the number of observations from Treatment group k as the standard deviation of the estimator, and if k is 1, then $\gamma_k = 0$.

---

[4] If there is more than one categorical variable, then they should each be analyzed separately. For instance, if one has both drug types and ethnicities, then z_i should take on only the drug types or ethnicity values in the subsequent inference methods.

TABLE 3.4
Results for a least significant difference test

Treatment	Mean	n_i	Groupings
4	18.2	19	A
2	13.6	17	B
1	12.8	21	B C
5	11.6	15	B C
3	10.4	18	C

Table 3.4 gives a mock-up for a typical result of a least significant difference analysis. The last column, "Groupings", is a multivalued label indicating treatments which are not significantly different in pairwise comparisons (i.e. treatments can be grouped into the same category). In this example Treatment 4 is significantly different from all other treatments, since Treatment 4 is in its own group A. Treatment 2, 1, and 5 are not significantly different from each other since they all are belong to group B. Similarly, Treatment 1, 5, and 3 can be grouped into group C; therefore, they are not significantly different from each other. There is however, a significant difference between Treatment 3 and 2 since they do not contain a mutual group label (Treatment 2 is in group B and Treatment 3 is in group C).

3.7.4 Nonlinear regression

It is often of importance to test for nonlinear relationships between a response and covariates. In many cases this relationship may be performed using the methods of standard linear regression by transforming the original variables into new linear terms. For instance if one wanted to test for a quadratic effect of the variable X_2, one could append a new variable, X_2^2 to the original data matrix X. Adding interactions is another way to incorporate non-linear effects into linear models. An interaction is made by multiplying terms in the original data matrix to create new linear terms. The interpretation of the interaction term differs greatly depending on whether the original terms are continuous or categorical.

> **Toddler weight vs height and gender continued**
> Looking at Figure 3.1, it may be unsettling that the lines for both the Male and Female cohorts are parallel. If one wanted to test for a difference in slope as a function of gender, including the interaction term $\mathbf{X}_3 = \mathbf{X}_1 * I(\text{gender} = \text{Male})$ would allow for an offset in the slope for the Male cohort. The results for the model including the interaction term are displayed in Table 3.5. As can be seen, the additional interaction term does not appear to be significant. Moreover, it

TABLE 3.5
Results of linear regression. The overall model has an F-statistic of 24.29 based on 3 and 8 *df* with a p-value of .0002.

Parameter	Estimate	St. Dev.	t-statistic	p-value
β_0 (Intercept)	−1.43	8.86	0.16	0.875
β_1 (Height)	0.70	0.26	2.66	0.029
β_2 (Gender)	−12.48	10.81	−1.15	0.282
β_3 (Height×Gender)	0.48	0.32	1.51	0.170

diminishes the significance of the original Gender term. If one only viewed Table 3.5, it would appear that there is no Gender effect on the weight of toddler.

While this addition of linear terms can help with many relationships, it cannot satisfy all models. For instance,

$$y_i = Ae^{-\beta^t x_i} + Be^{-\alpha^t x_i} + \epsilon_i,$$

is a common decay curve used within biology which cannot be converted into a linear form (8). Likewise, one may expect that the distribution of the error term depends on X and an unknown parameter θ which transform the error term $G(\theta, x_i, \epsilon_i)$. Nonlinear regression and generalized linear regression can be used to fit these models and perform inference on each (5; 9).

3.8 Conclusion

These methods are likely to be used in many applications when a continuous response is observed, yet they are very crude and incapable of performing all analyses. As you progress in your studies you are likely to find much more complicated tests which are better suited to a given analysis. Nonetheless, the more advanced tests tend to require more stringent modeling assumptions, higher mathematical maturity and/or have greater computational burdens. The simplicity and general ease of use with these methods will make them a valuable addition to your statistical toolbox, and understanding these fundamental analysis methods will help with more advanced studies.

The preceding has included a very limited introduction to linear regression. A more complete regression presentation is available in (10) or (1), among a plethora of other books on the subject. These books include more detailed treatment of the aforementioned tests along with much more elaborate ANCOVA methods, corrections for violations from the assumptions, computational methods, prediction, outlier diagnostics, variable selection and other topics which arise in practice.

Bibliography

[1] Ronald Christensen. *Plane answers to complex questions: The theory of linear models.* Springer texts in statistics. Springer, third edition, 2002.

[2] A. Dmitrienko, A.C. Tamhane, and F. Bretz. *Multiple Testing Problems in Pharmaceutical Statistics.* Chapman & Hall/CRC Biostatistics Series. CRC Press, 2009.

[3] Jaroslav Hájek, Zbyněk Šidák, and Pranab K. Sen. *Theory of Rank Tests.* Probability and Mathematical Statistics. Academic Press, second edition edition, 1999.

[4] E. L. Lehmann and Joseph P. Romano. *Testing statistical hypotheses.* Springer Texts in Statistics. Springer, New York, third edition, 2005.

[5] P. McCullagh and J. A. Nelder. *Generalized Linear Models.* Chapman & Hall / CRC, London, 2nd edition, 1997.

[6] David S. Moore. *The Basic Practice of Statistics*. W. H. Freeman & Co., New York, NY, USA, 2nd edition, 1999.

[7] B. Rosner. *Fundamentals of Biostatistics*. Cengage Learning, 8th edition, 2010.

[8] W. Rudin. *Principles of Mathematical Analysis*. International series in pure and applied mathematics. McGraw-Hill, 3rd edition, 1976.

[9] G.A.F. Seber and C.J. Wild. *Nonlinear Regression*. Wiley Series in Probability and Statistics. Wiley, 2003.

[10] George A.F. Seber and Alan J. Lee. *Linear regression analysis*. Wiley series in probability and statistics. Wiley, second edition, 2003.

[11] Z. Sidak, P.K. Sen, and J. Hajek. *Theory of Rank Tests*. ISSN. Elsevier Science, 1999.

[12] J.H. Zar. *Biostatistical Analysis*. Prentice Hall, 2010.

4

Time to Event Data

Daniel Scharfstein, Yuxin Zhu and Anastasios Tsiatis

CONTENTS

4.1 Introduction

In many randomized, controlled clinical trials, the primary outcome is often time from randomization until the occurrence of an event of interest (e.g., death, relapse). The primary outcome is commonly referred to as *time to event*, *survival time*, or *failure time*. In such trials, the major focus is on drawing inference about the distribution of time to event for competing treatments.

In most clinical trials, the time to event may not be observed as each participant is only followed over a finite time horizon and the event of the interest has not occurred before the end of that horizon. The follow-up time can vary from participant to participant. This variation can be due to staggered entry into the clinical trial, loss to follow-up or premature discontinuation of participation in the trial. Those participants who do not have observed failure times are referred to as *right censored*. For such participants, partial information is available about the time to event. Specifically, it is known that the failure time occurs after the follow-up time. Right censoring led to a whole new area of statistics called *Survival Analysis*.

This chapter is organized as follows. The chapter presumes familiarity with probability and basic calculus. In Section 4.2, we introduce a randomized trial that we use to motivate and illustrate the methods discussed in this chapter. Section 4.3 introduces the mathematical fundamentals of survival analysis. Section 4.3.1 introduces mathematical notation. Section 4.3.2 introduces the concept of the net hazard function and links it to the distribution of the time to event of interest. Section 4.3.3 discusses censoring and the distribution of the observable data. Section 4.4 discusses the so-called "one-sample" problem, which is focused on inference about the distribution of the time-to-event based on a sample of data subject to right censoring. The section introduces a key assumption, called non-informative censoring, that is usually imposed in order to estimate this distribution; it introduces the Kaplan-Meier estimator and presents Greenwood's variance formula. Section 4.5 discusses the so-called "two-sample" problem, which is focused on testing whether the distributions of the time to event is equal between independent samples of data subject to right censoring; it introduces the well-known log-rank test statistic. Section 4.6 discusses the Cox regression model. Section 4.7 discusses the one- and two-sample problem in the context of informative censoring. The last section provides key references for topics covered and not covered in this chapter as well as software.

4.2   ACTG 320

ACTG 320 was a randomized trial designed to evaluate whether indinavir sulfate is effective in treating participants with advanced HIV (human immunodeficiency virus) (i.e., CD4 counts less than 200) (24). Participants were randomized to receive open-label AZT and 3TC with or without indinavir sulfate for at least 48 weeks. Randomization was stratified according to CD4 count measured at the time of screening: greater than 50 versus less than or equal to 50. A total of 1156 participants were randomized between January 29, 1996 and January 27, 1997. Participants who developed intolerance to AZT or had progressive disease after 24 weeks on study were allowed to substitute d4T for AZT. Participants were scheduled to be followed at weeks 4, 8, 16, 24, 32, 40, and 48 and every 8 weeks thereafter up to week 96. The primary outcome was the time from randomization to the development of the AIDS or death.

Figure 4.1(a) presents a schematic representation of data for a random sample of 45 participants from ACTG 320. Each line represents data for a participant. The line starts at the calendar time of randomization. The line ends at the calendar time of end of follow-up. The symbol at the end of the line denotes the participant status on that calendar date. If it is an x, then the participant either developed AIDS or died at that time point. For these participants, the length of the line represents the failure time. If the symbol at the end of the line is a ∘, then follow-up has ended at that calendar time without the occurrence of AIDS or death. For these participants, the occurrence of the event of interest is known to occur after the last date of follow-up and the failure time is larger than the length of the line. Figure 4.1(b) presents the same data but on a study time scale (in days), i.e., time zero is the date of randomization. For the moment, ignore the treatment stop symbol on these figures; we will discuss the use of these data in Section 4.7.

4.3   Mathematical Fundamentals

4.3.1   Notation

Let T denote the time to event. Let $F(t) = P(T \leq t)$ and $S(t) = P(T > t)$ be the cumulative distribution function and survivor functions of the random variable T, respectively.

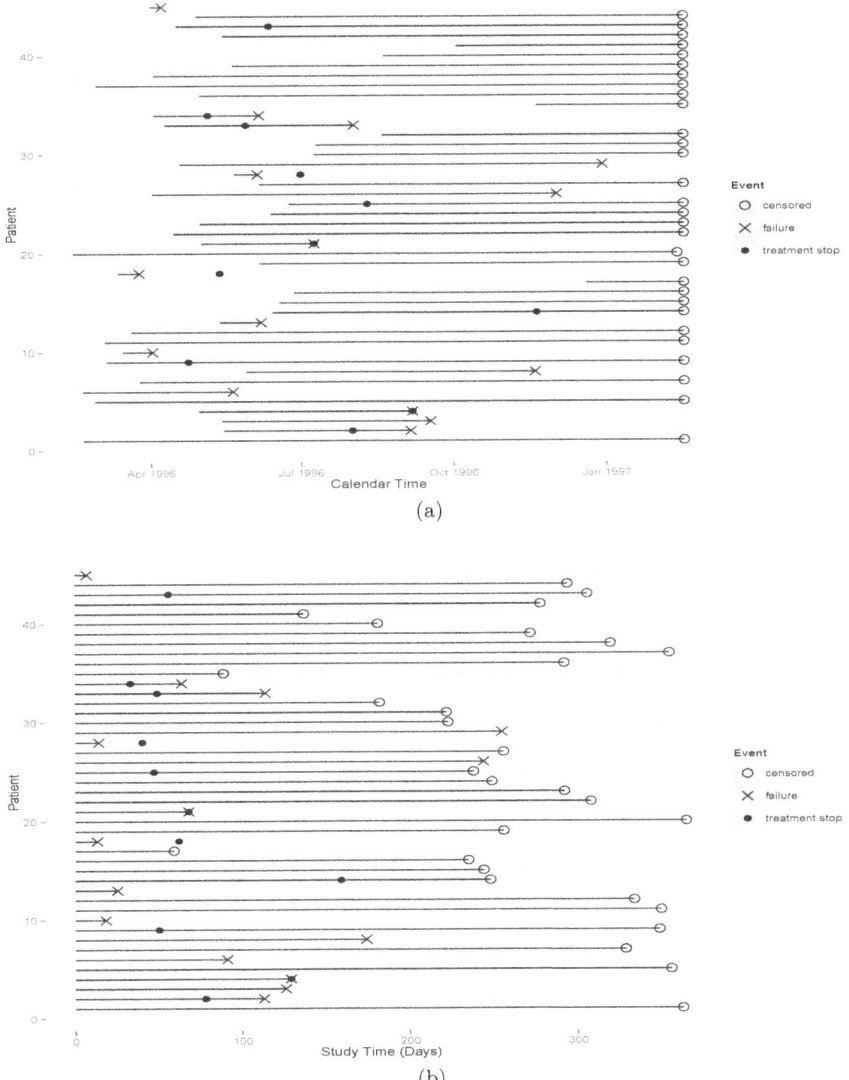

FIGURE 4.1
ACTG 320: Schematic representation of data for a random sample of 45 participants. (a) Calendar time scale; (b) Study time scale.

There are some key technical conditions on $F(\cdot)$ that underlie survival analysis methodology. Specifically, we assume that

1. $F(\cdot)$ can possibly have a countable number of jumps at finite times $0 \leq u_1 < u_2 < \ldots$,

2. $F(\cdot)$ is right continuous; this means that $\lim_{u \to t^+} F(u) = F(t)$ for all t, and

3. $F(\cdot)$ is right differentiable between the jumps; this means that, for $u_j < t < u_{j+1}, \lim_{dt \to 0^+} \frac{F(t+dt)-F(t)}{dt}$ is a finite real number.

Define $u_0 = 0$ and $\tau = \sup_j u_j$ (sup is smallest upper bound). Let $dF(t) = F'(t)dt$

if t is a continuity point of $F(\cdot)$ and $dF(t) = F(t) - F(t-)$ if t is a jump point of $F(\cdot)$, where $t-$ denotes time just prior to t and $F'(t)$ is the right derivative of $F(\cdot)$ at t. Formally, $F'(t) = \lim_{dt\to 0+} \frac{F(t+dt)-F(t)}{dt} = \lim_{dt\to 0+} \frac{P[t \leq T \leq t+dt]}{dt}$.

4.3.2 Hazard

The (net) hazard is a useful way of characterizing the distribution of T as it describes the changing risk of failure over time among those who remain at risk. Let

$$d\Lambda(t) = \frac{dF(t)}{S(t-)},$$

where $\Lambda(t)$ is the integrated or cumulative hazard function. When t is a continuity point of $F(\cdot)$, $d\Lambda(t) = \lambda(t)dt$, where

$$\lambda(t) = \frac{F'(t)}{S(t-)} = \lim_{dt\to 0+} \frac{P[t \leq T \leq t+dt \mid T \geq t]}{dt};$$

and when t is a jump point of $F(\cdot)$, $d\Lambda(t) = P[T = t | T \geq t]$. Notice that when t is a jump point of $F(\cdot)$, $d\Lambda(t)$ is the conditional probability of experiencing an event at time t given it occurs at or after t. If t is a continuity point of $F(\cdot)$, $d\Lambda(t)$ is approximately equal, for small dt, to the conditional probability of experiencing an event in the interval $[t, t + dt]$ given it occurs at or after t. The function $\lambda(t)$ is called the hazard rate, which is the instantaneous risk of an event at time t given it occurs at or after t. The hazard rate is *not* a probability.

In general, the survival function can be written in terms of the hazard as follows:

$$S(t) = \prod_{u_j \leq t} \{1 - d\Lambda(u_j)\} \times \exp\left\{-\sum_{j\geq 1} \int_{u_{j-1}}^{u_j} I(s \leq t)d\Lambda(s) - \int_{\tau}^{\infty} I(s \leq t)d\Lambda(s)\right\}, \quad (4.1)$$

where $I(s \leq t)$ is an indicator function which takes on the value 1 if $s \leq t$ and 0 if $s > t$.

4.3.3 Censoring and observed data

Let C denote the follow-up time defined in the hypothetical world in which the time to event does not pre-empt its observation (e.g., time from randomization until database lock). We consider the observed outcome data for a participant as (X, Δ), where $X = \min(T, C)$ and $\Delta = I(T \leq C)$. If $\Delta = 1$, then the time to event is observed (i.e., $T = X$). If $\Delta = 0$, then the time to event is known to occur after X (i.e., $T > X$) .

The distribution of the observed data for a participant can be characterized by the following quantities: $S_X(t) = P[X > t]$ and $F^\dagger(t) = P[X \leq t, \Delta = 1]$. The latter quantity is referred to as the sub-distribution or cumulative incidence function for failure. Note that $F^\dagger(t) \neq F(t)$ and $S_X(t) \neq S(t)$; the right hand side, not the left hand side, are the quantities that are primary interest. Notice that $P[\Delta = 1] = F^\dagger(\infty)$ and $P[X \leq t, \Delta = 0] = 1 - S_X(t) - F^\dagger(t)$.

Another characteristic of the distribution of the observed data is the cause-specific or observed hazard for failure defined as:

$$d\Lambda^\dagger(t) = \frac{dF^\dagger(t)}{S_X(t-)}.$$

When t is a continuity point of $F^\dagger(\cdot)$, $d\Lambda^\dagger(t) = \lambda^\dagger(t)dt$, where

$$\lambda^\dagger(t) = \lim_{dt\to 0+} \frac{P[t \leq X \leq t+dt, \Delta = 1 \mid X \geq t]}{dt};$$

and when t is a jump point of $F^\dagger(\cdot)$, $d\Lambda^\dagger(t) = P[X = t, \Delta = 1 | X \geq t]$. Notice that when t is a jump point of $F^\dagger(\cdot)$, $d\Lambda^\dagger(t)$ is the conditional probability of *observing* a failure event at time t given at risk for *observing* failure at time t. If t is a continuity point of $F^\dagger(\cdot)$, $d\Lambda^\dagger(t)$ is approximately equal, for small dt, to the conditional probability of *observing* a failure in the interval $[t, t + dt]$ given at risk for *observing* failure at time t.

4.4   Estimation of Survival Distribution

Assumptions are required in order to draw inference about the distribution of the time to event based on observed data from a random sample of n independent participants (below, subscript i will denote data for the ith participant). It is typically assumed that censoring is non-informative. Mathematically, non-informative censoring corresponds to assuming, for all t,

$$d\Lambda(t) = d\Lambda^\dagger(t), \tag{4.2}$$

i.e., the net hazard of failure is equal to the cause-specific hazard of failure. If T and C are independent (i.e., independent censoring), then the non-informative assumption will hold. Unless there are secular trends in enrollment, censoring arising due to study termination should, in principle, be non-informative. Censoring due to premature drop-out, competing risks or treatment termination may be informative. We will discuss how to address this issue in Section 4.7. The utility of non-informative censoring is that it allows identification of $S(\cdot)$ since $d\Lambda^\dagger(\cdot)$ depends on the distribution of the observed data and $S(\cdot)$ can be computed from $d\Lambda(\cdot)$ (see formula (4.1) above).

We can estimate $F^\dagger(t)$ by $\widehat{F}^\dagger(t) = N(t)/n$ and $S_X(t-)$ by $\widehat{S}_X(t-) = Y(t)/n$, where $N(t) = \sum_{i=1}^{n} I(X_i \leq t, \Delta_i = 1)$ is the called the counting process for failure and $Y(t) = \sum_{i=1}^{n} I(X_i \geq t)$ is called the "at-risk" process. Notice that $N(t)$ is a step function with jumps at the observed failure times (say, $t_1, \ldots, t_k$); the jump at a failure time t_j is $dN(t_j) = N(t_j) - N(t_j-)$. We estimate $d\Lambda(t)$, under non-informative censoring by

$$d\widehat{\Lambda}(t) = \frac{d\widehat{F}^\dagger(t)}{\widehat{S}_X(t-)} = \frac{dN(t)}{Y(t)}.$$

This estimator only takes positive values at the observed failure times; it is zero at all other times. Plugging this estimator for $d\Lambda(t)$ into the right hand side of (4.1), we obtain the Kaplan-Meier estimator (32):

$$\widehat{S}(t) = \prod_{t_j \leq t} \left\{ 1 - \frac{d\widehat{F}^\dagger(t_j)}{\widehat{S}_X(t_j-)} \right\} = \prod_{t_j \leq t} \left\{ 1 - \frac{dN(t_j)}{Y(t_j)} \right\}.$$

In Figures 4.2(a)–4.2(d), we display, for ACTG 320, the treatment-specific estimators of $F^\dagger(t)$, $S_X(t-)$, $\Lambda(t)$, and $S(t)$, respectively.

The estimated variance of $\widehat{S}(t)$ is given by Greenwood's formula (23):

$$\widehat{Var}[\widehat{S}(t)] = \widehat{S}(t)^2 \prod_{t_j \leq t} \left\{ \frac{dN(t_j)}{Y(t_j)(Y(t_j) - N(t_j))} \right\}.$$

This can be used to form a $(1 - \alpha)\%$ point-wise confidence interval for $S(t)$. To deal with the fact that $S(t)$ is bounded between 0 and 1, it is recommended that one develops a

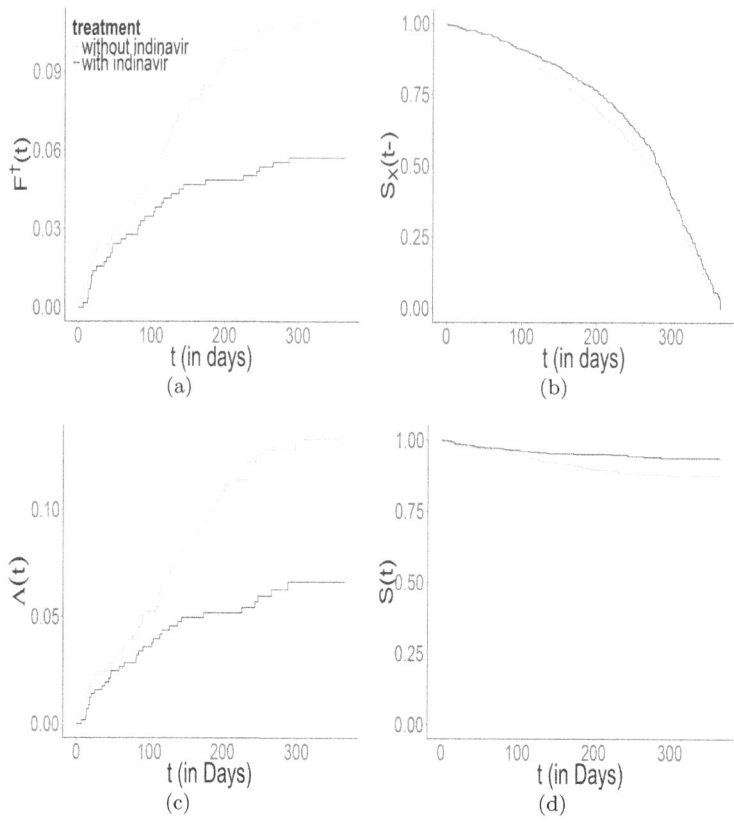

FIGURE 4.2
ACTG 320: Treatment-specific estimators of (a) $F^\dagger(t)$, (b) $S_X(t-)$, (c) $\Lambda(t)$, and (d) $S(t)$.

confidence interval for $\log(-\log\{S(t)\})$ and then back-transforms to a confidence interval for $S(t)$. Specifically, a confidence interval for $\log(-\log\{S(t)\})$ is of the form

$$\log(-\log\{\widehat{S}(t)\}) \pm z_{\alpha/2} \sqrt{\frac{\widehat{Var}[\widehat{S}(t)]}{(\widehat{S}(t)\log\{\widehat{S}(t)\})^2}},$$

where z_x is the $1-x$ quantile of the standard normal distribution.

4.5 Hypothesis Testing

The methods discussed in this section are adapted from (2; 18; 25; 43).

Suppose we are interested in comparing the survival curves of two randomized treatment groups. We assume non-informative censoring in both treatment groups. Let $S^{(0)}(t)$ ($\Lambda^{(0)}(t)$) and $S^{(1)}(t)$ ($\Lambda^{(1)}(t)$) denote the survival (cumulative hazard) functions for treatment groups 0 and 1, respectively. We wish to test the null hypothesis that $S^{(0)}(t) = S^{(1)}(t)$ for all t (or, equivalently $\Lambda^{(0)}(t) = \Lambda^{(1)}(t)$ for all t). Let $N^{(0)}(t)$ ($Y^{(0)}(t)$) and $N^{(1)}(t)$ ($Y^{(0)}(t)$) denote the counting process for failure (at-risk process) in groups 0 and 1, respectively. Let

$N(t) = N^{(0)}(t) + N^{(1)}(t)$ and $Y(t) = Y^{(0)}(t) + Y^{(1)}(t)$. Let $t_1, \ldots, t_k$ be the observed failure times for both groups combined.

Consider the integrated weighted difference between the hazard functions, defined as

$$\beta(w) = \int w(t)\{d\Lambda^{(1)}(t) - d\Lambda^{(0)}(t)\},$$

where $w(t)$ is a non-negative weight function. Under the null hypothesis, $\beta(w)$ will be zero. If $d\Lambda^{(1)}(t) > d\Lambda^{(0)}(t)$ for all t, $\beta(w) > 0$ and if $d\Lambda^{(1)}(t) < d\Lambda^{(0)}(t)$ for all t, $\beta(w) < 0$. We can estimate $\beta(w)$ by

$$\widehat{\beta}(w) = \int w(t)\{d\widehat{\Lambda}^{(1)}(t) - d\widehat{\Lambda}^{(0)}(t)\} = \int w(t)\left\{\frac{dN^{(1)}(t)}{Y^{(1)}(t)} - \frac{dN^{(0)}(t)}{Y^{(0)}(t)}\right\}.$$

Interestingly, $\widehat{\beta}(w)$ can be re-written as:

$$\int k(t)\left\{dN^{(1)}(t) - \frac{Y^{(1)}(t)}{Y(t)}dN(t)\right\} = \sum_{t_j} k(t_j)\left\{dN^{(1)}(t_j) - \frac{Y^{(1)}(t_j)}{Y(t_j)}dN(t_j)\right\},$$

where $k(t) = w(t)\frac{Y(t)}{Y^{(1)}(t)Y^{(0)}(t)}$. Notice that the term in brackets is the typical "observed minus expected" quantity computed from a two-by-two table constructed based on the set of participants who are at-risk for being observed to fail at or after time t_j (i.e., $\{i : X_i \geq t_j\}$), where the columns denote treatment assignment and the rows denote failure at time t_j. The two-by-two table has the following form:

Fail\Treatment	1	0	total
Yes	$dN^{(1)}(t_j)$	$dN^{(0)}(t_j)$	$dN(t_j)$
No	$Y^{(1)}(t_j) - dN^{(1)}(t_j)$	$Y^{(0)}(t_j) - dN^{(0)}(t_j)$	$Y(t_j) - dN(t_j)$
Total	$Y^{(1)}(t_j)$	$Y^{(0)}(t_j)$	$Y(t_j)$

In this table, the observed number of failures at time t_j for treatment 1 is $dN^{(1)}(t_j)$. Under the null hypothesis, the expected number of failures at time t_j for treatment 1 is $\frac{Y^{(1)}(t_j)}{Y(t_j)}dN(t_j)$, resulting in the term in brackets above. Thus, $\widehat{\beta}(w)$ is a weighted average of "observed-expected" terms from two-by-two tables constructed at each observed failure time.

The estimated variance of $\widehat{\beta}(w)$, under the null, is

$$\widehat{Var}[\widehat{\beta}(w)] = \sum_{t_j} w(t_j)^2 \left\{\frac{Y(t_j)}{Y^{(1)}(t_j)Y^{(0)}(t_j)}\right\} \frac{dN(t_j)}{Y(t_j)} \frac{Y(t_j) - dN(t_j)}{Y(t_j) - 1}.$$

Under the null,

$$T(w) = \frac{\widehat{\beta}(w)}{\sqrt{\widehat{Var}[\widehat{\beta}(w)]}} \approx N(0, 1).$$

The null is rejected at the 0.05 level if $|T(w)| > 1.96$.

With specific choices of $w(t)$, we can generate various test statistics that have been proposed for testing for treatment differences. For example, $w(t) = w_{LR}(t) = \frac{Y^{(1)}(t)Y^{(0)}(t)}{Y(t)}$ (or $k(t) = 1$) yields the log-rank statistic, $w(t) = w_{GB}(t) = Y^{(1)}(t)Y^{(0)}(t)$ (or $k(t) = Y(t)$) yields the Gehan-Breslow statistic, and $w(t) = w_{GW}(t) = \frac{Y^{(1)}(t)Y^{(0)}(t)}{Y(t)}\widehat{S}(t-)$ (or $k(t) = \widehat{S}(t-)$) yields the generalized Wilcoxon statistic, where $\widehat{S}$ is the Kaplan-Meier estimator of failure based on both treatment groups. Notice that the log-rank statistic places equal weight over time of the observed minus expected differences. The Gehan-Breslow and generalized Wilcoxon statistics place less weight over time.

In ACTG 320, the log-rank, Gehan-Breslow and generalized Wilcoxon statistics are -3.23, -3.09 and -3.20, respectively. The associated p-values are all less than 0.005, providing statistical evidence in favor of indinavir sulfate.

4.6 Cox Regression Model

Let $Z = (Z_1, \ldots, Z_k)$ be a k-dimensional vector of baseline covariates recorded on a participant. We assume non-informative censoring within levels of Z, i.e.,

$$d\Lambda(t|z) = d\Lambda^\dagger(t|z) \text{ for all } z,$$

where $d\Lambda(t|z)$ and $d\Lambda^\dagger(t|z)$ are the net and cause-specific hazards of failure for participants with covariates $Z = z$.

In 1972, Cox (8) proposed the following regression model

$$\frac{d\Lambda(t|z)}{1 - d\Lambda(t|z)} = \frac{d\Lambda_0(t)}{1 - d\Lambda_0(t)} \exp\{\gamma^T z\}, \tag{4.3}$$

where γ is a k-dimensional vector of unknown parameters and $d\Lambda_0(t)$ is the so-called baseline hazard function as it represents the hazard for participants with covariates $Z = 0$. In this model, the baseline function is left completely unspecified. At continuity points t, (4.3) reduces to

$$\lambda(t|z) = \lambda_0(t) \exp\{\gamma^T z\}; \tag{4.4}$$

at jump points t, (4.3) reduces to

$$\frac{P[T = t | T \geq t, Z = z]}{P[T > t | T \geq t, Z = z]} = \frac{P[T = t | T \geq t, Z = 0]}{P[T > t | T \geq t, Z = 0]} \exp\{\gamma^T z\}.$$

Here, $\exp\{\gamma_j\}$ quantifies the relative change in the risk associated with an increase of one unit in the covariate Z_j. For this model, the relative change is assumed to be the same throughout time. The parameter value $\gamma_j = 0$ corresponds to the case where the jth covariate has no effect on survival. When $\gamma_j > 0$ (< 0), the risk of failure at any point in time increases (decreases) as Z_j increases.

It is important to note that the regression model that is usually specified (i.e., (4.4) holds for all t) is for an underlying failure time that is assumed to have a continuous distribution. Because of inexact measurement (e.g., failure time measured to the level of days), the distribution of the underlying *measurable* failure time (even in the absence of censoring) is discrete. In this chapter, we specify a regression model (i.e., (4.3)) for the underlying *measurable* failure time. The impact of this distinction is on (a) the interpretation of γ and (b) how ties are handled when drawing inference about γ.

To estimate γ, Cox (8; 9) proposed the partial likelihood technique. The partial likelihood is constructed as the product of conditional likelihoods at each observed failure time t_j. Specifically, the contribution to the partial likelihood at t_j is the conditional likelihood of observing the participants who actually failed at t_j given information just prior to t_j *and* that there are $d_j = \sum_{i=1}^{n} dN_i(t_j)$ participants who fail at t_j (without specification of which participants). Let $\mathcal{Q}_j$ be all subsets of d_j participants from the set of $n_j = \sum_{i=1}^{n} Y_i(t_j)$

participants at risk at time t_j; there are $\begin{pmatrix} n_j \\ d_j \end{pmatrix}$ subsets in $\mathcal{Q}_j$. Let S_j be the sum of the covariate vectors for participants who are observed to fail at t_j. Let $S_{j,k}$ be the sum of the covariate vectors for participants in the kth subset in $\mathcal{Q}_j$. Then the conditional likelihood at t_j can be written as

$$L_{t_j}(\gamma) = \frac{\exp(\gamma^T S_j)}{\sum_{k \in \mathcal{Q}_j} \exp(\gamma^T S_{j,k})}. \tag{4.5}$$

The overall partial likelihood is $PL(\gamma) = \prod_{t_j} L_{t_j}(\gamma)$.[1] Since using this likelihood can be computationally intensive, various likelihood approximations have been proposed by (4; 12). (4) and (12) replace the denominator in (4.5) by $\begin{pmatrix} n_j \\ d_j \end{pmatrix} \{\frac{1}{n_j} \sum_{l \in \mathcal{R}_j} \exp(\gamma^T Z_l)\}^{d_j}$ and by $\prod_{k=0}^{d_j-1} \{\sum_{l \in \mathcal{R}_j} \exp(\gamma^T Z_l) - \frac{k}{d_j} \sum_{l \in \mathcal{D}_j} \exp(\gamma^T Z_l)\}$, respectively, where $\mathcal{R}_j$ is set of participants at risk at t_j and $\mathcal{D}_j$ is the set of participants who fail at t_j.[2] These approximated partial likelihoods are implemented in all the major software packages (i.e., R, Stata, SAS). These approximations have been shown to perform well when the ratio of d_j to n_j is small for most t_j (13).

The score function associated with $PL(\gamma)$ (or one of its approximations) is

$$\mathcal{S}(\gamma) = \sum_{t_j} \left\{ S_j - \frac{A'_j(\gamma)}{A_j(\gamma)} \right\},$$

where $A_j(\gamma)$ equals $\sum_{k \in \mathcal{Q}_j} \exp(\gamma^T S_{j,k})$ (or one of its approximations) and $A'_j(\gamma)$ is the derivative of $A_j(\gamma)$ with respect to γ. In the special case of Breslow's approximation (4), it can be shown that

$$\mathcal{S}(\gamma) = \sum_{i=1}^n \int \left\{ Z_i - \frac{\sum_{j=1}^n Y_j(t) Z_j \exp(\gamma^T Z_j)}{\sum_{j=1}^n Y_j(t) \exp(\gamma^T Z_j)} \right\} dN_i(t). \tag{4.6}$$

The parameter γ is estimated as the maximizer, $\widehat{\gamma}$, of $PL(\gamma)$ (or one of its approximations). The maximizer is found by solving $\mathcal{S}(\gamma) = 0$. The estimator $\widehat{\gamma}$ will be approximately normal with mean γ and the inverse of the Hessian (matrix of second derivatives) of log of $PL(\gamma)$ (or one its approximations) evaluated at $\widehat{\gamma}$. The variance of $\widehat{\gamma}_j$ is estimated by the jth diagonal component of the inverse of the aforementioned Hessian matrix. A 95% confidence interval for γ_j can be computed as $\widehat{\gamma}_j \pm 1.96\sqrt{\widehat{Var}[\widehat{\gamma}_j]}$. The null hypothesis that $\gamma_j = 0$ versus the alternative that $\gamma_j \neq 0$ can be tested at 0.05 level by rejecting the null if the 95% confidence interval does not contain zero.

In the special case where Z is the indicator of treatment of assignment, γ quantifies the log relative risk of failure between the two treatments. In this setting, it is interesting to note that

- $\mathcal{S}(0)$ in (4.6) reduces to $\widehat{\beta}(w_{LR})$

- Testing whether $\gamma = 0$ using the approach above is equivalent, in large samples, to testing for a treatment difference using the logrank test statistic.

---

[1] Under the model for assuming the underlying failure time has a continuous distribution (i.e., (4.4) holds for all t), $L_{t_j}(\gamma)$ would be computed differently. Specifically, in the presence of ties at t_j, the conditional likelihood at that time needs to incorporate all the possible ways of "untying" the tied failure times.

[2] The resulting approximated partial likelihoods are identical to the approximated partial likelihoods that are used when it is assumed that the underlying failure time has a continuous distribution.

Let $h_0(t) = \frac{d\Lambda_0(t)}{1 - d\Lambda_0(t)}$. A profile likelihood estimator of $h_0(t)$ puts mass only at failure times t_j. Let $h_j = h_0(t_j)$. The profile estimate for h_j is the unique non-negative solution, $\widehat{h}_j$, to the following equation:

$$\sum_{i \in \mathcal{R}_j} \frac{h_j \exp(\widehat{\gamma}^T Z_i)}{1 + h_j \exp(\widehat{\gamma}^T Z_i)} = d_j. \tag{4.7}$$

Further, an estimator of the conditional survivor function of T given $Z = z$, $S(t|z)$, is

$$\widehat{S}(t|z) = \prod_{t_j \leq t} \left\{ \frac{1}{1 + \widehat{h}_j \exp(\widehat{\gamma}^T z)} \right\}. \tag{4.8}$$

Above, we have focused on modeling the risk of failure as a function of covariates that do not depend on time. In many studies with time-to-event outcomes, the covariates of interest may also change with time. Such covariates are referred to as time-dependent covariates. In ACTG 320, for example, participants were clinically evaluated at multiple occasions after enrollment; at these evaluations, CD4 counts were measured. In evaluating whether CD4 is a potential surrogate marker for the development of AIDS or death (i.e., failure), it is natural to ask how the risk of failure at a given time t relates to the history of CD4 counts prior to time t.

Let $Z(t) = (Z_1(t), \ldots, Z_l(t))$ be a l-dimensional vector of covariates that is known at time t. Let $\overline{Z}(t)$ be the history of these covariates through time t, i.e., $\overline{Z}(t) = \{Z(u) : 0 \leq u \leq t\}$. A covariate that does not vary with time can be considered as a special case of a time-varying covariate. We assume non-informative censoring within covariate histories, i.e.,

$$d\Lambda(t|\overline{z}(t)) = d\Lambda^\dagger(t|\overline{z}(t)) \text{ for all } \overline{z}(t),$$

where $d\Lambda(t|\overline{z}(t))$ and $d\Lambda^\dagger(t|\overline{z}(t))$ are the net and cause-specific hazards of failure for participants with covariate history $\overline{Z}(t) = \overline{z}(t)$.

The Cox regression model posits that

$$\frac{d\Lambda(t|\overline{z}(t))}{1 - d\Lambda(t|\overline{z}(t))} = \frac{d\Lambda_0(t)}{1 - d\Lambda_0(t)} \exp\{\gamma^T g(t, \overline{z}(t))\}, \tag{4.9}$$

where $g(t, \overline{z}(t))$ is a k-dimensional known function of t and $\overline{z}(t)$, γ is a k-dimensional vector of unknown parameters and $d\Lambda_0(t)$ is the baseline hazard function. In the ACTG 320 example, suppose $l = 1$, $Z_1(t)$ is the most recently recorded CD4 at or prior to time t. We might consider $g(t, \overline{Z}(t)) = Z_1(t)$ in which γ represents the common (over time) change in the risk of failure at time t per unit increase in the CD4 count known at that time. Or, we might consider $g(t, \overline{Z}(t)) = (Z_1(t)I(t \leq 56), Z_1(t)I(56 < t \leq 112), \ldots, Z_1(t)I(168 < t \leq 224), Z_1(t)I(t > 224))$, in which case, the effect of CD4 count is allowed to vary over time according to how the time-axis is partitioned.

Estimation of γ and $h_0(t)$ proceeds as above. It is important to emphasize that, in the case of time-independent covariates, it makes sense to estimate $S(t|z)$ as above. In the case of time-dependent covariates, it may not make sense to estimate $S(t|\overline{z}(t)) = P[T > t|\overline{Z}(t) = \overline{z}(t)]$. This is because the very fact that $\overline{Z}(t)$ is measured can imply that the participants are alive or event-free. To distinguish settings where it makes sense to estimate $S(t|\overline{z}(t))$, it is important to distinguish between internal and external covariates. An external covariate is one that can affect a participant, but can be measured even if the participant is not on study, e.g., air pollution levels. In contrast, an internal covariate is one in which the change in the covariate depends on the participant, e.g., CD4 count.

In ACTG 320, investigators were interested in evaluating whether the effect of treatment varied by baseline CD4 status. Specifically, they wanted to know whether the treatment effect was different for participants with baseline CD4 less than or equal to 50 as compared to participants with baseline CD4 between 51 and 200. For each of these CD4 strata, we can fit a Cox regression model with a single treatment-indicator covariate (1 for indinavir, 0 otherwise) and then compare the strata-specific treatment effect estimators using a Wald-type test statistic. The estimated log relative risk in participants with low and high baseline CD4 are -0.67 (standard error $= 0.24$) and -0.70 (standard error $= 0.27$), respectively. The Wald test statistic for the difference in these log relative risks is 0.099, with an associated p-value of 0.92. Thus, there is very weak statistical evidence of effect modification by baseline CD4.

4.7 Informative Censoring

The methods described above for estimation of $S(t)$ and testing for differences in survival between treatment groups rely on the assumption of non-informative censoring, i.e., Equation (4.2) holds for each treatment group. It can be shown that Assumption (4.2) is equivalent to assuming, for t and $t' > t$,

$$d\Lambda_C^\dagger(t|T = t') = d\Lambda_C^\dagger(t|T > t), \tag{4.10}$$

where $d\Lambda_C^\dagger(t|T = t')$ is the cause-specific hazard of censoring at t given information that failure occurs at time t' and $d\Lambda_C^\dagger(t|T > t)$ is the cause-specific hazard of censoring at t given that failure occurs after t (not when it occurs) (38).

Robins and Finkelstein (38) developed a method that relaxes this assumption. Specifically, they assume that

$$d\Lambda_C^\dagger(t|\overline{z}(t), T = t') = d\Lambda_C^\dagger(t|\overline{z}(t), T > t), \tag{4.11}$$

where $d\Lambda_C^\dagger(t|\overline{z}(t), T = t')$ is the cause-specific hazard of censoring at t given information on $\overline{Z}(t) = \overline{z}(t)$ and that failure occurs at time t' and $d\Lambda_C^\dagger(t|\overline{z}(t), T > t)$ is the cause-specific hazard of censoring at t given information on $\overline{Z}(t) = \overline{z}(t)$ and that failure occurs after t (not when it occurs). They further assume a dimension-reduction model for $d\Lambda_C^\dagger(t|\overline{z}(t), T > t)$. Here, we consider a model of the form:

$$\frac{d\Lambda_C^\dagger(t|\overline{z}(t), T > t)}{1 - d\Lambda_C^\dagger(t|\overline{z}(t), T > t)} = \frac{d\Lambda_{C,0}^\dagger(t)}{1 - d\Lambda_{C,0}^\dagger(t)} \exp\{\eta^T g(t, \overline{z}(t))\}, \tag{4.12}$$

where $g(t, \overline{z}(t))$ is a k-dimensional known function of t and $\overline{z}(t)$, η is a k-dimensional vector of unknown parameters and $d\Lambda_{C,0}^\dagger(t)$ is the baseline hazard function. To estimate η and $h_{C,0}^\dagger(t) = \frac{d\Lambda_{C,0}^\dagger(t)}{1-d\Lambda_{C,0}^\dagger(t)}$, use the techniques described in Section 4.6 above with the following modifications: (1) reverse the roles of T and C and (2) break ties between failure and censoring times by assuming that failure precedes censoring. Let

$$\tilde{K}_i(t; \tilde{\eta}) = \prod_{c_k \leq t} \left\{ \frac{1}{1 + \tilde{h}_{C,0}^\dagger(c_k; \tilde{\eta}) \exp\{\tilde{\eta}^T g(t_k, \overline{Z}_i(c_k))\}} \right\},$$

where c_k are the unique ordered censoring times, $\tilde{\eta}$ is an estimator of η in Model (4.12),

$\tilde{h}_{C,0}^{\dagger}(t;\tilde{\eta})$ is the corresponding profile estimator of $h_{C,0}^{\dagger}(t)$, and $\overline{Z}_i(t)$ is the time varying co-variate vector associated with participant i. Here, $\tilde{K}_i(t;\tilde{\eta})$ is an estimator of the probability that participant i is uncensored at time t. The adjusted survival curve of (38) is of the form:

$$\tilde{S}(t) = \prod_{t_j \leq t} \left\{ 1 - \frac{\sum_{i=1}^n \left(dN_i(t_j)/\tilde{K}_i(t_j;\tilde{\eta}) \right)}{\sum_{i=1}^n \left(Y_i(t_j)/\tilde{K}_i(t_j;\tilde{\eta}) \right)} \right\},$$

where t_j are the unique ordered failure times.

The intuition of the adjusted estimator is as follows. The numerator and denominator of the ratio inside the curly brackets estimates, in the absence of censoring, the number of participants who are expected to fail at t_j and the number of participants expected to be at risk for failure at t_j, respectively. Why? In the numerator (denominator), each participant who fails (is at risk) at t_j is inverse weighted by the probability of being uncensored at that time. Inverse weighting (a technique derived from the survey sampling literature) serves to upweight the contribution of participants with observed data to account for themselves and others like them who were unobserved. For example, if the probability of being uncensored at time t_j for a participant is 0.25, then he/she accounts for him/her-self plus three other similar participants who were unobserved at that time. In the absence of censoring, the ratio in the curly brackets estimates the hazard of failure at t_j and one minus the ratio estimates the probability of being event free at t_j given at risk at t_j. Therefore, the product of the terms in curly brackets through time t estimates the probability of being event free at time t. It is important to note that if $\tilde{\gamma} = 0$ (i.e., $\overline{Z}(t)$ is not prognostic for censoring), the survival curve estimator of (38) reduces to the Kaplan-Meier estimator.

Robins and Finkelstein (38) also showed how to construct an adjusted estimator of γ in model (4.3) with treatment assignment indicator as the sole covariate by modifying $\mathcal{S}(\gamma)$ in (4.6) to ensure that it has mean zero (in large samples) when (4.11) and (4.12) are assumed for each treatment group *and* reducs to (4.6) when the covariates in model (4.12) are not prognostic for censoring. Towards this end, (38) defined the modified score function:

$$\tilde{S}(\gamma) = \sum_{i=1}^n \int \left\{ Z_i - \frac{\sum_{j=1}^n \tilde{W}_j(t)Y_j(t)Z_j \exp(\gamma^T Z_j)}{\sum_{j=1}^n \tilde{W}_j(t)Y_j(t) \exp(\gamma^T Z_j)} \right\} \tilde{W}_i(t)dN_i(t),$$

where

$$\tilde{W}_i(t) = \frac{Z_i\tilde{K}_i^{(1)}(t;0) + (1 - Z_i)\tilde{K}_i^{(0)}(t;0)}{Z_i\tilde{K}_i^{(1)}(t;\tilde{\eta}^{(1)}) + (1 - Z_i)\tilde{K}_i^{(0)}(t;\tilde{\eta}^{(0)})},$$

the superscripts are used to reference treatment groups, and $\tilde{K}_i^{(z)}(t;0)$ are the treatment-specific Kaplan-Meier estimators of the survival curve for censoring (with ties broken as above). The adjusted estimator $\tilde{\gamma}$ of γ is the solution to $\tilde{S}(\gamma) = 0$. (38) showed that $\tilde{\gamma}$ will be normally distributed in large samples. While it is possible to compute an analytic expression for the standard error of $\tilde{\gamma}$, it is easier to use bootstrapping procedures to estimate standard errors for $\hat{\gamma}$ and construct confidence intervals for γ.

To illustrate this method, we use data from ACTG 320 where we further censor participants at discontinuation of their assigned treatment if it occurs prior to failure. That is, we use the earlier of the open and solid circles in Figure 4.1. In the analyses discussed above, there were 96 failures and 1054 censored observations. Incorporating censoring at treatment stop reduces the number of failures to 66 (43 without indinavir, 23 with indinavir) and increases the number of censored observations to 1084. Specifically, 30 of the 96 failures occurred after treatment stop – 20 from the without indinavir arm and 10 from the indinavor arm. In addition, 207 of the original 1054 censored observations are moved to an earlier censoring time – 147 from the without indinavir arm and 60 from the indinavor arm.

TABLE 4.1

Treatment-specific censoring model fits: exponentiated regression coefficients and associated 95% confidence intervals.

Covariate	Without indinavir		With indinavir	
	Effect	95% CI	Effect	95% CI
CD4	0.9997	[0.9987, 1.001]	0.9984	[0.9976, 0.9992]
Karnofsky Score	0.9926	[0.9820, 1.003]	1.0075	[0.9961, 1.0190]
Hemoglobin	1.0290	[0.7388, 1.433]	1.4818	[0.9706, 2.2622]

It is plausible that such censoring may be informative as participants who are sicker or experiencing side effects may be more likely to discontinue their assigned therapy. In this analysis, we seek to estimate the effect of treatment in a world without non-compliance. This is considered a "hypothetical" estimand (27). It contrasts with the aforementioned analyses which was focused on estimating what is often referred to as the intention-to-treat effect or "policy" estimand (27).

For each treatment group, we fit Model (4.12) with Karnofsky score and hemoglobin at baseline as time-independent covariates and CD4 as a time-varying covariate. We used Breslow's approximation to the partial likelihood; the results based on Efron's approximation were similar. Table 4.1 displays the exponentiated regression coefficients and associated 95% confidence intervals from fitting these models. The table shows that participants with higher CD4 (i.e., healthier) are less likely to be censored. The effects of baseline Karnofsky score were close to null in each treatment group. Participants with higher baseline hemoglobin were more likely to be censored, with a larger effect in the indinavir arm; the effects were coupled with appreciable uncertainty. These results suggest that a non-informative censoring analysis may be optimistic. Figure 4.3 shows the unadjusted (i.e., Kaplan-Meier) and adjusted estimated survival curves. As expected, the figure shows that the adjusted curves tend to be lower than the unadjusted curves, although the shifts are negligible. The unadjusted and adjusted estimates of the relative risk are 0.49 (95% CI: 0.30–0.82) and 0.46 (95% CI: 0.25–0.77), respectively. In contrast, the estimate of the relative risk in the previous analysis (i.e., not censoring at treatment stop) was 0.51 (95% CI: 0.33-0.77). The estimate of the effect of indinavir under full compliance is slightly larger, although it does not appear to be clinically significant.

Scharfstein and Robins (40) developed methods for evaluating the sensitivity of survival curve estimation to deviations from Assumption (4.11). Rotnitzky *et al.*(39) extended the ideas of (40) to address competing causes of censoring (e.g., end-of-study censoring vs. censoring due to treatment stop). Zhang *et al.*(48) used the ideas of (38) to draw inference, in the presence of non-compliance, about the distribution of a time-to-event for treatment regimes with specified treatment stops.

4.8 Conclusion

In this chapter, we discussed the most commonly used survival analysis methods (survival curve estimation, treatment group comparsions, Cox regression) in clinical trials. We also reviewed a method for adjusting for informative censoring that we think should be more widely utilized. We emphasized the discrete Cox regression model because it has been our experience that event times are usually measured inexactly (e.g., at the level of days

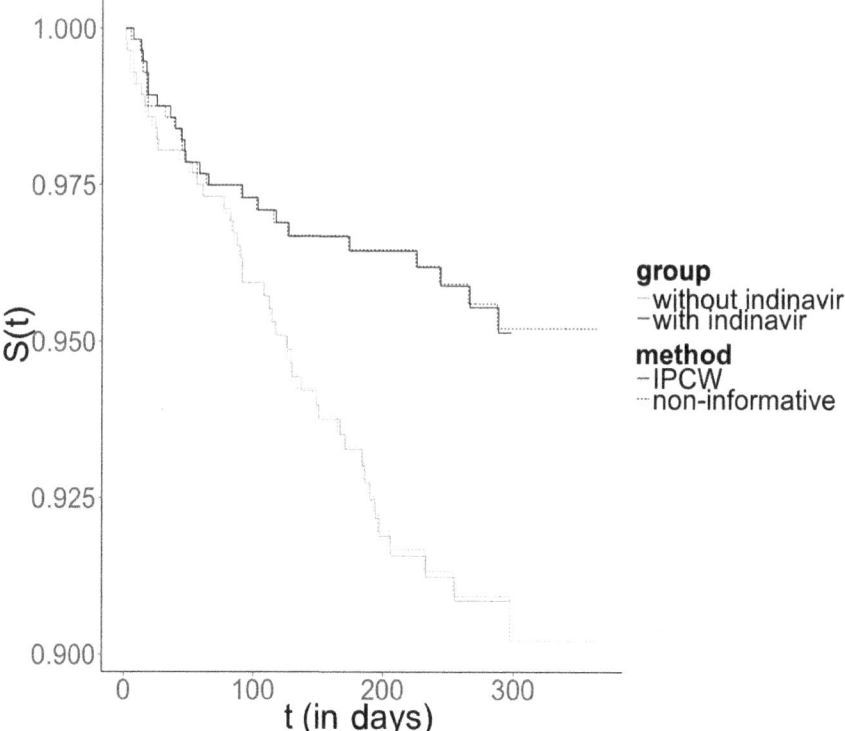

FIGURE 4.3
Treatment-specific unadjusted (Kaplan-Meier [32]) and adjusted (Robins and Finkestein [38]) estimated survival curves.

rather than hours or seconds). Such inexactness renders the events times to have discrete support. Importantly, the partial likelihood approximations under a discrete Cox model are identical to the approximations for handling ties in the continuous time Cox model. Thus, the estimators resulting from maximizing these approximated partial likelihoods can be considered as estimating the regression parameter in either the discrete or continuous time Cox model.

There is a wide body of survival analysis methods that we have not discussed. While we focused on time to event data that may be right censored, there are methods that handle event times that may also be interval censored (i.e., only known to fall into a finite time interval) (14–17; 20; 21; 34; 42; 45; 46). There is also a great deal of work on alternative regression models, including the accelerated failure time model (3; 30; 37; 47) and the semi-parametric transformation model (5; 6).

The issue of competing risks, whereby participants are at-risk for multiple *pre-emptive* causes of failure, is particularly challenging. If a particpant is observed to fail from one cause then he/she is no longer at risk for failure from another cause. Thus, when analyzing failure due to a given cause, it is not appropriate to simply consider participants who failed due to another cause as censored observations. This is why many analysts often work with a composite outcome which is the time of the first failure regardless of cause. Alternatively, some analysts report cause-specific hazards (36) or cause-specific sub-distribution functions (22).

Another important area is multivariate survival analysis, where multiple failure events are to be recorded on each participant (either in series or in parallel) or a single failure event on participants who are themselves clustered into groups. Methods are available that are similar in spirit to the marginal, copula and random effects models using in longitudinal data analysis, with the exception that in survival analysis random effects used are often referred to as frailty models. (26) provides a detailed review of methods for analyzing multivariate survival data.

Methods are also available for the design and analysis of clinical trials in which time-to-event data are to be analyzed at interim time points at which a decision can be made to prematurely stop the trial for efficacy or futility. Scharfstein, Tsiatis and Robins (41) and Jennison and Turnbull (28) developed a general framework, based on the concept of statistical information, for designing and monitoring such trials in which a type I error spending function (developed by (11)) along with a stopping boundary is utilized to preserve the overall operating characteristics of the trial. Jennison and Turnbull (29) is a great resource to learn more about what is often called group sequential clinical trials.

Survival analysis is a very well researched field. There are great reference books available, including but not limited to (2; 10; 19; 31; 33; 44). There are also great software routines available in SAS (1), R (35) and STATA (7) for analyzing survival data. There is a great deal of online material demonstrating how to use these routines.

Bibliography

[1] Paul D Allison. *Survival analysis using SAS: a practical guide.* Sas Institute, 2010.

[2] Per Kragh Andersen, Ornulf Borgan, Richard D Gill, and Niels Keiding. *Statistical models based on counting processes.* Springer Science & Business Media, 2012.

[3] Rebecca A Betensky, Daniel Rabinowitz, and Anastasios A Tsiatis. Computationally simple accelerated failure time regression for interval censored data. *Biometrika*, 88(3):703–711, 2001.

[4] Norman Breslow. Covariance analysis of censored survival data. *Biometrics*, pages 89–99, 1974.

[5] Kani Chen, Zhezhen Jin, and Zhiliang Ying. Semiparametric analysis of transformation models with censored data. *Biometrika*, 89(3):659–668, 2002.

[6] SC Cheng, LJ Wei, and Z Ying. Analysis of transformation models with censored data. *Biometrika*, 82(4):835–845, 1995.

[7] Mario Cleves. *An introduction to survival analysis using Stata.* Stata Press, 2008.

[8] David R Cox. Regression models and life-tables. *Journal of the Royal Statistical Society. Series B (Methodological)*, pages 187–220, 1972.

[9] David R Cox. Partial likelihood. *Biometrika*, 62(2):269–276, 1975.

[10] David Roxbee Cox and David Oakes. *Analysis of survival data*, volume 21. CRC Press, 1984.

[11] David L Demets and KK Lan. Interim analysis: the alpha spending function approach. *Statistics in medicine*, 13(13-14):1341–1352, 1994.

[12] Bradley Efron. The efficiency of cox's likelihood function for censored data. *Journal of the American statistical Association*, 72(359):557–565, 1977.

[13] VT Farewell and Ross L Prentice. The approximation of partial likelihood with emphasis on case-control studies. *Biometrika*, 67(2):273–278, 1980.

[14] Michael P Fay. Comparing several score tests for interval censored data. *Statistics in Medicine*, 18(3):273–285, 1999.

[15] Michael P Fay and Pamela A Shaw. Exact and asymptotic weighted logrank tests for interval censored data: the interval r package. *Journal of Statistical Software*, 36(2), 2010.

[16] Dianne M Finkelstein. A proportional hazards model for interval-censored failure time data. *Biometrics*, pages 845–854, 1986.

[17] Dianne M Finkelstein and Robert A Wolfe. A semiparametric model for regression analysis of interval-censored failure time data. *Biometrics*, pages 933–945, 1985.

[18] Thomas R Fleming and David P Harrington. A class of hypothesis tests for one and two sample censored survival data. *Communications in Statistics-Theory and Methods*, 10(8):763–794, 1981.

[19] Thomas R Fleming and David P Harrington. *Counting processes and survival analysis*, volume 169. John Wiley & Sons, 2011.

[20] Els Goetghebeur and Louise Ryan. Semiparametric regression analysis of interval-censored data. *Biometrics*, 56(4):1139–1144, 2000.

[21] William B Goggins, Dianne M Finkelstein, David A Schoenfeld, and Alan M Zaslavsky. A markov chain monte carlo em algorithm for analyzing interval-censored data under the cox proportional hazards model. *Biometrics*, pages 1498–1507, 1998.

[22] Robert J Gray. A class of k-sample tests for comparing the cumulative incidence of a competing risk. *The Annals of statistics*, pages 1141–1154, 1988.

[23] Major Greenwood et al. A report on the natural duration of cancer. *Reports on Public Health and Medical Subjects. Ministry of Health*, (33), 1926.

[24] Scott M Hammer, Kathleen E Squires, Michael D Hughes, Janet M Grimes, Lisa M Demeter, Judith S Currier, Joseph J Eron Jr, Judith E Feinberg, Henry H Balfour Jr, Lawrence R Deyton, et al. A controlled trial of two nucleoside analogues plus indinavir in persons with human immunodeficiency virus infection and cd4 cell counts of 200 per cubic millimeter or less. *New England Journal of Medicine*, 337(11):725–733, 1997.

[25] David P Harrington and Thomas R Fleming. A class of rank test procedures for censored survival data. *Biometrika*, 69(3):553–566, 1982.

[26] Philip Hougaard. *Analysis of multivariate survival data*. Springer Science & Business Media, 2012.

[27] CHMP ICH. E9 (R1) Addendum on estimands and sensitivity analysis in clinical trials to the guideline on statistical principles for clinical trials. In *Proceedings of the International Conference on Harmonisation of Technical Requirements for Registration of Pharmaceuticals for Human Use*, 2019.

[28] Christopher Jennison and Bruce W Turnbull. Group-sequential analysis incorporating covariate information. *Journal of the American Statistical Association*, 92(440):1330–1341, 1997.

[29] Christopher Jennison and Bruce W Turnbull. *Group sequential methods with applications to clinical trials*. CRC Press, 1999.

[30] Zhezhen Jin, DY Lin, LJ Wei, and Zhiliang Ying. Rank-based inference for the accelerated failure time model. *Biometrika*, 90(2):341–353, 2003.

[31] John D Kalbfleisch and Ross L Prentice. *The statistical analysis of failure time data*, volume 360. John Wiley & Sons, 2011.

[32] Edward L Kaplan and Paul Meier. Nonparametric estimation from incomplete observations. *Journal of the American statistical association*, 53(282):457–481, 1958.

[33] David G Kleinbaum and Mitchel Klein. *Survival analysis*. Springer, 1996.

[34] Jane C Lindsey and Louise M Ryan. Methods for interval-censored data. *Statistics in medicine*, 17(2):219–238, 1998.

[35] Melinda Mills. *Introducing survival and event history analysis*. Sage Publications, 2011.

[36] Ross L Prentice, John D Kalbfleisch, Arthur V Peterson Jr, Nancy Flournoy, VT Farewell, and NE Breslow. The analysis of failure times in the presence of competing risks. *Biometrics*, pages 541–554, 1978.

[37] James Robins and Anastasios A Tsiatis. Semiparametric estimation of an accelerated failure time model with time-dependent covariates. *Biometrika*, 79(2):311–319, 1992.

[38] James M Robins and Dianne M Finkelstein. Correcting for noncompliance and dependent censoring in an aids clinical trial with inverse probability of censoring weighted (ipcw) log-rank tests. *Biometrics*, pages 779–788, 2000.

[39] Andrea Rotnitzky, Andres Farall, Andrea Bergesio, and Daniel Scharfstein. Analysis of failure time data under competing censoring mechanisms. *Journal of the Royal Statistical Society: Series B (Statistical Methodology)*, 69(3):307–327, 2007.

[40] Daniel O Scharfstein and James M Robins. Estimation of the failure time distribution in the presence of informative censoring. *Biometrika*, 89(3):617–634, 2002.

[41] Daniel O Scharfstein, Anastasios A Tsiatis, and James M Robins. Semiparametric efficiency and its implication on the design and analysis of group-sequential studies. *Journal of the American Statistical Association*, 92(440):1342–1350, 1997.

[42] Jianguo Sun. *The statistical analysis of interval-censored failure time data*. Springer Science & Business Media, 2007.

[43] Robert E Tarone and James Ware. On distribution-free tests for equality of survival distributions. *Biometrika*, 64(1):156–160, 1977.

[44] Terry M Therneau and Patricia M Grambsch. *Modeling survival data: extending the Cox model*. Springer Science & Business Media, 2000.

[45] Bruce W Turnbull. The empirical distribution function with arbitrarily grouped, censored and truncated data. *Journal of the Royal Statistical Society. Series B (Methodological)*, pages 290–295, 1976.

[46] Lianming Wang, Christopher S McMahan, Michael G Hudgens, and Zaina P Qureshi. A flexible, computationally efficient method for fitting the proportional hazards model to interval-censored data. *Biometrics*, 2015.

[47] LJ Wei. The accelerated failure time model: a useful alternative to the cox regression model in survival analysis. *Statistics in medicine*, 11(14–15):1871–1879, 1992.

[48] Min Zhang, Anastasios A Tsiatis, Marie Davidian, Karen S Pieper, and Kenneth W Mahaffey. Inference on treatment effects from a randomized clinical trial in the presence of premature treatment discontinuation: the synergy trial. *Biostatistics*, 12(2):258–269, 2011.

5

Count Data

Xin He and Jianguo "Tony" Sun

CONTENTS

5.1   Introduction

In many randomized controlled trials, the outcome of interest is the count of certain events, such as the number of recurrent disease infections, repeated adverse events, or tumor occurrences experienced by a study subject. Such studies can be generally classified into two types. One is that all counts are independent of each other, which will be referred to as simple count data. A common example of such data is that each subject gives or records the total number of occurrences of the events of interest for the whole study period (Cameron and Trivedi, 1998). The other type is the clustered or correlated count data in which the observed counts may be related. A common situation that yields such data, the focus of this chapter, is that each study subject is observed at discrete time points and thus instead of only one total count of the events of interest during the whole study period, we observe several counts of the events at multiple time points during the whole follow-up period. It is apparent that the counts from the same subject will be correlated and such data are often referred to as panel count data in the literature (Kalbfleisch and Lawless, 1985; Sun, 2006; Sun and Zhao, 2013). This chapter will provide a brief overview for regression analysis of these two types of count data described above and some comments will be given below for other related count data.

Examples of simple count data can be found in almost every field such as agriculture, economics, medical studies, and social science. To give a specific example of correlated or panel count data, consider the data arising from the National Cooperative Gallstone Study (NCGS), which is a 10-year, multicenter, randomized, controlled, double-blinded clinical trial on the use of the natural bile acid chenodeoxycholic acid (chenodiol) for the dissolution of cholesterol gallstones (Schoenfield et al., 1981; Thall and Lachin, 1988). The original study consists of 916 patients who were randomly assigned into one of three treatments, and treated for up to two years: high dose (750 mg per day), low dose (375 mg per day), and placebo. A part of the resulting data was presented in Sun and Zhao (2013) including the successive visit times in study weeks and the associated counts of episodes of nausea between the visits over the first 52 weeks of follow-up for 113 patients with floating gallstones in high dose (65) and placebo (48) groups. In addition, patients were scheduled to return for clinical observations at 1, 2, 3, 6, 9, and 12 months during the first year follow-up, but the actual visit or observation times differed from patient to patient. One primary objective of the study was to test the difference of the two treatments with respect to the recurrence rate of nausea. This study gave panel count data on the recurrences of nausea. On the other hand, to make the analysis simple, one may get and only focus on simple count data given by just the total or cumulative number of occurrences of nausea from the whole follow-up period. This example will be used to illustrate the regression analysis techniques later on.

Extensive literature has been developed for the analysis of simple and panel count data. In particular, Cameron and Trivedi (1998) and Sun and Zhao (2013) provided comprehensive coverages of related statistical approaches for these two types of count data, respectively. For regression analysis of simple count data, Poisson and negative binomial models have been widely used. For regression analysis of panel count data, it is usually more convenient to regard the observations as realizations of some underlying counting processes and work directly on the mean function of the counting processes conditional on covariate processes due to the incomplete nature of panel count data. Some simple approaches include fitting the data to parametric Poisson processes (Breslow, 1984; Hinde, 1982) or mixed parametric Poisson processes (Thall, 1988), and treating the data as longitudinal count data for which the generalized estimating equation (GEE) approach can be applied (Diggle et al., 2002). However, a major drawback of such parametric methods is that it is usually difficult to find an appropriate parametric model for panel count data in a given problem. In this chapter, we will focus on the estimation of covariate effects on the mean function of the underlying recurrent event process using semiparametric approaches.

In the following, we start with a brief review of regression analysis of simple count data in Section 5.2 using the maximum likelihood approach under the Poisson and negative binomial regression models. In Section 5.3, we will discuss two likelihood-based approaches for regression analysis of correlated count data when the underlying counting processes can be regarded as non-homogeneous Poisson or negative binomial processes. Section 5.4 presents two estimating equation-based approaches which do not rely on any distribution assumption. An illustration of the presented estimation procedures using the NCGS data is provided in each of Sections 5.2, 5.3, and 5.4. Section 5.5 concludes with some discussion and remarks on issues not discussed in the preceding sections.

5.2　Regression Analysis of Simple Count Data

For regression analysis of simple count data, the most commonly employed method is Poisson regression (Cameron and Trivedi, 1998; Jorgensen, 1961; Kianifard and Gallo, 1995).

Under the Poisson regression model, the equality of conditional mean and conditional variance is a critical assumption for the statistical inference using the Poisson maximum likelihood estimator. In many empirical applications, this equidispersion assumption is too restrictive. Overdispersion occurs in the situation when the observed variance exceeds the theoretical variance under the assumed Poisson distribution. One simple approach to relax the equidispersion restriction of the Poisson regression model is the negative binomial regression model (Cameron and Trivedi, 1998). More discussion on this is given below.

In the following, the focus will be on the maximum likelihood estimation approaches.

5.2.1 Poisson regression for count

Consider a recurrent event study involving n independent subjects and the only information observable from each subject is the number of events that have occurred during a common follow-up time period. For subject i, define N_i to be the cumulative count of events and let $\mathbf{X}_i = (X_{i1}, \cdots, X_{ip})'$ be a $p \times 1$ covariate vector, $i = 1, \cdots, n$. For the covariate effect on N_i, we assume that given $\mathbf{X}_i$, the conditional mean of N_i has the form

$$E(N_i|\mathbf{X}_i) = \mu_i = \exp(\beta'\mathbf{X}_i) , \tag{5.1}$$

where β is a vector of unknown regression parameters. Suppose that the N_i's have a Poisson distribution with parameter μ_i and the probability mass function

$$f(N_i|\mu_i) = \frac{\exp(-\mu_i)\mu_i^{N_i}}{\Gamma(N_i + 1)} , \tag{5.2}$$

where $\mu_i > 0$ and the gamma function $\Gamma(x) = \int_0^\infty t^{x-1}\exp(-t)dt$. Under this Poisson assumption, it can be shown that the conditional mean and the conditional variance are equal, i.e., $Var(N_i|\mathbf{X}_i) = E(N_i|\mathbf{X}_i) = \mu_i$.

The likelihood function for the n independently and identically distributed Poisson observations is a product of probabilities given by (5.2). Then it is clear that the log likelihood function of β is proportional to

$$l(\beta) = \sum_{i=1}^n \{N_i \log(\mu_i) - \mu_i\} = \sum_{i=1}^n \{N_i\beta'\mathbf{X}_i - \exp(\beta'\mathbf{X}_i)\} . \tag{5.3}$$

Taking derivatives of the log likelihood function (5.3) with respect to β, it can be shown (Sun and Zhao, 2013) that the maximum likelihood estimator, denoted by $\hat{\beta}_P$, of β can be obtained by solving the estimating equation

$$\sum_{i=1}^n \mathbf{X}_i\{N_i - \exp(\beta'\mathbf{X}_i)\} = 0 .$$

Let β_0 denote the true value of β. It can be easily shown that $\sqrt{n}(\hat{\beta}_P - \beta_0)$ converges in distribution to a multivariate normal vector with mean zero and covariance matrix

$$Var(\beta_0) = \frac{1}{n}\left\{\sum_{i=1}^n \mathbf{X}_i\mathbf{X}_i'\exp(\beta_0'\mathbf{X}_i)\right\}^{-1} .$$

A consistent estimate of $Var(\beta_0)$ is given by $Var(\hat{\beta}_P)$.

5.2.2　Negative binomial regression for count

To relax the equidispersion assumption, the negative binomial regression model is often used in practice. Assume that there exists a latent variable, ν_i, representing nonnegative random unobserved heterogeneity, which is independently and identically distributed under a gamma distribution with the density function

$$g(\nu_i|\alpha) = \frac{1}{\Gamma(\alpha^{-1})\alpha^{1/\alpha}}\nu_i^{\alpha^{-1}-1}\exp\left(-\frac{\nu_i}{\alpha}\right) , \tag{5.4}$$

with a shape parameter $\alpha^{-1} > 0$, a scale parameter $\alpha > 0$, $E(\nu_i) = 1$, and $Var(\nu_i) = \alpha$.

Assume that the N_i's follows a mixed Poisson distribution with the conditional mean

$$E(N_i|\mathbf{X}_i, \nu_i) = \nu_i\mu_i = \nu_i\exp(\beta'\mathbf{X}_i) , \tag{5.5}$$

$i = 1, \cdots, n$. The latent variable ν_i induces the overdispersion while preserving the conditional mean as below

$$E(N_i|\mathbf{X}_i) = \mu_i = \exp(\beta'\mathbf{X}_i)$$

and

$$Var(N_i|\mathbf{X}_i) = \mu_i(1 + \alpha\mu_i) .$$

That is, the conditional variance of N_i is greater or equal to it's conditional mean.

Under (5.4) and (5.5), the marginal probability mass function of N_i is given by

$$f(N_i|\mu_i, \alpha) = \frac{\Gamma(N_i + \alpha^{-1})}{\Gamma(N_i + 1)\Gamma(\alpha^{-1})}\left(\frac{\mu_i}{\mu_i + \alpha^{-1}}\right)^{N_i}\left(\frac{\alpha^{-1}}{\mu_i + \alpha^{-1}}\right)^{\alpha^{-1}} . \tag{5.6}$$

That is, N_i follows a negative binomial distribution, which is also known as the gamma-Poisson mixture distribution (Greenwood and Yule, 1920). In fact, the Poisson distribution can be seen as a limiting case of the negative binomial distribution when α approaches zero.

The log likelihood function of β and α is proportional to

$$l(\beta, \alpha) = \sum_{i=1}^{n}\left\{\sum_{j=1}^{N_i-1}\log(j + \alpha^{-1}) - (N_i + \alpha^{-1})\log[\exp(\beta'\mathbf{X}_i) + \alpha^{-1}]\right.$$
$$\left. -\alpha^{-1}\log\alpha + N_i\beta'\mathbf{X}_i\right\} \tag{5.7}$$

(Sun and Zhao, 2013). Taking derivatives of the log likelihood function (5.7) with respect to β and α, the maximum likelihood estimators of β and α, denoted by $\hat{\beta}_{NB}$ and $\hat{\alpha}_{NB}$, are the solutions to the following estimating equations

$$\sum_{i=1}^{n}\frac{N_i - \exp(\beta'\mathbf{X}_i)}{\alpha\exp(\beta'\mathbf{X}_i) + 1}\mathbf{X}_i = 0$$

and

$$\sum_{i=1}^{n}\left\{\frac{1}{\alpha^2}\left[\log\{1 + \alpha\exp(\beta'\mathbf{X}_i)\} - \sum_{j=0}^{N_i-1}\frac{1}{j + \alpha^{-1}}\right] + \frac{N_i - \exp(\beta'\mathbf{X}_i)}{\alpha[\alpha\exp(\beta'\mathbf{X}_i) + 1]}\right\} = 0 .$$

It can be easily shown that $\hat{\beta}_{NB}$ and $\hat{\alpha}_{NB}$ are consistent estimators of the true values of β and α, denoted by β_0 and α_0, respectively. In addition, their joint distribution can be asymptotically approximated by the multivariate normal distribution with mean $(\beta_0', \alpha_0)'$ and the covariance matrix determined by

$$Var(\hat{\beta}_{NB}) = \frac{1}{n} \left\{ \sum_{i=1}^{n} \frac{\exp(\hat{\beta}_{NB}'\mathbf{X}_i)}{1 + \hat{\alpha}_{NB} \exp(\hat{\beta}_{NB}'\mathbf{X}_i)} \mathbf{X}_i \mathbf{X}_i' \right\}^{-1} ,$$

$$Var(\hat{\alpha}_{NB}) = \frac{1}{n} \left\{ \sum_{i=1}^{n} \left(\log\left[1 + \hat{\alpha}_{NB} \exp(\hat{\beta}_{NB}'\mathbf{X}_i)\right] - \sum_{j=0}^{N_i-1} \frac{1}{j + \hat{\alpha}_{NB}^{-1}} \right)^2 \right.$$
$$\left. + \frac{\exp(\hat{\beta}_{NB}'\mathbf{X}_i)}{\hat{\alpha}_{NB}^2 \left[1 + \hat{\alpha}_{NB} \exp(\hat{\beta}_{NB}'\mathbf{X}_i)\right]} \right\}^{-1},$$

and $Cov(\hat{\beta}_{NB}, \hat{\alpha}_{NB}) = 0$.

A test for overdispersion can be developed for $H_0 : \alpha = 0$ versus $H_1 : \alpha > 0$ to determine whether the Poisson regression model is adequate. However, the standard Wald or likelihood ratio test is not valid since the null hypothesis is on the boundary of the parameter space. Thus a modified likelihood ratio test is more appropriate (Sun and Zhao, 2013). In particular, the null distribution of the modified likelihood ratio test statistic is a $50 : 50$ mixture of chi-squared distributions with 0 and 1 degrees of freedom.

5.2.3 Poisson and negative binomial regression for rate

In Subsections 5.2.1 and 5.2.2, it is assumed that the follow-up time period (or exposure time) is the same for all the subjects. However, the follow-up time may vary greatly subject by subject in many randomized trials. For subject i, let T_i denote the follow-up time, $i = 1, \cdots, n$. Then the observed data are $\{(N_i, T_i, \mathbf{X_i}); i = 1, \cdots, n\}$.

The Poisson regression model for rate can be written as

$$E\left(\frac{N_i}{T_i} \middle| \mathbf{X}_i\right) = \exp(\beta'\mathbf{X}_i) . \tag{5.8}$$

Similarly, the negative binomial regression model for rate has the form

$$E\left(\frac{N_i}{T_i} \middle| \mathbf{X}_i, \nu_i\right) = \nu_i \exp(\beta'\mathbf{X}_i) . \tag{5.9}$$

Under models (5.8) and (5.9), the conditional means of N_i are given by

$$E(N_i | \mathbf{X}_i) = \exp\left\{\beta'\mathbf{X}_i + \log(T_i)\right\}$$

and

$$E(N_i | \mathbf{X}_i, \nu_i) = \nu_i \exp\left\{\beta'\mathbf{X}_i + \log(T_i)\right\} ,$$

respectively, which are special cases of models (5.1) and (5.5) by setting the first component of β as one. The term $\log(T_i)$ is often referred to as an "offset". The same maximum likelihood estimation procedures discussed above for β and α can be applied.

To illustrate the regression analysis procedures discussed above for simple count data, the NCGS data described in Section 5.1 and given in the data set I of Chapter 9 in Sun and Zhao (2013), are used to determine the treatment effect on the nausea recurrence rate. Let N_i denote the total or cumulative count of episodes of nausea over the maximum follow-up time T_i for subject i, $i = 1, \cdots, 113$. Define X_i to be equal to 1 if the patient was in the high dose group and 0 otherwise. To estimate the effect of the high dose treatment, we analyze the average occurrence rate of nausea using the negative binomial regression model (5.9). The application of the maximum likelihood estimation procedure described in Subsection 5.2.2 gives $\hat{\beta}_{NB} = -0.766$ with the estimated standard error of 0.538 and a p-value of 0.155 for testing no treatment effect on the average occurrence rate of nausea. This indicates that there seems no dose effect on the occurrence rate of nausea. In addition, $\hat{\alpha}_{NB} = 7.720$ with a p-value less than 0.001 for testing overdispersion using the modified likelihood ratio test in Subsection 5.2.2, indicating that a Poisson regression model is inappropriate for the nausea count data.

5.2.4 Other models for simple count data

Although the negative binomial regression model can be used to capture overdispersion, it may not be sufficient to analyze empirical count data with an excess number of zeros. Some alternative regression models for count have been proposed, including the hurdle regression model (Mullahy, 1986) and the zero-inflated Poisson regression model (Lambert, 1992). The hurdle model has an interpretation as a two-part model which consists of a binary outcome model (e.g., logit or probit) to predict zeros and a truncated-at-zero count model to predict positive counts. The zero-inflated Poisson regression model can be interpreted as a two-component mixture model combining a point mass at zero with a Poisson count distribution. Both models allow for two sources of overdispersion. The former one allows for extra (or too few) zeros, while the latter one allows for overdispersion induced by individual heterogeneity in the positive set. The standard maximum likelihood estimation procedure can be applied to both models and more details on these methods can be found in Cameron and Trivedi (1998). Of course, one could also apply the Poisson maximum pseudo-likelihood estimator (Gourieroux et al., 1984a, 1984b), the Poisson generalized linear model estimator (McCullagh and Nelder, 1989), and the Poisson generalized methods of moments estimator (Hansen, 1982). The first method is based on the correct specification of the conditional mean without assuming equidispersion, while the other two methods are based on correct specification of both the conditional mean and conditional variance.

5.3 Regression Analysis of Correlated Count Data: Likelihood-Based Approaches

In this and next sections, we will discuss regression analysis of correlated or panel count data. Consider a follow-up study that involves n independent subjects who may experience recurrent events of interest, and let $N_i(t)$ represent the cumulative number of events that have occurred prior to time t for the ith subject with $N_i(0) = 0$, $i = 1, \cdots, n$. Suppose that $N_i(\cdot)$ is observed only at finite time points $T_{i,1} < \cdots < T_{i,m_i}$, where m_i denotes the potential number of observation times for subject i, $i = 1, \cdots, n$. Let $\mathbf{X}_i$ be time-independent as defined in the preceding section. Then the observed panel count data have the form

$$\{(T_{i,j}, N_{i,j}, \mathbf{X}_i); \ j = 1, \cdots, m_i, \ i = 1, \cdots, n\} \ .$$

Given $\mathbf{X}_i$, a commonly used model for the mean function of $N_i(t)$ is the so-called proportional mean model

$$E(N_i(t)|\mathbf{X}_i) = \mu_0(t)\exp(\beta'\mathbf{X}_i) \ , \tag{5.10}$$

where $\mu_0(t)$ is an unspecified continuous baseline mean function and β represents a $p \times 1$ vector of unknown regression parameters.

In the following, we will consider two Poisson process-based approaches.

5.3.1 Maximum pseudo-likelihood estimation for the Poisson model

Given $\mathbf{X}_i$, by assuming that the $N_i(t)$'s are non-homogeneous Poisson processes with the mean function (5.10) and ignoring the dependence of $\{N_i(T_{i,j}), j = 1, \cdots, m_i\}$ for each i , Zhang (2002) derived a log pseudo-likelihood function of the form

$$l_p(\mu_0, \beta) = \sum_{i=1}^{n} \sum_{j=1}^{m_i} \{N_{i,j} \log \mu_0(T_{i,j}) + N_{i,j}\beta'\mathbf{X}_i - \mu_0(T_{i,j}) \exp(\beta'\mathbf{X}_i)\} \ . \tag{5.11}$$

To estimate $\mu_0(t)$ and β, it is natural to maximize the log pseudo-likelihood function (5.11).

Let $s_1 < \cdots < s_m$ denote the ordered distinct observation time points in the set of all observation time points $\{T_{i,j}, j = 1, \cdots, m_i, i = 1, \cdots, n\}$, and w_l and $\bar{N}_l$ be the number and mean value of the observations made at s_l, respectively, $l = 1, \cdots, m$. For a given β, define

$$\bar{a}_l(\beta) = \frac{1}{w_l} \sum_{i=1}^{n} \sum_{j=1}^{m_i} \exp(\beta'\mathbf{X}_i)I(T_{i,j} = s_l)$$

and

$$\bar{b}_l(\beta) = \frac{1}{w_l} \sum_{i=1}^{n} \sum_{j=1}^{m_i} N_{i,j}\beta'\mathbf{X}_iI(T_{i,j} = s_l)$$

for $l = 1, \cdots, m$. Then the log pseudo-likelihood function (5.11) can be rewritten as

$$l_p(\mu_0, \beta) = \sum_{l=1}^{m} w_l\{\bar{N}_l \log \mu_0(s_l) - \bar{a}_l(\beta)\mu_0(s_l) + \bar{b}_l(\beta)\} \ . \tag{5.12}$$

Let $\hat{\mu}_{PL}(t)$ and $\hat{\beta}_{PL}$ denote the estimators of $\mu_0(t)$ and β that maximize (5.12) with $\hat{\mu}_{PL}(t)$ being a non-decreasing step function with possible jumps only at the s_l's. Let $\mu = (\mu_0(s_1), \cdots, \mu_0(s_m))' = (\mu_1, \cdots, \mu_m)'$. Then $\hat{\mu}_{PL}(t)$ and $\hat{\beta}_{PL}$ can be determined by maximizing

$$l_p(\mu, \beta) = \sum_{l=1}^{m} w_l\{\bar{N}_l \log \mu_l - \bar{a}_l(\beta)\mu_l + \bar{b}_l(\beta)\} \ . \tag{5.13}$$

over the vectors of parameters μ and β under the restriction $\mu_1 \leq \cdots \leq \mu_m$.

Zhang (2002) proposed a two-step iterative algorithm which maximizes (5.13) over μ and β alternatively. For fixed β, the maximization of (5.13) over μ is equivalent to maximizing

$$\sum_{l=1}^{m} w_l\bar{a}_l(\beta) \left\{ \frac{\bar{N}_l}{\bar{a}_l(\beta)} \log \mu_l - \mu_l \right\} \ .$$

Therefore, for given β, the $\hat{\mu}_{PL}(s_l)$'s are the isotonic regression estimator (IRE) of $\{\bar{N}_1/\bar{a}_1(\beta), \cdots, \bar{N}_m/\bar{a}_m(\beta)\}$ with weights $\{w_1\bar{a}_1(\beta), \cdots, w_m\bar{a}_m(\beta)\}$. That is,

$$\hat{\mu}_{PL}(s_l; \beta) = \max_{r \leq l} \min_{s \geq l} \frac{\sum_{v=r}^{s} w_v\bar{N}_v}{\sum_{v=r}^{s} w_v\bar{a}_v(\beta)} = \min_{s \geq l} \max_{r \leq l} \frac{\sum_{v=r}^{s} w_v\bar{N}_v}{\sum_{v=r}^{s} w_v\bar{a}_v(\beta)}$$

given by the max-min formula of the IRE (Barlow et al., 1972; Robertson et al., 1988).

For given $\mu_0(t)$ or μ, one can apply the standard Newton-Raphson algorithm for the estimation of β. Zhang (2002) showed that the log pseudo-likelihood function (5.13) given μ is a concave function of β and its value increases after each iteration. For the convergence criterion, one may check the absolute relative change of the log pseudo-likelihood function (5.13) between two successive estimators of $\mu_0(t)$ and β. In general, the algorithm is shown to be robust and always converges regardless of the choice of the initial estimator of β. For the variance estimation of $\hat{\beta}_{PL}$, the simple bootstrap procedure can be applied.

As an illustration, we apply the method above to the NCGS data discussed before. Let $N_i(t)$ be the underlying recurrent event process for the occurrences of nausea for subject i, $i = 1, \cdots, 113$. Under the proportional mean model (5.10), the above maximum pseudo-likelihood estimation procedure based on 200 bootstrap samples, yields $\hat{\beta}_{PL} = -0.533$ with the estimated standard error of 0.543 and a p-value of 0.326 for testing no treatment effect on the mean function of nausea recurrences. A similar conclusion is reached as that in Subsection 5.2.3 regarding the treatment effect.

5.3.2 Maximum likelihood estimation for the Poisson model

Besides maximizing the log pseudo-likelihood function in Subsection 5.3.1, one may consider the maximization of the following full log likelihood function

$$l(\mu_0, \beta) = \sum_{l'=1}^{m-1} \sum_{l=l'+1}^{m} \tilde{n}_{l,l'} \log[\mu_0(s_l) - \mu_0(s_{l'})] - \sum_{l=1}^{m} b_l(\beta)\mu_0(s_l) + \sum_{i=1}^{n} N_{i,m_i}\beta'\mathbf{X}_i , \quad (5.14)$$

where

$$b_l(\beta) = \sum_{i=1}^{n} I(T_{i,m_i} = s_l) \exp(\beta'\mathbf{X}_i)$$

and

$$\tilde{n}_{l,l'} = \sum_{i=1}^{n} \sum_{j=1}^{m_i} (N_{i,j} - N_{i,j-1})I(T_{i,j} = s_l, T_{i,j-1} = s_{l'})$$

for $0 \leq l' < l \leq m$.

Let $\hat{\mu}_L(t)$ and $\hat{\beta}_L$ denote the estimators of $\mu_0(t)$ and β that maximize (5.14). As shown in Wellner et al. (2004), the maximum likelihood estimator $\hat{\beta}_L$ is more efficient compared to the maximum pseudo-likelihood estimator $\hat{\beta}_{PL}$ in the preceding subsection, but the pseudo-likelihood estimation method has advantages in terms of computational simplicity. Wellner and Zhang (2007) further studied the asymptotic properties of the estimators $(\hat{\beta}_{PL}, \hat{\mu}_{PL}(t))$ and $(\hat{\beta}_L, \hat{\mu}_L(t))$, proved the strong consistency, derived the rate of convergence of both estimators in some L-metrics related to the observation scheme, and established the asymptotic normality of both $\hat{\beta}_{PL}$ and $\hat{\beta}_L$ under appropriate hypotheses. To improve computing efficiency, Lu et al. (2009) developed another semiparametric maximum likelihood estimation procedure for $\mu_0(t)$ and β in which $\mu_0(t)$ was approximated by monotone B-splines and the generalized Rosen algorithm in Jamshidian (2004) was applied for the parameter estimation.

5.3.3 Maximum likelihood estimation for the negative binomial model

As discussed in Section 5.2, the Poisson process may be too restrictive in practice, thus one may consider a mixed Poisson process. In particular, assume that the $N_i(t)$'s are non-homogeneous Poisson processes with the mean function

$$E(N_i(t)|\mathbf{X}_i, \nu_i) = \nu_i\mu_0(t) \exp(\beta'\mathbf{X}_i) ,$$

given the covariate vector $\mathbf{X}_i$ and a latent variable ν_i which follows a gamma distribution with the density function (5.4). Then it can be shown that the $N_{i,j}$'s follow a negative binomial distribution and the resulting full likelihood function is given by

$$L_{NB}(\mu_0, \beta) = \prod_{i=1}^{n} \prod_{j=1}^{m_i} \left\{ \frac{\Gamma(N_{i,j} + \alpha^{-1})}{\Gamma(N_{i,j} + 1)\Gamma(\alpha^{-1})} \left[\frac{\mu_0(T_{i,j}) \exp(\beta' \mathbf{X}_i)}{\mu_0(T_{i,j}) \exp(\beta' \mathbf{X}_i) + \alpha^{-1}} \right]^{N_{i,j}} \right.$$
$$\left. \left[\frac{\alpha^{-1}}{\mu_0(T_{i,j}) \exp(\beta' \mathbf{X}_i) + \alpha^{-1}} \right]^{\alpha^{-1}} \right\} . \tag{5.15}$$

The same maximum likelihood estimation procedure discussed in Subsection 5.3.2 can be applied to estimate β under (5.15).

5.4 Regression Analysis of Correlated Count Data: Distribution-Free Approaches

In the previous section, we have focused on the situation where the underlying counting processes can be regarded as non-homogeneous Poisson processes and as mentioned before, this may not be true in practice. Also note that the implementation of the estimation procedures above is relatively expensive in computation due to the involved estimation of the unknown baseline mean function $\mu_0(t)$ in the proportional mean model (5.10). In this section, we will present two estimating equation-based approaches that do not require any distribution assumption and not involve the estimation of the unknown baseline mean function.

5.4.1 Conditional estimating equation method

Let the $N_i(t)$'s, $T_{i,j}$'s, $N_{i,j}$'s, $\mathbf{X}_i$'s, and s_l's be defined as before and suppose that the mean function $E(N_i(t)|\mathbf{X}_i)$ is given by model (5.10). For subject i, let the random variable C_i denote the follow-up time associated with the subject. Define $O_i(t) = \sum_{i=1}^{m_i} I(T_{i,j} \leq t)$ which is a counting process representing the accumulated number of observations on subject i up to time t, $i = 1, \cdots, n$. Let $o_i(t) = O_i(t) - O_i(t^-)$ indicating whether subject i has an observation at time t. In the following, we assume that $N_i(t)$, $O_i(t)$ and C_i may be dependent, but given $\mathbf{X}_i$, $N_i(t)$, $O_i(t)$, and C_i are independent of each other. The goal is to make inference about β.

For subject i, define a new process

$$\tilde{N}_i(t) = \int_0^t N_i(s) dO_i(s) \ , t \geq 0 \ ,$$

which has possible jumps only at the observation time points $T_{i,j}$'s with respective jump size $N_{i,j}$'s. It can seen that recurrent event data are available for the $\tilde{N}_i(t)$'s and that

$$E\{d\tilde{N}_i(t)|O_i(s), 0 < s \leq t; \mathbf{X}_i\} = \mu_0(t) \exp(\beta' \mathbf{X}_i) dO_i(t) \ .$$

In the following, let τ be the longest follow-up time. Define

$$S_C^{(j)}(\beta; t) = \frac{\sum_{i=1}^{n} I(C_i \geq t) \mathbf{X}_i^{\otimes j} \exp(\beta' \mathbf{X}_i) o_i(t)}{\sum_{i=1}^{n} o_i(t)}$$

for t with $\sum_{i=1}^{n} o_i(t) > 0$ and $j = 0, 1, 2$, where $a^{\otimes j} = 1, a, aa'$, for $j = 0, 1, 2$. Then for the estimation of β, motivated by the Cox partial score function in Lawless and Nadeau (1995), Hu et al. (2003) constructed the following estimating function

$$U_n^C(\beta; W) = \sum_{i=1}^{n} \int_0^{\tau} W(t)I(C_i \geq t)\{\mathbf{X}_i - \bar{\mathbf{X}}_C(t; \beta)\}d\tilde{N}_i(t) , \qquad (5.16)$$

where $W(\cdot)$ is a known weight function and $\bar{\mathbf{X}}_C(t; \beta) = S_C^{(1)}(\beta; t)/S_C^{(0)}(\beta; t)$.

For the simple situation where all the subjects have only one observation at the same time point $T_0 < \tau$ with $C_i = \tau$ for all i and a constant weight function $W(t) = 1$, the estimating function (5.16) reduces to

$$U_n^C(\beta; 1) = \sum_{i=1}^{n} \mathbf{X}_i N_i(T_0) - \left\{ \sum_{i=1}^{n} \int_0^{T_0} \frac{1}{\sum_{j=1}^{n} \exp(\beta'\mathbf{X}_j)} dN_i(t) \right\}$$

$$\times \left\{ \sum_{i=1}^{n} \mathbf{X}_i \exp(\beta'\mathbf{X}_i) \right\} .$$

Let β_0 be the true value of β and $\hat{\beta}_n^C$ be the solution to $U_n^C(\beta; W) = 0$. Define

$$\hat{\mu}_0^C(t; \beta) = \frac{\sum_{i=1}^{n} I(C_i \geq t)N_i(t)o_i(t)}{\sum_{i=1}^{n} I(C_i \geq t)\exp(\beta'\mathbf{X}_i)o_i(t)}$$

and

$$\hat{M}_i^C(t; \beta) = \int_0^t I(C_i \geq s)\{N_i(s) - \hat{\mu}_0^C(s; \beta)\exp(\beta'\mathbf{X}_i)\}dO_i(s)$$

for $t \in [0, \tau]$. Hu et al. (2003) showed that the estimator $\hat{\beta}_n^C$ is consistent and the distribution of $\sqrt{n}(\hat{\beta}_n^C - \beta_0)$ can be asymptotically approximated by the multivariate normal distribution with mean zero and the covariance matrix $\hat{\Sigma}^C = A_C^{-1}(\hat{\beta}_n^C)B_C(\hat{\beta}_n^C)A_C^{-1}(\hat{\beta}_n^C)$, where

$$A_C(\beta) = -\frac{1}{n} \sum_{i=1}^{n} \int_0^{\tau} I(C_i \geq t) \left[\frac{S_C^{(2)}(\beta; t)}{S_C^{(0)}(\beta; t)} - \bar{\mathbf{X}}_C(t; \beta)^{\otimes 2} \right] d\tilde{N}_i(t)$$

and

$$B_C(\beta) = \frac{1}{n} \left[\sum_{i=1}^{n} \int_0^{\tau} \{\mathbf{X}_i - \bar{\mathbf{X}}_C(t; \beta)\}d\hat{M}_i^C(t; \beta) \right]^{\otimes 2} .$$

5.4.2 Unconditional estimating equation method

In Subsection 5.4.1, the conditional estimating equation method considers fixed observation time points $T_{i,1} < \cdots < T_{i,m_i}$, $i = 1, \cdots, n$, and it is assumed that the covariates $\mathbf{X}_i$ have no effect on the observation process $O_i(t)$. Now we discuss an unconditional estimating equation method by modeling the observation process.

Assume that the observation process $O_i(t)$ follows the proportional rate model

$$E\{dO_i(t)|\mathbf{X}_i\} = \exp(\gamma'\mathbf{X}_i)d\tilde{\mu}_0(t) , \qquad (5.17)$$

where $\tilde{\mu}_0(.)$ is either discrete or absolutely continuous with respect to Lebesgue measure and γ is a $p \times 1$ vector of regression parameters representing the covariate effects on the observation process. Combining models (5.10) and (5.17), we have

$$E\{d\tilde{N}_i(t)|\mathbf{X}_i\} = \exp(\tilde{\beta}'\mathbf{X}_i)d\tilde{\mu}_0^*(t) \ ,$$

where $\tilde{\beta} = \beta + \gamma$ and $\tilde{\mu}_0^*(t) = \int_0^t \mu_0(s)d\tilde{\mu}_0(s)$.

Define

$$S_M^{(j)}(\tilde{\beta};t) = \frac{1}{n}\sum_{i=1}^n I(C_i \geq t)\mathbf{X}_i^{\otimes j}\exp(\tilde{\beta}'\mathbf{X}_i)$$

for $j = 0, 1, 2$ and $\bar{\mathbf{X}}_M(t;\tilde{\beta}) = S_M^{(1)}(\tilde{\beta};t)/S_M^{(0)}(\tilde{\beta};t)$. Similar to the estimating function (5.16), an unbiased estimating function for $\tilde{\beta}$ is given by

$$U_n^M(\tilde{\beta};W) = \sum_{i=1}^n \int_0^\tau W(t)I(C_i \geq t)\{\mathbf{X}_i - \bar{\mathbf{X}}_M(t;\tilde{\beta})\}d\tilde{N}_i(t) \ , \tag{5.18}$$

where $W(\cdot)$ is a known weight function as before. Assume a constant weight function $W(t) = 1$ and $C_i = \tau$ for all i, the estimating function (5.18) reduces to

$$U_n^M(\tilde{\beta};1) = \sum_{i=1}^n \mathbf{X}_i \int_0^\tau N_i(t)dO_i(t) - \left\{\sum_{i=1}^n \mathbf{X}_i\exp(\tilde{\beta}'\mathbf{X}_i)\right\} \times \int_0^\tau \frac{\sum_{l=1}^n N_l(t)dO_l(t)}{\sum_{j=1}^n \exp(\tilde{\beta}'\mathbf{X}_j)} \ .$$

Let $\hat{\gamma}$ be the estimator for γ using the following estimating function

$$L(\gamma) = \sum_{i=1}^n \int_0^\tau \left\{\gamma'\mathbf{X}_i - \log\left[\sum_{l=1}^n I(C_l \geq t)\exp(\gamma'\mathbf{X}_i)\right]\right\}dO_i(t) \ ,$$

and $\hat{\tilde{\beta}}$ be the estimator of $\tilde{\beta}$ given by the solution to the estimating equation $U_n^M(\tilde{\beta};W) = 0$ for a given $W(t)$. Then an estimator for β can be obtained as $\hat{\beta}_n^M = \hat{\tilde{\beta}} - \hat{\gamma}$.

Define

$$a_{M,i}^{11}(\tilde{\beta},\gamma) = \int_0^\tau I(C_i \geq t)\left[\frac{S_M^{(2)}(\tilde{\beta};t)}{S_M^{(0)}(\tilde{\beta};t)} - \bar{\mathbf{X}}_M(t;\tilde{\beta})^{\otimes 2}\right]d\tilde{N}_i(t) \ ,$$

$$a_{M,i}^{22}(\tilde{\beta},\gamma) = \int_0^\tau I(C_i \geq t)\left[\frac{S_M^{(2)}(\gamma;t)}{S_M^{(0)}(\gamma;t)} - \bar{\mathbf{X}}_M(t;\gamma)^{\otimes 2}\right]dO_i(t) \ ,$$

$$\hat{M}_i^M(t;\tilde{\beta}) = \int_0^t I(C_i \geq s)\left\{d\tilde{N}_i(s) - \exp(\tilde{\beta}'\mathbf{X}_i)d\hat{\tilde{\mu}}_0^*(s;\tilde{\beta})\right\} \ ,$$

and

$$\hat{M}_i^O(t;\gamma) = \int_0^t I(C_i \geq s)\left\{dO_i(s) - \exp(\gamma'\mathbf{X}_i)d\hat{\tilde{\mu}}_0(s;\gamma)\right\}$$

for $i = 1, \cdots, n$. Here $\hat{\tilde{\mu}}_0^*(t;\tilde{\beta})$ and $\hat{\tilde{\mu}}_0(t;\gamma)$ are the Breslow estimators for $\tilde{\mu}_0^*(t)$ and $\tilde{\mu}_0(t)$, respectively.

Hu et al. (2003) proved that the estimator $\hat{\beta}_n^M$ is consistent and the distribution of $\sqrt{n}(\hat{\beta}_n^M - \beta_0)$ can be asymptotically approximated by the multivariate normal distribution with mean zero and the covariance matrix

$$\hat{\Sigma}^M = (\mathbf{I}_p, -\mathbf{I}_p)A_M^{-1}(\hat{\tilde{\beta}},\hat{\gamma})B_M(\hat{\tilde{\beta}},\hat{\gamma})A_M^{-1}(\hat{\tilde{\beta}},\hat{\gamma})(\mathbf{I}_p, -\mathbf{I}_p)'$$

TABLE 5.1
Estimated treatment effects under the proportional mean model for panel count data.

Method	Estimate	Standard error	p-value for $\beta = 0$
$\hat{\beta}_{PL}$	-0.533	0.543	0.326
$\hat{\beta}_n^C$	-0.419	0.537	0.435
$\hat{\beta}_n^M$	-0.527	0.628	0.401

where $\mathbf{I}_p$ is the $p \times p$ identity matrix,

$$A_M(\tilde{\beta}, \gamma) = -\frac{1}{n} \sum_{i=1}^{n} \mathrm{diag}\left(a_{M,i}^{11}(\tilde{\beta}, \gamma), a_{M,i}^{22}(\tilde{\beta}, \gamma)\right) ,$$

and

$$B_M(\tilde{\beta}, \gamma) = \frac{1}{n} \sum_{i=1}^{n} \left[\begin{array}{c} \int_0^\tau \{\mathbf{X}_i - \bar{\mathbf{X}}_M(t; \tilde{\beta})\} d\hat{M}_i^M(t; \tilde{\beta}) \\ \int_0^\tau \{\mathbf{X}_i - \bar{\mathbf{X}}_M(t; \gamma)\} d\hat{M}_i^O(t; \gamma) \end{array} \right]^{\otimes 2} .$$

5.4.3　Analysis of the National Cooperative Gallstone Study

In this subsection, we illustrate the estimation procedures described above through the NCGS data on the nausea occurrence. Table 5.1 summarizes the results by applying the maximum pseudo-likelihood estimation procedure in Subsection 5.3.1, the conditional estimating equation procedure in Subsection 5.4.1, and the unconditional estimating equation procedure in Subsection 5.4.2. Note that the first likelihood-based procedure relies on the Poisson process assumption, while the other two estimating equation-based approaches do not require any distribution assumption. One can see from the table that all three methods give the same conclusion that the high dose treatment did not significantly reduce the recurrence rate of nausea for patients with floating gallstones. Furthermore, the unconditional estimating equation procedure also yields $\hat{\gamma} = -0.024$ with the estimated standard error of 0.040 and a p-value of 0.549 for testing $H_0 : \gamma = 0$ in model (5.17). This suggests that there was no significant treatment effect on the patient observation process too.

5.5　Discussion and Concluding Remarks

In this chapter, likelihood-based approaches have been reviewed for regression analysis of simple count data under the Poisson and negative binomial models. For regression analysis of correlated or panel count data, several likelihood-based and distribution-free approaches have been discussed under the proportional mean model (5.10). In addition to these methods, many other authors have investigated the inference procedures for the proportional mean model (e.g., Cheng and Wei, 2000; Sun and Wei, 2000) and the proportional rate model (e.g., Lawless and Zhan, 1998; Nielsen and Dean, 2008; Staniswalls et al., 1997). In particular, Cheng and Wei (2000) developed an estimating equation approach to incorporate time-dependent covariates. Lawless and Zhan (1998) discussed a Poisson-based likelihood procedure and extended generalized estimating equations by using a piecewise constant rate function. Nielsen and Dean (2008) and Staniswalls et al. (1997) employed some smoothing techniques along with Poisson process-based assumptions.

In all of the previous discussions, a basic assumption is that the observation and censoring times are independent of the underlying recurrent event process of interest completely or given covariates. In practice, however, this may not be true and actually some methods have been developed for regression analysis of panel count data with dependent observation and follow-up processes. These include the joint modeling inference procedures (e.g., He et al., 2009; Huang et al., 2006; Kim, 2006; Sun et al., 2007; Zhao and Tong, 2011; Zhao et al., 2013) and the marginal approaches based on, for example, semiparametric transformation models (e.g., Li et al., 2010; Li et al., 2013; Zhao et al., 2013).

An additional type of count data that is closely related to panel count data is the so-called recurrent event data, a topic that will also be discussed in this book. Recurrent event data arise when the subjects in a study concerning some recurrent events are observed or followed continuously and thus instead of only counts of the occurrences of the event, they give all exact occurrence times of these occurrences. From the view of the underlying counting process that controls the occurrence of the event, panel count data give us incomplete data, while recurrent event data provide complete data. For regression analysis of recurrent event data, the readers are referred to Chapter 7 of this book or Cook and Lawless (2007).

Another important issue related to the panel count data discussed above is regression analysis of multivariate panel count data, in which multi-type recurrent event processes of interest may be correlated. Among others, Chen et al. (2005) proposed a mixed Poisson process approach with piecewise constant baseline intensities and multivariate log-normal random effects. He et al. (2008) developed a marginal approach using estimating equations. Li et al. (2011) studied the regression analysis of multivariate panel count data with dependent observation times. A complete and thorough review of literature on analysis of panel count data can be found in Sun and Zhao (2013).

Bibliography

Barlow, R., Bartholomew, D., Bremner, J. and Brunk, H. (1972). *Statistical inference under order restrictions*. New York: John Wiley & Sons.

Breslow, N. E. (1984). Extra-Poisson variation in log-linear models. *Applied Statistics*, 33, 38–44.

Cameron, A. C. and Trivedi, P. K. (1998). *Regression analysis of count data*. Econometric Society Monograph, No. 30, Cambridge University Press.

Chen, B. E., Cook, R. J., Lawless, J. F. and Zhan, M. (2005). Statistical methods for multivariate interval-censored recurrent events. *Statistics in Medicine*, 24, 671–691.

Cheng, S. C. and Wei, L. J. (2000). Inferences for a semiparametric model with panel data. *Biometrika*, 87, 89–97.

Cook, R. J. and Lawless, J. F. (2007). *The statistical analysis of recurrent events*. New York: Springer-Verlag.

Diggle, P. J., Heagerty, P., Liang, K.-Y. and Zeger, S. L. (2002). *Analysis of Longitudinal Data*, 2nd Edition. New York: Oxford University Press.

Gourieroux, C., Monfort, A. and Trognon, A. (1984a). Pseudo maximum likelihood methods: Theory. *Econometrica*, 52, 681–700.

Gourieroux, C., Monfort, A. and Trognon, A. (1984b). Pseudo maximum likelihood methods: Applications to Poisson models. *Econometrica*, 52, 701–720.

Greenwood, M. and Yule, G. U. (1920). An inquiry into the nature of frequency distributions of multiple happenings, with particular reference to the occurrence of multiple attacks of disease or repeated accidents. *Journal of the Royal Statistical Society, Series A*, 83, 255–279.

Hansen, L. P. (1982). Large sample properties of generalized method of moments estimators. *Econometrica*, 50, 1029–1054.

He, X., Tong, X. and Sun, J. (2009). Semiparametric analysis of panel count data with correlated observation and follow-up times. *Lifetime Data Analysis*, 15, 177–196.

He, X., Tong, X., Sun, J. and Cook, R. J. (2008). Regression analysis of multivariate panel count data. *Biostatistics*, 9, 234–248.

Hinde, J. (1982). Compound Poisson regression models. In GLIM 82: *Proceedings of the International Conference in Generalized Linear Models*, R. Gilchrist, (ed.), Berlin: Springer-Verlag, 109–121.

Hu, X. J., Sun, J. and Wei, L. J. (2003). Regression parameter estimation from panel counts. *Scandinavian Journal of Statistics*, 30, 25–43.

Huang, C. Y., Wang, M. C. and Zhang, Y. (2006). Analyzing panel count data with informative observation times. *Biometrika*, 93, 763–775.

Jamshidian, M. (2004). On algorithms for restricted maximum likelihood estimation. *Computational Statistics and Data Analysis*, 45, 137–157.

Jorgensen, D. W. (1961). Multiple regression analysis of a Poisson process. *Journal of the American Statistical Association*, 56, 235–245.

Kalbfleisch, J. D. and Lawless, J. F. (1985). The analysis of panel data under a Markov assumption. *Journal of the American Statistical Association*, 80, 863–871.

Kianifard, F. and Gallo P. P. (1995). Poisson regression analysis in clinical research. *Journal of Biopharmaceutical Statistics*, 5, 115–129.

Kim, Y-J. (2006). Analysis of panel count data with dependent observation times. *Communications in Statistics – Simulation and Computation*, 35, 983–990.

Lamber, D. (1992). Zero-inflated Poisson regression with an application to defects in manufacturing. *Technometrics*, 34, 1–14.

Lawless, J. F. and Nadeau, J. C. (1995). Some simple robust methods for the analysis of recurrent events. *Technometrics*, 37, 158–168.

Lawless, J. F. and Zhan, M. (1998). Analysis of interval-grouped recurrent event data using piecewise constant rate functions. *Canadian Journal of Statistics*, 26, 549–565.

Li, N., Park, D-H., Sun, J. and Kim, K. (2011). Semiparametric transformation models for multivariate panel count data with dependent observation process. *Canadian Journal of Statistics*, 39, 458–474.

Li, N., Sun, L. and Sun, J. (2010). Semiparametric transformation models for panel count data with dependent observation processes. *Statistics in Biosciences*, 2, 191–210.

Li, N., Zhao, H. and Sun, J. (2013). Semiparametric transformation models for panel count data with correlated observation and follow-up times. *Statistics in Medicine*, 32, 3039–3054.

Lu, M., Zhang, Y. and Huang, J. (2009). Semiparametric estimation methods for panel count data using monotone B-splines. *Journal of the American Statistical Association*, 104, 1060–1070.

McCullagh, P. and Nelder, J. A. (1989). *Generalized linear models*, 2nd Edition. London: Chapman and Hall.

Mullahy, J. (1986). Specification and testing of some modified count data models. *Journal of Econometrics*, 33, 341–365.

Nielsen, J. D. and Dean, C. B. (2008). Clustered mixed nonhomogeneous Poisson process spline models for the analysis of recurrent event panel data. *Biometrics*, 64, 751–761.

Robertson, T., Wright, F. T. and Dykstra, R. (1988). *Order restricted statistical inference*. New York: John Wiley & Sons.

Schoenfield, L. J., Lachin, J. M., the Steering Committee and the NCGS Group (1981). Chenodiol (chenodeoxycholic acid) for dissolution of gallstones: The National Cooperative Gallstone Study. *Annals of Internal Medicine*, 95, 257–282.

Staniswalls, J. G., Thall, P. F. and Salch, J. (1997). Semiparametric regression analysis for recurrent event interval counts. *Biometrics*, 53, 1334–1353.

Sun, J. (2006). *The statistical analysis of interval-censored failure time data*. New York: Springer.

Sun, J., Tong, X. and He, X. (2007). Regression analysis of panel count data with dependent observation times. *Biometrics*, 63, 1053–1059.

Sun, J. and Wei, L. J. (2000). Regression analysis of panel count data with covariate-dependent observation and censoring times. *Journal of the Royal Statistical Society, Series B*, 62, 293–302.

Sun, J. and Zhao, X. (2013). *The statistical analysis of panel count data*. New York: Springer.

Thall, P. F. (1988). Mixed Poisson likelihood regression models for longitudinal interval count data. *Biometrics*, 44, 197–209.

Thall, P. F. and Lachin, J. M. (1988). Analysis of recurrent events: Nonparametric methods for random-interval count data. *Journal of the American Statistical Association*, 83, 339–347.

Wellner, J. A. and Zhang, Y. (2007). Two likelihood-based semiparametric estimation methods for panel count data with covariates. *Annals of Statistics*, 35, 2106–2142.

Wellner, J. A., Zhang, Y. and Liu, H. (2004). A semiparametric regression model for panel count data: when do pseudo-likelihood estimators become badly inefficient? *Proceedings of the Second Seattle Symposium in Biostatistics*, New York: Springer, 143–174.

Zhang, Y. (2002). A semiparametric pseudolikelihood estimation method for panel count data. *Biometrika*, 89, 39–48.

Zhao, H., Li, Y. and Sun, J. (2013). Analysing panel count data with dependent observation process and a terminal event. *Canadian Journal of Statistics*, 41, 174–191.

Zhao, X. and Tong, X. (2011). Semiparametric regression analysis of panel count data with informative observation times. *Computational Statistics and Data Analysis*, 55, 291–300.

Zhao, X., Tong, X. and Sun, J. (2013). Robust estimation for panel count data with informative observation times. *Computational Statistics and Data Analysis*, 57, 33–40.

6

Longitudinal Data

Soeun Kim and Myunghee Cho Paik

CONTENTS

6.1 Introduction

In longitudinal studies, outcomes are obtained by following patients over time. In many cases, outcomes consist of repeatedly measured values of participants. Repeatedly measured outcomes are not independent and the main caveat of longitudinal studies is to take care of such dependence. In some studies, outcome is time to a certain event or a value at the end of the follow-up, which we will not deal with in this chapter.

The goal of this chapter is to review the methods and address the topics of longitudinal analysis especially in randomized clinical trials. Although analysis ignoring all covariates yields a valid inference due to randomization, we discuss modeling approaches for following two reasons. First, we can improve efficiency of the estimator of the main effect of interest through modeling. Second, score test statistics obtained through modeling can be shown to be robust with respect to model mispecification due to randomization.

Longitudinal studies often have a vector of outcomes. Generalized Estimating Equations (GEE) (1; 2) and Generalized Linear Mixed Models (GLMMs) (3) are two different extensions of Generalized Linear Models (GLM) (4). As in GLM, they cover both continuous and discrete outcomes, but the GEE is a marginal approach, and the GLMM is a conditional approach. Specifically, the regression coefficients in GEE features *population average effects* that compare the mean outcome for groups of subjects in the population. On the other hand, the regression coefficients in GLMMs can be interpreted as *subject-specific effects* that compare the mean outcome when the same subject is in one group versus the other (5).

In the following sections, GEE and GLMMs are reviewed. Section 1.2 starts with a review of GEE method describing asymptotic properties, efficiency of the GEE estimator, and model selection criterion in GEE. In section 1.3, GLMMs are presented as the subject-specific models, and covers a review of estimation procedures. Section 1.4 is new and devoted to special features under randomization, and section 1.5 outlines the methods dealing with missing data in GEE and in GLMMs. Finally, a case study of a clinical trial in antihypertensive drug in isolated hypertension is given in section 1.6.

6.2   Generalized Linear Models

Before going into GEE and GLMMs, we briefly introduce generalized linear models (4) in this section. GLMs are the class of models that extend linear regression models to accommodate continuous or categorical response variables. In other words, GLMs offer a flexibility by allowing different types of response variables, including continuous, ordinal, count, categorical responses. Logistic regression models, Poission regression models as well as analysis of variance are some of the most widely used GLMs that can be used for analyzing data with categorical response, counts, or for continuous responses, respectively (6). GLMs are specified with a distributional assumption, a systematic component, and a link function. Canonical link function for the Normal distribution is Identity function, whereas for Poisson distribution it is log function, and for Bernoulli distribution logit is the canonical link function. By taking the systematic component and link function, GLMs have the following form.

$$g(\mu_i) = \beta_1 X_{i1} + \beta_2 X_{i2} + ... + \beta_p X_{ip}$$

where $g(.)$ is a known link function, i denotes subject, and p is the number of covariates.

6.3   Generalized Estimating Equations

The GEE method is an inferential procedure concerning the marginal mean of a multivariate outcome through regression models. The GEE can be viewed as a natural extension of GLM to multivariate outcomes while keeping each element's mean structure the same. Liang and Zeger (1) extended the GLM to accomodate multivariate distributions by borrowing the idea of quasi-likelihood (7). Instead of making a full distributional assumption, the GEE method makes assumptions only on the mean and variance of an outcome as in the quasi-likelihood models. The GEE method can be viewed as a multivariate extension of quasi-likelihood allowing nuisance parameters in the variance other than the constant scale factor.

Various extensions of GEE have been proposed by many authors. Ziegler (8) gives an extensive annotated bibliography on GEE.

6.3.1 Notations

The GEE resembles the score equation from GLMs. Let $Y_i = (Y_{i1}, Y_{i2}, Y_{i3}, \cdots, Y_{in_i})^T$, $(i = 1, 2, \cdots, K)$ be a vector of outcomes for subject i, and $X_i = (X_{i1}^T, X_{i2}^T, \cdots X_{in_i}^T)$ be a $p \times n_i$ covariate matrix where X_{ij} is a $1 \times p$ vector of covariates. Conditioning on X_i, the mean of Y_i is denoted by $E(Y_i|X_i) = \mu_i$, where $\mu_i = (\mu_{i1}, \mu_{i2}, \cdots, \mu_{in_i})^T$. The outcome Y_i could represent measurements taken at n_i different occasions, or time points, for subject i. While the elements in Y_i may be correlated, $Y_1, Y_2, \cdots, Y_K$ are assumed to be independent vectors. The mean of Y_{ij} is $E(Y_{ij}|X_{ij}) = \mu_{ij} = h^{-1}(X_{ij}\beta)$, where $h(.)$ is a link function which relates the linear predictor $X_{ij}\beta$ to the mean μ_{ij}, and β is an unknown parameter of interest. For example, in logistic regression with binary outcomes, $log\frac{\mu_{ij}}{1-\mu_{ij}} = X_{ij}\beta$, the link function is $h(a) = log\frac{a}{(1-a)}$, and the parameter of interest, β has interpretation of log odds ratio. We assume the variance of Y_{ij} has a form $v_{ij}(\mu_{ij})/\phi$, where ϕ is a scalar dispersion parameter. For example, if Y_{ij} is binary, $E(Y_{ij}) = \mu_{ij}$, $v_{ij}(\mu_{ij}) = \mu_{ij}(1 - \mu_{ij})$, and $\phi = 1$. If Y_{ij} is distributed as normal with mean μ_{ij} and variance σ^2, then $v_{ij}(\mu_{ij}) = 1$, and $\phi = 1/\sigma^2$. We denote the variance-covariance matrix of a random vector Y_i by V_i where

$$V_i = A_i(\mu_i)^{1/2} R_i(\alpha) A_i(\mu_i)^{1/2}/\phi,$$

A_i is a $n_i \times n_i$ diagonal matrix with the jth diagonal element $v_{ij}(\mu_{ij})$, and $R_i(\alpha)$ is a "working" correlation matrix. The qualifier "working" implies that $R_i(\alpha)$ is not necessarily the true correlation, thus V_i is not the true variance of Y_i but only a specified variance. The GEE is defined as

$$U(\beta, \alpha) = \sum_{i=1}^{K} U_i(\beta, \alpha) = \sum_{i=1}^{K} \frac{\partial \mu_i^T}{\partial \beta} V_i^{-1}(Y_i - \mu_i) = 0. \tag{6.1}$$

Although V_i depends on ϕ, the solution of (6.1) does not, and we drop ϕ in the arguments of $U(\beta, \alpha)$. If α is known, the estimate of β, say $\hat{\beta}$, can be obtained by solving equation (6.1). The estimating function $U(\beta, \alpha)$ has a comparable form to the score function of a GLM, and importantly, $E(U) = 0$, as long as the conditional mean of Y is $E(Y_{ij}|X_{i1}, ..., X_{in_i}) = E(Y_{ij}|X_{ij})$ (9).

6.3.2 Asymptotic properties

Let the solution of (6.1) be $\hat{\beta}(\alpha)$. When α is known, the consistency and asymptotic normality of $\hat{\beta}(\alpha)$ can be obtained using the same argument of quasi-likelihood theory. In deriving the asymptotic properties of $\hat{\beta}(\alpha)$, uncertainty of the estimator of α should be taken into account, which in turn, depends on ϕ. A consistent estimator of ϕ given β can be obtained by

$$\hat{\phi}^{-1}(\beta) = \sum_{i=1}^{K} \sum_{j=1}^{n_i} \left(\frac{Y_{ij} - \mu_{ij}(\beta)}{\sqrt{v(\mu_{ij})}} \right)^2 \bigg/ \{\sum_{i=1}^{K} n_i - p\}.$$

Different forms of $R(\alpha)$ results in different estimators of α. In general, the estimators of α involve β and ϕ. For example, when all off-diagonal elements of R_i are the same, the estimator of α given β and ϕ can be obtained by

$$\hat{\alpha}(\beta, \phi) = \phi \left[\sum_{i=1}^{K} \sum_{j>k} \frac{(Y_{ij} - \mu_{ij})}{\sqrt{v(\mu_{ij})}} \frac{(Y_{ik} - \mu_{ik})}{\sqrt{v(\mu_{ik})}} \right] \bigg/ \{\sum_{i=1}^{K} n_i(n_i - 1)/2 - p\}.$$

Note that the estimation of α depends on β and ϕ, and in turn, the estimation of β depends on α. Each step should be alternated until both parameters converge. It turns out that estimating α does not affect inference on β given some assumptions (1). A crucial assumption is that the estimate of α is $\sqrt{K}$–consistent.

Let $W_0(\beta_0, \alpha) = E(-K^{-1} \frac{\partial U}{\partial \beta^T})$, $Var\{K^{-\frac{1}{2}} U(\beta_0, \alpha)\}$ be $W_1(\beta_0, \alpha)$, and

$$W_1(\beta_0, \alpha) = E\left(K^{-1} \sum_{i=1}^{K} \frac{\partial \mu_i^T}{\partial \beta} V_i^{-1} (Y_i - \mu_i)(Y_i - \mu_i)^T V_i^{-1} \frac{\partial \mu_i}{\partial \beta^T}\right).$$

Then $K^{\frac{1}{2}}(\hat{\beta} - \beta_0)$ can be shown to be asymptotically normal with mean 0 and variance $W_0^{-1} W_1 W_0^{-1}$. If $R(\alpha)$ is correctly specified, $W_1 = W_0$. Even when $R(\alpha)$ is misspecified, $W_1(\beta_0, \alpha)$, can be consistently estimated by

$$K^{-1} \sum_{i=1}^{K} \frac{\partial \mu_i^T}{\partial \beta} V_i^{-1} (Y_i - \mu_i)(Y_i - \mu_i)^T V_i^{-1} \frac{\partial \mu_i}{\partial \beta^T},$$

evaluated at $\beta = \hat{\beta}$.

Alternatively the sandwich variance, $W_0^{-1} W_1 W_0^{-1}$, can be consistently estimated by the Jackknife variance, $(K - p) \sum_{i=1}^{K} (\hat{\beta}_{-i} - \hat{\beta})(\hat{\beta}_{-i} - \hat{\beta})^T$, where $\hat{\beta}_{-i}$ is the GEE estimate obtained after deleting the ith independent unit or subject (10; 11).

Note that the proof relies on the $\sqrt{K}$-consistency of $\hat{\alpha}$. Since $R(\alpha)$ is merely a working correlation, it is not clear what $\hat{\alpha}$ is estimating. Crowder (12) cautioned that in some cases, the estimand α is subject to uncertainty of definition, leading to a breakdown of the asymptotic properties of $\hat{\beta}$. In the case of independent working correlation models, this problem can be avoided. Especially in the case of canonical link functions, e.g. logit link for binary, log link for Poisson, and identity link for normal, the GEE estimates are guaranteed to exist under some regularity conditions. It is a good practice to fit an independent working correlation model first before fitting more complicated working correlation structures.

We should note that the asymptotic properties of GEE depend on large K and fixed n_i. Asymptotic results when either K or n_i, or both increase are given in Xie and Yang (13). More recently asymptotic results are given by Wang (14) when the number of covariates grows to infinity with the number of clusters.

6.3.3 Efficiency

The misspecified working correlation yields consistent estimates, but advantage of correctly specifying the correlation structure is gain in efficiency. Let the correct correlation matrix be R_i^* and corresponding W_0 be W_0^*. The variance of $\hat{\beta}$ is smallest when R is correctly specified, and has the form $K^{-1} W_0^{*-1}$, whereas the variance of $\hat{\beta}$ is $K^{-1} W_0^{-1} W_1 W_0^{-1}$ when $R_i \neq R_i^*$. The asymptotic relative efficiency (ARE) of the estimator with misspecified correlation structure is

$$\text{ARE} = W_0^{-1} W_1 W_0^{-1} W_0^*.$$

When the dimension of β is one, Figure 6.1 shows the ARE of the GEE estimator under independent working correlation structure. As anticipated, the ARE decreases as the correlation increases, but also decreases as the number of time point increases. Rotnitzky and Jewell (15) provide the upper and lower bounds of the ARE of a linear combination of $\hat{\beta}$ under some special correlation structures. Efficiency of GEE when $R(\alpha)$ is misspecified has been examined analytically (16) and numerically (17; 18).

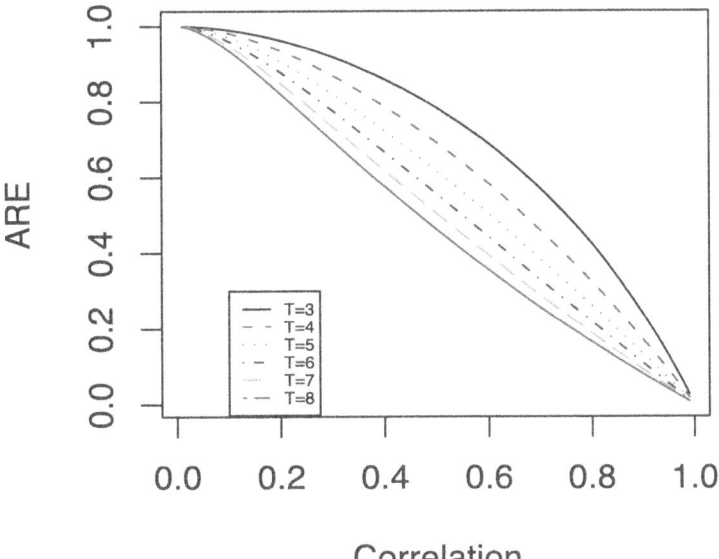

FIGURE 6.1
Asymptotic relative efficiency.

Efficiency of $\hat{\beta}$ can be improved by modeling the variance-covariance matrix. Prentice (19) considered modeling the correlation as a function of covariates as well as the mean. The estimating equation for the correlation is stacked with the original GEE for β to form a joint estimating equation. Liang, Zeger, and Qaqish (20) modelled jointly the mean and the odds ratio between the repeated binary outcomes. Paik (21) considered joint modeling of mean and scale parameter, ϕ. A joint modeling of mean, scale and correlation parameter by separate link functions was considered by Yan and Fine (22) and Lee, Paik and Lee (23). For multi-level multivariate data, for example, data with multiple measurements per each family member and for multiple family members, correlation structure becomes complicated. For such problems, a joint modeling of mean, scale, and canonical correlation is proposed (24).

6.3.4 Model selection criterion in GEE

When data are independent, Akaike's Information Criterion (AIC) by Akaike (25) is widely used for model selection criterion. In GEE, the AIC cannot be used due to lack of likelihood function. To handle this problem, one solution is to use a class of multivariate distributions called partly exponential family proposed by Zhao, Prentice, and Self (26). They showed that the score function of the partly exponential family has the same form as GEE if the GEE variance is correctly specified. Prentice and Zhao (27) showed joint modeling using the partly exponential family for binary outcomes. Disadvantage is that parameters in partly exponential family distributions are implicitly related to means and variance. Nonetheless, existence of such family of distributions provides a basis of data generating mechanisms useful for research in GEE.

An alternative method is proposed by Pan (28) extending the AIC to GEE utilizing the likelihood function under working independence model. The working independence likelihood can be viewed as quasi-likelihood by Wedderburn (7),

$$Q(\beta; y) = \int_y^\mu \frac{y - t}{\phi V(t)} dt,$$

with matching mean and variance. Pan (28) proposed a model selection method in GEE and referred to as quasi-likelihood under independence model criterion (QIC), where

$$\text{QIC} = -2Q(\hat{\beta}; y) + 2\text{trace}(\hat{\Omega}\hat{V}_K),$$

$\hat{\Omega} = -\partial^2 Q(\beta)/\partial\beta\partial\beta'|_{\beta=\hat{\beta}}$ and $\hat{V}_K = K^{-1}\hat{W}_0^{-1}\hat{W}_1\hat{W}_0^{-1}$. Note that $\hat{\Omega}^{-1}$ and $\hat{V}_K$ are asymptotically equivalent when all models are correctly specified. When outcomes are independent, QIC reduces to AIC. As in the model selection procedure with AIC, the model with the smallest QIC is selected as the preferred GEE model.

In the presence of missing data, there are a few model selection methods in GEE. When responses are subject to missing and missing mechanism is missing at random, Shen and Chen (29) proposed new criteria for selecting mean model and correlation structure with missing outcomes in GEE.

6.4 Generalized Linear Mixed Models

6.4.1 Notations

GLMMs are an extension of GLMs where correlations are induced by sharing an unobservable common factor, called the random effect. Let $Y_i = (Y_{i1}, Y_{i2}, \cdots, Y_{in_i})^T$ be the response for the ith unit ($i = 1, \cdots, K$). We assume that Y_{ij} arises independently from an exponential family distribution given b_i, say $f(Y_{ij}|b_i; \beta, \phi)$, where

$$\log f(Y_{ij}|b_i; \beta, \phi) = [Y_{ij}\psi_{ij}(\beta) - c\{\psi_{ij}(\beta)\}]/\phi + d(Y_{ij}, \phi),$$

ψ_{ij} denotes the canonical parameter, and ϕ is the dispersion parameter. Let b_i be a $q \times 1$ random effect and distributed as $g(b; D)$ indexed by unknown D. Let (X_i, Z_i) be $n_i \times p$, $n_i \times q$ matrices of explanatory variables associated with fixed and random effects, and β be a $p \times 1$ vector associated with the fixed effect. We denote the conditional mean by $\mu_{ij} \equiv E(Y_{ij}|b_i) = c'(\psi_{ij}) = \mu_{ij}$, and $\eta_{ij} = h(\mu_{ij})$, with $h(\cdot)$ as the link function and $\eta_{ij} = X_{ij}\beta + Z_{ij}b_i$. The conditional variance is $var(Y_{ij} \mid b_i) = \phi V(\mu_{ij}(b_i))$. Typically, b_i is assumed to be distributed as multivariate normal, although it is not necessary. When $q = 1$, instead of imposing distributional assumption on b_i, one can impose distributional assumption on a function of that random effect, leading to hierarchical generalized linear models (30).

6.4.2 Population average versus subject-specific model

The coefficients in GLMMs have a subject-specific interpretation whereas the GEE counterpart can be interpreted as a population average effect (5; 31). These two approaches answer different questions about change over time. GEE are marginal models that takes into account population average and describe the correlation among repeated measurements over time. On the other hand GLMMs can model subject specific effects over time. For example, in a logistic model, if the response probability given the random effect b_i is

$$P(Y_{ij} = 1|X_{ij}, b_i) = \frac{\exp(\beta_0 + \beta_1 X_{ij} + b_i)}{1 + \exp(\beta_0 + \beta_1 X_{ij} + b_i)},$$

the odds ratio is

$$\exp(\beta_1) = \frac{P(Y_{ij} = 1 | X_{ij} = 1, b_i)P(Y_{ij} = 0 | X_{ij} = 0, b_i)}{P(Y_{ij} = 1 | X_{ij} = 0, b_i)P(Y_{ij} = 0 | X_{ij} = 1, b_i)},$$

and is different from the GEE counterpart, $\frac{P(Y_{ij}=1|X_{ij}=1)P(Y_{ij}=0|X_{ij}=0)}{P(Y_{ij}=1|X_{ij}=0)P(Y_{ij}=0|X_{ij}=1)}$. Note that the marginal probability, $\int P(Y_{ij} = 1 | X_{ij}, b_i)g(b_i)db_i$, is not of logit form. Wang and Louis (32) showed that the Bridge distribution for a random effect preserves the link function in the marginal scale as in the conditional scale.

6.4.3   Estimation procedures

Two main inferential procedures can be used for GLMMs. Since b_i is not observed, b_i can be either integrated out to obtain a marginal likelihood or eliminated by conditioning to obtain a conditional likelihood.

6.4.3.1   Marginal likelihood

If b_i is assumed to be multivariate normally distributed, the marginal likelihood is

$$
\begin{aligned}
e^{l(\beta, \phi, D)} &= \prod_{i=1}^{K} \int \left[\prod_{j=1}^{n_i} f(Y_{ij} \mid b_i; \beta, \phi) \right] g(b_i; D)db_i \\
&\propto \prod_{i=1}^{K} |D|^{-\frac{1}{2}} \int \exp\left[\sum_{j=1}^{n_i} l_{ij}(\beta, \phi; b_i) - \frac{1}{2} b_i^T D^{-1} b_i \right] db_i,
\end{aligned}
$$

where $l_{ij}(\beta, \phi; b_i) = \log f(Y_{ij}|b_i; \beta, \phi)$. Usually the marginal likelihood does not have a closed form, and evaluating the marginal likelihood is a challenging computational task. Most built-in functions for GLMMs in popular software provide options for computational methods, namely, Gaussian quadrature, Laplace approximation, and penalized quasi-likelihood. Each method maximizes differently the approximated marginal likelihood. Gaussian quadrature method by numerically integrating the marginal likelihood is the most accurate among the three, but fails to converge often in practice. Laplace approximation of the marginal likelihood is

$$l(\beta, \phi, D) \approx \sum_{i=1}^{K} \left[\sum_{j=1}^{n_i} l_{ij}(\beta, \phi; \tilde{b}_i) - \frac{1}{2} \tilde{b}_i^T D^{-1} \tilde{b}_i - \frac{1}{2} log|I + Z_i^T W_i Z_i D| + \frac{q}{2} \log(2\pi) \right],$$

where $\tilde{b}_i$ is the solution of $\frac{\partial}{\partial b_i} \{ \sum_{j=1}^{n_i} l_{ij}(\beta, \phi; b_i) - \frac{1}{2} b_i^T D^{-1} b_i \} = 0$, and W_i is an $n_i \times n_i$ diagonal matrix with j^{th} diagonal element as $[\phi V(\mu_{ij}(b_i))\{h^{'}(\mu_{ij}(b_i))\}^2]^{-1}$. This approximation becomes accurate for large n_i and is exact when the conditional distribution of b_i given Y_i is multivariate normal. Although Laplace approximation is computationally less challenging than the Gaussian quadrature method, it sometimes fails to converge in practice.

The most commonly used method in practice is to maximize the penalized quasi-likelihood (PQL) (3),

$$\sum_{i=1}^{K} \left[\sum_{j=1}^{n_i} l_{ij}(\beta, \phi; \tilde{b}_i) - \frac{1}{2} \tilde{b}_i^T D^{-1} \tilde{b}_i + \frac{q}{2} \log(2\pi) - \frac{1}{2} \log |D| \right].$$

The PQL function ignores the correction term in the Laplace approximation, $-\sum_{i=1}^{K} \frac{1}{2} log|D + Z_i^T W_i Z_i|$. If $\frac{\partial}{\partial \beta} W_i \approx 0$, ignoring this term has negligible effect, but

$\frac{\partial}{\partial\beta}W_i \neq 0$ in general except in the normal case. In fact the PQL estimator maximizes $\prod_{i=1}^{K}\{\prod_{j=1}^{n_i} f(Y_{ij}|b_i;\beta,\phi)g(b_i;D)\}$ jointly over $(\beta^T, b_1, b_2, \cdots, b_K)$. The qualifier "penalized" comes from the fact that this method maximizes $\prod_{i=1}^{K}\{\prod_{j=1}^{n_i} f(Y_{ij}|b_i;\beta,\phi)\}$ over $(\beta^T, b_1, b_2, \cdots, b_K)$ with penalty $g(b_i;D)$, so that "estimator" of b_i obtained as if they are fixed can be pulled toward zero. The maximum PQL estimator can be computed via iteratively re-weighted least square method as in the GLM, using pseudo-outcome $Y_i^* = X_i\alpha + Z_i b_i + \frac{\partial\eta_i}{\partial\mu_i(b_i)}(Y_i - \mu_i(b_i))$, so that $E(Y_i^*|b_i) = X_i\alpha + Z_i b_i$ and $Var(Y_i^*|b_i) = W_i^{-1}$.

6.4.3.2 Conditional likelihood

Consider the case of binary Y_{ij} with logit link and $\eta_{ij} = X_{ij}\beta_1 + Z_i\beta_2 + b_i$, and assume that X_{ij} can change over time, called a time varying covariate. When the interest is only in the regression coefficient for time varying covariate, β_1, a conditional likelihood given $\sum_{j=1}^{K} Y_{ij} = y_{i+}$ is

$$\prod_{i=1}^{K} \frac{\prod_{j=1}^{n_i} P(Y_{ij}=1|X_{ij},b_i)^{Y_{ij}} P(Y_{ij}=0|X_{ij},b_i)^{1-Y_{ij}} g(b_i;D)}{\sum \prod_{j=1}^{n_i} P(Y_{ij}=1|X_{ij},b_i)^{Y_{ij}} P(Y_{ij}=0|X_{ij},b_i)^{1-Y_{ij}} g(b_i;D)}$$
$$= \prod_{i=1}^{K} \frac{\prod_{j=1}^{n_i} \exp(Y_{ij}\beta_1 X_{ij})}{\sum \prod_{j=1}^{n_i} \exp(Y_{ij}\beta_1 X_{ij})}.$$

where summation is over $(Y_{i1}, Y_{i2}, \cdots, Y_{in_i})$ that satisfy $\sum_{j=1}^{n_i} Y_{ij} = y_{i+}$. Note that β_2 and b_i are eliminated by conditioning. Even when Y_{ij} is not binary, conditioning on order statistics $Y_{(i1)}, Y_{(i2)}, \cdots, Y_{(in_i)}$, a conditional likelihood with canonical link functions (33) can be written

$$\prod \frac{\prod_{j=1}^{n_i} f(Y_{ij}|b_i;\beta,\phi)g(b_i;D)}{\sum \prod_{j=1}^{n_i} f(Y_{ij}|b_i;\beta,\phi)g(b_i;D)} = \prod \frac{\prod_{j=1}^{n_i} \exp(Y_{ij}\beta_1 X_{ij})}{\sum \prod_{j=1}^{n_i} \exp(Y_{(ij)}\beta_1 X_{ij})}.$$

We can verify that $g(b_i;D)$ is canceled in the conditional likelihood, which implies β_1 can be estimated without specifying $g(b_i;D)$.

Evaluating the denominator of the conditional likelihood can be computationally prohibitive. Denoting $f(Y_{ij}|X_{ik},b_i;\beta,\phi)$ by f_{ijk}, pairwise conditioning leads to a pseudo likelihood (34), which is defined by

$$\prod_{i=1}^{K}\prod_{j>k} \frac{f_{ijj}f_{ikk}}{f_{ijj}f_{ikk} + f_{ijk}f_{ikj}}$$
$$= \prod_{i=1}^{K}\prod_{j>k} \frac{\exp(Y_{ij}\beta_1 X_{ij})\exp(Y_{ik}\beta_1 X_{ik})}{\exp(Y_{ij}\beta_1 X_{ij})\exp(Y_{ik}\beta_1 X_{ik}) + \exp(Y_{ij}\beta_1 X_{ik})\exp(Y_{ik}\beta_1 X_{ij})}.$$

This approach is useful in handling missing data (35).

6.5 Test Statistics Under Randomization

In clinical trials, it is common that testing hypothesis is of primary interest. In this section, we examine the test statistics derived from GEE and GLMMs, and discuss their special features resulting from randomization. For GLMMs, there are three test statistics, Wald, score, and likelihood-ratio tests based on the likelihood function. For GEE, there is no likelihood, but Wald-type, score-type, and the likelihood-ratio-type tests under working independence (15) can be constructed. Wald and Wald-type tests for GLMMs or GEE require all estimates of the specified model, and are not the focus of this section. The rest of the section is devoted to the score and score-type tests for GLMMs and GEE, respectively, and their special features due to randomization.

6.5.1 Notations

Hypothesis testing through modeling typically involves parameters of interest and nuisance parameters. Consider a partition, $\beta^T = (\beta_1^T, \beta_2^T)$, where β_1 is the parameter of interest of length q, related to the primary hypothesis in the trial, and β_2 is nuisance parameter of length $p - q$ pertaining to adjusting variables. We are interested in testing hypothesis $H_0 : \beta_1 = 0$. We also partition the GEE estimating function, $U(\beta, \alpha) = \begin{pmatrix} U_1(\beta_1, \beta_2, \alpha) \\ U_2(\beta_1, \beta_2, \alpha) \end{pmatrix}$, where $U_1(\beta_1, \beta_2, \alpha)$ is a q-variate sub-vector of $U(\beta, \alpha)$ corresponding to β_1. Similarly we partition the matrices W_0 and W_1:

$$W_0 = K^{-1} E \begin{pmatrix} -\frac{\partial U_1}{\partial \beta_1^T} & -\frac{\partial U_1}{\partial \beta_2^T} \\ -\frac{\partial U_2}{\partial \beta_1^T} & -\frac{\partial U_2}{\partial \beta_2^T} \end{pmatrix} = \begin{pmatrix} B_{11}, B_{12} \\ B_{21}, B_{22} \end{pmatrix},$$

and $W_1 = \begin{pmatrix} M_{11}, M_{12} \\ M_{21}, M_{22} \end{pmatrix}$. It can be verified that $B_{21} = B_{12}^T$ and $M_{21} = M_{12}^T$.

6.5.2 Score-type test for GEE under randomization

For correlated binary outcomes, Lefkopoulou, Moore and Ryan (36) examined the score-type statistics for inference of GEE under the independent working correlation structure, which assumes independence between time points so that correlation is zero between different time points. The score-type statistics can be constructed in the same way for general structure of $R(\alpha)$ as follows:

$$T_S = U_1(0, \tilde{\beta}_2, \tilde{\alpha})^T var\{U_1(0, \tilde{\beta}_2, \tilde{\alpha})\}^{-1} U_1(0, \tilde{\beta}_2, \tilde{\alpha}),$$

where $\tilde{\beta}_2$ and $\tilde{\alpha}$ are the GEE estimators under the restriction $\beta_1 = 0$, and replacing α with $\tilde{\alpha}$ does not affect the asymptotic distribution of the test statistic. In general,

$$var\{K^{-\frac{1}{2}} U_1(0, \tilde{\beta}_2, \alpha)\}$$
$$= M_{11} + B_{12} B_{22}^{-1} M_{22} B_{22}^{-1} B_{21} - B_{12} B_{22}^{-1} M_{21} - M_{12} B_{22}^{-1} B_{21}.$$

To facilitate the discussion, let X_1 represent the group assignment, and X_2 represent all other covariates to be adjusted for. Without loss of generality we assume $E(X_1) = 0$.

Now we show that $E(B_{12}) = 0$ under the null hypothesis $H_0 : \beta_1 = 0$, due to randomization, where the expectation is over X. Note that the score for β_1 under the null has a form

$$U_1(0, \tilde{\beta}_2, \tilde{\alpha}) = \sum_{i=1}^{K} X_{1i} \Delta_i V_i^{-1}(\tilde{\beta}_2, \tilde{\alpha})\{Y_i - \mu_i(\tilde{\beta}_2)\}, \tag{6.2}$$

where Δ_i is an $n_i \times n_i$ matrix with diagonal element $\frac{\partial \mu_{ij}}{\partial \eta_{ij}}$. Note that μ_i and V_i in (6.2) do not depend on X_{1i} since they are evaluated under the null. Then

$$K^{-\frac{1}{2}} U_1(0, \tilde{\beta}_2, \alpha) = K^{-\frac{1}{2}} U_1(0, \beta_2, \alpha) + B_{12} B_{22}^{-1} K^{-\frac{1}{2}} U_2(0, \beta_2, \alpha) + o_p(1).$$

Now we can show that $E(B_{12}) = 0$ as follows:

$$E(B_{12}) = E\left[-\frac{\partial U_1(0, \beta_2, \alpha)}{\partial \beta_2^T} \right] = E \sum_{i=1}^{K} X_{1i} \frac{\partial}{\partial \beta_2} \left[\Delta_i V_i^{-1}(\tilde{\beta}_2, \tilde{\alpha})\{Y_i - \mu_i(\tilde{\beta}_2)\} \right]$$

$$= E \sum_{i=1}^{K} X_{1i} \frac{\partial}{\partial \beta_2} \left[\Delta_i V_i^{-1}(\tilde{\beta}_2, \tilde{\alpha}) \right] \{Y_i - \mu_i(\tilde{\beta}_2)\} \tag{6.3}$$

$$- E \sum_{i=1}^{K} X_{1i} \Delta_i V_i^{-1}(\tilde{\beta}_2, \tilde{\alpha}) \frac{\partial \mu_i(\tilde{\beta}_2)}{\partial \beta_2}.$$

The first term of (6.3) is zero since $E(Y_i|X_i) = \mu_i$, and the second term is zero due to randomization and $EX_{1i}=0$. Under randomization, the middle term of the test statistic T_S, $var\{K^{-\frac{1}{2}} U_1(0, \tilde{\beta}_2, \alpha)\}$ simplifies to M_{11}. This statistic requires computation of $\tilde{\beta}_2$, but not its variance, and allows to ignore the effect of estimating β_2.

Further inspection of the score-type statistic reveals robustness of the test based on T_S with respect to model misspecification. Specifically, consider the case where the mean model for Y_i is misspecified, so that $E(Y_i|X_i) \neq \mu_i$. The second term of (6.3) is still zero due to randomization and $EX_{1i}=0$, while the first term of (6.3) also becomes zero as follows:

$$E \sum_{i=1}^{K} X_{1i} \frac{\partial}{\partial \beta_2} \left[\Delta_i V_i^{-1}(\tilde{\beta}_2, \tilde{\alpha}) \right] \{E(Y_i|X_{2i}) - \mu_i(\tilde{\beta}_2)\}$$

$$= E \sum_{i=1}^{K} X_{1i} E \frac{\partial}{\partial \beta_2} \left[\Delta_i V_i^{-1}(\tilde{\beta}_2, \tilde{\alpha}) \right] \{E(Y_i|X_{2i}) - \mu_i(\tilde{\beta}_2)\} = 0.$$

The first equality is because $\frac{\partial}{\partial \beta_2}\left[\Delta_i V_i^{-1}(\tilde{\beta}_2, \tilde{\alpha})\right]\{E(Y_i|X_{2i})-\mu_i(\tilde{\beta}_2)\}$ is a function of X_2 only, and X_1 and X_2 are independent. The second equality is due to $E(X_1) = 0$. This implies that even when the mean μ_i is misspecified, i.e., $E(Y_i|X_i) \neq \mu_i$, we still have $EU_1(0, \beta_2, \alpha) = 0$ mainly due to independence between X_1 and X_2. Therefore the score-type test is robust with respect to misspecification of the model.

Another simplified test statistic can be contemplated due to randomization, and provides insight in efficiency gains via modeling. Note that the second term of (6.2), $\sum_{i=1}^{K} X_{1i}^T \Delta_i V_i^{-1}(\beta_2, \tilde{\alpha})\mu_i(\beta_2)$, has zero expectation due to independence between X_1 and X_2. This implies $A(\tilde{\beta}_2, \tilde{\alpha})var(A(\tilde{\beta}_2, \tilde{\alpha}))^{-1}A(\tilde{\beta}_2, \tilde{\alpha})$, where $A(\tilde{\beta}_2, \tilde{\alpha}) = \sum_{i=1}^{K} X_{1i}^T \Delta_i V_i^{-1}(\tilde{\beta}_2, \tilde{\alpha})Y_i$, can serve as a valid test statistic. In the case of cannonical link, $A(\tilde{\beta}_2, \tilde{\alpha})$ reduces to $\sum_{i=1}^{K} X_{1i}^T R_i^{-1}(\tilde{\alpha})Y_i$, and does not require computing $\tilde{\beta}_2$. If independent working correlation structure is used, $A(\tilde{\beta}_2, \tilde{\alpha})$ is the group difference in Y. However, the variance term in the score test features the conditional variance given X_2, and thus the resulting test is more efficient than the test based on a simple difference without modeling. The efficiency gained due to modeling will depend on the size of β_2.

6.5.3　Score test for GLMMs under randomization

We denote score functions and information for GLMMs by U_{GL} and B_{GL} and use similar partitioning as in the previous section. For GLMMs, usual Wald, score and likelihood ratio tests can be constructed from the marginal likelihood. In the case of canonical link function, the score function for β_1 evaluated under the null hypothesis $H_0 : \beta_1 = 0$ has a form

$$U_{1,GL}(0, \tilde{\beta}_2, \tilde{D}) = \sum_{i=1}^{K} \int \sum_{j=1}^{n_i} X_{1ij}\{Y_{ij} - \mu_{ij}(\tilde{\beta}_2, b_i)\}k(b_i|Y_i)db_i,$$

where $k(b_i|Y_i) = \frac{\{\prod_{j=1}^{n_i} f(Y_{ij}|b_i;\beta,\phi)\}g(b_i;D)}{\int \{\prod_{j=1}^{n_i} f(Y_{ij}|b_i;\beta,\phi)\}g(b_i;D)db_i}$, $\tilde{\beta}_2$ is the solution of

$$U_{2,GL}(0,\tilde{\beta}_2,\tilde{D}) = \sum_{i=1}^{K} \int \sum_{j=1}^{n_i} X_{2ij}\{Y_{ij} - \mu_{ij}(\tilde{\beta}_2,b_i)\}k(b_i|Y_i)db_i = 0,$$

and $\tilde{D}$ is the maximum likelihood estimator under the restriction $\beta_1 = 0$. Using similar arguments in GEE, we can verify $EB_{12,GL} = 0$, where the expectation is over X. The test statistic is

$$T_{S,GL} = U_{1,GL}(0,\tilde{\beta}_2,\tilde{D})^T B_{11,GL}(0,\tilde{\beta}_2,\tilde{D})^{-1} U_{1,GL}(0,\tilde{\beta}_2,\tilde{D}).$$

As in the GEE case, we can check $EU_{1,GL}(0,\beta_2,D) = 0$ due to randomization even if $E(Y_{ij}|X_{ij},b_i)$ is misspecified. Therefore the test based on $T_{S,GL}$ is robust against the misspecification of the mean model.

Furthermore, consider the case $n_i = 2$ and treatment is assigned within person and $X_{2ij} \equiv X_{2i}$ for all j. For example, in ophthalmologic studies, the treatment is given to one eye, and placebo to the other. Then the second term of $U_{1,GL}(0,\tilde{\beta}_2,\tilde{D})$, $\sum_{i=1}^{K} \sum_{j=1}^{n_i} \int -X_{1ij}\mu_{ij}(X_{2i},\tilde{\beta}_2,b_i)h(b_i|Y_i)db_i$, is exactly zero. Then the score function evaluated under the null, $U_{1,GL}(0,\tilde{\beta}_2,\tilde{D})$ reduces to $\sum_{i=1}^{K} \sum_{j=1}^{n_i} X_{1i}Y_{ij}$. In this case, the numerator of the score test statistic is the group difference, but the variance is $B_{11,GL}(0,\tilde{\beta}_2,\tilde{D})$, conditioning on X_2 following the GLMMs. As in the GEE case, test based on $T_{S,GL}$ is more efficient than the test based on a simple difference without modeling.

6.6   Handling Missing Data in Clinical Trials

When outcomes are repeatedly measured over time in longitudinal studies, data are more prone to missingness than in studies collecting data at a single time point. When outcomes are missing, intent-to-treat analysis cannot be conducted, and the inference needs to resort to modeling. Although the same principles apply for handling missing Y and missing X, the case of missing X is more challenging than the case of missing Y. In clinical trials, most missingness occurs in outcomes, and we restrict our attention to the case where outcomes are missing and covariates are completely observed.

Rubin (37) introduced the notion of missing data mechanism and its impact on inference. In the longitudinal setting, we use the same definition as in Liang and Zeger (1) and define data to be missing completely at random (MCAR) if the missingness does not depend on Y. Under MCAR, the GEE method by Liang and Zeger (1) is valid. We define data to be missing at random (MAR) if missingness depends on observed Y given X. The data are defined to be nonignorably missing when missingness depends on unobserved data. The most popular way to handle missing data is a marginal likelihood approach based on the observed data, which can be obtained by integrating a full likelihood over the missing data. We can use this approach in GLMMs. Under the MAR assumption, the GLMM is valid, but GEE is not generally valid. With non-likelihood inference as in GEE, two approaches have commonly been used to mitigate bias under the MAR assumption, namely, inverse probability weighting (IPW) and imputation. We briefly review the two methods to handle missing outcomes in GEE focusing on the case of MAR, and discuss principles of the likelihood method for GLMMs.

6.6.1 Missing data in GEE

The GEE naturally handles varying numbers of measurements across subjects resulting from dropout or skipping planned visits, and is valid when data are MCAR. If the probability of missingness depends on observed or unobserved Y, the GEE estimator is not consistent. Fitzmaurice et al. (38) examined the bias when the missingness mechanism is not MCAR. To fix this problem, Robins, Rotnizky, and Zhao (39) proposed IPW GEE in which each observation is weighted by the inverse of probability of observation. Specifically, let δ_{ij} be the observation indicator for Y_{ij} for individual i at time j, where $i = 1, \cdots, N$, $j = 1, \cdots, M$. Data are called to have monotone missing pattern when $\mathrm{P}(\delta_{ij} = 1|\delta_{i,j-1} = 0) = 0$. Also let L_{ij} be a vector of all the data collected on individual i at time j, $\overline{L}_{ij} = (L_{i1}^T, \ldots, L_{ij}^T)^T$, and $\underline{L}_{ij} = (L_{ij+1}^T, \ldots, L_{iM}^T)^T$. Under monotone missing pattern the IPW GEE estimator is the solution of

$$\sum_{i=1}^{N} \frac{\partial \mu_i^T}{\partial \beta} V_i(\rho)^{-1} \Delta_i(\alpha)(Y_i - \mu_i) = 0,$$

where $\Delta_i(\alpha)$ is a diagonal matrix with j^{th} diagonal element being δ_{ij}/π_{ij}, $\pi_{ij} = \pi_{ij}(\alpha) = \mathrm{P}(\delta_{ij} = 1|\overline{L}_{i,j-1})$, $\pi_{ij} = \lambda_{i1} \cdots \lambda_{ij}$, and $\lambda_{ij}(\alpha) = \mathrm{P}(\delta_{ij} = 1|\overline{L}_{i,j-1}, R_{i,j-1} = 1)$. The weights attempt to restore the full data by properly representing the missing records. In imputed GEE, missing outcomes are replaced with the estimates of the conditional expectation given observed data (40). The IPW GEE and imputation methods are consistent under correct specification of probability of observation model, or under correct imputation model, respectively. Robins and colleagues also proposed a modified IPW GEE to improve the efficiency of the weighted GEE (41). This method is referred to as doubly-robust (DR) method since it requires both auxiliary models but yield consistent results when either the dropout or the imputation model is correctly specified (42–44). Copas and Seaman (45) presented a DR estimator for the GEE and provided a thorough examination of GEE methods with incomplete outcome data under MAR dropout mechanism. With minor coding effort we can implement IPW GEE, which we will use in the case study in the later section.

6.6.2 Missing data in GLMMs

In GLMMs, a joint distribution of (Y, b) is assumed along with the distribution of b. Let the observed part of Y_i be Y_i^o and missing part be Y_i^m. For example, if there are three time points for i, $n_i = 3$, and $\delta_{i1} = \delta_{i2} = 1$ and $\delta_{i3} = 0$, $Y_i^o = (Y_{i1}, Y_{i2})^T$ and $Y_i^m = Y_{i3}$. Let $\delta_i = (\delta_{i1}, \cdots, \delta_{iM})^T$ and $f(Y_i|b_i; \beta, \phi) = \prod_{j=1}^{M} f(Y_{ij}|b_i; \beta, \phi)$. The marginal likelihood given the observed data can be written as

$$\prod_{i=1}^{K} \int P(\delta_i|Y_i, X_i, b_i; \theta) f(Y_i|b_i; \beta, \phi) g(b_i; D) dY_i^m db_i. \tag{6.4}$$

Partitioning the joint distribution of (δ_i, Y_i, b_i) as in (6.4) enables us to directly examine the effect of missing mechanism on inference. If missingness depends on Y^o but not on Y^m nor b, that is, $P(\delta_i|Y_i, X_i, b_i; \theta) = P(\delta_i|Y_i^o, X_i; \theta)$, the likelihood function can be factored into

$$\prod_{i=1}^{K} \int P(\delta_i|Y_i^o, X_i; \theta) f(Y_i|b_i; \beta, \phi) g(b_i; D) dY_i^m db_i$$

$$= \prod_{i=1}^{K} P(\delta_i|Y_i^o, X_i; \theta) \int f(Y_i|b_i; \beta, \phi) g(b_i; D) dY_i^m db_i.$$

In this case, data are missing at random and the estimator maximising the marginal likelihood ignoring missingness probability is valid. Similar factorization occurs when data are MCAR, that is, $P(\delta_i|Y_i, X_i, b_i; \theta) = P(\delta_i|X_i; \theta)$. This implies that the likelihood approach is valid without explicitly modeling the missingness mechanism when data are missing at random or missing completely at random.

There is a variety of literature that investigate GLMMs under nonignorable missingness (46) (47). To handle nonignorably missing data, assumptions on both missingness probability and imputation models are required, and all the specified models should be correctly specified for the results to be valid. Sensitivity analysis can be accompanied to evaluate robustness of the results.

A subclass of nonignorable missingness received attention recently. The case where $P(\delta_i|Y_i, X_i, b_i; \theta) = P(\delta_i|X_i, b_i; \theta)$ constitutes nonignorable missingness since the probability of missingness depends on unobserved b_i, but this mechanism is less harmful than the case where missingness depends on Y_i^m. Yuan and Little (48) referred to this case as cluster specific nonignorable missingness (CSNI) and proposed a simple solution. In regression setting, the pairwise pseudo-likelihood inference by Liang and Qin (34) for the GLMMs under CSNI becomes free of the missingness probability and $g(b_i; D)$. Details can be found in Zhang and Paik (35).

6.7   Case Study

We analyzed the longitudinal data from a randomized double blind trial in a study of isolated systolic hypertension in the elderly (49; 50). This study was conducted across multiple centers in the United States, where a total of 4736 participants were randomized to antihypertensive drug treatment ($n = 2365$) and placebo ($n = 2371$). The primary purpose of the study was to assess the effect of antihypertensive drug on the risk of stroke among people aged 60 or more with isolated systolic hypertension.

The objective of our analysis is to test the difference in treatment effect on systolic blood pressure (SBP) and diastolic blood pressure (DBP) annually measured for five years. We restricted our attention to those who showed up for the first follow-up. We also excluded the measurements after the first missed visit to monotonize the missing pattern. The analyses are based on a sample of 4244 subjects, with 2165 in treatment group and 2079 subjects in the placebo group. Change in SBP (CSBP) and in DBP (CDBP) from baseline are the primary outcomes of interest. We denote by CSBP_{ij} the change in SBP from baseline to year j for subject i. Figures 6.2 and 6.3 show changes of the mean of the CSBP and CDBP over time. While SBP had a large initial drop during the first year, then maintained at a stable level, DBP declined gradually over time.

Table 6.1 shows the baseline characteristics of the subjects in this analysis. In both placebo and treatment group, approximately 44% were males, approximately 80% of subjects were white and 13% were black, and the mean age of subjects at baseline was a little above 71 in both groups.

We fitted the GEE and GLMMs with outcome CSBP or CDBP, and time, age, and treatment as predictors. Based on QIC for GEE and AIC for GLMMs, we chose linear time effect over categorical and decided not to include any examined interaction terms. We draw inference based on the following models:

$$\text{M0 : CSBP}_{ij} = \beta_0 + \beta_1 Time_i + \beta_2 Age_i + \beta_3 Treatment_i + e_{ij}$$

$$\text{M1 : CSBP}_{ij} = (\beta_0 + b_{i0}) + \beta_1 Time_{ij} + \beta_2 Age_i + \beta_3 Treatment_i + e_{ij}$$

TABLE 6.1
Baseline characteristics by treatment group.

		Placebo (N=2079)		Treatment (N=2165)	
		N	%	N	%
Sex	Male	904	43.5	946	43.7
	Female	1175	56.5	1219	56.3
Race	White	1664	80.0	1723	79.6
	Black	269	12.9	288	13.3
	Asian	96	4.6	97	4.5
	Hispanic	32	1.5	37	1.7
	Other	18	0.9	20	0.9
		N	Mean(SD)	N	Mean(SD)
Age		2079	71.3 (6.6)	2165	71.6 (6.6)
Weight(lbs.)		2063	162.9 (34.1)	2145	162.9 (33.7)
Height(ins.)		2060	64.3 (4.0)	2135	64.3 (3.9)
Body Mass Index		2058	27.6 (5.1)	2134	27.6 (5.0)
Systolic BP		2079	170.0 (9.2)	2165	170.4 (9.4)
Diastolic BP		2079	76.4 (9.8)	2165	76.8 (9.6)

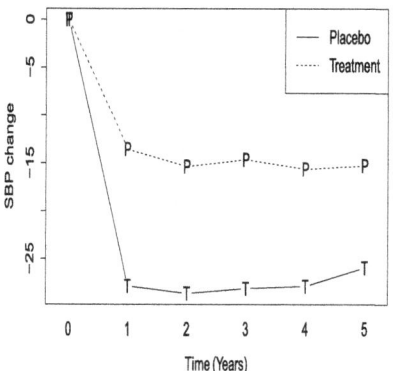
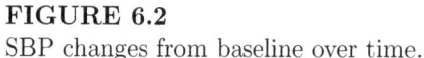

FIGURE 6.2
SBP changes from baseline over time.

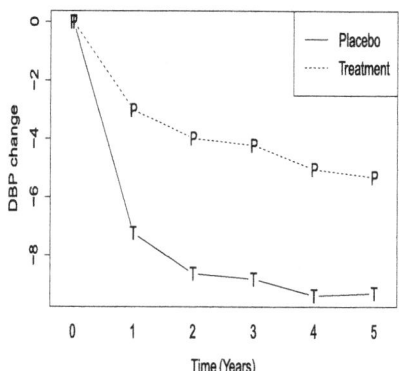

FIGURE 6.3
DBP changes from baseline over time.

M2 : $\text{CSBP}_{ij} = (\beta_0 + b_{i0}) + (\beta_1 + b_{i1})Time_{ij} + \beta_2 Age_i + \beta_3 Treatment_i + e_{ij}$.

The model (M0) is GEE while (M1) and (M2) are GLMMs. For (M1) and (M2), it is assumed that $e_{ij} \sim N(0, \sigma^2)$, and $(b_{i0}, b_{i1})^T \sim N\left(\begin{pmatrix} 0 \\ 0 \end{pmatrix}, \begin{pmatrix} D_{00} & D_{01} \\ D_{01} & D_{11} \end{pmatrix}\right)$. For GEE models, we fitted independence, exchangeable, and unstructured working correlation structures. Similarly, the following are models for CDBP.

M0 : $\text{CDBP}_{ij} = \beta_0 + \beta_1 Time_i + \beta_2 Age_i + \beta_3 Treatment_i + e_{ij}$

M1 : $\text{CDBP}_{ij} = (\beta_0 + b_{i0}) + \beta_1 Time_{ij} + \beta_2 Age_i + \beta_3 Treatment_i + e_{ij}$

M2 : $\text{CDBP}_{ij} = (\beta_0 + b_{i0}) + (\beta_1 + b_{i1})Time_{ij} + \beta_2 Age_i + \beta_3 Treatment_i + e_{ij}$.

Under missing at random assumption, IPW GEE was fitted for the three working correlation structures.

TABLE 6.2
Fitted GEE models from (M0).

CSBP	Intercept	Treatment	Age	Time
GEE Independence	$-19.1(2.37)^{***}$	$-13.3(0.42)^{***}$	$0.07(0.03)^*$	$-0.101(0.11)$
GEE Exchangeable	$-18.8(2.32)^{***}$	$-13.3(0.40)^{***}$	$0.06(0.03)$.	$0.008(0.10)$
GEE Unstructured	$-18.7(2.29)^{***}$	$-13.4(0.40)^{***}$	$0.06(0.03)$.	$0.040(0.10)$
IPW Independence	$-20.2(2.72)^{***}$	$-13.0(0.48)^{***}$	$0.07(0.04)$.	$0.17(0.14)$
IPW Exchangeable	$-19.7(2.41)^{***}$	$-13.3(0.42)^{***}$	$0.07(0.03)^*$	$0.12(0.13)$
IPW Unstructured	$-19.7(2.41)^{***}$	$-13.4(0.42)^{***}$	$0.07(0.03)^*$	$0.13(0.13)$
CDBP	**Intercept**	**Treatment**	**Age**	**Time**
GEE Independence	$-0.95(1.47)$	$-4.43(0.27)^{***}$	$-0.02(0.02)$	$-0.58(0.06)^{***}$
GEE Exchangeable	$-1.21(1.44)$	$-4.41(0.26)^{***}$	$-0.02(0.02)$	$-0.55(0.06)^{***}$
GEE Unstructured	$-1.12(1.41)$	$-4.40(0.26)^{***}$	$-0.02(0.02)$	$-0.54(0.06)^{***}$
IPW Independence	$-1.17(1.59)$	$-4.43(0.30)^{***}$	$-0.02(0.02)$	$-0.46(0.08)^{***}$
IPW Exchangeable	$-0.80(1.48)$	$-4.42(0.28)^{***}$	$-0.03(0.02)$	$-0.47(0.08)^{***}$
IPW Unstructured	$-0.60(1.49)$	$-4.41(0.29)^{***}$	$-0.03(0.02)$	$-0.46(0.08)^{***}$

*** $p < 0.001$, ** $p < 0.01$, * $p < 0.05$, and $p < 0.10$

Table 6.2 presents the results from GEE analysis for CSBP and CDBP. The average difference in CSBP between the treatment group and the placebo group was approximately 13 units, with treatment group having a significantly larger decline than the placebo group. The treatment effect was also significant for CDBP with the treatment group having approximately 4.4 units larger decline than the placebo group. On the other hand, the time effect was nonsignificant for CSBP, but significant for CDBP, suggesting that the SBP declined within a year and stabilized thereafter, but the DBP declined gradually over five years. Overall results were similar for all three working correlation structures. The estimated working correlation parameters for exchangeable working correlation structure were 0.39 and 0.45 for CSBP and CDBP, respectively, and similar estimates were obtained for the IPW GEE models. Table 6.3 reports the estimated unstructured working correlation from IPW GEE, which are similar to the estimates from GEE and not reported.

Table 6.4 shows the fitted λ_{ij}, $j = 2, \cdots, 5$. We fitted logistic models for binary observation indicators at time points from 2 to 5 with treatment, age, and SBP (or DBP)

TABLE 6.3

Estimated unstructured working correlation under missing at random assumption.

			CSBP					CDBP		
IPW	1	0.38	0.35	0.30	0.30	1	0.42	0.41	0.43	0.37
		1	0.42	0.37	0.31		1	0.48	0.47	0.39
			1	0.47	0.44			1	0.53	0.53
				1	0.54				1	0.42
					1					1

measured at previous time points as predictors, given that the previous time point is observed. Interaction terms were examined but none were significant. For some models, the probability of observation depended on completely observed covariates such as treatment or age, implying data are missing completely at random. However, in models for λ_{i3} and λ_{i5}, SBP_1, and DBP_1 were significant, respectively, hinting data may be missing at random. Since the study is not designed to detect significance of the observation probability model, we need to interpret statistical significance with caution. With small coefficient estimates for SBP_1 and DBP_1, IPW GEE results resemble those of GEE as shown in Table 6.2.

For mixed models, we fitted the models (M1) and (M2). Model (M2) has a random intercept and slope allowing the slope to be different for each individual. Table 6.5 presents the estimates and standard errors from (M1) and (M2) for CSBP and CDBP. The results were coherent with those of GEE models. Although we anticipated similar results between

TABLE 6.4

Fitted conditional observation probabilities.

CSBP	Treatment	Age	SBP_1	SBP_2	SBP_3	SBP_4
λ_2	0.51***	−0.03***	0.005			
	(0.14)	(0.01)	(0.003)			
λ_3	−0.22	−0.04***	−0.01*	0.007.		
	(0.15)	(0.01)	(0.004)	(0.004)		
λ_4	−0.06	0.01*	−0.003	−0.004.	−0.002	
	(0.08)	(0.005)	(0.002)	(0.002)	(0.002)	
λ_5	−0.04	−0.002	−0.001	−0.001	−0.003	0.001
	(0.09)	(0.006)	(0.003)	(0.003)	(0.003)	(0.003)
CDBP	**Treatment**	**Age**	DBP_1	DBP_2	DBP_3	DBP_4
λ_2	0.49***	−0.03***	0.01.			
	(0.13)	(0.01)	(0.01)			
λ_3	−0.15	−0.04***	0.001	0.005		
	(0.13)	(0.01)	(0.007)	(0.007)		
λ_4	0.06	0.01*	0.006	−0.007.	−0.004	
	(0.07)	(0.005)	(0.004)	(0.004)	(0.004)	
λ_5	−0.02	−0.004	−0.01*	0.001	0.006	−0.006
	(0.08)	(0.006)	(0.004)	(0.005)	(0.004)	(0.004)

Values are Estimate(SD). *** $p < 0.001$, ** $p < 0.01$, * $p < 0.05$, and $p < 0.10$

(M1) and the GEE model with exchangeable working correlation, the estimates did not change much when we added random slope in model (M2).

In summary, results from GEE and GLMMs demonstrated that the group difference was significant for CSBP and CDBP, but the change occurred within the first year for CSBP, thus showing no time effect, whereas the change occurred gradually over five-year period for CDBP. This trend was coherent in all the GEE models with independence, exchangeable, unstructured working correlations as well as in random intercept linear mixed models with or without random slope. The treatment group difference was highly significant for both CSBP and CDBP, but CDBP yielded smaller magnitude compared to CSBP.

TABLE 6.5
Fitted GLMMs from (M1) and (M2).

CSBP	Intercept	Group	Age	Time	σ^2	D_{00}	D_{01}	D_{11}
M1	-18.8***	-13.3 ***	0.06.	0.01	186.1	119.6		
	(2.24)	(0.40)	(0.03)	(0.09)	(2.5)	(3.8)		
M2	-18.7***	-13.6***	0.06.	0.009	171.6	108.2	-7.04	8.1
	(2.20)	(0.40)	(0.03)	(0.01)	(2.7)	(8.5)	(2.5)	(0.95)

CDBP	Intercept	Group	Age	Time	σ^2	D_{00}	D_{01}	D_{11}
M1	-1.21	-4.41***	-0.02	-0.55***	65.4	52.3		
	(1.43)	(0.26)	(0.02)	(0.06)	(0.87)	(1.6)		
M2	-0.90	-4.41***	-0.02	-0.55***	63.6	36.7	1.9	1.1
	(1.41)	(0.26)	(0.02)	(0.06)	(1.0)	(3.0)	(0.8)	(1.0)

Bibliography

[1] Liang K.Y. and Zeger S. L. Longitudinal data analysis using generalized linear models. *Biometrika*, 73:13–22, 1986.

[2] Zeger S. L. and Liang K. Y. Longitudinal data analysis for discrete and continuous outcomes. *Biometrics*, 42:121–130, 1986.

[3] Breslow N. E. and Clayton D. G. Approximate inference in Generalized Linear Mixed Models. *Journal of the American Statistical Association*, 88:9–25, 1993.

[4] McCullagh P. and Nelder J. A. *Generalized Linear Models*. 1989.

[5] Liang K. Y. Zeger S. L. and Albert P. S. Models for longitudinal data: A Generalized Estimating Equation approach. *Biometrics*, 44:1046–1060, 1988.

[6] Laird N.M. Fitzmaurice G.M. and Ware J.H. *Applied Longitudinal Analysis*. 2011.

[7] Wedderburn R. W. M. Quasi-likelihood functions, Generalized Linear Models, and the Gauss-Newton method. *Biometrika*, 61:439–447, 1974.

[8] Kastner C. Ziegler A. and Blettner M. The Generalised Estimating Equations: An Annotated Bibliography. *Biometrical Journal*, 40(40):115–139, 1988.

[9] Pepe M.S. and Anderson G.L. A cautionary note on inference for marginal regression models with longitudinal data and general correlated response data. *Communications in Statistics-Simulation and Computation*, 23:939–951, 1994.

[10] Ziegler A. Practical considerations of the Jackknife estimator of variance for Generalized Estimating Equations. *Statistical Papers*, 38:363–369, 1997.

[11] Dear K. B. G. Lipsitz S. R. and Zhao L. Jackknife estimators of variance for parameter estimates from estimating equations with applications to clustered survival data. *Biometrics*, 50:842–846, 1994.

[12] Crowder M. On the use of a working correlation matrix in using Generalized Linear Models for repeated measures. *Biometrika*, 82:407–410, 1995.

[13] Xie M. and Yang Y. Asymptotics for generalized estimating equations with large cluster sizes. *The Annals of Statistics*, 31:310–347, 2003.

[14] Wang L. GEE analysis of clustered binary data with diverging number of covariates. *The Annals of Statistics*, 39:389–417, 2011.

[15] Rotnitzky A. and Jewell N. P. Hypothesis testing of regression parameters in semiparametric generalized linear models for cluster correlated data. *Biometrika*, 77(77):485–497, 1990.

[16] Mancl L. A. and Leroux B. G. Efficiency of regression estimates for clustered data. *Biometrics*, 52:500–511, 1996.

[17] Paik M. C. Repeated measurement analysis for nonnormal data in small samples. *Communications in Statistics*, B17:1155–1171, 1988.

[18] Park T. A comparison of the Generalized Estimating Equation approach with the maximum likelihood approach for repeated measurements. *Statistics in Medicine*, 12:1723–1732, 1993.

[19] Prentice R.L. Correlated binary regression with covariates specific to each binary observation. *Biometrics*, 44:1033–1048, 1988.

[20] Qaqish B. Liang K. Y., Zeger S. L. Multivariate regression analyses for categorical data. *Journal of the Royal Statistical Society Series B-Methodological*, 54(1):3–40, 1992.

[21] Paik M. C. Parametric variance function estimation for nonnormal repeated measurement data. *Biometrics*, 48:19–30, 1992.

[22] Yan J. and Fine J. Estimating equations for association structures. *Statistics in Medicine*, 23:859–874, 2004.

[23] Paik M.C. Lee H-S and Lee J.H. Genotype adjusted familial correlation analysis using three generalized estimating equations. *Statistics in Medicine*, 27(26):5471–5483, 2008.

[24] Paik M.C. Lee H-S and Lee J.H. Estimating a multivariate familial correlation using Joint models for canonical correlations: Application to memory score analysis from familial Hispanic Alzheimer's disease study. *Biometrics*, 65:463–469, 2009.

[25] Akaike H. Information theory as an extension of the maximum likelihood. *Proceeding of IEEE International Symposium on Information Theory*, 2:0, 1973.

[26] Prentice R. L. Zhao L. P. and Self S. G. Multivariate mean parameter estimation by using a partly exponential model. *Journal of the Royal Statistical Society Series B-Methodological*, 54(3):805–811, 1992.

[27] Prentice R. L. and Zhao L. P. Estimating equations for parameters in mean and covariances of multivariate discrete and continuous response. *Biometrics*, 47(47):825–839, 1991.

[28] Pan W. Akaike's information criterion in generalized estimating equations. *Biometrics*, 57:120–125, 2001.

[29] Shen C.W. and Chen Y.H. Model selection for Generalized Estimating Equations accommodating dropout missingness. *Biometrics*, 68:1046–1054, 2012.

[30] Lee Y. and Nelder J. A. Hierarchical Generalized Linear Models. *Journal of the Royal Statistical Society, Series B*, 58:619–656, 1996.

[31] Stiratelli R., Laird N., and Ware J.H. Random effects models for serial observations with binary response. *Biometrics*, 40:961–71, 1984.

[32] Wang Z. and Louis T.A. Matching conditional and marginal shapes in binary random intercept models using a bridge distribution function. *Biometrika*, 90:765–775, 2003.

[33] Kalbfleisch J.D. Non-parametric bayesian analysis of survival time data. *Journal of the Royal Statistical Society. Series B*, 40:214–221, 1978.

[34] Liang K.Y. and Qin J. Regression analysis under non-standard situations: a pairwise pseudolikelihood approach. *Journal of the Royal Statistical Society: Series B*, 62:773–786, 2000.

[35] Zhang H. and Paik M.C. Handling missing responses in generalized linear mixed model without specifying missing mechanism. *Journal of Biopharmaceutical Statistics*, 19:1001–1017, 2009.

[36] Moore D. Lefkopoulou M. and Ryan L. The Analysis of multiple correlated binary outcomes: Application to rodent teratology experiments. *Journal of the American Statistical Association*, 84:810–815, 1989.

[37] Rubin D.B. Inference and missing data. *Biometrika*, 63:581–592, 1976.

[38] Ishii H. Fitzmaurice G.M. and W.A.S. Buxton. In *Proceedings of the SIGCHI conference on Human factors in computing systems - CHI '95*, pages 442–449, New York, New York, USA, 1995. ACM Press.

[39] Rotnitzky A. Robins J.M. and Zhao L.P. Analysis of semiparametric regression models for repeated outcomes in the presence of missing data. *Journal of the American Statistical Association*, 90:106–121, 1995.

[40] Paik M. C. The Generalized Estimating Equation approach when data are not missing completely at random. *Journal of American Statistical Association*, 92:1320–1329, 1997.

[41] Robins J.M. and Rotnitzky A. Semiparametric efficiency in multivariate regression models with missing data. *Journal of the American Statistical Association*, 90:122–129, 1995.

[42] Kenward M.G. Carpenter J.R. and Vansteelandt S. A comparison of multiple imputation and doubly robust estimation for analyses with missing data. *Journal of the Royal Statistical Society: Series A*, 169:571–584, 2006.

[43] Tsiatis A. *Semiparametric Theory and Missing Data*. Springer, New York, 2006.

[44] Seaman S. and Copas A. Doubly robust generalized estimating equations for longitudinal data. *Statistics in Medicine*, 28:937–55, 2009.

[45] Copas A. and Seaman S.R. Bias from the use of generalized estimating equations to analyze incomplete longitudinal binary data. *Journal of applied statistics*, 37:911–922, 2010.

[46] Wu M.C. and Carroll R.J. Estimation and comparison of changes in the presence of informative right censoring by modeling the censoring process. *Biometrics*, 44:175–188, 1988.

[47] Chen M.H. Ibrahim J.G. and Lipsitz S. R. Missing responses in generalized linear mixed models when the missing data mechanism is nonignorable. *Biometrika*, 88:551–564, 2001.

[48] Yuan Y. and Little R.J.A. Parametric and semiparametric model-based estimates of the finite population mean for two-stage cluster samples with item nonresponse. *Biometrics*, 63:1172–1180, 2007.

[49] Davis B.R. Data source: personal communication.

[50] SHEP Cooperative Research Group. Prevention of Stroke by Antihypertensive Drug Treatment in Older Persons With Isolated Systolic Hypertension. *The Journal of the American Medical Association*, 265:3255–64, 1991.

7

Recurrent Events

Yujie Zhong and Richard Cook

CONTENTS

7.1 Introduction

7.1.1 Recurrent event data

Recurrent event data arise in a diverse array of chronic diseases where affected individuals experience repeated acute manifestations of disease. In respirology, for example, individuals with asthma (Verona et al., 2003) or chronic obstructive pulmonary disease (Grossman et al., 1998) are at risk of recurrent attacks or exacerbations of symptoms possibly due to infection. In neurology, individuals with epilepsy (Pledger et al., 1994) are at risk of recurrent seizures. In hematology individuals with thrombocytopenia are at risk of recurrent episodes of bleeding (Heddle et al., 2003). In each case the events of interest are transient but they can impair quality of life and functional ability, as well as lead to an accumulation of damage. The goal in many prophylactic trials is to reduce the occurrence of these events.

There has been considerable statistical research conducted over the last twenty years on methods for the analysis of recurrent event data. Models and methods can be broadly classified as based on intensity functions (Andersen et al., 1993), marginal rate functions (Lawless and Nadeau, 1995), and random effect models (Lawless, 1987). Which of the various approaches to analysis is most suitable for a given setting depends on the precise scientific objectives. Intensity-based methods are well suited to developing a deep understanding of the risks for event occurrence based on fixed covariates and time-dependent covariates which may summarize exogenous or endogenous factors (Kalbfleisch and Prentice, 2002). Marginal analyses based on rate functions, however, offer a natural basis for treatment comparisons in randomized studies; robust variance estimates are recommended in this case to ensure valid inference. Likelihood rate-based analysis are also possible based on mixed Poisson models. Key considerations when selecting an approach to analysis in randomized trials include the ability to make causal statements, efficiency, robustness of inferences to model-misspecification, and the facility to carry out sample size calculations. More subtle issues such as sensitivity to event-dependent withdrawal, and complications arising when mortality rates are appreciable, also warrant consideration.

The remainder of this chapter is organized as follows. In Section 7.2, we introduce notation for censored recurrent event data and define intensity and rate functions. The form of the corresponding likelihoods are provided along with estimating functions supporting robust inference. The impact of history-dependent censoring is discussed for likelihood and robust analyses, along with methods for adjusting for this. An application to a trial of individuals with cystic fibrosis is also considered in Section 7.2 to illustrate how to construct data frames, carry out analyses and interpret results. In Section 7.3, we develop sample size criteria for standard parallel group trials under a mixed Poisson (negative binomial) model. Alternative methods of analysis are critically discussed in Section 7.4, in which issues of causal inference and model misspecification are raised. Adaptive methods of sample size estimation are also reviewed, as are methods for dealing with recurrent and terminal events. Further remarks are provided in Section 7.5. We next describe a study we use for illustrative purpose.

7.1.2   Data from a cystic fibrosis Trial

Fuchs et al. (1994) report on a clinical trial of individuals with cystic fibrosis. Such individuals are at substantial risk of an accumulation of mucus in the lungs leading to pulmonary exacerbations and, in the longer term, a deterioration of their lung function. In a randomized clinical trial, an experimental treatment at the time, called rhDNase, is a purified recombinant form of the human enzyme DNase I. For individuals randomized to the experimental arm they were to receive it through daily administration whereas those randomised to standard care were administered a placebo in a blinded fashion. The occurrences of pulmonary exacerbations were recorded over the course of follow-up, which was approximately 169 days. Figure 7.1 contains a plot of event occurrence over the course of follow-up for 10 individuals receiving rhDNase and 10 receiving the placebo treatment. The length of the timelines conveys the duration of follow-up and it can be seen that there is some variation across individuals. The onset of exacerbations are denoted by dots and one can likewise see that some individuals experience none (e.g. individual 1 in the placebo group) and some experience several (e.g. individual 7 in the placebo group). A primary objective of the study is to compare the two treatment groups in terms of exacerbation occurrence.

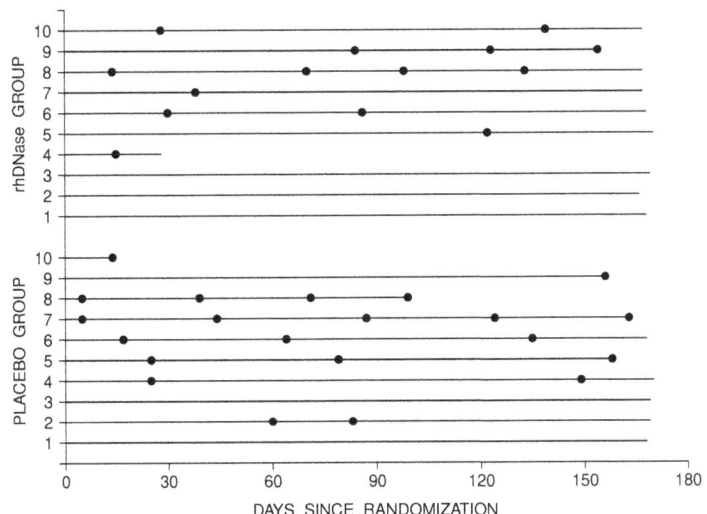

FIGURE 7.1

Plot of event occurrence and duration of follow-up time for selected individuals from the cystic fibrosis trial (Fuchs et al., 1994).

7.2 Notation and Model Formulation

7.2.1 Analysis considerations with recurrent event data

Let $0 < T_{i1} < T_{i2} < \cdots$ denote the times of the events of interest for individual i in a sample of size m, $i = 1, \ldots, m$. The random variable $N_i(t) = \sum_{j=1}^{\infty} I(T_{ij} \leq t)$ counts the number of events occurring over $(0, t]$ for individual i, where $I(\cdot)$ is an indicator function such that $I(A) = 1$ if A is true and $I(A) = 0$ otherwise. When viewed as a stochastic process over time, $\{N_i(s), 0 \leq s\}$ is a right-continuous counting process. We let $\Delta N_i(t) = N_i(t, t + \Delta t)$ denote the number of events experienced by individual i over $[t, t + \Delta t)$ and $dN_i(t) = \lim_{\Delta t \downarrow 0} \Delta N_i(t)$ indicate whether an event occurred at time t for individual i. In a one sample problem, the mean function for a recurrent event process is defined as

$$\mu(t) = E\{N_i(t)\}, \tag{7.1}$$

which gives the expected number of events over $(0, t]$ per individual.

If X_i is a $p \times 1$ vector of fixed covariates, let $\mathcal{H}_i(t) = \{N_i(s), 0 < s < t, X_i\}$ denote the history of the process for individual i at time t, which records the number and times of events from 0 to t^- along with the covariate values. The intensity function of the recurrent event process is the instantaneous conditional probability of an event occurring at t conditional on the history at t, and is written as

$$\lambda(t | \mathcal{H}_i(t)) = \lim_{\Delta t \to 0} \frac{P\left(\Delta N_i(t) = 1 | \mathcal{H}_i(t)\right)}{\Delta t}. \tag{7.2}$$

In the continuous time setting here it is assumed that two events cannot occur at the same time, in which case the intensity function completely specifies the event process (Cook and

Lawless, 2007). The intensity function is analogous to the hazard function used in survival analysis in that it conveys the instantaneous risk of an event. It is a more general function however, in that events can recur in this setting, and the history includes information on the number and timing of events in the past; in the survival setting the history of the response simply includes the fact that the individual in mind has not failed previously. We also note that in survival analysis involving time-varying covariates, models are typically based on intensity functions in a strict sense, given they usually involve conditioning on the path of a covariate process.

Consider a randomized trial in which individuals are to be followed over time in order to record the occurrence of clinical events of interest. There is typically a planned period of observation specified in a trial protocol where $t = 0$ denotes the time of randomization and A represents an administrative censoring time. To accommodate loss to follow-up we define a random variable R_i to represent the time individual i withdraws from the study, giving a random observation period of $(0, C_i]$ where $C_i = \min(A, R_i)$; we also let $Y_i(t) = I(t \leq C_i)$.

We let $d\bar{N}_i(t) = Y_i(t)dN_i(t)$ indicate that an event was observed at time t for individual i. Under the assumption of conditionally independent and non-informative censoring (Cook and Lawless, 2007), the probability that "n_i events occur at times $t_{i1} < \cdots < t_{in_i}$" over $(0, C_i]$ for individual i conditional on X_i is

$$\prod_{j=1}^{n_i} \lambda(t_{ij}|\mathcal{H}_i(t_{ij})) \cdot \exp\left\{-\int_0^\infty Y_i(u)\lambda(u|\mathcal{H}_i(u))du\right\}, \tag{7.3}$$

which also constitutes the contribution to the partial likelihood.

7.2.2 Methods based on rate and mean functions

While intensity-based models are useful when interest lies in understanding the dynamic aspects of a point process, marginal features such as mean or rate functions are more appealing when analysing data from clinical trials. The primary reason for this is that conditioning on events realized post-randomization induces confounding which prevents one from making causal inferences about the treatment effects. As a consequence there is considerable appeal in the use of rate and mean functions when assessing treatment effects on recurrent event outcomes.

The rate function $\rho(t)$ is the derivative of the mean function and hence reflects how the marginal risk of events changes over time; since $\rho(t)dt = d\mu(t) = E\{dN_i(t)\}$ the mean function can be expressed as

$$\mu(t) = \int_0^t \rho(s)ds. \tag{7.4}$$

For Poisson processes, the rate function is in fact the intensity function and rate-based likelihood analyses can be carried based on (7.3) under the assumption that the events are generated according to a Poisson process.

There have been many developments in recent years on ways of formulating regression models but the proportional rate function models of the form $\rho_i(t) = \rho(t|X_i) = \rho_0(t)g(X_i; \beta)$ remain the most common with $\rho_0(s)$ a baseline rate function applicable when $X_i = 0$, $g(X_i; \beta) = \exp(X_i'\beta)$, and β is a vector of regression coefficients. For a Poisson process the likelihood (7.3) can then be written as

$$\left[\prod_{j=1}^{n_i} \rho(t_{ij}|X_i)\right] \exp\left(-\int_0^\infty Y_i(s)\rho(s|X_i)ds\right). \tag{7.5}$$

While Poisson processes are determined entirely in terms of the conditional rate function the model is highly restrictive in that $E\{N_i(t)|X_i\} = \text{var}\{N_i(t)|X_i\} = \mu_i(t)$. This can be addressed by using the Poisson formulation as a working model for estimation and then deriving robust variance estimates to provide protection against misspecification.

Suppose $\rho_0(t;\alpha)$ in (7.5) is indexed by an $r \times 1$ parameter α and β is a $p \times 1$ vector of regression coefficients. Let $C = \max(C_1, \ldots, C_m)$ and $\theta = (\alpha', \beta')'$. The log-likelihood from (7.5) is

$$\ell(\theta) = \sum_{i=1}^{m} \int_0^C Y_i(s)[\log \rho_i(s;\theta)dN_i(s) - \rho_i(s;\theta)ds] \ . \tag{7.6}$$

which gives the score equations

$$U_\alpha(\theta) = \frac{\partial \ell(\theta)}{\partial \alpha} = \sum_{i=1}^{m} \int_0^C Y_i(s)\frac{\partial \log \rho_0(s;\alpha)}{\partial \alpha}\{dN_i(s) - \rho_i(s;\theta)ds\} = 0 \ , \tag{7.7}$$

$$U_\beta(\theta) = \frac{\partial \ell(\theta)}{\partial \beta} = \sum_{i=1}^{m} \int_0^C Y_i(s)\{dN_i(s) - \rho_i(s;\theta)ds\}X_i = 0 \ , \tag{7.8}$$

and we let $U(\theta) = (U_\alpha'(\theta), U_\beta'(\theta))'$ denote the full score vector.

The semiparametric multiplicative model in which $\rho_0(t)$ is not assumed to have any particular parametric form is called the Andersen-Gill Model (Andersen and Gill, 1982). In this framework, we treat $d\mu_0(t) = \rho_0(t)dt$ as a parameter giving the score equation

$$\sum_{i=1}^{m} Y_i(s)\{dN_i(s) - \exp(X_i'\beta)d\mu_0(s)\} = 0 \ , \qquad 0 \le s \ , \tag{7.9}$$

from (7.7). Solving for $d\mu_0(s)$ gives

$$d\widetilde{\mu}_0(s;\beta) = \frac{d\bar{N}.(s)}{\sum_{i=1}^{m} Y_i(s)\exp(X_i'\beta)} \ ,$$

where $d\bar{N}.(s) = \sum_{i=1}^{m} Y_i(s)dN_i(s)$. Substituting this into (7.8) gives an estimating equation $U(\beta) = \sum_{i=1}^{m} U_i(\beta) = 0$ where

$$U_i(\beta) = \int_0^C Y_i(s)X_i\left[dN_i(s) - \frac{d\bar{N}.(s)}{\sum_{\ell=1}^{m} Y_\ell(s)\exp(X_\ell'\beta)}\exp(X_i'\beta)\right] \ . \tag{7.10}$$

Solving $U(\beta) = 0$ yields the estimate $\widehat{\beta}$. Note that (7.10) has the form of the Cox partial likelihood score equation and hence software for fitting Cox models can be used to obtain $\widehat{\beta}$.

A robust asymptotic covariance matrix for $\widehat{\beta}$ is given by

$$\text{asvar}\{\sqrt{m}(\widehat{\beta} - \beta)\} = \mathcal{A}^{-1}(\beta)\mathcal{B}(\beta)[\mathcal{A}^{-1}(\beta)]' \ , \tag{7.11}$$

where $\mathcal{A}(\beta) = m^{-1}E\{-\partial U(\beta)/\partial \beta'\}$ and $\mathcal{B}(\beta) = m^{-1}E\{U(\beta)U'(\beta)\}$. A robust covariance matrix for $\widehat{\beta}$ is estimated by using an empirical expectation of these quantities and inserting $\widehat{\beta}$ in place of β to give

$$\widehat{\text{asvar}}\{\sqrt{m}(\widehat{\beta} - \beta)\} = \widehat{A}^{-1}(\beta)\widehat{B}(\beta)[\widehat{A}^{-1}(\beta)]'\big|_{\beta=\widehat{\beta}} \ , \tag{7.12}$$

where $\widehat{A}(\beta) = -m^{-1}\partial U(\beta)/\partial \beta'$ and

$$\widehat{B}(\beta) = \frac{1}{m}\sum_{i=1}^{m} \int_0^C \int_0^C Y_i(u)Y_i(v)W_i(u;\beta)W_i'(v;\beta)dM_i(u)dM_i(v) \ ,$$

with

$$W_i(s; \beta) = X_i - \frac{\sum_{l=1}^{m} Y_l(s) \exp(X_l' \beta) X_l}{\sum_{l=1}^{m} Y_l(s) \exp(X_l' \beta)} , \tag{7.13}$$

and $dM_i(u) = dN_i(u) - d\mu_i(u)$. Lin et al. (2000) provide a rigorous derivation of the large sample theory based on martingales. Therneau and Grambsch (2000) and Cook and Lawless (2007) show how this model can be fitted using R (R Core Team, 2014) or Splus (Venables and Ripley, 2002). This variance estimate is recommended for general use as it is valid under a Poisson model but offers protection against departures from the Poisson process. We show how this semiparametric Andersen-Gill model can be fitted to the cystic fibrosis data and the robust variance is obtained in the first analysis of Section 7.2.4

Flexible fully-specified rate-based models can be obtained by introducing multiplicative random effects which reflect heterogeneity in the event rates across individuals. In this setting each individual is still presumed to experience events according to a Poisson process, but they do so with their own rate of events expressible as a multiple of other individuals rate functions. That is suppose that given U_i, $\{N_i(t), 0 \le t\}$ follows a Poisson process with rate function $\rho(s|U_i, X_i) = U_i \rho_0(s) \exp(X_i' \beta)$, where $U_i > 0$ is a subject-specific random effect. We assume that $U_i \perp X_i$ and U_i, $i = 1, \ldots, m$ are independent and identically distributed non-negative random variables with $E(U_i) = 1$ and $\text{var}(U_i) = \phi$.

The random effects are of course conceptualized and unobserved and it is natural to derive properties of the model after marginalizing over these quantities. Because the random effects act multiplicatively, the simple mean structure is retained in the marginal models: $E\{N_i(t)|X_i\} = E(U_i|X_i)\mu_0(t) \exp(X_i'\beta) = \mu_i(t)$ and hence $E\{dN_i(t)|X_i\} = \rho_0(t)dt \exp(X_i'\beta)$. Moreover by the conditional variance formula we find $\text{var}(N_i(t)|X_i) = \mu_i(t) + \phi\mu_i^2(t)$, and $\text{cov}(N_i(a, b), N_i(s, t)|X_i) = \phi\mu_i(a, b)\mu_i(s, t)$ for any $a < b < s < t$, where $N_i(s, t)$ is the number of event occurring over $(s, t]$ and $\mu_i(s, t) = E[N_i(s, t)]$. Thus the mixed Poisson model accommodates both extra-Poisson variation arising due to unexplained (multiplicative) variation in risk across individuals, and a serial dependence in the counts over disjoint intervals within individuals. This latter phenomenon is often of concern when it is apparent that event occurrence is associated with an increase in the risk of future events.

So far we have not considered a specific distribution for U_i, but if we let $G(\cdot; \phi)$ denote the cumulative distribution function for U_i then the marginal likelihood becomes

$$\prod_{i=1}^{m} \int_0^\infty \left\{ \left[\prod_{j=1}^{n_i} u_i \rho(t_{ij}|X_i) \right] \exp\left(-\int_0^\infty Y_i(s) u_i \rho(s|X_i) ds \right) \right\} dG(u_i) , \tag{7.14}$$

Any distribution for a non-negative random variable can be adopted (e.g log-normal, positive stable) but the most common is the gamma distribution which is conjugate to the Poisson model (Lawless, 1987). The resulting likelihood takes the explicit form

$$\prod_{i=1}^{m} \left\{ \left[\prod_{j=1}^{n_i} \frac{\rho_0(t_{ij})}{\mu_0(C_i)} \right] \frac{\Gamma(n_i + \phi^{-1})}{\Gamma(\phi^{-1})} \frac{(\phi\mu_i(C_i))^{n_i}}{(1 + \phi\mu_i(C_i))^{n_i + \phi^{-1}}} \right\} . \tag{7.15}$$

The intensity for a mixed Poisson process with a multiplicative random effect is $\lambda(t|\mathcal{H}_i(t)) = E(U_i|\mathcal{H}_i(t))\rho_0(t) \exp(X_i'\beta)$. For a negative binomial process it has the form

$$\lambda(t|\mathcal{H}_i(t)) = \frac{1 + \phi N_i(t^-)}{1 + \phi\mu_i(t)} \cdot \rho_0(t) \exp(X_i'\beta) , \tag{7.16}$$

from which it can be seen that there will be a transient spike in the risk of events arising upon the occurrence of each event. In many medical settings this phenomenon arises when the events occur as a consequence of a change in an underlying condition.

7.2.3   Censoring, Likelihood, and Marginal Methods

Here we examine assumptions regarding the censoring process more closely. We let $\bar{\mathcal{H}}_i(t) = \{(\bar{N}_i(u), Y_i(u)), 0 < u < t, X_i\}$ denote the full history including the information on the censoring process. The intensity of the censoring time is defined by

$$\lambda_i^c(t|\bar{\mathcal{H}}_i(t)) = \lim_{\Delta t \downarrow 0} \frac{P(t \le C_i \le t + \Delta t|\bar{\mathcal{H}}_i(t))}{\Delta t} , \qquad (7.17)$$

which shows that the risk of study withdrawal can in general depend on the previous record of event occurrence. Event-dependent censoring does not cause problems with likelihood-based analyses because the full likelihood for the event and censoring process written in product integral (Cook and Lawless, 2007) notation,

$$\prod_{[0,\infty)} P\left(d\bar{N}_i(u), dY_i(u)|\bar{\mathcal{H}}_i(u)\right)$$

can be factored as

$$\prod_{[0,\infty)} P\left(d\bar{N}_i(u)|\bar{\mathcal{H}}_i(u)\right) \prod_{[0,\infty)} P\left(dY_i(u)|d\bar{N}_i(u), \bar{\mathcal{H}}_i(u)\right) . \qquad (7.18)$$

If the censoring process is non-informative, meaning (7.17) does not share any parameters with the event intensity (7.2), then the second term in (7.18) can be ignored. Moreover if the censoring and event processes are conditionally independent given the history, then the first term of (7.18) is the likelihood in (7.3).

There are two distinct settings in which one may consider marginal rate-based models. If one believes that the events are generated by a Poisson process then the likelihood arguments may apply and consistent estimation is achieved even with event-dependent censoring. If one believes that a mixed-Poisson model is appropriate, rate-based inference is still possible based on such a formulation. However, if the multiplicative rate model is adopted for ease of interpretation of treatment effects, and estimating functions are to be used with robust sandwich type variance estimates, analyses are not likelihood-based, the factorization in (7.18) is not applicable, and we must re-examine things.

To do so we consider the simplest case of a one-sample problem where the goal is to estimate the rate and mean functions. The estimating equation for $d\mu(s)$ under the Poisson assumption is

$$\sum_{i=1}^m Y_i(s)\{dN_i(s) - d\mu(s)\} = 0 \qquad (7.19)$$

which has solution $d\hat{\mu}(s) = d\bar{N}.(s)/Y.(s)$ with $Y.(s) = \sum_{i=1}^m Y_i(s)$. This in turn gives the Nelson-Aalen estimate of the mean function

$$\hat{\mu}(t) = \int_0^t d\hat{\mu}(s) . \qquad (7.20)$$

Under completely independent censoring (7.19) yields an unbiased estimate of the mean function for any point process. More generally, if $Y_i(s)$ is not independent of $N_i(t)$ given $\bar{\mathcal{H}}_i(t)$ then $d\hat{\mu}(s)$ is consistent for the solution to

$$E\{Y_i(s)(dN_i(s) - d\mu(s))\} = P(Y_i(s) = 1)[E(dN_i(s)|Y_i(s) = 1) - d\mu(s))] = 0 ,$$

which we denote as $d\mu^*(s)$ and we obtain $\mu^*(t)$ as the limiting value of the mean function estimate.

Much interest has developed in recent years in inverse probability weighted (IPW) estimating functions and estimates which correct for dependent censoring. This is achieved by modeling the censoring process and formulating IPW estimating equations of the form $\sum_{i=1}^{m} U_i^w(s) = 0$ where

$$U_i^w(s) = \frac{Y_i(s)}{P(C_i \geq s | \bar{\mathcal{H}}_i(s))} \{ dN_i(s) - d\mu(s) \} . \tag{7.21}$$

Of course in any given analysis the censoring intensities must be estimated; consistent estimation of $\mu(t)$ is possible if the censoring model is correctly specified and unknown parameters are replaced by $\sqrt{m}$-consistent estimators. The discussion above is in the context of a one sample problem but the same issues arise when assessing treatment effects via marginal rate-based models. Further details on the formulation of censoring models for recurrent event processes are beyond the scope of this article but we refer readers to Cook and Lawless (2007) for further discussion.

The key point is that when the recurrent event model is intensity-based and fully specified, the factorization of the likelihood means that history-dependent censoring not problematic. For reasons discussed earlier, and which we revisit in Section 7.4, in clinical trials it is undesirable to condition on the process history when evaluating treatment effects. This is fine if the events are generated by a Poisson process as in this setting likelihood analyses do not condition on the process history. It is a very restrictive model, however, so typically one adopts rate-based models in a semiparametric setting and use robust variance estimation. In this case however, censoring must be completely independent of the event history for consistent estimation based on standard estimating equations. Inverse probability of censoring weighted equations such as (7.21) provide protection against history-dependent censoring but require specification of a model which characterizes how the risk of censoring depends on the process history.

7.2.4　Assessment based on exacerbations in cystic fibrosis

A select number of lines of the data frame are included below for four individuals using the counting process format (Therneau and Grambsch, 2000). The R function `coxph` was written for the analysis of survival data and because it can handle left-truncation we can use it to compute the likelihood function by noting that the contribution to (7.5) from individual i is

$$\prod_{j=1}^{n_i} \left\{ \rho(t_{ij}|X_i) \exp \left(- \int_{t_{i,j-1}}^{t_{ij}} Y_i(s) \rho(s|X_i) ds \right) \right\} \cdot \exp \left(- \int_{t_{i,n_i}}^{C_i} Y_i(s) \rho(s|X_i) ds \right) .$$

If individual i has $n_i > 0$ events then we have $n_i + 1$ sub-intervals $[0, t_{i1}), [t_{i1}, t_{i2}), \ldots, [t_{i,n_i-1}, t_{in_i})$ and $[t_{in_i}, C_i]$ and thus we can express the contribution from individual i as a product of $n_i + 1$ contributions with n_i contributions taking the same form as contributions from left-truncated survival times (the left-truncation time being 0 for the first interval) and one left-truncated and right-censored observation corresponding to $[t_{in_i}, C_i]$; the `start` times are analogous to the left-truncation times and the `stop` times are the recurrent event times if `status = 1` or otherwise it is the final censoring time. The individual with id=493305 had an exacerbation which occurred 65 days after randomization and they contribute two lines to the data frame. Note that the start time on their second line is 75; during intervening days this individual was experiencing an exacerbation and therefore was not at risk so this interval was removed from the at risk period.

```
> rhDNase[1:5,]
```

```
     id start stop status enum trt
1 493301    0  168      0    1   1
2 493303    0  169      0    1   1
3 493305    0   65      1    1   0
4 493305   75  168      0    2   0
5 493309    0  168      0    1   1
```

The function `coxph` in R can be called to fit the Andersen-Gill model and with the specification `cluster(id)` a robust variance estimate based on (7.12) is obtained.

```
> library(survival)
> fitr <- coxph(Surv(start, stop, status) ~ trt + cluster(id),
               data=rhDNase, method="breslow")
```

The result of applying this function is an object `fitr`, the contents of which are provided through the call to the `summary` function.

```
> summary(fitr)
  n= 965, number of events= 360

        coef exp(coef) se(coef) robust se       z Pr(>|z|)
trt -0.2865    0.7509   0.1064    0.1309  -2.188   0.0287

     exp(coef) exp(-coef) lower .95 upper .95
trt    0.7509      1.332    0.5809    0.9706

Likelihood ratio test= 7.32 on 1 df, p=0.006836
Wald test            = 4.79 on 1 df, p=0.02867
Score (logrank) test = 7.29 on 1 df, p=0.006921, Robust = 4.8  p=0.02853
```

Here we see that there is an approximately 25% lower rate of exacerbations in the rhDNase group ($RR = 0.75$; 95% CI: 0.58, 0.97; $p = 0.029$). The likelihood ratio test and the score test are based on the assumption that the working Poisson model is correct but the Wald and robust pseudo-score tests accommodate extra-Poisson variation.

If a semiparametric negative binomial analysis is of interest the likelihood in (7.15) can be maximized by specification of the `frailty(id)` option instead of the `cluster(id)` specification, but the call to `coxph` is otherwise identical.

```
> fitf <- coxph(Surv(start, stop, status) ~ trt + frailty(id),
               data=rhDNase, method="breslow")
```

Here the output is slightly different with the introduction of an estimate of the variance parameter of the frailty distribution, ϕ. This function uses an approach based on penalized spline functions to approximate the maximum likelihood estimates, when the gamma distribution is used for the random effect distribution (the default).

```
> summary(fitf)
  n= 965, number of events= 360
             coef se(coef)  se2   Chisq   DF      p
trt         -0.302  0.1385 0.1068   4.75  1.0 2.9e-02
frailty(id)                       386.29 270.7 4.8e-06
      exp(coef) exp(-coef) lower .95 upper .95
trt      0.7394      1.353    0.5636      0.97

   Variance of random effect= 1.229903   I-likelihood = -2263.9
Likelihood ratio test= 530.4  on 271.3 df,   p=0
```

An almost identical estimated relative rate is obtained at 0.74 (95% CI: 0.56, 0.97; $p = 0.029$). The estimate $\hat{\phi} = 1.23$ suggests quite substantial heterogeneity in the rate of events; as we will see this will play a key role in sample size calculations for future studies; additional output from the call to the summary function are omitted.

It is also helpful to plot the nonparametric Nelson-Aalen estimates (7.20) of the cumulative mean functions. There is only modest variation in the duration of follow-up and so little cause for concern regarding the effects of dependent censoring which we discussed in Section 7.2.3. The code for obtaining the standard unweighted Nelson-Aalen estimates is based on the code for estimating the cumulative hazard function in survival analysis. The following code provides the estimate for the placebo group and gives the dashed line in Figure 7.2.

```
> na0 <- survfit(fitr, type="aalen", newdata=data.frame(trt = 0))

> plot(0, 0, type="n", axes=F, xlim=c(0,180), ylim=c(0,0.7), xlab="", ylab="")
> lines(c(0, na0$time), -log(c(1, na0$surv)), type="s", lty=2)
```

The plot reveals a roughly constant risk of events and a consistently lower mean function for individuals receiving rhDNase.

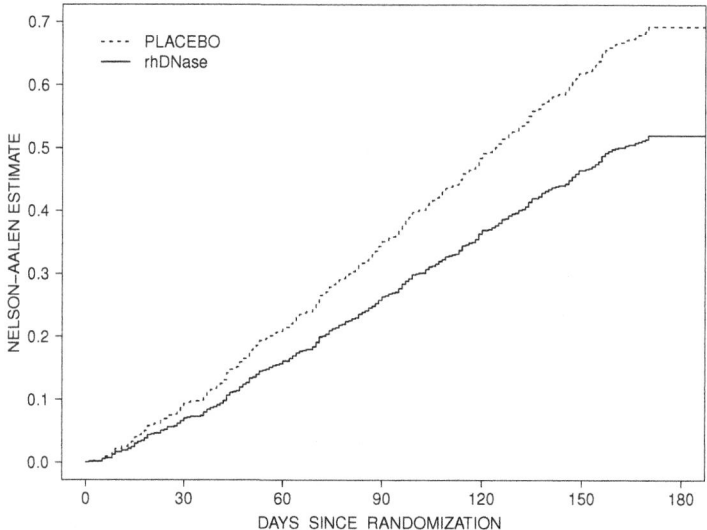

FIGURE 7.2
Nelson-Aalen estimate (7.20) of the cumulative mean functions for the rhDNase and placebo groups for the study by Fuchs et al. (1994).

7.3   Sample Size Based on Proportional Rate Functions

7.3.1   Derivations under a negative binomial model

Sample size formula for a trial with recurrent events generated according to a Poisson model were developed by Cook (1995). Bernardo and Harrington (2001) develop general results based on large sample theory which coincide with these of Cook (1995) in special cases. Note that any model can be specified for the recurrent events and provided one can compute the solution to $E_T\{U(\beta)\} = 0$ and the matrices $\mathcal{A}(\beta)$ and $\mathcal{B}(\beta)$ in (7.11) sample size requirements can be derived. The general challenge is to find such models that render the multiplicative rate function model correct and the limiting value β^* interpretable. We focus on estimation and consider the framework of a mixed Poisson process for sample size calculations.

We consider the setting of a two-armed randomized trial in which individuals are assigned to receive an experimental treatment with probability 0.5 and receive standard care otherwise; we let $X_i = 1$ indicate assignment to the experimental arm and set $X_i = 0$ otherwise. To simplify the derivations we consider a constant (time-homogeneous) event rate and let $\rho_0(s) = \rho_0$. The log-likelihood contribution from individual i denoted $\ell_i(\theta)$ is then

$$\sum_{j=0}^{n_i-1} \log(1 + j\phi) + n_i(\beta_0 + \beta_1 X_i) - (\phi^{-1} + n_i) \log(1 + \phi \exp(\beta_0 + \beta_1 x_i) C_i), \qquad (7.22)$$

where $\beta_0 = \log \rho_0$, $\beta = (\beta_0, \beta_1)'$, and $\theta = (\beta', \phi)'$. Notice that (7.22) depends only on the event counts, the duration of follow-up, and the treatment assignment. The score vector is denoted $U_\theta(\theta) = \partial \ell(\theta)/\partial \theta = (U_\beta'(\theta), U_\phi(\theta))'$, where $U_\beta(\theta) = \partial \ell(\theta)/\partial \beta$ and $U_\phi(\theta) = \partial \ell(\theta)/\partial \phi$, and we let $I(\theta) = -\partial^2 \ell(\theta)/\partial \theta \partial \theta'$ and $\mathcal{I}(\theta) = E\{I(\theta)\}$ denote the observed and expected information matrices, respectively. Moreover β is orthogonal to ϕ (i.e. $\mathcal{I}_{\beta\phi}(\theta) = 0$) and hence an asymptotic covariance matrix for $\widehat{\beta}$ is $\mathcal{I}_{\beta\beta}^{-1}(\theta)$, where $\mathcal{I}_{\beta\beta}(\theta)$ is the 2×2 submatrix of $\mathcal{I}(\theta)$ conformable with β (Lawless, 1987). This can be well-approximated by

$$\text{asvar}\{\sqrt{m}(\widehat{\beta}_0 - \beta_0)\} = \left\{0.5 * E\left[\frac{\exp(\beta_0) C}{1 + \phi \exp(\beta_0) C} \Big| X = 0\right]\right\}^{-1}, \qquad (7.23)$$

$$\text{asvar}\{\sqrt{m}(\widehat{\beta}_1 - \beta_1)\} = \sum_{x=0}^{1} \left\{0.5 * E\left[\frac{\exp(\beta_0 + \beta_1 x) C}{1 + \phi \exp(\beta_0 + \beta_1 x) C} \Big| X = x\right]\right\}^{-1}, \qquad (7.24)$$

where C represents the net follow-up duration. When planning studies we typically assume that study withdrawal times (R) are independent and nondifferential (i.e. independent of treatment) and adopt convenient distributions for R (e.g. exponential) to facilitate calculation of the expectations in (7.23) and (7.24). In this case if $F_r(r; \psi)$ is the c.d.f. of R indexed by parameter ψ and $h(\cdot)$ is a function of C, we simply have

$$E(h(C)) = \int_0^A h(r) \cdot dF_r(r; \psi) + h(A) \cdot [1 - F_r(A; \psi)] .$$

Of course when follow-up is over a common time period $(0, A]$ (i.e. there is no loss to follow-up) then an analysis based simply on counts is sufficient. We can replace $\mu_i(C_i)$ with $\exp(\log \mu_0(A) + X_i\beta)$ in (7.15) and simply concern ourselves with the expected number of events in the two arms.

If γ_1 and γ_2 denote the desired type I and type II error rates of a two-sided test of $H_0 : \beta_1 = \beta_{10}$, where the alternative value is $\beta_1 = \beta_{1A}$, then one needs to find the minimum sample size m that satisfies

$$m > \frac{\left(\sqrt{\text{asvar}_0(\sqrt{m}(\widehat{\beta}_1 - \beta_{10}))} \, Z_{\gamma_1/2} + \sqrt{\text{asvar}_A(\sqrt{m}(\widehat{\beta}_1 - \beta_{1A}))} \, Z_{\gamma_2}\right)^2}{(\beta_{10} - \beta_{1A})^2}, \qquad (7.25)$$

where $\text{asvar}_0(\sqrt{m}(\widehat{\beta}_1 - \beta_{10}))$ and $\text{asvar}_A(\sqrt{m}(\widehat{\beta}_1 - \beta_{1A}))$ denote the expressions for the asymptotic variance given by (7.24) under the null and alternative hypotheses, respectively, and Z_p represents the $100(1-p)$ percentile for a standard normal distribution. Values for β_{10} and β_{1A} are needed, as well as provisional values for the other parameters in the response model and the censoring distribution. Often $\beta_{10} = 0$ so that the null hypothesis is that the events occur at the same rate in the treatment and control groups.

Finally, although sample size calculations are often based on Wald-type test statistics, as here, actual analyses may be based on other approaches such as the robust pseudo-score tests (Cook et al., 1996). The frequency properties of different tests are generally comparable for large samples, so sample size choices based on (7.25) can still be employed.

7.3.2 Illustrative sample size calculation

Here we consider a setting of an asthma trial in which individuals are randomized in a balanced fashion to an experimental treatment or standard care. Suppose follow-up is planned over the period $(0, 1]$ and on average there are two events experienced over this period under standard care with events generated by a time homogeneous model. We suppose individuals withdraw from the study with (possibly latent) withdrawal times (R_i) generated by an exponential model with hazard ψ such that 85% of individuals complete follow-up, so $\psi = -\log 0.85$. Prior studies suggest a modest degree of heterogeneity with $\phi = 0.15$. The goal is to ensure 90% power to detect a 25% reduction in the rate of events based on a two-sided test of $H_0 : \beta = \beta_{10} = 0$ versus $H_A : \beta \neq 0$ at the 5% level, so $\beta_{1A} = \log 0.75$, $\gamma_1 = 0.05$, $Z_{\gamma_1/2} = 1.96$, $\gamma_2 = 0.10$, and $Z_{\gamma_2} = 1.28$. Plugging these parameters into (7.24) gives $\text{asvar}_0(\sqrt{m}(\hat{\beta}_1 - \beta_{10})) = 2.797$ and $\text{asvar}_A(\sqrt{m}(\hat{\beta}_1 - \beta_{1A})) = 3.159$. This gives the minimum sample size of $m = 373$ individuals in the trial. Table 7.1 contains the sample size for additional values of ϕ including 0.10, 0.20, and 0.25. for 80% as well as 90% power. We can see the intuitive results that the sample size requirement increase as the effect size gets smaller and as the extent of heterogeneity increases.

TABLE 7.1

Sample size calculation for a balanced randomized asthma trial under negative binomial model based on proportional rate function.

	80% Power				90% Power			
$\exp(\beta_{1A})$	$\phi = 0.10$	0.15	0.20	0.25	$\phi = 0.10$	0.15	0.20	0.25
0.90	1855	2004	2152	2299	2494	2693	2891	3089
0.85	786	849	911	973	1060	1143	1226	1309
0.80	421	454	487	520	569	613	657	701
0.75	256	276	296	316	347	373	400	427
0.70	169	181	194	207	229	246	264	281

7.4 Other Considerations in Recurrent Event Analyses

7.4.1 Issues regarding causal inference

Rate functions characterize a marginal feature of the event process but are not presumed to fully characterize the model except for the special case of Poisson processes. In general settings there may be a history dependence which is ignored, or even simply fixed prognostic baseline variables which are not accounted for. In what follows we omit subscripts indexing individuals for convenience and consider the case of a single baseline variable Z which is associated with the event occurrence through a Poisson model $E(dN(t)|Z) = \rho_0(t)\exp(Z\zeta)$; let $F_Z(z)$ denote the cumulative distribution function for Z. When we recruit individuals into a clinical trial and randomize the treatment allocation ($X = 1$ with probability 0.5 and $X = 0$ otherwise), we may suppose that events are governed by a Poisson process with rate function

$$E(dN(t)|X, Z) = \rho_0^*(t)dt\exp(X\beta + Z\zeta) . \tag{7.26}$$

We can, and typically do, consider the model

$$E(dN(t)|X) = E_{Z|X}\{E(dN(t)|X, Z)\} = \rho_0(t)dt\exp(X\beta) , \tag{7.27}$$

where $\rho_0(t) = \rho^*(t)M_z(\zeta)$ with $M_Z(\zeta)$ the moment generating function of Z evaluated at ζ. The fact that $X \perp Z$ due to randomization ensures that X still has a multiplicative effect in the marginal model. Thus omitting or conditioning on baseline variables is perfectly acceptable in marginal rate-based models since the effect of X in both (7.26) and (7.27) has a causal meaning.

The "stratified Andersen-Gill model" is a partially conditional model in which one conditions on the cumulative event count and so focus on $E\{dN(t)|X, N(t^-) = n\}$. This situation is quite different than the previous one since $N(t^-)$ is observed post-randomization and is responsive to treatment. It is well-known in survival analysis that one should not evaluate treatment effects in regression models which control for features observed post-randomization (Kalbfleisch and Prentice, 2002). In the recurrent event setting we have

$$dF_Z(Z|X, N(t^-) = n) = \frac{\mu(t|X, Z)^n \exp(-\mu(t|X, Z))dF_Z(z)}{E_Z\{\mu(t|X, Z)^n \exp(-\mu(t|X, Z))\}} . \tag{7.28}$$

Since this probability cannot be factored it is apparent that $X \not\perp Z|N(t^-) = n$ despite the randomization at $t = 0$. Use of a partially conditional model precludes causal inference about treatment effects in a formal sense, since conditioning on the cumulative event count induces confounding. Examination of partially conditional rate functions can be helpful when exploring the extent of risk on past event occurrence of course, and if the event process is Markov this offers a valid strategy for analysis. If the goal is to carry out robust inferences regarding treatment effects in clinical trials, conditioning on the event history is not recommended; see Zhong and Cook (2019) for further details.

In some settings the event process is most naturally conceived of in terms of gap time models. In this case intensity functions may take the form of a simple model

$$\lambda(t|\mathcal{H}(t)) = h_n(B(t)|X) , \tag{7.29}$$

where $N(t^-) = n$ and $B(t) = t - t_{N(t^-)}$ is the time since the most recent event. Like stratified Andersen-Gill model this conditions on $N(t^-)$ and $T_{N(t^-)}$, and moreover this resets the time origin for the event risk at the time of each successive event. Any dependence between gap times that remains unaccounted for also induces dependent censoring (Lin et al., 1999) which is more challenging to deal with than the sort discussed in Section 7.2.3. Use of gap time models is therefore not recommended for use when evaluating treatment effects in clinical trials.

7.4.2 Marginal multivariate failure times models

Another framework for analysis is to consider the times from randomization to the first, second, and subsequent events using the approach of Wei et al. (1989) developed for multivariate failure time data. Estimating in this approach can proceed by fitting separate marginal Cox regression models for the time from randomization to the first event (T_{i1}), the time from randomization to the second event (T_{i2}), etc. until the the time from randomization to the Kth event (T_{iK}), where K is specified in the protocol. Robust variance estimates are computed to address the fact that the event times are not independent, and the result does ensure valid inferences for tests of the null hypothesis of no treatment. More generally, however, there are limitations when applying this approach. First there are no point process models which are compatible with the proportional hazards assumption simultaneously holding for $T_{i1}, T_{i2}, \ldots, T_{iK}$ and so the interpretation of the individual β_k estimates are difficult. An alternative approach of stratifying on the event number and constraining the coefficients to be the same yields a single measure β which is likewise impossible to assign meaning to. Moreover under misspecified models the distribution of the censoring time influences the limiting value of the estimator so different studies with identical recurrent event processes but different administrative and/or random censoring distributions will yield estimates of different limiting quantities. Finally it is undesirable to have to specify the number of events to be analysed; the Andersen-Gill analysis does not require one ignore time of events after any particular number.

7.4.3 Adaptive two-stage sample size estimation

In clinical trials involving recurrent events sample size calculations require specification of event rates and measures of variability of the resulting estimators. Published trials will routinely report estimates of the average number of events in each treatment arm often with accompanying plots of Nelson-Aalen estimates. The robust methods of Section 7.2.2 are routinely adopted for inference however, which means that estimates of ϕ may not be available from the published literature. As a result in the absence of patient-level data it can be difficult to select suitable values for ϕ for sample size calculations based on (7.25). If ϕ is larger than the value specified in (7.25), then the actual power will be less than the desired level. When there is considerable uncertainty in the value of ϕ or other parameters, it is advisable to specify a range of values as reported in Table 7.1, and then select the largest sample size resulting from the parameter value configurations. Alternatively, one can consider adaptive designs which carry out sample size re-estimation periodically throughout the course of the study; Wittes and Brittain (1990) use the term internal-pilot studies. We focus on the setting of a mixed-Poisson process.

The approach essentially involves using preliminary estimates of the design parameters to obtain an initial target sample size. After some individuals have been recruited and randomized and have provided responses the preliminary data can be used to update the estimates of the design parameters through parameter estimation under blinded or unblinded procedures. It is generally well-recognized that retaining the blind is desirable in the course of a randomized trial and so blinded procedures are generally preferred. There are of course identifiability issues which may arise depending on which parameter is to be estimated. Cook et al. (2009a) develop expectation-maximization algorithm (Dempster et al., 1977) in which the treatment indicator is considered a missing covariate; the baseline probability of treatment assignment is known of course by the randomization process. Cook et al. (2009a) highlight the identifiability issue for the estimate of the baseline rate in the control arm by providing contour plots of the profile observed data likelihood surface, which suggests constraints may be necessary to select the suitable estimate of the control event rate. The variance parameter ϕ does not suffer from this so can be more easily estimated based on blinded data.

The use of an expectation-maximization algorithm for blinded sample size estimation is controversial and its utility will often be context dependent. Friede and Schmidli (2010) adopts a different approach by pooling data across the treatment arms and estimating an overall event rate which is a function of the control event rate and the treatment effect expressible as a relative rate. If one specifies the treatment effect as the clinically important effect one wishes to detect, then one can easily solve for the control event rate. A quasi-likelihood approach is used to model and estimate the variance inflation factor when heterogeneity is a concern. This approach has appeal in that the imposition of additional assumptions eliminates the convergence and identifiability issues pointed out in Cook et al. (2009a) and it has been developed and studied further in related papers including Schneider et al. (2013).

7.4.4 Recurrent and terminal events

Recurrent events frequently arise in settings where individuals are at high risk of a terminating event after which no further events are possible. In transplant studies, graft transient rejection episodes are indicative of the recipients immune reaction to the transplanted organ and their occurrence affects individuals' quality of life and incur health care costs. Total graft rejection terminates this recurrent event process (Cole et al., 1994). In studies of individuals with cancer metastatic to bone, the weakened bone puts people at risk of fractures and other skeletal related events which new therapies may aim to prevent. In individual with metastatic cancer mortality rates are appreciable and so both recurrent skeletal complications and death are observed post-randomization (Hortobagyi et al., 1996). The problem is similar to a semi-competing risks problem in that the recurrent events do not preclude the competing event of death but following death no recurrent events can arise.

To consider the statistical issues we consider first a one-sample problem and let $\{N_i(s), 0 < s\}$ and $\{N_i^d(t), 0 < t\}$ denote the counting process for recurrent events and death where T_i^d is the time of death and $N_i^d(t) = I(T_i^d \le t)$. A partially conditional rate function can be defined as $d\mu_i^\dagger(u) = E\{dN_i(u)|N_i^d(u^-) = 0\}$ which conditions on being alive at time u^- and we let $\mathcal{F}^d(u) = P(T_i^d \ge u)$ be the survivor function for T_i^d. A marginal mean function can then be defined as

$$E\{N_i(t)\} = \int_0^t \mathcal{F}^d(u) \, d\mu_i^\dagger(u) , \qquad (7.30)$$

which gives the marginal expected number of recurrent events over $(0, t]$, while accounting for the possible realization of the terminating event.

In a sample of size m, if C_i is a right censoring time and $Y_i(t) = I(t \le C_i)$, $i = 1, \ldots, m$, as before, we introduce $Y_i^d(t) = I(t \le T_i^d)$ which indicates whether individual i is alive at t and let $\bar{Y}_i(t) = Y_i(t)Y_i^d(t)$. The quantities $d\bar{N}_i(t) = \bar{Y}_i(t)dN_i(t)$ and $d\bar{N}_i^d(t) = \bar{Y}_i(t)dN_i^d(t)$ are observable and the history of the observable process is $\bar{\mathcal{H}}_i(t) = \{(\bar{N}_i(u), \bar{N}_i^d(u), Y_i(u)), 0 < u < t\}$. An estimate of (7.30) is obtained by computing $d\widehat{\mu}_i^\dagger(t) = \sum_{i=1}^m \bar{Y}_i(t)dN_i(t) / \sum_{i=1}^m \bar{Y}_i(t)$, and if $\widehat{\mathcal{F}}^d(t)$ is the Kaplan-Meier estimate for the survival distribution then $\widehat{\mu}_i(t) = \int_0^t \widehat{\mathcal{F}}^d(u) \, d\widehat{\mu}_i^\dagger(u)$. Examination of this quantity alone is insufficient to gain insight into the nature of the treatment effect since the survival-adjusted marginal mean can be affected by a treatment effect on the conditional rate function or an effect on the survival distribution. Therefore, as is the case in settings with semi-competing risks it is best to examine these plots in concert with estimates of the survival distributions in treatment groups.

Large sample properties of this estimator proposed by Cook and Lawless (1997) were developed by Ghosh and Lin (2000), who also considered two-sample tests of hypotheses incorporating both the recurrent event and survival processes. Methods for regression analyses in this setting were developed by Ghosh and Lin (2003). Much work has been done in the development of fully specified models accommodating a dependence between the recurrent event and death processes through shared or correlated random effects (Liu et al., 2004). But such methods, like many related joint models for longitudinal and survival data, enable one to predict recurrent events following death. An alternative framework can be adopted by casting this problem into a multistate setting where event occurrence leads to a transition between transient states and death results in entry to an absorbing state (Cook et al., 2009b). This has the advantage of enabling one to summarize event histories by assigning utilities to different states and making causal comparisons between mean cumulative utility functions for each treatment group.

7.5   Discussion

When contemplating an approach to analyse recurrent event data a paramount concern should be to ensure that the estimated treatment effect is interpretable, tests are valid, and causal inference can be drawn. In addition in clinical trials we typically aim to make as few assumptions as possible and ensure inferences are robust to model misspecification. Boher and Cook (2006) use large sample theory and empirical studies to critique the commonly used methods and based on the above criteria, rate based analyses are recommended. These can be carried out by fitting a semiparametric Poisson proportional rate function model (Andersen and Gill, 1982; Lawless, 1987) and using a robust variance estimate (Lawless and Nadeau, 1995; Lin et al., 2000) to ensure valid tests of the null hypothesis of no treatment effect, and valid frequency statements if the proportionality assumption is valid. This is a partially specified model, however, analogous to the marginal models fitted based on generalized estimation equations for longitudinal data. As such it cannot be used as a basis for prediction following model fit. The mixed Poisson model does facilitate prediction but requires specification of a distribution for the random effect. Empirical studies suggest that inferences about regression coefficients are relatively insensitive to misspecification of $G(u; \phi)$ and so this is not a major consideration for treatment comparisons but estimation of ϕ and prediction are sensitive to misspecification.

When interest lies in carrying out periodic interim monitoring of data from trials involving recurrent events the overall risk of falsely rejecting the null hypothesis can be controlled by specification of an error spending function (Demets and Lan, 1994) which mimics the properties of the common Pocock (1977) or O'Brien and Fleming (1979) group sequential designs while allowing flexibility in the ultimate timing of analyses. The key requirement in such settings is the derivation of the asymptotic joint distribution of successive point estimates or test statistics. Cook and Lawless (1996) develop such methods and Jiang (1999) consider similar issues in the setting of mixed Poisson processes.

Acknowledgments

This research was supported by a grant from National Natural Science Foundation of China to Yujie Zhong (NSFC-11901376), sponsored by Shanghai Pujiang Program (2019PJC051), a Discovery Grant and Supplement Award from the Natural Science and Engineering Research Council of Canada to Richard J Cook (RGPIN 155849 and RGPIN 04207), and a grant from the Canadian Institutes for Health Research (FRN 13887). Richard J Cook is a Faculty of Mathematics Research Chair, University of Waterloo.

Bibliography

Andersen, P., Borgan, O., Gill, R., and Keiding, N. (1993). *Statistical Models Based on Counting Processes*. Springer Verlag, New York.

Andersen, P. and Gill, R. (1982). Cox's regression model for counting processes: a large sample study. *The Annals of Statistics*, 10(4):1100–1120.

Bernardo, M. and Harrington, D. (2001). Sample size calculations for the two-sample problem using the multiplicative intensity model. *Statistics in Medicine*, 20(4):557–579.

Boher, J. and Cook, R. (2006). Implications of model misspecification in robust tests for recurrent events. *Lifetime Data Analysis*, 12(1):69–95.

Cole, E., Cattran, D., Farewell, V., Aprile, M., Bear, R., Pei, Y., Fenton, S., Tober, J., and Cardella, C. (1994). A comparison of rabbit antithymocyte serum and OKT3 as prophylaxis against renal allograft rejection. *Transplantation*, 57(1):60–67.

Cook, R. (1995). The design and analysis of randomized trials with recurrent events. *Statistics in Medicine*, 14(19):2081–2098.

Cook, R., Bergeron, P.-J., Boher, J.-M., and Liu, Y. (2009a). Two-stage design of clinical trials involving recurrent events. *Statistics in Medicine*, 28(21):2617–2638.

Cook, R. and Lawless, J. (1996). Interim monitoring of longitudinal comparative studies with recurrent event responses. *Biometrics*, 52(4):1311–1323.

Cook, R. and Lawless, J. (1997). Marginal analysis of recurrent events and a terminating event. *Statistics in Medicine*, 16(8):911–924.

Cook, R. and Lawless, J. (2007). *The Statistical Analysis of Recurrent Events*. Springer, New York.

Cook, R., Lawless, J., Lakhal-Chaieb, L., and Lee, K.-A. (2009b). Robust estimation of mean functions and treatment effects for recurrent events under event-dependent censoring and termination: application to skeletal complications in cancer metastatic to bone. *Journal of the American Statistical Association*, 104(485):60–75.

Cook, R., Lawless, J., and Nadeau, C. (1996). Robust tests for treatment comparisons based on recurrent event responses. *Biometrics*, 52(2):557–571.

Demets, D. and Lan, K. (1994). Interim analysis: the alpha spending function approach. *Statistics in Medicine*, 13(13-14):1341–1352.

Dempster, A., Laird, N., and Rubin, D. (1977). Maximum likelihood from incomplete data via the em algorithm. *Journal of the Royal Statistical Society. Series B (methodological)*, 39(1):1–38.

Friede, T. and Schmidli, H. (2010). Blinded sample size reestimation with count data: methods and applications in multiple sclerosis. *Statistics in Medicine*, 29(10):1145–1156.

Fuchs, H., Borowitz, D., Christiansen, D., Morris, E., Nash, M., Ramsey, B., Rosenstein, B., Smith, A., and Wohl, M. (1994). Effect of aerosolized recombinant human DNase on exacerbations of respiratory symptoms and on pulmonary function in patients with cystic fibrosis. The Pulmozyme Study Group. *New England Journal of Medicine*, 331:637–642.

Ghosh, D. and Lin, D. (2000). Nonparametric analysis of recurrent events and death. *Biometrics*, 56(2):554–562.

Ghosh, D. and Lin, D. (2003). Semiparametric analysis of recurrent events data in the presence of dependent censoring. *Biometrics*, 59(4):877–885.

Grossman, R., Mukherjee, J., Vaughan, D., Cook, R., LaForge, J., Lampron, N., and Eastwood, C. (1998). A 1-year community-based health economic study of ciprofloxacin vs usual antibiotic treatment in acute exacerbations of chronic bronchitis: the canadian ciprofloxacin health economic study group. *CHEST Journal*, 113(1):131–141.

Heddle, N., Cook, R., Webert, K., Sigouin, C., and Rebulla, P. (2003). Methodologic issues in the use of bleeding as an outcome in transfusion medicine studies. *Transfusion*, 43(6):742–752.

Hortobagyi, G., Theriault, R., Porter, L., Blayney, D., Lipton, A., Sinoff, C., Wheeler, H., Simeone, J., Seaman, J., Knight, R., Heffernan, M., Reitsma, D., Kennedy, I., Allan, S., and Mellars, K. f. t. P. . A. B. C. S. G. (1996). Efficacy of pamidronate in reducing skeletal complications in patients with breast cancer and lytic bone metastases. *New England Journal of Medicine*, 335(24):1785–1792.

Jiang, W. (1999). Group sequential procedures for repeated events data with frailty. *Journal of Biopharmaceutical Statistics*, 9(3):379–399.

Kalbfleisch, J. and Prentice, R. (2002). *The Statistical Analysis of Failure Time Data*. John Wiley and Sons, New York, 2nd Edition.

Lawless, J. (1987). Negative binomial and mixed poisson regression. *Canadian Journal of Statistics*, 15(3):209–225.

Lawless, J. and Nadeau, C. (1995). Some simple robust methods for the analysis of recurrent events. *Technometrics*, 37(2):158–168.

Lin, D., Sun, W., and Ying, Z. (1999). Nonparametric estimation of the gap time distribution for serial events with censored data. *Biometrika*, 86(1):59–70.

Lin, D., Wei, L., Yang, I., and Ying, Z. (2000). Semiparametric regression for the mean and rate functions of recurrent events. *Journal of the Royal Statistical Society. Series B, Statistical Methodology*, 62(4):711–730.

Liu, L., Wolfe, R., and Huang, X. (2004). Shared frailty models for recurrent events and a terminal event. *Biometrics*, 60(3):747–756.

O'Brien, P. and Fleming, T. (1979). A multiple testing procedure for clinical trials. *Biometrics*, 35(3):549–556.

Pledger, G., Sackellares, J., Treiman, D., Pellock, J., Wright, F., Mikati, M., Sahlroot, J., Tsay, J., Drake, M., Olson, L., Handforth, C., Garnett, W., Schachter, S., Kupferberg, H., Ashworth, M., McCormick, C., Leiderman, D., Kapetanovic, I., Driscoll, S., O'Hara, K., Torchin, C., Gentile, J., Kay, A., and Cereghino, J. (1994). Flunarizine for treatment of partial seizures. *Neurology*, 44(10):1830–1836.

Pocock, S. (1977). Group sequential methods in the design and analysis of clinical trials. *Biometrika*, 64(2):191–199.

R Core Team (2014). *R: A Language and Environment for Statistical Computing*. R Foundation for Statistical Computing, Vienna, Austria.

Schneider, S., Schmidli, H., and Friede, T. (2013). Blinded sample size re-estimation for recurrent event data with time trends. *Statistics in Medicine*, 32:5448–5457.

Therneau, T. and Grambsch, P. (2000). *Modeling Survival Data: Extending The Cox Model*. Springer Science & Business Media, New York.

Venables, W. and Ripley, B. (2002). *Modern Applied Statistics with S*. Springer Science + Business Media, LLC, New York, Fourth Edition.

Verona, E., Petrov, D., Cserhati, E., Hofman, J., Geppe, N., Medley, H., and Hughes, S. (2003). Fluticasone propionate in asthma: A long term dose comparison study. *Archives of Disease in Childhood*, 88(6):503–509.

Wei, L., Lin, D., and Weissfeld, L. (1989). Regression analysis of multivariate incomplete failure time data by modeling marginal distributions. *Journal of the American Statistical Association*, 84(408):1065–1073.

Wittes, J. and Brittain, E. (1990). The role of internal pilot studies in increasing the efficiency of clinical trials. *Statistics in Medicine*, 9(1–2):65–72.

Zhong, Y. and Cook, R. J. (2019). The effect of omitted covariates in marginal and partially conditional recurrent event analyses. *Lifetime Data Analysis*, 25(2):280–300.

Part III

Design of Randomized Controlled Trials

8

Cross-Over Designs

Stephen Senn

CONTENTS

8.1 Introduction

Cross-over designs, or perhaps more commonly cross-over *trials*, are designs in which patients are allocated to sequences of treatments with the purpose of comparing different elements of these sequences(1). In other words, it is not sufficient that patients be allocated to sequences; this allocation has to be for the purpose of comparing the treatments that make up the sequences rather than (say) evaluating comparatively the combined total effect of all the elements in a sequence with that of another.

The fact that each patient can act as his or her own control means that the influence of an important component of variation, namely the difference between patients, can be eliminated. This can have important consequences as regards reducing the variability of treatment effect estimates and hence also in reducing the number of patients that need to be studied. This can make them an attractive option on occasion.

However, although cross-over trials are common in drug development, they are rarely used in phase III. They are often employed for dose-finding, especially in chronic diseases and so commonly encountered in phase II. They may also be used for the purpose of studying pharmacokinetics in phase I. A special type of cross-over trial that is very common is one in which formulations of a treatment are compared as regards the concentration of the drug in the blood, the objective being to show that the two formulations are equivalent. Such studies are called *bioequivalence* studies and they are commonly carried out by generic drug manufacturers seeking a license once the patent of the innovator product is expired. A successful bioequivalence study will obviate the need for a full development of the generic product and lead to considerable savings.

The nature and purpose of cross-over trials is perhaps best understood by giving some examples. However, before doing so, it is worth noting that a minority term for cross-over trials is *changeover* trials and that also where one has sets of so-called *n-of-1 trials*, these taken as a whole have the structure of a cross-over trial. This point will be discussed in due course.

One minor point regarding presentation of sequences is worth making. In the medical statistics literature it is usual to write sequences with periods going across the page. Successive rows will then represent further sequences. In the optimal design literature, which has been much concerned with cross-over trials, it is usual to write periods going down the page and sequences across. Thus in the medical literature treatment A followed by treatment B will be written AB, whereas in the design literature it will be written:

$$A$$
$$B$$

Here we shall follow the former convention but instead of always writing sequences below each other will sometimes use the/sign to indicate the start of a new sequence.

8.2 Some Examples

We now consider some example of differing complexity that illustrates some features of cross-over trials. In so doing we shall raise without discussion in depth some points, in particular to do with analysis, that will be dealt with more completely later.

TABLE 8.1
Sequences for an 'AB/BA' cross-over trial in asthma.

Sequence	Period 1	Wash-out	Period 2
SF	salbutamol		formoterol
FS	formoterol		salbutamol

8.2.1   Example 1 : An AB/BA design

In a double blind cross-over study in paediatric asthma, 13 children (mean age about 11 years) were allocated to receive two sequences of two treatments: a single dose of 200 μg salbutamol (S) or a single dose of 12 μg formoterol (F). Thus, each child was randomised to receive either S followed by F or F followed by S. The first treatment (S or F depending on allocated sequence) was delivered by physicians in the clinic in the morning and the children had their peak expiratory flow (PEF) measured over regular intervals for 8 hours. They then travelled home with their parents and had their PEF measured again at home 10, 11 and 12 hours after treatment. After a suitable wash-out period of at least one day, the children returned and were administered their second treatment (F or S depending on allocated sequence) and the measurement procedure was repeated. The trial is described in detail in(2) and the data are reported in(3) and also[1].

The design of the trial is represented in Table 8.1. We shall refer to designs like this one as an AB/BA cross-over where in general terms, we suppose that a treatment A is compared to another treatment B. For example, A might be salbutamol and B might be formoterol, although we have used the more specific labels S and F here. Such a design is sometimes referred to as a two-period design but such terminology is best avoided since there are also two period designs in which all four sequences are used, that is to say not only AB and BA but also AA and BB. Such a design is referred to as Balaam's design(4) but we shall not consider it further here.

Remark

Note that the patients were treated in two *periods*, which we can refer to as period 1 and period 2. However, this labelling does not imply that all the patients recruited were treated simultaneously in period 1 and then again simultaneously in period 2. This point is worth labouring since there has been some confusion in the methodological literature regarding this(5; 6). As with any parallel group trial the patients were recruited sequentially. Some of them will have completed period 2 before others started period 1. For this reason, when considering possible differences there might be in measurement conditions from period to period one sometimes speaks of *order effects*. This terminology will not be used here but the reader should be warned that periods are not calendar periods and as regards date, a given period may and often will differ from patient to patient.

8.2.2   Example 2: A design in three treatments, three periods, and six sequences

In a double-blind cross-over trial run in Sweden and Finland(1; 7), two doses (50 mg and 100 mg) of the potassium salt of diclofenac were compared to placebo as regards efficacy in dealing with migraine. The main outcome variable was pain 2 hours after treatment measured using a so-called *visual analogue scale* (VAS). Here the patient is asked to mark the degree of pain on a line of length 100 mm, with 0 mm representing no pain and 100 mm representing unbearable pain. Blinding was achieved by giving patients either two 50mg diclofenac tablets two matching placebo tablets or one of each.

If we label the treatments as P (placebo), D_{50} (diclofenac 50 mg), and D_{100} (diclofenac 100 mg), then patients were randomised to one of the six possible sequences for which each patient receives each treatment once, namely:

$$D_{50}D_{100}P/P\, D_{100}D_{50}/D_{100}D_{50}P/D_{100}P\, D_{50}/D_{50}PD_{100}/P\, D_{50}D_{100}.$$

Remark

Note that over the set of six sequences each treatment appears twice in every period. This has advantages in that should there be an effect of period, and should the patients have been allocated in equal number to the sequences, then this will be orthogonal to the effect of treatment should one wish to fit both in a model. This orthogonality means that by putting period effects in the model they can be efficiently eliminated, not only from the estimate of the treatment effect but also from the estimate of its variance. However, to achieve this it would not have been necessary to use six sequences. Three would have sufficed with the three forming a so-called Latin square as in Table 8.2.

TABLE 8.2
The six sequences used in Example 1 represented as two Latin squares.

	Latin Squares				
I			**II**		
Period			**Period**		
1	**2**	**3**	**1**	**2**	**3**
P	D_{50}	D_{100}	P	D_{100}	D_{50}
D_{100}	P	D_{50}	D_{100}	D_{50}	P
D_{50}	D_{100}	P	D_{50}	P	D_{100}

A design using either of the two Latin squares shown would be adequate for the purpose of eliminating the period effect.

Remark

An interesting feature of this design is that the period between treatments is necessarily variable. In Example 1, it could vary by choice since a minimum wash-out period of one day was set. In Example 2, however, it is the unpredictable onset of migraine that triggers the resort to treatment and this will lead to a variable interval.

8.2.3 Example 3: An incomplete blocks design with fewer periods than treatments

This example, a trial known as MTA02, is included to illustrate the complexity that cross-over trials can sometimes have. The trial concerned was a parallel assay in asthma comparing three-doses (6 μg, 12 μg and 24 μg) of a new formulation (MTA) to the same doses of an existing formulation (ISF) of formoterol. Also included was a placebo. There were thus seven treatments in all. However, practical considerations dictated that patients could be treated on five occasions only. The design chosen was a so-called *balanced incomplete block* design. *Balanced* in this context means that every possible pair of treatments, of which there were $(7 \times 6)/2 = 21$, was intended to be equally frequently studied within patients. Since each patient afforded the possibility of making $(5 \times 4)/2 = 10$ pairwise comparisons, this

TABLE 8.3
Sequences of treatments for patients used in *Example 3* MTA02.

Sequence	Period 1	Period 2	Period 3	Period 4	Period 5
A	MTA 6μg	Placebo	ISF 24μg	MTA 24μg	ISF 6μg
B	MTA 6μg	ISF 24μg	ISF 6μg	ISF 12μg	Placebo
C	MTA 6μg	MTA 24μg	ISF 12μg	ISF 24μg	MTA 12μg
D	MTA 12μg	ISF 12μg	MTA 6μg	Placebo	ISF 24μg
E	MTA 12μg	MTA 6μg	ISF 24μg	ISF 6μg	ISF 12μg
F	MTA 12μg	Placebo	ISF 6μg	MTA 6μg	MTA 24μg
G	MTA 24μg	ISF 6μg	MTA 12μg	ISF 12μg	MTA 6μg
H	MTA 24μg	MTA 12μg	MTA 6μg	ISF 24μg	ISF 6μg
I	MTA 24μg	ISF 12μg	ISF 24μg	MTA 12μg	Placebo
J	ISF 6μg	ISF 12μg	MTA 12μg	MTA 6μg	Placebo
K	ISF 6μg	ISF 12μg	Placebo	MTA 24μg	MTA 12μg
L	ISF 6μg	MTA 6μg	MTA 24μg	ISF 12μg	ISF 24μg
M	ISF 12μg	MTA 6μg	Placebo	ISF 24μg	MTA 24μg
N	ISF 12μg	Placebo	MTA 24μg	MTA 12μg	MTA 6μg
O	ISF 12μg	ISF 24μg	MTA 12μg	Placebo	ISF 6μg
P	ISF 24μg	MTA 24μg	MTA 12μg	ISF 6μg	ISF 12μg
Q	ISF 24μg	ISF 6μg	ISF 12μg	Placebo	MTA 24μg
R	ISF 24μg	MTA 12μg	Placebo	ISF 6μg	MTA 6μg
S	Placebo	ISF 24μg	MTA 24μg	ISF 6μg	MTA 12μg
T	Placebo	MTA 24μg	MTA 12μg	MTA 6μg	ISF 24μg
U	Placebo	ISF 6μg	MTA 6μg	MTA 24μg	ISF 12μg

was achieved by choosing a design with 21 sequences. This meant that for every replication of this basic design each pair of treatments appeared 10 times within patients. The design chosen is given in Table 8.3.

The design was also chosen to be *uniform on the periods*, i.e. each treatment appeared equally often in each period. Since there were 5 periods and 21 sequences for a single replicate this meant that each treatment was used in $(5 \times 21)/7 = 15$ patients and appeared 3 times in every period. (See, for example, period 1 of Table 8.3) The design was planned to be replicated 6 times (126 patients in total) but in the end 161 patients were recruited although not all of them completed the sequence of 5 treatments and only 148 provided information on each of five treatments.

The treatments were delivered in the clinic as a single dose only and frequent measurements over 12 hours of FEV_1 were taken. The purpose of the trial was to show that the two formulations of formoterol were equipotent. Unfortunately the trial failed in this respect and the new formulation was shown to have at most $1/4$ the potency of the existing one, even though all doses of both formulations gave effective bronchodilation compared to placebo at 12 h.

The trial is described in great detail in(8).

Remark

Great efforts were made for this trial to try and balance as much as possible. The design was balanced in the incomplete blocks sense in that given pairs of treatments were intended to be studied equally often. It was also uniform on the periods. However, it should be appreciated that although, other things being equal, balance is generally desirable, it is not the be all and end all of design. For example, suppose that a 4 period design had been used. Each patient would have provided the means of comparing $(4 \times 3)/2 = 6$ pairs of treatments. A design with 7 sequences, much simpler than that employed, would have been balanced in the sense that each of the 21 possible treatment pairs would have appeared in two sequences (since $6 \times 7 = 42$). Suppose that it had been decided that 7 sequences would be used, more being too complicated to arrange, but that 5 periods could be employed. Then since the division of $7 \times [(5 \times 4)/2] = 70$ by 21 is not an integer it would be impossible to have a design in which each pair of treatments was compared equally frequently within patients. However, $(7 \times 4) + (14 \times 3) = 70$ and a design can be found in which 7 pairs are represented 4 times and the other 14 are represented 3 times and thus more information could be obtained for each and every pairwise comparison. This shows that balance is not the be all and end all of design: it is desirable other things being equal.

In fact the number of patients eventually recruited was greater than the anticipated five-fold replicate of the basic design of 21 sequences and the numbers of pairs were as given in Table 8.4.

TABLE 8.4

Frequency with which pairs of treatment were given to patients in the MTA02 trial. The diagonal gives the frequency with which treatments were given and the off-diagonals the frequency of the pairs.

	MTA 6 μg	MTA 12 μg	MTA 24 μg	ISF 6 μg	ISF 12 μg	ISF 24 μg	Placebo
MTA 6 μg	115						
MTA 12 μg	76	114					
MTA 24 μg	77	77	116				
ISF 6 μg	77	78	77	116			
ISF 12 μg	78	75	80	78	116		
ISF 24 μg	75	74	75	77	76	113	
Placebo	77	76	78	77	77	75	115

8.2.4　Example 4: A replicate cross-over design with more periods than treatments

Senn, Rolfe, and Julious(9) describe a cross-over trial in Parkinsonism (EASE-PD) reported by Stocchi et al.,(10) comparing two formulations of ropinirole, an immediate release form (I) and a prolonged release form (P). After an initial titration period of 12 weeks, patients then entered three 8 week periods. At the end of the first 8 week period they were either maintained on the current formulation or switched. At the end of the second 8 week period those who had been maintained at the end of the first week period were switched and those who been switched were maintained. The net result was that the sequences used were I I P/ P P I/ I P P/ P I I.

Remark

This design was chosen so as to avoid having to switch the treatment the patients were given more than once during the trial. Note also that this trial is different from the examples discussed so far in that patients were given regular treatment during several weeks whereas in the previous three examples each treatment was given on one occasion only. Unlike the previous three examples there was no wash-out between periods. This point will be picked up when discussing the phenomenon of carry-over subsequently.

8.2.5 Example 5: A replicate bioequivalence study comparing two formulations in four periods

Shumaker and Metzler(11) report a four-period bioequivalence study in which 26 healthy volunteers were given two formulations of phenytoin, each being administered as a single dose on two occasions. As is common in such studies there was a test (T) and reference (R) formulation, the purpose being to show that the test formulation had the same pharmaco-dynamic characteristics as the reference and hence could be considered safe and efficacious. The subjects were allocated at random to one of two sequences TRRT/RTTR. Regular blood samples were taken and the concentration in the blood measured over a period of 100 hours.

The concentration measures were, as is usual, used to calculate area under the concentration time curve (AUC) and the maximum measured concentration (C_{max}). The summary measures are provided by the authors in their article as well as SAS® code so that this is a good example for those wishing to try their hand at analysis.

Remark

Each pair of periods 1 and 2 or 3 and 4 produces either the subsequence RT or TR. Thus the design can be regarded as a replicate standard RT/TR cross-over with the subject being switched to the alternative sequence for the second replicate. The authors used the fact that additional periods were available to see if there was any evidence of a subject by formulation interaction (there was not).

It is hoped that these five examples have given some flavour of the variety of design that is possible for cross-over trials. We have already briefly covered some examples of uses of cross-over trials in the introduction. We now look at some of the circumstances under which they are used in more detail.

8.3 General Considerations

8.3.1 Phase of drug development

As already indicated cross-over trials are rarely used in Phase III. Amongst the many reasons is the important one that precisely because they are efficient, fewer patients will be recruited. However, for many indications the sample size required to provide enough information to support judgements of safety exceeds that needed for efficacy, hence the savings in number that cross-over trial would offer are not required.

Furthermore, phase III trials are far more likely to use so-called hard outcomes, of which mortality would be an extreme example, and these are rarely suited for cross-over trials. However, by the same token, it is often extremely difficult to do dose-finding using hard endpoints. Here it is quite usual to use pharmacodynamic alternatives. For example, a phase III programme in hypertension-lowering might target cardiovascular outcomes or strokes. However, the phase II dose-finding programme might use blood pressure itself as the outcome and for such a purpose a cross-over trial might be envisaged.

8.3.2 Suitable indications

Cross-over trials are easiest to interpret if the effect of treatment is reversible (at least roughly). This means that they are far more suited to chronic diseases where the condition

TABLE 8.5
Possible fixed effect parameters for Example 1.

Sequence	Period 1	Period 2
SF	μ	$\mu + \pi + \tau$
FS	$\mu + \tau$	$\mu + \pi$

of the patient is relatively stable and the purpose of treatment is to deal with symptoms rather than effecting a cure. Pain control(12), as for example in rheumatism or migraine (as in Example 3) are thus suitable indications. Asthma is another (see Example 1 and Example 3) where cross-over trials are often used. They are sometimes also used in epilepsy or in Parkinsonism(10).

Cross-over trials are rarely used where 'cure' is an outcome. One exception is trials in infertility. This may at first sight seem to be an unsuitable indication since if conception is achieved with the first treatment (for example) the couple seeking treatment will not need to be given the second. However, the reason that the second data are missing is then known so that this is an example of a missing at random problem(13) and it turns out that using a random effects model or a suitable stratification and the Mantel-Haenszel(14) approach as an analysis is possible(15). Similar situations can occur in other indications(16). It is arguable, however, as to whether such designs are really cross-over designs. They are perhaps best regarded as augmented parallel group trials. They will not be covered further here.

8.4 Issues in Analysis

8.4.1 Models for cross-over trials

A common approach to analysing cross-over trials involves fitting patient and period effects in addition to treatment effects as predictors of the outcome variable of interest. Period effects and treatment effects will very naturally appear as fixed in such models but patient effects are sometimes fitted as fixed and sometimes as random.

We may model the response Y_{ij} for the period j for patient i of a cross-over trial like this

$$Y_{ij} = \nu_{ij} + \xi_i + \epsilon_{ij} \dots I.$$

Here ν_{ij} is the expected response given the treatment and the period concerned, ξ_i represents the patient effect, and ϵ_{ij} is a so-called within-patient error and can be regarded as the random difference from the true value we would see from occasion to occasion given that all other effects (patient, period, and treatment) were fixed. Typically, or at least most simply, we assume all such terms to be independent and with the same variance, say σ^2. Note that it is awkward to replace the term ν_{ij} in $Y_{ij} = \nu_{ij} + \xi_i + \epsilon_{ij} \dots 1$ by specific terms involving the treatment and period effects because a given patient is (of necessity) only treated with a given treatment in a given period. This means that once the patient and the period are indexed the treatment is known. It is often easier to represent the fixed effects in a table. Table 8.5 presents a possible parameterization for Example 1.

Here μ is the expected response under salbutamol in period 1, π is the expected difference period 2 – period 1, other things being equal, and τ is the expected difference formoterol – salbutamol, other things being equal.

8.4.2 Patient effects and variance structures

We now pick up the issue of modelling the patient effects. In favour of fitting the patient effect as random is the aesthetic point that for parallel group trials patient effects are implicitly treated as being random since if they were declared to be fixed they would be confounded with treatment and no estimate of the treatment effect would be possible. Patient effects have to be explicitly handled in a cross-over trial and so a natural approach is to make the implicit parallel group model the starting point.

However, it is also the case that for many cross-over trials it makes little difference to any inferences made (and sometimes no difference at all) whether the patient effect is fitted as fixed or random. Such is, for example, the situation for the common AB/BA design provided there are no missing data. The reason is that the appropriate analysis involves within-patient differences and such differences eliminate the patient effect anyway, so that it is moot as to whether it needs to be regarded as fixed or random. An advantage of treating them as fixed is that it avoids having to use software for mixed models and simpler algorithms based on ordinary least squares can be employed.

However, for a design such as that given by Example 3, it is the case that not only the differences within patients carry information about treatments but also the totals (or equivalently averages) for the patients. This is because different patients receive different treatments. Thus there is information available not only within but between patients. However, the latter requires that we should model the patient effects as random and this is typically done by treating the terms as random realizations from a Normal distribution so that we have $\xi_i \sim N\left(0, \gamma^2\right)$ where γ^2 is the between-patient variance.

However, even for Example 3, the gain in efficiency by treating the patient effect as random is small. What the gain will be depends on the ratio γ^2/σ^2. A ratio of 0 leads to the maximal gain and it can be shown in this case that the fixed effects model has an efficiency of 93% compared to the random effects model. As the ratio approaches infinity, the two solutions converge and hence the fixed effects model approaches 100% efficiency. In this specific example, for the outcome variable of mean FEV_1 over 12 hours, the ratio γ^2/σ^2 is more than 18 and the efficiency of the fixed effects approach, compared to the random effects approach, is nearly 99% and, indeed, the two solutions were virtually indistinguishable.

Note that the combination of the between patient and within patient models implies the following covariance for two observations Y_{ij}, $Y_{i'j'}$:

$$cov\left(Y_{ij},\ Y_{i'j'}\right) = \begin{array}{l} 0\ if\ i' \neq i \\ \gamma^2\ if\ i' = i\ \ j' \neq j \\ \gamma^2 + \sigma^2\ if\ i' = i\ \ j' = j \end{array} \tag{8.1}$$

which is to say, the covariance between observations on different patients is 0, that between different observations on the same patient is γ^2, whereas the variance of individual observations is $\gamma^2 + \sigma^2$.

An equivalent form of the covariances is to use the intra-class correlation coefficient $\rho = \gamma^2/\left(\gamma^2 + \sigma^2\right)$ and write $\phi^2 = \gamma^2 + \sigma^2$ we can thus re-write 2 as

$$cov\left(Y_{ij},\ Y_{i'j'}\right) = \begin{array}{l} 0\ if\ i' \neq i \\ \rho\phi^2\ if\ i' = i\ \ \ j' \neq j \\ \phi^2\ if\ i' = i\ \ \ j' = j \end{array} \tag{8.2}$$

with the proviso that $0 \leq \rho \leq 1$. However 3 also suggests a slightly more general form in which the condition that $\rho \geq 0$ is dropped. Such a model cannot arise as an additive error model but is theoretically possible. For a general discussion of the issues in the context of the linear mixed models (but not specifically cross-over trials) the reader should consult Verbeke and Molenberghs(17).

8.4.3　Carry-over effects

Carry-over has been defined as, 'the persistence (whether physically or in terms of effect) of a treatment applied in one period in a subsequent period of treatment'(1) (p. 8). The problem that the phenomenon of carry-over poses is that if it occurs we may fail to estimate the difference between treatments correctly, mistaking what is in fact a difference of complex mixtures of effects for the sort of simple difference we would see in a parallel group trial. The potential presence of carry-over is seen both as being the biggest difficulty in running and analysing cross-over trials and also as the biggest technical challenge in producing more complicated designs and appropriate analyses to deal with it. In fact there is a huge technical literature on the subject.

A simple precaution in design is to institute a wash-out period. This can take two forms: it can be either passive or active. In the former case, no treatment is given during the wash-out period. In the latter case, a treatment is given but the measurement of the effect of that treatment is limited to a later part of the treatment period by which time the effect of the previous treatment is assumed to have disappeared, i.e., washed out. Examples 1–3 all used passive wash-out periods. However, Example 4 used an active wash-out period thus avoiding having to leave the patients without treatment.

Modelling of carry-over by statisticians has tended to be formal, unimaginative and unrealistic(1; 6; 18). Typically it is assumed to last for one period only if present and to reflect only the engendering and not the perturbed treatment. This will be referred to as the *simple carry-over model*. For example, if applied to Example 3 MTA02, it would be assumed that the carry-over in period 5 of sequence B, in which placebo was given, would be the same as the carry-over in period 4 of sequence C, in which the 24 μg dose of ISF was given, simply because the preceding treatment in both cases was the 12 μg dose of ISF.

For the simple AB/BA design however, there is only one period, carry-over can only appear in period 2 and since as soon as one knows the engendering treatment one knows the perturbed treatment, more complicated models are unnecessary. A simple carry-over model suffices. The problem is that it can be shown that no efficient unbiased estimate of the treatment effect is possible in the presence of carry-over. This has led some to propose more complicated designs involving further period or other sequences. As soon as this happens, however, assumptions become critical.

Although extremely popular in the theoretical literature, this model has been much criticised by statisticians working on clinical trials as being unrealistic and overlooking, amongst other things, that if a drug has reached steady state there can be no carry-over into itself(19; 20). It can also be shown that adjusting for carry-over for such a model can make matters worse where carry-over applies(21). A rather different tradition has been followed by pharmacometricians, who have used the same pharmacokinetic and pharmacodynamic models for carry-over that they have employed for the main effect of treatment(22).

The issue of carry-over will be taken up again when discussing design in the following sections.

8.4.4　Residual degrees of freedom and error estimation

For the AB/BA cross-over design, the residual degrees of freedom for error estimation do not exceed the number of patients. The analysis of variance decomposition is as in Table 8.6 below.

Thus there are n patients and $n-2$ degrees of freedom for error. However, if we look at the design of Example 4, we find that the corresponding table would be as in Table 8.7 below.

TABLE 8.6
Degrees of freedom for an AB/BA cross-over design.

Source	Degrees of freedom
Patients	$n-1$
Periods	1
Treatments	1
Residual	$n-2$
Total	$2n-1$

TABLE 8.7
Degrees of freedom for a trial of the type illustrated by *Example 4*.

Source	Degrees of freedom
Patients	$n-1$
Periods	2
Treatments	1
Residual	$2n-3$
Total	$3n-1$

Now we see that degrees of freedom for error exceed the number of patients. Where this happens it is a both an opportunity and a warning[23]. The opportunity is to study treatment-by-patient interaction. In fact we could extend the analysis of variance represented in TABLE 8.7 by including a term for treatment-by-patient interaction. Such a term would have $n-1$ degrees of freedom, leaving $n-2$ for error as before. Thus replicate cross-overs provide an excellent design for the purpose of establishing individual response to treatment[24].

The danger is that although one could argue that a model that ignores interaction could be valid for this simple purpose of testing a strict null hypothesis that treatments are equal (since if equal there can be no interaction), for other inferential purposes, naively pooling within-patient and patient-by-interaction degrees of freedom will mislead. It may be appropriate to set up a random effects model with a random term for treatment-by-patient interaction. It is surprising how much discussion there has been of the analogous issue in meta-analysis and how little for cross-over trials.

For cross-over trials with more than two periods, even if no treatments are repeated, a similar issue with degrees of freedom will arise although identification of interaction terms in the same simple way is not possible. Here the issue is rather that heteroscedasticity may apply. This issue is not unique to cross-over trials since in three armed parallel group trials, for example, one may worry about pooling degrees of freedom from all three arms when comparing only two. Note that in a sense such heteroscedasticity must also reflect treatment-by-patient interaction since given randomization and under a strict null that all three treatments are identical the variances must be identical. However, the point is that this is for the global null hypothesis that all treatments are identical and when comparing A to B, say, the equality of C to either does not form part of the hypothesis.

A way of directly constructing variance estimates that avoids these difficulties will be discussed under analysis below.

8.5 Examples of Analysis

The data for the PEF readings in L/min 8 hours after treatment for Example 1 are given in Table 8.8.

TABLE 8.8
Readings of PEF(L/min) 8 hours after treatment for Example 1.

Sequence	Patient	Salbutamol	Formoterol
SF	1	270	310
	4	260	310
	6	300	370
	7	390	410
	10	210	250
	11	350	380
	14	365	330
FS	2	370	385
	3	310	400
	5	380	410
	9	290	320
	12	260	340
	13	90	220

Before beginning any analysis it is always wise to inspect the data. Here two features are noticeable. The first is that the precision appears to be to the nearest 5 L/min but if that is so there is an excess of values ending in 0. In fact out of 26 values only 2 end in 5. The second point is that the patient numbers run from 1 to 14 but there are only 13 patients, as patient 8 is missing. We shall assume that neither of these matters causes a problem for analysis. In fact patient 8 dropped out after the first period and we shall make the assumption that this is for reasons unconnected to treatment.

8.5.1 Basic estimator approach

A possible analysis is to reduce the data to a within-patient contrast measuring the effect of interest. This contrast has been referred to as a 'basic estimator' and was used as an approach to analysis in a famous paper by Hills and Armitage on the AB/BA design[25]. This contrast, as well as two other possible contrasts of interest are given in Table 8.9.

Here the period semi-difference is the period 2 – period 1 PEF reading divided by two. It is thus half the basic estimator for the SF sequence and minus one half the basic estimator for the FS sequence. The mean is just the sum of the two PEF values for each patient divided by two. Some useful intermediate calculations on these three summary statistics are given in Table 8.10 .

Using the basic estimators, we have two (possibly biased) estimates of the treatment effect from the two sequences 30.71 L/min and 62.50/L/min. If there is a period effect then each of these is confounded with it but this effect will be eliminated on averaging them. Thus an estimate that eliminates not only patient effects but also period effects is given by $(30.71 + 62.50)/2 = 46.6$ L/min. If the variance of a basic estimator is σ^2 then the variance of the mean from the SF sequence is $\sigma^2/7$ and for the FS sequence is $\sigma^2/6$ and the variance

TABLE 8.9
Three statistics calculated for PEF 8 hours after treatment for *Example 1*: the difference formoterol-salbutamol (basic estimator), half the difference period 2 – period 1, the mean of both readings.

Patient	Sequence	Basic estimator	Period semi-difference	Mean
1	SF	40	20.0	290.0
4	SF	50	25.0	285.0
6	SF	70	35.0	335.0
7	SF	20	10.0	400.0
10	SF	40	20.0	230.0
11	SF	30	15.0	365.0
14	SF	−35	−17.5	347.5
2	FS	15	−7.5	377.5
3	FS	90	−45.0	355.0
5	FS	30	−15.0	395.0
9	FS	30	−15.0	305.0
12	FS	80	−40.0	300.0
13	FS	130	−65.0	155.0

of the mean of them both will be

$$\frac{1}{4}\left(\frac{\sigma^2}{7} + \frac{\sigma^2}{6}\right).$$

To estimate σ^2 we use the variation within sequences of the basic estimator since, just as we have eliminated the effect of periods from our estimate of the treatment effect, we need to eliminate it from the estimate of its variance. We thus sum the two corrected sums of squares together and divide by the sum of their degrees of freedom to obtain

$$\hat{\sigma}^2 = \frac{6521.43 + 9987.50}{(7-1) + (6-1)} = \frac{16508.43}{11} = 1500.8\left(\frac{L}{min}\right)^2$$

Substituting this for σ^2 we obtain an estimated variance of the treatment estimate of 116.1

TABLE 8.10
Calculation on the by-patient statistics.

		Sequence					
		SF (n=7)			FS (n=6)		
		Summary Statistic			Summary statistic		
Statistic	Basic estimator	Period semi-difference	Mean	Basic estimator	Period semi-difference	Mean
Mean, L/min	30.71	15.36	321.79	62.50	−31.25	314.58
Corrected sum of squares, L²/min²	6521.43	1630.36	19608.93	9987.50	2496.88	37830.21

$L^2/\min^2$ and taking the square root an estimated standard error of 10.8 L/min. The ratio of the estimate 46.6 L/min to its standard error gives us a t-statistic of 4.3 which is clearly significant. Alternatively we can note that the 97.5% quantile of the t-distribution with 11 degrees of freedom is 2.201 and by multiplying this by the standard error and adding and subtracting it from the estimate we get lower and upper confidence intervals of 23 L/min and 70 L/min, respectively.

In order to discuss the analysis of AB/BA cross-over designs more easily it will be useful to have a simple label for this estimator and, following a previous convention, we shall refer to it as *CROS*.

The general approach of (i) estimating a contrast of interest for each patient, (ii) estimating the variance of such contrasts directly by seeing how they vary within sequences and pooling over sequences, (iii) forming an unweighted mean as the overall estimate, and (iv) then estimating the variance of the mean using the appropriate multipliers (in this case $1/6 + 1/7$ divided by 4) can be applied to any cross-over trial in which (a) every treatment is given to every patient and (b) the treatments are uniform on the periods.

This particular form of analysis has a number of advantages; (a) it is fairly robust, (b) it can be applied to many cross-over designs, and (c) it estimates the variance directly in one stage using the extent to which a direct estimate of the treatment effect calculated for every patient varies.

The disadvantage is that it is not fully efficient for more complex designs where numbers in the sequences are unequal and that it can only be applied with difficulty to incomplete block designs.

8.5.2   Two-sample t-test approach

The same results can be achieved by using the semi-period differences and comparing these using a two-sample *t*-test. Irrespective of the sequence, the period one value is subtracted from the period two values (and then divided by two). Assuming no carry-over, these differences will differ in expectation only as a consequence of the treatments given: in one case the salbutamol value is subtracted from the formoterol value and in the other case vice-versa. Thus, the difference between the two means will estimate the treatment effect. This can be seen here by subtracting −31.25 from 15.36 to give 46.6 L/min as before. If the corrected sums of squares for the semi-period difference are studied it will be seen that they are Œ the corresponding value for the basic estimators. This must be so since in terms of absolute values the semi-period differences are half the basic estimators. This explains why their sum only has to be multiplied by $(1/7+1/6)$ to get the estimate of the treatment variance as in the standard two-sample t test. The confidence intervals and P-value are, of course, the same as before.

8.5.3   Linear and mixed models

Two alternative analyses here give identical results to that above. For these analyses the outcomes are represented as a single vector with 26 values rather than as two columns with 13. Three further columns contain (1) the identity of the patient for whom the values were obtained as a factor with 13 levels (in this example), (2) the period in which the data were recorded as a factor with two levels, and (3) the treatment given prior to the recorded outcome as a factor with two levels. For both approaches treatment and period are included as fixed effects. For both approaches patient is also included in the model but each handles it differently.

For the first method, the linear models approach, the patient factor is treated as fixed. The second is a mixed model approach which treats patient as random and period and treatment as fixed. The inferences are identical to those already found by working with summary statistics.

8.5.4   Testing for carry-over

So far we have not used the means over the two periods, the third of the three statistics calculated for each patient given in Table 8.9. Some statisticians like to use these to test for carry-over. The argument is that since effects of both periods and of both treatments are reflected in the mean they cannot be the cause of any difference, which should therefore either be random or due to the order in which treatments were given. Differential carry-over from one treatment into another could cause means to differ by sequence. For instance, in Example 1, it might be the case that the duration of action of formoterol is longer than salbutamol (such is, indeed, known to be the case) so that where salbutamol was taken after formoterol values might be boosted, whereas formoterol taken after salbutamol would not have values boosted.

In such a case the parameters in Table 8.5 could be modified to look like Table 8.11 below.

TABLE 8.11
Possible parameters for *Example 1* when carry-over applies.

Sequence	Period 1	Period 2
SF	μ	$\mu + \pi + \tau + \lambda_S$
FS	$\mu + \tau$	$\mu + \pi + \lambda_F$

In that case, the standard within-patient estimator would have an expectation equal to $\tau + (\lambda_S - \lambda_F)/2$. We can thus see that the treatment effect will be biased by half the differential carry-over, $(\lambda_S - \lambda_F)$.

However, we can also see in that the expected difference of the total over both periods) for the sequence SF minus that for sequence FS is simply $(\lambda_S - \lambda_F)$ the differential carry-over effect. Thus a suitable statistic for testing carry-over is the observed total over both periods averaged over patients in the first sequence minus the corresponding total difference in the second sequence. We shall refer to this statistic as CARRY.

In fact, since the bias in CROS is CARRY/2 it becomes more useful to work with the semi-carry-over effect and this can be done by working with the means rather than the totals. For this purpose the means given in the last column of Table 8.9 or the summary statistics in Table 8.10 can be used to compare the two sequences. For instance we may estimate CARRY/2 as

$$321.79 - 314.55 = 7.24 \; L/min.$$

The variance of the patient means may be estimated from the corrected sums of squares as

$$\widehat{\sigma}_m^2 = \frac{19608.93 + 37830.21}{(7-1) + (6-1)} = \frac{57439.14}{11} = 5221.74 \left(\frac{L}{min}\right)^2$$

and from this the variance of the estimate of the semi- carry-over difference as

$$\left(\frac{1}{7} + \frac{1}{6}\right) 5221.74 = 1616.25 \left(\frac{L}{min}\right)^2$$

and hence a standard error of $40.2 L/min$ from which, since the quantiles of the t-distribution on 11 degrees of freedom corresponding to 2.5% and 97.5% are -2.201 and 2.201 so that the confidence limits are $7.24 \pm 40.2 \times 2.201 = (-81.28, 95.69)$.

The confidence intervals straddle zero, so conventionally, there is no evidence of carry-over. This is, or at least seems to be reassuring. On the other hand the limits are very wide and the possible bias in the CROS estimate in either direction indicated by the confidence limits in absolute terms is larger (at least twice as much) than the estimate of the treatment effect itself. This points to a problem with the test for carry-over. Since it is based on between-patient information, its power to detect carry-over is low and the associated estimate is very unreliable.

8.5.5   8.5.5 An unbiased estimate of the treatment effect

If the second period values are discarded, then the data that remain have the structure of a parallel group trial and, being obviously incapable of being affected by carry-over, will provide the means of providing an unbiased estimate of the treatment effect. We shall refer to the estimate of the difference in first period means as PAR.

If we carry out a two-sample t analysis of the first period data in Example 1 then we obtain an estimate of 53.81 L/min with a standard error of 45.28 L/min and 95% confidence limits of $(-46, 153)$L/min. The t-statistic is 1.19 on 11 degrees of freedom and the two-sided P-value is 0.13.

This illustrates a problem with the analysis. Although PAR is unbiased by carry-over, it is extremely inefficient with, in this case, a standard error that is more than four times that for CROS. Of course, if one intends to use PAR, there is no point in carrying out a cross-over trial, since the second period values, obtained at considerable inconvenience to investigator and patient are simply thrown away. For many years, however, a strategy of using CARRY, CROS and PAR to analyse cross-over trials was recommended, which we now consider.

8.5.6   The two-stage procedure

For many years a popular approach to analysing cross-over trials was the so-called two-stage procedure(25; 26). This involved a preliminary test for carry-over using CARRY. If this was not-significant then CROS would be used to estimate and test the treatment effect. If CARRY was significant then PAR was used instead. Because of the low power of CARRY, it was commonly recommended that this should be carried out at the 10% level.

Many seemed to find the logic of this attractive. The possibility of carry-over could never be completely dismissed and CARRY provided the means for testing for it. If carry-over was detected then PAR provided an unbiased back-up, if not CROS provided an efficient test.

However, in a key paper(27) in 1989, Peter Freeman examined the procedure as a whole and showed that it led to an inflation of the type I error. The problem is that CARRY and PAR are not independent but highly correlated, first because the first period difference in means enters into both statistics, it is 100% of PAR and 50% of CARRY, and second because the second period difference in means, which forms the remaining 50% of CARRY, is correlated with the first period difference anyway.

The consequence of this is that the probability that PAR is significant when CARRY is significant is much higher under the null hypothesis of no difference between treatments than the nominal level of the test. There is no such problem with CROS which is orthogonal to CARRY by construction but the net result is that the two stage procedure gives the trialist 'two bites of the cherry' and the overall type I error rate is not 5% but between 7% and 9.5%.

Furthermore the three statistics satisfy the relationship PAR=CROS+CARRY/2 which is illustrated by the three estimates here since we have 53.8 L/min = 46.6 L/min + 7.2

L/min. The consequence of this is that conditionally (on CARRY) PAR and CROS do not differ in terms of variance, since given that you know CROS you know PAR and vice versa. They differ in terms of their conditional bias and quite plausibly the bias of PAR is far worse(28). The net consequence of this is that the conditional type I error rate of PAR is between 25% and 50% when there is no carry-over but CARRY is significant. Thus, the problem with the two-stage procedure is that either it is irrelevant or it leads to a highly inflated type I error rate(1; 29).

A way of understanding the problem is the following. Since CARRY/2 is based on the mean values over two periods but is a contrast between sequences, it will tend to differ not only when carry-over is appreciable but also when randomisation has allocated patients who differ considerably according to their level of illness between the two sequences. However, these are precisely the circumstances under which we would mistrust a between-patient estimator and prefer a within-patient one. Thus choosing PAR under these circumstances is perverse.

The advice is clear. Despite the fact that it was popular at one time and continued to be recommended in general texts on medical statistics for some years after Freeman's paper(30), the two-stage procedure should not be used.

8.6   Issues in Design

8.6.1   Choosing sequences

There is a voluminous literature dealing with the choice of optimal sequences for a cross-over trial, much of it unfortunately, based on the unrealistic simple carry-over model. The issues will be illustrated using an example of Dawid and Senn's(31) to illustrate problems in choosing models.

TABLE 8.12
Weights to be used in combining cell means to provide an unbiased treatment estimate given different models for carry-over.

	Sequence								
	AAPP				PPAA				
	Period (Treatment)				Period(Treatment)				
Model/Estimator	1(A)	2(A)	3(P)	4(P)	1(P)	2(P)	3(A)	4(A)	q
1 No	1/4	1/4	−1/4	−1/4	−1/4	−1/4	1/4	1/4	1
2 Simple	6/20	4/20	−7/20	−3/20	−6/20	−4/20	7/20	3/20	1.1
3 Steady	1/4	1/4	0	−1/2	−1/4	−1/4	0	1/2	1.5
4 General	1	−1/2	0	−1/2	−1	1/2	0	1/2	6

We suppose that we have a cross-over design in two treatments (A for active and P for placebo) in four periods and two-sequences. The two sequences chosen are AAPP/PPAA. This design is one of two known to be optimal given the restriction to four periods and two sequences and if a simple carry-over model applies.

The design and some possible schemes for estimation are illustrated in TABLE 8.12, which is taken from Dawid & Senn(31). The fractions within the body of the table are the weights with which the eight cell means, corresponding to the combination of sequence group and period, would be combined to provide an unbiased estimate of the treatment effect depending on the model assumed for carry-over. The first column gives four kinds of carry-over model. *No* is a model for which it is assumed no-carry-over occurs. *Simple* corresponds to carry-over lasting for one period and being the same from active into placebo as from active into active. *Steady* is a steady-state model proposed by Fleiss(19; 20) who pointed out that if the periods of treatment were long enough to study the steady-state effects of treatment there could be no carry-over from a treatment into itself. *General* allows for any arbitrary mixture of *Simple* & *Steady* and includes them, and indeed *No* as special cases. The final column q is a measure of the variance of the contrast and is obtained by summing the squares of the weights and multiplying by 2 (this in order to make the sum for *No* equal 1). Provided that within patient errors are independent and with equal variance, the variance for a particular estimate will be proportional to q.

If the weights are studied, it will be seen that in addition to summing to 1 over the cells labelled A and to -1 over the cells labelled P, thus estimating the contrast of interest, they sum to 0 over periods and sequences, thus eliminating period and patient effects. In addition the weights for *Simple* sum to 0 over cells following an A (and similarly for all cells following a P), thus eliminating simple carry-over. The value of q is as small as it can be consistent with that aim. The weights for *Steady* are equal to 0 in any cell in which the treatment has just been switched from A to P (and also from P to A). These are the only cells in which carry-over can occur under the steady-state model, thus that form of carry-over is eliminated. If we look at the weights for G*eneral,* we shall see that they satisfy both the simple carry-over requirement of summing to 0 over any cell following A and also of summing to 0 when the treatment has been switched from A to P (and vice versa).

Note that *General* is unbiased for the other two kinds of carry-over model. However, no applied statistician would pay the price of having a variance that was six times that for *No* by adopting this model. The situation for *Simple* might be rather different if it could be believed that this was a realistic model. However, the arguments of Fleiss and others suggest that this is implausible[1,18−20,22]. That being so, the model associated with *No* becomes attractive.

However, as soon as this is admitted, the advantages of particular designs, such as the AAPP/PPAA compared to others disappear. Any design in which A appears as often as P in any sequence or period is fully efficient for the purpose of estimating the treatment effect.

8.6.2   Other issues

This does not mean, however, that there are no issues for the applied statistician to address. For example, one that is often overlooked is the length of a period. Consider the AAPP/PPAA design. Either the effects of single doses are being studied, in which case adequate wash-out is easy to arrange, or a course of treatment will involve repeated administration in which case the purpose the design will be to study the steady state of treatment. In that case, carry-over from the drug into the drug cannot occur. If it does occur then the period chosen is too short. Understood in these terms it can be seen that a key task of those designing the trial, including the statistician, is the choice of the length of the period and that this choice would also have to be made if a parallel group trial were used. This can be seen even more clearly if one realises that an AAPP/PPAA design is no more than an AP/PA design with an extra measurement taken at mid-period.

In fact, authors contributing to the mathematical statistical literature have sometimes failed to realise that not only is period not a calendar period but the number of periods itself is not necessarily a primitive constraint(32). It might be in a single-dose pharmacodynamic study such as Example 3, but in a repeated dose design such as Example 4 the constraints were the total time available for study and the number of switches of therapy.

A further issue regarding planning is that replicate periods are sometimes added to reduce the sample size. Such could have been the function of the extra periods in Example 5. Here the practical planning issue is (usually) whether the extra inconvenience to the subject of the extra periods outweighs the convenience to the trialist that a reduced number of subjects will bring. Note that it is often the case that time to recruit subjects outweighs time to treat them so that extending the number of periods in a cross-over trial often, paradoxically, can make the trial shorter. Healthy volunteer studies, however, are an exception. Here a pool of healthy volunteers may be available and so recruitment time is not a problem. In this example, Shumaker and Metzler indicate that the ability to study treatment-by-subject interaction was the main advantage of the design. This was related to concerns that drugs that could be equivalent on average might not be so individually. At the time the trial was run, so-called 'individual bioequivalence' was a hot topic(33; 34).

Where a number of treatments are being compared, the choice of sequences is clearly important. A popular choice is designs based on Latin squares. It is not uncommon when three treatments are being compared to use all six sequences for which each treatment appears once. This means that a balanced design will have to have a multiple of six subjects. For larger numbers of treatments it seems to be more usual to use a single Latin square. So-called Williams squares(35) are also popular. These are particular Latin Squares in which each treatment follows every other. They have advantages if there is carry-over and if it has the 'simple carry-over' form. However, for reasons already discussed, this model is unrealistic and this makes the design less useful than is often supposed.

Finding suitable sequences for balanced incomplete block designs can also be a design challenge but that is a technical matter beyond the scope of this chapter.

8.6.3   Planning the sample size

Consider an AB/BA cross-over trial with equal numbers, n, allocated to each of the two sequences and suppose that a continuous outcome is being measured. There will be $2n$ basic estimators for comparing A to B and if each of these differences has variance $2\varphi^2$, where φ^2 is the variance of the within-patient error, the variance of the overall estimate will be $2\varphi^2/(2n) = \varphi^2/n$ whether we perform the average in one step or first within sequences and then over sequences. Suppose we have a clinically relevant difference of Δ, and require a standardised value of this of δ, where, in a frequentist framework, the value of δ would reflect the type I and II errors we would tolerate, then we have the (approximate) requirement

$$\frac{\Delta}{\varphi/\sqrt{n}} = \delta, \quad n = \frac{\delta^2\varphi^2}{\Delta^2} \tag{4}$$

For a rough calculation, with δ taken from tables (or a computer program) for the Normal distribution, equation (4) is usually adequate. For instance, suppose we seek a type I error probability of 0.025 one-sided and a power of 0.80. The values of the standardised Normal distribution are 1.96 and 0.84 and adding these together we get a value of $\delta = 2.8$. Suppose that we have a trial in asthma with FEV_1 as the target variable and that we have a clinically relevant difference of 200 ml and a within patient standard deviation of 300 ml. We then have

$$n = 2.8^2 \left(\frac{300}{200}\right)^2 = 18$$

(rounded up to the nearest whole number).

If we use proprietary software such as nQuery® to do the calculation, then the following statement is issued:

When the sample size in each sequence group is 19 (a total sample size of 38), a 2 × 2 cross-over design will have 80% power to detect a difference in means of 200.000 (the difference between a Treatment 1 mean, μ_1, of and a Treatment 2 mean, μ_2, of) assuming that the Cross-over ANOVA $\sqrt{\text{MSE}}$ is 300.000 (the Standard deviation of differences, σ_d, is 424.264) using a two group t-test (Cross-over ANOVA) with a 0.025 one-sided significance level.

So that the more sophisticated calculation, based on the non-central t-distribution has added one patient per sequence, a not particularly important difference. More important, however, is to heed the implicit warning in the statement:

'the Cross-over ANOVA $\sqrt{\text{MSE}}$ is 300.000 (the Standard deviation of differences, σ_d, is 424.264).'

Two different conventions are commonly used within the cross-over trials literature for within-patient error. One is the variance of the difference between the measured value and the 'true' value, the other is the variance of the difference between two measured values. The latter is twice the former and hence the standard deviation involved is $\sqrt{2}$ times as large and in this example $\sqrt{2} \times 300 = 424$. The important point to note here is that *it is essential to make sure that input and formula are compatible.* The danger is that a value will be picked up from the literature and will be falsely assumed to be one when it is in fact the other. Such confusion is of far greater importance than whether a crude formula or the more sophisticated version is used.

For more complex cross-over designs, the trick is to relate them to the AB/BA design. For designs that are uniform on the period and balanced in sequences and for which the number of treatment equals the number of periods, the result for the total number of patients for an AB/BA design may be used. For example, if a four treatment designs were being run in a single Latin square of four sequences, with similar design parameters to the example above then the requirement of 38 patients for an AB/B design could be rounded up to 40 thus indicating ten patients per sequences. Although there are some slight issues to do with degrees of freedom, here they are unimportant. Any adjustment for multiplicity would have to have been separately dealt with, say by adjusting the type I error probability.

Issues are more complicated for more complex studies, but treatment of these is again beyond the scope of this chapter.

8.7   N-of-1 trials

N of 1 trials are trials in which a single patient is repeatedly randomised to treatments being compared. For example the patients might be treated in several pairs of periods and be treated with treatment A in one period and B in another, the order being random. Where a number of patients have been treated in this way, the set of results together have the structure of a cross-over trial with more periods than treatment.

Such trials are ideally suited to identifying various components of variation and may be suitable analysed using a mixed model, although the approach to be used will depend very much on the purpose, which may range from establishing whether there is a discernable difference between treatments, to predicting what the effect will be for a given patient.

There is no space to discuss this topic in detail. The reader is referred to the references cited here(36)ˉ(39).

8.8 Conclusion

Where cross-over trials can be applied, they can bring considerable gains in efficiency and valuable insight on components of variation. They are not suitable for all indications and they are hardly ever useful in phase III. Nevertheless for a number of purposes, most notably bioequivalence and dose-finding for chronic diseases they can be very useful. They bring with them many challenges, however, and care is needed in design and analysis.

Further reading

Much of the material of this chapter is covered in *Cross-over Trials in Clinical Research*. A rival text by Jones and Kenward, *Design and Analysis of Cross-over Trials*(40) has a rather different perspective as regards carry-over, with which this author does not agree, but is highly recommended nonetheless for its treatment of background theory and in particular for its extensive coverage of binary data. A review of cross-over trials in the first 25 years of Statistics in Medicine6 may also be of interest as may an encyclopaedia article on the subject (41).

Acknowledgement

Support from the European Union's FP7 programme through the IDEAL project, grant number 602552 is gratefully acknowledged.

Bibliography

[1] Senn SJ. Cross-over Trials in Clinical Research. Second ed. Chichester: Wiley 2002.

[2] Graff-Lonnevig V, Browaldh L. Twelve hours bronchodilating effect of inhaled formoterol in children with asthma: a double-blind cross-over study versus salbutamol. Clinical and Experimental Allergy 1990;20:429–32.

[3] Senn SJ, Auclair P. The graphical representation of clinical trials with particular reference to measurements over time [published erratum appears in Statistics in Medicine 1991 Mar;10(3):487]. Statistics in Medicine 1990;9(11):1287–302.

[4] Balaam LN. A two period design with t2 experimental units. Biometrics 1968;24:61–73.

[5] Senn S. Misunderstandings regarding clinical cross-over trials. Stat Med 2005;24(23):3675-8. doi: 10.1002/sim.2155 [published Online First: 2005/11/02]

[6] Senn SJ. Cross-over trials in Statistics in Medicine: the first '25' years. Stat Med 2006;25(20):3430–42.

[7] Dahlof C, Bjorkman R. Diclofenac-K (50 and 100 mg) and placebo in the acute treatment of migraine. Cephalalgia : an international journal of headache 1993;13(2):117–23.

[8] Senn SJ, Lillienthal J, Patalano F, et al. An incomplete blocks cross-over in asthma: a case study in collaboration. In: Vollmar J, Hothorn LA, eds. Cross-over Clinical Trials. Stuttgart: Fischer 1997:3–26.

[9] Senn S, Rolfe K, Julious SA. Investigating variability in patient response to treatment–a case study from a replicate cross-over study. Stat Methods Med Res 2011;20(6):657–66. doi: 10.1177/0962280210379174 [published Online First: 2010/08/27]

[10] Stocchi F, Hersh BP, Scott BL, et al. Ropinirole 24-hour prolonged release and ropinirole immediate release in early Parkinson's disease: a randomized, double-blind, non-inferiority crossover study. Curr Med Res Opin 2008;24(10):2883-95. doi: 10.1185/03007990802387130 [published Online First: 2008/09/05]

[11] Shumaker RC, Metzler CM. The phenytoin trial is a case study of "individual bioequivalence". Drug Information Journal 1998;32:1063–72.

[12] Dworkin RH, McDermott MP, Farrar JT, et al. Interpreting patient treatment response in analgesic clinical trials: implications for genotyping, phenotyping, and personalized pain treatment. Pain 2014;155(3):457–60. doi: 10.1016/j.pain.2013.09.019

[13] Rubin D. Inference and missing data (with discussion). Biometrika 1976;63:581–92.

[14] Mantel N, Haenszel W. Statistical aspects of the analysis of data from retrospective studies. Journal of the National Cancer Institute 1959;22(4):719–48.

[15] Makubate B, Senn S. Planning and analysis of cross-over trials in infertility. Stat Med 2010;29(30):3203–10. doi: 10.1002/sim.3981 [published Online First: 2010/12/21]

[16] Nason M, Follmann D. Design and analysis of crossover trials for absorbing binary endpoints. Biometrics 2010;66(3):958–65.

[17] Verbeke G, Molenberghs G. The use of score tests for inference on variance components. Biometrics 2003;59(2):254–62.

[18] Senn SJ. Is the 'simple carry-over' model useful? [published erratum appears in Statistics in Medicine 1992 Sep 15;11(12):1619]. Statistics in Medicine 1992;11(6):715–26.

[19] Fleiss JL. Letter to the editor. Biometrics 1986;42:449–50.

[20] Fleiss JL. A critique of recent research on the two-treatment cross-over design. Controlled clinical trials 1989;10:237–43.

[21] Senn SJ, Lambrou D. Robust and realistic approaches to carry-over. Statistics in Medicine 1998;17(24):2849–64.

[22] Sheiner LB, Hashimoto Y, Beal SL. A simulation study comparing designs for dose-ranging. Statistics in Medicine 1991;10(3):303–21.

[23] Senn SJ, Hildebrand H. Crossover trials, degrees of freedom, the carryover problem and its dual. Statistics in Medicine 1991;10(9):1361–74.

[24] Senn SJ. Individual response to treatment: is it a valid assumption? BMJ 2004;329(7472):966–68.

[25] Hills M, Armitage P. The two-period cross-over clinical trial. Br J Clin Pharmacol 1979;8:7–20.

[26] Grizzle JE. The two-period change over design and its use in clinical trials. Biometrics 1965;21:467-80.

[27] Freeman P. The performance of the two-stage analysis of two-treatment, two-period cross-over trials. Statistics in Medicine 1989;8:1421—32.

[28] Senn SJ. The AB/BA cross-over: how to perform the two-stage analysis if you can't be persuaded that you shouldn't. In: Hansen B, de Ridder M, eds. Liber Amicorum Roel van Strik. Rotterdam: Erasmus University 1996:93–100.

[29] Senn SJ. The AB/BA crossover: past, present and future? Statistical methods in medical research 1994;3(4):303–24.

[30] Senn SJ, Lee S. The analysis of the AB/BA cross-over trial in the medical literature. Pharm Stat 2004;3(2):123—31.

[31] Dawid AP, Senn S. Statistical model selection. In: Christie M, Cliffe A, Dawid AP, et al., eds. Simplicity, Complexity and Modelling. Chichester: Wiley 2011:11–31.

[32] Senn SJ. Letter to the editor: Misunderstandings regarding clinical cross-over trials. Statistics in Medicine 2005;24(23):3675–78.

[33] Hauck WW, Anderson S. Individual Bioequivalence - What Matters to the Patient. Statistics in Medicine 1991;10(6):959–60.

[34] Senn SJ. Statistical issues in bioequivalence. Statistics in Medicine 2001;20(17-18):2785–99.

[35] Williams EJ. Experimental designs balanced for the estimation of residual effects. Australian Journal of Scientific Research 1949;2:149–68.

[36] Zucker DR, Schmid CH, McIntosh MW, et al. Combining single patient (N-of-1) trials to estimate population treatment effects and to evaluate individual patient responses to treatment. Journal of clinical epidemiology 1997;50(4):401–10. doi: S0895-4356(96)00429-5 [pii] [published Online First: 1997/04/01]

[37] Zucker DR, Ruthazer R, Schmid CH. Individual (N-of-1) trials can be combined to give population comparative treatment effect estimates: methodologic considerations. Journal of clinical epidemiology 2010;63(12):1312–23. doi: S0895-4356(10)00194-0 [pii]10.1016/j.jclinepi.2010.04.020 [published Online First: 2010/09/25]

[38] Araujo A, Julious S, Senn SJ. Understanding variation in sets of n-of-1 trials. PloS one 2016;11(12):e0167167.

[39] Senn S. Sample size considerations for n-of-1 trials. Statistical methods in medical research 2017:962280217726801. doi: 10.1177/0962280217726801 [published Online First: 2017/09/09]

[40] Jones B, Kenward MG. Design and analysis of cross-over trials: CRC Press 2014.

[41] Senn SJ. Cross-over Trials. In: Everitt BS, Palmer CR, eds. Encyclopaedic Companion to Medical Statistics,: Hodder Arnold 2005:88–91.

9

Factorial Designs

Bibhas Chakraborty and Palash Ghosh

CONTENTS

9.1 Introduction

Clinical trials, by their very nature, are quite challenging to conduct even with the simplest of experimental designs such as two-arm randomized controlled trials. Factorial designs bring about additional challenges in conduct because of multiple treatment arms and in statistical analysis because of potential interactions between factors and accompanying issues. In this chapter, we will systematically study these designs.

Factorial designs, a class of randomized experimental designs, constitute a mature topic in the design of experiments literature within statistics. These designs were pioneered by Fisher (1942) and Yates (1937) in the first half of the twentieth century in England. Early applications of factorial designs were in agricultural experiments; subsequently, they also found extensive applications in engineering, especially in industrial experiments (Box et al., 1978; Wu and Hamada, 2000). Their usage in clinical trials followed much later; the earliest published factorial clinical trial that we found was in the mid-1970s (Motolese et al., 1975).

These designs were developed to simultaneously study more than one treatments (or treatment components) – historically called *factors* – within the same experiment, thereby offering data efficiency. A factor can have more than one *levels*. For example, in the context of clinical or intervention trials, a factor can be defined in terms of the presence or absence of a single drug, or a single surgical intervention, or a single behavioral intervention. Different doses of the same drug, or different versions of a behavioral intervention, can constitute levels of a factor. The simplest kind of factorial design is a 2×2 or 2^2 factorial, which involves 2 factors each having 2 levels, resulting in $2 \times 2 = 4$ treatment combinations, or *cells*. Likewise, a design with 3 factors, each having 2 levels, has $2 \times 2 \times 2 = 8$ cells, and is called a $2 \times 2 \times 2$ or 2^3 factorial, and so on. A more complex factorial trial can have factors with varying number of levels. For example, the early study by Motolese et al. (1975) for treatment of hypertension employed a $5 \times 2 \times 2$ factorial design with $5 \times 2 \times 2 = 20$ cells; each patient was randomized to receive one of five doses of oxprenolol (placebo, or 20, 40, 60, or 80 mg daily), one of two doses of dihydralazine (placebo or 30 mg daily) and one of two doses of hydrochlorothiazide (placebo or 30 mg daily). However this kind of complex factorials are relatively uncommon in clinical trials. In the following, we briefly discuss a few large and well-known factorial clinical trials; all of them were two-level factorials.

The relatively less usage of factorial designs in medical research can be partially attributed to certain logistical issues associated with such designs when applied in a clinical setting, as well as certain general misconceptions. For example, a common perception is that a factorial design may not be a feasible option when the number of components is large. While there is some truth to it, it is often possible to employ *fractional* factorial designs rather than *full* factorial designs to address logistical challenges. The objective of this chapter is to give a general overview of factorial designs and to discuss their various roles in the context of clinical trials.

The remainder of the chapter is organized as follows. In Section 9.2, we discuss various usages of factorial designs as well as situations where they are not suitable. Section 9.3 provides a basic theoretical exposure of full factorial designs, explaining the key concepts. This is further expanded in Section 9.4 where fractional factorial designs are presented in detail. We present analysis plan for factorial designs in Section 9.5. Next, follow-up studies in case of developing multicomponent interventions are discussed in brief in Section 9.6. Power and sample size considerations are presented in Section 9.7. Finally, we conclude with an overall discussion in Section 9.8.

9.2   Different Usages of Factorial Designs

In clinical research, like any other research, it is of paramount importance to design a study properly. Ill-designed studies are often not able to answer the questions researchers want to answer, no matter how sophisticated statistical analysis tools are applied in the end. Hence it is critically important to understand the roles of various study designs – when to, and when not to, apply them. In the present section, we discuss various situations where factorial designs are useful, as well as situations where they are not.

In the current chapter, we consider three classes of trial designs, each with a somewhat different purpose. In all these types of trials, factorial designs may prove useful, as discussed here. First, we consider what are often called *confirmatory trials*. In such trials, the intervention(s) under consideration have already been developed and/or fine-tuned through either earlier phases of experimentation or observational data analysis, but needs a confirmatory evaluation of its efficacy in a large, fully powered study. In the regulatory context of drug development, such confirmatory trials are often known as *phase III trials*. However, we prefer the term confirmatory trial as it encompasses other types of interventions (e.g. behavioral) beyond the formal regulatory framework.

The second type of trials that we focus on in the current chapter are sometimes called *screening trials* in the context of developing multicomponent interventions; here the term *screening* refers to screening for active intervention components from a large pool of potential components, and thus is not to be confused with screening for subjects. Finally, while developing multicomponent interventions, screening trials are often followed by what are known as *refining* or *follow-up studies*; such studies aim to resolve any remaining questions about the multicomponent interventions not addressed by the screening trials. Screening and follow-up studies, followed by a confirmatory randomized trial, constitute the various phases of the *multiphase optimization strategy (MOST)* for developing, optimizing, and evaluating multicomponent behavioral or biobehavioral interventions (Collins et al., 2005). Both screening and follow-up studies are essentially developmental trials, as opposed to confirmatory trials. Such developmental trials are somewhat analogous to *early phase trials* conducted prior to phase III trials for drug development. While the use of factorial designs in confirmatory and screening trials are discussed in detail in the current section, the corresponding discussion for follow-up studies is deferred to Section 9.6 in order to achieve better flow in description.

9.2.1 Efficiency of confirmatory trials: Evaluation of more than one Intervention in a single study

Randomized controlled trials are generally deemed to be the best source of confirmatory evidence in medical research (Grimes and Schulz, 2002). While informative, these typically large confirmatory trials are also expensive and time-consuming in general. So it is natural for investigators to consider the option of evaluating more than one intervention in the same study, if possible (Montgomery et al., 2003), because that way one can answer more than one confirmatory research questions at the cost of one. In fact, evaluating more than one interventions in the same study can be achieved by employing a factorial design. In particular, when the two or more interventions employed in the factorial design are independent, i.e., they do not have an interaction (positive or negative synergy), the interpretation of the effects are straightforward and the estimation is efficient. However, the estimated effects of individual interventions and the interpretations thereof become problematic if there is an active interaction, and so extra care must be taken for proper inference and interpretation (Byar and Piantadosi, 1985; Byar et al., 1976); see Section 9.3 for more precise discussion of this issue. In case of medication trials, the assumption of no interaction is plausible when the two (or more) medications under study operate via different biological pathways. Below we discuss a few famous examples of the use of factorial design with the purpose of data efficiency (e.g., answering more than one questions at the cost of one):

1. Physicians' Health Study I (Hennekens and Eberlein, 1985): This was a 2×2 factorial prevention trial conducted in the United States involving 22,071 male physicians as trial subjects, and was designed to test the effects of: (1) aspirin (vs. placebo) on reducing cardiovascular mortality, and (2) β-carotene (vs. placebo) on reducing cancer incidence. Note that the study aimed to test two interventions in unrelated diseases, and hence it may be reasonable to assume that the two interventions operate independently without any interaction. This assumption allowed the investigators to independently assess the effect of aspirin on cardiovascular mortality and the effect of β-carotene on reducing cancer incidence, as if these were two separate trials superimposed on one another.

2. Clinical trial component of the Women's Health Initiative (Women's Health Initiative Study Group, 1998): This was a 2^3 factorial trial involving 64,500 women and three factors, viz., dietary modification (low-fat eating pattern vs. self-selected dietary behavior), hormone replacement therapy (vs. matching placebo), and calcium plus vitamin D supplementation (vs. matching placebo). The outcomes were coronary disease, breast cancer, and osteoporosis. Again, the different factors targeted different outcomes, and thus it may be reasonable to assume that they operate independently.

3. Physicians' Health Study II (Christen et al., 2000). This was a 2^4 factorial trial for cancer prevention, and involved 15,000 participants. The four factors were β-carotene, vitamin E, vitamin C, and multivitamin. The two levels were presence or absence (matching palcebo) of the corresponding factor. Note that all four factors targeted a single outcome (incidence of cancer) in this study, and thus they may not be independent.

4. α-tocopherol β-carotene Lung Cancer Prevention Trial (The ATBC Cancer Prevention Study Group, 1994): This was a 2×2 factorial design involving 29,133 male smokers in Finland between 1987 and 1994, with the incidence of lung cancer as the outcome. The two factors were α-tocopherol and β-carotene, and the two levels were presence or absence (matching palcebo) of the corresponding factor. Note that both factors targeted a single outcome (incidence of lung cancer) in this study, and thus they may not be independent.

Thus, factorial designs offer a time- and resource-efficient option for designing trials to assess more than one research questions simultaneously, particularly when the interventions under study do not have an interaction. Even when the intervention factors have some active interactions, the factorial designs can still be used, but the efficiency gain may be compromised due to the correlation between estimated effects. Historically, the above fact created a misconception in the clinical research community that factorial designs cannot be used in presence of interaction between factors. This is anything but true. There are other kinds of clinical research situations where factorial designs are used precisely for the purpose of estimating interactions among multiple factors. In fact, factorial designs are the only scientifically and statistically sound (based on randomization) design for systematically studying interactions. Thus, while efficiency is one common reason for using factorial designs, studying interactions is another important reason. This second reason plays a key role in the development of multicomponent interventions (see Section 9.2.2).

Historically there has been some debate about whether one needs to adjust the Type I error rate for multiple testing in factorial designs in the current context of efficiency of confirmatory trials. Byar and Piantadosi (1985) argued that such adjustment is not necessary when the intervention factors target different outcomes (biological pathways) and so the corresponding factorial trials can be viewed as two (or more) separate, independent trials (for which we do not normally adjust the Type I error). Examples 1 and 2 fall under this category. On the other hand, in examples 3 and 4, where various factors target the same outcome (biological pathway), they may not be viewed as completely independent trials; in such cases, adjustment for multiple testing may be reasonable. Lin et al. (2016) discussed a factorial trial where the regulatory agency specifically asked for multiplicity adjustment.

9.2.2　Screening trials: Developing multicomponent interventions

Multicomponent or multi-factor interventions are increasingly common in various health domains, e.g. AIDS (Golin et al., 2006), cardiovascular diseases (Cuffe, 2006), depression (Williams et al., 2007), diabetes (Paul et al., 2007), drug abuse (Riggs et al., 2006), gerontology (Allore et al., 2005), obesity (Bluford et al., 2007), and smoking cessation (Strecher et al., 2008). While some components may involve a medication, many components are behavioral, implementation, or delivery factors (Cuffe, 2006; Strecher et al., 2008; Williams et al., 2007). It has been recognized in the literature (Allore et al., 2005; Campbell et al., 2000; Friedli and King, 1998; Stephenson and Imrie, 1998; West et al., 1993) that development and evaluation of these multicomponent interventions pose additional design challenges over those of single-component interventions, and these challenges are inadequately addressed by the standard two-group randomized controlled trials. A particular challenge is to screening out the inactive or less active components from a large list of potential components. Trials geared towards screening out inactive components to eventually build high quality multicomponent interventions are often known as *screening trials* (Chakraborty et al., 2009; Collins et al., 2005; Nair et al., 2008); it is important to note that here the term *screening* refers to screening of intervention components, not screening of study participants. It has been argued that full and fractional factorial designs are ideally suited to this endeavor (Chakraborty et al., 2009).

Project Quit (Strecher et al., 2008) is a recent example of a screening trial that aimed to develop an effective web-based multicomponent behavioral intervention for smoking cessation and that utilized fractional factorial designs. For illustrative purposes, we present a slightly modified version of *Project Quit*, following Nair et al. (2008). The investigators decided to study six behavioral components, viz., *depth of outcome expectations, depth of efficacy expectations, depth of success stories, personalization of message source, mode of message framing,* and *exposure schedule* (depth refers to the degree to which the communication was tailored to the baseline information on each individual). Since varying all six components across all possible levels in a single study was logistically prohibitive, the investigators decided to move forward in phases, where results of the research conducted in the first phase would inform the second, and so on (Collins et al., 2005; Nair et al., 2008). The goal of the first phase was to identify the active components and screen out inactive components. Each component was varied at two levels as is common in screening studies. In addition, all individuals were provided a 10-week free supply of nicotine patches, to address the pharmacological aspects of smoking cessation. The investigators decided to use a 16-cell fractional factorial design (FFD); see Section 9.4 for details. The primary outcome was self-reported seven-day point-prevalence abstinence at the 6-month follow-up from the date of randomization. More information on this study can be found in Strecher et al. (2008), Nair et al. (2008), and Chakraborty et al. (2009).

Factorial Design Based Approach vs. Alternatives

To get a more complete perspective, it may be useful for the reader to see a discussion on potential alternatives to factorial designs when the goal is the development of a multicomponent intervention. The traditional approach of empirically developing multicomponent interventions, sometimes called the *treatment package strategy* (Kazdin, 1986; West and Aiken, 1997; West et al., 1993), is to formulate a "likely best" intervention based on existing literature, theory, clinical experience, and potentially some information – if at all available – from limited prior experimentation with some of the components in stand-alone trials. Implicitly one often assumes that more treatment is always better so the "likely best" intervention includes many components. An additional implicit assumption is that any ill effects due to including inactive components are minor. The developed multicomponent intervention is then evaluated in a standard two-arm randomized trial. These two-arm trials are confirmatory in that they are designed to provide high quality information on whether the multicomponent intervention performs better than the standard care; they are not designed to provide direct information on which components are active, whether they have been set at optimal levels, and whether there is any interaction between the components (Stephenson and Imrie, 1998). To address the latter questions, investigators may use observational analyses, such as a dose-response analysis with the subject adherence level to the treatment as the dose (Allore et al., 2005; Riggs et al., 2006; Tinetti et al., 1994, 1996), or theory-based mediational analysis (West et al., 1993; Wolchik et al., 1993). The intervention is often refined based on the findings of these analyses, and then the refined version is tested in another two-arm randomized trial. Sometimes several such iterations are performed to refine the multicomponent intervention.

The main problem with this approach is that it depends heavily on the non-experimental, observational analyses. As is well-known (Holland, 1986; Rosenbaum and Rubin, 1984; Rubin, 1974), findings that are not based on randomization are hard to replicate due to likely presence of unknown *confounders* (i.e., the variables that affect both the receipt of an intervention component and the outcome, resulting in a mixing of treatment effect with the effect of these variables). As a consequence, the effects of individual components and interactions may be misinterpreted resulting in a suboptimal intervention. Collins et al. (2009a) provided a head-to-head comparison between the above approach and an experimental procedure using FFDs in an extensive simulation study. The simulation results showed that the FFD-based experimental approach outperformed the traditional approach (two-arm randomized trial followed by observational analyses) in terms of various criteria, e.g., optimizing the mean outcome of the final intervention and the rate of identifying the best multicomponent intervention. Of course the relative merit of the FFD-based experimental approach depends on the degree of confounding; using observational analyses to investigate interactions might work well when the unknown confounder is only weakly related to the receipt of the components or the outcome.

Another alternative to FFDs is to conduct a series of *dismantling* or *subtractive* trials (West et al., 1993) where a "more complete" version of the multicomponent intervention is compared with a reduced version with one or more components eliminated. A close variant of this is known as the *constructive strategy* (West et al., 1993) or *treatment augmentation design* (Hosking et al., 2005), where a base intervention is compared with an augmented version in which one or more components are added to the base intervention. Yet another alternative is known as the *comparative treatment strategy* (West et al., 1993), where several versions of the intervention are directly compared. For example, if there are k components under consideration, a comparative strategy would compare $(k + 1)$ experimental arms: k arms, each setting a single component at the high level and the rest at the low level, plus a control arm where all components are set at the low level. The above three approaches (i.e., dismantling, constructive, and comparative strategies) sometimes come under the umbrella term of *single-factor designs* (Collins et al., 2009b) whenever the experimental arms under comparison differ by manipulating a single factor.

Note that there are several problems with using a series of single-factor experimental designs to construct a multicomponent intervention. First, as discussed by Box et al. (1978, pp. 510–513), the use of single-factor designs often tacitly assumes that the effect of one component is independent of the levels of other components. This is not true in general, e.g., when there is a sizeable qualitative interaction between the components. Thus adopting a single-factor design often implicitly assumes that there is no interaction. Because of this limitation, using a series of single-factor designs to construct a multicomponent intervention may fail to achieve the best intervention.

The second problem regarding single-factor designs arises in designing the trials, e.g., deciding which factor to add (in constructive strategy) or subtract (in dismantling strategy), or which two versions of the multicomponent interventions to compare (in comparative strategy). These decisions are often driven by theory, cost, burden, or the results of observational analyses. To the extent that the results are driven by observational dose-response analyses of the amount of treatment received, they are vulnerable to confounding bias. As a consequence, in the sequence of single-factor trials conducted to find the best multicomponent intervention, active components may be accidentally eliminated in a dismantling strategy, and less active components may be erroneously added earlier than more active components in a constructive strategy.

A third problem with single-factor designs is that they often require many more subjects than comparable factorial designs to achieve similar power, rendering factorial designs a more efficient choice (Collins et al., 2009b).

To summarize, in contrast to the treatment package strategy or the single-factor designs, inference about individual components in FFDs are strictly based on randomization, and hence less vulnerable to confounding bias. Furthermore, single-factor experiments are not equipped to take care of interactions, and often have higher sample size requirement. Although some *aliasing* of effects happens in FFDs, the investigator can still control this based on prior substantive knowledge (see Section 9.4 for details). Thus by using FFDs, one often trades uncontrolled confounding for controlled aliasing. Thus full and fractional factorial experimental designs offer a gold standard for developing multicomponent interventions.

9.2.3 Situations where factorial designs are not suitable

Factorial design is not suitable for treatment factors or components that cannot be "crossed", i.e., that cannot be employed in conjunction with one another without compromising the meanings of factor levels. In other words, to use factorial designs, all treatment combinations arising out of the factorial structure should be implementable. This has been a fundamental concern regarding the use of factorial designs in medication trials. In medication trials, toxicity often precludes the combined use of multiple factors (e.g. drugs) unless the dosage is altered (Byar and Piantadosi, 1985; Piantadosi, 2005). That is, the combination of drug A and drug B uses lower doses of both A and B, compared to the case when either drug A or drug B is used alone. So a high level of drug A in the presence (high level) of drug B does not mean the same thing as a high level of drug A in the absence (low level) of drug B. In such cases, the factors lose their meanings, and factorial designs become inappropriate. Another example situation where factorial design is not suitable is where two factors are two different surgical procedures for the same medical problem but cannot be administered in conjunction with each other. Thus before employing a factorial design in a clinical trial, an investigator should consider whether the factors of interest, when crossed, can retain their meanings. Most behavioral, delivery, or implementation factors satisfy this criterion; and so do multiple medication factors as long as they use different biological pathways.

When some factors cannot be crossed, the clinical trials literature provides some approaches. Byar et al. (1993) discussed *incomplete factorial* designs along with analysis strategies to take care of such cases. These designs are full or fractional factorial designs, minus some unpermitted combinations. Although these designs are not balanced (see Section 9.3 for definition of balance), one can still estimate many of the relevant factorial effects.

In case of developing multi-factor interventions, when the number of factors (k) is moderately large, full factorial designs may be impractical due to cost of designing and implementing too many cells, i.e., making each treatment combination work together and ensuring implementation fidelity by staff (West et al., 1993). This criticism has been the main motivation behind the development of fractional factorial designs (FFDs). It is possible to select an FFD with substantially fewer cells, but still estimate the main effects (and sometimes important two-way interactions) without bias and with the same precision as in a full factorial design under plausible assumptions. A full factorial design allows the estimation of every individual factorial effect, including all higher order interactions. However, in the absence of compelling prior theory or evidence to the contrary, third- and higher-order interactions are likely negligible in size in most multicomponent interventions (Box et al., 1978; Nair et al., 2008). FFDs sacrifice the ability to estimate some of these higher order interactions, and in return, enable the study to have fewer cells. The choice of interactions to be sacrificed is informed by scientific theory, past studies, and investigator's experience. The practical price paid to buy the economy offered by an FFD is that the effects of interest, such as the main effects and two-way interactions, are *aliased* with some higher order interactions. When two or more effects are aliased, one can estimate only the sum of the aliased effects.

To overcome this problem, ideally an FFD is chosen when each "aliased bundle" includes only one effect that is a priori believed to be active, with any other effects included in the bundle likely negligible in size. If this is not possible, follow-up experiments (Collins et al., 2005; Meyer et al., 1996; Wu and Hamada, 2000) can be conducted to settle any ambiguity about which effects are most important in the aliased bundle of effects. The above ideas were used by both *Project Quit* (Strecher et al., 2008) and *Guide to Decide* (Nair et al., 2008) to design FFD trials. See Section 9.4 for further technical details about FFDs.

The strong use of theory and investigators' experience in determining which interactions to alias in an FFD is often initially disconcerting to scientists. Note however that in a two-arm randomized trial of a multi-factor intervention vs. control, the multi-factor intervention must be determined completely by theory and investigator's experience, and furthermore in these two-arm trials every factorial effect (main effects and interactions) is aliased with every other effect. Thus all analyses concerning individual factors hinge on the use of a correct model; if the model is too simple then finding out what each effect is estimating is often difficult or even impossible. In this regard, FFDs offer a clearly better option in that the entire aliasing pattern is under the investigator's control, and there are principled ways (e.g. follow-up experiments) to disentangle any aliased effect. Moreover in non-experimental studies (that often follow the two-group comparisons) in which often the receipt of treatment depends on adherence to or availability of certain components, staff decisions as to who to offer what treatment etc., the resulting confounding is uncontrolled.

Often concerns about feasibility are intertwined with a perceived need to include many subjects in each cell of the design; this may occur because investigators erroneously think that comparisons between individual cells will be required. This however is not the case; see Section 9.7 for a discussion of this along with power considerations. None-the-less there are some situations in which investigators are unable to hire sufficient staff so as to implement multiple multi-factor interventions or are unable to train the staff to implement multiple multi-factor interventions simultaneously. In these settings even FFDs are not feasible.

9.3 Full Factorial Designs: A Theoretical Background

As discussed earlier, a factorial experiment is an experiment consisting of two or more factors each of them having two or more levels; the subjects are allocated randomly to the various cells (ideally an equal number of subjects should be allocated to each cell). In general, a factorial experiment can involve k (> 1) factors, having $l_1, \ldots, l_k$ levels respectively, in which case it results in $l_1 \times \ldots \times l_k$ cells; note that for a particular experiment, $k, l_1, \ldots, l_k$ are fixed numbers. For simplicity, in this chapter we consider the theoretical framework of two-level factorials only, i.e., when $l_1 = \ldots = l_k = 2$; this results in 2^k cells, and thus the resulting design is called a 2^k *full* or *complete* factorial.

For simpler illustration, let us discuss 2^2 and 2^3 full factorials in some more detail. In a 2^2 design with two factors, say A_1 and A_2, the two levels usually represent "high vs. low" or "presence vs. absence" or "active vs. placebo" of treatment factors, and are denoted by "+" and "−"; in total there are four cells, as shown in the left panel of Figure 9.1. A subject allocated to the $(+, -)$ cell (second row of the left panel of Figure 9.1) receives the treatment factor A_1 at "+" level and A_2 at "−" level. Likewise, the cells of a 2^3 full factorial design are shown in the right panel of Figure 9.1.

One important property of a factorial design is *balance*. Balance means that each level of each factor in a design appears in the same number of cells, and is assigned to the same number of subjects. Full factorial and appropriately chosen *fractional* factorial designs (defined in Section 9.4) satisfy the balance property.

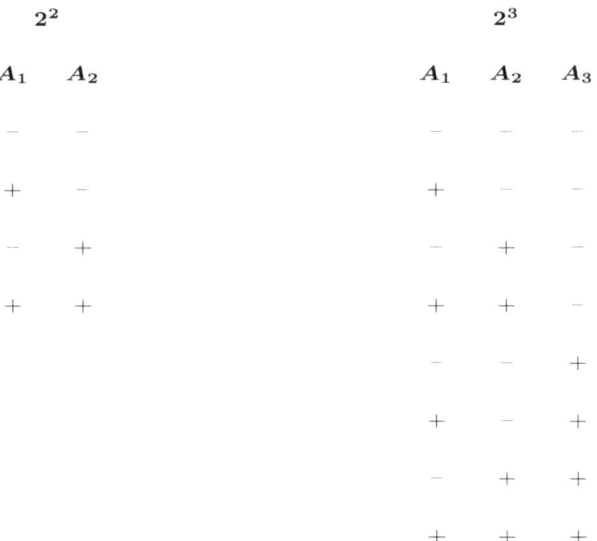

FIGURE 9.1
Different treatment combinations (cells) for 2^2 and 2^3 factorial design.

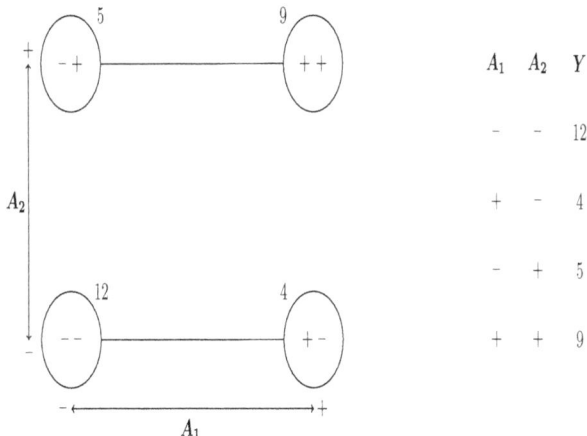

FIGURE 9.2
Computing main effect of the factor A_1.

Key concepts underlying factorial designs are those of *main effect, simple effect* and *interaction effect*, which are often generically referred to as *factorial effects*. Because of the underlying balance, all main effects and interaction effects are *orthogonal* to each other, i.e., they correspond to independent contrasts of the outcome data, which in practice leads to these factorial effect estimates being uncorrelated. Here we illustrate the concepts of various factorial effects using a 2^2 design and a continuous outcome Y. Let $\mu_{(-,-)}$, $\mu_{(-,+)}$, $\mu_{(+,-)}$ and $\mu_{(+,+)}$ be the mean outcomes corresponding to the $(-,-)$, $(-,+)$, $(+,-)$ and $(+,+)$ cells of the design, respectively. At the population level, the main effect of the component A_1, say $m(A_1)$, is defined as $m(A_1) = \frac{1}{2}[s(A_1|A_2+)+s(A_1|A_2-)]$, where $s(A_1|A_2+) = (\mu_{(+,+)} - \mu_{(-,+)})$ and $s(A_1|A_2-) = (\mu_{(+,-)} - \mu_{(-,-)})$ are two *simple effects*, denoting the effect of A_1

when A_2 is fixed at "+" and "−" level, respectively. Simple effects are alternatively known as *conditional main effects* (Wu and Hamada, 2000) because they condition on certain values of other factors. Thus the main effect of A_1 is defined as the average of the two simple effects, and hence can be interpreted as the effect of A_1 when half the subjects in the population are exposed to (the high level of) A_2 and the remaining half are not. Note that the main effect can be re-written as, $m(A_1) = \frac{1}{2}(\mu_{(+,+)} + \mu_{(+,-)}) - \frac{1}{2}(\mu_{(-,+)} + \mu_{(-,-)}) = \mu_{(+,\cdot)} - \mu_{(-,\cdot)}$, say. One can analogously define main effects and simple effects at the sample level. From the last expression, the main effect of A_1 is the difference between the average of Y values corresponding to all the observations in the study at '+' (high) level of A_1 and the average of Y values corresponding to all the observations in the study at '−' (low) level of A_1. Figure 9.2 gives an illustration of how to calculate the main effect of a factor A_1 from sample data. According to the above definition, and using the values in the figure, the main effect of A_1 is $m(A_1) = (\frac{9+4}{2}) - (\frac{12+5}{2}) = -2$, which is also the average of two simple effects, e.g., $s(A_1|A_2+) = (9 - 5) = 4$ and $s(A_1|A_2-) = (4 - 12) = -8$. Similarly, one can calculate simple effects and main effect of A_2.

For three two-level factors A_1, A_2 and A_3, the main effect of A_1 is defined as $m(A_1) = \frac{1}{4}[s(A_1|A_2+, A_3+) + s(A_1|A_2+, A_3-) + s(A_1|A_2-, A_3+) + s(A_1|A_2-, A_3-)]$, where $s(A_1|A_2+, A_3+) = (\mu_{+,+,+} - \mu_{(-,+,+)})$, $s(A_1|A_2+, A_3-) = (\mu_{(+,+,-)} - \mu_{(-,+,-)})$, $s(A_1|A_2-, A_3+) = (\mu_{(+,-,+)} - \mu_{(-,-,+)})$, and $s(A_1|A_2-, A_3-) = (\mu_{(+,-,-)} - \mu_{(-,-,-)})$ are the four simple effects. In general, for a design involving k two-level factors, there are 2^{k-1} simple effects; and the main effect is the average of these 2^{k-1} simple effects.

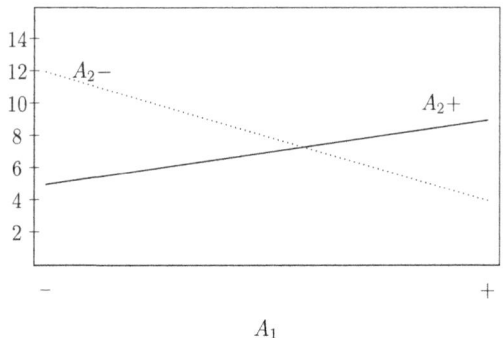

FIGURE 9.3
Interaction effect plot of $A_1 \times A_2$.

One of the main advantages of a factorial design is its ability to find out the interaction effects due to different combination of factors. By interaction, we mean the effect which is due to the joint effect of two or more factors but is not reflected in the main effect of any particular factor. Formally, the interaction effect of A_1 and A_2, sometimes denoted by $int(A_1, A_2)$ or $A_1 \times A_2$, can be defined in three equivalent but different ways (Wu and Hamada, 2000) as follows

$$
\begin{aligned}
int(A_1, A_2) &= \frac{1}{2}\underbrace{[\mu_{(+,+)} - \mu_{(+,-)}]}_{s(A_2|A_1+)} - \frac{1}{2}\underbrace{[\mu_{(-,+)} - \mu_{(-,-)}]}_{s(A_2|A_1-)} \\
&= \frac{1}{2}\underbrace{[\mu_{(+,+)} - \mu_{(-,+)}]}_{s(A_1|A_2+)} - \frac{1}{2}\underbrace{[\mu_{(+,-)} - \mu_{(-,-)}]}_{s(A_1|A_2-)} \\
&= \frac{1}{2}[\mu_{(+,+)} + \mu_{(-,-)}] - \frac{1}{2}[\mu_{(+,-)} + \mu_{(-,+)}].
\end{aligned}
\tag{9.1}
$$

Clearly, from the first two lines of the above display, the interaction between A_1 and A_2 is half the difference between two simple effects either of A_1 or of A_2. A large value of the interaction effect means that the two simple effects or conditional main effects differ by a large value, i.e., the effect of one factor depends on the presence or absence of the other factor, thereby justifying the usage of the term "interaction". In connection with Figure 9.2, $int(A_1, A_2) = (\frac{9-4}{2}) - (\frac{5-12}{2}) = 6$. An interaction plot is shown in Figure 9.3; the effects of A_1 are in opposite directions when interacting with high ('+') and low ('−') levels of the factor A_2.

In a factorial experiment, higher-order interactions can be analogously defined. For example, in a factorial design involving a total of k factors, a p-factor interaction or p-th order interaction $(1 \le p \le k)$, say $A_{i_1} A_{i_2} \ldots A_{i_p}$ $(1 \le i_1, \ldots, i_p \le k)$, is defined as half the difference between the two corresponding "conditional interaction effects" of order $(p-1)$; see Wu and Hamada (2000) for more details.

Before concluding the discussion about various factorial effects, we want to highlight an important issue about the definition of effects in a factorial design. Note that the most common simple effect of a factor is obtained when all the other factors are at '−' (low or absent), i.e., $s(A_1|A_2-)$; this is often perceived by clinical investigators as the "treatment effect" of the factor of interest (Piantadosi, 2005, p. 506). In absence of interaction between the factors, the two simple effects are equal, and hence either one of them is equal to the main effect. In this case, the conceptual mismatch between the formal definition of main effect and the often-perceived "treatment effect" of a factor (which is a simple effect, $s(A_1|A_2-)$) does not cause any problem. However, the main effect could be very different from a simple effect in presence of a sizeable interaction. Historically this issue had caused much confusion and resulting criticism of factorial designs in the clinical trial literature. See Byar and Piantadosi (1985), Piantadosi (2005), and Chakraborty et al. (2009) for further discussion on this topic.

Fundamental principles about factorial effects

It is important to understand how to prioritize the various factorial effects as well as to understand their inter-relationships. When there is strong substantive theory or prior results, this prioritization is straightforward. However, in the absence of such external and/or a priori information, some design principles are necessary to guide an investigator or analyst. Listed below are three such fundamental principles available in the classical design of experiment literature (Wu and Hamada, 2000).

1. *Hierarchical Ordering Principle:* Lower order effects are more likely to be important than higher order effects, and effects of the same order are equally likely to be important.

2. *Effect Sparsity Principle:* The number of relatively important effects in a factorial experiment is small.

3. *Effect Heredity Principle:* In order for an interaction to be significant, at least one of the parent factors should be significant.

These principles are employed in various aspects of factorial and other related designs (Wu and Hamada, 2000); in particular, the hierarchical ordering principle is often used in the design and analysis of *fractional factorial designs* (FFD) that we will discuss in more detail in the next section.

9.4   Fractional Factorial Designs

When the number of factors (k) in a study is moderately large, a full factorial design, even with two levels for all k factors, resulting in 2^k treatment combinations (cells) is often practically infeasible due to the cost of designing and implementing too many cells. In such settings, fractional factorial designs (FFDs) (Box and Hunter, 1961) offer an appealing alternative, since it is possible to select an FFD with substantially fewer cells, but still estimate the main effects (and sometimes important two-way interactions) without bias and with the same precision as in a full factorial design under plausible assumptions. A two-level fractional factorial design is denoted by 2^{k-p}, where k is the total number of factors and 2^{-p} fraction of the total number of cells are discarded.

Project Quit (Strecher et al., 2008) is a recent example of a web-based smoking cessation trial that successfully employed an FFD, and we will use this example in detail to illustrate various notions underlying FFDs. As discussed in Section 9.2.2, this study involved six behavioral and communication factors ($k = 6$) that were meant to address various psychosocial aspects of smoking cessation, namely: *depth of outcome expectations, depth of efficacy expectations, depth of success stories, personalization of message source, mode of message framing*, and *exposure schedule* (depth refers to the degree to which the communication was tailored to the baseline information on each individual). A full factorial design with the above six factors would need $2^6 = 64$ cells. But by using an FFD, it was possible to restrict the study to only 16 cells by removing 2^{-2} fraction of the total number of cells, and still be able to estimate all the main effects and some two-way interactions under reasonable assumptions. Thus, the use of an FFD offered substantial efficiency.

Aliasing, Defining Relation and Resolution

A full factorial design allows the estimation of all individual main effects and interactions. However, in the absence of compelling prior theory or evidence to the contrary, third- and higher-order interactions are likely negligible in size (Box et al., 1978; Nair et al., 2008). FFDs sacrifice the ability to estimate some of these higher order interactions, and in return, enable the study to have fewer cells. The choice of interactions to be sacrificed is informed by scientific theory, past studies, and investigator's experience. The practical price paid to buy the economy offered by an FFD is that the effects of interest, such as the main effects and two-way interactions, are *aliased* (i.e., mixed) with some higher order interactions. When two or more effects are aliased, one can estimate only the sum of the aliased effects. To overcome this problem, ideally an FFD is chosen in which each "aliased bundle" includes only one effect that is a priori believed to be active, with any other effects included in the bundle likely negligible in size. If this is not possible, follow-up experiments (Chakraborty et al., 2009; Collins et al., 2005; Meyer et al., 1996; Wu and Hamada, 2000) can be conducted to settle any ambiguity about which effects are most important in the aliased bundle of effects. The above ideas were used by both *Project Quit* (Strecher et al., 2008) and *Guide to Decide* (Nair et al., 2008) teams to design FFD trials.

To better understand the phenomenon of aliasing, the concepts of *defining relation* and *resolution* are useful. An FFD is completely characterized by its *defining relation* (Wu and Hamada, 2000), a rule from which the aliasing pattern of the FFD can be obtained. Suppose a study involving five factors, say $A_1, \ldots, A_5$, is restricted to 16 cells, as in the *Guide to Decide* study described in Nair et al. (2008). Then a 2^{-1} fraction of the 2^5 full factorial design should be used. With 16 cells, one can construct a full factorial design with four factors, say with $A_1, \ldots, A_4$. The strategy then is to alias the fifth factor, say A_5, with the 4-way interaction $A_1A_2A_3A_4$. This means, the column (in the design matrix) of A_5

is identical to that of the element-wise product of the columns of A_1, A_2, A_3, and A_4, i.e., $A_5 = A_1 A_2 A_3 A_4$. Note that all the elements in any of the columns are either $+1$ or -1. So element-wise product of any column with itself leads to the identity column, say I (with all its entries $+1$). In particular, $A_5 A_5 = I$. Multiplying both sides of the equation $A_5 = A_1 A_2 A_3 A_4$ by A_5 gives

$$I = A_1 A_2 A_3 A_4 A_5. \tag{9.2}$$

The condition (9.2) completely specifies the aliasing pattern of the 2^{5-1} FFD under consideration, and hence called its *defining relation*. The alias of any factorial effect can be found by multiplying both sides of (9.2) by that effect and then using the facts that $A_j I = A_j$ and $A_j A_j = I$ for all j. The word $A_1 A_2 A_3 A_4 A_5$ is called the *defining word*. The length (i.e., number of elements or letters) of the *defining word* is called the *resolution* of the design. So the design specified by (9.2) is a resolution V design. In a resolution V design, the main effects are aliased with four-way interactions, and the two-way interactions are aliased with three-way interactions. In a setting where the third- or higher-order interactions are negligible, resolution V FFDs are almost as good as the full factorial designs in that the main effects and two-way interactions are estimable without bias.

However, due to cost and feasibility constraints, one may have to use smaller (than 2^{-1}) fraction of full factorial designs, leading to lower resolution. Loosely speaking, the higher the resolution, the better is the design. The *Project Quit* study described before used a resolution IV FFD. In the following, we illustrate resolution IV designs with an example. Suppose there are six factors, say $A_1, \ldots, A_6$, in a study that is restricted to 16 cells (as in *Project Quit*). This means constructing a 2^{-2} fraction of the 2^6 ($= 64$ cells) full factorial design. With 16 cells, one can construct a full factorial with four factors, say with $A_1, \ldots, A_4$. Now, the strategy is to make the columns of the remaining two components A_5 and A_6 identical to some higher-order interactions. One such choice is to set $A_5 = A_1 A_3 A_4$ and $A_6 = A_2 A_3 A_4$. Using the same rules as before, one gets $I = A_1 A_3 A_4 A_5$ from the first aliasing relation, and $I = A_2 A_3 A_4 A_6$ from the second aliasing relation. Multiplying these two, a third equation $I = A_1 A_2 A_5 A_6$ follows. Thus the defining relation of this FFD is

$$I = A_1 A_3 A_4 A_5 = A_2 A_3 A_4 A_6 = A_1 A_2 A_5 A_6. \tag{9.3}$$

By definition, (9.3) is a resolution IV design, since the length of each defining word is 4. In a resolution IV design, the main effects are aliased with three-way or higher-order interactions, but the two-way interactions are aliased with other two-ways. In the following, we first discuss the FFD used in the *Project Quit* study. Next, we discuss a general approach to construct appropriate FFD trials.

FFD used in the Project Quit study

Denote the six factors of the *Project Quit* study by A_1, A_2, A_3, A_4, A_5, and A_6. In this study, prior knowledge suggested that the interactions between outcome expectations and efficacy expectations ($A_1 A_2$), outcome expectations and success stories ($A_1 A_3$), outcome expectations and message framing ($A_1 A_5$), and efficacy expectations and message framing ($A_2 A_5$) were likely active (let us call them *anticipated* interactions), and that all other interactions should be negligibly small in size. So a design was constructed such that one could estimate the $A_1 A_2$, $A_1 A_3$, $A_1 A_5$, and $A_2 A_5$ interactions, assuming all others to be small. Due to cost constraints, 16 cells were used in the design. So the design used was a 16-cell FFD with the defining relation

$$I = A_1 A_2 A_4 A_5 = A_1 A_3 A_4 A_6 = A_2 A_3 A_5 A_6. \tag{9.4}$$

This is a resolution IV design where some of the two-way interactions are aliased with other two-way interactions. The anticipated two-way interactions are listed on the left-hand side of the following aliasing equations (obtained from the defining relation (9.4)):

$$A_1 A_2 = A_4 A_5$$
$$A_1 A_3 = A_4 A_6$$
$$A_1 A_5 = A_2 A_4$$
$$A_2 A_5 = A_1 A_4$$

Note that the anticipated interactions were aliased with other two-way interactions that were considered negligible, and hence were estimable without bias. The defining relation $I = A_1 A_2 A_4 A_5 = A_1 A_3 A_4 A_6 = A_2 A_3 A_5 A_6$ was "cleverly" chosen to accomplish this goal. Of course, the investigator's assumption about the interactions could be wrong, but one can verify any critical working assumptions using follow-up studies (Collins et al., 2005).

General strategy for constructing FFD trials

For the purpose of illustration, we assume that regardless of the number of factors studied, the number of cells used can be at most 16 (equal to the number of cells used in the *Project Quit* study). Of course this number can vary from one setting to another. If four or fewer factors are to be studied, a full factorial design can be used. If five factors, say $A_1, \ldots, A_5$ are to be studied, then one should use the resolution V FFD with the defining relation $I = A_1 A_2 A_3 A_4 A_5$; this is the case in the *Guide to Decide* project described in Nair et al. (2008). If six factors, say $A_1, \ldots, A_6$ are to be studied, resolution IV designs are generally recommended. If prior knowledge suggests a few anticipated two-way interactions, an FFD can be chosen carefully so that the anticipated two-way interactions are not aliased with each other (this consideration often drives the construction of the design). Assuming the unanticipated interactions to be negligible, this ensures that each anticipated interaction can be estimated without bias. When there is only one anticipated interaction, any 16-cell resolution IV FFD can be used. However, for two or more anticipated interactions, choices are limited. Some software programs (e.g., SAS PROC FACTEX, JMP, Minitab) can be used to generate the designs in such cases; however, they only provide one possible design that satisfies the constraints of resolution and/or anticipated interactions, instead of giving the complete list of possible designs. For two or three anticipated interactions, the complete set of recommended designs were given by Chakraborty et al. (2009), and are reproduced here in Table 9.1.

9.5 Analysis Strategies

Let Y be the continuous outcome of interest, and $\mathbf{Y}$ denote the $n \times 1$ vector of observations on Y, where n is the total sample size in the trial. Consider a linear regression model $\mathbf{Y} = \mathbf{X}\beta + \epsilon$, where $\mathbf{X}$ is the $n \times m$ design matrix, β is an $m \times 1$ vector of unknown parameters and ϵ denotes the $n \times 1$ vector of errors. It is assumed that $E(\epsilon) = 0$ and $Cov(\epsilon) = \sigma^2 \mathbf{I}$. The least-squares estimate of β, denoted $\hat{\beta}$, is $(\mathbf{X}^T \mathbf{X})^{-1} \mathbf{X}^T \mathbf{Y}$ and the corresponding covariance matrix, $Cov(\hat{\beta}) = \sigma^2 (\mathbf{X}^T \mathbf{X})^{-1}$. Generally, the elements of the design matrix $\mathbf{X}$ are coded as $-1/1$, where 1 denotes the higher level of a factor and -1 denotes the lower level of a factor. In a 2^k factorial design, a particular row of a particular column of $\mathbf{X}$ denotes level status $(-1/1)$ of the corresponding column variable for the subject corresponding to the row. A

TABLE 9.1
Recommended resolution IV FFDs under varying anticipated interactions.

Case	Anticipated interactions of the form	Recommended designs (defining relations)
1	A_1A_2, A_3A_4 (no component shared)	$I = A_1A_2A_3A_5 = A_1A_3A_4A_6 = A_2A_4A_5A_6$
		$I = A_1A_2A_3A_5 = A_2A_3A_4A_6 = A_1A_4A_5A_6$
		$I = A_1A_2A_4A_5 = A_1A_3A_4A_6 = A_2A_3A_5A_6$
		$I = A_1A_2A_4A_5 = A_2A_3A_4A_6 = A_1A_3A_5A_6$
		$I = A_1A_2A_3A_6 = A_1A_3A_4A_5 = A_2A_4A_5A_6$
		$I = A_1A_2A_3A_6 = A_2A_3A_4A_5 = A_1A_4A_5A_6$
		$I = A_1A_2A_4A_6 = A_1A_3A_4A_5 = A_2A_3A_5A_6$
		$I = A_1A_2A_4A_6 = A_2A_3A_4A_5 = A_1A_3A_5A_6$
2	A_1A_2, A_1A_3 (one component shared)	$I = A_1A_2A_4A_5 = A_1A_3A_4A_6 = A_2A_3A_5A_6$
		$I = A_1A_2A_4A_6 = A_1A_3A_4A_5 = A_2A_3A_5A_6$
		$I = A_1A_2A_4A_5 = A_1A_3A_5A_6 = A_2A_3A_4A_6$
		$I = A_1A_2A_5A_6 = A_1A_3A_4A_5 = A_2A_3A_4A_6$
		$I = A_1A_2A_4A_6 = A_1A_3A_5A_6 = A_2A_3A_4A_5$
		$I = A_1A_2A_5A_6 = A_1A_3A_4A_6 = A_2A_3A_4A_5$
3	A_1A_2, A_3A_4, A_5A_6	same as case 1
4	A_1A_2, A_1A_3, A_4A_5	$I = A_1A_2A_5A_6 = A_1A_3A_4A_6 = A_2A_3A_4A_5$
		$I = A_1A_2A_4A_6 = A_1A_3A_5A_6 = A_2A_3A_4A_5$
5	A_1A_2, A_1A_3, A_2A_4	$I = A_1A_2A_5A_6 = A_1A_3A_4A_5 = A_2A_3A_4A_6$
		$I = A_1A_2A_5A_6 = A_1A_3A_4A_6 = A_2A_3A_4A_5$
6	A_1A_2, A_1A_3, A_1A_4	$I = A_1A_2A_5A_6 = A_1A_3A_4A_5 = A_2A_3A_4A_6$
7	A_1A_2, A_1A_3, A_2A_3	same as case 2

design matrix contains columns corresponding to the main effects of all the factors and their interaction effects, in addition to a column for the grand mean. The length m of parameter vector β is the number of columns in the design matrix X, i.e., the total number of main effects and interaction effects plus one (for the grand mean or intercept); it turns out that $m = 2^k$ for full factorials and $m = 2^{k-p}$ for fractional factorials. More generally, if one also wants to include baseline variables in the analysis model (as in analysis of covariance), more parameters can be incorporated into the vector β. The least-squares estimates of the model parameters β can be easily obtained from popular statistical softwares like R and SAS. However, these estimates are not exactly equal to the main effect of a factor or interaction effect between two factors, but of course they are directly related, as long as the design matrix is coded in $-1/1$. As discussed in Byar et al. (1993), the main effect of a factor A_1 is estimated by $2\hat{\beta}_{A_1}$, A_1A_2 interaction is estimated by $4\hat{\beta}_{A_1A_2}$, and so on. In general, an l-factor ($1 \leq l \leq k$) interaction, say $A_{i_1} \ldots A_{i_l}$ (with $1 \leq i_1, \ldots, i_l \leq k$), is estimated by $2^l \hat{\beta}_{A_{i_1} \ldots A_{i_l}}$. If variance heterogeneity across different cells is anticipated in a study, one can use a robust estimator, e.g. *sandwich estimator* (White, 1980) of the covariance matrix given by

$$(\mathbf{X}^T\mathbf{X})^{-1}\mathbf{X}^T(diag(\mathbf{Y} - \mathbf{X}\hat{\beta}))^2\mathbf{X}(\mathbf{X}^T\mathbf{X})^{-1}$$

in the linear model; however sample sizes should be at least moderately large for this estimator to work well. Wu and Hamada (2000) provide alternative methods to deal with variance heterogeneity. As discussed by Montgomery et al. (2003), the regression approach can be used for *unbalanced* data (in this context, balance means that each level of each factor appears in same number of cells and is assigned to the same number of subjects), and can estimate the factorial effects controlling for baseline or stratification variables. In case of binary (more generally, categorical) outcomes, the regression approach can be generalized via a generalized linear model. For example, if the outcome is binary, a logistic regression model can be used to analyze the data from a factorial design (Wu and Hamada, 2000, Ch. 13). See Nair et al. (2008) and Strecher et al. (2008) for examples of such analyses.

Analytic considerations in screening studies

In general, the data analysis in screening studies employs a linear model (in case of continuous outcome) or a generalized linear model (in case of binary or categorical outcome). A few considerations to be made during the analysis are:

1. The level of significance α for testing the effects in the screening study might be set higher than 0.05 to achieve greater power for detecting effects. α can be viewed as a tuning parameter of the procedure. One possible choice is to use $\alpha = 0.1$ for the main effects and *anticipated* two-way interactions, and a Bonferroni-corrected 0.1 level for the *unanticipated* interactions.

2. As an alternative (or augmentation) to performing significance tests at the screening study, one can rank-order the absolute values of the test statistic corresponding to the factorial effects (or equivalently p-values) and move to follow-up studies with the largest m. Then this m becomes a tuning parameter of the procedure. This approach should work better in case all individual effects are small, but together they produce some effect (significance test often accepts the null hypothesis of no effect in such cases, and hence perform poorly). To be resistant to the noise in the data, one may choose to rank-order only the main effects and *anticipated* interactions. This strategy with $m = 3$ was followed in the simulation study described in Collins et al. (2009a).

Examples of the screening analysis in the *Guide to Decide* and *Project Quit* studies can be found in Nair et al. (2008) and Strecher et al. (2008). Based on the screening analysis of the *Project Quit* study, the investigators decided to move to the follow-up study with the components having the highest two p-values (e.g. *success stories* and *message source*). Furthermore, since three of the components (outcome expectations, efficacy expectations, and success stories) were set at levels corresponding to high depth of tailoring versus low depth of tailoring, the investigators considered a regression of overall depth of tailoring (over all components) and found that as the depth of tailoring increased the smoking cessation rate increased. Hence the investigators decided to use a high depth of tailoring in the follow-up study.

Time-to-event analysis in factorial trials

Apart from continuous or binary outcomes, clinical trials often have survival or other time-to-event outcomes. While a detailed exposition of time-to-event analysis in the context of factorial clinical trials is beyond the scope of the current chapter, there exist various methods in the literature to handle such settings. In particular, Slud (1994) considered time-to-event analysis, including asymptotic theory, under a Cox proportional hazards model in the context of 2×2 factorial designs. Within his analysis framework, he examined the possibilities of using log-rank, adjusted log-rank, stratified log-rank and standardized maximum partial likelihood estimator based testing procedures, and even provided some guidelines about group-sequential analysis and early termination of one treatment factor due to ethical considerations. In a similar setting, Lin et al. (2016) considered simultaneous inference about the factorial effects, controlling for overall Type I error for regulatory purposes. Moser and Halabi (2015) considered time-to-event analysis in general factorial designs with any number of factors and any number of levels per factor, but assumed the time-to-event to follow an exponential distribution.

9.6 Follow-up Studies: Developing Multicomponent Interventions

In the process of developing a multicomponent intervention, an investigator often conducts follow-up studies involving the *significant* (in this section, we will use the term *significant* loosely to mean any effects that come out important according to the screening analysis strategy outlined in Section 9.5) factorial effects from the screening study to fine-tune the results, e.g., finding the best level (or dose) of a significant component which is either continuous or has more than two levels by a dose-response experiment (where the subjects are randomized to ethically acceptable doses of the component), or de-aliasing significant aliased interactions by a smaller factorial experiment. In this section, first we provide a hypothetical example of a follow-up study to provide some general intuition, and then briefly describe the follow-up phase of the *Project Quit* study. Further strategies for conducting follow-up studies can be found, for example, in Meyer et al. (1996) and Wu and Hamada (2000). Also, in case there is at least one component with more than two levels (e.g. a continuous component), dose-response experiments (Box and Draper, 1987; Myers and Montgomery, 1995) where subjects are randomized to ethical doses should be used to find the optimal dose of these components.

Hypothetical example

In this example, we assume that there are six components in the study, e.g., $A_1, \ldots, A_6$, out of which only A_1 is a 3-level component (say, high, medium, low levels – only high and low levels are studied at the screening trial) and the rest are binary (high and low). High values of the outcome are preferred. A 16-cell resolution IV FFD is used as the screening design (see Section 9.4 for details). Further, we assume that three-way (or higher-order) interactions are negligible in size compared to the noise in the data; hence even though main effects are aliased with three-way interactions, we assign the estimated effect to the main effect.

Suppose the significant effects along with their signs based on screening analysis are:

$$A_1(+), A_2(+), A_3(+), A_5(-), A_2 A_3 = A_4 A_5(-),$$

where both the interactions $A_2 A_3$ and $A_4 A_5$ involved in the aliased bundle are unanticipated. Since the main effect of A_4 is insignificant, the main effect of A_5 is negative, and the aliased $A_4 A_5$ interaction is negative (even though we are not sure if the observed effect is really due to $A_4 A_5$), one reasonable step would be to set A_4 at its high level (provided the high level of A_4 is not very expensive or burdensome) and A_5 at its low level (note that our decision about the optimum levels of A_4 and A_5 would be same when $A_4 A_5$ effect is really negative as when $A_4 A_5$ is null). Also, we would set A_6 to the low level. If there is concern about the potential $A_2 A_3$ interaction then the follow-up study could be a 8-group trial, where medium and high levels of A_1 are crossed with the two levels of A_2 and A_3 each to form the 8 groups (setting A_4, A_5, and A_6 at high, low, and low levels respectively).

Follow-up study design of Project Quit

An alternative to the follow-up studies outlined above is provided by *Project Quit* study, in which all components were two-level (hence a dose-response experiment was unnecessary) and no (unanticipated) aliased interaction were found significant (hence no de-aliasing experiment was necessary). The investigators decided to study different aspects (not studied in the screening trial) of the two important components (e.g., *success stories* and *message source*). The decision was to vary message source at two levels (high/low) of additional personalization, and to vary success stories in terms of the archetype (language and picture) of the hypothetical character in the story at three levels (e.g., a rebel, care-giver, or self-made character). Two new two-level components, e.g., *order* (of appearance on the web site: *success stories* first vs. *health advice* first) and *email quit status request* (yes/no) were added to the follow-up study. Subjects randomized to the "yes" level of *email quit status request* were contacted by the study staff at regular intervals about their quit status. The follow-up study consisted of 25 groups in total: 24 groups from the $2 \times 3 \times 2 \times 2$ factorial structure of the above four components, plus a control group. In all groups, the original components from the screening trial not studied in the follow-up study were set as follows: deeply-tailored efficacy expectation and outcome expectation messages, gain framing, and multiple exposures. All three levels of the success stories were also deeply-tailored. The control group received the best intervention according to the results of the screening study (e.g., highly personalized source at the first session only, deeply-tailored story with fixed archetype as in the screening study, deeply-tailored efficacy expectation and outcome expectation messages, gain framing, and multiple exposures) – they did not receive any email about their quit status.

9.7 Power and Sample Size Considerations

In this section, we will discuss several issues and considerations regarding power and sample size that come up in the context of factorial designs.

Concern about large sample size

There is often a general perception in the clinical research community that full or fractional factorial designs require larger sample size to maintain the same power compared to corresponding conventional trials. This perception stems from the fact that investigators sometimes use factorial designs to evaluate or compare a few multicomponent interventions, e.g., compare one cell against another cell (Green et al., 2002), or otherwise assess simple effects. This approach does not really exploit the factorial structure of the data *per se*; instead it is actually a parallel-groups comparison, where the groups are constructed by crossing more than one factor. This naturally leads to a large sample size requirement since each cell (group of subjects) must be large. However in case of screening trials (for developing multicomponent interventions), people primarily focus on main effects and sometimes also a few anticipated two-way interactions to be able to successfully screen components. The focus on main effects and lower order interactions for the purpose of screening can be partially justified by the *Hierarchical Ordering Principle* (Wu and Hamada, 2000), which says that main effects and lower order interactions are likely more important than higher order interactions. Recall that the main effect of a factor is an average of all the 2^{k-1} simple effects. Thus even though several components are studied, the total sample size required for assessing the significance of a main effect is the same as that for a two-group trial (for example in a linear model, the estimator of the main effect is proportional to the difference between the means of two groups of cells; all cells in the FFD belong to one or the other group). Furthermore, in the multiphase approach to intervention development (Collins et al., 2005;

Nair et al., 2008), ascertaining the best treatment combination is done through follow-up studies, in which one usually focuses on only a few combinations of components while holding the levels of the remaining components constant. See Section 9.6 for a discussion of follow-up studies.

Formulating the test statistic in screening studies

One important question is how one should formulate the test statistics to detect the effects of treatment components in a screening study. Note that in a screening study the goal is to screen out inactive components, and not to estimate either a simple effect or a main effect *per se*. Below we show that even when the data are generated using non-zero simple effects, often the power to detect the resulting main effect is higher than the power to detect the original simple effect. Hence in a screening study, formulating the test statistics based on main effects is in general better than formulating test statistics based on simple effects. To discuss this, consider the 2×2 factorial design with two components, say A_1 and A_2, r subjects per cell, and the continuous outcome Y. The true data-generating model is specified in terms of simple effects, which is consistent with an investigator's conceptualization. Thus the true data-generating model is given by

$$Y = b_0 + b_1 A_1 + b_2 A_2 + b_{12} A_1 A_2 + \varepsilon, \tag{9.5}$$

where the components A_1, A_2 are coded in 0/1. In the following, we show that by basing the test statistic on main effects, we can in general screen non-zero simple effects with greater power.

For simplicity, assume $\mathrm{Var}(\varepsilon) = \sigma^2$ is known (and homogeneous across cells). If a linear regression model with 0/1 coding is used as in Piantadosi (Piantadosi, 2005, pp. 508–509) then the following model is fit:

$$\beta_0 + \beta_1 A_1 + \beta_2 A_2 + \beta_{12} A_1 A_2. \tag{9.6}$$

In 0/1 coding, β_1, the coefficient of A_1, is a simple effect representing the comparison of the $(1,0)$ cell with the $(0,0)$ cell, i.e., $\beta_1 = \mu_{(1,0)} - \mu_{(0,0)} = b_1$, where $\mu_{(1,0)}$ is the population mean of Y in the $(1,0)$ cell, and so on. Now β_1 is estimated by $\hat{\beta}_1 = \bar{Y}_{(1,0)} - \bar{Y}_{(0,0)}$, where $\bar{Y}_{(1,0)}$ is the sample mean of Y in the $(1,0)$ cell, and so on. Clearly, $E(\hat{\beta}_1) = b_1$, and

$$Var(\hat{\beta}_1) = Var(\bar{Y}_{(1,0)}) + Var(\bar{Y}_{(0,0)}) = \frac{\sigma^2}{r} + \frac{\sigma^2}{r} = \frac{2\sigma^2}{r}.$$

So the signal-to-noise ratio (SNR), or standardized effect size, governing the power to screen A_1 with 0/1 coding in the analysis model (i.e., basing the test statistics on simple effects) is

$$SNR_{[0/1]} = \frac{|E(\hat{\beta}_1)|}{\sqrt{\mathrm{Var}(\hat{\beta}_1)}} = \frac{|b_1|\sqrt{r}}{\sqrt{2}\sigma}.$$

On the other hand, if we use the analysis model (9.6) with $-1/1$ coding, it follows that

$$\begin{aligned}
\beta_1 &= \frac{1}{4}\left[(\mu_{(+,+)} - \mu_{(-,+)}) + (\mu_{(+,-)} - \mu_{(-,-)})\right] \\
&= \frac{1}{2} \times (\text{the main effect of } A_1) \\
&= \frac{1}{2} \times (\text{the average of 2 simple effects}),
\end{aligned}$$

and is estimated by the sample version $\hat{\beta}_1$ (where μ is replaced by $\bar{Y}$). Then,

$$E(\hat{\beta}_1) = \beta_1 = \left(\frac{b_1}{2} + \frac{b_{12}}{4}\right),$$

$$Var(\hat{\beta}_1) = \frac{1}{4} \times \frac{1}{2} \times \text{(variance of an estimated simple effect)}$$

$$= \frac{1}{4} \times \frac{1}{2} \times \frac{2\sigma^2}{r} = \frac{\sigma^2}{4r}.$$

So the signal-to-noise ratio governing the power to screen A_1 with $-1/1$ coding in the analysis model (i.e., basing the test statistics on main effects) is

$$SNR_{[-1/1]} = \frac{|E(\hat{\beta}_1)|}{\sqrt{Var(\hat{\beta}_1)}} = \left|b_1 + \frac{b_{12}}{2}\right| \frac{\sqrt{r}}{\sigma}.$$

A measure of relative efficiency of the two coding schemes (equivalently, two ways of forming the test statistics) in screening A_1 is given by

$$\eta = \frac{SNR_{[-1/1]}}{SNR_{[0/1]}} = \sqrt{2}\left|b_1 + \frac{b_{12}}{2}\right| / |b_1| = \sqrt{2}\left|1 + \frac{b_{12}}{2b_1}\right|.$$

In absence of an interaction (i.e., $b_{12} = 0$), $\eta = \sqrt{2} > 1$, and hence the $-1/1$ coding gives higher power for screening components. In case of synergistic interaction (i.e., b_1 and b_{12} are of same sign), η is even larger, so the $-1/1$ coding gives higher power. Even in case of antagonistic interaction (i.e., b_1 and b_{12} are of opposite sign), the $-1/1$ coding gives higher power in screening components (i.e., $\eta > 1$) if $b_1 < 0$ and $0 < b_{12} < -(2+\sqrt{2})b_1$, or if $b_1 > 0$ and $0 > b_{12} > -(2 - \sqrt{2})b_1$. If we have k (≥ 2) components in a factorial experiment, and there may be a two-way but no higher-order interaction in the true data-generating model, then the relative efficiency of the two coding schemes is given by

$$\eta = 2^{(k-1)/2}\left|1 + \frac{b_{12}}{2b_1}\right|,$$

which is an increasing function of k; see Chakraborty et al. (2009) for a derivation of this.

To illustrate the power implications of basing the test statistics on main effects rather than simple effects in a regression analysis, we consider a small simulation study originally presented in Chakraborty et al. (2009). The data-generating model is $Y|A_1, A_2 \sim N(\mu = b_0 + b_1A_1 + b_2A_2 + b_{12}A_1A_2, \sigma = 1)$, where A_1, A_2 are coded in $0/1$. That is, the data-generating model is specified in terms of simple effects (as is usually conceptualized by an investigator). The coefficients b_1, b_2 are set according to Cohen's (1988) small or medium effect size (i.e., $b_1 = b_2 = 0.2, 0.5$). The coefficient b_{12} of the interaction term is varied: $b_{12} = b_1, b_1/2, 0, -b_1/2, -b_1$ (i.e., same size and sign as b_1, half the size of and same sign as b_1, absent, half the size of b_1 but of opposite sign, same size as b_1 but of opposite sign). A 0.05 level of significance is used throughout, while varying the sample size: $n = 100, 200, 500$. The goal of this simulation is to illustrate that even when the data-generating model is specified in terms of simple effects, basing the test statistics on main effects (using $-1/1$ coding) leads to higher power in most settings than basing the test statistics on simple effects (using $0/1$ coding). Note that the signal-to-noise ratios of the coding schemes govern the corresponding powers. In the following, we consider the power to screen A_1 both in presence and absence of an interaction term A_1A_2 (synergistic as well as antagonistic). Table 9.2 contains a Monte Carlo estimate (using 1000 iterations) of the power for screening A_1 under different scenarios.

TABLE 9.2
Power to screen A_1 in absence and presence of an interaction.

n	Interaction size (b_{12})	Interaction type	$b_1 = 0.2$ Analysis model in 0/1 coding	$b_1 = 0.2$ Analysis model in $-1/1$ coding	$b_1 = 0.5$ Analysis model in 0/1 coding	$b_1 = 0.5$ Analysis model in $-1/1$ coding
	Same ($= b_1$)	synergistic	0.1030	0.2910	0.4150	0.9550
	Half ($= \frac{b_1}{2}$)	synergistic	0.1030	0.2290	0.4150	0.8730
100	Absent ($= 0$)	none	0.1030	0.1720	0.4150	0.6830
	Half ($= -\frac{b_1}{2}$)	antagonistic	0.1030	0.1110	0.4150	0.4420
	Same ($= -b_1$)	antagonistic	0.1030	0.0820	0.4150	0.2290
	Same ($= b_1$)	synergistic	0.1690	0.5440	0.6920	1.0000
	Half ($= \frac{b_1}{2}$)	synergistic	0.1690	0.3940	0.6920	0.9870
200	Absent ($= 0$)	none	0.1690	0.2840	0.6920	0.9430
	Half ($= -\frac{b_1}{2}$)	antagonistic	0.1690	0.1720	0.6920	0.7510
	Same ($= -b_1$)	antagonistic	0.1690	0.1040	0.6920	0.3940
	Same ($= b_1$)	synergistic	0.3460	0.9210	0.9740	1.0000
	Half ($= \frac{b_1}{2}$)	synergistic	0.3460	0.8040	0.9740	1.0000
500	Absent ($= 0$)	none	0.3460	0.6050	0.9740	1.0000
	Half ($= -\frac{b_1}{2}$)	antagonistic	0.3460	0.3730	0.9740	0.9870
	Same ($= -b_1$)	antagonistic	0.3460	0.1890	0.9740	0.8040

Note that in Table 9.2, the power to screen A_1 is higher in general when the analysis model is coded in $-1/1$ compared to when the analysis model is coded in 0/1 (e.g. comparing the 4th vs. 5th column, and comparing the 6th vs. 7th column), except when the interaction is of same size and opposite in sign as the simple effect of A_1 (as expected from the above discussion). However according to the *Hierarchical Ordering Principle* (Wu and Hamada, 2000), interactions are usually of smaller order of magnitude than the main effects (absent strong scientific theory to the contrary), and hence this is a fairly unlikely scenario. A secondary point to note is that when the data-analysis model uses the $-1/1$ coding, there is a decrease in power to screen A_1 as the interaction term b_{12} decreases from highly synergistic to highly antagonistic (moving down the 5th and 7th columns). However, when the data-analysis model uses the 0/1 coding, the power for screening A_1 is independent of the size of the interaction term b_{12} (moving down the 4th and 6th columns). But the decrease in power in the 5th and 7th columns due to interaction often does not pose a serious threat (as compared to the loss of power from using 0/1 coding) if the goal is to screen components, since in most settings $-1/1$ coding gives better power anyway.

Power to detect interactions

Factorial designs are often criticized on the ground that the power to detect an interaction is much lower than the power to detect a main effect of the same size (Montgomery et al., 2003; Piantadosi, 2005). However, it is also recognized that factorial designs are the only experimental designs that can systematically investigate interactions. To overcome the low power for detecting interactions in a confirmatory (not screening) trial, the general recommendation in the literature (Byar and Piantadosi, 1985) is that if an interaction is strongly anticipated based on the investigator's prior knowledge, the study should be powered with larger sample size. When criticizing factorial designs on the ground of low power for interactions in the context of screening trials for developing multicomponent interventions, it is useful to consider the pros and cons of the possible alternatives. The natural alternative is to conduct non-experimental analyses using treatment adherence or other post-randomization outcomes as doses or factor levels from a randomized trial or to use observational data sets. As discussed previously, the relative merit of FFDs over the above strategy depends on the degree of confounding in the data. The crux is that the low power to detect interactions in a factorial design can be offset by its ability to perform valid estimation and inference, and its ability to control (by design) aliasing in a principled manner, in comparison to observational analyses.

Power and sample size in screening trials

In a screening trial using a factorial design, the power calculation used to size the trial focuses on main effects of each component. Thus the power calculation is similar to that of a two-arm randomized trial in that the two levels of a single component (averaged over the levels of all other components) serve as the two arms. Below we provide the power calculation for the *Project Quit* study as an example. For *Project Quit*, the planned initial recruitment size was 2000; this number was chosen to achieve a total sample size of 1500 for the analysis, anticipating a 75% response rate at the 6-month follow-up. Assuming no differential attrition across cells, this meant roughly 750 subjects per level of each intervention component. The primary outcome was binary, e.g., seven-day point-prevalence smoking cessation at the 6-month follow-up. So the power analysis involved binomial calculations (using a normal approximation) assuming a baseline average cessation rate of 10% found in a previous study (Dijkstra et al., 1998). For each main effect, the sample size of 750 per level provides approximately 80% power for detecting a 4.5% difference in cessation rates. The same power characteristics exist for each of the six components. Note that to achieve the same power to detect the same difference in cessation rates, one would need the same sample size in a usual two-arm study (so the sample size requirement is not increased by using a factorial design). The formula for calculating power in the present set-up is given by

$$\Phi\left(\frac{\sqrt{\frac{n}{2}}|\Delta| - z_{\alpha/2}\sqrt{2p(1-p)}}{\sqrt{p(1-p) + (p+\Delta)(1-p-\Delta)}}\right),$$

where n is the total sample size, p is the baseline cessation rate, Δ is the change in cessation rate to be detected, α is the Type I error rate, $z_{\alpha/2}$ is the upper $100(\frac{\alpha}{2})\%$ cutoff point of a standard normal distribution, and Φ is the standard normal distribution function. For more details on power calculation and a SAS macro, see Dziak et al. (2013).

Power and sample size for time-to-event analysis in factorial trials

For time-to-event analysis in 2×2 factorial trials, Slud (1994) derived theoretical expressions for power under a Cox proportional hazards model. On the other hand, Moser and Halabi (2015) considered time-to-event analysis in general factorial designs with any number of factors and any number of levels per factor, but assumed the time-to-event to follow an exponential distribution. They looked at the issues of power and sample size in more detail; specifically, they provided closed-form solutions for required sample size and study duration for given power and other relevant parameters when comparing hazard rates across main effects or interactions. They further extended their work to the setting of incomplete factorial designs (Byar et al., 1993).

9.8   Discussion

In this chapter, we have discussed the use of full and fractional factorial designs in various clinical research scenarios. Specifically, we have discussed their usage in classical confirmatory medication trials where such designs are used predominantly for efficiency, e.g., to study more than one medications simultaneously at the cost of running one trial. We have also discussed in great detail the role of factorial designs in developing multicomponent interventions that are becoming increasingly common in health sciences; in this context, it is critical to study interactions among factors.

In our discussion of FFDs, we assumed that third- and higher-order interactions are negligible (Box et al., 1978; Nair et al., 2008). This is not a binding constraint. Suppose prior knowledge suggests that interactions of up to order three involving a certain component are likely important, whereas even two-way interactions involving some other components are negligible. One can still use a carefully chosen FFD (Wu and Hamada, 2000).

Previously in Section 9.2.3, we have discussed some situations of clinical research where factorial designs are not applicable, e.g., when some of the factors cannot be crossed or when implementing all the cells is not feasible. In addition, one setting in which factorial designs are not particularly well suited is when the main effects of all the individual components are weak, but there are some high-order interactions in the data-generating model that produce a strong effect on the outcome (i.e., a setting where the *Hierarchical Ordering Principle* is violated).

Another important caveat regarding the use of factorial designs for developing multi-component interventions is the presence of nested components (e.g., levels of component B are nested within the levels of component A). Generalization of the usual factorial designs called *nested factorials* (Ankerman et al., 2003; Smith and Beverly, 1981) can incorporate nested components. Analysis of such designs can employ mixed-effects models (Searle et al., 2002), as in cluster randomized trials. A somewhat similar issue is when some intervention components are applied most naturally in a grouped setting. For example, some intervention components are provided to all patients at a clinic (Daunica et al., 2006) or to all children in a classroom or school (Flay and Collins, 2005). See Dziak et al. (2012) for a factorial-based experimental framework tailored to such settings. As pointed out by Couper et al. (2005), sometimes factorial designs are also criticized for slower recruitment rate (since subjects need to meet the inclusion criteria for all the factors) and potential lower compliance (due to a more complicated treatment protocol) than single-factor trials. However, these are common to any studies of multi-factor interventions, and not problems specific to factorial designs.

More recently, there has been some work that expanded the scope of factorial designs from their classical context. For example, Murphy and Bingham (2009) developed a framework of experimental designs for developing dynamic treatment regimes using concepts of factorial designs. Another important recent development is the work by Dasgupta et al. (2015) in which the authors have developed a framework for causal inference using potential outcomes, where the data come from 2^k factorial studies that are not randomized trials (e.g. quasi-experimental studies). These recent works provide new opportunities for research along these lines of thought.

Bibliography

Allore, H., M. Tinettia, T. Gill, and P. Peduzzi (2005). Experimental designs for multi-component interventions among persons with multifactorial geriatric syndromes. *Clinical Trials* 2(1), 13–21.

Ankerman, B., A. Aviles, and J. Pinheiro (2003). Optimal designs for mixed-effects models with two random nested factors. *Statistica Sinica 13*, 385–401.

Bluford, D., B. Sherry, and K. Scanlon (2007). Interventions to prevent or treat obesity in preschool children: A review of evaluated programs. *Obesity 15*, 1356–72.

Box, G. and N. Draper (1987). *Empirical Model-building and Response Surfaces*. New York: Wiley.

Box, G. and J. Hunter (1961). The 2^{k-p} fractional factorial designs. *Technometrics 3*, 311–351 and 449–458.

Box, G., W. Hunter, and J. Hunter (1978). *Statistics for Experimenters: An Introduction to Design, Data Analysis, and Model building*. New York: Wiley.

Byar, D., A. Herzberg, and W. Tan (1993). Incomplete factorial designs for randomized clinical trials. *Statistics in Medicine 12*, 1629–1641.

Byar, D. and S. Piantadosi (1985). Factorial designs for randomized clinical trials. *Cancer Treatment Reports 69*, 1055–1063.

Byar, D., R. Simon, W. Friedewald, J. Schlesselman, D. DeMets, J. Ellenberg, M. Gail, and J. Ware (1976). Randomized clinical trials — perspectives on some recent ideas. *New England Journal of Medicine 295*, 74–80.

Campbell, M., R. Fitzpatrick, A. Haines, A. Kinmonth, P. Sandercock, D. Spiegelhalter, and P. Tyrer (2000). Framework for design and evaluation of complex interventions to improve health. *British Medical Journal 321*, 694–696.

Chakraborty, B., L. Collins, V. Strecher, and S. Murphy (2009). Developing multicomponent interventions using fractional factorial designs. *Stat Med 28*(21), 2687–708.

Christen, W., J. Gaziano, and C. Hennekens (2000). Design of physicians' health study ii: A randomized trial of beta-carotene, vitamins e and c, and multivitamins, in prevention of cancer, cardiovascular disease, and eye disease, and review of results of completed trials. *Annals of Epidemiology 10*, 125–134.

Cohen, J. (1988). *Statistical Power for the Behavioral Sciences* (2nd ed.). Hillsdale, NJ: Erlbaum.

Collins, L., B. Chakraborty, S. Murphy, and V. Strecher (2009a). Comparison of a phased experimental approach and a single randomized clinical trial for developing multicomponent behavioral interventions. *Clinical Trials 6*, 5–15.

Collins, L., J. Dziak, and R. Li (2009b). Design of experiments with multiple independent variables: A resource management perspective on complete and reduced factorial designs. *Psychological Methods 14*, 202–224.

Collins, L., S. Murphy, V. Nair, and V. Strecher (2005). A strategy for optimizing and evaluating behavioral interventions. *Annals of Behavioral Medicine 30*, 65–73.

Couper, D., J. Hosking, R. Cisler, D. Gastfriend, and D. Kivlahan (2005). Factorial designs in clinical trials: Options for combination treatment studies. *Journal of Studies on Alcohol S15*, 24–32.

Cuffe, M. (2006). The patient with cardiovascular disease: treatment strategies for preventing major events. *Clinical Cardiology 29*, II4–12.

Dasgupta, T., N. Pillai, and D. Rubin (2015). Causal inference from 2^k factorial designs by using potential outcomes. *Journal of the Royal Statistical Society: Series B (Statistical Methodology)* 77(4), 727–753.

Daunica, A., S. Smitha, E. Branka, and R. Penfield (2006). Classroom-based cognitivebehavioral intervention to prevent aggression: Efficacy and social validity. *Journal of School Psychology* 44(2), 123–139.

Dijkstra, A., H. DeVries, J. Roijackers, and G. V. Breukelen (1998). Tailored interventions to communicate stage-matched information to smokers in different motivational stages. *Journal of Consulting and Clinical Psychology* 66(3), 549–557.

Dziak, J., L. Collins, and A. Wagner (2013). FactorialPowerPlan users' guide (Version 1.0). *University Park: The Methodology Center, Penn State. Retrieved from http://methodology.psu.edu.*

Dziak, J., I. Nahum-Shani, and L. Collins (2012). Multilevel factorial experiments for developing behavioral interventions: power, sample size, and resource considerations. *Psychological Methods* 17(2), 153–175.

Fisher, R. (1942). *The Design of Experiments, 3rd ed.* Edinburgh: Oliver & Boyd.

Flay, B. and L. Collins (2005). Historical review of school-based randomized trials for evaluating problem behavior prevention programs. *Annals of the American Academy of Political and Social Science* 599, 115–146.

Friedli, K. and M. King (1998). Psychological treatments and their evaluation. *International Review of Psychiatry* 10, 123 – 126.

Golin, C., J. Earp, H. Tien, P. Stewart, C. Porter, and L. Howie (2006). A 2-arm, randomized, controlled trial of a motivational interviewing-based intervention to improve adherence to antiretroviral therapy (art) among patients failing or initiating art. *Journal of Acquired Immune Deficiency Syndrome* 42, 42–51.

Green, S., P. Liu, and J. O'Sullivan (2002). Factorial design considerations. *Journal of Clinical Oncology* 20(16), 3424 – 3430.

Grimes, D. and K. Schulz (2002). An overview of clinical research: the lay of the land. *Lancet* 359, 57–61.

Hennekens, C. and K. Eberlein (1985). A randomized trial of aspirin and beta-carotene among u.s. physicians. *Preventive Medicine* 14, 165–168.

Holland, P. (1986). Statistics and causal inference. *Journal of the American Statistical Association* 81, 945–970.

Hosking, J., R. Cisler, D. Couper, D. Gastfriend, D. Kivlahan, and R. Anton (2005). Design and analysis of trials of combination therapies. *Journal of Studies on Alcohol S15*, 34–42.

Kazdin, A. (1986). The evaluation of psychotherapy: Research design and methodology. In S. Garfield and A. Bergin (Eds.), *Handbook of psychotherapy and behavior change* (3rd ed.)., New York, pp. 23 – 68. Wiley.

Lin, D., J. Gong, P. Gallo, P. Bunn, and D. Couper (2016). Simultaneous inference on treatment effects in survival studies with factorial designs. *Biometrics* 72(4), 1078 – 1085.

Meyer, R., D. Steinberg, and G. Box (1996). Follow-up designs to resolve confounding in multifactor experiments. *Technometrics 38*, 303–332.

Montgomery, A., T. Peters, and P. Little (2003). Design, analysis and presentation of factorial randomised controlled trials. *BMC Medical Research Methodology 3*(26).

Moser, B. and S. Halabi (2015). Sample size requirements and study duration for testing main effects and interactions in completely randomized factorial designs when time to event is the outcome. *Communications in Statistics - Theory and Methods 44*(2), 275–285.

Motolese, M., G. Muiesan, and A. Colombi (1975). Hypotensive effect of oxprenolol in mild to moderate hypertension: A multicentre controlled study. *European Journal of Clinical Pharmacology 8*(21–31).

Murphy, S. and D. Bingham (2009). Screening experiments for developing dynamic treatment regimes. *Journal of the American Statistical Association 104*(485), 391–408.

Myers, R. and D. Montgomery (1995). *Response Surface Methodology*. New York: Wiley.

Nair, V., V. Strecher, A. Fagerlin, P. Ubel, K. Resnicow, S. Murphy, R. Little, B. Chakraborty, and A. Zhang (2008). Screening experiments and fractional factorial designs in behavioral intervention research. *American Journal of Public Health 98*, 1354–1359.

Paul, G., S. Smith, D. Whitford, F. O'Kelly, and T. O'Dowd (2007). Development of a complex intervention to test the effectiveness of peer support in type 2 diabetes. *BMC Health Services Research 7*, 136.

Piantadosi, S. (2005). *Clinical Trials: A Methodologic Perspective*. New York: Wiley.

Riggs, N., P. Elfenbaum, and M. Pentz (2006). Parent program component analysis in a drug abuse prevention trial. *Journal of Adolescent Health 39*, 66–72.

Rosenbaum, P. and D. Rubin (1984). Reducing bias in observational studies using subclassification on the propensity score. *Journal of the American Statistical Association 79*, 516–524.

Rubin, D. (1974). Estimating causal effects of treatments in randomized and nonrandomized studies. *Journal of Educational Psychology 66*, 688–701.

Searle, S., G. Casella, and C. McCulloch (2002). *Variance Components*. New York: Wiley.

Slud, E. (1994). Analysis of factorial survival experiments. *Biometrics 50*(1), 25 – 38.

Smith, J. and J. Beverly (1981). The use and analysis of staggered nested factorial designs. *Journal of Quality Technology 13*, 166–173.

Stephenson, J. and J. Imrie (1998). Why do we need randomised controlled trials to assess behavioural interventions? *British Medical Journal 316*, 611 – 613.

Strecher, V., J. McClure, G. Alexander, B. Chakraborty, V. Nair, J. Konkel, S. Greene, L. Collins, C. Carlier, C. Wiese, R. Little, C. Pomerleau, and O. Pomerleau (2008). Web-based smoking cessation components and tailoring depth: Results of a randomized trial. *American Journal of Preventive Medicine 34*(5), 373 – 381.

The ATBC Cancer Prevention Study Group (1994). The alpha-tocopherol, beta-carotene lung cancer prevention study: Design, methods, participant characteristics, and compliance. *Annals of Epidemiology 4*, 1–10.

Tinetti, M., D. Baker, G. McAvary, and et al. (1994). A multifactorial intervention to reduce the risk of falling among elderly people living in the community. *New England Journal of Medicine 331*, 821–827.

Tinetti, M., G. McAvary, and E. Claus (1996). Does multiple risk factor reduction explain the reduction in fall rate in the yale ficsit trial? *American Journal of Epidemiology 144*, 389–399.

West, S. and L. Aiken (1997). Toward understanding individual effects in multicomponent prevention programs: Design and analysis strategies. In K. Bryant, M. Windle, and S. West (Eds.), *The Science of Prevention: Methodological Advances From Alcohol and Substance Use Research*, Washington, DC. American Psychological Association.

West, S., L. Aiken, and M. Todd (1993). Probing the effects of individual components in multiple component prevention programs. *American Journal of Community Psychology 21*(5), 571–605.

White, H. (1980). A heteroskedasticity-consistent covariance matrix estimator and a direct test for heteroskedasticity. *Econometrica 48*, 817–838.

Williams, J., M. Gerrity, T. Holsinger, S. Dobscha, B. Gaynes, and A. Dietrich (2007). Systematic review of multifaceted interventions to improve depression care. *General Hospital Psychiatry 29*, 91–116.

Wolchik, S., S. West, S. Westover, I. Sandler, A. Martin, J. Lustig, J. Tein, and J. Fisher (1993). The children of divorce intervention project: Outcome evaluation of an empirically based parenting program. *American Journal of Community Psychology 21*, 293–331.

Women's Health Initiative Study Group (1998). Design of the women's health initiative clinical trial and observational study. *Controlled Clinical Trials 19*, 61–109.

Wu, C. and M. Hamada (2000). *Experiments: Planning, Analysis, and Parameter Design Optimization*. New York: Wiley.

Yates, F. (1937). The design and analysis of factorial experiments. *Imperial Bureau of Soil Sciences - Technical communication No. 35*.

10

Cluster Randomized Designs

Martin Bland and Mona Kanaan

CONTENTS

10.1   What is a Cluster Randomized Trial?

In a cluster randomized trial, research participants are not sampled independently, but as a group or cluster. For example, we may allocate to treatments, not individual people, but primary care centres, hospital wards, schools, or patient intake over different weeks. In this chapter we shall explain why randomizing groups or clusters of participants rather than individuals leads to difficulties in design, execution, and analysis. We explain why clustering leads to problems of analysis and describe several different possible approaches, including the use of summary statistics for the cluster, robust standard errors, multilevel modeling, generalized estimating equations, and the stepped wedge design. We describe the need for an increased sample size and the role of the intracluster correlation coefficient in estimating this. Finally, we describe why cluster randomization leads to practical difficulties and possible biases in recruitment and data collection.

Why might we want to do a cluster randomized trial? One reason is that the intervention takes place at the level of a cluster. A typical example is a trial of clinical guidelines (2), where the intervention is given to a care provider, such as a physician. For example, The Early Labour Study in Scotland (TELSiS) was a trial of an algorithm for diagnosis of onset of labour in women having their first birth, a diagnosis which can be quite difficult. Fourteen delivery units in Scotland were allocated either to have all the midwives offered training in the new algorithm for diagnosis or to be left untreated, to provide care as usual (10). The proportions of women experiencing augmentation of labour were compared. The intervention was training given to midwives, but we were interested in the effect on women in labour.

It would be impossible for midwives trained in the algorithm to use it for some patients and to forget it and return to their previous practice for others. Educational interventions are often like this, where a whole class is taught together. In the Roots of Empathy trial, ongoing at the time of writing, 74 primary schools have been allocated to usual teaching or to the Roots of Empathy programme, which is taught to an entire school class, and the children in one Year 5 class (8–9 years old) form the cluster.

Another reason for allocating clusters is the possibility of contamination. Members of a cluster may influence other members of the cluster. Brooks et al. (7) reported a trial of incentives to improve attendance at adult literacy classes, where students in half the classes, chosen at random, were offered money part way through the course if they continued attending, the outcome variable being future attendance. If some students were offered an incentive and others in the same class were not, this might cause resentment in those who did not get the offer. In the CADET (Collaborative Depression Trial) trial of collaborative care for depression in primary care (32; 33), 50 UK general practices were allocated to either treatment as usual or collaborative care, where a mental health worker was attached to the practice to provide support for patients presenting with depression as part of their problem. The cluster was all patients presenting with depression to the same practice. This was done because a feasibility study (34) had found that the difference between collaborative care and usual care patients was smaller in individually randomized participants within the same practice than the difference between patients allocated to collaborative care and those in practices where all were allocated to care as usual. This was interpreted as suggesting that the presence of the collaborative care worker in the practice could influence the care of all depressed patients.

We might have action of an individual intervention upon other members of the cluster, where a successful intervention on an individual may affect surrounding people. For example, in a trial of the use of helminthicides in children infected with intestinal parasites (11), there was concern that children who have been treated are less likely to infect others, but that children who were treated but were then surrounded by infectious children were likely to be reinfected rapidly.

Another possible reason for cluster allocation is administrative convenience. In the trial of helminthicides in children (11), it was much easier to treat everyone in a school than to treat a randomly selected half.

10.2   The Problem of Clustering

What makes a cluster randomized trial different from individually randomized trials is that members of a cluster will be more like one another than they are like members of other clusters. We need to take this into account in both the design and analysis (12). Observations within the same cluster are correlated. Methods which ignore clustering may mislead, because they assume that all subjects are independent observations. This may lead to standard errors which are too small, confidence intervals which are too narrow, and P values which are too small.

A little simulation will illustrate this. Four random cluster means were generated, two in each of two groups, from a normal distribution with mean 10.0 and standard deviation 2.0. Then 10 members of each cluster were generated by adding a random number from a normal distribution with mean 0.0 and standard deviation 1.0. The null hypothesis that there is no difference between the means in the two populations from which these groups are drawn is true. We can carry out a two-sample t test comparing the means, ignoring the

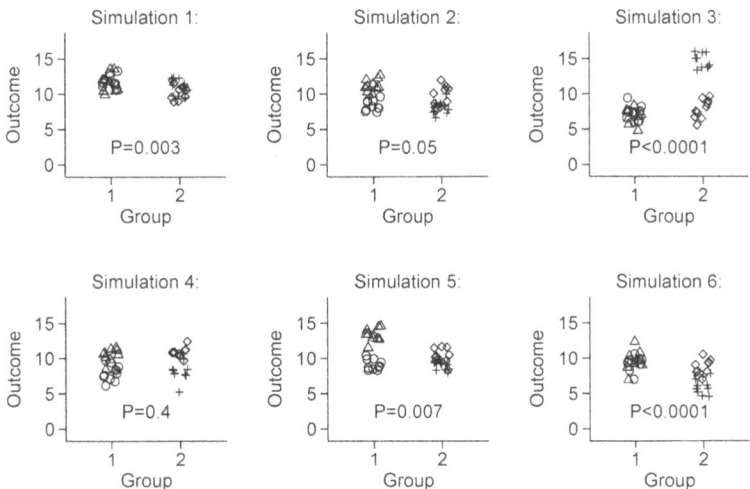

FIGURE 10.1
Six simulated studies with a large cluster effect but no real difference between groups.

clustering. If the t test were valid, this would give us a significant difference, $P< 0.05$, in one in 20 or 5% of possible samples.

Figure 10.1 shows the results of six runs of this simulation. Five of the six are statistically significant, though which group has the higher mean varies. When this was repeated 1 000 times, there were 600 significant differences, with $P< 0.05$, 60.0%. Of these, 502 were highly significant, with $P< 0.01$, 50.2%. If a t test ignoring the clustering were valid, we would expect 50 significant differences, 5%, and 10 highly significant ones, 1%.

What went wrong? The analysis using the two sample t test assumes that we have 20 independent observations in each group. This is not true. We have two independent clusters of observations, but the observations in those clusters are really the same thing repeated ten times.

One possible valid statistical analysis would be to find the means for the four clusters and to carry out a two-sample t test using these four means only. If we do this for the simulated data in Figure 10.1, the P values are 0.09, 0.6, 0.3, 0.8, 0.5, and 0.3. None of them are significant. When this was repeated over the 1 000 simulation runs, there were 53 (5.3%) tests which were significant at $P< 0.05$ and 14 (1.4%) which were highly significant at $P< 0.01$.

This simulation is very extreme, because there are only two groups of two clusters and a very large cluster effect. (One of us has actually reviewed a proposed study with two groups of two clusters.) A smaller cluster effect would reduce the shrinking of the P values, but it would not remove it. The simulation shows that spurious significant differences can occur if we ignore the clustering.

There are several approaches which can be used to analyse data randomized in clusters, including using a summary statistic for each cluster, robust variance estimates, multilevel modeling, generalized estimating equation (GEE) models, and Bayesian hierarchical models. Any method which takes into account the clustering should be an improvement compared to methods which do not. Bland (3) reported that up to 2003 the number of cluster randomized trials had been increasing and that there was much greater awareness of the importance

TABLE 10.1
Artificial rupture of membranes (ARM) in the TELSiS trial, ignoring clustering.

Group	ARM	No ARM	Total
Intervention	401 (45.0%)	491 (55.0%)	892
Control	500 (39.1%)	779 (60.9%)	1 279
Total	1 270	901	2,171

of taking the data structure into account in the analysis, something which had frequently been ignored in the past.

Similar problems arise in other types of research. In clinical studies, there are trials which treat body parts as the unit of analysis; limbs, eyes, teeth, warts, and moles have all had this treatment. For example, in sub-fertility trials we might have multiple eggs, multiple embryos, multiple implants per treatment cycle, or multiple treatment cycles per woman. The unit of analysis in such trials should be the woman, but we might have egg, embryo, implant, or treatment cycle as the unit of analysis. Vail and Gardener (40) reported that 32 out of 39 trials, 82%, did not use the woman as the unit of analysis when they should.

10.3　Summary Statistics

In the TELSiS trial of diagnosis of onset of labour, one of the secondary outcome variables was artificial rupture of membranes (ARM). If we tabulate this by treatment group, we get Table 10.1. We could carry out a test of significance on this table using a simple contingency table chi-squared test: $\chi^2 = 7.44, d.f. = 1, P = 0.006$. This test would not be valid, however, because it ignores the clustering.

Table 10.2 shows the percentage of women experiencing ARM in each hospital in TELSiS. Data are shown for the period leading up to randomization as well as the period following randomization and intervention. Table 10.1 shows the post-randomization data only.

One way to deal with clustering is to calculate a summary statistic for each cluster, such as the percentage of women who experienced ARM in each hospital. We can then compare the two groups using a two sample t test, as we did for the simulation in §10.2. If we do this, we get $t = -0.24, d.f. = 12, P = 0.8$, not significant and very different from the $P = 0.006$ given by the chi-squared test for Table 10.1. We can also get an estimate of the difference between the randomized groups and a confidence interval. The mean percentage of women with ARM is 41.1 in the Intervention Group and 39.9 in the Control Group, difference = 1.2 percentage points, 95% confidence interval = -9.7 to 12.0 percentage points. This analysis takes into account the variation between clusters in a way which the crude chi-squared test does not. It suggests that the chi-squared test was as misleading as the t test using individual observations in §10.2.

The t test analysis does not take into account the variation between the hospitals in the number of women giving birth. We can do this by carrying out a weighted analysis, weighting the percentages in the t test by the number of women in the hospital. The estimated weighted difference is 5.9 percentage points, 95% CI = -3.9 to 15.6, $t = 1.31$, $P = 0.2$. As this calculation takes the varying cluster size into account, it is preferable.

Figure 10.2 shows that, in this particular example, in the control group the hospital with the highest percentage is relatively small, whereas in the intervention group the two smallest percentages are from small hospitals. The small difference apparent in Figure 10.2

TABLE 10.2
Artificial rupture of membranes (ARM) in the TELSiS trial, by hospital.

Hospital code	Randomized group	Before randomization		After randomization	
		Number	% ARM	Number	% ARM
1	Control	201	44.8	200	40.5
2	Intervention	198	42.9	200	58.5
3	Control	199	40.2	199	41.7
4	Intervention	48	52.1	64	46.9
5	Control	199	56.8	200	45.5
6	Control	200	41.5	200	36.5
7	Intervention	83	15.7	56	25.0
8	Control	199	27.1	200	38.0
9	Intervention	202	44.6	199	48.2
10	Intervention	200	39.5	60	31.7
11	Control	197	26.4	200	28.5
12	Intervention	162	38.3	200	43.5
13	Control	96	43.8	80	48.8
14	Intervention	136	22.1	113	34.5

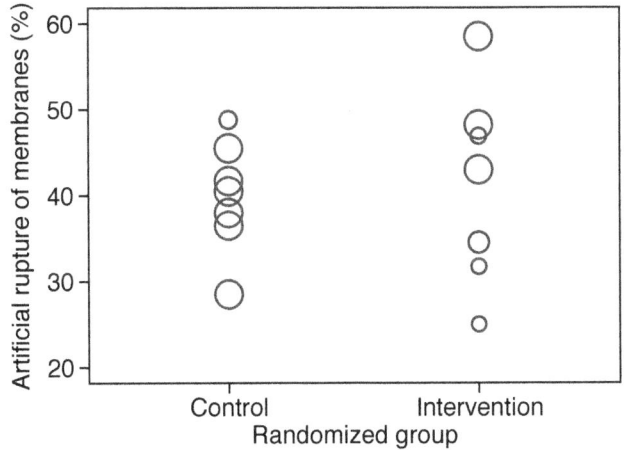

FIGURE 10.2
Percentage of women per hospital with ARM in TELSiS, by randomized group. (Area of symbol is proportional to the number of women.)

is minimized if all the centres are given equal weight. Weighting by hospital size is an improvement.

This analysis always works, for any number of clusters, and the analysis is valid, provided the assumptions of the t test are met. However, as we shall see, it is limited in its ability to accommodate adjustment for other sources of variation.

10.4   The Intra-Cluster Correlation Coefficient and the Design Effect

Intra-cluster correlation coefficient or ICC is the correlation between pairs of subjects where each pair is chosen at random from different cluster. To see how this works, first consider a set of pairs of observations such as pulse rate made on a group of people, one pair of pulses per person. We can calculate a correlation coefficient between the two pulse measurements, taking the first measurement as Variable X and the second measurement as Variable Y. This is often done in the study of measurement error (5). Suppose we now reverse the order of some of the pairs, say for all the females. We will get a slightly different value for the correlation coefficient. There are many possible ways we could order the pairs, so many possible correlation coefficients. The order of measurement should not be important. It should not matter which observation came first. We need a central value. One possibility is the intra-class correlation coefficient, also denoted by ICC.

We can split the variance of the pulse measurement into two parts:

$$\sigma^2 = \sigma_b^2 + \sigma_w^2$$

where σ_b^2 = variance between subjects "true" or average values over all possible measurements and σ_w^2 = variance of repeated measurements on the same subject, the measurement error, which we shall assume to be the same for everybody. Then the expected value of the sample correlation coefficient, which must be positive, is

$$\rho = \frac{\sigma_b^2}{\sigma_b^2 + \sigma_w^2}$$

We can estimate σ_b^2 and σ_w^2 from the data and use these to estimate the correlation coefficient. This estimate, the ICC, does not depend on the order of the pairs. The people who provide the pairs of measurements are the classes.

Given this formulation, ICC can be used for any number of repeated observations within the class. It estimates the correlation between pairs of observations, each pair drawn from a different class. Early applications were for twin studies, for litters of pigs, etc. Thus we can apply it to a set of data where clusters form the classes. The ICC measures the correlation between pairs of observations, where each pair is drawn from a different cluster. In this application we call it the intra-cluster correlation coefficient, also denoted by ICC. The ICC is also the proportion of the variance of observations, all made from different clusters, which is due to membership of a cluster.

We can estimate σ_b^2 and σ_w^2 easily from an analysis of variance table. Table 10.3 shows this for the TELSiS ARM data, using the observations before randomization, where no intervention has been applied to increase variability. The outcome variable in Table 10.3 is binary, 1 for ARM and 0 for no ARM, so we have not given the F test for the variance ratio. We require only the mean squares, which provide estimates of the variances.

The mean square within hospitals provides a simple estimate of σ_w^2, 0.2372513. The mean

TABLE 10.3
One way analysis of variance for ARM before intervention, variable = 1 if ARM, 0 if not ARM.

Source	Sum of squares	D.f.	Mean square	Variance ratio
Between hospitals	23.32086	13	1.7939124	7.85
Within hospitals	526.86491	2306	0.2284757	
Total	550.18578	2319	0.2372513	

square between hospitals is more complicated and includes both the variability within and between hospitals. If we had the same number of observations, m, for each hospital it would be easy, it would estimate $m\sigma_b^2 + \sigma_w^2$. If the numbers within the n hospitals vary, say m_i for the ith hospital, we need to calculate

$$m_0 = \frac{1}{n-1}\left(\Sigma m_i - \frac{\Sigma m_i^2}{\Sigma m_i}\right)$$

then the mean square between hospitals would estimate $m_0\sigma_b^2 + \sigma_w^2$. Hence we get estimates of the two variances and hence estimate the ICC. Here $m_0 = 164.49703$ so our estimate of σ_b^2 is

$$\frac{1.7939124 - 0.2284757}{164.4970} = 0.00951651$$

Should this produce a negative estimate, we set it to zero, the minimum possible value. The ICC is given by

$$\rho = \frac{0.00951651}{0.00951651 + 0.2284757} = 0.03998664$$

which we can round to 0.040. This might seem like a very small correlation coefficient, but it is typical of cluster randomized trials in medicine.

The ratio of the variance obtained from a specific sampling scheme, such as sampling in clusters, to the variance which would be obtained by a simple random sample of the same total size drawn from the same population is called the design effect of the sample, often denoted by *Deff*. For a cluster sample with m units per cluster, $Deff = 1 + (m-1)\rho$. If we have one unit per cluster, $Deff = 1$, and this would be the same as a simple random sample. If the numbers in the clusters vary, we replace m by

$$\tilde{m} = \frac{\Sigma m_i^2}{\Sigma m_i}$$

It is easy to see that if all the m_i have the same value, m, then $\tilde{m} = m$. It is possible to show that, if the clusters vary in size, $\tilde{m}$ is always greater than the mean cluster size and that, for a given total sample size, the design effect is a minimum when the clusters are of equal size.

If we estimate a sample size for an individually randomized trial, as described in Chapter 12, it would be too small if the trial were cluster-randomized. We can estimate the size of a cluster randomized trial by calculating the size as if the trial were individually randomized and then multiplying this by the design effect. For example, in CADET the primary outcome measure was depression reported on the PHQ-9 depression scale, which gives an integer score between 0 and 27. This was to be treated as a quantitative measure and mean scores would be compared. The trial was designed to detect an effect size of 0.4 standard deviations. With 90% power and significance level alpha = 0.05 we would require 132 patients per group in a two armed individually randomized trial. The proposed cluster-randomized trial would

TABLE 10.4
The effect of increasing cluster size in TELSiS.

Cluster size	Number of women per group	Design effect	Effective sample size per group
100	600	5.059	119
200	1200	9.159	131
300	1800	13.259	136
400	2400	17.359	138

have 14 patients per cluster. In a feasibility study (34) we had found ICC = 0.06, so the design effect would be $1 + (14 - 1) \times 0.06 = 1.65$. The required cluster trial sample size is, therefore, $132 \times 1.65 = 220$ per group. In order to follow up 220 patients, we decided to try to recruit 275 patients per group to anticipate a loss to follow up of 20% (33).

In the TELSiS trial, clusters would be maternity hospitals and the primary outcome variable would be whether women had augmentation of labour using oxytocin. We had data on the numbers of women for whom this had happened in all the eligible hospitals and from these estimated ICC = 0.041. The planned difference sought was 10 percentage points and, from the pilot data, the control proportion = 34%. To detect a difference between 34% and 24% with power 0.90 would need 431 women per group in an individually randomized study. Suppose we recruit 100 women per cluster. Then $Deff = 1 + (m - 1) \times ICC = 1 + (100 - 1) \times 0.041 = 5.059$. An individually randomized trial would need 431 women per group. We would need $431 \times 5.059 = 2180$ women $= 22$ hospitals per group. There were only 17 maternity hospitals in Scotland, we hoped to recruit at least 12 to TELSiS. The study would not be big enough. If we did a study with 12 clusters of 100 women each, what would be the effective sample size, i.e. the size of an individually allocated study of the same power? We can divide our cluster trial size by the design effect to get this. For 12 hospitals we would have 600 women per group, with effective sample size $= 600/5.059 = 119$. The difference detectable would be a reduction from 34% to 16%, much bigger than we anticipated.

Could we use bigger clusters? Table 10.4 shows the effect of increasing cluster size on the design effect and effective sample size for a 12-hospital trial. We were never going to reach 431 women per group and decided that we must redesign the trial.

10.5   Baseline and Other Adjustments

Hospitals vary in their usual clinical practice and it was hoped that in TELSiS we could improve the power of the study by adjusting for a baseline variable which measured this. We opted for the proportion of women observed to have the same event, whatever it was, before the intervention or before an equivalent time in the control group. (The women providing the before and after data were different, of course; this is a cluster level variable.) For ARM, these are also shown in Table 10.2. Figure 10.3 shows the relationship between the percentage ARM pre-intervention and post-intervention; clearly there is a hospital effect.

We can adjust for the baseline level of ARM by regression of post-intervention ARM on pre-intervention ARM and intervention group. We weighted this by total number of women

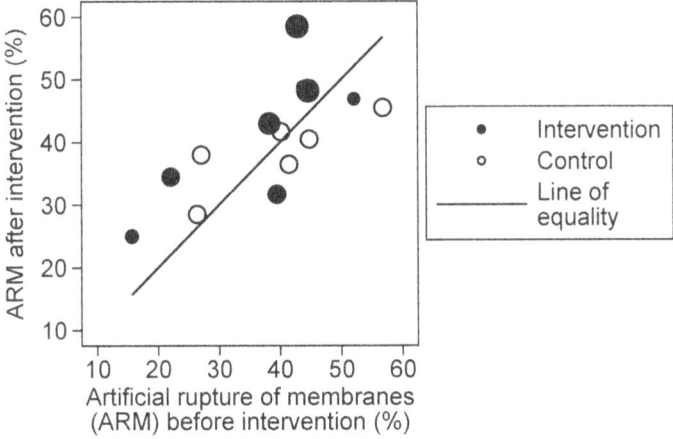

FIGURE 10.3
Percentage of women per hospital with ARM in TELSiS, post-intervention against pre-intervention, by randomized group. (Area of symbol is proportional to the total number of women.)

for the hospital. For the pre-intervention percentage the regression coefficient was 0.56 (95% CI 0.17 to 0.94, P = 0.008) and for the intervention 5.53 (−2.23 to 13.29, P = 0.1).

There are limitations to this kind of adjustment at the level of the cluster. Even two predictor variables, baseline and treatment, is a lot for multiple regression with only 14 observations. We could not include another variable. Also, percentage with ARM pre-treatment is a cluster level variable, we do not have the same women before and after. We could deal with a woman-level variable by computing the cluster summary statistic for it and using that, but we might lose much of the information. We need one of the individual-level methods of analysis described here. However, these all need more clusters than TELSiS provides.

10.6 Robust Standard Errors

The CADET trial of collaborative care of depression in primary care, where a mental health worker was attached to a general practice to support patients presenting with depression, had 50 clusters allocated to two groups. Patients identified as presenting with depression were recruited and asked to complete the PHQ-9 depression scale. It was thought that the baseline depression score would be a predictor of the outcome score, with which we would adjust the estimated treatment effect on PHQ-9.

We can carry out the same analyses using summary statistics for CADET as were done for TELSiS. Figure 10.4 shows mean PHQ-9 after four month against mean at baseline, for participants for whom there was a PHQ-9 after four months. There is a noticeable fall in both groups, a positive association between PHQ-9 after four months and at baseline, and a slightly lower mean score for the collaborative care group. If we carry out a weighted t test comparing treatment groups, we get difference in mean PHQ-9, collaborative care minus treatment as usual, = −1.58 (95% CI −2.82 to −0.33, P = 0.014). Hence we had evidence that collaborative care reduces depression score. If we adjust this for the mean baseline

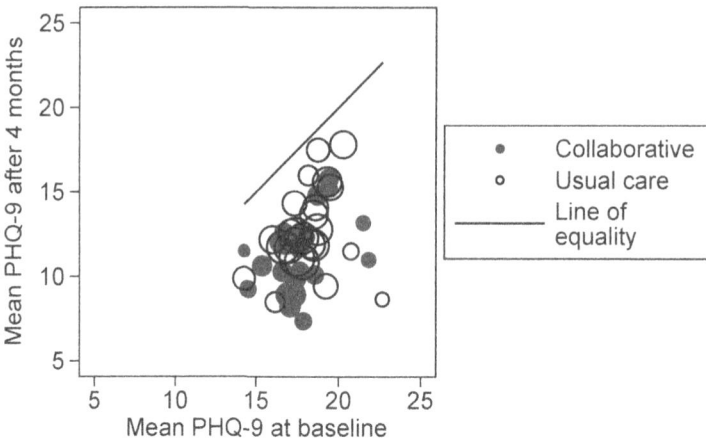

FIGURE 10.4
Mean PHQ-9 depression score per practice in CADET, score after 4 months against baseline, by randomized group. (Area of symbol is proportional to the total number of participants observed at 4 months.)

PHQ-9 for the cluster, we get a significant effect of baseline, the adjusted difference is -1.31 (95% CI -2.47 to -0.15, P = 0.028), the coefficient for baseline PHQ-9 being 0.63 (95% CI 0.22 to 1.03, P = 0.003). The estimate treatment effect is reduced slightly, with a slightly narrower confidence interval. It remains statistically significant, but with a larger P value.

Baseline PHQ-9 score is measured on the same people as the PHQ-9 at four months, so we would like to adjust at the level of the individual participant, if we can. One method to do this makes use of the Huber-White-sandwich estimator, which uses robust standard errors to adjust the standard error for the estimate while keeping the estimate itself the same. The application was suggested by Williams (43). If we carry out regression of PHQ-9 at four months on treatment group, with practice defined as the cluster, the estimated difference is -1.58, as before, with 95% CI -2.80 to -0.35, P = 0.013). If we carry out the incorrect individual data two sample test, ignoring the clustering, we get difference = -1.58 (95% CI -2.81 to -0.34, P = 0.013), so the only effect of allowing for clustering in this way is to widen the confidence interval slightly.

If we adjust for individual baseline PHQ-9 using the robust standard error approach, we get adjusted difference = -1.32 (95% CI -2.39 to -0.25, P = 0.017), the coefficient for baseline PHQ-9 being 0.60 (95% CI 0.47 to 0.72, P < 0.001). Being analysed at the level the data were collected, the person, the relationship between PHQ-9 after four months and at baseline is better estimated here than in the summary statistics analysis, but is remarkably similar.

In CADET, practices were allocated by minimization (§10.11) using the practice list size, number of full time equivalent GPs, and index of multiple deprivation (IMD) for the geographical location of the practice, and this was done separately for each of three cities. The primary analysis for the trial therefore included these variables as predictors. If we stratify or minimise the allocation, it should be because we think that these variables have some effect on the outcome. In practice, list size and full time equivalents were highly correlated, so full time equivalents was not included in the regression. This left us with five cluster-level predictors, intervention group, list size, IMD, and two variables for the three cities, which seemed acceptable for 50 clusters. It was also thought that age of the participant might affect the PHQ-9 score, younger people possibly being quicker to recover from depression. This was therefore included in the regression, giving two participant-level variables: baseline PHQ-9 and age. The estimated effect of collaborative care after this adjustment was virtually unchanged at -1.33 (95% CI -2.31 to -0.35, P $= 0.009$). List size was also a significant predictor, larger practices having lower PHQ-9 in their patients, though by only 0.18 PHQ-9 points per 1 000 patients, suggesting that larger practices may be slightly more effective in managing depression.

IMD had a negative coefficient, not significant in this analysis but significant at the 12 month follow-up. This meant that high deprivation was associated with lower depression scores, which several collaborators found counterintuitive. A possible explanation is that if depression has been provoked by adverse material circumstances, it may be easier to recover than when depression arises without such external causes.

In CADET, the ICC was small and the cluster sizes were small, hence the design effect was small and so the robust standard error confidence interval was very similar to the incorrect estimate ignoring clustering. In the trial of financial incentives in education (7), members of the cluster were in the same adult literacy class and so were interacting directly. This increased the ICC considerably, so the design effect was substantial, despite the fairly small clusters. This trial had 152 participants in 28 clusters. The data were discussed in detail by Bland (4). The mean number of sessions attended post incentive was 5.28 for the incentive group and 6.69 for the control group. A crude t test, ignoring the clustering, gave -1.41 (95% CI -2.40 to -0.41, P $= 0.006$), which was potentially highly misleading. A weighted t test using cluster means gave -1.41 (95% CI -3.10 to 0.29, P $= 0.1$). Analysis using robust standard errors produced a difference $= -1.41$ (95% CI -3.13 to 0.32, P $= 0.1$). As in TELSiS, an analysis which takes the structure of the data into account has produced no significant difference when the incorrect analysis was significant. Unlike TELSiS, this study has an individual level covariate, a literacy score at randomization. We might hypothesise that students with higher scores may continue with the class for a shorter time, as they will achieve satisfactory literacy sooner. If we add this score as a second predictor, we get the estimated adjusted difference $= -1.53$ (95% CI -2.79 to -0.028, P $= 0.019$). The coefficient for literacy score was -0.049 (95% CI -0.071 to -0.028), a negative coefficient, as hypothesised. The main finding is the rather counter-intuitive one that a financial incentive discouraged students from attending.

Robust standard errors are a large sample approximation which breaks down when the number of clusters is small. A guideline for its validity is 30 or more clusters. (Brooks et al. (7) considered 28 clusters to be close enough; we agree.) As the number of clusters falls below this, the modification of the standard error becomes less good and the standard errors tend to be too small, producing confidence intervals which are too narrow and P values which are too small.

10.7 Multilevel Modeling

Summary statistics methods are very straightforward and easy to use, robust standard errors are easy to use provided we have enough clusters and have access to suitable software. They are both rather limited in the extension to more complex designs which might be found in cluster-randomized trials. For example, we might have more than one level of clustering, such as children within school classes and classes within schools. We might have a factorial trial where we allocate schools to one treatment comparison and within schools we allocate classes to another comparison. Such designs are very rare in medical applications and are more likely to be found in educational research, the area where a lot of development of multilevel methodology took place. Multilevel modeling (20) is a good approach for such designs, and we can also use this method for a simpler trial, such as CADET.

In a multilevel model, we define different levels for the data. In CADET, the individual trial participants form the first level of the data, the practices form the second, higher level of the data. The simplest multilevel model is:

$$y_{ik} = \mu + \beta_1 x_{1,i} + u_i + e_{ik} \tag{10.1}$$

where y_{ik} is the observed outcome variable for participant k in cluster i, μ is the overall mean, β_1 is the coefficient for the intervention group variable x_1, $x_{1,i}$ is the value of x_1 for the ith cluster, u_i is a random variable with mean zero and variance σ_b^2 representing the variation between clusters, and e_{ik} is a random variable with mean zero and variance σ_w^2 for each i, representing the variation between participants within cluster i. The model is that the clusters vary in the mean y_{ik} within them, but the variability is the same.

If we fit this model for the CADET data, the estimates are $\hat{\beta}_1 = -1.58$ (95% CI -2.81 to -0.34, P $= 0.012$), $\hat{\sigma}_b = 7.20 \times 10^{-08}$, $\hat{\sigma}_w = 7.03$. The cluster effect for CADET was vanishingly small, hence the minute $\hat{\sigma}_b$. The same model fitted for the incentives trial produced $\hat{\beta}_1 = -1.40$ (95% CI -3.14 to 0.17, P $= 0.078$), $\hat{\sigma}_b = 1.94$, $\hat{\sigma}_w = 2.49$. We can estimate the ICC from these estimates by

$$\hat{\rho} = \frac{\hat{\sigma}_b^2}{\hat{\sigma}_b^2 + \hat{\sigma}_w^2} = \frac{1.94^2}{1.94^2 + 2.49^2} = 0.38$$

which is a very large ICC.

We can add covariates measured at the level of the cluster, such as list size, IMD, and the three-level factor city in CADET. The model becomes

$$y_{ik} = \mu + \beta_1 x_{1,i} + \beta_2 x_{2,i} + \beta_3 x_{3,i} + \beta_4 x_{4,i} + \beta_5 x_{5,i} + u_i + e_{ik} \tag{10.2}$$

where $x_{2,i}$ represents list size, $x_{3,i}$ is IMD, and $x_{4,i}$ and $x_{5,i}$ are indicator variables for the second and third cities. These are all fixed effects and fitted with one value for all clusters. We can also add the two participant-level variables, baseline depression score and age, which we will denote by x_6 and x_7. As each participant has their own values for these, the observed values for the kth person in the ith cluster will be $x_{6,ik}$ and $x_{7,ik}$. The model becomes:

$$y_{ik} = \mu + \beta_1 x_{1,i} + \beta_2 x_{2,i} + \beta_3 x_{3,i} + \beta_4 x_{4,i} + \beta_5 x_{5,i} + \beta_6 x_{6,ik} + \beta_7 x_{7,ik} + u_i + e_{ik} \tag{10.3}$$

The effects are still fixed, with only a single parameter for baseline PHQ-9 and a single parameter for age. If we fit this model, the estimated value of β_1, the treatment effect, is $\hat{\beta}_1 = -1.33$ (95% CI -2.44 to -0.22, P $= 0.019$) and the error standard deviations are $\hat{\sigma}_b = 3.33 \times 10^{-11}$ and $\hat{\sigma}_w = 6.26$. The estimate is similar to that without the covariates and the errors are slightly smaller, suggesting that the covariates have explained some of the variation between clusters and between people within the clusters.

If we add the one available individual-level covariate, literacy score, to the model for the incentives trial, $\hat{\beta}_1 = -1.56$ (95% CI -2.87 to -0.25, P $= 0.020$), $\hat{\sigma}_b = 1.37$, $\hat{\sigma}_w = 2.53$. We can estimate the ICC adjusted for literacy score from these estimates by

$$\hat{\rho} = \frac{\hat{\sigma}_b^2}{\hat{\sigma}_b^2 + \hat{\sigma}_w^2} = \frac{1.37^2}{1.37^2 + 2.53^2} = 0.23$$

which is a large ICC, but less than the unadjusted value. The adjustment has removed some of the variation between groups.

The estimates from the multilevel analyses are similar to those obtained by the robust standard errors. We need quite a lot of clusters for multilevel modeling, 40 is sometimes recommended. Maas and Hox (27) recommended 50. The incentives trial, with 28 clusters, is just about big enough to meet the guideline for robust standard errors, a bit too small for multilevel modeling.

Similar approaches in the Bayesian framework are described by Spiegelhalter (36) and Turner et al. (39).

10.8   Generalized Estimating Equations (GEE) Models

Generalized or general estimating equation models were originally designed for analysing longitudinal or panel data (13; 26), when we have a group of participants whom we follow over time. They can also be used for analysing cluster-randomized trials and make use of robust standard errors as part of the method. They are called marginal models, though for the application to cluster randomized trials this is not particularly important. In the marginal model the treatment effect is the expected value of the difference between the outcomes for the trial treatment and for the comparison treatment for the average cluster. We model the marginal expectation, the average response for observations sharing the same values of covariates (such as baseline depression or literacy score) across all clusters. In conditional models, such as the multilevel model or regression model with robust standard errors described above, the treatment effect is the expected value of the difference between the response of participants in the cluster to the given treatment and to the opposite treatment. If the covariate values were different in the two treatment groups, we might get substantially different estimates for multilevel and GEE models. In a cluster-randomized trial, this should not be the case, so we should get similar answers.

When fitting a GEE model, we first have to choose an appropriate correlation structure, describing the correlation within a cluster. In a cluster-randomized trial, it is reasonable to choose what is called an exchangeable correlation matrix, where all the correlations are assumed to be the same. This might not be the case in the original application of GEE models where we would be following people with multiple observations made over time. We are not using this potential advantage of the GEE model.

The way a GEE model works is that we fit a standard regression model, then estimate residuals from this regression model, the differences between the observed outcome and the value predicted by the regression. We use the empirical distribution of these residuals to estimate the correlation parameters. We refit the model incorporating the correlation matrix estimated. We re-estimate residuals and repeat until the estimates converge. Of course, the computer program does all this for us, we just sit and watch.

For a simple GEE model applied to the CADET data, where only the treatment group is included and an exchangeable correlation matrix is assumed, the estimates are $\hat{\beta}_1 = -1.57$ (95% CI -2.76 to -0.39, P $= 0.009$). The confidence interval is very slightly narrower than that given by the multilevel model. If we include all the covariates, both cluster-level and individual-level, we get $\hat{\beta}_1 = -1.28$ (95% CI -2.19 to -0.36, P $= 0.006$). The confidence interval is slightly narrower than the corresponding multilevel model.

If we apply the simple GEE model to the incentives trial data, only the treatment group included, the estimated effect is $\hat{\beta}_1 = -1.48$ (95% CI -3.00 to 0.04, P $= 0.057$). Here, too, the confidence interval is slightly narrower than that given by the multilevel model. If we include the covariate, literacy score, we get $\hat{\beta}_1 = -1.56$ (95% CI -2.80 to -0.33, P $= 0.013$). Again, the confidence interval is slightly narrower than the corresponding multilevel model.

Like the multilevel and robust standard errors methods, GEE models need a substantial amount of data for reliable estimation. Twenty clusters per treatment is recommended. Bland (4) reported a simulation study of a trial having only 14 clusters per intervention group, like the incentives trial. A simulation of 5000 cluster randomized trials with two groups of 14 clusters of varying sizes, mean cluster size = 5.4, standard deviation 1.5 as in the incentives trial, and ICC = 0.33, and where the null hypothesis was true, produced 7.9% of trials with significant treatment effect in a GEE model, which was highly significantly greater than the 5% we should have, P< 0.0001. This also happened with 6.9% of multilevel models, P< 0.0001 compared to the nominal 5%. Robust standard errors, for which the number of clusters was close to the recommended 30, performed better. Only 5.4% of trials gave significant treatment differences, P = 0.2 compared to 5%. Weighted summary statistics also produced 5.4% of trials with significant differences.

GEE models do not extend easily to more than two levels, but this is not going to be a problem in most cluster randomized trials. They do not give estimates of the between-clusters and within-cluster variances, either, so if these are important, for example to estimate an adjusted ICC, a multilevel model might be preferred. Omar and Thompson (30) compare GEE and multilevel models with analyses by summary statistics for a cluster randomized trial with binary outcome data.

10.9   Stepped Wedge Designs

In the above, cluster randomized controlled trials where clusters were randomized simultaneously to receive one of two (or more) interventions were discussed. In some instances, however, it is desirable that all clusters being randomized receive the intervention by the end of the study. This situation could arise where it is believed that the intervention will do more good than harm or where a public policy intervention is being rolled-out before effectiveness has been demonstrated. In such cases, one could employ the "stepped wedge" randomized controlled trial design. For example, Pearson *et al.* (31), used the stepped wedge design to evaluate a Risk-Need-Responsivity probation supervision program, Citizenship, using six office units under the supervision of the Teesside (UK) probation area. It was considered not ethical to withhold an intervention that had been shown to be potentially effective in an observational study.

In the stepped wedge design, the intervention is introduced to all clusters in random order over a number of sequential time points or steps, as they are sometimes known. At each step, one or more randomly selected clusters receive the intervention. Typically, the study measurements are then taken for all clusters before the next step commences. The procedure is continued until all clusters have been allocated the treatment. The time interval between each two successive steps is constant and predetermined, primarily by the nature of the intervention. It is worth mentioning that the design can also be used to randomize individuals rather than clusters, however, the trials that use the design tend to randomize clusters.

Table 10.5 is a graphical representation illustrating the roll-out of an intervention (a training package) to 105 mental health therapists in a stepped wedge design. In this particular instance, 15 therapists are randomized to the intervention at each time point; therefore a total of 7 time points is required to introduce the intervention to all therapists. The intervention is rolled out to a new group of therapists every three months, after a baseline measurement has been taken for all clusters (Time 0). The PHQ-9 score for a patient was the primary outcome for this study and it was believed that the intervention will lead to

TABLE 10.5
Illustration of the stepped wedge design.

		T0	T1	T2	T3	T4	T5	T6	T7
	Month	0	3	6	9	12	15	18	21
Group	**Therapist**								
Group 1	1–15	C	I	I	I	I	I	I	I
Group 2	16–30	C	C	I	I	I	I	I	I
Group 3	31–45	C	C	C	I	I	I	I	I
Group 4	46–60	C	C	C	C	I	I	I	I
Group 5	61–75	C	C	C	C	C	I	I	I
Group 6	76–90	C	C	C	C	C	C	I	I
Group 7	91–105	C	C	C	C	C	C	C	I

Therapists are randomized in groups of 15 and the intervention rolled out at 7 time points after baseline measurements at time 0. Here C indicates that therapists were in Control periods, whereas I indicates they were in Intervention periods. The intervention points are separated by 3 months.

a reduction on this score, which is considered a positive outcome. To achieve a power of 80%, 5 patients are to be recruited by each therapist at each time point; therefore, each therapist is considered as a cluster. Hence, measurements for a total of 4200 ($5 \times 8 \times 105$) patients will be included in the study. Table 10.5 shows when therapists are in the control or intervention periods of the study.

The number of trials using this design is on the increase. In 2006 a systematic review identified 12 studies (8); this had increased to 25 in a second systematic review in 2011 (28). Ethical, financial, and logistic reasons are usually cited for employing the design. Financial constraints, limited availability of the vaccine, and logistic difficulties were incentives to use the stepped wedge design in the Gambia Hepatitis Intervention Study (GHIS) (21) which assessed the effectiveness of infant hepatitis B immunization in the prevention of chronic liver disease in adult life. The units of randomization were the 17 immunization teams, each responsible for administering vaccinations within a defined region of the Gambia. Every 10–12 weeks one of the teams was randomly selected to start the programme in its area, so it took almost four years to introduce the study intervention in all regions.

Ethical considerations and practical issues were cited by Bashour et al. (1) for using the design. The study investigated whether training doctors in communication skills in four Syrian hospitals has an effect on women's satisfaction with doctor-woman relationship during labour and delivery. One hundred expectant mothers were recruited from maternal wards in each of four hospitals at five periods; this resulted in a total of 2000 women who were later surveyed about their satisfaction with the doctor-woman relationship in labour and delivery rooms, measured on a modified Medical Interview Satisfaction Scale (MMISS).

As is the case with parallel randomized clustered trials, when determining sample size for a stepped wedge design the treatment effect to be detected, its variability, the cluster size, the ICC, the power and the significance level must be taken into account. In addition, the number of time periods (steps) and the design of how the randomization is rolled out play a rôle in the sample size calculations. Hussey and Hughes (24) derive a formula to estimate the sample size for a stepped-wedge cluster randomized trial with a continuous outcome. (The design where the number of randomization steps is equal to the number of clusters has optimal power (24).) Furthermore, for high ICC the stepped wedge design is more powerful than a parallel design cluster randomized trial of similar size.

Recently, there has been an interest in developing the methodological aspects of the design; to this effect Hemming et al. (23) suggested a generic framework to encompass variations of the design, including cases where not all clusters contribute measurements at each time period. Work is in progress to extend the CONSORT statement to stepped wedge cluster randomized trials (22).

In order to determine the effectiveness of the intervention, the control and intervention periods of the stepped wedge design are compared using methods that account for clustering and the time periods, such as Generalized Linear Mixed Models. In addition to comparing the intervention to the control, the design allows for the evaluation of whether the intervention's effect changes over time.

As it is the case with parallel cluster RCTs, analysis of stepped wedge cluster RCTs requires using complex models that take into account clustering of the data and the roll-out design. To this effect, Equation 10.1 can be modified to model observations from a stepped wedge cluster RCT as follows:

$$y_{ijk} = \mu + \alpha_j + \beta_1 x_{1,ij} + u_i + e_{ijk} \tag{10.4}$$

where y_{ijk} is the observed outcome variable for participant k at time j in cluster i, μ is the overall mean, α_j is a fixed effect corresponding to time interval j (step j), β_1 is the coefficient for the indicator variable $x_{1,ij}$ ($x_{1,ij}$ is set equal to 1 when cluster i at time j is in the intervention steps and to 0 when it is in the control steps), u_i is as defined above, and e_{ijk} is a random variable with mean zero and variance σ_w^2 for each i,j, representing the variation between participants within cluster i at time j.

If we fit model 10.4 to the data on women's satisfaction with doctor-woman relationship during labour and delivery MMISS scores (1), we obtain the following estimates. For the intervention effect, $\hat{\beta}_1 = 0.03$ (95% CI -0.08 to 0.15, P $= 0.583$). For the first step effect, $\hat{\alpha}_0 = 0$ for identifiability reasons, (here time 0 corresponds to the step where all the hospitals have not received the intervention). For the estimate corresponding to time 1, when the first randomly selected hospital receives the intervention, $\hat{\alpha}_1 = 0.02$ (95% CI -0.08 to 0.12, P $= 0.680$). For the subsequent steps, $\hat{\alpha}_2 = -0.03$ (95% CI -0.14 to 0.08, P $= 0.605$), $\hat{\alpha}_3 = 0.05$ (95% CI -0.08 to 0.18, P $= 0.457$), and $\hat{\alpha}_4 = 0.01$ (95% CI -0.15 to 0.16, P $= 0.943$). There might be an effect of time arising through learning or through some other, unrelated changes happening outside the study. The estimates for the time periods indicate that we do not have evidence for a time effect on the outcome. The estimates for the random effects are $\hat{\sigma}_w^2 = 0.51$ (standard error (SE) 0.016) and $\hat{\sigma}_b^2 = 0.02$ (SE $= 0.016$).

10.10 Sample Size Estimation

When we plan a clinical trial, we have to decide how many research participants we want to recruit. To do this, we think of the primary analysis we want to do, then we choose the sample size to enable us to do this with the precision we want. For example, we might plan a significance test between two group means, or the estimation of an odds ratio. In the first case, we would decide how big a difference we want to be able to detect, then choose the sample size to give the target power to detect this. In the second, we decide how wide or how narrow we want the confidence interval to be. To do these things, we must estimate the likely variability of quantitative variables or the likely size of underlying proportions which we wish to compare. Methods to do this are described in Chapter 12.

In a cluster design, as we saw in Section 10.4, to get the sample size for a cluster randomized desing, we then multiply the individually randomized sample size by the design effect. To estimate the sample size for a cluster design, therefore, we must also estimate the design effect. This will require the cluster sizes and the variability between clusters or the ICC. Cluster sizes are usually built into the research population. For example, we may have little control over the numbers of suitable patients in a primary care practice or children in a school class. Sometimes, they are rather an unknown quantity. For example, in CADET, the clusters were patients attending a primary practice for whom the physician noted depression as a possible factor in their presentation and who were willing to take part in the study when subsequently approached by a researcher. The cluster size was planned to be 14, to be made up of the first 14 suitable participants who agreed to take part. However, for many practices we did not manage to recruit this number and for some we over-recruited as partial compensation for this. The range of cluster sizes was 4 to 20 participants in a practice, mean 11.9. This variability of the cluster size inflated the design effect. Fortunately, in the main trial the ICC was less than in the feasibility study and the trial was adequately powered.

For CADET, we had carried out a feasibility study, which enabled us to estimate the ICC. A feasibility study or a pilot study for a cluster randomized trial may be a major undertaking and is often omitted. Instead, authors rely on published data from other studies in the same field. As noted above, the CONSORT statement (9) reminds authors to include an estimate of the ICC. This helps us to see more about the trial in question but is also of great value to other researchers in the same area. The corresponding information about the distribution of cluster size would also be useful.

Having obtained estimates of the standard deviations, proportions, or whatever else would be needed for a sample size calculation in an individually allocated study, we then estimate the sample size as described in Chapter 12. We then multiply this by the design effect to get the number of participants required. Division of this total number of participants by the mean cluster size gives us the required number of clusters.

Clustering may have a large effect if the ICC is large or if the cluster size is large. For example, if ICC = 0.001, cluster size = 500, the design effect will be $1 + (500 - 1) \times 0.001 = 1.5$. We would need to increase the sample size by 50% to achieve the same power as an individually randomized trial.

What makes the ICC large? ICC is usually quite small when the clusters are convenience clusters and members have little contact with one another. For example, in a study of screening for atrial fibrillation, Fitzmaurice et al. (18) reported that "The intraclass correlation was 0.0027 across all practices". ICC is increased when the intervention is to the care provider. In a trial where 72 practices received interventions to implement guidelines for urinary tract infection and 70 practices received interventions to implement guidelines for sore throat (19), the main outcome measures were changes in rates of use of antibiotics, laboratory tests, and telephone consultations. Flottorp et al. (19) reported that "The intracluster correlation coefficients for the primary outcomes varied from 0.05 to 0.21". ICC is also increased when the members of a cluster interact closely. For example in a study of exercise classes for older women to promote continence, Sackley et al. (35) used as the outcome answers to the question "Do you ever leak any urine when you don't mean to?". They reported ICC = 0.37. When the trial intervention increases the ICC, as for the exercise classes where women could compare experiences, ICCs estimated before treatment or with no treatment may be far too small.

To estimate the sample size for a cluster randomized trial, we must have an estimate of the ICC. We might have pilot data which we can use or we might find a trial with similar clusters and outcome variable. Many trial reports do not give the ICC, however. Authors are encouraged to publish ICCs to facilitate future trials. Eldridge et al. (17) reviewed trials in primary care and reported the median ICC = 0.04. This figure is often quoted in the design of other trials.

It is useful to think about the planned primary analysis for a proposed cluster randomized trial, to ensure that the trial will be big enough for this analysis to be valid. For example, if we want to adjust for individual level covariates, we need sufficient clusters for the proposed regression-based method, whether robust standard errors, multilevel modeling, or generalized linear modeling, to be valid. Such covariates might be a baseline measurement of the outcome measure, for example, such as PHQ-9 in CADET, or age and gender of the participant.

If we want to adjust for cluster level covariates, we need sufficient clusters to do this. It is usually considered that we need 10 observations per degree of freedom in a least squares regression model and this guideline applies to the clusters. It is good practice in randomized trials to adjust for any stratification or minimization variables, if possible, because these variables should explain some of the variability in the outcome variable. If they don't, why stratify for them? In CADET, the clusters were primary care practices and they were allocated by minimization on three variables: number of patients registered with the practice, number of whole time equivalent primary care physicians, and the index of multiple deprivation (IMD) derived from census data for the practice location. This allocation was done separately for each of three geographical sites. If we were to adjust for all of these we would have six degrees of freedom for the model, one each for intervention, number of registered patients, number of physicians, and IMD, plus two for the three sites. As we had only 51 clusters, the number of physicians was omitted from the model. This was considered acceptable because it was highly correlated ($r = 0.80$) with the number of registered patients, which was included.

As with all sample size estimates, there is often a lot of uncertainty involved in the estimate for a cluster-randomized study. For a cluster randomized study, there is additional uncertainty in the ICC and, sometimes, in the achievable cluster sizes. If possible, we might want to increase the sample size to take account of these, in the same way as we might inflate the sample size in many designs to allow for participants who withdraw or for whom we have some missing data.

10.11 Practical Problems of Cluster Randomized Trials

Cluster randomized trials present several problems, including obtaining the consent of trial participants, allocation of clusters to treatments, and recruitment of cluster members.

There is a well-established principle, particularly in health research, that research participants should give informed consent before being recruited into trials. Potential participants should be able to refuse to join the trial. Participants should be able to withdraw at any time. This is complicated in cluster randomized trials. If the treatment is given directly to individuals, e.g. collaborative care for depression, we could seek consent from individual patients. The problem is they have already been randomized, because their general practice is either collaborative care or control. They can consent to treatment, but this is different in the two groups. We might get differential acceptance. Patients can opt out of treatment, but this, too, may be different in the two groups. Participants can consent to provide data, which would be similar in the two groups, indeed, data collection should appear as close to identical as we can make it. We usually ask participants for permission to use their data in a research project. Intention to treat should be very important in analysis, but if they have refused to provide data, this will be difficult. When a treatment is given to care providers, such as treatment guidelines or training, as in TELSiS, the provider must give consent, before randomization. How do we get consent from patients? They cannot opt out, but can consent only for data. In some trials there are people who act as gatekeepers, for example the head of a school. Gatekeepers can give consent, but they are not participants in the trial themselves. In a trial of a school curriculum intervention where the outcome variable was a national assessment, such as SATS or GCSE tests, neither the children or their parents can realistically withhold consent, even for data collection. Because of the problems of consent, it has been alleged that cluster randomized trials are not ethical at all. See Hutton (25) and Edwards et al. (15) for discussions of the special ethical problems of cluster randomized trials.

How many clusters should we have in a trial? We need enough clusters to allow differences between clusters to balance and to estimate the variation between clusters. Murray (29) recommends at least 4 clusters per group, Donner and Klar (14) recommend at least 7. We definitely need more than one cluster per group, as comparing just two clusters, one per treatment, gives no way to estimate the variability between clusters and so to separate treatment and cluster effects. We have seen it done. The best design from the statistical viewpoint has as many clusters as possible. Clusters should be as small as possible and as uniform as possible. The ideal has clusters of size one, i.e. individually randomized. As always, trial design is a compromise between statistical concerns and practical concerns. Modeling techniques require a larger number of clusters, 20 or more per treatment group.

How should we allocate clusters to treatment groups? Clusters may vary. For example, schools may vary systematically, in total size, religious orientation, number of children receiving free school meals, academic ability of the intake, mixed sex or single sex, etc. Primary care practices may vary in practice size, socioeconomic structure of the catchment population, and urban or rural location. Researchers may worry that simple random allocation will not balance these varying characteristics.

In an individually randomized trial we can usually solve the problem of concerns about variability between participants by stratification, randomizing in blocks within strata defined by the characteristics that concern us. In a cluster randomized trial, there may be too few clusters to do this. For example, in the Together4All educational trial, researchers wanted to allocate to two groups schools which varied in total size, religious orientation, the percentage of children receiving free school meals, mixed sex/single sex, and percentage of children falling below the "expected" level in English and in Maths. If we could classify each of these variables into just two strata, the total number of possible strata would be $2 \times 2 \times 2 \times 2 \times 2 \times 2 = 64$, yet there were only 13 available schools. Many strata might have only one school. We would not be able to adjust at the analysis stage, there will be insufficient clusters.

One approach might be to match clusters in pairs then allocate one of each pair to each group. Matching has several problems: we might not be able to match all clusters, some pairs may be more closely matched than others, if a cluster drops out we would have to drop its matched pair, one of an odd number of clusters would be unpaired. If we were then to carry out a paired t test on summary statistics, we would have very few degrees of freedom.

It might be possible to achieve approximate balance by minimization (6; 16; 37; 38). This is based on the idea that the next cluster to enter the trial is given whichever treatment would minimize the overall imbalance between the groups at that stage of the trial. The Together4All trial, TELSiS, CADET, and the incentives trial were all allocated by minimization. Minimization does not solve all problems and may not produce balance for unrecorded variables, which we expect random allocation to achieve. Minimization is not random, though we can introduce a random element by allocating clusters to the group indicated by minimization with a probability greater than 0.5, rather than always. Minimization may not be accepted for regulatory purposes, but this is not usually the reason for a cluster randomized trial. In individually allocated trials, it may be possible for a potential partcipant's characteristics to influence the investigator's decision about recruiting that partcipant to the trial, as the investigator might know what treatment the partcipant would receive using the minimization algorithm. This is less likely in a cluster trial. Variables which are used in minimization should be taken into account in the analysis if possible, as was done for CADET, as this will reduce the variability within the groups. However, this may not be possible due to small numbers of clusters, as for TELSiS or Together4All.

Just as knowledge of the treatment group should not influence whether a cluster becomes part of a trial, it should not influence membership of clusters. Ideally, recruitment to clusters should be done before randomization. We should know who will be in each cluster. For example, in a school study, the pupils of the school or of a school class form the cluster. If a single class will form the cluster, it should be identified when the school joins the study and before allocation. In a trial of small group education of GPs to improve the outcome of chronic asthma in general practice (42), for each GP 30 patients were found who agreed to provide symptom questionnaire data before randozization.

What if we do not know who is in the cluster? In some primary care trials, researchers can trawl practice lists for cluster members. In some trials, we are limited to new potential participants presenting, as in TELSiS or CADET. In CADET, practice databases were searched weekly for new presenters with depressive symptoms. We should not rely on clinicians or other service providers who know the allocation of the cluster to identify and recruit those they consider to be eligible participants. This gives great potential for bias. Torgerson and Torgerson (38) give a good discussion.

Refusal by members of clusters, as certainly happened in CADET, may occur when not all cluster members get the treatment to which they were allocated or refuse to provide outcome data. Typically, this happens in the intervention arm more frequently than in the control arm, which is often treatment as usual. In CADET, participants could cease contact with both the collaborative care worker, whom the participant may simply dislike, and the GP providing usual care. We can, and should, analyse the trial according to the intention to treat where the estimated treatment effect is likely to be less than if all participants had received the allocated treatment. Torgerson and Torgerson (38) argue that this bias may be more than contamination would cause. We can also allow for missing data by a method such as multiple imputation (41). This relies on the assumption that we can predict the missing observation, apart from random variation, from the other information we have about the participant.

In reporting cluster randomized trials, as in other trials, the CONSORT statement is a valuable tool, which helps us to ensure that important information is not left out. The CONSORT statement has been extended to cluster randomized trials (9). For example, it reminds us of the value of including an estimate of the ICC.

Should we do a cluster randomized trial? If we can answer the research question with an individually randomized trial, we would suggest that is the preferable option. However, this may not be possible and a cluster randomized trial may be the only way. It is not to be undertaken lightly.

Bibliography

[1] H.N. Bashour, M. Kanaan, M.H. Kharouf, A.A. Abdulsalam, M. Tabbaa, and S.A. Cheikha. The effect of training doctors in communication skills on women's satisfaction with labour and delivery: a stepped wedge cluster randomized trial in Damascus. *BMJ Open*, 3:e002674, 2013.

[2] J.M. Bland. Sample size in guidelines trials. *Family Practice*, 17:S17–S20, 2000.

[3] J.M. Bland. Cluster randomised trials in the medical literature: two bibliometric surveys. *BMC Medical Research Methodology*, 4(21), 2004.

[4] J.M. Bland. The analysis of cluster-randomised trials in education. *Effective Education*, 2:165–180, 2015.

[5] J.M. Bland and D.G. Altman. Measurement error and correlation coefficients. *British Medical Journal*, 313:41–2, 1996.

[6] M. Bland. *An Introduction to Medical Statistics*. University Press, Oxford, 2015.

[7] G. Brooks, M. Burton, P. Cole, J. Miles, C. Torgerson, and D. Torgerson. Randomised controlled trial of incentives to improve attendance at adult literacy classes. *Oxford Review of Education*, 34:493–504, 2008.

[8] C.A. Brown and R. J. Lilford. The stepped wedge trial design: a systematic review. *BMC Medical Research Methodology*, 6:54, 2006.

[9] M.K. Campbell, D.R. Elbourne, and D.G. Altman. CONSORT statement: extension to cluster randomised trials. *British Medical Journal*, 328:702–708, 2004.

[10] H. Cheyne, V. Hundley, D. Dowding, J.M. Bland, P. McNamee, I. Greer, M. Styles, C.A. Barnett, G. Scotland, and C. Niven. Effects of algorithm for diagnosis of active labour: cluster randomised trial. *British Medical Journal*, 337:a2396, 2008.

[11] P.J. Cooper, A.L. Moncayo M.E. Chico, M.G. Vaca, J.M. Bland, E. Mafla, F. Sanchez, L.C. Rodrigues, D.P. Strachan, and G.E. Griffin. Effect of albendazole treatments on the prevalence of atopy in children living in communities endemic for geohelminth parasites: a cluster-randomised trial. *Lancet*, 367:1598–1603, 2006.

[12] J. Cornfield. Randomization by group: a formal analysis. *American Journal of Epidemiology*, 108:100–102, 1978.

[13] P. Diggle, P. Heagerty, K. Liang, and S. Zeger. *Analysis of Longitudinal Data, 2nd Ed.* Oxford University Press, 2002.

[14] A. Donner and N. Klar. *Design and Analysis of Cluster Randomised Trials in Health Research*. Arnold, London, 2000.

[15] S.J.L. Edwards, D.A. Braunholtz, R.J. Lilford, and A.J. Stevens. Ethical issues in the design and conduct of cluster randomised controlled trials. *British Medical Journal*, 318:1407–1409, 1999.

[16] S. Eldridge and S. Kerry. *A Practical Guide to Cluster Randomised Trials in Health Services Research*. Wiley, Chichester, 2012.

[17] S.M. Eldridge, D. Ashby, G.S. Feder, A.R. Rudnicka, and O.C. Ukoumunne. Lessons for cluster randomised trials in the 21st century: a systematic review of trials in primary care. *Clinical Trials*, 1:80–90, 2004.

[18] D.A. Fitzmaurice, F.D.R. Hobbs, S. Jowett, J. Mant, E.T. Murray, R. Holder, J.P. Raftery, S. Bryan, M. Davies, G.Y.H. Lip, and T.F. Allan. Screening versus routine practice in detection of atrial fibrillation in patients aged 65 or over: cluster randomised controlled trial. *British Medical Journal*, 335:383, 2007.

[19] S. Flottorp, A.D. Oxman, K. Hovelsrud, S. Treweek, and J. Herrin. Cluster randomised controlled trial of tailored interventions to improve the management of urinary tract infections in women and sore throat. *British Medical Journal*, 325:367, 2002.

[20] H. Goldstein. *Multilevel Statistical Models, 4th Ed.* Wiley, Chichester, 2011.

[21] The Gambia Hepatitis Study Group. The Gambia Hepatitis Intervention Study. *Cancer Research*, 47:5782–7, 1987.

[22] K. Hemming, A. Girling, T. Haines, and R. Lilford. Protocol: CON-SORT extension to stepped wedge cluster randomised controlled trial. Available from: http://www.equator-network.org/wpcontent/uploads/2009/02/consort-sw-protocol-v1.pdf [accessed 10 march 2015b].

[23] K. Hemming, R. Lilford, and A. Girling. Stepped-wedge cluster randomised controlled trials: a generic framework including parallel and multiple-level design. *Statistics in Medicine*, 34:181–196, 2015.

[24] A. Hussey and J.P. Hughes. Design and analysis of stepped wedge cluster randomized trials. *Contemporary Clinical Trials*, 28:182–191, 2007.

[25] J. Hutton. Are distinctive ethical principles required for cluster randomised controlled trials? *Statistics in Medicine*, 20:473–488, 2001.

[26] K. Liang and S. Zeger. Longitudinal data analysis using generalized linear models. *Biometrika*, 73:13–22, 1986.

[27] C.J.M. Maas and J.J. Hox. Sufficient sample sizes for multilevel modeling. *Methodology*, 1:86–92, 2005.

[28] N. Mdege, M. Man, C.A. Taylor (née Brown), and D.J. Torgerson. Systematic review of stepped wedge cluster randomized trials shows that design is particularly used to evaluate interventions during routine implementation. *Journal of Clinical Epidemiology*, 64:936–948, 2011.

[29] D.M. Murray. *The Design and Analysis of Group-Randomized Trials*. University Press, Oxford, 1998.

[30] R.Z. Omar and S.G. Thompson. Analysis of a cluster randomized trial with binary outcome data using a multi-level model. *Statistics in Medicine*, 19:2675–2688, 2000.

[31] D. Pearson, C. McDougall, M. Kanaan, D.J. Torgerson, and R. Bowles. Evaluation of the citizenship evidence-based probation supervision program using a stepped wedge cluster randomized controlled trial. *Crime & Deliquency*, 60:1–26, 2014.

[32] D.A. Richards, J.J. Hill, L. Gask, K. Lovell, C. Chew-Graham, P. Bower, J. Cape, S. Pilling, R. Araya, D. Kessler, J.M. Bland, C. Green, G. Gilbody, G. Lewis, C. Manning, A. Hughes-Morley, and M. Barkham. Clinical effectiveness of collaborative care for depression in UK primary care (CADET): cluster randomised controlled trial. *BMJ*, 347:f4913, 2013.

[33] D.A. Richards, A. Hughes-Morley, R.A. Hayes, R. Araya, M. Barkham, J.M. Bland, P. Bower, J. Cape, C. Chew-Graham, L. Gask, S. Gilbody, C. Green, D. Kessler, G. Lewis, K. Lovell, C. Manning, and S. Pilling. Collaborative depression trial (CADET): multi-centre randomised controlled trial of collaborative care for depression - study protocol. *BMC Health Services Research*, 9:188, 2009.

[34] D.A. Richards, K. Lovell, S. Gilbody, L. Gask, D. Torgerson, M. Barkham, M. Bland, P. Bower, A.J. Lankshear, A. Simpson, J. Fletcher, D. Escott, S. Hennessy, and R. Richardson. Collaborative care for depression in UK primary care: a randomized controlled trial. *Psychological Medicine*, 38:279–287, 2008.

[35] C.M. Sackley, N.A. Rodriguez, F. Badger, C. Wright, J. Besemer, and K.T.V. van Reeuwijk. A phase II exploratory cluster randomized controlled trial of a group mobility training and staff education intervention to promote urinary continence in UK care homes. *Clinical Rehabilitation*, 22:714–721, 2008.

[36] D.J. Spiegelhalter. Bayesian methods for cluster randomized trials with continuous responses. *Statistics in Medicine*, 20:435–452, 2001.

[37] D.R. Taves. Minimization - new method of assigning patients to treatment and control groups. *Clinical Pharmacology & Therapeutics*, 15:443–453, 1974.

[38] D.J. Torgerson and C.J. Torgerson. *Designing Randomised Trials in Health, Education, and the Social Sciences: An Introduction.* Palgrave Macmillan, Basingstoke, 2008.

[39] R.M. Turner, R.Z. Omar, and S.G. Thompson. Bayesian methods of analysis for cluster randomized trials with binary outcome data. *Statistics in Medicine*, 20:453–472, 2001.

[40] A. Vail and E. Gardener. Common statistical errors in the design and analysis of subfertility trials. *Human Reproduction*, 18:1000–1004, 2003.

[41] I.R. White, P. Royston, and A.M. Wood. Multiple imputation using chained equations: issues and guidance for practice. *Statistics in Medicine*, 30:377–399, 2011.

[42] P.T. White, C.A. Pharoah, H.R. Anderson, and P. Freeling. Improving the outcome of chronic asthma in general practice: a randomized controlled trial of small group education. *Journal of the Royal College of General Practitioners*, 39:182–186, 1989.

[43] R.L. Williams. A note on robust variance estimation for cluster-correlated data. *Biometrics*, 56:645–646, 2000.

11

Randomization, Stratification, and Outcome-Adaptive Allocation

Oleksandr Sverdlov and Yevgen Ryeznik

CONTENTS

11.1   Introduction

Randomized controlled trials (RCTs) play an indispensable role in advancing clinical practice and delivering best medicines to patients with difficult conditions [31]. The most common objective of RCTs (in superiority trials) is to test the hypothesis that an experimental treatment is better than the standard of care in the population of patients with the disease of interest. To achieve this goal, eligible patients are enrolled into the trial and are randomized into two or more groups to receive either the experimental treatment(s) or the control (standard of care). After each study patient is treated and prospectively followed up for a well-defined period while contributing important clinical data, the treatment effects are compared with respect to the primary outcome using a statistical test. Based on the RCT results, a conclusion may be drawn/reached that the experimental treatment is safe and efficacious in treating the disease and can lead to an improved quality of life—this will have an important implication for patients with the disease.

Randomization refers to generation of a sequence of treatment assignments by means of some known random mechanism and the implementation of this sequence in the trial. When properly implemented, randomization can contribute to the validity and credibility of the trial results by mitigating various experimental biases and promoting comparability of treatment groups. However, to achieve these goals, randomization should be complemented by some other important design techniques [64].

Let us explore the role of randomization and other methodological tools in mitigating the experimental biases that can be introduced during the conduct of an RCT. To fix ideas, we shall focus on a two-arm RCT and assume that the randomization list (a sequence of treatment assignments, E or C, where E stands for the experimental treatment and C stands for the control treatment) has been generated in advance, so that every patient

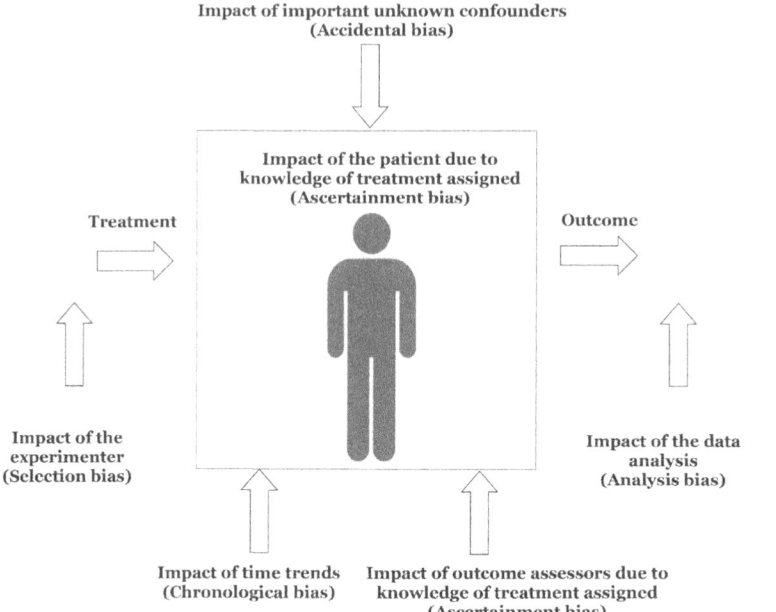

FIGURE 11.1
Major types of potential experimental biases in an RCT.

who enrolls into the study is assigned to one of the treatments (E or C) according to the randomization list. The major types of experimental bias are displayed in Figure 11.1.

Selection bias [11] can occur when an investigator knows or is able to guess with high probability which treatment is to be assigned to an upcoming patient. The advance knowledge of the treatment assignment can motivate an investigator to selectively enroll a particular type of patients who are thought to benefit most from the given treatment. This may lead to systematic covariate imbalances and confounding of the true treatment effect. For instance, if the experimental group consists of more patients with better prognosis than the control group, the analysis not accounting for this imbalance may suggest that the experimental treatment is more efficacious while in reality the two treatments are equally efficacious. The best solution to mitigate selection bias in an RCT is to combine randomization with *double-blinding* (neither the study subjects nor the investigator or the staff involved in the treatment or clinical evaluation should be aware of the treatments received). When double-blinding is not possible, one should use *allocation concealment*—every treatment allocation should be kept confidential (until after it is executed) to the investigator involved in the patient enrollment. This can be done, for instance, by having a third party (e.g. independent Contract Research Organization) to implement randomization. Also it is important to avoid using allocation schemes that are too restrictive, i.e. where treatment assignments can be predicted with high probability [11, 12].

Ascertainment bias can occur due to the knowledge of which treatment the patient is receiving. For instance, patients may be more likely to drop out from the study if they know that they have been assigned to the placebo; outcome assessors may be biased in their assessment of the outcome based on their previous experience and expectations of treatment effects. To minimize potential for ascertainment bias, double-blinding is regarded as the optimal approach. As noted by Friedman, Furberg and DeMets [26]: *A clinical trial should, ideally, have a double-blind design in order to avoid potential problems of bias during data collection and assessment. In studies where such a design is impossible, other measures to reduce potential bias are advocated.*

Accidental bias [24] is the bias due to unknown factors that affect the primary outcome. For instance, if there is an unknown important prognostic factor and the statistical analysis fails to adjust for it, the estimated treatment effect can be biased. Randomization tends to mitigate accidental bias, including bias due to unmeasured confounders. By law of large numbers, randomization ensures that, on average, the two groups are comparable with respect to important patient baseline characteristics and important unknown and unmeasured confounders. The more random a procedure is, the less susceptible it is to accidental bias. In practice it is important to avoid using randomization schemes that are likely to produce a sequence of treatment assignments with some particular periodicity, such as the truncated binomial design with a large sample size [52].

Chronological bias [11, 46] is the bias due to time trends. It can arise in situations when some patient characteristics change over the recruitment process and a randomization sequence has too many consecutive treatment allocations of the same type, thereby causing confounding of the time and treatment effects. Chronological bias can be mitigated by placing restrictions on randomization to ensure that treatment numbers are approximately the same throughout the trial. It is important to note that the requirements of treatment balance and allocation randomness are in conflict. Restricted randomization procedures (see Section 11.2) can be used to provide tradeoff between balance and randomness.

Analysis bias can arise due to selective inclusion/exclusion of subjects in the analysis datasets and/or due to a failure to apply appropriate covariate adjustments. This type of bias can occur even in the well-conducted RCTs with proper blinding and allocation concealment and it can have devastating impact on the trial results. To avoid analysis bias, one important principle is to use the pre-specified analysis datasets. The "intent-to-treat" (ITT) principle posits that all randomized subjects should be followed so long as alive, able, and consenting. The primary analysis should use the "all randomized" patients dataset to ensure alignment with the ITT principle. In addition, the statistical methodology to analyze data, the important covariates to be adjusted for, the ways of checking model assumptions (and alternative ways to analyze data if these assumptions are violated), and the strategies for handling missing data should be pre-specified in the study protocol or in a separate statistical analysis plan.

Overall, the three key merits of randomization in the RCT context can be summarized as follows:

1. Randomization helps mitigate selection bias due to an investigator's attempt to guess the treatment assignment and selectively enroll study patients.

2. Randomization helps promote comparability of treatment groups with respect to both important patient baseline characteristics and important unknown and thus unmeasured confounders.

3. Randomization adds to the validity of the model-based analysis by making the assumption of independence of experimental errors more feasible and it can be also used as a basis for randomization-based inference.

Now that the importance of randomization has been acknowledged, one may wonder which randomization procedures should be used in practice. To facilitate the discussion, we shall first describe some mathematics of randomization.

Let E and C denote, respectively, the experimental and the control treatments. For a trial with n patients, a randomization sequence is a random vector $\mathbf{\Delta}_n = (\delta_1, \ldots, \delta_n)$, where $\delta_i = 1$ if the ith patient is assigned to E and $\delta_i = 0$ if the ith patient is assigned to C. Suppose there are q important baseline covariates such that for the ith patient one observes the covariate vector $\mathbf{z}_i = (z_{1i}, \ldots, z_{qi})'$. Let Y_i be the ith patient's outcome which is modeled as $E(Y_i|\delta_i, \mathbf{z}_i) = g(\delta_i, \mathbf{z}_i, \boldsymbol{\theta})$, where $g(\cdot)$ is some regression function and $\boldsymbol{\theta}$ is a vector of model parameters including the effects of treatments, covariates, and, when appropriate, treatment-by-covariate interactions. Eligible and consenting patients enter the trial sequentially and are randomly assigned to either E or C. Suppose the target allocation ratio is 1:1, that is we want to have an equal number of patients assigned to E and C. Let ϕ_{i+1} denote the probability that the $(i+1)$st patient is randomized to E. Then we have the following classification of randomization designs [52]:

- *Simple randomization*, if $\phi_{i+1} = \Pr(\delta_{i+1} = 1) = 1/2$, $i \geq 0$. In other words, each patient's treatment assignment is made with probability $1/2$ and all treatment assignments are mutually independent.

- *Restricted randomization*, if $\phi_{i+1} = \Pr(\delta_{i+1} = 1|\mathbf{\Delta}_i)$, where $i \geq 1$ and $\phi_1 = 1/2$. In other words, the randomization probability of the $(i+1)$st patient is conditional on treatment assignments of the previous i patients.

- *Covariate-adaptive randomization* (which includes *stratified randomization* as a special case), if $\phi_{i+1} = \Pr(\delta_{i+1} = 1|\mathbf{\Delta}_i, \mathbf{z}_1, \ldots, \mathbf{z}_i, \mathbf{z}_{i+1})$, where $i \geq 1$ and $\phi_1 = 1/2$. In other words, the randomization probability of the $(i+1)$st patient is conditional on treatment assignments and covariates of the previous i patients, and the covariate vector of the $(i+1)$st patient.

If the study goal is to achieve unequal allocation (e.g. skewed in favor of and empirically better treatment), then one has

- *Outcome-adaptive randomization*, if $\phi_{i+1} = \Pr(\delta_{i+1} = 1|\mathbf{\Delta}_i, \mathbf{Y}_i)$, where $\mathbf{Y}_i = (Y_1, \ldots, Y_i)$, $i \geq 1$ and $\phi_1 = 1/2$. In other words, the randomization probability of the $(i+1)$st patient is conditional on treatment assignments and outcomes of the previous i patients.

In the following sections we shall discuss in detail the different types of randomization designs: simple and restricted randomization (Section 11.2), stratified and covariate-adaptive randomization (Section 11.3) and outcome-adaptive randomization (Section 11.4).

11.2 Simple and Restricted Randomization

With the simple randomization (completely randomized design, CRD), treatment assignments form a sequence of independent and identically distributed Bernoulli random variables: $\delta_1, \ldots, \delta_n \sim \text{Bernoulli}(1/2)$, where $\delta_i = 1(0)$ for treatment $E(C)$. A major advantage of CRD is that it is completely unpredictable and eliminates selection bias. A major disadvantage is that in small studies it can result, with non-negligible probability, in treatment imbalances that can lead to loss of statistical efficiency. Balanced (1:1) allocation is desirable

TABLE 11.1
Probability that $N_E(n)/n$ deviates from 0.5 by more than 0.1 with the CRD.

n	Sample allocation $N_E(n) : N_C(n)$	Imbalance w	$\mathbf{Pr}(|N_E(n) - N_C(n)| > w)$
20	(12:8) or (8:12)	4	0.263
40	(24:16) or (16:24)	8	0.154
100	(60:40) or (40:60)	20	0.035
200	(120:80) or (80:120)	40	0.004

as it frequently maximizes power of statistical tests. However, in theory, CRD can result in all subjects being assigned to the same treatment group! For a trial with n patients, $N_E(n) = \sum_{i=1}^{n} \delta_i$ is the number of patients assigned to E and $N_C(n) = \sum_{i=1}^{n} (1 - \delta_i)$ is the number of patients assigned to C. Let $D_n = N_E(n) - N_C(n) = 2N_E(n) - n$ denote the treatment imbalance after n allocations. Since $N_E(n) \sim \text{Binomial}(n, 1/2)$, we can compute $\Pr(|D_n| > w) = 1 - \Pr(|D_n| \leq w)$, the probability of imbalance more extreme than w, where w is a positive integer. This can be evaluated exactly using binomial probabilities or using normal approximation: $\Pr(|D_n| > w) \approx 2(1 - \Phi(w/\sqrt{n}))$.

Table 11.1 shows some values of probability that the sample proportion $N_E(n)/n$ for the CRD deviates from the target value of 0.5 by more than 0.1. For instance, if $n = 20$, there is a 26% chance that the trial will result is more unbalanced allocation than 12:8 or 8:12. As n gets larger, imbalance becomes less of a concern. For $n = 100$ and $n = 200$, the probability that $N_E(n)/n$ deviates from 0.5 by more than 0.1 is 3.5% and 0.4%, respectively.

In practice, one may wish to achieve balanced (or nearly balanced) allocation throughout the trial. Various restricted randomization procedures have been proposed for this purpose [52].

The most common restricted randomization procedure is the *permuted block design*. The treatment assignments are made at random in blocks such that for every $2b$ subjects (where b is a positive integer), exactly b subjects are assigned to each of the two treatments E or C. Balance within the block can be ensured by either the *random allocation rule* or the *truncated binomial design*. For the random allocation rule, the randomization probability for the $(i+1)$st subject within a given block can be written as $\phi_{i+1} = \{b - b_E(i)\}/\{2b - i\}$, $i = 0, \ldots, b-1$, where $b_E(i)$ is the number of treatment E assignments among i allocations within the block (with the convention $b_E(0) = 0$). For the truncated binomial design, subjects are randomized with probability $1/2$ to E or C until one of the treatments has b assignments, in which case the remaining assignments are made deterministically to the opposite treatment. Importantly, while these two approaches equalize treatment numbers in every block of size $2b$, they have quite different probability structures. For instance, for a block of size 4 there are 6 possible random sequences of treatment assignments: $EECC$, $ECEC$, $CECE$, $ECCE$, $CEEC$, and $CCEE$. For the random allocation rule the probability of each sequence is $1/6$, whereas for the truncated binomial design each of the sequences $EECC$ and $CCEE$ occurs with probability $1/4$ and each of the remaining sequences occurs with probability $1/8$. In practice the random allocation rule is more common than the truncated binomial design.

One challenge for the permuted block design is the choice of the block size. If block size is 2, then every pair of subjects results in the balanced allocation; however, every even allocation is deterministic which opens a potential for selection bias. In practice, blocks of size 4 to 8 are common, and variable block sizes are also possible. However, even with moderately large block size, allocations in the tail of each block can be guessed with high probability, which is undesirable if the trial is not double-blinded [11, 52]. Furthermore, the variable block design does not necessarily reduce predictability compared to the fixed block design as one might expect [59].

A requirement of exact balance can be replaced with a less stringent requirement of near-balance at any stage of the trial. Berger, Ivanova and Knoll [14] proposed the *maximal procedure* which minimizes predictability of treatment assignments while maintaining imbalance below a pre-specified threshold. For a given sample size and a given value of the maximal tolerated imbalance, a randomization sequence from the maximal procedure can be generated using an efficient algorithm [56]. An important feature of this procedure is that all randomization sequences in its allocation space (in the case of 1:1 randomization) are equally likely and therefore the construction of a randomization test should be straightforward. One disadvantage of the maximal procedure is the lack of the close form expression for the allocation probability $\phi_{i+1} = \Pr(\delta_{i+1} = 1|\boldsymbol{\Delta}_i)$; thus the randomization schedule from the maximal procedure can only be generated in advance for a pre-specified value of the sample size.

A major advancement in the design of clinical trials was made by Efron [24] who introduced the *biased coin design* (BCD) to balance treatment assignments while maintaining randomness. With BCD, the $(i+1)$st patient is randomized to treatment E with probability

$$\Pr(\delta_{i+1} = 1|D_i) = \begin{cases} p, & \text{if } D_i < 0; \\ 1/2, & \text{if } D_i = 0; \\ 1-p, & \text{if } D_i > 0. \end{cases} \tag{11.1}$$

In (11.1), $p \in (1/2, 1]$ is a user-defined coin bias probability (Efron's favorite choice is $p = 2/3$). The BCD can be thought of as a middle ground between CRD and the permuted block design with block size 2. Several valuable extensions of Efron's BCD have been developed. Soares and Wu [63] proposed the *big stick design* for which treatment allocations are made with probability $1/2$ as long as $|D_i| < c$ for some pre-specified positive integer c; once $|D_i| = c$, the allocation is made with probability 1 to reduce imbalance. Chen [20] proposed the *biased coin design with imbalance tolerance* as a hybrid of Efron's [24] and Soares and Wu's [63] designs.

In practice, BCD is frequently applied with stratified or covariate-adaptive randomization (see Section 11.3); e.g. within each stratum a separate BCD procedure is run to promote balance. Reference [9] provides a recent example of a randomized open-label trial comparing closed loop insulin delivery versus conventional subcutaneous insulin delivery in subjects with type 2 diabetes, which used biased coin randomization stratified according to glycated hemoglobin level, body mass index, and pre-trial total daily insulin dose.

Baldi Antognini and Giovagnoli [7] proposed a class of *adjustable biased coin designs* (ABCDs) for which the coin bias is tuned according to the value of current imbalance. The $(i + 1)$st subject is randomized to E with probability $\Pr(\delta_{i+1} = 1|D_i) = F(D_i)$, where $F : \mathbb{Z} \to [0, 1]$ is a non-decreasing function of imbalance defined on the set of integers satisfying $F(D_i) + F(-D_i) = 1$. Baldi Antognini and Giovagnoli [7] advocate the class of functions

$$F_a(D_i) = \begin{cases} \frac{|D_i|^a}{|D_i|^a+1}, & \text{if } D_i \leq -1; \\ 1/2, & \text{if } D_i = 0; \\ \frac{1}{|D_i|^a+1}, & \text{if } D_i \geq 1, \end{cases} \tag{11.2}$$

where $a \geq 0$ is a user-defined parameter controlling the degree of randomness ($a = 0$ is simple randomization and as a increases the procedure becomes more deterministic). Baldi Antognini [6] proved that ABCD converges to balance faster than any other BCD thereby resulting in more powerful Z- and t-tests. Another useful approach to formally optimize the value of the coin bias probability p in (11.1) can be found in [8].

Wei [75] and Smith [62] introduced a class of *generalized biased coin designs* (GBCDs) for which randomization probabilities change according to the value of imbalance measured as the difference in sample allocation proportions. One advantage of this approach is that the randomization procedure takes into account the current size of the trial: it forces balance when the trial is small and it becomes more random as the trial gets larger. Let $d_i = i^{-1}(N_E(i) - N_C(i))$. Then the $(i+1)$st subject is randomized to treatment E with probability $\Pr(\delta_{i+1} = 1|d_i) = f(d_i)$, where $f : [-1, 1] \to [0, 1]$ is a continuous non-increasing function satisfying $f(d_i) + f(-d_i) = 1$. Smith [62] proposed the following family of allocation functions:

$$f_\rho(d_i) = \frac{(1-d_i)^\rho}{(1-d_i)^\rho + (1+d_i)^\rho} = \frac{\{N_C(i)\}^\rho}{\{N_E(i)\}^\rho + \{N_C(i)\}^\rho}, \tag{11.3}$$

where $\rho \geq 0$ is a user-defined parameter controlling the degree of randomness. If $\rho = 0$ we have CRD; if $\rho = 1$ we have the D-optimum procedure of Wei [75]; if $\rho = 2$ we have the D_A-optimum procedure of Atkinson [1]; and if $\rho \to \infty$ the procedure becomes equivalent to the permuted block design with block size 2. Wei [75] and Smith [62] established asymptotic properties of treatment imbalance following the procedure (11.3). As $n \to \infty$, $n^{-1/2}D_n$ converges to a normal distribution with mean 0 and asymptotic variance $\{1 - 4f'_\rho(0)\}^{-1}$, where $f'_\rho(0)$ is the derivative of $f_\rho(x)$ at $x = 0$. Therefore, procedure (11.3) is asymptotically balanced at the rate of convergence to balance which is of order $n^{-1/2}$.

Since various restricted randomization procedures targeting 1:1 allocation are available, one may wonder how to select the "best" one in practice. A randomization procedure should provide a good tradeoff between balance and randomness, it should be simple to implement and it should have established statistical properties and lead to valid statistical inference.

One simple approach to assess the goodness of a randomization procedure is to examine simultaneously two criteria: *bias* and *loss* [1, 3, 4]. In this context, bias refers to lack of randomness of a randomization procedure. One measure of lack of randomness was proposed by Heritier, Gebski and Pillai [33]. For a trial of size n, they define *forcing index* at the ith allocation step as

$$FI_i = \frac{1}{i} \sum_{m=1}^{i} \frac{|p_m - 0.5|}{0.5}, \quad i = 1, \ldots, n, \tag{11.4}$$

where p_m is the treatment E allocation probability for the mth patient. Clearly, $FI_i \equiv 0$ for the CRD. On the other hand, the most restrictive permuted block design with block size 2, PBD(2), has 50% of assignments to E made with probability 0.5, and other 50% of assignments to E made with probability 0 or 1. Therefore, for PBD(2) we have $F_i = 0.5$ (if i is even) or $F_i = (i-1)/2i$ (if i is odd). For any other randomization procedure, we have $0 \leq F_i \leq 0.5$ for $i \geq 1$, and lower values of F_i indicate greater randomness.

The measure of loss quantifies statistical efficiency of a randomization procedure. It refers to the effective number of patients on whom information is lost due to imbalance in the design. For a trial with n patients, the loss at the ith step is expressed as

$$L_i = \frac{\{N_E(i) - N_C(i)\}^2}{i}, \quad i = 1, \ldots, n. \tag{11.5}$$

If $N_E(i) - N_C(i) = 0$, the groups are balanced and the loss is zero. In general, L_n can take values $0, \ldots, n$ according to the probability distribution induced by a randomization design. Importantly, $E(L_n) = \text{Var}(D(n))/n$, which directly relates loss and variability: the less variable the randomization procedure is, the lower loss it has. Ideally, a randomization procedure should have low values of both forcing index and loss throughout the trial. Various randomization procedures can be compared graphically to identify ones that provide suitable tradeoff between the two criteria [2–4].

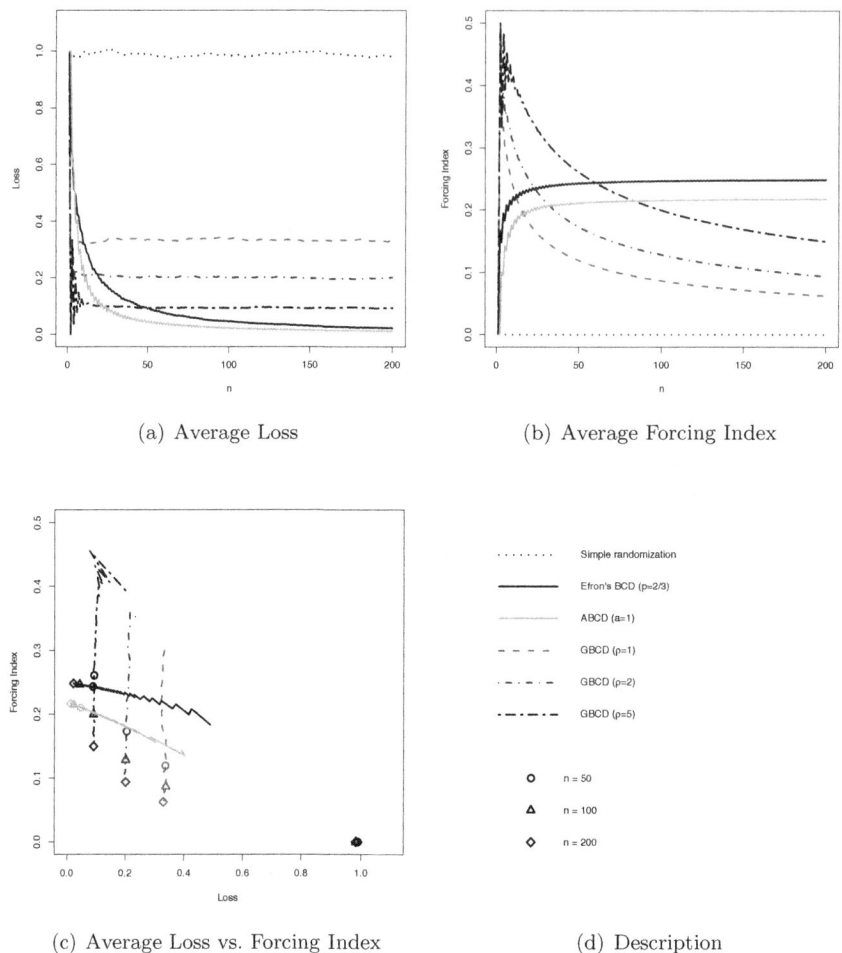

(a) Average Loss

(b) Average Forcing Index

(c) Average Loss vs. Forcing Index

(d) Description

FIGURE 11.2
Average loss and average forcing index of six randomization designs, based on 10,000 simulation runs.

To illustrate these ideas, we performed a simulation study comparing six randomization procedures: (i) CRD; (ii) BCD($p = 2/3$); (iii) ABCD($a = 2$); (iv) GBCD($\rho = 1$); (v) GBCD($\rho = 2$); and (iv) GBCD($\rho = 5$). The six procedures are compared with respect to forcing index (11.4) and loss (11.5) in a hypothetical trial with up to $n = 200$ patients. Each procedure was simulated 10,000 times and the average values of forcing index and loss were derived for each intermediate value of the sample size from 1 to 200.

Figure 11.2 shows a summary of the simulation results. Examining simultaneously Figures 11.2(a) and 11.2(b), one can see that for CRD average loss is 1 (worst among the designs) and average forcing index is 0 (best among the designs). For BCD($p = 2/3$) and ABCD($a = 2$), loss $\to 0$ as n increases, and forcing index increases as the trial progresses. Notably, ABCD($a = 2$) is uniformly better than BCD($p = 2/3$) both in terms of loss and forcing index. For three GBCD designs, average loss is constant, but average forcing index $\to 0$ as $n \to \infty$. This is a fundamental difference between the procedures that consider imbalance as difference in treatment counts (BCD and ABCD) and those that consider

TABLE 11.2
Distance to the origin (0,0) in the loss–forcing index space for six randomization procedures.

Procedure	$n = 50$	$n = 100$	$n = 200$
CRD	0.99	0.97	0.99
BCD($p = 2/3$)	0.26	0.25	0.25
ABCD($a = 2$)	0.21	0.22	0.22
GBCD($\rho = 1$)	0.35	0.35	0.34
GBCD($\rho = 2$)	0.25	0.24	0.24
GBCD($\rho = 5$)	0.28	0.22	0.18

imbalance as difference in treatment proportions (GBCD). For the former, the allocation proportion converges to a degenerate distribution at $1/2$, whereas for the latter, the allocation proportion converges to a normal distribution with mean $1/2$ and nonzero variance. Figure 11.2(c) is a plot of average loss versus average forcing index. The origin (0,0) is the "most desirable" point where both loss and forcing index are zero. While this point is impossible to achieve for any randomization procedure, the goodness of a procedure can be quantified by a distance to (0,0). Table 11.2 shows the values of this distance for six randomization designs for $n = 50, 100, 200$. One can see that ABCD($a = 2$) provides best tradeoff between loss and forcing index for $n = 50$; ABCD($a = 2$) and GBCD($\rho = 5$) are best for $n = 100$; and GBCD($\rho = 5$) is best for $n = 200$.

Another approach to evaluate merits of different randomization procedures is to study their inferential characteristics, such as Type I error rate, power, estimation efficiency, etc. These measures are more reflective of the primary objective of an RCT—to provide an unbiased treatment comparison.

There are two distinct approaches to facilitate statistical inference following an RCT— an *invoked population model* approach and a *randomization-based* approach [52]. The former approach is more common in clinical research. It assumes that there is a target population from which a random sample of patients is drawn for participation in a clinical trial, and randomization is applied to form treatment groups. Individual outcomes are assumed to satisfy some statistical model and are analyzed accordingly; e.g. using two-sample t-test for continuous data. Unfortunately, such an approach is subject to certain assumptions that may be quite difficult to verify in practice.

Alternatively, one can use a randomization-based approach which makes no model assumptions on the outcome data. Under the null hypothesis of the equality of treatment effects, an individual outcome is not affected by the treatment assignment and is regarded as fixed. The treatment assignments are permuted in all possible ways consistent with the randomization procedure implemented in the trial. The randomization-based p-value is the sum of probabilities of those randomization sequences that yield the treatment difference the same or more extreme than the one actually observed in the trial. For a more detailed exposition to randomization-based inference, see recent papers by Proschan and Dodd [50] and Rosenberger, Uschner and Wang [54].

In the following example, we illustrate by simulation utility of the two approaches to statistical inference in several plausible RCT settings. The purpose is to compare the performance of various randomization procedures in situations when the model assumptions are satisfied and when they are violated. Consistent with some previous works [55, 65], we generate individual outcome data from the following model:

$$Y_i = \mu_E \delta_i + \mu_C (1 - \delta_i) + u_i + \varepsilon_i, \quad i = 1, \ldots, n, \tag{11.6}$$

where μ_E and μ_C are the mean effects for treatments E and C, respectively, $\delta_i = 1(0)$ for E

(C), u_i is an unknown term associated with the ith patient, and $\varepsilon_i \sim N(0,1)$ is a random error term, independent of δ_i and u_i.

We consider six procedures: CRD, BCD($p = 2/3$), ABCD($a = 2$), and GBCD with $\rho = 1$, 2, and 5. Two different scenarios for the mean treatment effects are explored: Null ($\mu_E = \mu_C = 0$) and Alternative ($\mu_E = 0.5$; $\mu_C = 0$). Three different scenarios for the subject effects u_i's are explored:

- "Idealized": $u_i \equiv 0$ for $i = 1, \ldots, n$.

  In this case, we have a standard setup for a two-sample t-test with no violations in the model assumptions.

- "Linear trend": $u_i = \frac{5i}{n+1}$, $i = 1, \ldots, n$.

  In this case, the outcomes are affected by a linear time trend.

- "Selection bias": $u_{i+1} = \nu \cdot \mathbf{1}\{N_E(i) < N_C(i)\} - \nu \cdot \mathbf{1}\{N_E(i) > N_C(i)\}$, $i = 0, \ldots, n-1$, with the convention $N_E(0) = N_C(0) = 0$. Here $\mathbf{1}\{\cdot\}$ stands for the indicator function and $\nu > 0$ is the "bias effect" (we set $\nu = 0.5$).

  In this case we have a setting when the investigator prefers the experimental treatment (E) and is interested in demonstrating its superiority over the control (C). We assume that higher values of the outcome signify treatment efficacy, and the investigator is aware of the past assignments and wants to "beat" randomization by guessing the next treatment assignment and selectively enrolling a "suitable" patient.

  In other words, at the $(i+1)$st allocation step, the investigator knows the numbers $N_E(i)$ and $N_C(i)$ and guesses that the next assignment will be to the treatment group that is currently underrepresented (the *convergence strategy* [17]). The biasing mechanism is as follows. Enroll a "healthier" patient (with higher expected outcome, $u_{i+1} = 0.5$), if $N_E(i) < N_C(i)$; enroll a "sicker" patient (with lower expected outcome, $u_{i+1} = -0.5$), if $N_E(i) > N_C(i)$; or enroll a "regular" patient ($u_{i+1} = 0$), if $N_E(i) = N_C(i)$.

  Note that the described selection bias model can arise not only in open-label single-center studies where investigators can have access to randomization lists, but also in randomized double-blinded studies using very restrictive and, therefore, highly predictable randomization procedures, such as the permuted block design with a small block size [11].

For the analysis of data generated according to model (11.6), we consider a two-sample t-test and a randomization-based test. For the former, the test statistic is

$$T = \frac{\bar{Y}_E - \bar{Y}_C}{\sqrt{n(S_E^2 + S_C^2)}} \sqrt{(n-2)N_E(n)N_C(n)},$$

where $n = N_E(n) + N_C(n)$, $\bar{Y}_E = \frac{\sum_{i=1}^n \delta_i Y_i}{N_E(n)}$, $\bar{Y}_C = \frac{\sum_{i=1}^n (1-\delta_i)Y_i}{N_C(n)}$, $S_E^2 = \sum_{i=1}^n \delta_i(Y_i - \bar{Y}_E)^2$, and $S_C^2 = \sum_{i=1}^n (1-\delta_i)(Y_i - \bar{Y}_C)^2$. Then $H_0 : \mu_E = \mu_C$ is rejected in favor of $H_1 : \mu_E \neq \mu_C$ at significance level α, if $|T| > t_{1-\alpha/2,n-2}$, where $t_{1-\alpha/2,n-2}$ is the $100(1 - \alpha/2)$th percentile of the t-distribution with $n - 2$ degrees of freedom.

For the randomization-based test, we first compute the observed mean difference $S_{obs} = \bar{Y}_E - \bar{Y}_C$. The randomization based p-value is $P_{rando} = \sum_{\ell \in \Omega} \mathbf{1}\{|S_\ell| \geq |S_{obs}|\} \Pr(\ell)$, where Ω is a set of all possible randomization sequences, and ℓ is a particular randomization sequence from Ω. In practice, Ω is very large and challenging to enumerate, especially when n is large. However, P_{rando} can be estimated using Monte-Carlo simulation as follows: generate B randomization sequences of length n from a given randomization algorithm, and for the mth sequence compute the test statistic, $S_m = \bar{Y}_{E,m} - \bar{Y}_{C,m}$, $m = 1, ..., B$. Then the proportion of sequences for which the test statistic is at least as extreme as the observed one is a consistent estimator of the randomization-based p-value: $\hat{P}_{rando} = B^{-1} \sum_{m=1}^{B} \mathbf{1}\{|S_m| \geq |S_{obs}|\}$. Statistical significance is declared, if $\hat{P}_{rando} < \alpha$.

We ran 10,000 simulations of a trial with $n = 200$ patients, using six different randomization procedures, with data generated under two scenarios for the mean effects (Null or Alternative), three scenarios for the subject effects("idealized", "linear trend", or "selection bias"), and data analysis using either two-sample t-test or a randomization-based test. The two-sided significance level was set to $\alpha = 0.05$. The proportion of simulation runs with statistically significant results was taken as an estimate of size (Type I error rate) under H_0, or power under H_1. For the randomization-based test we used $B = 15,000$.

Table 11.3 shows the simulation results. In the "idealized" setting, all randomization procedures maintain the Type I error rate and have similar power, both with the t-test and randomization-based test.

In the "linear trend" scenario, with the t-test, only CRD maintains the nominal 0.05 Type I error rate, whereas five other procedures are quite conservative, with Type I error rates in the range 0.001–0.018. Also, power of the t-test in this case is much reduced compared to the "idealized" case. By contrast, the randomization-based test for all six procedures yields the 0.05 Type I error rate and higher power than the t-test. These results are in good concordance with some earlier findings [27, 55]. ABCD($a = 2$) seems to have highest power of the randomization-based test in the presence of a linear trend.

Finally, in the "selection bias" scenario, all designs have degraded performance. For both the t-test and the randomization-based test, only CRD does maintain the Type I error rate at the 0.05 level, whereas five other procedures have inflated Type I error. Interestingly, the designs that privilege balance over randomness, e.g. BCD($p = 2/3$) and ABCD($a = 2$), also have greater Type I error inflation in the presence of selection bias.

The presented example highlights several important points. The choice of a randomization procedure does matter as far as the quality of statistical tests is concerned. While achieving a good tradeoff between treatment balance and allocation randomness seems like an important consideration, one should also be mindful about statistical properties of a randomization procedure, such as Type I error and power, as they are direct measures of the trial design quality. The choice of a statistical test following the randomization design is important as well. For normally distributed outcomes, both the t-test and the randomization-based test yield valid results under a population model assumptions; however, when the model assumptions are violated, the randomization test can be more robust than the t-test. Furthermore, some randomization designs (especially those that privilege balance over randomness) may be quite vulnerable to selection bias. Overall, a judicious assessment of different candidate randomization procedures for selecting one for use in a given trial should be routinely performed. A useful template, Evaluation of Randomization procedures for Design Optimization (ERDO), was recently proposed by Hilgers et al. [34]. A statistical software package `randomizeR` [73] can be also useful for this purpose.

11.3   Stratified and Covariate-Adaptive Randomization

A major prerequisite for the valid treatment comparison is similarity of the comparison groups with respect to important prognostic factors that impact the study outcome. For example, in trials of heart disease the important prognostic factors are the cholesterol level, blood pressure, age and gender, to name a few. In multi-center trials, different centers can be heterogeneous in terms of demography of the patients and adherence to the

TABLE 11.3

Type I error rate and power of a two-sample t-test and a randomization test for six randomization procedures with $n = 200$.

	t-test		Randomization test	
	Type I error rate	power	Type I error rate	power
Procedure	*Idealized scenario*			
CRD	0.052	0.938	0.052	0.939
BCD($p = 2/3$)	0.048	0.943	0.049	0.943
ABCD($a = 2$)	0.048	0.938	0.049	0.937
GBCD($\rho = 1$)	0.047	0.940	0.046	0.940
GBCD($\rho = 2$)	0.051	0.938	0.051	0.938
GBCD($\rho = 5$)	0.053	0.939	0.053	0.938
	Linear trend scenario			
CRD	0.049	0.516	0.049	0.514
BCD($p = 2/3$)	0.002	0.529	0.049	0.905
ABCD($a = 2$)	0.001	0.530	0.051	0.922
GBCD($\rho = 1$)	0.018	0.523	0.048	0.673
GBCD($\rho = 2$)	0.011	0.520	0.049	0.737
GBCD($\rho = 5$)	0.005	0.528	0.051	0.822
	Selection bias scenario			
CRD	0.054	0.948	0.053	0.948
BCD($p = 2/3$)	0.380	0.999	0.396	0.999
ABCD($a = 2$)	0.279	0.998	0.300	0.999
GBCD($\rho = 1$)	0.069	0.968	0.070	0.970
GBCD($\rho = 2$)	0.089	0.976	0.094	0.979
GBCD($\rho = 5$)	0.158	0.991	0.168	0.992

protocol procedures; therefore, the center can be an important covariate as well. While simple randomization balances treatment assignments and distributions of important covariates on average, in a single trial covariate imbalances can occur by chance, especially in small studies. Observed covariate imbalances, even though they are random by definition in a well-conducted RCT, can lead to criticism of the trial results [45]. Severe covariate imbalances may be indicative of selection bias in unblinded and open-label studies [11, 12]. Restricted randomization procedures described in Section 11.2 attempt to equalize treatment assignments throughout the trial, but they are not designed explicitly to promote balance with respect to selected covariates.

In principle, observed covariate imbalances can be adjusted for in the analysis, e.g. using regression modeling. However, it may be also desirable to avoid covariate imbalances by applying appropriate design techniques [53]. A class of randomization designs which attempt to prospectively balance treatment assignments across important selected covariates is referred to as *covariate-adaptive randomization*. The simplest covariate-adaptive randomization design is the *stratified randomization*. The International Conference on Harmonization E9 guidance [37] contains the following statement: *"Stratification by important prognostic factors measured at baseline (e.g., severity of disease, age, sex, etc.) may sometimes be valuable in order to promote balanced allocation within strata; this has greater potential benefit in small trials"*.

A stratified randomization design can be described as follows. Suppose there are M discrete covariates (stratification factors), the jth of which has ℓ_j levels. The choice of these covariates must be justified, preferably based on the literature, and these covariates should

Pros	Cons
• Ensures similarity of distributions of selected prognostic factors in the experimental and control groups, to achieve an unbiased estimate of the treatment effect. This is particularly relevant is small trials.	• If one or more stratification factors has low prevalence of some stratification levels, some strata will be small or empty.
• Decreases variability between the experimental and control groups and may potentially increase power, especially in small trials.	• When using permuted block within strata, small strata might lead to imbalances in treatment allocation due to incomplete blocks.
• Enhances interpretability of the trial results. An imbalance for a key prognostic variable may be considered a design flaw by regulatory agencies or the scientific community, even if it is adjusted by the statistical model.	• If the value of a stratification variable needs to be specifically assessed for the trial (e.g. certain biomarkers), this might prolong time to randomization.
• Provides a basis for an unbiased subgroup analysis on the basis of the corresponding stratification variables.	• The potential gain of stratification decreases as the trial size increases.
• Can facilitate important comparisons of safety outcomes.	• Using a stratification variable that is not prognostic will cause a loss of power, even with large strata.

FIGURE 11.3
Pros and cons of stratified randomization.

have strong relationship with the primary outcome. A group of patients with a particular combination of covariate levels is said to form a stratum. The total number of strata in the trial is $s = \prod_{j=1}^{M} \ell_j$. Within each stratum, a separate randomization procedure is applied using some restricted randomization design to balance treatment assignments. The most common approach is to use a permuted block design within strata. Since in practice the exact number of patients in each stratum is unknown at the outset, sufficiently many blocks of treatment assignments should be pre-generated, acknowledging that some of these blocks will not be utilized in the trial. Some other restricted randomization procedures to balance treatment assignments within strata can be used [82, 83].

Stratified randomization has both advantages and disadvantages. Some of these are summarized in Figure 11.3; see also Kernan et al. [40]. One important advantage is that stratified randomization followed by a stratified analysis can improve efficiency of the treatment effect estimation compared to post-stratification alone, especially in small trials [29].

If the number of strata s is small, stratified randomization works well. However, if s is large, some strata will contain very few patients or may even be empty for one or both treatments. This is because in a stratum with few patients only initial allocations in the first block will be utilized, and consequently imbalance may add up across strata.

Hallstrom and Davis [30] give an insightful model to evaluate probability of the overall trial imbalance with stratified randomization due to unfilled blocks. Consider a trial with n patients and s strata. Let b_r denote the block size in the rth stratum, where $r = 1, \ldots, s$. Let N_r denote the number of subjects assigned in the last block of the rth stratum and E_r denote the number of subjects assigned to treatment E among the N_r subjects ($E_r \leq N_r \leq b_r$). Conditional on N_r, E_r follows a hypergeometric distribution with mean $N_r/2$ and variance $\frac{N_r}{4} \frac{b_r - N_r}{b_r - 1}$, and if we let $D_r = N_r - 2E_r$ denote the treatment imbalance in the rth stratum, we have $E(D_r) = 0$ and $\text{Var}(D_r) = \frac{E(N_r(b_r - N_r))}{b_r - 1}$. Suppose $N_r \sim Uniform\{1, 2, \ldots, b_r\}$. Then $\text{Var}(D_r) = \frac{b_r + 1}{6}$ and summing over independent strata, the total imbalance in the trial is $D = \sum_{r=1}^{s} D_r$. Using Central Limit Theorem, D follows a normal distribution with zero mean and variance $\text{Var}(D) = \frac{\sum_{r=1}^{s} b_r + s}{6}$. Thus we can evaluate the probability of imbalance greater than w as $\Pr(|D| > w) \approx 2\left(1 - \Phi(w/\sqrt{\text{Var}(D)})\right)$. Some calculations from [30] are instructive to present here. The Cardiac Arrhythmia Suppression Trial (CAST) was planned with a total of $n = 4200$ patients. A multicenter trial design with 27 centers and 2 additional stratification variables, Ejection Fraction (5 levels) and Time Since MI Infarction (2 levels), results in $27 \times 5 \times 2 = 270$ strata. Assuming blocks of four ($b_r = 4$) and assuming that $N_r \sim Uniform\{1, 2, 3, 4\}$, we have $\text{Var}(D) = (\sum_{r=1}^{270} b_r + 270)/6 = 225$ and $\Pr(|D| > 30) \approx 2(1 - \Phi(30/15)) = 0.05$. In other words, there is 5% chance that the final treatment imbalance is 30 or more—this may not be a concern. However, if in addition stratification is made by individual hospital and there are 4 hospitals per center, the number of strata becomes $27 \times 5 \times 2 \times 4 = 1080$ with $\text{Var}(D) = (\sum_{r=1}^{1080} b_r + 1080)/6 = 900$ and $\Pr(|D| > 30) \approx 2(1 - \Phi(30/30)) = 0.317$, which means that there is 32% chance of the final imbalance of 30 or more—this may be of a concern in practice.

The above example shows that the merit of stratified randomization decreases as the number of stratification variables increases. However, in many clinical trials (e.g. in oncology) investigators may wish to achieve balance over a very large number of prognostic factors simultaneously. Instead of balancing treatment numbers within each of the s strata, treatment assignments can be balanced marginally, within each of the $\ell = \sum_{j=1}^{M} \ell_j$ levels of given covariates. Such marginal balance is sufficient if only main covariate effects but not their interactions affect the primary outcome [76]. The dynamic allocation procedures, also known as *covariate-adaptive randomization*, are alternative methods to stratified randomization for clinical trials where balance with respect to a large number of covariates is sought.

The most famous covariate-adaptive randomization procedure is the *minimization* procedure developed independently by Taves [66] and Pocock and Simon [48]. For a trial with treatments E and C and discrete covariates $Z_1, \ldots, Z_M$, suppose the $(i+1)$st subject enters the trial and has a covariate vector $z_{i+1} = (z_1, \ldots, z_M)'$. Let $N_{E,z_j}(i)$ denote the current number of treatment E assignments within level z_j ($j = 1, \ldots, M$), and $N_{C,z_j}(i)$ denote the similar quantity for treatment C. Take $D(i) = \sum_{j=1}^{M}(N_{E,z_j}(i) - N_{C,z_j}(i))$ as a measure of an overall covariate imbalance corresponding to the observed covariate profile of the $(i+1)$st subject. Then the $(i+1)$st subject is randomized to treatment E with probability

$$\phi_{i+1} = \Pr(\delta_{i+1} = 1 | D(i)) = \begin{cases} p, & \text{if } D(i) < 0; \\ 1/2, & \text{if } D(i) = 0; \\ 1 - p, & \text{if } D(i) > 0. \end{cases} \tag{11.7}$$

where $p \in (1/2, 1]$ is a user-specified coin bias probability. The minimization procedure can be viewed as an extension of Efron's BCD [24] with important covariate data. If $p = 1$, we have Taves's [66] procedure which is almost deterministic. Pocock and Simon [48] favored $p = 3/4$. Various extensions of the minimization procedure have been proposed [57]. Simulation studies comparing minimization with simple or stratified randomization show that minimization improves balance for a large number of covariates and may slightly improve power if covariates balanced in the design are also adjusted for in the analysis [2, 70, 78]. Some recent reviews show that minimization is an increasingly popular method in clinical trial practice [49, 66].

A less frequently used covariate-adaptive randomization procedure is the *marginal urn design* proposed by Wei [76]. For each covariate level, an urn containing equal number of balls of two types E and C is designated. When an eligible patient enters the trial, the urns corresponding to the patient's observed covariate levels are selected. The urn with the largest absolute difference between proportions of type E balls and type C balls is chosen to generate the treatment assignment (a random choice is made in case of a tie). A ball is drawn from the selected urn and the patient is assigned according to the type of the ball drawn. The ball is then replaced into the urn and additionally $\theta \geq 0$ balls of the same type and $\xi > \theta$ balls of the opposite type are placed into each of the urns corresponding to the patient's covariate levels. A simple approach is to set $\theta = 0$ and $\xi = 1$. The procedure is repeated for each new patient entering the trial. As with the minimization procedure, Wei's urn design can be extended in a number of ways. For instance, instead of adding balls to the urn, one can use Ivanova's *drop-the loser rule* [38] that sequentially removes balls to maintain balance over covariates. Such a procedure is less variable and is potentially more powerful than Wei's original proposal.

An alternative approach to sequentially minimize covariate imbalances was proposed by Atkinson [1] and Smith [62]. This approach is based on optimal design theory. Consider a 2-arm trial comparing E versus C with responses following a homoscedastic linear model:

$$Y_i = \alpha \delta_i + z_i'\beta + \varepsilon_i, \tag{11.8}$$

where $\delta_i = 1(-1)$ for $E(C)$, $z_i' = (1, z_{1i}, \ldots, z_{qi})$, α is the treatment effect, β is the vector of covariate effects and $\varepsilon_i \sim N(0, \sigma^2)$ are independent error terms, $i = 1, \ldots, n$. Let $\Delta_n' = (\delta_1, \ldots, \delta_n)$ denote the vector of treatment assignments and $\mathbf{Z}_n$ denote the $n \times (q+1)$ matrix of covariate values for the n patients. The variance of the estimated treatment effect is

$$\mathrm{Var}(\hat{\alpha}) = \sigma^2 \{n - \Delta_n' \mathbf{Z}_n (\mathbf{Z}_n' \mathbf{Z}_n)^{-1} \mathbf{Z}_n' \Delta_n\}^{-1}.$$

The *loss* $L_n = \Delta_n' \mathbf{Z}_n (\mathbf{Z}_n' \mathbf{Z}_n)^{-1} \mathbf{Z}_n' \Delta_n$ represents the effective number of patients from whom information is lost due to imbalance in the design. Note that L_n is a random variable because both $\mathbf{Z}_n$ and Δ_n are random, and $\mathrm{Var}(\hat{\alpha})$ is minimized if $L_n = 0$. In a trial with covariate-adaptive randomization, a sequence of observed patient covariate profiles impacts the allocation sequence Δ_n. Suppose i patients have been enrolled into the trial and the $(i+1)$st patient with covariate vector z_{i+1} is to be randomized. Based on Δ_i, $\mathbf{Z}_i$ and z_{i+1}, we want to determine δ_{i+1} that would minimize the value of L_{i+1}. Atkinson [1] and Smith [62] showed that minimizing L_{i+1} is equivalent to choosing δ_{i+1} which maximizes the value of $\{\delta_{i+1} - z_{i+1}'(\mathbf{Z}_i'\mathbf{Z}_i)^{-1}\mathbf{Z}_i'\Delta_i\}^2$. Such a rule is referred to as the *sequential D_A-optimal design*. Atkinson [1] also suggested randomizing the $(i+1)$st patient to treatment E with probability

$$\phi_{i+1} = \frac{\{1 - z_{i+1}'(\mathbf{Z}_i'\mathbf{Z}_i)^{-1}\mathbf{Z}_i'\Delta_i\}^2}{\{1 - z_{i+1}'(\mathbf{Z}_i'\mathbf{Z}_i)^{-1}\mathbf{Z}_i'\Delta_i\}^2 + \{1 + z_{i+1}'(\mathbf{Z}_i'\mathbf{Z}_i)^{-1}\mathbf{Z}_i'\Delta_i\}^2}. \tag{11.9}$$

Both the minimization and the D_A-optimal design approach have pros and cons which are summarized in Figure 11.4.

To illustrate utility of covariate-adaptive randomization, we used simulation to investigate the performance of the following seven designs:

(i) Simple randomization (CRD);

(ii) Stratified permuted block design with block size 4;

(iii) The marginal urn design of Wei [76];

(iv) The minimization procedure with $p = 3/4$ (Pocock and Simon [48]);

(v) The minimization procedure with $p = 1$ (Taves [66]);

(vi) The randomized D_A-optimal design (Atkinson [1]);

(vii) The non-randomized D_A-optimal design (Atkinson [1], Smith [62]).

	Minimization	**D_A-optimal design**
Pros	• Very simple to implement, requires no matrix inversion.	• Incorporates formal optimality into the trial design.
		• Can naturally handle continuous covariates and their interactions.
	• Both approaches have good balancing properties (as shown theoretically and by numerous simulation studies in the literature).	
	• Both approaches can improve power and efficiency for linear models, especially in small and moderate sample sizes.	
	• Theoretical properties of both approaches have been established; in particular, valid statistical tests are available (Hasegawa & Tango 2009; Shao et al. 2010).	
Cons	• Continuous covariates must be discretized before implementation.	• Relies on a correctly specified linear model; may be subject to modeling bias.
	• Has many ad-hoc features: user-defined measure of covariate imbalance, the value of coin bias probability, etc.	

FIGURE 11.4

Pros and cons of the minimization procedure and the D_A-optimal design.

We consider a two-arm trial with the total of 200 patients and a continuous response which follows a normal linear model with 3 covariates:

$$Y_i = \mu + \alpha \delta_i + \beta_1 z_{1i} + \beta_2 z_{2i} + \beta_3 z_{3i} + \varepsilon_i, \quad \varepsilon_i \sim N(0,1), \quad i = 1, \ldots, 200.$$

In the above model, $\delta_i = \pm 1$ (treatment indicator) and the covariates are as follows: $z_{1i} \sim Bernoulli(0.5)$ (gender), $z_{2i} \sim Discrete\ Uniform\{18, \ldots, 75\}$ (age), $z_{3i} \sim N(\mu = 200, \sigma = 30)$ (cholesterol level). To facilitate a comparison among the randomization designs, the covariate values for the 200 patients are simulated once and are fixed constant throughout simulations. For implementation of the stratified permuted block design (ii), Wei's marginal urn design (iii) and the minimization procedures (iv) and (v), covariates z_2 (age) and z_3 (cholesterol level) are discretized into 4 and 3 levels, respectively:

$$\widetilde{z}_2 = \begin{cases} 1, & \text{if } z_2 \in [18, 30]; \\ 2, & \text{if } z_2 \in [31, 45]; \\ 3, & \text{if } z_2 \in [46, 59]; \\ 4, & \text{if } z_2 \in [60, 75], \end{cases}$$

and

$$\widetilde{z}_3 = \begin{cases} 1 \text{ (normal level)}, & \text{if } z_3 < 200; \\ 2 \text{ (borderline risk)}, & \text{if } z_3 \in [200, 240); \\ 3 \text{ (high hisk)}, & \text{if } z_3 \geq 240. \end{cases}$$

Table 11.4 shows the distribution of covariates in the 24 strata after discretization of z_2 and z_3. Note that in some strata the numbers are not multiples of 4 and therefore the stratified permuted block design (ii) cannot achieve perfectly balanced treatment numbers due to incomplete blocks in these strata.

For the D_A-optimal design procedures (vi) and (vii) no discretization of the covariates is required. These procedures are applied directly to the observed covariate values; however we require that initial 10 allocations be performed using a permuted block design, to ensure the matrix $\mathbf{Z}_i' \mathbf{Z}_i$ can be inverted afterwards. The parameter values are set as follows

TABLE 11.4
Distribution of covariate values for a hypothetical trial with 200 patients and 3 covariates (gender, age, cholesterol level), after discretization of age and cholesterol level into 4 and 3 levels, respectively, for the total of 24 strata.

| | Male | | | Female | | |
| | Cholesterol (mg/dL) | | | Cholesterol (mg/dL) | | |
Age (years)	< 200	200–240	≥ 240	< 200	200–240	≥ 240
18–30	12	10	2	10	10	2
31–45	18	14	3	10	6	2
46–59	10	10	1	11	11	4
60–75	13	9	2	9	17	4

throughout simulations: $\mu = 1$, $\alpha = 0.5$, $\beta_1 = 0.5$, $\beta_2 = 1$, $\beta_3 = 1$. With such a specification, assuming 1 : 1 allocation, the ANCOVA t-test has approximately 94% power for rejecting the hypothesis of no treatment effect. Each procedure was simulated 10,000 times. The operating characteristics included:

- Distributions of treatment imbalance, overall in the trial and within covariate margins.

- Loss $L_n = \mathbf{\Delta}'_n \mathbf{Z}_n (\mathbf{Z}'_n \mathbf{Z}_n)^{-1} \mathbf{Z}'_n \mathbf{\Delta}_n$ as a measure of estimation precision (lower values of loss are desirable).

- Forcing index $FI_n = \frac{1}{n} \sum_{i=1}^{n} \frac{|p_i - 0.5|}{0.5}$ as a measure of lack of randomness (lower values of forcing index are desirable).

- Power of the analysis of covariance (ANCOVA) t-test of $H_0 : \alpha = 0$ (zero treatment effect).

Let us examine the balancing properties of the designs. Figure 11.5 shows simulated distributions of treatment imbalance (number on E minus number on C), overall and within covariate margins, when $n = 200$. As expected, all imbalance distributions are symmetric and centered around zero. CRD is most variable and its range of the overall trial imbalance is from -50 to 50. Wei's marginal urn design (iii) and the randomized D_A-optimal design (vi) are less variable than simple randomization and have imbalance ranges roughly from -28 to 28. For the stratified permuted block design (ii) the overall imbalance ranges from -18 to 18 (due to unfilled blocks in some strata). The two minimization designs (iv) and (v) provide very tight balance, both overall and within covariate margins—for instance, an imbalance of more than 4 is impossible for the minimization with $p = 1$ and it is very unlikely for the minimization with $p = 0.75$. Interestingly, the non-randomized D_A-optimal design (vii) achieves very good treatment balance overall in the trial and within each gender; yet within the discrete levels of age and cholesterol, the balance is worse compared to the minimization with $p = 1$. This is because the D_A-optimal design operates on the original (non-discretized) values of the covariates.

Of course, balance and improved statistical efficiency comes at the expense of randomness. Table 11.5 compares the seven designs in terms of forcing index and loss (displayed are the mean, minimum, and maximum values of the simulated distributions), and average power of ANCOVA t-test, for $n = 50, 100, 200$. Simple randomization is best in terms of randomness (forcing index is zero) but is worst in terms of loss (average loss is ~ 4, and maximum loss is > 20). The non-randomized D_A-optimal design has lowest loss but highest forcing index. Other designs provide some tradeoff between loss and forcing index. The randomized D_A-optimal design seems to have best performance overall; in particular, it has

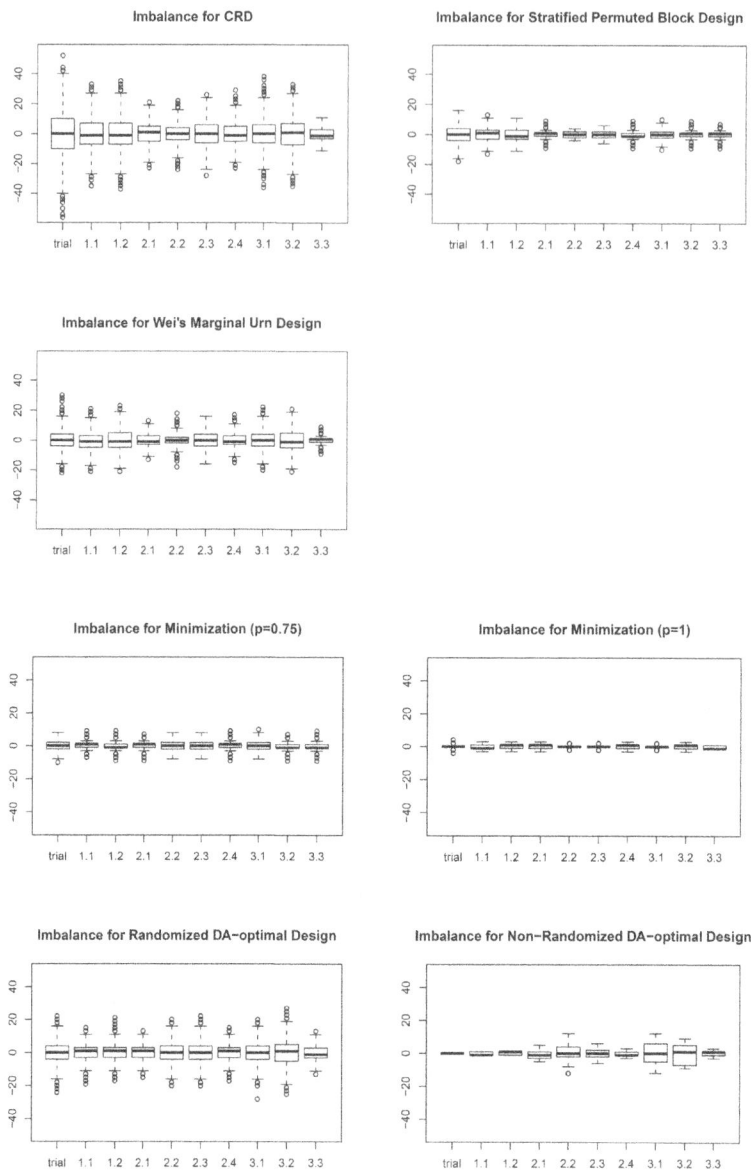

FIGURE 11.5
Simulated distributions of imbalance (number on E minus number on C), overall in the trial and within covariate margins, for seven randomization designs.

both lower forcing index and lower loss compared to the stratified permuted block design. In terms of power, covariate-adaptive randomization methods are slightly more powerful than CRD when sample size is small ($n = 50$) or moderate ($n = 100$); however, when sample size is large ($n = 200$) all designs have the same power.

In conclusion, covariate-adaptive randomization procedures do provide some advantages over complete randomization, especially for small to moderate sample sizes. The benefits include better balance, both overall in the trial and within covariate margins, better tradeoff

TABLE 11.5
Comparison of seven randomization designs in terms of Forcing Index, Loss, and Power of an ANCOVA t-test for $n = 50, 100, 200$.

Design	Forcing Index Mean (Min – Max)	Loss Mean (Min – Max)	Average Power
	$n = 50$		
(i)	0 (0 – 0)	3.97 (0.08 – 24.72)	0.38
(ii)	0.27 (0.21 – 0.37)	1.77 (0.08 – 9.08)	0.40
(iii)	0.25 (0.15 – 0.38)	1.94 (0.08 – 10.21)	0.40
(iv)	0.41 (0.32 – 0.48)	1.16 (0.08 – 8.28)	0.40
(v)	0.74 (0.62 – 0.86)	0.60 (0.08 – 3.96)	0.40
(vi)	0.40 (0.28 – 0.54)	0.88 (0.08 – 4.71)	0.41
(vii)	0.93 (0.90 – 0.94)	0.16 (0.10 – 0.21)	0.40
	$n = 100$		
(i)	0 (0 – 0)	3.94 (0.05 – 19.90)	0.69
(ii)	0.33 (0.27 – 0.41)	1.18 (0.04 – 6.24)	0.69
(iii)	0.21 (0.13 – 0.30)	1.83 (0.04 – 10.42)	0.68
(iv)	0.42 (0.36 – 0.47)	0.79 (0.04 – 6.23)	0.70
(v)	0.75 (0.65 – 0.86)	0.45 (0.04 – 3.89)	0.70
(vi)	0.28 (0.20 – 0.40)	0.82 (0.04 – 4.24)	0.71
(vii)	0.97 (0.95 – 0.97)	0.05 (0.04 – 0.06)	0.71
	$n = 200$		
(i)	0 (0 – 0)	3.96 (0.07 – 22.48)	0.94
(ii)	0.37 (0.31 – 0.42)	0.80 (0.00 – 7.16)	0.94
(iii)	0.16 (0.11 – 0.25)	1.80 (0.00 – 10.40)	0.94
(iv)	0.43 (0.39 – 0.47)	0.53 (0.00 – 5.97)	0.94
(v)	0.78 (0.70 – 0.85)	0.35 (0.00 – 3.13)	0.94
(vi)	0.20 (0.14 – 0.30)	0.84 (0.00 – 6.27)	0.94
(vii)	0.98 (0.97 – 0.99)	0.01 (0.00 – 0.06)	0.94

Designs: (i) Simple randomization (CRD); (ii) Stratified Permuted Block Design; (iii) Marginal Urn Design; (iv) Minimization with $p = 0.75$; (v) Minimization with $p = 1$; (vi) Randomized D_A-optimal Design; (vii) Non-Randomized D_A-optimal Design.

between estimation efficiency and randomness, and slight increase in power. Importantly, the covariates used in the design should also be adjusted for in the analysis via regression modeling. As noted by Senn, Anisimov and Fedorov [58]: "...*The gain that can be made through including covariates in a model is often much greater than the gain that can be made by merely balancing for them.*"

11.4 Outcome-Adaptive Randomization

While the primary goal of the RCT is the valid treatment comparison and 1:1 randomization provides a mechanism for treatment allocation, in some disease settings there is a strong ethical imperative to increase the chance of trial participants to receive an empirically better treatment. Such settings include trials for rare and severe or life threatening diseases [47] and trials for highly contagious diseases (such as Ebola) where the hope is that the disease might be eradicated by the treatment or vaccine under study [13]. Outcome-adaptive randomization (also known as *response-adaptive randomization* [36]) is a class of randomization procedures for which the treatment allocation probability is sequentially modified based on past treatment assignments and outcomes to skew allocation to the better performing treatment and achieve possibly other goals. Outcome-adaptive randomization has a long history in the biostatistics literature [77, 81]. Refer to recent monographs by Hu and Rosenberger [36] and Atkinson and Biswas [5] for details on this subject.

In this section we shall discuss Bayesian adaptive randomization (BAR), a subclass of outcome-adaptive randomization for which treatment allocation probability is sequentially modified according to the value of posterior probability that one treatment is better than the other(s). The motivation is to treat the trial patients as effectively as possible while maintaining the randomized nature of the experiment. Therefore, BAR attempts to provide a tradeoff between 1:1 randomization which is optimal for a definitive treatment comparison and a deterministic treatment assignment based on a physician's reference. Although BAR has not found broad use in practice, some interesting applications can be found in [21, 28, 68, 72].

To describe BAR, it is instructive to refer to the paper by Thompson [71]. Consider an allocation problem for a randomized trial comparing treatments E and C with respect to binary outcomes. Let p_E and p_C denote, respectively, the success probabilities for treatments E and C, which are assumed to have independent beta prior distributions. The posterior distributions of p_E and p_C are updated continuously based on accumulating binary outcome data in the trial. These posteriors are also beta distributions with parameters updated according to the observed number of "successes" among the patients assigned to E and C. The new patient is randomized between treatments E and C with probabilities

$$\lambda_E = \Pr(p_E > p_C | data) \quad \text{and} \quad \lambda_C = 1 - \lambda_E.$$

In other words, the current patient is assigned with higher probability to the treatment that is viewed more successful given data observed thus far. While this approach has an intuitive appeal, it is not optimal in any formal sense. In fact, treatment allocation probabilities (λ_E, λ_C) are highly variable and the risk of skewing allocation towards an inferior treatment may be nontrivial. To stabilize the procedure, Thall and Wathen [69] proposed modifying Thompson's [71] procedure as follows:

$$\lambda_E = \frac{\{\Pr(p_E > p_C | data)\}^\gamma}{\{\Pr(p_E > p_C | data)\}^\gamma + \{\Pr(p_E < p_C | data)\}^\gamma} \quad \text{and} \quad \lambda_C = 1 - \lambda_E, \quad (11.10)$$

where $\gamma \geq 0$ is a tuning parameter. Based on empirical evidence, Thall and Wathen [69] recommended setting $\gamma = 1/2$ or $\gamma = n/2N$, where n is the current sample size and N is the maximum sample size for the study.

To appreciate the impact of the parameter γ, we performed a simulation study. Four designs were considered: the design with $\gamma = 0$ corresponds to equal randomization (ER); the design with $\gamma = 1$ is Thompson's [71] procedure; and the designs with $\gamma = 1/2$ and $\gamma = n/(2N)$ are practical modifications of Thompson's procedure proposed by Thall and Wathen [69]. The true values of success rates are set as $p_E = 0.4$ and $p_C = 0.2$, and independent beta(1,1) priors are assumed for p_E and p_C. The maximum sample size is chosen to be $N = 50$.

Figure 11.6 shows simulated distributions of the allocation proportion to treatment E after 20, 30, and 50 patients. For the BAR procedures, the median values of the allocation proportion are greater than 0.5, in favor of the superior arm (treatment E). At the same time BAR procedures are appreciably more variable than ER, and there is a nontrivial probability that with BAR the allocation is unbalanced in the wrong direction. Smaller

Allocation proportion for Treatment E

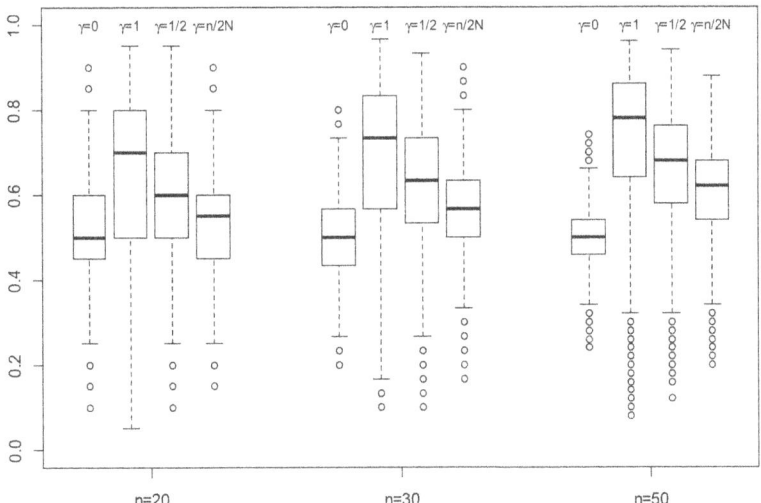

FIGURE 11.6
Simulated distributions of the allocation proportion to treatment E after 20, 30, and 50 patients for the equal randomization ($\gamma = 0$) and three BAR procedures ($\gamma = 1$, $\gamma = 1/2$, $\gamma = n/(2N)$).

values of γ (e.g. $\gamma = 1/2$ or $\gamma = n/2N$) make the randomization procedure less variable and less skewed towards the superior treatment arm.

Korn and Freidlin [42] found via simulation that BAR provides modest-to-no benefits in terms of reducing the number of non-responders in the trial compared to 1:1 and 2:1 randomization. They argue that even these modest benefits may be outweighed by additional limitations of BAR such as susceptibility to estimation bias in the presence of time trends, the need to observe outcomes fast to enable meaningful adaptation, and logistical complexities. Yuan and Yin [80] quantified the maximum percentage reduction in the expected number of non-responders that any response-adaptive design can achieve relative to 1:1 and 2:1 allocations. Such reductions are not dramatic, typically $< 3\%$ but can be as large as 10%. Berry [15] commented that BAR indeed has somewhat limited usefulness in a two-arm setting; yet its potential is much greater in multi-arm multi-objective trials with important predictive biomarkers.

Several authors extended Thompson's [42] approach to multi-armed RCTs. Let p_0 denote the response rate for the control arm and $p_1, \ldots, p_K$ denote the response rates for $K \geq 1$ experimental arms. Assume independent beta priors for p_k, $k = 0, 1, \ldots, K$. Conditional on the observed outcome data, the posterior beta distributions of p_k's are derived. Yin, Chen and Lee [79] suggested taking the average of the posterior samples of the response rates, $\bar{p} = \sum_{k=1}^{K} p_k / K$, computing the posterior probabilities $\pi_k = \Pr(p_k > \bar{p} | data)$, $k = 1, \ldots, K$, and setting the randomization probability to the kth arm as $\lambda_k = \pi_k^\gamma / \sum_{j=0}^{K} \pi_j^\gamma$. Clearly, such an approach increases allocation frequency to more successful treatment arms.

In multi-armed RCTs it is important to have sufficient number of patients in the control group. Trippa et al. [72] and Wason and Trippa [74] proposed a BAR procedure which favors more successful experimental treatments, but at the same time maintains approximate balance between the most frequently assigned experimental treatment and the control arm. Let $n_0, n_1, \ldots, n_K$ denote the current numbers of patients assigned to treatments $0, 1, \ldots, K$. Define the randomization probabilities of enrolling the new (ith) patient to treatment k as follows:

$$\pi_{ik} \propto \begin{cases} \frac{\{\Pr(p_k > p_0 | data)\}^{\gamma(i)}}{\sum_{j=1}^{K} \{\Pr(p_j > p_0 | data)\}^{\gamma(i)}}, & k = 1, \ldots, K; \\ \{\exp(\max(n_1, \ldots, n_K) - n_0)\}^{\eta(i)} / K, & k = 0. \end{cases} \quad (11.11)$$

In (11.11), $\gamma(i) \geq 0$ and $\eta(i) \geq 0$ are tuning parameters (the index i signifies that $\gamma(i)$ and $\eta(i)$ may vary across patients in the study). If $\gamma(i) = 0$, the randomization probabilities are equal across all experimental treatment arms. If $\gamma(i) \to \infty$, the allocation is concentrated at the most successful experimental treatment. Intermediate values of $\gamma(i)$ yield the allocation probabilities within these two extreme cases. The parameter $\eta(i)$ determines the degree of skewing towards the control arm: the larger $\eta(i)$ is, the higher is the probability of assigning the control treatment. Trippa et al. [72] and Wason and Trippa [74] recommend choosing $\eta(i)$ as a linear increasing function with η(overall sample size) = 0.25. After proper calibration, the procedure (11.11) can balance the sample sizes for the control and the better performing experimental treatment. In fact, if there is a single "best" experimental treatment, then as the size of the trial increases, the design will result in approximately 50% of the patients assigned to this experimental treatment and 50% of the patients assigned to the control.

A useful application of BAR is in dose-ranging studies where the objective is to determine "most informative" dose levels such as the ED95 (the dose producing 95% of the maximum treatment effect), or stop the study early if the dose–response is flat. One famous example is the ASTIN trial [16]. This trial implemented a BAR design in which randomization probability to any dose level was set proportional to the posterior probability that the given dose is the ED95. The dose–response curve was found to be flat and study was terminated early for futility, thereby achieving substantial savings in the study cost and patient resources.

Finally, it should be noted that BAR is becoming an increasingly useful approach in targeted agent development where predictive covariates (biomarkers) define subgroups of patients who are most likely to benefit from the investigational treatment [44]. The objectives of targeted agent trials are multi-fold and may include the following: (i) to test significance of treatment-biomarker interactions; (ii) to test significance of the treatment effect in certain biomarker subsets; (iii) to assign more trial patients to the treatments that are more effective given the patients' biomarker profiles. BAR attempts to address these objectives simultaneously. A few examples of BAR in targeted agent trials include the Biomarker-Integrated Approaches of Targeted Therapy of Lung Cancer Elimination (BATTLE) trial [41, 84], the Investigation of Serial Studies to Predict Your Therapeutic Response with Imaging and Molecular Analysis 2 (I-SPY 2) trial [10], and a collaborative development strategy in Alzheimer's disease [43].

11.5   Concluding Remarks

In contrast to many other experimental areas, the RCT has two unique features. First, the eligible subjects enroll into the study sequentially and the recruitment period may take several years. Second, the study subjects are humans, often diagnosed with a fatal disease, and therefore the principle of medical ethics prescribes that each study patient should receive the best available treatment as part of the trial protocol.

Randomization is a foundational element of the RCT. When properly implemented, randomization can promote various experimental objectives and contribute to the validity of the trial results. Depending on the study goals, several types of randomization designs are available. Simple randomization eliminates selection bias and balances treatment assignments asymptotically; however it can result, with non-negligible probability, in deviations from the target allocation ratio in small samples. Restricted randomization it is often used to achieve balanced or nearly balanced allocation throughout the trial.

If there is a small number of important prognostic factors correlated with the primary outcome, one can use stratified randomization to balance treatment assignments within strata formed by crossing the levels of these factors. If a number of important prognostic factors is large, one can consider using covariate-adaptive randomization designs such as the minimization procedure [48, 66] or a randomized D_A-optimal design [1, 62]. It is important that stratified and covariate-adaptive randomization designs should always be followed by a covariate-adjusted analysis for the inference to be valid [58, 60]. An alternative to the model-based inference is the randomization-based inference for which the experimental randomization forms the basis for statistical tests [32]. While dynamic allocation procedures raised controversy and criticism from the European regulators in 2003 [22], the follow-up discussions and commentaries [18, 23, 51], as well as recent methodological advances [35] have led to a revised guidance on adjustment for baseline covariates [25].

In some clinical trials there is a strong need to increase the chance of a study patient to be assigned to the empirically better treatment. These settings include, for instance, trials for rare diseases such as childhood cancers or trials for highly contagious diseases such as Ebola. Outcome-adaptive randomization can be used to provide a practical solution to the "treatment versus experimentation" dilemma. Bayesian adaptive randomization is a subclass of outcome-adaptive randomization for which treatment allocation probability is modified according to the value of posterior probability that one treatment is better than the other(s). Other types of outcome-adaptive randomization are also available [5, 36]. While outcome-adaptive randomization can be attractive from the individual ethics perspectives, it is operationally more complex than other types of randomization; in particular, the outcome data must be observed quickly and it must be of excellent quality to enable accurate estimation and adaptation throughout the trial. One major criticism of outcome-adaptive randomization is that it can lead to biased estimates and inflated probability of Type I error in the presence of time trends. To address this issue, some researchers suggested updating randomization probabilities after blocks of patients rather than after every patient, and adjust for the blocking factor through regression modeling [19, 39]. Alternatively, randomization tests that ensure control of the Type I error rate in the presence of time trends in trials with outcome-adaptive randomization can be used [50, 61].

Bibliography

[1] Atkinson, A. C. (1982). Optimal biased coin designs for sequential clinical trials with prognostic factors. *Biometrika* **69(1)**, 61–67.

[2] Atkinson, A. C. (2002). The comparison of designs for sequential clinical trials with covariate information. *Journal of the Royal Statistical Society Series A* **165(2)**, 349–373.

[3] Atkinson, A. C. (2012). Bias and loss: the two sides of a biased coin. *Statistics in Medicine* **31**, 3494–3503.

[4] Atkinson, A. C. (2014). Selecting a biased-coin design. *Statistical Science* **29(1)**, 144–163.

[5] Atkinson, A. C. and Biswas, A. (2014). *Randomised Response-Adaptive Designs in Clinical Trials*. Chapman and Hall/CRC Press, Boca Raton, FL.

[6] Baldi Antognini, A. (2008). A theoretical analysis of power of biased coin designs. *Journal of Statistical Planning and Inference* **138**, 1792–1798.

[7] Baldi Antognini, A. and Giovagnoli, A. (2004). A new 'biased coin design' for the sequential allocation of two treatments. *Applied Statistics* **53**, 651–664.

[8] Baldi Antognini, A., Rosenberger, W. F., Wang, Y., and Zagoraiou, M. (2015). Exact optimum coin bias in Efron's randomization procedure. *Statistics in Medicine* **34**, 3760–3768.

[9] Bally, L., Thabit, H., Hartnell, S., Andereggen, E., Ruan, Y., Wilinska, M. E., Evans, M. L., Wertli, M. M., Coll, A. P., Stettler, C., and Hovorka, A. R. (2018). Closed-loop insulin delivery for glycemic control in noncritical care. *The New England Journal of Medicine* **379**, 547–556.

[10] Barker, A. D., Sigman, C. C., Kelloff, G. J., Hylton, N. M., Berry, D. A., and Esserman, L. J. (2009). I-SPY 2: an adaptive breast cancer trial design in the setting of neoadjuvant chemotherapy. *Clinical Pharmacology and Therapeutics* **86(1)**, 97–100.

[11] Berger, V. W. (2005a). *Selection Bias and Covariate Imbalances in Randomized Clinical Trials*. Wiley, New York.

[12] Berger, V. W. (2005b). Quantifying the magnitude of baseline covariate imbalances resulting from selection bias in randomized clinical trials. *Biometrical Journal* **47(2)**, 119–127.

[13] Berger, V. W. (2015). Letter to the Editor: A note on response-adaptive randomization. *Contemporary Clinical Trials* **40**, 240.

[14] Berger, V. W., Ivanova, A., and Knoll, M. D. (2003). Minimizing predictability while retaining balance through the use of less restrictive randomization procedures. *Statistics in Medicine* **22**, 3017–3028.

[15] Berry, D. A. (2011). Adaptive clinical trials: the promise and the caution. *Journal of Clinical Oncology* **29(6)**, 606–609.

[16] Berry, D. A., Müller, P., Grieve, A. P., Smith, M. K., Parke, T., Blazek, R., Mitchard, N., and Krams, M. (2002). Adaptive Bayesian designs for dose-ranging drug trials. In B. Carlin, A. Carriquiry, C. Gatsonis, A. Gelman, R. E. Kass, I. Verdinelli, M. West, editors, *Case Studies in Bayesian Statistics V*, pages 99–181, Springer, Berlin.

[17] Blackwell, D. and Hodges, J. L. (1957). Design for the control of selection bias. *The Annals of Mathematical Statistics* **28(2)**, 449–460.

[18] Buyse, M. and McEntegart, D. (2004). Achieving balance in clinical trials: An unbalanced view from EU regulators. *Applied Clinical Trials* **13(5)**, 36–40.

[19] Chappell, R. and Karrison, T. (2006). Continuous Bayesian adaptive randomization based on event times with covariates by Cheung et al. (with Authors' reply). *Statistics in Medicine* **25**, 3050–3054.

[20] Chen, Y. P. (1999). Biased coin design with imbalance tolerance. *Communications in Statistics—Stochastic Models* **15(5)**, 953–975.

[21] Cheung, Y. K., Inoue, L. Y. T., Wathen, J. K., and Thall, P. F. (2006). Continuous Bayesian adaptive randomization based on event times with covariates. *Statistics in Medicine* **25**, 55–70.

[22] Committee for Proprietary Medicinal Products (CPMP). (2003). Points to Consider on Adjustment for Baseline Covariates. CPMP/EWP/2863/99.

[23] Day, S., Grouin, J. M., and Lewis, J. (2005). Achieving balance in clinical trials. *Applied Clinical Trials* **13**, 41–43.

[24] Efron, B. (1971). Forcing sequential experiments to be balanced. *Biometrika* **58**, 403–417.

[25] European Medicines Agency (EMA) Committee for Medicinal Products for Human Use. Guideline on adjustment for baseline covariates in clinical trials. 26 February, 2015.

[26] Friedman, L. M., Furberg, C. D., and DeMets, D. L. (2010). *Fundamentals of Clinical Trials*, 4th edition. Springer, New York.

[27] Galbete, A., and Rosenberger, W. F. (2016). On the use of randomization tests following adaptive designs. *Journal of Biopharmaceutical Statistics* **26(3)**, 466–474.

[28] Giles, F. J., Kantarjan, H. M., Cortes, J. E., Garcia-Manero, G., Verstovsek, S., Faderl, S., Thomas, D. A., Ferrajoli, A., O'Brien, S., Wathen, J. K., Xiao, L. C., Berry, D. A., and Estey, E. H. (2003). Adaptive randomized study of Idarubicin and Cytarabine versus Troxacitabine and Cytarabine versus Troxacitabine and Idarubicin in untreated patients 50 years or older with adverse karyotype acute myeloid leukemia. *Journal of Clinical Oncology* **21(9)**, 1722–1727.

[29] Grizzle, J. E. (1982). A note on stratifying versus complete random assignment in clinical trials. *Controlled Clinical Trials* **3**, 365–368.

[30] Hallstrom, A. and Davis, K. (1988). Imbalance in treatment assignments in stratified blocked randomization. *Controlled Clinical Trials* **9**, 375–382.

[31] Harrington, D. P. (2000). The randomized clinical trial. *Journal of the American Statistical Association* **95(449)**, 312–315.

[32] Hasegawa, T. and Tango, T. (2009). Permutation test following covariate-adaptive randomization in randomized controlled trials. *Journal of Biopharmaceutical Statistics* **19(1)**, 106–119.

[33] Heritier, S., Gebski, V., and Pillai, A. (2005). Dynamic balancing randomization in controlled clinical trials. *Statistics in Medicine* **24**, 3729–3741.

[34] Hilgers, R-D., Uschner, D., Rosenberger, W. F., and Heussen, N. (2017). ERDO – a framework to select an appropriate randomization procedure for clinical trials. *BMC Medical Research Methodology* **17**:159.

[35] Hu, F., Hu, Y., Ma, Z., and Rosenberger, W. F. (2014). Adaptive randomization for balancing over covariates. *Wiley Interdisciplinary Reviews: Computational Statistics* **6(4)**, 288–303.

[36] Hu, F. and Rosenberger, W. F. (2006). *The Theory of Response-Adaptive Randomization in Clinical Trials*. Wiley, New York.

[37] International Conference on Harmonization of Technical Requirements for Registration of Pharmaceuticals for Human Use. Topic E9: Statistical principles for clinical trials. (1998).

[38] Ivanova, A. (2003) A play-the-winner-type urn design with reduced variability. *Metrika* **58**, 1–13.

[39] Karrison, T. G., Huo, D., and Chappell, R. (2003). A group sequential, response-adaptive design for randomized clinical trials. *Controlled Clinical Trials* **24**, 506–522.

[40] Kernan, W. N., Viscoli, C. M., Makuch, R. W., Brass, L. M., and Horwitz, R. I. (1999). Stratified randomization for clinical trials. *Journal of Clinical Epidemiology* **52(1)**, 19–26.

[41] Kim, E. S., Herbst, R. S., Wistuba, I. I., Lee, J. J., Blumenschein, G. R. Jr., Tsao, A., Stewart, D. J., Hicks, M. E., Erasmus, J. Jr., Gupta, S., Alden, C. M., Liu, S., Tang, X., Khuri, F. R., Tran, H. T., Johnson, B. E., Heymach, J. V., Mao, L., Fossella, F., Kies, M. S., Papadimitrakopoulou, V., Davis, S. E., Lippman, S. M., and Hong, W. K. (2011). The BATTLE trial: personalizing therapy for lung cancer. *Cancer Discovery* **1(1)**, 44–53.

[42] Korn, E. L. and Freidlin, B. (2011). Outcome-adaptive randomization: is it useful? *Journal of Clinical Oncology* **29(6)**, 771–776.

[43] Krams, M. and Dragalin, V. (2014). Considerations and optimization of adaptive trial design in clinical development programs. In Weile He, Jose Pinheiro and Olga M. Kuznetsova, editors. *Practical Considerations for Adaptive Trial Design and Implementation*: pages 69–90. Springer, New York.

[44] Lee, J. J., Gu, X., and Liu, S. (2010). Bayesian adaptive randomization for targeted agent development. *Clinical Trials* **7**, 584–596.

[45] Leyland-Jones B. on behalf of the BEST Investigators and Study Group. (2003). Breast cancer trial with erythropoietin terminated unexpectedly. *Lancet Oncology* **4(8)**, 459–460.

[46] Matts, J. P. and McHugh, R. B. (1983). Conditional Markov chain design for accrual clinical trials. *Biometrical Journal* **25**, 563–577.

[47] Palmer, C. R. and Rosenberger, W. F. (1999). Ethics and practice: Alternative designs for phase III randomized clinical trials. *Controlled Clinical Trials* **20**, 172–186.

[48] Pocock, S. J. and Simon, R. (1975). Sequential treatment assignment with balancing for prognostic factors in the controlled clinical trial. *Biometrics* **31**, 103–115.

[49] Pond, G. R. (2010). Trends in the application of dynamic allocation methods in multi-arm cancer clinical trials *Clinical Trials* **7(3)**, 227–234.

[50] Proschan, M. A. and Dodd, L. E. (2019). Re-randomization tests in clinical trials *Statistics in Medicine* **38**, 2292–2302.

[51] Roes, K. C. B. (2004). Regulatory perspectives: dynamic allocation as a balancing act. *Pharmaceutical Statistics* **3**, 187–191.

[52] Rosenberger, W. F. and Lachin, J. L. (2015). *Randomization in Clinical Trials: Theory and Practice*, 2nd edition. Wiley, New York.

[53] Rosenberger, W. F. and Sverdlov, O. (2008). Handling covariates in the design of clinical trials. *Statistical Science* **23(3)**, 404–419.

[54] Rosenberger, W. F., Uschner, D., and Wang, Y. (2019). Randomization: The forgotten component of the randomized clinical trial. *Statistics in Medicine* **38**, 1–12.

[55] Rosenkranz, G. K. (2011). The impact of randomization on the analysis of clinical trials. *Statistics in Medicine* **30**, 3475–3487.

[56] Salama, I., Ivanova, A., and Quaqish, B. (2008). Efficient generation of constrained block allocation sequences. *Statistics in Medicine* **27**, 1421–1428.

[57] Scott, N. W., McPherson, G. C., Ramsay, C. R., and Campbell, M. K. (2002). The method of minimization for allocation to clinical trials: a review. *Controlled Clinical Trials* **23**, 662–674.

[58] Senn, S., Anisimov, V. V., and Fedorov, V. V. (2010). Comparisons of minimization and Atkinson's algorithm. *Statistics in Medicine* **29**, 721–730.

[59] Shao, H. and Rosenberger, W. F. (2016). Properties of the random block design for clinical trials. In J. Kunert, C. H. Müller, A. C. Atkinson, editors, *mODa 11—Advances in Model-Oriented Design and Analysis*, pages 225–233, Springer International Publishing Switzerland.

[60] Shao, J., Yu, X., and Zhong, B. (2010). A theory of testing hypotheses under covariate adaptive randomization. *Biometrika* **97**, 347–360.

[61] Simon, R. and Simon, N. (2011). Using randomization tests to preserve Type I error with response-adaptive and covariate-adaptive randomization. *Statistics and Probability Letters* **81**, 767–772.

[62] Smith, R. L. (1984). Sequential treatment allocation using biased coin designs. *Journal of the Royal Statistical Society Series B* **46**, 519–543.

[63] Soares, J. F. and Wu, C. F. J. (1983). Some restricted randomization rules in sequential designs. *Communications in Statistics—Theory and Methods* **12(17)**, 2017–2034.

[64] Sverdlov, O. and Rosenberger, W. F. (2013). Randomization in clinical trials: can we eliminate bias? *Clinical Investigation* **3(1)**, 37–47.

[65] Sverdlov, O. and Ryeznik, Y. (2019). Implementing unequal randomization in clinical trials with heterogeneous treatment costs. *Statistics in Medicine* **38**, 2905–2927.

[66] Taves, D. R. (1974). Minimization: a new method of assigning subjects to treatment and control groups. *Clinical Pharmacology and Therapeutics* **15**, 443–453.

[67] Taves, D. R. (2010). The use of minimization in clinical trials. *Contemporary Clinical Trials* **31**, 180–184.

[68] Thall, P. F. and Wathen, J. K. (2005). Covariate-adjusted adaptive randomization in a sarcoma trial with multi-stage treatments. *Statistics in Medicine* **24**, 1947–1964.

[69] Thall, P. F. and Wathen, J. K. (2007). Practical Bayesian adaptive randomization in clinical trials. *European Journal of Cancer* **43**, 860–867.

[70] Therneau, T. M. (1993). How many stratification factors are 'too many' to use in a randomization plan? *Controlled Clinical Trials* **14(2)**, 98–108.

[71] Thompson, W. R. (1933). On the likelihood that one unknown probability exceeds another in the view of the evidence of the two samples. *Biometrika* **25**, 275–294.

[72] Trippa, L., Lee, E. Q., Wen, P. Y., Batchelor, T. T., Cloughesy, T., Parmigiani, G., and Alexander, B. M. (2012). Bayesian adaptive randomized trial design for patients with recurrent glioblastoma. *Journal of Clinical Oncology* **30(26)**, 3258–3263.

[73] Uschner, D., Schindler, D., Heussen, N., and Hilgers, R-D. (2018). randomizeR: An R package for the assessment and implementation of randomization in clinical trials. *Journal of Statistical Software* **85**:8.

[74] Wason, J. M. S. and Trippa, L. (2014). A comparison of Bayesian adaptive randomization and multi-stage designs for multi-arm clinical trials. *Statistics in Medicine* **33**, 2206–2221.

[75] Wei, L. J. (1978a). The adaptive biased coin design for sequential experiments. *The Annals of Statistics* **6(1)**, 92–100.

[76] Wei, L. J. (1978b). An application of an urn model to the design of sequential controlled clinical trials. *Journal of the American Statistical Association* **73**, 559–563.

[77] Wei, L. J. and Durham, S. (1978). The randomized play-the-winner rule in medical trials. *Journal of the American Statistical Association* **73**, 840–843.

[78] Weir, C. J. and Lees, K. R. (2003). Comparison of stratification and adaptive methods for treatment allocation in an acute stroke clinical trial. *Statistics in Medicine* **22**, 705–726.

[79] Yin, G., Chen, N., and Lee, J. J. (2012). Phase II trial design with Bayesian adaptive randomization and predictive probability *Applied Statistics* **61(2)**, 219–235.

[80] Yuan, Y. and Yin, G. (2011). On the usefulness of outcome-adaptive randomization. *Journal of Clinical Oncology* **29(13)**, e390–e392.

[81] Zelen, M. (1969). Play the winner rule and the controlled clinical trial. *Journal of the American Statistical Association* **64**, 131–146.

[82] Zelen, M. (1974). The randomization and stratification of patients to clinical trials. *Journal of Chronic Diseases* **28**, 365–375.

[83] Zhao, W. (2014). A better alternative to stratified permuted block design for subject randomization in clinical trials. *Statistics in Medicine* **33**, 5239–5248.

[84] Zhou, X., Liu, S., Kim, E. S., Herbst, R. S., and Lee, J. J. (2008). Bayesian adaptive design for targeted therapy development in lung cancer—a step towards personalized medicine. *Clinical Trials* **5**, 181–193.

12

Background to Sample Size Calculations

Jo Rothwell, Cindy Cooper, Steven Julious and Mike Campbell

CONTENTS

12.1   Introduction

Chapters 8-11 highlighted the importance of design in clinical trials, as well as the fundamental components of a typical sample size calculation. This chapter will discuss the various sample size formulae used for different trial designs and outcome measures, as well as discussing the sensitivity of the calculation to the anticipated effect.

The sample size calculation is an important part of a trial protocol. The calculation estimates the minimum number of patients needed in the trial for a given power and significance level for a pre-specified clinically meaningful difference between the two treatments (d).

The clinically meaningful difference is important for a number of reasons, both clinical and statistical. The difference in treatments could be, for example, the difference in the proportion of patients surviving on an experimental treatment compared to the standard treatment, or the expected change in systolic blood pressure on an experimental treatment compared to an active control treatment. It could be argued that the sample size calculation is most sensitive to the target difference. If the difference is halved then the sample size quadruples (Fayers and Machin, 1995). This is a serious consideration since very large sample sizes may not be achievable due to cost, time or other resources. If it is too small, it may be clinically irrelevant; whereas if it is too large, it may be unrealistic.

This chapter will describe the sample size calculations for parallel group trials and cross-over trials (discussed in Chapter 8) of each of the following types of study: superiority studies, equivalence studies, and non-inferiority studies. These calculations will be shown for both continuous outcomes and binary outcomes, and worked examples for each trial and outcome type will be provided.

12.2   Types of Trials

For the purpose of this chapter we shall be focusing on trials comparing two treatments in the form of a parallel group trial and a cross-over trial.

12.2.1   Parallel group trials

In a parallel group trial, patients are randomly assigned to one of two treatment groups. The ideal scenario is that all other baseline characteristics of the patients are roughly similar in each group (i.e. equal number of males and females, location of hospital, patient ages). If we consider a diagram of the trial, it would look like Figure 12.1.

Each group is given a different intervention (e.g. treatment versus placebo, experimental treatment versus current standard treatment) then the two groups are compared directly with each other.

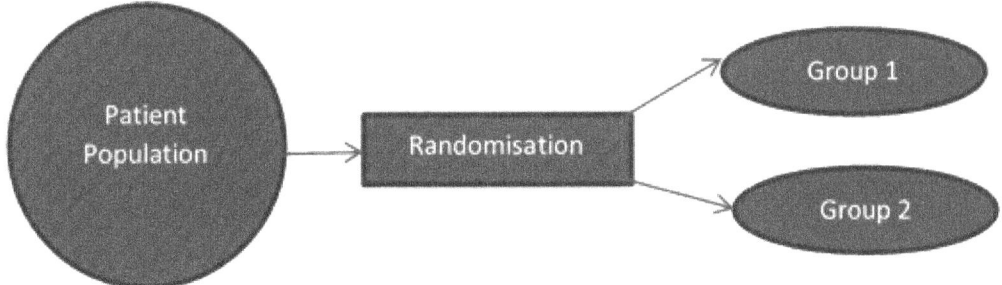

FIGURE 12.1
Illustration of a typical randomisation for a parallel group trial.

12.2.2 Cross-over trials

The theory of cross-over design has been discussed previously in Chapter 8, however a summary is included here. The aim for this type of trial is that all patients experience all treatments under investigation. This is done by randomly allocating the patients to different treatment sequences, for two period cross-over studies these would be A followed by B or B followed by A. Thus, each group has a different order of treatments assigned to it. This is illustrated in Figure 12.2.

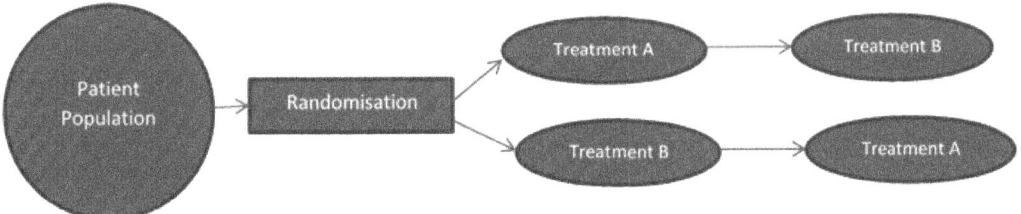

FIGURE 12.2
Illustration of a typical randomisation for a cross-over trial

The key point for this trial is that all patients get all treatments; the order in which they are given is random. Once the first treatment has been given, the groups usually have a washout period in order for a patient to clinically return to baseline, therefore providing a constant base on which to test each of the subsequent treatments.

The concentration in this chapter will be on two period cross-over studies. The key point for this trial is that all patients get all treatments; the order in which they are given is random. Once the first treatment has been given, the groups are usually given a washout period in order for key clinical measurements to return to baseline, therefore providing a constant base to test each of the subsequent treatments on. These trials enable within-patient analysis since each patient has experienced both treatments.

There are two main assumptions when performing a cross-over trial, namely that the order the treatments are received does not affect the patients' response to the treatment, i.e. lack of period effects, and that all patients return to baseline prior to the second treatment, i.e. lack of carryover effects. This type of trial is best suited to long-term stable conditions such as eczema or asthma, less so for conditions where the patient is likely to get worse over time (degenerative conditions).

TABLE 12.1
Type I and Type II errors and power.

Decide to	Null hypothesis is actually	
Reject Null Hypothesis	Correct Decision Power of test $(1 - \beta)$	Type I Error (α)
Not Reject Null Hypothesis	Type II Error (β)	Correct Decision

12.3 Continuous Outcomes

This section will focus on trials with continuous outcomes. Within each outcome type, the trial designs will be discussed separately.

12.3.1 Superiority trials

Superiority trials focus on providing statistically significant evidence against the null hypothesis that the two treatments in question are the same with respect to the comparison of interest (e.g. mean response time). The null and alternate hypotheses are as follows:

- H_0: $\mu_A = \mu_B$ (The two treatments are the same)

- H_1: $\mu_A \neq \mu_B$ (The two treatments are different)

The two types of errors which can occur when testing the null hypothesis are Type I and Type II errors (Julious and Walters, 2014; Neyman and Pearson, 1928a,b, 1933a,b). These errors are shown in Table 12.1 in terms of the decision made at the end of the trial and whether the null hypothesis is really true or not, but are also summarised below.

- Type I Error : Rejecting H_0 when it is true

- Type II Error : Not rejecting H_0 when it is false

The sample size calculation takes these errors into consideration and aims to estimate the sample size whilst minimising them. The Type I error is conventionally fixed at a two-sided level of 0.05 which is also the significance level of the trial. The Type II error is usually fixed at 0.1 but can be as high as 0.2.

A Type I error is deemed more serious in both medical and financial terms. From a medical perspective, a Type I error would mean an ineffective treatment being shown to be effective when it is not the case. Giving a patient an ineffective treatment would be unethical, as it may be preventing the patient from receiving an effective treatment, and could result in the condition of the patient deteriorating. From a financial perspective, money may be spent to change treatments and medical practice unnecessarily if the new treatment were not more effective than the current treatment.

A Type II error is less costly both clinically and financially. A Type II error would not result in a change in medical practice, although the patients may be deprived of an effective intervention, and is therefore seen as the lesser of two evils. It is because a Type II error is deemed not as costly as a Type I error that it can be set at a higher level. In a regulated study setting by health authorities, it is referred to as the Sponsor's risk.

It is more common to think not in terms of the Type II error but in terms of the power of a trial (power= $1 - \beta = 1-$Type II Error). The power of a trial can be understood as the probability of rejecting H_0 when a specific alternative is true, so getting the result correct. Therefore the higher the power of a trial is, the more likely that a significant result will be declared if there is truly a difference. It is the chance that the study will provide a conclusive result.

If we consider the null hypothesis for a superiority trial, it can be seen that there are two ways in which the null can be rejected. This can occur if $\mu_A < \mu_B$ or if $\mu_A > \mu_B$. It is due to these two chances that this type of test is referred to as a two-tailed test. This simply means that the Type I error is split equally between the two instances, so each tail has a 0.025 probability of occurring under the null hypothesis. The remainder of this chapter will assume $\sigma_A^2 = \sigma_B^2 = \sigma^2$ (homoscedasticity).

Superiority trials are normally used when comparing a treatment to a control. This control could be a placebo (negative control) or a current treatment (active control). Superiority trials are preferred for this when new drugs have been formulated since you don't want the new treatment to be only as good as the control; you specifically want to demonstrate that the new treatment is superior.

12.3.1.1 Parallel group trials

As mentioned earlier in the chapter, a parallel group trial consists of two groups with the sample size for group B being able to be written as a multiple of the sample size for group A ($rn_B = n_A$) where r is the allocation ratio. A preliminary sample size calculation for the population would be (Brush, 1988)

$$n_A = \frac{(r+1)(Z_{1-\beta} + Z_{1-\alpha/2})^2\sigma^2}{rd_S^2}, \tag{12.1}$$

where r is the allocation ratio, d_S is the target difference between the treatments, $Z_{1-\beta}$ and $Z_{1-\alpha/2}$ are the Normal quantiles for power $(1 - \beta)$ and significance (α), and σ^2 is the population variance.

This calculation is based on the population variance estimate; therefore the Normal distribution values can be used. However, when a trial has been conducted and the data collected, the population variance σ^2 is considered to be unknown and the sample variance estimate from the trial, s^2, is used instead. As a result of this, the Normal distribution and by extension the Z-quantiles cannot be used so the t-distribution and t-quantile

$$n_A = \frac{(r+1)(Z_{1-\beta} + Z_{1-\alpha/2})^2\sigma^2}{rd_S^2}, \tag{12.2}$$

are used (Brush, 1988; Chow et al., 2002; Senn, 1993). This gives us the following equation, where n_A is the smallest integer value that satisfies it

$$n_A \geq \frac{(r+1)(Z_{1-\beta} + t_{1-\alpha/2, n_A(r+1)-2})^2\sigma^2}{rd_S^2}. \tag{12.3}$$

However, it is noticeable that this equation does not give a direct estimate since n_A appears on both sides of the equation. The best method to deal with this is to re-write the equation in terms of the power and solve using an iterative technique

$$1 - \beta = \Phi\left(\sqrt{\frac{rn_A d_S^2}{(r+1)\sigma^2}} - t_{1-\alpha/2, n_A(r+1)-2}\right). \tag{12.4}$$

Here, $\Phi(.)$ is the cumulative density of a Normal distribution. When the sample variance is being used instead of the population variance, Senn describes how instead of using the Normal distribution, the power should be estimated from the non-central t-distribution with $n_A(r+1) - 2$ degrees of freedom and a non-centrality parameter $\sqrt{\frac{rn_A}{(r+1)}}$ (Senn, 1993). This is due to the power being estimated under the alternative hypothesis, which states $d \neq 0$ therefore the corresponding t-distribution would be non-central. The two distributions (t- and *Normal-*) are very similar, with the t-distribution being slightly fatter than the Normal distribution (Julious, 2010). The equation above can then therefore be rewritten as follows, using the non-central t-distribution

$$1 - \beta = 1 - T^{-1}\left(t_{1-\frac{\alpha}{2}, n_A(r+1)-2}, n_A(r+1) - 2, \sqrt{\frac{rn_A d_S^2}{(r+1)\sigma^2}}\right), \qquad (12.5)$$

where $T^{-1}(\dots)$ is the cumulative distribution function of a non-central t-distribution. To allow for the Normal approximation to the non-central t-distribution a small correction factor can be added to (12.2) so it better approximates to (12.5) as follows (Guenther, 1981; Julious et al., 1999):

$$n_A = \frac{(r+1)(Z_{1-\beta} + Z_{1-\alpha/2})^2 \sigma^2}{rd_S^2} + \frac{Z_{1-\alpha/2}}{4}. \qquad (12.6)$$

12.3.1.2   Quick results

If a quick estimate is required for a sample size, we can use the quick results in each section of this chapter. It simply estimates the $(r+1)(Z_{1-\beta} + Z_{1-\alpha/2})^2$ part of the equation with their values from the Normal tables for the Type I and Type II errors.

For 0.90 power, equal allocation, 2-sided significance level the sample size per arm can be calculated by (Julious, 2004)

$$n_A = \frac{21\sigma^2}{d_S^2} \qquad (12.7)$$

where d_S is the target difference between the treatments, σ^2 is the population variance. This equation comes from the following equation when $r = 1$:

$$n_A = \frac{10.5(r+1)\sigma^2}{rd_S^2}. \qquad (12.8)$$

If there was not equal allocation, ($n_B = rn_A$), then this would be the quick result to use. The value 10.5 arises from the $(Z_{1-\beta} + Z_{1-\alpha/2})^2$ part of Equation 12.2, it is an estimate as oppose to the exact value when $Z_{1-\beta} = Z_{0.9} = 1.282$ and $Z_{1-\alpha/2} = Z_{0.975} = 1.96$.

For 0.80 power, equal allocation, 2-sided significance level (equal allocation) we can use

$$n_A = \frac{16\sigma^2}{d_S^2} \qquad (12.9)$$

where d_S is the target difference between the treatments and σ^2 is the population variance. Again, the value of 16 arises from $r = 1$, $Z_{1-\beta} = Z_{0.8} = 0.842$, and $Z_{0.975} = 1.96$. A more accurate result would have 15.7, not 16, so this result would be a little more conservative.

Both the quick results give reasonable estimates of the sample size required and could be used as initial sample size estimates.

Another consideration for parallel group trials is whether there is a cluster effect. This only affects cluster parallel group trials and results in a slightly different sample size formula and calculation. This topic is discussed in further detail at the end of the sectionin Section 12.4 (see also Chapter 10). Note 1: The sample size tables used in the examples throughout this chapter provide a more accurate result than the hand calculations because they are taken using the truncated t-distribution as oppose to the Normal approximation. Note 2: Any sample sizes which result in a decimal shall be rounded up to account for the extra participant needed.

12.3.1.3 Worked example 1

Let us consider a trial of blood pressure medication which is expected to reduce systolic blood pressure by 10 mmHg. The allocation the investigator has chosen is equal between the treatment and control groups, and the expected standard deviation in the population which the trial is aimed at is 50 mmHg. Using the formal equation, the number of subjects per arm is

$$n_A = \frac{(r+1)(Z_{1-\beta} + Z_{1-\alpha/2})^2 \sigma^2}{r d_S^2} + \frac{Z_{1-\alpha/2}}{4} = \frac{2 \times (1.282 + 1.96)^2 \times 50^2}{10^2} + \frac{1.96}{4} = 526.02 \approx 527$$
(12.10)

Using the information provided by the researcher and the quick result above for 0.90 power and a significance level of 0.05, we get the sample size to be 525 per group.

Using 12.2, it can be seen that for a given $d = 10$, $\sigma = 50$, and allocation rate $r = 1$ we get a standardised difference of $\delta = 0.2$ which results in a sample size of 527 per group.

It is worth noting that there is a difference of 2 patients per group depending on which method is used; the full calculation, the quick calculation or the tables, however in the grand scheme of things these differences are negligible.

12.3.1.4 Cross-over trials

In order for us to be able to estimate a sample size for a cross-over trial, we need to first estimate the within-subject standard deviation, σ_w. This can be extracted from the residual line of the analysis of variance (ANOVA) model; it evaluates the variation which occurs through repeated measures on the same patient. The within-subject variability from the ANOVA model is directly related to the variability around the difference, σ_d, from a paired t-test as $\sigma_d^2 = 2\sigma_w^2$. In this chapter we shall assume an equal allocation (i.e. $r = 1$).

The sample size can be calculated using the within-group standard deviation and the effect size, the sample size calculation can be attained by a similar method as in the parallel group study (Guenther, 1981),

$$n = \frac{2(Z_{1-\beta} + Z_{1-\alpha/2})^2 \sigma_w^2)}{d_S^2} + \frac{Z_{1-\alpha/2}}{2},$$
(12.11)

where n is the total sample size. Similar to parallel group trials, an additional factor of $\frac{Z_{(1-\alpha/2)}}{2}$ can be added to allow for the approximation to the Normal distribution. The non-central t-distribution result with $n - 2$ degrees of freedom and non-centrality parameter $\sqrt{\frac{nd_S^2}{2\sigma_w^2}}$ is given by (Senn, 1993)

$$1 - \beta = 1 - T^{-1}\left(t_{1-\alpha/2, n-2}, n-2, \sqrt{\frac{nd_S^2}{2\sigma_w^2}}\right).$$
(12.12)

TABLE 12.2

Sample size requirements for one group, n_A (where $n_B = rn_A$), with various standardised differences ($\delta = \frac{d}{\sigma}$) and allocation ratios, r, for a parallel group trial. This table has been calculated for a 0.90 power and a two sided Type I error probability of 0.05. These sample sizes are calculated from the non-central t-distribution.

	Allocation ratios			
δ	1	2	3	4
0.05	8407	6306	5605	5255
0.10	2103	1577	1402	1314
0.15	935	702	624	585
0.20	527	395	351	329
0.25	338	253	225	211
0.30	235	176	157	147
0.35	173	130	115	108
0.40	133	100	89	83
0.45	105	79	70	66
0.50	86	64	57	53
0.55	71	53	47	44
0.60	60	45	40	37
0.65	51	38	34	32
0.70	44	33	30	28
0.75	39	29	26	24
0.80	34	26	23	21
0.85	31	23	20	19
0.90	27	21	18	17
0.95	25	19	17	15
1.00	23	17	15	14
1.05	21	15	14	13
1.10	19	14	13	12
1.15	17	13	12	11
1.20	16	12	11	10
1.25	15	11	10	9
1.30	14	11	9	9
1.35	13	10	9	8
1.40	12	9	8	8
1.45	12	9	8	7
1.50	11	8	7	7

12.3.1.5 Quick results

For a 0.90 power and a two-sided 0.05 Type I error:

$$n = \frac{21\sigma_w^2)}{d^2} + \frac{Z_{1-\alpha/2}}{2}, \tag{12.13}$$

where σ_w^2 is the within-subject variance and d is the expected difference, n is the total sample size.

12.3.1.6 Worked example 2

For this example let us consider two different inhalers for asthma sufferers, one being the standard and the other being a new inhaler. The expected difference in the number of asthma incidences is 2 (units) with an expected within-subject population standard deviation of 4 (units). Note that the expected within-subject standard deviation (σ_w) is expected to be half the between-subject standard deviation (σ_d). The patients are randomised to either have the standard inhaler first, or the new inhaler first. Using the formula to calculate the total subjects for the study (which will be split between the treatment sequences):

$$n = \frac{2(Z_{1-\beta} + Z_{1-\alpha/2})^2 \sigma_w^2)}{d_S^2} + \frac{Z_{1-\alpha/2}}{2} = \frac{2 \times (1.282 + 1.96)^2 \times 4^2)}{2^2} + \frac{1.96}{2} = 85.06 \approx 86$$

$$(12.14)$$

Using Table 12.3, we can compare the results gained from the worked example with those from the table. With a standardised difference of $\delta = \frac{d}{\sigma_w} = \frac{2}{4} = 0.5$, the total sample size is 87 based on the table. This is one more than the calculated result.

12.3.2 Equivalence trials

Equivalence trials are carried out to determine whether two interventions produce the same results for the patients. An example of an equivalence trial is comparing dihydrocodeine and methadone in the treatment of heroin addiction. These are two different substances which are being tested to determine if they have the same effect on the patient. It could be that one method of treatment is cheaper than the other, or easier to formulate. The hypotheses being tested for equivalence trials are (Julious, 2004):

$$H_0 : \mu_A \neq \mu_B$$

$$H_1 : \mu_A = \mu_B$$

This is usually written in terms of the clinical difference, d_e

$$H_0 : \mu_A - \mu_B \leq -d_e \text{ or } \mu_A - \mu_B \geq d_e$$

$$H_1 : -d_e \leq \mu_A - \mu_B \leq d_e$$

Both parts in the null hypothesis need to be rejected in order for a complete rejection of the null hypothesis. This is an example of an intersect-union test. In these tests, each component is tested at an α level and this gives a composite test which is also of significance level, α (Berger and Hsu, 1996). Usually we do the two one-sided tests (TOST) which tests each component of the null hypothesis. This is operationally the same as a $(1 - 2\alpha)100\%$ confidence interval, where equivalence is established if each end of the confidence interval falls within the region $(-d_e, d_e)$. In other words the 95% confidence interval (CI) must lie within $(-d_e, d_e)$ in order for the treatments to be deemed equivalent at a 0.025 significance level (Jones et al., 1996).

One consideration for both equivalence and non-inferiority trials is the equivalence or non-inferiority margin. The setting of this can be rather controversial and has been defined as the largest difference that is clinically acceptable, so that a difference bigger than this would matter in practice (for the Evaluation of Medicinal Products, 2000). Commonly used methods to establish the margin are clinical judgement and statistical reasoning. Usually the margin is set at a fraction of the limit of the placebo effect (Kaul and Diamond, 2006; Rothmann et al., 2003).

TABLE 12.3
The total sample size (n) for a cross-over study for various standardised differences ($\delta = \frac{d}{\sigma_w}$) with 0.90 power and a two sided Type I error probability of 0.05. These sample sizes are calculated from the non-central t-distribution.

δ	n
0.05	8408
0.10	2104
0.15	936
0.20	528
0.25	340
0.30	236
0.35	174
0.40	134
0.45	106
0.50	88
0.55	72
0.60	62
0.65	52
0.70	46
0.75	40
0.80	36
0.85	32
0.90	30
0.95	26
1.00	24
1.05	22
1.10	20
1.15	20
1.20	18
1.25	16
1.30	16
1.35	14
1.40	14
1.45	14
1.50	12

Often the decision on the equivalence limit is based on some comparison to placebo. The following steps (Agostino et al., 2003; Wiens, 2002) should be considered when determining the limit:

1. We must be confident that the active control would have been different from placebo had one been employed.

2. We should be able to determine that there is no clinically meaningful difference between investigative treatment and control.

3. Through comparing the investigative treatment to control we should indirectly be able to determine that it is superior to placebo.

This limit needs to be established on a study-by-study basis with advice from the relevant agencies involved in the trial. The issues raised for equivalence limits are the same as for non-inferiority limits discussed earlier in the chapter. Figure 12.3 is a diagram of how confidence intervals can be used to test the different hypotheses of superiority, equivalence, and non-inferiority trials (Julious, 2004).

Figure 12.3 represents the area which the 95% confidence interval needs to lie in order to reject the null hypothesis. Notice for the equivalent and non-inferiority trials there is a pre-defined region or value which is close to zero, this provides a little bit of leeway when testing the treatments. This is referred to as the equivalence limit or the non-inferiority margin.

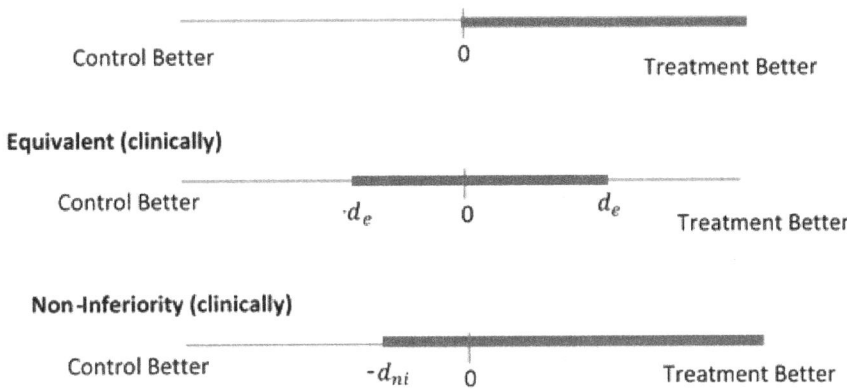

FIGURE 12.3
Illustration of the aims for each trial design.

12.3.2.1 Parallel group trials

General case

For equivalence trials the sample size cannot be derived directly for the general case where the expected true mean difference is not zero. This is due to there being two one-sided tests being performed, as mentioned above. This results in two chances of making a Type II error. Hence, the sample size cannot be derived directly for the case where the expected true mean difference is not zero ($\mu_A - \mu_B \neq 0$) since the Type II error must be split between the two tests.

The power can therefore be written as (Julious, 2004)

$$
1 - \beta = \Phi\left(\sqrt{\frac{rn_A\left((\mu_A - \mu_B) - d_e\right)^2}{(r+1)\sigma^2}} - Z_{1-\alpha}\right)
$$

$$
+ \Phi\left(\sqrt{\frac{rn_A\left((\mu_A - \mu_B) + d_e\right)^2}{(r+1)\sigma^2}} - Z_{1-\alpha}\right) - 1 \quad (12.15)
$$

This equation is then iterated until the desired power is achieved. As for the superiority trials, when the variance is unknown the $Z-$ values can no longer be used, so the equation is rewritten as (Julious, 2004)

$$1 - \beta = \Phi\left(\sqrt{\frac{rn_A\left((\mu_A - \mu_B) - d_e\right)^2}{(r+1)\sigma^2}} - t_{1-\alpha, n_A(r+1)-2}\right)$$
$$+ \Phi\left(\sqrt{\frac{rn_A\left((\mu_A - \mu_B) + d_e\right)^2}{(r+1)\sigma^2}} - t_{1-\alpha, n_A(r+1)-2}\right) - 1 \quad (12.16)$$

Again this uses the non-central Normal-distribution to calculate the Type II error probability and power. For the sample size calculations, the following approximation equation is used. The non-central t-distribution gives (Chow et al., 2002; Hauschke et al., 1992; Owen, 1965)

$$1 - \beta = T^{-1}\left(-t_{1-\alpha, n_A(r+1)-2}, n_A(r+1) - 2, \tau_2\right) - T^{-1}\left(t_{1-\alpha, n_A(r+1)-2}, n_A(r+1) - 2, \tau_1\right)$$
$$(12.17)$$

Where

$$\tau_1 = \frac{\left((\mu_A - \mu_B) + d_e\right)\sqrt{rn_A}}{(r+1)\sigma^2} \quad (12.18)$$

and

$$\tau_2 = \frac{\left((\mu_A - \mu_B) - d_e\right)\sqrt{rn_A}}{(r+1)\sigma^2} \quad (12.19)$$

are the non-centrality parameters.

If we simplify this calculation for an initial sample size for the iterations, we get (Julious, 2004)

$$n_A = \frac{(r+1)\sigma^2(Z_{1-\beta} + Z_{1-\alpha})^2}{r((\mu_A - \mu_B) - d_e)^2} \quad (12.20)$$

Special case (no treatment difference)

If there is no treatment difference ($\mu_A - \mu_B = 0$), then Equation 12.20 becomes (S. A. Julious, 2004)

$$n_A = \frac{(r+1)\sigma^2(Z_{1-\beta/2} + Z_{1-\alpha})^2}{rd_e^2} \quad (12.21)$$

The reason that $1 - \beta/2$ is now used as oppose to $1 - \beta$ is that the Type II error is split between the two one-sided tests. However, as the mean difference is now zero, the Type II error is split equally so we can have a direct estimate of the sample size. The non-central t-distribution result for the power is (Julious, 2004)

$$1 - \beta = 2 \times T^{-1}\left(-t_{1-\alpha, n_A(r+1)-2}, n_A(r+1) - 2, \tau\right) - 1 \quad (12.22)$$

where

$$\tau = \frac{-\sqrt{n_A} \times rd_e}{\sqrt{(r+1)\sigma^2}} \quad (12.23)$$

is the non-centrality parameter.

12.3.2.2 Worked example 3

An effect used to detect equivalence is a change on the visual analogue scale (VAS) of 10 mm (this would be d). The VAS is often used in clinical environments to assess pain. If the allocation to the two groups was equal in a hypothetical trial investigating the equivalence of two pain treatments, $r = 1$.

First, consider that the true mean difference is zero ($\mu_A - \mu_B = 0$) and the expected population standard deviation for the population in which the trial will be run is 40 mm (σ). Calculating the standardised equivalence limits in the usual way

$$\pm\delta = \pm\frac{d}{\sigma} = \pm\frac{10}{40} = \pm 0.25 \tag{12.24}$$

The $\pm$ sign needs to be included because, if you recall from the diagram above, the equivalence limit lays either side of 0. The Type I and Type II errors are fixed at 0.025 and 0.10 (meaning the power is 0.90), respectively. Let us first calculate the sample size using the formal equation:

$$n_A = \frac{(r+1)\sigma^2(Z_{1-\beta} + Z_{1-\alpha})^2}{r((\mu_A - \mu_B) - d_e)^2} = \frac{2 \times 40^2 \times 1.645 + 1.96)^2}{10^2} = 415.87 \approx 416 \tag{12.25}$$

With the quick formula we get

$$n_A = \frac{13\sigma^2(r+1)}{rd^2} = \frac{13 \times 40^2 \times 2}{10^2} = 416 \tag{12.26}$$

If we now compare these results with Table 12.4, looking at the 0% mean difference and $\delta = 0.25$, we get a sample size per arm of 417.

12.3.2.3 Cross-over trials

"nobreak

General case

As in previous sections, we will use an equal allocation ($r = 1$) and the power can be estimated using (Julious, 2004)

$$1 - \beta = \Phi\left(\sqrt{\frac{n((\mu_A - \mu_B) - d_e)^2}{2\sigma_w^2}} - Z_{1-\alpha}\right)$$
$$+ \Phi\left(\sqrt{\frac{n((\mu_A - \mu_B) + d_e)^2}{2\sigma_w^2}} - Z_{1-\alpha}\right) - 1 \tag{12.27}$$

When the population variance is unknown, this equation can be rewritten as

$$1 - \beta = \Phi\left(\sqrt{\frac{n((\mu_A - \mu_B) - d_e)^2}{2\sigma_w^2}} - -t_{1-\alpha,n-2}\right)$$
$$+ \Phi\left(\sqrt{\frac{n((\mu_A - \mu_B) + d_e)^2}{2\sigma_w^2}} - t_{1-\alpha,n-2}\right) - 1 \tag{12.28}$$

This shows that the methodology for cross-over trials is similar to that for parallel group

TABLE 12.4
Various sample sizes (n_A) for each arm of a parallel group equivalence study with allocation ratio ($r = 1$) for differing standardised equivalence limits ($\delta = \frac{d}{\sigma}$) and true mean differences as a percentage of δ for 0.90 power and 0.025 Type I error probability. These sample sizes are calculated from the non-central t-distribution.

δ	0%	10%	15%	20%	25%
0.05	10398	11042	11916	13218	14960
0.10	2600	2762	2980	3306	3742
0.15	1158	1228	1326	1470	1664
0.20	652	692	746	828	936
0.25	418	444	478	530	600
0.30	290	308	332	370	418
0.35	214	228	246	272	308
0.40	164	174	188	208	236
0.45	130	138	150	166	186
0.50	106	112	122	134	152
0.55	88	94	100	112	126
0.60	74	78	84	94	106
0.65	64	68	72	80	90
0.70	56	58	62	70	78
0.75	48	52	54	60	68
0.80	42	46	48	54	60
0.85	38	40	44	48	54
0.90	34	36	38	42	48
0.95	30	34	36	40	44
1.00	28	30	32	36	40
1.05	26	28	30	34	36
1.10	22	24	26	30	32
1.15	22	22	24	26	30
1.20	20	22	22	24	28
1.25	18	20	22	24	26
1.30	18	18	20	22	24
1.35	16	18	18	20	22
1.40	16	16	18	18	22
1.45	14	16	16	18	20
1.50	14	14	16	16	18

trials for equivalence studies (Chow et al., 2002; Hauschke et al., 1992; Owen, 1965). The non-central t-distribution result for the power is given by (Julious, 2004)

$$1 - \beta = \Phi(\sqrt{\frac{n((\mu_A = \mu_B) - d)^2}{2\sigma_w^2}} - Z_{1-\alpha}) + \Phi(\sqrt{\frac{n((\mu_A = \mu_B) - d)^2}{2\sigma_w^2}} - Z_{1-\alpha}) - 1 \quad (12.29)$$

When the population variance is uknown, this equation can be rewritten as

$$1 - \beta = \Phi(\sqrt{\frac{n((\mu_A = \mu_B) - d)^2}{2\sigma_w^2}} - t_{1-\alpha,n-2}) + \Phi(\sqrt{\frac{n((\mu_A = \mu_B) - d)^2}{2\sigma_w^2}} - t_{1-\alpha,n-2}) - 1$$

$$(12.30)$$

This shows that the methodology for cross-over trials is similar to that for parallel group trials for equivalence studies.

For a quick calculation (for $\mu_A - \mu_B > 0$)(Julious, 2004)

$$n = \frac{2\sigma_w^2 (Z_{1-\beta} + Z_{1-\alpha})^2}{\left((\mu_A - \mu_B) - d_e\right)^2} \tag{12.31}$$

For quick calculations at a 0.90 power and 0.025 Type I error probability (Julious, 2004):

$$n = \frac{21\sigma_w^2}{\left((\mu_A - \mu_B) - d_e\right)^2} \tag{12.32}$$

Special case of no treatment difference

For the special case where there is no treatment difference ($\mu_A - \mu_B = 0$) the direct estimate of the sample size is (Julious, 2004)

$$n = \frac{2\sigma_w^2 (Z_{1-\beta} + Z_{1-\alpha})^2}{d_e^2} \tag{12.33}$$

The quick calculation is (for a 0.90 power and a 0.025 Type I error) (Julious, 2004)

$$n = \frac{26\sigma_w^2}{d_e^2} \tag{12.34}$$

12.3.2.4   Worked example 4

Let us use a similar example as in the parallel group equivalence section. Consider an investigator who wants to assess two different pain treatments using a cross-over design and using the VAS for assessment. It is clinically accepted that a change of 10mm on the VAS is the largest effect for treatments to be equivalent. This means that if the difference is greater than 10mm, then equivalence can't be assumed. If the difference is less than 10mm then equivalence can be accepted. Consider the true mean difference being zero ($mu_A - \mu_B = 0$) and the expected within-subject standard deviation of 20mm (σ_w). This gives a standardised equivalence limit of

$$\pm\delta = \pm\frac{d}{\sigma} = \pm\frac{10}{20} = \pm0.50 \tag{12.35}$$

The Type I and Type II errors are fixed at 0.025 and 0.10 (meaning the power is 0.90), respectively. Let us first calculate the sample size using the formal equation:

$$n = \frac{2\sigma_w^2 (Z_{1-\beta} + Z_{1-\alpha})^2}{d_e^2} = \frac{2 \times 20^2 \times (1.645 + 1.96)^2}{10^2} = 103.97 \approx 104 \tag{12.36}$$

With the quick forumla we get

$$n = \frac{26\sigma_w^2}{d_e^2} = \frac{26 \times 20^2}{10^2} = 104 \tag{12.37}$$

If we use Table 12.5 and recall that $\delta = 0.5$ and the percentage mean difference is 0% we get a sample size of 106 in total for the trial. Now let us consider the case where the true mean difference is believed to be 2mm ($\mu_A - \mu_B = 2$). This equates to 20% of the standardised equivalence limits, the sample size would be 135 according to Table 12.5.

TABLE 12.5

Various total sample sizes (n) for equivalence studies using a cross-over design for different standardised equivalence limits ($\delta = \frac{d}{\sigma}$) and different true mean differences as a percentage of δ. These are for a 0.90 power and a Type I error probability of 0.025. These sample sizes are calculated from the non-central t-distribution.

δ	0%	10%	15%	20%	25%
0.05	10398	11044	11916	13220	14962
0.10	2602	2764	2982	3308	3742
0.15	1158	1230	1326	1472	1666
0.20	652	692	748	828	938
0.25	418	444	480	532	602
0.30	292	310	334	370	418
0.35	216	228	246	272	308
0.40	166	176	190	210	236
0.45	132	140	150	166	188
0.50	106	114	122	136	152
0.55	88	94	102	112	126
0.60	76	80	86	94	106
0.65	64	68	74	82	92
0.70	56	60	64	70	80
0.75	50	52	56	62	70
0.80	44	46	50	54	62
0.85	40	42	44	48	54
0.90	36	38	40	44	50
0.95	32	34	36	40	44
1.00	30	30	32	36	40
1.05	26	28	30	34	36
1.10	24	26	28	30	34
1.15	22	24	26	28	32
1.20	22	22	24	26	30
1.25	20	20	22	24	28
1.30	18	20	20	22	26
1.35	18	18	20	22	24
1.40	16	18	18	20	22
1.45	16	16	18	18	20
1.50	14	16	16	18	20

We can also calculate this by hand. Assume all other variables remain the same ($d = 10$, equal allocation ratio). Then using the formula we get a sample size of:

$$n = \frac{2\sigma_w^2 (Z_{1-\beta} + Z_{1-\alpha})^2}{((\mu_A - \mu_B) - d)^2} = \frac{2 \times 20^2 \times (1.282 + 1.96)^2}{(2 - 10)^2} = 131.4 \approx 132 \qquad (12.38)$$

$$n = \frac{21\sigma_w^2}{((\mu_A - \mu_B) - d)^2} = \frac{21 \times 20^2}{(2 - 10)^2} = 131.25 \approx 132 \qquad (12.39)$$

12.3.3   Non-inferiority trials

Non-inferiority trials are done when there is an active control, such as a current treatment on the market. The aim for this trial is to show that the new treatment is as good as the current one or better than the current one. This could be useful when the side effects are less with the new treatment, or the cost is less. An example of this type of trial being useful would be to investigate whether doctors could be replaced with nurses for performing a specific therapy or treatment. The main focus is to show that nurses are as good as doctors at providing the therapy. A bonus from this would be if it were shown that nurses were better than doctors at providing that therapy.

The hypothesis for non-inferiority trials can be stated as follows: H_0: The given treatment is inferior with respect to the mean response. H_1: The given treatment is non-inferior with respect to the mean response.

This is usually written in terms of the clinical difference, d $H_0 : \mu_A - \mu_B \leq -d$
$H_1 : \mu_A - \mu_B > -d$
In this contet, $-d$ is the non-inferiority margin. Non-inferiority trials are tested using a one-sided hypothesis test, equivalent to testing just one part of the two parts of the TOST procedure in the equivalence trials. Realistically this is the same principle as getting a $(1 - 2\alpha)100$ confidence interval and concluding non-inferiority if the entire interval is greater than $-d$.

12.3.3.1   Parallel group trials

The methods required to arrive at the sample size calculation follow closely with those from previous sections and have not been included here. These formulae are very similar to the formulae for the equivalence trials. We can once again assume that there is equal allocation between groups ($r = 1$) in this section. The sample size for a parallel group non-inferiority trial is

$$n = \frac{2\sigma^2(Z_{1-\beta} + Z_{1-\alpha})^2}{((\mu_A - \mu_B) - d)^2} \tag{12.40}$$

The quick calculation at a 0.90 power and a 0.025 Type I error probability:

$$n = \frac{10.5\sigma^2(r + 1)}{(r(\mu_A - \mu_B) - d)^2} = \frac{21\sigma^2}{((\mu_A - \mu_B) - d)^2} \tag{12.41}$$

The equations for the sample sizes when the differences in means are zero are the same as those for the equivalence trials.

12.3.3.2   Worked example 5

Let us consider the following example. A clinical investigator is doing a trial on two different treatments for hypertension, one being the standard therapy and the other being the experimental therapy. The largest clinically acceptable effect to be able to declare non-inferiority is a blood pressure change of 10 mmHg, this will be d. This means that if the mean difference is greater than 10 mmHg non-inferiority cannot be assumed. Let's first investigate the case where the true mean difference between the treatments is zero ($\mu_A - \mu_B = 0$) and the expected standard deviation of the trial population is 40 mmHg (σ). There is an equal allocation ratio ($r = 1$).

Recall that a visual representation of this trial is shown in Figure 12.3. If the difference is greater than d in a negative direction, non-inferiority cannot be assumed.

Using the information provided by the researcher, the standardised non-inferiority limit would be

$$-\delta = -\frac{d}{\sigma} = frac1040 = -0.25 \tag{12.42}$$

It is negative because that is the sign of d in Fgure 12.3. The Type I and Type II errors are fixed at 0.025 and 0.80, respectively. Using the formulae we get a sample size of

$$n = \frac{(r+1)\sigma^2(Z_{1-\beta} + Z_{1-\alpha})^2}{((\mu_A - \mu_B) - d)^2} = \frac{2 \times 40^2(1.282 + 1.96)^2}{(0 - 10)^2} = 336.34 \approx 337 \; per \; arm \quad (12.43)$$

Using the quick formula we get a sample size of

$$n = \frac{10.5\sigma^2(r+1)}{(r(\mu_A - \mu_B) - d)^2} = \frac{10.5 \times 40^2 \times 2}{(0 - 10)^2} = 336 \; per \; arm \quad\quad\quad (12.44)$$

From Table 12.6, with a percentage mean difference of 0% and $\delta = 0.25$, we get a sample size of 338 per arm. If we compare the results from these three methods, we get a difference in sample size of up to 2 patients per arm.

Let us now consider a scenario in which the investigator now thinks that the experimental therapy (μ_B) is slightly superior to the standard therapy, so that the true mean difference is believed to be 2 mmHg ($\mu_A - \mu_B = -2$). We can think about this in terms of Figure 12.3, if the new treatment is deemed to be a difference of 2 better than the old treatment, the treatment difference will be 12 away from the non-inferiority margin so fewer participants will be needed in order to assume non-inferiority. We can use the formulae above to recalculate the sample size for each arm of the trial.

$$n = \frac{2\sigma^2(Z_{1-\beta} + Z_{1-\alpha})^2}{((\mu_A - \mu_B) - d)^2} = \frac{2 \times 40^2 \times (1.282 + 1.96)^2}{(-2 - 10)^2} = 233.57 \approx 234 \; per \; arm \quad (12.45)$$

The quick formula gives the following result

$$n = \frac{10.5\sigma^2(r+1)}{(r(\mu_A - \mu_B) - d)^2} = \frac{10.5 \times 40^2 \times 2}{(-2 - 10)^2} = 233.33 \approx 234 \; per \; arm \quad\quad (12.46)$$

From Table 12.6 with a percentage mean difference of $\frac{(\mu_A - \mu_B)}{d} = \frac{-2}{10} = -20\%$ and $\delta = 0.25$ the sample size is 235 per arm. The difference between the sample sizes using each method is one patient per arm, which is not a considerably large difference so would not make a vast difference in the long run for the trial.

12.3.3.3   Cross-over trials

As for the cross-over trials in equivalence trials, we need to include the within-subject standard deviation for the population (σ_w). The general formula for sample sizes per arm in cross-over trials is

$$n = \frac{2\sigma_w^2(Z_{1-\beta} + Z_{1-\alpha})^2}{((\mu_A - \mu_B) - d)^2} \quad\quad\quad\quad\quad (12.47)$$

The quick formula (at 0.90 power and 0.025 Type I error probability) is

$$n = \frac{21\sigma_w^2}{((\mu_A - \mu_B) - d)^2} \qu\quad\quad\quad\quad\quad (12.48)$$

12.3.3.4   Worked example 6

Let us again use the example from the parallel group non-inferiority trial example regarding hypertension. Now the investigator would like to design a cross-over trial for the two treatments, with the same clinically acceptable non-inferiority effect margin of 10 mmHg (d). First, we consider that the true mean difference between the two treatments is zero

TABLE 12.6

This table shows different sample sizes (n_A) for one arm of a parallel group design for a non-inferiority trial with an allocation of $r = 1$ (equal allocation) for various standardised equivalence limits ($\delta = d/\sigma$). It shows the sample sizes for different true mean differences as a percentage of δ for a 0.90 power and Type I error probability of 0.025. These sample sizes are calculated from the non-central t-distribution.

δ	−25%	−20%	−15%	−10%	−5%	0%	0%	5%	15%	20%	25%
0.05	5381	5839	6358	6949	7626	8407	9316	10379	11636	13136	14945
0.10	1346	1461	1590	1738	1908	2103	2330	2596	2910	3285	3737
0.15	599	650	708	773	849	935	1036	1155	1294	1461	1662
0.20	338	366	399	436	478	527	584	650	729	822	935
0.25	217	235	256	279	306	338	374	417	467	527	599
0.30	151	164	178	194	213	235	260	290	325	366	417
0.35	111	121	131	143	157	173	192	213	239	270	306
0.40	86	93	101	110	121	133	147	164	183	207	235
0.45	68	74	80	87	96	105	116	130	145	164	186
0.50	55	60	65	71	78	86	95	105	118	133	151
0.55	46	50	54	59	64	71	78	87	98	110	125
0.60	39	42	46	50	54	60	66	74	82	93	105
0.65	33	36	39	43	47	51	57	63	70	79	90
0.70	29	31	34	37	40	44	49	54	61	68	78
0.75	25	27	30	32	35	39	43	48	53	60	68
0.80	23	24	26	29	31	34	38	42	47	53	60
0.85	20	22	23	26	28	31	34	37	42	47	53
0.90	18	20	21	23	25	27	30	34	37	42	48
0.95	16	18	19	21	23	25	27	30	34	38	43
1.00	15	16	17	19	21	23	25	27	31	34	39
1.05	14	15	16	17	19	21	23	25	28	31	35
1.10	13	14	15	16	17	19	21	23	26	29	32
1.15	12	13	14	15	16	17	19	21	23	26	30
1.20	11	12	13	14	15	16	18	20	22	24	27
1.25	10	11	12	13	14	15	16	18	20	23	25
1.30	10	10	11	12	13	14	15	17	19	21	24
1.35	9	10	10	11	12	13	14	16	17	20	22
1.40	8	9	10	10	11	12	13	15	16	18	21
1.45	8	9	9	10	11	12	13	14	15	17	19
1.50	8	8	9	9	10	11	12	13	14	16	18

($\mu_A - \mu_B = 0$) and the expected within-subject population standard deviation is 20 mmHg (σ_w). This gives a non-inferiority limit of

$$-\delta = -\frac{d}{\sigma_w} = -\frac{10}{20} = -0.50 \tag{12.49}$$

If the Type I and Type II error probabilities are fixed at 0.025 and 0.10, respectively, one can calculate the sample size required for the trial using the formulae above.

$$n = \frac{2\sigma_w^2 (Z_{1-\beta} + Z_{1-\alpha})^2}{((\mu_A - \mu_B) - d)^2} = \frac{2 \times 20^2 \times (1.282 + 1.96)^2}{(0 - 10)^2} = 84.08 \approx 84 \ in \ total \tag{12.50}$$

Note that we can round this number down because the total sample size needs to be an even number for a two-arm cross-over trial so that equal subjects are in each sequence group.

Using the quick formula, the total sample size required is

$$n = \frac{21\sigma_w^2}{((\mu_A - \mu_B) - d)^2} = \frac{21 \times 20^2}{(0 - 10)^2} = 84 \tag{12.51}$$

If we use Table 12.7 with a percentage mean difference of 0% ($\mu_A - \mu_B = 0$) and $\delta = 0.50$, we get a sample size of 88 patients in total. This shows that for smaller sample sizes, the variation between the methods is more pronounced, here with a difference of up to four patients. Once more, this difference in total patients is not overly large. Let us now consider if the true mean difference is believed to be 2 mmHg ($\mu_A - \mu_B = -2$). In terms of the example, this means that the new treatment is supposedly 2 better than the old treatment. The sample sizes using the formulae above are

$$n = \frac{2\sigma_w^2 (Z_{1-\beta} + Z_{1-\alpha})^2}{((\mu_A - \mu_B) - d)^2} = \frac{2 \times 20^2 \times (1.282 + 1.96)^2}{(-2 - 10)^2} = 58.4 \approx 58 \; in \; total \tag{12.52}$$

Note that we can round this number down because the total sample size needs to be an even number for a two-arm cross-over trial so that equal subjects are in each sequence group.

Using the quick formula, the total sample size required is

$$n = \frac{21\sigma_w^2}{((\mu_A - \mu_B) - d)^2} = \frac{21 \times 20^2}{(-2 - 10)^2} = 58.33 \approx 58 \tag{12.53}$$

The percentage mean difference would now be $\frac{(\mu_A - \mu_B)}{d} = \frac{-2}{10} = -20\%$, with a standardised non-inferiority limit of $\delta = 0.50$. Using Table 12.7 we get a total sample size of 61 which is 2 patients greater than the calculated sample sizes.

12.4 Binary Outcomes

12.4.1 Superiority trials

12.4.1.1 Parallel group trials

The parallel group superiority trial is investigating whether two different treatments/interventions are different in terms of their proportion of patients with a particular outcome. Let p_A and p_B be the proportion of adverse events in groups A and B, respectively. The two hypotheses of interest would be H_0: There is no difference between the two treatment effects in terms of the risk ratio ($p_A = p_B$) H_1: There is no difference between the two treatment effects in terms of the risk ratio ($p_A \neq_B$)

Consider Table 12.8, a summary table for atypical clinical trial with a binary end-point or outcome.

Let p_A be the proportion of responses in group A and p_B be the response in group B, n_A and n_B be the total number of patients in groups A and B, respectively, $n = n_A + n_B$ be the total number of patients in the study, $\bar{p} = \frac{n_A p_A + n_B p_B}{n_A + n_B}$ be the average response across the treatments.

There are two methods for getting a sample size calculation for this type of outcome, one of which is using the anticipated responses under the alternative hypothesis and the other is using the responses under the null and alternative hypotheses.

TABLE 12.7

This table shows different total sample sizes (n) of a cross-over design for a non-inferiority trial with an allocation of $r = 1$ (equal allocation) for various standardised equivalence limits ($\delta = \frac{d}{\sigma}$). It shows the sample sizes for different true mean differences as a percentage of δ for a 0.90 power and Type I error probability of 0.025. These sample sizes are calculated from the non-central t-distribution.

δ	-25%	-20%	-15%	-10%	-5%	0%	0%	5%	15%	20%	25%
0.05	5382	5840	6360	6950	7628	8408	9316	10380	11638	13138	14946
0.10	1348	1462	1592	1740	1910	2104	2332	2598	2912	3286	3738
0.15	600	652	710	774	850	936	1038	1156	1296	1462	1664
0.20	340	368	400	438	480	528	586	652	730	824	936
0.25	218	236	258	280	308	340	376	418	468	528	600
0.30	152	166	180	196	214	236	262	292	326	368	418
0.35	112	122	132	144	158	174	194	214	240	270	308
0.40	88	94	102	112	122	134	148	166	184	208	236
0.45	70	76	82	88	98	106	118	132	146	166	188
0.50	56	62	66	72	80	88	96	106	120	134	152
0.55	48	52	56	60	66	72	80	88	100	112	126
0.60	40	44	48	52	56	62	68	76	84	94	106
0.65	34	38	40	44	48	52	58	64	72	80	92
0.70	30	32	36	38	42	46	50	56	62	70	80
0.75	26	30	32	34	36	40	44	50	54	62	70
0.80	24	26	28	30	32	36	40	44	48	54	62
0.85	22	24	26	28	30	32	36	38	44	48	54
0.90	20	22	22	24	26	30	32	36	38	44	50
0.95	18	20	20	22	24	26	28	32	36	40	44
1.00	16	18	20	20	22	24	26	30	32	36	40
1.05	16	16	18	18	20	22	24	26	30	32	36
1.10	14	16	16	18	18	20	22	24	28	30	34
1.15	14	14	16	16	18	20	20	22	26	28	32
1.20	12	14	14	16	16	18	20	22	24	26	30
1.25	12	12	14	14	16	16	18	20	22	24	26
1.30	12	12	12	14	14	16	16	18	20	22	26
1.35	10	10	12	12	14	14	16	18	20	22	24
1.40	10	10	12	12	14	14	16	16	18	20	22
1.45	10	10	10	12	12	14	14	16	18	18	20
1.50	8	8	10	12	12	12	14	14	16	18	20

Method 1

This method is a simple calculation to get an approximate sample size relatively quickly. With this method we get a sample size calculation of

$$n_A = \frac{(Z_{1-\beta} + Z + 1 - \alpha/2)^2(p_A(1 - p_A) + p_B(1 - p_B))}{(p_A - p_B)^2} \tag{12.54}$$

This formula is for the case of equally sized groups (i.e. $n_A = n_B$)

There are two formulae to quickly calculate a sample size estimate for a superiority trial with a binary outcome. For a 0.90 power and Type I error probability of 0.05 we get

$$n_A = \frac{5.25}{(p_A - p_B)^2} \tag{12.55}$$

TABLE 12.8

	Outcome		
Treatment	1	0	Total
A	p_A	$1 - p_A$	n_A
B	p_B	$1 - p_B$	n_B
Overall Response	$\bar{p}$	$1 - \bar{p}$	$n = n_A + n_B$

For an 0.80 power and a two-sided Type I error probability of 0.05 we can estimate a sample size using

$$n_A = \frac{4}{(p_A - p_B)^2} \qquad (12.56)$$

Both these quick formulae will provide conservative "maximum" estimates of the sample size per arm.

12.4.1.2 Method 2

The alternative method used for calculating the sample size uses the responses under the null and alternative hypotheses. This method arose because in method 1 the variances under each hypothesis are assumed to be equal, when in practice this is unlikely to be the case since under the null we have $p_A = p_B$ and under the alternative we have $p_A \neq p_B$. This method tends to be executed using software packages. The adjusted sample size calculation is as follows

$$n_A = \frac{(Z_{1-\alpha/2}\sqrt{variance under null} + Z_{1-\beta}\sqrt{variance under alternative})^2}{(p_A - p_B)^2} \qquad (12.57)$$

This means the sample size can be estimated using the following formula

$$n_A = \frac{(Z_{1-\alpha/2}\sqrt{2\bar{p}_1(1-\bar{p}_1)} + Z_{1-\beta}\sqrt{p_A(1-p_A) + p_B(1-p_B)})^2}{(p_A - p_B)^2} \qquad (12.58)$$

where $\bar{p}_1 = \frac{p_A + p_B}{2}$.

One final consideration for parallel group trials with a binary end-point is whether a continuity correction needs to be implemented. This is a more traditional method which was used when computing power wasn't as strong as it is currently. It was included if the chi-squared assumptions didn't hold, for example if there were small or zero cell counts. Nowadays the more conventional method is to use Fisher's exact test. The formula for this sample size is used once the initial sample size has been calculated using one of the previous formulae, the corrected sample size is

$$n_{cc} = \frac{n_A}{4}[1 + \sqrt{1 + \frac{4}{n_A \times (p_A - p_B)}}]^2 \qquad (12.59)$$

This result can also be used to provide an estimate for the sample size using Fisher's exact test.

12.4.1.3 Worked example 7

An investigator wishes to compare two treatments for nausea, one being placebo and the other being a new experimental drug. The absolute risk of nausea on placebo is predicted

TABLE 12.9
This table shows various sample size estimates using method 1 described above for each arm of a parallel group trial.

p_A	\multicolumn{9}{c}{p_B}								
	0.05	**0.10**	**0.15**	**0.20**	**0.25**	**0.30**	**0.35**	**0.40**	**0.45**
0.10	578								
0.15	184	915							
0.20	97	263	1209						
0.25	63	120	331	1461					
0.30	44	79	158	389	1671				
0.35	33	54	94	182	437	1839			
0.40	25	39	62	106	200	473	1965		
0.45	20	29	44	69	115	214	500	2048	
0.50	16	23	33	48	74	121	223	515	2091

to be 50% and it is thought that the new treatment would be worth using if it reduced the absolute risk of nausea to 30%, meaning that the treatment effect would have an absolute risk reduction of 20%. The investigator wishes to design a trial which has 0.90 power and a two-sided significance level of 0.05. Using Method 1, we get the following sample size per arm using the simplified calculation

$$n_A = \frac{(Z_{1-\beta} + Z_{1-\alpha/2})^2 \left(p_A(1 - p_A) + p_B(1 - p_B)\right)}{(p_A - p_B)^2} \tag{12.60}$$

$$n_A = \frac{(1.282 + 1.96)^2 (0.30 \times 0.70 + 0.50 \times 0.50)}{(0.30 - 0.50)^2} = 120.87 \approx 121 \tag{12.61}$$

Using the quick formula for 0.90 power and 0.05 Type I error probability, we get

$$n_A = \frac{5.25}{(p_A - p_B)^2} = \frac{5.25}{(0.30 - 0.50)^2} = 131.25 \approx 132 \tag{12.62}$$

If we use Table 12.9, we get a sample size of 121 per arm. The tables are symmetrical so one is able to interchange p_A and p_B.

Using method 2, we get

$$\bar{p}_1 = \frac{p_A + p_B}{2} = \frac{0.30 + 0.50}{2} = 0.40 \tag{12.63}$$

$$n_A = \frac{(Z_{1-\alpha/2}\sqrt{2\bar{p}_1(1 - \bar{p}_1)} + Z_{1-\beta}\sqrt{p_A(1 - p_A) + p_B(1 - p_B)})^2}{(p_A - p_B)^2} \tag{12.64}$$

$$n_A = \frac{(1.96 \times \sqrt{2 \times 0.40 \times 0.60} + 1.282 \times \sqrt{0.30 \times 0.70 + 0.50 \times 0.50})^2}{(0.3 - 0.5)^2} = 124.04 \approx 125 per arm \tag{12.65}$$

If we use Table 12.10 we get a sample size of 124 per arm. The tables are symmetrical so one is able to interchange p_A and p_B.

Using the value of 121 patients per arm, we could use the continuity correction formulae to get a corrected sample size of $130.80 \approx 131$ patients per arm.

TABLE 12.10

A table showing the sample size estimates using method 2 for each arm of a parallel group trial. These sample sizes are based on different expected outcome responses for a given treatment (p_A) and control (p_B) for a trial with a two sided Type I error probability of 0.05 and 0.90 power.

					p_B				
p_A	0.05	0.10	0.15	0.20	0.25	0.30	0.35	0.40	0.45
0.10	582								
0.15	188	918							
0.20	101	266	1212						
0.25	65	133	335	1464					
0.30	47	82	161	392	1674				
0.35	36	57	97	185	440	1842			
0.40	28	42	65	109	203	477	1969		
0.45	23	33	47	72	118	217	503	2053	
0.50	19	26	36	52	77	124	227	519	2095
0.55	16	21	28	39	54	81	128	231	524
0.60	14	17	23	30	40	56	82	130	231
0.65	12	15	19	24	31	41	57	82	128
0.70	10	12	15	19	24	31	41	56	81
0.75	8	10	13	16	19	24	31	40	54
0.80	7	9	11	13	16	19	24	30	39
0.85	6	7	9	11	13	15	19	23	28
0.90	5	6	7	9	10	12	15	17	21
0.95	4	5	6	7	8	10	12	14	16

12.4.1.4 Cross-over trials

Cross-over trials with binary data are very different to anything that has been discussed up to this point. However, they are very similar for each of the three trial types (superiority, equivalence, and non-inferiority). The hypotheses for this type of trial are H_0: There is no difference between the two treatments ($p_A = p_B$) H_1: There is a difference between the two treatments ($p_A \neq p_B$)

If we first consider Table 12.11, for a cross-over trial the only cells of interest are the discordant cells (the cells '01' and '10'). The reason for this is that the concordant cells in a superiority trial agree with the null hypothesis that there is no difference between the treatments, whereas the alternative hypothesis is stating that one treatment is preferable.

TABLE 12.11

An example summary of a hypothetical cross-over trial, where n_{xy} are the number of responses in cell xy. The end row and column give the total responses for each treatment. These overall responses are the numbers expected to be seen if it were a parallel group trial.

	Treatment B		
Treatment A	1	0	Total
1	n_{11}	n_{10}	n_{A1}
0	n_{01}	n_{00}	n_{A0}
	n_{B1}	n_{1B}	n

TABLE 12.12

A summary table of a hypothetical cross-over trial where $\lambda_{xy} = \frac{n_{xy}}{n}$, $p_A = \frac{n_{A1}}{n}$ and $p_B = \frac{n_{B1}}{n}$. The marginal totals are found in the end row and column and each cell shows the proportion of responses for each treatment combination/outcome.

	Treatment B		
Treatment A	1	0	Total
1	λ_{11}	λ_{10}	p_A
0	λ_{01}	λ_{00}	$1 - p_A$
	p_B	$1 - p_B$	1

Table 12.11 can be re-written in terms of the proportions of responses, giving Table 12.12.

The trial can then be summarised in an odds ratio using the discordant cells by $\Psi = \frac{\lambda_{10}}{\lambda_{01}}$. This can sometimes be difficult to interpret, therefore an approximate odds ratio can be gathered from the marginal totals of Table 12.12 where (Royston, 1993).

$$\Psi = \frac{p_A(1 - p_B)}{p_B(1 - p_A)} \tag{12.66}$$

The discordant sample size can be estimated using the odds ratio, then from this value the total sample size required for the trial can be calculated. The discordant sample size, n_d, is deemed useful since it does not contain any "unknown" values, it is purely based on the odds ratio (Ψ), $Z_{1-\alpha/2}$ and $Z_{1-\beta}$ (Connett et al., 1987; Fleiss and Levin, 1988; Julious et al., 1999; Royston, 1993; Schlesselman and Schneiderman, 1982)

$$n_d = \frac{Z_{1-\alpha/2}(\Psi + 1) + 2 \times Z_{1-\beta} \times \sqrt{\Psi})^2}{(\Psi - 1)^2} \tag{12.67}$$

The total sample size can then be estimated using (Connett et al., 1987; Julious et al., 1999; Royston, 1993)

$$N_{total} = \frac{n_d}{\lambda_{01} + \lambda_{10}} \tag{12.68}$$

Table 12.13 gives the total sample sizes required for various odds ratios.

12.4.1.5 Worked example 8

Consider the scenario that an investigator wishes to test two treatments for migraine, one being the experimental treatment and the other being the standard treatment currently being prescribed. A summary of the predicted responses to treatment is shown in Table 12.14.

Using this table we can calculate the odds ratio

$$\Psi = \frac{\lambda_{10}}{\lambda_{01}} = \frac{0.32}{0.16} = 2 \tag{12.69}$$

or

$$\Psi = \frac{p_A(1 - p_B)}{p_B(1 - p_A)} = \frac{0.72 \times 0.44}{0.56 \times 0.28} = 2.02 \approx 2 \tag{12.70}$$

The investigator would like to have a 0.90 power and a two-sided level of significance of

TABLE 12.13

This table gives the discordant sample sizes for different odds ratios for a cross-over trial with 0.90 power and a two-sided significance level of 0.05. The total sample sizes can be calculated from this, they can't be given here since they depend on the discordant proportions and each proportion could be made in numerous ways.

Odds Ratio (Ψ)	Discordant Sample Size (n_d)
0.05	8
0.10	12
0.15	16
0.20	20
0.25	25
0.30	32
0.35	42
0.40	53
0.45	69
0.50	91
0.55	121
0.60	164
0.65	230
0.70	334
0.75	511
0.80	848
0.85	1595
0.90	3791
0.95	15983

TABLE 12.14

This table shows a summary of the expected responses to two migraine treatments.

	Experiment Treatment		
Standard Treatment	1	0	**Total**
1	0.40	0.32	0.72(p_A)
0	0.16	0.12	0.28
	0.56(p_B)	0.44	1

0.05. If we substitute the odds ratio $\Psi = 2$ into the equation to calculate an estimate of the discordant sample size for the trial we get

$$n_d = \frac{(Z_{1-\alpha/2}(\Psi + 1) + 2 \times Z_{1-\beta} \times \sqrt{Psi})^2}{(Psi - 1)^2} = \frac{(1.96 \times (2 + 1) + 2 \times 1.282 \times \sqrt{2})^2}{(2 - 1)^2} = 90.36 \approx 91$$

$$(12.71)$$

Therefore the discordant sample size needed for the trial is 91 subjects. This value could be used in the recruitment stage so that patients were recruited until there were 91 discordant patients. This would only be possible for short term studies, for example, studies which take two weeks to complete. The patients could be recruited sequentially until there were 91 discordant pairs.

Alternatively, Table 12.13 could be used since an odds ratio of 2 is equivalent to an odds ratio of 0.5, it is just inverted. This results in discordant sample size required being 91 subjects also.

For completeness, the investigator should also determine the total sample size required given that 91 subjects are needed. Using the total sample size equation and the value for the discordant sample size ($n_d = 91$) we get

$$N_{total} = \frac{n_d}{\lambda_{01} + \lambda_{10}} = \frac{91}{(0.32 + 0.16)} = 189.58 \approx 190 \; subjects \tag{12.72}$$

12.4.2 Equivalence trials

12.4.2.1 Parallel group trials

The two hypotheses for this trial are (Julious and Campbell, 2012)

- H_0: The two treatments are different in terms of their risk difference. ($p_A \neq p_B$)

- H_1: The two treatments are not different in terms of their risk difference. ($p_A = p_B$)

Normally these hypotheses are written not in terms of the risk difference, but in terms of the clinical difference, d_e. They become

- H_0: $p_A - p_B \leq -d_e$ or $p_A - p_B \geq +d_e$

- H_1: $-d_e < p_A - p_B < +d_e$

These hypotheses, as with the continuous outcome equivalence and non-inferiority trials, are intersection-union tests. With these tests, as previous discussed, each component of the null hypothesis is tested at the level α (Berger and Hsu, 1996; Julious, 2004, 2010). The sample size for a parallel group equivalence trial (with equal allocation, $r = 1$) is (Julious and Campbell, 2012)

$$n_A = \frac{(Z_{1-\beta} + Z_{1-\alpha/2})^2 \Big(p_A(1 - p_A) + p_B(1 - p_B)\Big)}{(|p_A - p_B| - d_e)^2} \tag{12.73}$$

If there is no anticipated treatment difference ($p_A - p_B = 0$), a direct sample size estimate can be obtained from the following formula (Julious and Campbell, 2012)

$$n_A = \frac{2 \times (Z_{1-\beta} + Z_{1-\alpha/2})^2 \times \bar{p}(1 - \bar{p})}{d_e^2} \tag{12.74}$$

12.4.2.2 Worked example 9

An investigator is interested in designing a trial in which the anticipated response rate probability on the active control is 0.70 and the expected response rate probability on the therapy under investigation is 0.73. The equivalence limit is set to be 0.10, with a 0.90 power and one-sided 0.025 Type I error probability.

Using the formula, the sample size can be calculated

$$n_A = \frac{(Z_{1-\beta} + Z_{1-\alpha/2})^2 \Big(p_A(1 - p_A) + p_B(1 - p_B)\Big)}{(|p_A - p_B| - d_e)^2} \tag{12.75}$$

$$n_A = \frac{(1.282 + 1.96)^2 \Big(0.70 \times 0.30 + 0.73 \times 0.27\Big)}{(|0.70 - 073| - 0.10)^2} = 873.23 \approx 874 \; per \, arm \tag{12.76}$$

Using Table 12.15, where $p_A = 0.70$, $p_B - p_A = 0.03$ and the equivalence limit is 10% ($d = 0.10$), we get a sample size of 874 per group.

TABLE 12.15

A table of sample sizes per arm for a parallel group equivalence study. These values were estimated for 0.90 power and a Type I error probability of 0.025.

p_A	Limit	$p_B - p_A$										
		−0.05	−0.04	−0.03	−0.02	−0.01	0	0.01	0.02	0.03	0.04	0.05
0.70	0.05	-	45645	11325	4993	2802	2184	2749	4806	10694	42282	-
0.70	0.10	1839	1268	925	707	585	546	574	680	874	1175	1671
0.70	0.15	460	378	317	275	252	243	247	265	299	350	418
0.70	0.20	205	180	161	148	140	137	138	143	152	167	186
0.75	0.05	-	41337	10222	4491	2511	1950	2445	4257	9434	37134	-
0.75	0.10	1671	1149	835	636	525	488	511	603	771	1032	1461
0.75	0.15	418	342	286	248	226	217	220	235	264	308	366
0.75	0.20	186	163	145	133	126	122	122	126	134	146	163
0.80	0.05	-	35978	8856	3872	2154	1664	2075	3592	7910	30934	-
0.80	0.10	1461	1000	723	548	450	416	434	509	646	860	1209
0.80	0.15	366	298	248	214	194	185	187	198	222	256	303
0.80	0.20	163	142	126	115	108	104	104	107	113	122	135
0.85	0.05	-	29568	7227	3136	1731	1326	1639	2809	6124	23684	-
0.85	0.10	1209	822	590	444	362	332	343	398	500	658	915
0.85	0.15	303	245	202	173	156	148	148	155	172	196	229
0.85	0.20	135	117	103	93	87	83	82	84	87	94	102
0.90	0.05	-	22108	5336	2284	1242	936	1136	1911	4075	15383	-
0.90	0.10	915	615	436	324	260	234	238	271	333	428	578
0.90	0.15	229	183	150	126	112	104	102	106	114	128	145
0.90	0.20	102	87	76	68	62	59	57	57	58	61	65

12.4.2.3 Cross-over trials

A number of materials have covered this issue, the methods used are just extensions of the methods for superiority cross-over trials and parallel group non-inferiority trials (Julious, 2010). It is recommended that the same methodologies are used as for the parallel group trials to form an estimate of the total sample size for a non-inferiority cross-over trial. This is done by taking the sample size per arm to be the total sample size.

12.4.3 Non-inferiority trials

12.4.3.1 Parallel group trials

The two hypotheses for this type of trial can be rewritten in terms of a pre-specified clinical difference, d_{NI} (Chan, 2003; Chen et al., 2000; for the Evaluation of Medicinal Products, 2000).

$$H_0 : p_A - p_B \leq d_{NI}$$
$$H_1 : p_A - p_B > d_{NI}$$

where d_{NI} is the non-inferiority margin. This means that the null hypothesis, H_0, is that a given treatment is deemed inferior and the alternate hypothesis, H_1, states that the given treatment is not inferior. The setting of the non-inferiority limit is not easy, however it is defined as the largest difference that is clinically acceptable such that a larger difference than this would matter in clinical practice (for the Evaluation of Medicinal Products, 2000).

This type of study can be treated as a one-tailed study; therefore the α value we use is 0.025.

One way of thiking about it is

$$n_A = \frac{(Z_{1-\alpha}\sqrt{variance\ under\ null} + Z_{1-\beta}\sqrt{variance\ under\ alternative})^2}{((p_A - p_B) - d)^2} \tag{12.77}$$

We can write this as:

$$n_A = \frac{(Z_{1-\alpha}\sqrt{p_A(1 - p_A) + p_B(1 - p_B)} + Z_{1-\beta}\sqrt{\bar{p}_A(1 - \bar{p_A}) + \bar{p}_B(1 - \bar{p}_B)})^2}{((p_A - p_B) - d)^2} \tag{12.78}$$

where $\bar{p}_A$ and $\bar{p}_B$ are the estimates of the responses on treatment under the null hypothesis. If $n_A = n_B$ then a direct estimate of the sample size is

$$n_A = \frac{(p_A(1 - p_A) + p_B(1 - p_B))(Z_{1-\beta} + Z_{1-\alpha})^2}{((p_A - p_B) - d)^2} \tag{12.79}$$

where p_A is the assumed proportion of responses is expected in subjects on treatment A and p_B is the assumed proportion of responses expected on treatment B.

A quick method, using 0.90 power and two-sided significance level of 0.05, is

$$n_A = \frac{5.25}{((p_A - p_B) - d)^2} \tag{12.80}$$

A quick method, using 0.80 power and two-sided significance level of 0.05, is

$$n_A = \frac{4}{((p_A - p_B) - d)^2} \tag{12.81}$$

Both these quick methods will provide a conservative "maximum" sample size estimate. This is based on the equal-sized-groups equation above, as the maximum sample size would be when $\bar{p} = 0.5$ (where $\bar{p} = \frac{p_A + p_B}{2}$). Note these last two results are very conservative outside the range of $(0.3, 0.7)$ for $\bar{p}$.

12.4.3.2   Worked example 10

An investigator wishes to design a trial which compares two treatments, where the anticipated response probability on the active control is 0.85 and the anticipated response probability on the therapy to be investigated is also 0.85 (there is no true difference between the treatments). The non-inferiority margin is set to $d = 0.1$, the power is set to be 0.90 and the one-sided Type I error probability is 0.025. The sample size per arm can be calculated

$$n_A = \frac{(p_A(1 - p_A) + p_B(1 - p_B))(Z_{1-\beta} + Z_{1-\alpha})^2}{((p_A - p_B) - d)^2} \tag{12.82}$$

$$n_A = \frac{(0.85 \times 0.15 + 0.85 \times 0.15)(1.282 + 1.96)^2}{(0 - 0.1)^2} = 268.02 \approx 269 \tag{12.83}$$

Using Table 12.16 with $p_A = 0.85$, $p_B = 0.85$ (therefore $p_B - p_A = 0$) and the non-inferiority margin of $d = 0.1$ we get a sample size of 268 per arm.

If the anticipated investigative treatment response rate is actually $p_B = 0.82$ then the formula gives a sample size of

$$n_A = \frac{(0.85 \times 0.15 + 0.82 \times 0.18)(1.282 + 1.96)^2}{((0.85 - 0.82) - 0.1)^2} = 590.09 \approx 591 \tag{12.84}$$

Again using Table 12.16, with $p_A = 0.85$, $p_B = 0.82$ (therefore $p_B - p_A = -0.03$) and the non-inferiority margin of $d = 0.1$ we get a sample size of 590 per arm.

TABLE 12.16

A table of various sample sizes for a parallel group non-inferiority trial at 0.90 power and a 0.025 Type I error probability.

p_A	Limit	$p_B - p_A$										
		−0.05	**−0.04**	**−0.03**	**−0.02**	**−0.01**	**0**	**0.01**	**0.02**	**0.03**	**0.04**	**0.05**
0.70	0.05		45845	11325	4993	2784	1766	1214	883	669	522	418
0.70	0.10	1839	1268	925	703	550	442	362	301	254	216	186
0.70	0.15	460	378	315	266	228	197	171	150	133	118	105
0.70	0.20	205	179	157	139	124	111	100	90	81	74	67
0.75	0.05		41537	10222	4491	2495	1577	1080	782	590	459	366
0.75	0.10	1671	1149	835	632	493	395	322	267	224	190	163
0.75	0.15	418	342	284	240	204	176	152	133	117	103	92
0.75	0.20	186	162	142	125	111	99	89	80	72	65	59
0.80	0.05		36178	8856	3872	2141	1345	917	660	495	382	303
0.80	0.10	1461	1000	723	545	423	337	273	225	188	158	135
0.80	0.15	366	298	246	207	175	150	129	112	98	86	76
0.80	0.20	163	141	123	108	95	85	75	67	60	54	49
0.85	0.05		29768	7227	3136	1720	1072	724	516	383	293	229
0.85	0.10	1209	822	590	441	340	268	216	176	145	121	102
0.85	0.15	303	245	201	167	141	120	102	88	76	66	58
0.85	0.20	135	116	101	88	77	67	60	53	47	42	37
0.90	0.05		22308	5336	2284	1234	757	502	351	255	190	145
0.90	0.10	915	615	436	322	244	190	150	120	97	79	65
0.90	0.15	229	183	149	122	101	85	71	60	51	43	37
0.90	0.20	102	87	74	64	55	48	41	36	31	27	24

12.5   Final Remarks

This chapter has provided a summary of the various methods used to calculate sample sizes required for trials of parallel group or cross-over trials. The methods are similar for superiority, equivalence and non-inferiority designs. We have provided worked examples to assist with working through the calculations.

There are a number of software packages which also assist with performing sample size calculations illustrated in this chapter

Bibliography

Agostino, R. B., Massaro, J. M., and Sullivan, L. M. (2003). Non-inferiority trials: design concepts and issues – the encounters of academic consultants in statistics. *Statistics in Medicine*, 22(2):169–186.

Berger, R. L. and Hsu, J. C. (1996). Bioequivalence trials, intersection-union tests and equivalence confidence sets. *Statistical Science*, 11(4):283–302.

Brush, G. G. (1988). *How to Choose the Proper Sample Size*, volume 12 of *ASQC Statistical How-To Series*. ASQC Quality Press, Milwaukee, WI.

Chan, I. (2003). Proving non-inferiority or equivalence of two treatments with dichotomous endpoints using exact methods. *Statistical Methods In Medical Research*, 12(1):37–58.

Chen, J. J., Tsong, Y., and Kang, S.-H. (2000). Tests for equivalence or noninferiority between two proportions. *Drug Information Journal*, 34(2):569–578.

Chow, S.-C., Shao, J., and Wang, H. (2002). A note on sample size calculation for mean comparisons based on noncentral t -statistics. *Journal of Biopharmaceutical Statistics*, 12(4):441–456.

Connett, J. E., Smith, J. A., and McHugh, R. B. (1987). Sample size and power for pair-matched case-control studies. *Stat Med*, 6(1):53–9.

Fayers, P. M. and Machin, D. (1995). Sample-size - how many patients are necessary. *British Journal of Cancer*, 72(1):1–9.

Fleiss, J. L. and Levin, B. (1988). Sample size determination in studies with matched pairs. *Journal of Clinical Epidemiology*, 41(8):727–730.

European Agency for the Evaluation of Medicinal Products (2000). Points to consider on switching between supriority and non-inferiority. http://www.ema.europa.eu/docs/en_GB/document_library/Scientific_guideline/2009/09/WC500003658.pdf.

Guenther, W. (1981). Sample size formulas for normal theory t tests. *The American Statistician*, 35(4):243–244.

Hauschke, D., Steinijans, V., Diletti, E., and Burke, M. (1992). Sample size determination for bioequivalence assessment using a multiplicative model. *Journal of Pharmacokinetics and Biopharmaceutics*, 20(5):557–561.

Jones, E., Jarvis, P., Lewis, J., and Ebbutt, A. (1996). Trials to assess equivalence: The importance of rigorous methods. *British Medical Journal*, 313(7048):36–39.

Julious, S. A. (2004). Sample sizes for clinical trials with normal data. *Statistics in Medicine*, 23(12):1921–1986.

Julious, S. A. (2010). *Sample sizes for clinical trials [electronic resource]*. Boca Raton : CRC Press/Taylor & Francis, c2010, Boca Raton.

Julious, S. A. and Campbell, M. J. (2012). Tutorial in biostatistics: sample sizes for parallel group clinical trials with binary data. *Statistics in Medicine*, 31(24):2904–2936.

Julious, S. A., Campbell, M. J., and Altman, D. G. (1999). Estimating sample sizes for continuous, binary, and ordinal outcomes in paired comparisons: Practical hints. *Journal of Biopharmaceutical Statistics*, 9(2):241–251.

Julious, S. A. and Walters, S. J. (2014). Estimating effect sizes for health-related quality of life outcomes. *Stat. Methods Med. Res.*, 23(5):430–439.

Kaul, S. and Diamond, G. A. (2006). Good enough: A primer on the analysis and interpretation of noninferiority trials. *Annals of Internal Medicine*, 145(1):62–69.

Neyman, J. and Pearson, E. S. (1928a). On the use and interpretation of certain test criteria for purposes of statistical inference: Part i. *Biometrika*, 20A(1/2):175–240.

Neyman, J. and Pearson, E. S. (1928b). On the use and interpretation of certain test criteria for purposes of statistical inference: Part ii. *Biometrika*, 20A(3/4):263–294.

Neyman, J. and Pearson, E. S. (1933a). On the problem of the most efficient tests of statistical hypotheses. *Philosophical Transactions of the Royal Society of London. Series A, Containing Papers of a Mathematical or Physical Character*, 231:289–337.

Neyman, J. and Pearson, E. S. (1933b). The testing of statistical hypotheses in relation to probabilities a priori. *Math. Proc. Camb. Phil. Soc.*, 29(4):492–510.

Owen, D. B. (1965). A special case of a bivariate non-central *t*-distribution. *Biometrika*, 52(3/4):437–446.

Rothmann, M., Li, N., Chen, G., Chi, G. Y. H., Temple, R., and Tsou, H.-H. (2003). Design and analysis of non-inferiority mortality trials in oncology. *Statistics in Medicine*, 22(2):239–264.

Royston, P. (1993). Exact conditional and unconditional sample size for pair-matched studies with binary outcome: a practical guide. *Stat Med*, 12(7):699–712.

Schlesselman, J. J. and Schneiderman, M. A. (1982). Case control studies: Design, conduct, analysis. *Journal of Occupational and Environmental Medicine*, 24(11):879.

Senn, S. (1993). *Cross-over trials in clinical research*. Chichester : Wiley, c1993, Chichester.

Wiens, B. L. (2002). Choosing an equivalence limit for noninferiority or equivalence studies. *Control Clin Trials*, 23(1):2–14.

13

Sample Size Estimation and Power Analysis: Time to Event Data

Oliver Bautista and Keaven Anderson

CONTENTS

13.1 Introduction

Estimation of sample size (N) and power analysis for time-to-event analysis in randomized controlled trials poses higher complexity and considerations as compared to the routine sample size and power calculations for continuous or binary outcomes. In addition to the usual considerations for Type I (α) and Type II (β) error probabilities, expected treatment effect (θ), losses-to-follow-up, and event probability distribution (posited through hazard rate, cumulative event distribution, or median time-to-event), factors such as duration of recruitment (R), total study duration (T), staggered entry distribution, treatment non-compliance, competing risks, and possibly non-proportional hazards are important components of sample size and power analysis for randomized controlled trials with time-to-event outcomes, as extensively discussed by [43]. Evaluation of sample size and power for time-to-event outcomes was also discussed by [21]. This chapter focuses on strategies and considerations for different situations likely to be encountered in randomized controlled trials that generate time-to-event outcomes. For example, the strategy for sample size estimation in rare event situations typical of vaccine studies may be different compared to the strategy for drug studies where the endpoint event rate is not low. This chapter will also touch on recent novel developments in the area of time-to-event analysis.

13.2   Methods for Sample Size Estimation and Power Analysis

The logrank test is commonly used in testing a hypothesis on the ratio of hazard rates in a two-sample study with a time-to-event endpoint. It is asymptotically fully efficient when the hazards are proportional [21]. Thus, the development and evolution of sample size estimation and power analysis for time-to-event studies revolved around the logrank test and the proportional hazards assumption. Many of the innovations made in methods for computing sample size and power for a time-to-event endpoint were predicated on the assumption that the logrank test will be used for testing a hypothesis on the hazard ratio, and that the innovation is robust and preserves power even under violations of conditions that make the logrank test asymptotically fully efficient. Under a proportional hazards (PH) model, the power of the logrank test is asymptotically dependent on the expected total number of events E. When the significance level and power are fixed, methods of sample size estimation for time-to-event endpoints are intrinsically two-step procedures, where the total event count E required to reach a prespecified power is determined first, and then the total number of study participants N needed to acquire E events is determined next. The various components such as losses-to-follow-up, event probability distribution, duration of recruitment R, total study duration T, staggered entry distribution, treatment non-compliance, competing risks, and possibly non-proportional hazards all affect the estimation of the required number of participants.

13.2.1   Approaches relating to acquisition of events

Time-to-event study designs can be classified as maximum information [19, 22], maximum duration [19, 27], or fixed duration designs [20], depending on the temporal approach chosen to acquire E. A maximum information design is one where the foremost goal is to ensure that the power is at least $1 - \beta$ by acquiring a minimum of E events. Note that E is proportional to the statistical information associated with the logrank test for testing a hypothesis on the ratio of hazards of two independent groups under a PH model, so that power is related directly to E. In maximum information studies, one or more of the parameters R, T, and N are random and are adjusted until E events are acquired. Study designs characterized as event-driven are maximum information designs. The study described by Chen, Gesser and Luxembourg [8], particularly the phase III portion of the study, is event-driven where a fixed N was recruited and followed until such time as when E events were acquired. A maximum duration design is one where T is fixed. Typically N participants will be recruited over a period R, such that over the entire study duration T, it is expected but not guaranteed that E events will be observed. Thus, the power $1 - \beta$ of a maximum duration study is random. In addition, since participants enter the study over a duration R and the study ends at fixed time T, the i-th and j-th participants among the total of N who entered the study at different time points over R will have a total study follow-up times t_i, and t_j respectively, where $t_i \neq t_j$. A fixed duration design is a study where $t_i = t_j = t$ for each of the N

participants. In a fixed duration design, if R, T, and N are fixed such that it is expected but not guaranteed that E events will be observed, the power $1 - \beta$ is random, similar to a maximum duration design. Proponents of methods of sample size estimation and power analysis for time-to-event studies developed methods that are fitted to one of these three categorizations of time-to-event study designs. In actual practice of randomized controlled trials however, these three categorizations of time-to-event study designs are not necessarily mutually exclusive. The study reported by Fowler [16] is a hybrid fixed duration-maximum information design. In this study, follow-up time for each subject for observing the primary endpoint event is limited to a fixed duration of time t. The study was also event-driven, ensuring a study power equal to $1 - \beta$. Consequently, R, T, and N were random in this study.

13.2.2 Estimation of required number of events: no accounting of other design parameters

Early proponents of sample size estimation methods for time-to-event studies provided formulas that do not account for time and the various components described by Wittes [43] and Lachin [22]. Freedman [10] gave simple formulas for estimating the required number of events and participants in a time-to-event study where the hazard ratio of two treatment groups will be tested using the logrank test under the assumption that every participant in a given treatment group has approximately the same probability of experiencing an event and no one drops out from the study. George and Desu [12] and Bernstein and Lagakos [4] provided formulas for the special case when the time to event follows an exponential distribution, which became the basis of subsequent innovations. A more general form was given by Schoenfeld [36] that was derived for the logrank test without assuming an exponential event time model. The Schoenfeld method yields slightly smaller sample size than those derived from the Freedman formula.

Chan and Bohidar [6] (from hereon referred to as CB) provided exact methods of estimating the required number of events in order to sufficiently power a study where the event rate is expected to be very low. Their method was framed in the context of a vaccine efficacy study where the event rate in a placebo group is expected to be very low because study participants enter the study as healthy participants. If the vaccine is highly efficacious, the expected event rate in the vaccinated group is expected to be even lower. A hypothesis on the ratio of event rates in the vaccine and placebo groups is transformed into a hypothesis on a one-sample binomial proportion. They then proposed a method of calculating the required number of events needed to be observed corresponding to a pre-specified power to test a hypothesis on the binomial proportion. Their method focused on estimating the required number of events needed to be observed in order to attain a pre-specified power. Thus, it provides an estimate of the amount of information required for a maximum information design. Their calculation of the number of participants needed to be recruited into the study in order to accumulate the required number of events was rudimentary, and took into account only the expected event rates. While the CB method was not framed in the context of designing a time-to-event study, factors such as losses to follow-up, staggered subject accrual, total study duration, etc., that are normally incorporated in sample size estimation in time-to-event studies are just as applicable to vaccine efficacy studies. The next section will provide some possible strategies for estimating the number of participants needed to be recruited into a vaccine efficacy study that can be paired with the CB method of estimating the required number of events.

13.2.3   Estimation of required number of events: with accounting of other design parameters

Lachin and Foulkes [23] (from hereon referred to as LF) provided equations with closed-form solutions for estimating the required number of events and participants in a two-group time-to-event study where the time to event is assumed to follow an exponential distribution, accounting for design features such as staggered subject accrual over a recruitment period R, total study duration T, losses-to-follow-up, non-compliance, and stratification. They provided an equation from which one can derive estimates of the total number of events and total number of participants needed to appropriately power a test of hypothesis on the hazard ratio using the logrank test. The LF method provides innovations on the method of Rubinstein, Gail and Santner [34] that preceded it. Among several innovations, the LF method allows for non-uniform participant accrual whereas the Rubinstein, Gail and Santner method assumes a uniform participant accrual. Lachin and Foulkes also provided an equation from which one can derive estimates of the total number of events and total number of participants needed to appropriately power a test of hypothesis on the risk difference based on an asymptotic Wald statistic. The LF method is suited for maximum duration designs. It is an ideal starting point for sample size estimation and power analysis for such study designs because estimates are based on closed-form expressions and the method accomodates flexibility in incorporating various design parameters. An implementation of the LF method may be found in the R package gsDesign [1]. Using the LF calculation methods, gsDesign allows variations such as (piecewise) exponential event and dropout rates, piecewise constant enrollment rates, stratified populations, superiority or noninferiority designs, and fixing any two of the following three considerations: recruitment period R, enrollment rate, and minimum follow-up duration $(T - R)$. Provided later in this chapter are examples of applications of these methods to a hypothetical oncology trial and a hypothetical diabetes non-inferiority study to rule out excess cardiovascular risk that may be associated with a new diabetes medication.

Kim and Tsiatis [20] (from hereon referred to as KT), in the context of designing a group sequential controlled trial, proposed a procedure where the required number of events corresponding to a prespecified power is calculated first. Assuming a fixed rate at which participants are recruited into the study, the total sample size N is estimated by calculating the size of N with an associated expected number of events equal to calculated required events for varying participant recruitment duration R, total study duration T, and event-time distribution assumed to be exponentially distributed. Similar to the LF method, the KT method is suited for maximum duration designs. Unlike the LF method, there is no readily available closed-form expression for sample size associated with the KT.

Lakatos [24] and Lakatos and Lan [25] provided a method for sample size estimation for designs more complex than can be accomodated by the Lachin and Foulkes method. The method is also suited for designing maximum duration studies. The Lakatos method can account for design parameters such as duration of recruitment R, total study duration T, staggered participant entry distribution, treatment non-compliance, drop-in, lag in treatment effect, competing risks, non-proportional hazards, and stratification. The method assumes that the logrank test will be used to test a hypothesis on the ratio of hazards of two groups, but it does not require the typical assumptions of proportional hazards or exponential event-time distribution. The method allows for any probability distribution governing design parameters such as participant entry into the study, event-time distribution, treatment non-compliance, drop-ins, losses to follow-up, and competing risks, among many. A unique and innovative feature of the method is the way in which the design elements are put into a Markov chain framework. Elements such as non-compliance, drop-ins, losses to follow-up, competing risks, and ultimately becoming an event case, represent states of the

Markov chain and time factors such as recruitment period R and total study duration T are represented by the periods of the Markov chain. The arbitrary probability distributions that govern the different design elements form the basis of the transition probabilities from one state to another across the time periods. Because of these, there is no closed-form expression for sample size. However, an advantage of the method is sample size can be evaluated for complex situations without simulation. Barthel et al. [3] provided an extension of the Lakatos method for comparing more than two independent groups. The robustness of the Lakatos method with respect to deviations from the assumptions governing the logrank test may make it conservative in the sense of producing a larger sample size estimate compared to other methods, such as the Lachin and Foulkes method. In addition, as Wittes [43] commented, the method requires the designer of the randomized controlled trial to project in considerable detail the course of the study under both the null and alternative hypothesis.

13.3 Case Studies

13.3.1 Rare events with non-proportional hazard ratio

Many efficacy studies of prophylactic vaccines are typically studies on rare events because it involves waiting for the event of interest to develop among study participants who are healthy at the time of vaccination. The phase III portion of the randomized controlled trial reported by Chen et al. [8] is an example of such a study. In this chapter, it is used as a case study for developing a strategy on designing a two-sample randomized controlled trial with a rare-event primary outcome and potentially non-proportional hazards.

13.3.1.1 The study as designed

The phase III portion of the study reported by Chen et. al. [8] compared a novel nine-valent human papillomavirus (9vHPV) vaccine with a four-valent human papillomavirus (4vHPV) vaccine that served as an active control. The study was designed to demonstrate that the efficacy of 9vHPV relative to 4vHPV with respect to the primary endpoint of high grade disease is at least 25%, using a one-sided test of hypothesis that has at least 90% power and at most 0.025 Type I error. Assuming that the true vaccine efficacy is at least 83%, it was determined using the CB method that at least 30 primary endpoint events are needed to be observed in order to test the hypothesis within the desired design parameters. It was estimated that 7000 participants per group would be needed to be recruited into the study and that the primary efficacy analysis would be conducted after approximately 50% of participants had at least 24 months of follow-up and when at least 30 primary endpoint events had been observed [8]. Based on a KS-like calculation shown in Table 13.1, it was estimated that 7000 participants in the control group would yield approximately 32 events when each evaluable participant had completed 2 years of follow-up. Thus, 7000 participants per group is expected to provide sufficient statistical information to test the primary efficacy hypothesis within the desired parameters.

13.3.1.2 The study as it unfolded

Table 13.2 shows the life-table estimate of the primary endpoint event distribution through 3.5 years of follow-up post vaccination. In each vaccination group, 85% of vaccinated participants were eligible for efficacy follow-up in the primary analysis population, higher than the 69% projected during the study design stage. The dropout rate during the follow-up

TABLE 13.1
Calculation of expected events in the active control group.

		Inclusion rate	N
Participants vaccinated	Screened		7000
	Inclusion criteria	0.77	5390
	Vaccinated	0.9	**4851**

	Dropout rate	N
Dropped out during follow-up (5%/yr)	0.0975	**473**

		Calculation	Person years
Expected follow-up	Completers	$(4851 - 473) \times 2$	8756
	Dropouts	$\sim 473 \times 1$	473
	Total		**9229**

		Calculation	Events
Expected events	Person yrs $\times$ rate	9229×0.0035	**32**

period was less than the 5% per year projected during the design stage. The observed hazard rate in the active control group did not follow the projected constant rate of 0.35 cases per 100 person-years. It was 0.10 to 0.17 cases per 100 person-years during the first 24 months of post-vaccination follow-up, and then increased thereafter. A total of 10 primary endpoint events were accrued after 24 months median follow-up time post-vaccination; 31 were accrued after 40 months median follow-up post-vaccination. The study was successful in demonstrating the primary efficacy hypothesis, although the accrual of the required number of primary endpoint events occurred more than 1 year after the accrual time projected during the study design stage.

13.3.1.3 Insights gleaned from the study

The required number of participants needed to be recruited into the study was calculated based on an approach suitable for a maximum duration design, in the sense that a fixed study duration (2 years of post-vaccination) was assumed and then the number of participants that will yield the required number of events was determined. The study was executed as a maximum information design in the sense that the study did not stop after 2 years of post-vaccination follow-up. Rather, the study continued until the required number of events were accrued. Mathematically, the 9vHPV to 4vHPV hazard ratio was nearly constant over time only because of the 0.0 hazard rate in the 9vHPV group. In practical terms, the hazard ratio was non-proportional over time in the sense that the 9vHPV hazard was nearly constant over time while the 4vHPV hazard rate was increasing over time. The non-constant hazard rate in the 4vHPV vaccine, which was initially lower than expected and then increased later in the study, was the single major factor that contributed to the accrual of the required number of events that was 1 year later than the projected accrual time. The nearly 100% efficacy (not 83% as projected) of the 9vHPV relative to 4vHPV resulted in the 9vHPV vaccine group not contributing to event accrual.

TABLE 13.2
Observed event rate of primary endpoints over time.

Study time (Months)	At-risk	Endpoint cases	Hazard[1]	Cumulative event rate (%)
		9vHPV vaccine group		
0–6[2]	7,099	—	—	—
6–12	6,016	1	0.03	0.02
12–18	5,941	0	0.00	0.02
18–24	5,826	0	0.00	0.02
24–30	5,715	0	0.00	0.02
30–36	5,579	0	0.00	0.02
36–42	5,404	0	0.00	0.02
42–48	4,267	0	0.00	0.02
		4vHPV vaccine group		
0–6[2]	7,105	—	—	—
6–12	6,017	0	0.00	0.00
12–18	5,926	3	0.10	0.05
18–24	5,816	5	0.17	0.14
24–30	5,698	4	0.14	0.21
30–36	5,572	10	0.36	0.39
36–42	5,400	7	0.29	0.53
42–48	4,294	1	0.07	0.56

[1]Endpoint cases per 100 person-years.
[2]Vaccination period was month 1–6; follow-up period started at end of month 6.

13.3.1.4 Alternative strategies

Whenever feasible, practical, and resources permitting, conduct a study with low event rate and potentially non-proportional hazards as a maximum information study. When a maximum information design is not feasible to implement, designing the study will involve trade-offs and balancing of the following considerations.

1. *Use worst-case conservative assumptions relating to endpoint event rates.*

Lachin and Foulkes[23] gave a cautionary note that in reality, hazards likely will not be constant nor exactly proportional over time. They recommended using worst case assumptions such as using the lowest plausible event rates for the control group, assuming the smallest clinically relevant treatment effect, and appropriately adjusting the usual assumptions used in time-to-event sample size calculations.

Assuming the lowest plausible event rate in the control group has an accompanying disadvantage and advantage. The clear disadvantage is its effect on increasing the sample size. The advantage is it can guard against the risk of an underpowered maximum duration study, or against the risk of a maximum information study that has a longer than expected total study time. The unfolding of the study reported in Chen et. al. [8] demonstrated the resulting effect of having a longer than expected study duration when the assumed control group event rate is larger than that which occurred during the study.

A strategy of using a worst-case conservative assumption relating to endpoint event rates may also mean discarding the usual assumption of constant event rate over time. For a study where the control group event rate initially starts low and increases over time, such as the study reported by Chen et al. [8], assuming a Poisson-distributed case accrual with a low event rate that is constant over time may not completely guard against an underpowered maximum duration study or a longer than expected maximum information study. Thus, it may be prudent to design a study with a low event rate with potentially non-proportional hazards by using a control group event rate that initially starts low and increases over time. The Lakatos method can be useful for designing such a study. However, as Wittes [43] commented, the Lakatos method requires a considerably detailed projection of how the study will unfold.

2. Use conservatively low clinically relevant treatment effect for estimating the total events required to sufficiently power the study; use high treatment effect for the purpose of estimating the time course of case accrual.

Assuming the smallest clinically relevant treatment effect also has an accompanying advantage and disadvantage. A clear advantage is an increase in power if the observed treatment effect is greater than that assumed during the design stage. One disadvantage is it may have a consequent effect of underestimating the study time needed to accrue the required total number of events. Assuming a small treatment effect may result in overestimation of the number of events that will accrue in the experimental group, consequently leading to an underestimation of the total study time required to accumulate the required total number of events. The study reported by Chen et. al. [8] did the opposite and assumed the largest treatment effect (i.e., 100% efficacy) for the purpose of projecting the study time needed to accumulate the required number of events. They calculated the accrual of all the required number of events from the control group only (see Table 13.1).

There is no single sample size and power calculation method for time-to-event studies that will allow an analyst to simultaneously specify a low treatment effect for the purpose of estimating the total events required to sufficiently power the study and a high treatment effect for the purpose of estimating the time course of event accrual. This particular strategy requires sample size calculations for multiple settings and then selecting a sample size that provides a reasonable compromise among conflicting considerations.

13.3.1.5 Alternative strategy example

The following example shows how the study reported by Chen et al. [8] could be designed using the CB method to estimate the required number of events and the Lakatos method to project the time course of event accrual. The parameters used in the Lakatos method calculations are based on how the Chen et al. [8] study unfolded, having the benefit of hindsight. Designing a future study will have no such hindsight, although the mechanics illustrated here are applicable.

In the study reported by Chen et al. [8], a regimen of 3-dose vaccination was administered over a period of 6 months. Primary endpoint events were counted starting 4 weeks after the third vaccination dose at Month 6 (i.e., after Month 7). The primary analysis population was per-protocol, primarily consisting of participants who received all 3 doses of vaccination and uninfected with HPV during the vaccination period and through Month 7. Participants lost-to-follow-up or became infected with HPV at any time from Day 1 through Month 7 were excluded from the per-protocol population.

Determining the required total number of events: The CB method was used in the determination that a total of 30 primary endpoint events is sufficient statistical information in order to conduct a test of the null hypothesis that the efficacy of 9vHPV vaccine relative to qHPV vaccine in preventing the primary endpoint event is less than 25% against the alternative hypothesis that the vaccine efficacy is greater than 25%, with Type I error of at most 0.025 (1-sided) and power that is at least 90% when the real vaccine efficacy is at least 83%.

Determining the time course of event accrual: The Lakatos method was used to estimate the steady-state probability, say $\pi(\tau)$, of being in the primary endpoint event state at the time interval $(0, \tau ; \tau = 1, 2, \ldots 54$ months). The steady-state probability incorporates the transition probabilities from- and to- one of the following states:

a. Entry into the study (recruitment pattern): Recruitment in the Chen et al. [8] study unfolded with the following distribution of proportions of the total sample size recruited over a duration of 16 months: 0.0049, 0.086, 0.0785, 0.0372, 0.0025 0.0199, 0.0571, 0.1103, 0.1311, 0.1530, 0.0751, 0.0708, 0.0813, 0.0546, 0.0283, 0.0093.

b. Loss-to-follow-up: A total of 15% of the participants randomized into the study were excluded from the per-protocol analysis population due either to loss-to-follow-up or infection with HPV at any time from Day 1 through Month 7. Losses-to-follow-up during the post vaccination period (i.e., efficacy follow-up phase) occurred at a rate of 3% per year. Thus in the computation of the steady-state probability, the following loss-to-follow-up distribution was used:

   i. From 0 to 7 months: Exponential($\lambda = 15\%$ over 7 months) where λ is the Exponential distribution hazard rate;

   ii. From Month 7 onwards: Exponential($\lambda = 3\%$ per year).

c. Event distribution in the 9vHPV vaccine group: The following event distribution was used based on the design that primary endpoint events were counted after Month 7 and on the life-table estimate of the primary endpoint event rate shown in Table 13.2:

   i. From 0 to 7 months: Zero probability;

   ii. Months 8 to 12: Exponential (λ=0.03% per year);

   iii. Month 13 onwards: Zero probability.

d. Event distribution in the 4vHPV vaccine group: The following event distribution was used based on the design that primary endpoint events were counted after Month 7 and on the life-table estimate of the primary endpoint event rate shown in Table 13.2:

   i. From 0 to 12 months: Zero probability;

   ii. Months 13 to18: Exponential (λ=0.10% per year);

   iii. Months 19 to 24: Exponential (λ=0.17% per year);

   iv. Months 25 to 30: Exponential (λ=0.14% per year);

   v. Months 31 to 36: Exponential (λ=0.36% per year);

   vi. Months 37 to 42: Exponential (λ=0.29% per year);

   vii. Month 43 onwards: Exponential (λ=0.07% per year).

TABLE 13.3
Time course of event accrual.

Time (τ)	$\pi(\tau)$	Expected $(N \times \pi(\tau))$	Observed
		Cumulative events	
9vHPV Vaccine Group $N = 7,000$			
12	0.0001	1	1
24	0.0001	1	1
30	0.0001	1	1
36	0.0001	1	1
42	0.0001	1	1
48	0.0001	1	1
4vHPV Vaccine Group $N = 7,000$			
12	0.0000	0	0
24	0.0011	8	8
30	0.0017	12	12
36	0.0032	22	22
42	0.0043	30	29
48	0.0044	31	30

Let N_{9v} and N_{4v} denote the number of participants randomized into the 9vHPV and 4vHPV vaccine groups respectively; and let $\pi(\tau)_{9v}$ and $\pi(\tau)_{4v}$ denote the steady-state probability of being a primary endpoint event in the time interval $(0, \tau]$ in the 9vHPV and 4vHPV vaccine groups respectively; then the total primary endpoint events in the 9vHPV and 4vHPV vaccine groups during the time interval $(0, \tau]$ may be estimated by $N_{9v} \times \pi(\tau)_{9v}$ and $N_{4v} \times \pi(\tau)_{4v}$ respectively. Thus, the time course of event accrual may be assessed for $\tau = 1, 2, \ldots 54$ months for different values of N_{9v} and N_{4v}. Table 13.3 shows the estimated time course of event accrual for a sample size of $N = 7,000$ participants recruited in each of the 9vHPV and 4vHPV vaccine groups, together with the cumulative count of events actually observed during the study.

Table 13.3 shows that if one can very accurately predict and specify during the study design stage what will actually transpire during the study interms of recruitment pattern, losses-to-follow-up, and event cumulative distribution, then one can very accurately predict the time course of event accrual using the Lakatos method. The use of piecewise-exponential distribution to model the primary endpoint event distribution and losses-to-follow-up provides a convenient way of calculating from hazard rates the time interval-specific event probabilities that are needed to be specified to implement the Lakatos method.

13.3.2 An oncology study

In oncology trials, it is common to perform multiple analyses (interim and final) where both overall survival (OS) and progression-free survival (PFS) are analyzed. In this example, proportional hazards assumption is used for both PFS and OS in comparing a standard therapy with an experimental therapy, powering the study to detect a hazard ratio of 0.75 for OS and 0.6 for PFS. R program codes for sample size calculations using the gsDesign R package [1] are provided in this example. Help files from the gsDesign package provide additional information if needed. Splitting Type I error $\alpha = 0.025$ equally between OS and PFS, one would need 600=2*(ceiling(nEvents(hr=.75, alpha=.0125)/2)) OS events and 192=2*(ceiling(nEvents(hr=.6,alpha=.0125)/2)) PFS events in order for the trial to be powered at 90% for each endpoint based on the Schoenfeld formula. OS events generally accrue much more slowly than PFS events, which suggests three strategies. First, higher Type I error can be allocated to the OS endpoint and lower to PFS. Second, the power for testing the OS hypothesis can be set lower than the power for testing the PFs hypothesis. Finally, the testing of the PFS hypothesis can occur well before the testing of the OS hypothesis. The overall sample size is typically driven by the assumptions relating to OS. To illustrate the sample size calculations using gsDesign, the following design parameters are assumed:

- The time scale is months.

- One-sided $\alpha_{PFS} = 0.0025$ allocated to PFS.

- One-sided $\alpha_{OS} = 0.0225$ allocated to OS.

- Power=99% for testing the PFS hypothesis.

- Power=85% for testing the OS hypothesis.

- Exponential time-to-PFS-events with a median of 6 months in the control group and a hazard ratio of 0.6 (median of 6/0.6=10 months in the experimental group).

- Exponential time-to-OS-events with with a median of 14 months in the control group and a hazard ratio of 0.75 (median of 14/0.75=18.67 in the experimental group).

- Enrollment ramp-up to a steady state after 6 months, with enrollment rate for the first 2 months equal to 25% of the steady state enrollment rate; enrollment rate of 50% of steady state for months 2-4; and 75% of steady state for months 4-6.

- Enrollment is over 30 months; minimum follow-up of the last participant enrolled is 18 months (minfup) for a total study duration of 48 months (T).

- 2% per year loss-to-follow-up for OS (exponential dropout rate) in each treatmment group.

- 7.5% per year loss-to-follow-up for PFS (exponential dropout rate) in each treatment group.

- The specified power for testing the OS hypothesis will determine the total study sample size.

- The total study sample size will determine the time to accrual of PFS events and time of testing of the PFS hypothesis.

- An interim analysis for OS will be performed at the time of the final testing of the PFS hypothesis with efficacy bounds determined using an O'Brien-Fleming-like spending function [26] and futility bounds determined using a Hwang-Shih-DeCani spending function [13] with parameter $\gamma = -2$.

Given these design parameters, the sample size calculations using gsDesign starts as follows:

```
# one-sided Type I error
alphaPFS <- .0025
alphaOS <- .0225
# Type II error (1-power)
betaPFS <- .01
betaOS <- .15
# hazard rates determined by medians
lambdaPFS <- log(2)/6
lambdaOS <- log(2)/14
# treatment effect
hrPFS <- 0.6
hrOS <- 0.75
# enrollment period durations for ramp-up
# final enrollment period will be extended in function call
R <- c(2,2,2,6)
# relative enrollment rates during ramp-up and steady state
gamma <- c(.25,.5,.75,1)
# study duration in months
T <- 48
# minimum follow-up duration
minfup <- 18
# (note that enrollment duration is implicitly T-minfup)

# loss-to-follow-up hazard rates determined by
# 1-year probability of loss-to-follow-up
etaOS   <- -log(1-.02)/12
etaPFS <- -log(1-.075)/12
```

The derivation of the sample size needed for testing the OS hypothesis assuming no interim analysis (as a first approximation) follows here. The goal of this first approximation is to estimate the sample size needed for testing the OS hypothesis, which will then be used to estimate the time to accrual of the required total PFS events. The participant enrollment (accrual) rates have been set in order to accrue the required OS events within the desired study duration. The duration of the steady state enrollment rate has been adjusted based on the total study duration (T) and minimum follow-up (`minfup`).

```
xOS <- nSurv(lambdaC = lambdaOS, hr = hrOS, eta = etaOS,
             gamma = gamma, R = R, T = T , minfup = minfup,
             alpha = alphaOS, beta = betaOS)
xOS

## Fixed design, two-arm trial with time-to-event
## outcome (Lachin and Foulkes, 1986).
## Solving for:   Accrual rate
## Hazard ratio                   H1/H0=0.75/1
## Study duration:                   T=48
## Accrual duration:                  30
## Min. end-of-study follow-up: minfup=18
```

```
## Expected events (total, H1):        445.8257
## Expected sample size (total):       627.2417
## Accrual rates:
##        Stratum 1
## 0-2      5.8078
## 2-4     11.6156
## 4-6     17.4234
## 6-30    23.2312
## Control event rates (H1):
##        Stratum 1
## 0-Inf    0.0495
## Censoring rates:
##        Stratum 1
## 0-Inf    0.0017
## Power:                    100*(1-beta)=85%
## Type I error (1-sided):   100*alpha=2.25%
## Equal randomization:          ratio=1
```

Note that the total number of events corresponds to 71% = round(ceiling(xOS$eDC+ xOS$eDE)/ceiling(2*xOS$eNC)*100) of the total sample size, which would generally be considered a relatively mature follow-up for OS. While the expected total number of OS events shown above was calculated under the alternative hypothesis event rates, the LF calculations uses the null hypothesis event rates. For this example, the calculated expected total numbers of events based on the null and alternative hypothesis events rates are approximately similar since the assumed hazard ratio under the alternative hypothesis is 0.75.

The total sampe size $N=628$ calculated for testing the OS hypothesis is now used to derive the time to accrual of the required total events for testing the PFS hypothesis at the desired type I error and power. This is done by using the enrollment rates (xOS$gamma) and enrollment periods (xOS$R) generated in the calculation of the OS sample size along with the other assumed parameters relating to PFS. Note that for this calculation, the total study duration T and minimum follow-up minfup are set to NULL since these are the values that are being calculated. The calculated T corresponds to the estimated time to accrual of the required total PFS events. For this example, the power for testing the PFS hypothesis was selected to ensure that the study was fully enrolled prior to reaching the time to accrual of the required total PFS events, as indicated by the calculated value of T shown below.

```
xPFS <- nSurv(lambdaC = lambdaPFS, hr = hrPFS, eta = etaPFS,
             gamma = xOS$gamma, R = xOS$R, T = NULL,
             minfup = NULL, alpha = alphaPFS, beta = betaPFS)
xPFS

## Fixed design, two-arm trial with time-to-event
## outcome (Lachin and Foulkes, 1986).
## Solving for:  Follow-up duration
## Hazard ratio                  H1/H0=0.6/1
## Study duration:                  T=31.4742
## Accrual duration:                   30
## Min. end-of-study follow-up: minfup=1.4742
## Expected events (total, H1):      402.4135
## Expected sample size (total):     627.2417
## Accrual rates:
```

```
##         Stratum 1
## 0-2      5.8078
## 2-4     11.6156
## 4-6     17.4234
## 6-30    23.2312
## Control event rates (H1):
##         Stratum 1
## 0-Inf    0.1155
## Censoring rates:
##         Stratum 1
## 0-Inf    0.0065
## Power:                   100*(1-beta)=99%
## Type I error (1-sided):  100*alpha=0.25%
## Equal randomization:          ratio=1
```

Having estimated the time to accrual of the required total PFS events, the next step is to build a group sequential design for testing the OS hypothesis with 1 interim analysis at the time of accrual of the required PFS events. This building process is iterative. For this example, only a few iterations is needed.

```
OSevents <- xOS$eDC+xOS$eDE
for(i in 1:3){
# use OS events at current PFS design analysis to
# approximate portion at interim analysis
  timing <- nEventsIA(tIA=xPFS$T, x=xOS)/OSevents
# k=2 indicates 2 analyses, 1 interim + 1 final
# sfu and sfl indicate  desired efficacy and futility
# spending functions, respectively
# sflpar indicates spending function parameter for
# futility spending function
   xOS <- gsSurv(k=2, timing=timing, sfu=sfLDOF,
            sfl=sfHSD, sflpar=-2,
            lambdaC = lambdaOS, hr = hrOS, eta = etaOS,
            gamma = gamma, R = R, T = T, minfup = minfup,
            alpha=alphaOS, beta=betaOS)
# recalculate PFS design based on updated enrollment
  xPFS <- nSurv(lambdaC = lambdaPFS, hr = hrPFS, eta = etaPFS,
            gamma = xOS$gamma, R = xOS$R, T = NULL,
            minfup = NULL, alpha = alphaPFS, beta = betaPFS)
# update total OS events based on updated design
  OSevents <- xOS$n.I[2]
}
```

A tabular display of the resulting group sequential design is shown (via the summary(xOS) command) in Table 13.4. Some notable summaries that are useful for interpreting the tabular display are as follows: (1) It is estimated that a hazard ratio of at least 0.91 at the interim analysis would stop the study for futility; (2) A hazard ratio of at most 0.83 in the final analysis corresponds to successful demonstration of the OS hypothesis, and translates to an increase of at least 2.9 months in median OS compared to the 14 months median OS in the control group.

TABLE 13.4
Oncology OS group sequential design bounds.

Analysis	Value	Futility	Efficacy
IA 1: 58.7%	Z-value	0.76	2.76
N: 656	HR	0.91	0.72
Events: 274	p (1-sided)	0.2233	0.0029
30.5 months	P{Cross} if HR=1	0.7767	0.0029
	P{Cross} if HR=0.75	0.0525	0.3532
Final analysis	Z-value	2.02	2.02
N: 656	HR	0.83	0.83
Events: 467	p (1-sided)	0.0216	0.0216
48 months	P{Cross} if HR=1	0.9788	0.0212
	P{Cross} if HR=0.75	0.15	0.85

In a real study, the total OS events accrued at the time of testing of the PFS hypothesis (which is the time of OS interim analysis) will not be exactly equal to the count estimated during the time of building of the group sequential design. Thus in a real study, it is typically necessary to update the futility and efficacy boundaries corresponding to the total OS events actually available at the time of testing of the PFS hypothesis. In this particular example, suppose that 240 OS events (not the calculated 274 shown in the tabular display above) are accrued at the time of PFS hypothesis testing. The calculation of the updated futility and efficacy boundaries for conducting the OS interim analysis at 240 OS events are shown below.

```
OSupdate <- gsDesign(k=2, delta=xOS$delta, n.I=c(240,xOS$n.I[2]),
                     maxn.IPlan=xOS$n.I[2],
                     alpha=alphaOS, beta=betaOS,
                     sfu=sfLDOF, sfl=sfHSD, sflpar=-2)
gsBoundSummary(OSupdate)

## Analysis                 Value Efficacy Futility
## IA 1: 51%                    Z   2.9735   0.5063
##     N: 240        p (1-sided)   0.0015   0.3063
##                ~delta at bound   1.3327   0.2269
##             P(Cross) if delta=0   0.0015   0.6937
##             P(Cross) if delta=1   0.2290   0.0423
##     Final                    Z   2.0138   2.0138
##     N: 467        p (1-sided)   0.0220   0.0220
##                ~delta at bound   0.6476   0.6476
##             P(Cross) if delta=0   0.0214   0.9786
##             P(Cross) if delta=1   0.8523   0.1477
```

13.3.3   A diabetes noninferiority study

This example illustrates designing a noninferiority study to demonstrate a hazard ratio of less than 1.3 for a new treatment compared to a standard, corresponding a 2008 FDA guidance for evaluating cardiovascular (CV) risk in new diabetes therapies [5]. This example illustrates designing a meta analysis of two studies with an objective to demonstrate noninferiority; one is a low-risk study (LRS) population and the other is a high-risk study (HRS) population. For this example, the following design parameters are assumed:

- Exponential time to *CV* event with rate of 1.5% per year in the LRS; 3% per year in the HRS.

- Exponential loss-to-follow-up with event rate of 2% per year in each of the LRS and HRS.

- Hazard ratio (treatment/control) of 1.3 under the null hypothesis; 1.0 under the alternative hypothesis.

- Power = 90% for testing the hypotheses.

- Type I error $\alpha = 0.025$ (one-sided).

- LRS sample size = 2x HRS sample size.

- LRS begins enrollment 4 months before the HRS.

- LRS will enroll at 75% as fast as the HRS at steady state enrollment.

- In each study, enrollment ramps up over the course of about 6 months as in the oncology example provided previously.

- Total enrollment duration of 2 years (24 months)

- Total study time of 6 years.

- Minimum follow-up of last participant enrolled equal to 4 years

The total number of *CV* events required from the 2 studies to rule out the null hazard ratio of 1.3 is

```
ceiling(nEvents(hr=1,hr0=1.3,alpha=.025,beta=.1))
```

```
## [1] 611
```

The expected combined event rate per year in the 2 studies is 0.02 (calculated as 0.015x(2/3)+0.03x(1/3)) under the assumption of no difference in CV event rates between the treatment and control groups in each of the 2 studies. This translates to needing to accrue a total of

```
nEvents(hr=1,hr0=1.3,alpha=.025,beta=.1)/.02
```

```
## [1] 30529.3
```

participant-years of follow-up in the 2 studies to accrue the required CV events. The derivation of sample size for a design without interim analysis is as follows. Unlike in the oncology example, the time scale used in the derivations shown here is years, not months.

```
# enrollment rate period durations in years
R <- c(2,2,2,2,2,2,12)/12
# relative enrollment rates in low-risk population
glo <- c(.25,.5,.75,1,1,1,1)
# relative enrollment rates in high-risk population
ghi <- c(0,0,0,.25,.5,.75,1)*.75
# combine relative enrollment rates into a matrix
```

```
gamma <- matrix(c(glo,ghi),ncol=2)
# exponential event rates in the two studies
lambdaC <- matrix(c(.015,.03), ncol=2)
eta <- .02
# sample size
x <- nSurv(lambdaC=lambdaC, hr=1, hr0=1.3, eta=eta,
       alpha=.025, beta=.1, R=R, gamma=gamma, minfup=4, T=6)
# sample size per stratum (add the 2 treatment groups)
ceiling(x$eNC+x$eNE)
```

```
## [1] 4610 2470
```

```
# approximate event count per stratum; again add treatment groups
ceiling(x$eDC+x$eDE)
```

```
## [1] 311 307
```

Considering that a new diabetes drug may have excess cardiovascular risk or a substantial benefit, a group sequential design to monitor treatment effect during the course of the study is desirable. Provided below is a design with 3 interim analyses and 1 final analysis (k=4), with the 4 analyses equally spaced in terms of information fraction (number of events analyzed/total events at final analysis). The exponential spending function [2] of the form $f(t) = \alpha^{t^{-\nu}}$ is used in this example, where $\nu > 0$ for superiority since it completely orders the sample space which is helpful for sequential inference [29, 30, 40]. Specifically, $\nu = 0.75$ is chosen to approximate an O'Brien-Fleming efficacy (successful demonstration of noninferiority in this example) boundary. The Hwang-Shih-DeCani spending function [13] with $\gamma = -2$ is used to set the futility boundary.

The meta analysis design with interim analyses is provided in Table 13.5. Notable information to verify include the estimated timing of analyses; event counts at analyses; nominal α-levels at bounds; and hazard ratios at bounds, potentially to consider making a decision to discontinue the study at any interim analysis for either futility or efficacy. The design calculations are performed such that all stopping boundaries are non-binding, so that the study may be continued even if stopping boundaries are crossed at interim analyses and be legitimately analyzed at a following analysis.

Shown here is the calculation of the upper bounds of repeated confidence intervals for the hazard ratio given an observed hazard ratio of 1 at each analysis. The calculated hazard ratio upper bounds do not rule out a hazard ratio of 1.3 during the 3 interim analysis, but does so at the final analysis. Thus, noninferiority of the new treatment was established at the final analysis.

```
# Schoenfeld approximation of standard error
# of log(HR) at each analysis
se <- nEvents(n=ceiling(xgs$n.I),tbl=TRUE)$se
# compute upper repeated confidence bounds
# assuming observed HR=1.1
round(exp(log(1.1)+xgs$upper$bound*se),2)
```

```
## [1] 2.04 1.51 1.36 1.29
```

The 2008 FDA guidance [5] allows for the possibility of early conditional approval of a drug once a hazard ratio of 1.8 is ruled out. This can be assessed with a 2-sided 95% confidence interval for the hazard ratio that excludes 1.8, which could be evaluated a single

TABLE 13.5
Diabetes non-inferiority group sequential design bounds for CV endpoints.

Analysis	Value	Efficacy	Futility
IA 1: 25%	Z	4.02	-0.63
N: 7708	p (1-sided)	0.00	0.74
Events: 168	HR at bound	0.70	1.43
Year: 2.3	P(Cross) if HR=1.3	0.00	0.26
	P(Cross) if HR=1	0.01	0.01
IA 2: 50%	Z	2.88	0.36
N: 7708	p (1-sided)	0.00	0.36
Events: 336	HR at bound	0.95	1.25
Year: 3.5	P(Cross) if HR=1.3	0.00	0.65
	P(Cross) if HR=1	0.31	0.03
IA 3: 75%	Z	2.34	1.20
N: 7708	p (1-sided)	0.01	0.11
Events: 504	HR at bound	1.05	1.17
Year: 4.7	P(Cross) if HR=1.3	0.01	0.89
	P(Cross) if HR=1	0.72	0.05
Final	Z	2.02	2.02
N: 7708	p (1-sided)	0.02	0.02
Events: 672	HR at bound	1.11	1.11
Year: 6	P(Cross) if HR=1.3	0.02	0.98
	P(Cross) if HR=1	0.90	0.10

time at the first interim analysis. The calculations below show the approximate hazard ratio needed to be achieved to satisfy the requirement and the corresponding power under the assumption that event rates are equal in the two treatment groups.

```
# event counts at IA1
events <- ceiling(xgs$n.I[1])
# Schoenfeld approximation of standard error
# of log(HR) at IA1
se <- nEvents(n=events,tbl=TRUE)$se
# approximate observed HR that would rule out HR=1.8
# at IA1 using 95 percent CI
hr <- exp(log(1.8)-qnorm(.975)*se)
hr
```

```
## [1] 1.330238
```

```
# compute z-statistic to rule out HR=1
# corresponding to above hazard ratio
# done by translating observed HR and
# number of events
z <- -hrn2z(hr=hr, n=events[1], hr0=1, hr1=1.3)
# now compute cumulative normal probability for this
pnorm(z)
```

```
## [1] 0.9677948
```

13.4 Special Topics and Recent Developments

13.4.1 Treatment effects beyond hazard ratios

The Cox proportional hazards model and the logrank test are the most common statistical model and test used in analyzing time-to-event outcomes. Consequently, the hazard ratio is the default treatment effect by which treatment groups are being compared in studies with time-to-event outcomes, and for which analysis methods have been extensively developed. There are other treatment effect statistics by which one can compare treatment groups with respect to time-to-event outcomes, although analysis methods for these are not as extensively developed. Uno et. al.[39] provided a summary of some of these other treatment effect statistics. In this section, survival curve or survival rate is assumed to be derived from the survival function defined as one minus the cumulative distribution function for a time-to-event. When specifically referring to a mortality outcome, overall survival (OS) is used to refer to the survival function.

Difference of survival rates at time τ (τ-year survival): Comparison of treatment groups with respect to the treatment group-specific probability of survival at a fixed time point τ, or the τ-year survival S_τ when the unit of time is years, has a straightforward and clear interpretation. Point and confidence interval estimates of the difference or ratio of S_τ can be easily calculated based on Kaplan-Meier (KM) estimates of survival [39]. However, what is not clear and not easily implemented is the mechanics of calculating the total number of events needed to be accumulated in a time-to-event study that will test a hypothesis on the difference or ratio of S_τ at level of significance α with power ρ. Lin, et al. [28] showed how to implement a group sequential testing of a one-sample KM survival estimate S_τ when estimated at multiple times $t_1, t_2, \ldots, t_k$. The group sequential testing of the null hypothesis $H_0: S_\tau = S_0$ can be carried out by conducting a group sequential testing of the hypothesis $H_0: \Lambda_\tau = \Lambda_0$ based on the identity $S_\tau = e^{-\Lambda_\tau}$ where Λ_τ is the KM estimate of the cumulative hazard at τ. Calculation of the required number of participants N needed to conduct the test of hypothesis $H_0: \Lambda_\tau = \Lambda_0$ with power ρ at α level of significance is carried out based on the asymptotic standard normal distribution of a Z-score statistic resulting from a suitable normalization of $\log(\Lambda_\tau)$. For testing a null hypothesis $H_0: S_1(\tau) = S_2(\tau) = \Delta_0$ on the difference of τ-year survival between two treatment groups, Lin et. al. noted that it is straightforward to extend the results they showed for the one-sample case to the two-sample setting. There is currently no method available that provides an explicit way of calculating the total number of events needed to be accumulated in a time-to-event study that will test a hypothesis on the difference or ratio of S_τ at significance level α with power ρ. This can be an area of future research.

Difference of quantiles of time-to-event distribution functions (e.g., median survival): In designing time-to-event studies, median survival is most commonly used in conjunction with the Exponential distribution to specify the underlying probability distribution of the event time of interest. Once the time-to-event distribution has been determined, the logrank test is used to test hypothesis on the hazard ratio. Directly testing a hypothesis on equality of median survival between two treatment groups is uncommon. For time-to-event studies that have short follow-up duration or low event rates, the median survival may not be estimable from the observed data, and thus a direct test of hypothesis on the equality of median survival between two treatment groups is not feasible. Point estimate of median survival is easily obtained from the KM estimate. Interval estimation of median survival in the one-sample setting has been extensively studied.[9, 41]. An R package called bpcp implements the method of Fay et. al.[9]. Wang and Hettmansperger[41] also provided a method for interval estimation of the difference of median survival in the two-sample setting. There is currently no method available that provides an explicit way of calculating the total number of events needed to be accumulated in a time-to-event study that will test a hypothesis on the difference of median survival. This can be an area of future research.

Difference of restricted mean survival time (area under the survival curve): The restricted mean survival time (RMST) up to time τ is defined as $\mu_\tau = \int_0^\tau S(t)dt$, representing the area under the survival curve $S(t)$ up to time τ. It is interpreted as the expected event-free survival time associated with a planned follow-up duration of τ time units. The concept was pioneered by Irwin[14], whose work has been built upon by many others.[17, 18, 32, 33, 44]. Royston and Parmar[33] provided a simulation-based approach for estimating the required number of participants needed to be enrolled in a time-to-event study that will test a null hypothesis H_0: $\Delta_\tau = \mu_{1\tau} - \mu_{2\tau} = \Delta_0$ on the difference of RMSTs between two treatment groups, under a setting where the event time distribution in each treatment group is piecewise Exponential. They also provided an analogue to the concept of maximum statistical information, which they called data mature enough for final analysis. For a PH model setting that tests a hypothesis on the hazard ratio using the logrank test, the required total number of events needed to conduct an α-level test with power ρ corresponds to the maximum statistical information, and acquisition of that total number of events corresponds to the state when data are mature enough for final analysis. For a study that will test a hypothesis on the RMST difference Δ_τ, maximum information is reached, or data is mature enough for final analysis, when the variance of the point estimate of Δ_τ, *i.e.*, $\text{Var}(\Delta_\tau)$, becomes less than a threshold value that is a function of the level of significance α, the power ρ, and the hypothesized value Δ_1 of Δ_τ under the alternative hypothesis. They provided illustrations of how one can translate the maximum statistical information in terms of $\text{Var}(\Delta_\tau)$ into required total number of events by first formulating a putative PH model that will test hypotheses on the hazard ratio using the logrank test. Such an approach for estimating the required statistical information in terms of total number of events can expose one to a double jeopardy situation of gross misspecification of the treatment effects Δ_1 and the hazard ratio under the alternative hypothesis in the putative PH model. Given the complexity of Royston and Parmar's method, estimation of the total statistical information in terms of number of events needed to conduct an α-level test with power ρ in a time-to-event study that will test a hypothesis on the difference of RMST is still a wide open area for future research.

13.4.2 Sample size re-estimation

A maximum duration time-to-event study may become under powered if the event rates that are used in designing the study over estimate the emerging event rates during the study. At a scheduled interim analysis, the treatment group-blinded study team may be able to discern from the overall, treatment group-blinded event rate if the emerging event rate is lower than anticipated, and thus may eventually lead to an under powered test of the primary study hypothesis. Given such a situation, the choice of an appropriate course of action can be complicated and varied for a time-to-event study. For instance, unlike studies with normally distributed continuous outcome endpoint where each subject contributes 1 unit of statistical information and therefore increasing the sample size N correspondingly increases the statistical information, increasing the sample size N of a time-to-event study generally does not proportionately increase the statistical information due to reduced follow-up for patients enrolled at the end of the recruitment period.

A long time-course endpoint: For time-to-event endpoint that requires a long time-course to develop, such as the one in the study reported by Chen et. al.[8], increasing the sample size N by recruiting more participants late into the study may not result in an increase

: number of endpoint cases. participants recruited late into the study may not have
;h follow-up time in order for the study to observed the development of the endpoint of
st. For a time-to-event study with such an endpoint, the appropriate course of action
observing at an interim analysis an emerging event rate that is lower than anticipated
extend the total duration of the study to increase to follow-up time of each subject.
nay require converting a maximum duration study into a maximum information study
ler to maintain the pre-specified study power.

fixed-duration endpoint: The study reported by Fowler et al. [16] is one where the
primary endpoint is observable only for a limited fixed duration of time. In this situ-
the appropriate course of action upon observing, at an interim analysis, an emerging
rate that is lower than anticipated is to recruit more participants and increase the
Je size N. Extending each subject's follow-up time will not increase the likelihood of
ving an outcome event on a particular subject. The study reported by Fowler et al.
lesigned as event-driven, such that subject enrollment will continue until the the re-
1 number of events are accrued. Sample size re-estimation is not relevant for this kind
dy design as it is buit-in to the sampling procedure.

'hen sample size re-estimation is a viable option: In between the situations of a fixed-
ion endpoint such as the Fowler et al. study and a long time-course endpoint such as
hen et al. study is one where a sample size re-estimation may be a viable option for a
to-event study. Sample size re-estimation procedures generally come in two flavors with
ct to information used in the re-estimation procedure: unblinded procedures, where
ledge of emerging treatment effect or some treatment group-specific nuisance parameter
ded; and blinded procedures, where such knowledge is not required. Blinded sample
e-estimation procedures are preferred, firstly based on the scientific consideration that
ich procedures, there is minimal inflation of Type I error and power is preserved [38];
econdly based on perception that study integrity is maintained if only blinded looks
he data are conducted. Todd et al. [38] noted that sample size re-estimation methods
me-to-event studies tended to be focused in the setting of adaptive designs, such as
ntial, group sequential, or self-designing trials [15, 35, 37]. Mehta and Pocock [31] noted
idaptive methods of sample size re-estimation require non-standard final analysis, such
data from participants enrolled prior to the interim analysis when sample size was
imated are treated or weighted differently compared to those participants enrolled
the interim analysis. Such characteristic of this class of sample size re-estimation
dures makes it unattractive to practitioners for a variety of reasons, one of which
ieed to report to journals and regulatory submissions conventional hypothesis tests
-values uncluttered by complex statistical adjustments. Consequently, a new class of
dures emerged, wherein conventional hypothesis tests and p-values may be used in
nal analysis even after sample size re-estimation at an interim analysis, provided that
e-estimation is conducted only when the interim results are promising. The crux of
class of procedures is the definition of promising, [7, 11, 31] which typically involves
lation of conditional power based on the point estimate of treatment effect at the
m analysis, and thus requires someone to be unblinded to the emerging treatment
. Presented in this section is a strategy for sample size re-estimation for time-to-event
es along the lines proposed by Todd et al. [38] which is based only on blinded data at
ime of sample size review.

he Todd et. al. approach, which is an adaptation of the method by Whitehead et al. [42]
sed on the following framework. A PH model is assumed and a logrank test will be used
st a hypothesis on the log hazard ratio, so that the total number of events determines
otal statistical information. The study is a fixed sample size trial, where hypothesis
ig on the primary endpoint will be conducted only at the end of the study. A sample
eview, not an interim analysis on the primary hypothesis, will be conducted at some

time point during the study when there are event outcomes already observed so that a KM estimate of the survival function on the combined treatment groups can be calculated; not too early when there are scarce endpoint cases; or too late when the study is near its end. It can be at the time of accrual of 50% of required total number events. Implementation of the sample size re-estimation procedure requires calculation of the following:

1. A survival distribution function that represents the average of the survival distribution functions of the treatment and control groups assumed during the study design stage, denoted as $S_{old}(t)$;

2. An estimate of the emerging survival distribution function in the combined treament groups, calculated from available data during the time of sample size review, denoted as $S_{new}(t)$.

The idea behind the procedure is to get an estimate of deviation of the emerging survival distribution function $S_{new}(t)$ from the survival distribution function $S_{old}(t)$ assumed during the study design stage on the complementary log-log scale denoted by

$$\phi = -\log(-\log S_{new}(t)) + \log(-\log S_{old}(t)).$$

Note that ϕ is not denoted as a function of t as it is estimated either at a single time-point or averaged over some set of timepoints t_i where $S_{new}(t_i)$ can be estimated. This estimated deviation ϕ is projected to or through the time of end of study T, when at such time, it is assumed that the observed survival probability $\log(-\log S_{new}(T))$ will differ from $\log(-\log S_{old}(T))$ also by ϕ. What is the purpose of estimating $S_{new}(t)$ for $t \leq T$? To illustrate, assume a simplistic case where all study participants planned for during the study design stage, denoted by N_{old}, have a total follow-up duration of T time units. Denote $p_T = 1 - S_{new}(T)$ to be an estimate of the probability of being an endpoint case through time T. Then $D_{old} = N_{old} \times p_T$ is the total events at the end of the study expected to be accumulated from the N_{old} participants. If D_{old} is less than D_0, the total required statistical information calculated during the study design stage, then a new sample size N_{new} can be calculated such that $D_0 = N_{new}p_T$. Thus, estimation of $S_{new}(T)$ is a stepping stone that eventually leads to estimation of N_{new}. Described later is a strategy for estimating N_{new} based not on p_T, but on the steady-state probability of becoming an endpoint case through time T derived through the Markov process approach of Lakatos.

Let S denote the time of sample size review; $t_1, t_2, \ldots, t_q \leq S$ denote the sequence of the event times through time S; and $S_{new}(t_i): i = 1, \ldots, q$ the KM estimate of the emerging survival distribution function in the combined treament groups through time S. [Note: t_i-values are not what Todd used. For the 1st equation, they used equally-spaced points, presuming [implicitly] that the recruitment rate was constant over the recruitment period and, thus, $1/q$ of the patients would have t_i represented approximately by ri/q where r is the total recruitment duration and therefore for that portion of the population approximately $1 - S(t_i)$]would have events by time T.] Todd et al. [38] provided the following two distinct ways of estimating ϕ in the complementary log-log scale:

$$\phi = \frac{1}{q}\sum_{i=1}^{q}[-\log(-\log S_{new}(t_i)) + \log(-\log S_{old}(t_i))] \qquad (13.1)$$

$$\phi = -\log(-\log S_{new}(S)) + \log(-\log S_{old}(S)). \qquad (13.2)$$

The first represents the average deviation over the observed event times. The second is the most recent observed deviation just prior to the time of sample size review. Neither one perior to the other in terms of preserving the planned study Type I and II error rates

The estimate of $S_{new}(t)$ for $t > S$ through T can be obtained from

$$-\log(-\log S_{new}(t)) = -\log(-\log S_{old}(t)) + \phi. \qquad (13.3)$$

Given the projected survival function $S_{new}(t)$ for $t > S$ in equation (13.3), sample size stimation for any complex time-to-event study that can be designed in the framework he Lakatos method can be executed as follows:

Design the study using the Lakatos method. Let N_{old} and D_0 denote the total study sample size and total number of events, respectively, that are determined will be accumulated.

Run the Lakatos method a second time, assigning the total sample size N_{old} to one treatment group, say the control group, and using $S_{old}(t)$ to specify the event time probability in this group. Other design parameters such a recruitment pattern, losses-to-follow-up, etc. are as specified in the original design used in step 1. The design parameters for the other treatment group are immaterial because the desired information to be obtained in this step are the steady-state event probabilities π_{oldt} in the group that carries the total sample size N_{oldt} and event time probabilities derived from $S_{old}(t)$. Calculate $D_0 = N_{old}\pi_{old}T$, representing the expected number of events at end of study from the N_{old} participants based on the average survival distribution in the combined treatment groups, determined at the design stage.

At time S when sample size review is conducted, calculate $S_{new}(t)$ for $t_1, \ldots, t_q \le S$ and for $t > S$ through T. Repeat step 2 above, replacing $S_{old}(t)$ with $S_{new}(t)$. This will produce a new steady-state event probabilities π_{newT}. The re-estimated sample size can be obtained from $D_0 = N_{new}\pi_{newT}$.

This strategy is predicated on the assumption that the emerging treatment effect is at t as good, or better, than the treatment effect assumed during the design stage because strategy implicitly assumes that D_0 events based on the average survival distribution tion in the combined treatment groups are enough to sufficiently power the test of hyesis that will be conducted. If this assumption proves to be inaccurate, the re-estimated ple size may not provide enough events to maintain the desired study power. Such is price of a blinded sample size re-estimation procedure.

liography

Keaven M. Anderson. *gsDesign: Group Sequential Design*, 2015. R package version 2.9-4.

Keaven M. Anderson and Jason B. Clark. Fitting spending functions. *Statistics in Medicine*, 29:321–327, 2010.

FM-S Barthel, A Babiker, P Royston, and MKB Parmar. Evaluation of sample size and power for multi-arm survival trials allowing for non-uniform accrual, non-proportional hazards, loss to follow-up and cross-over. *Statistics in Medicine*, 25(15):2521–2542, 2006.

[4] D. Bernstein and S. Lagakos. Sample size and power determination for stratified clinical trials. *Journal of Statistical Computation and Simulation*, 8:65–73, 1978.

[5] Center for Drug Evaluation and Research. *Guidance for Industry: Diabetes Mellitus: Evaluating Cardiovascular Risk in New Antidiabetic Therapies to Treat Type 2 Diabetes.* United States Department of Health and Human Services, Food and Drug Administration, 2008.

[6] Ivan S.F. Chan and Norman R. Bohidar. Exact power and sample size for vaccine efficacy studies. *Communications in Statistics - Theory and Methods*, 27:1305–1322, 1998.

[7] Y. H. Joshua Chen, David L. DeMets, and K. K. Gordon Lan. Increasing the sample size when the unblinded interim result is promising. *Statistics in Medicine*, 23:1023–1038, 2004.

[8] Y.H. Chen, R. Gesser, and A. Luxembourg. A seamless phase IIB/III adaptive outcome trial: Design rationale and implementation challenges. *Clinical Trials*, pages 1–7, 2014.

[9] M.P. Fay, E.H. Brittain, and M.A. Proschan. Pointwise confidence intervals for a survival distribution with small samples or heavy censoring. *Biostatistics*, 14:723–736, 2013.

[10] L. S. Freedman. Tables of the numbers of patients required in clinical trials using the log-rank test. *Statistics in Medicine*, 1:121–129, 1982.

[11] Ping Gao, James H. Ware, and Cyrus Mehta. Sample size re-estimation for adaptive sequential design in clinical trials. *Journal of Biopharmaceutical Statistics*, 18:1184–1196, 2008.

[12] Stephen L. George and M. M. Desu. Planning the size and duration of a clinical trial studying the time to some critical event. *Journal of Chronic Diseases*, 27:15–24, 1974.

[13] I. K. Hwang, W. J. Shih, and J. S. DeCani. Group sequential designs using a family of type 1 error probability spending functions. *Statistics in Medicine*, 9:1439–1445, 1990.

[14] J.O. Irwin. The standard error of an estimate of expectation of life, with special reference to expectation of tumourless life in experiments with mice. *Journal of Hygiene*, 47:188–189, 1949.

[15] A. Jahn-Eimermacher and K. Ingel. Adaptive trial design: general methodology for censored time to event data. 30:171–177, 2009.

[16] V.G. Fowler Jr., K.B. Allen, E.D. Moreira Jr., et. al. Effect of an investigational vaccine for preventing staphylococcus aureus infections after cardiothoracic surgery. *Journal of the American Medical Association*, 309:1368–1379, 2013.

[17] T. Karrison. Restricted mean life with adjustment for covariates. *Journal of the American Statistical Association*, 82:1169–1171, 1987.

[18] T. Karrison. Use of irwin's restricted mean as an index for comparing survival in different treatment groups - interpretation and power considerations. *Journal of Controlled Clinical Trials*, 18:151–167, 1997.

[19] Kyungmann Kim, H. Boucher, and Anastasios A. Tsiatis. Design and analysis of group sequential logrank tests in maximum duration versus information trials. *Biometrics*, 51:988–1000, 1995.

Kyungmann Kim and Anastasios A. Tsiatis. Study duration for clinical trials with survival response and early stopping rule. *Biometrics*, 46:81–92, 1990.

John M. Lachin. *Biostatistical Methods. The Assessment of Relative Risks*. Wiley, New York, NY, 2000.

John M. Lachin. Maximum information designs. *Clinical Trials*, 2:453–464, 2005.

John M. Lachin and Mary A. Foulkes. Evaluation of sample size and power for analyses of survival with allowance for nonuniform patient entry, losses to follow-up, noncompliance, and stratification. *Biometrics*, 42:507–519, 1986.

Edward Lakatos. Sample size based on the log-rank statistic in complex clinical trials. *Biometrics*, 44:229–241, 1988.

Edward Lakatos and K.K.G. Lan. A comparison of sample size methods for the log rank statistic. *Statististics in Medicine*, 11:179–191, 1992.

K. K. G. Lan and David L. DeMets. Discrete sequential boundaries for clinical trials. *Biometrika*, 70:659–663, 1983.

K.K.G. Lan and John M. Lachin. Implementation of group sequential logrank tests in a maximum duration trial. *Biometrics*, 46:759–770, 1990.

D.Y. Lin, L. Shen, Z. Ying, and N.E. Breslow. Group sequential designs for monitoring survival probabilities. *Biometrics*, 52:1033–1041, 1996.

Qing Liu and Keaven M. Anderson. On adaptive extensions of group sequential trials for clinical investigations. *Journal of the American Statistical Association*, 103:1621–1630, 2008.

Willi Maurer and Frank Bretz. Multiple testing in group sequential trials using graphical approaches. *Statistics in Biopharmaceutical Research*, 5:311–320, 2013.

Cyrus Mehta and Stuart Pocock. Adaptive increase in sample size when interim results are promising: A practical guide with examples. *Statistics in Medicine*, 30:3267–3284, 2011.

P. Royston and M.K.B. Parmar. The use of restricted mean survival time to estimate the treatment effect in randomized clinical trials when the proportional hazards assumption is in doubt. *Statistics in Medicine*, 30:2409–2421, 2011.

P. Royston and M.K.B. Parmar. Restricted mean survival time: An alternative to the hazard ratio for the design and analysis of randomized trials with a time-to-event outcome. *BMC Medical Research Methodology*, 13:152, 2013.

L.V. Rubinstein, Mitchell H. Gail, and T.J. Santner. Planning the duration of a comparative clinical trial with loss to follow-up and a period of continued observation. *Journal of Chronic Diseases*, 34:469–479, 1981.

H. Schafer and H.H. Muller. Modification of the sample size and the schedule of interim analysis in survival trials based on data inspections. *Statistics in Medicine*, 20:3741–3751, 2001.

David Schoenfeld. The asymptotic properties of nonparametric tests for comparing survival distributions. *Biometrika*, 68:316–319, 1981.

[37] Y. Shen and J.W. Cai. Sample size reestimation for clinical trials with censored survival data. *Journal of the American Statistical Association*, 98:418–426, 2003.

[38] S. Todd, E. Valdés-Marquez, and J. West. A practical comparison of blinded methods for sample size reviews in survival data clinical trials. *Pharmaceutical Statistics*, 11:141–148, 2012.

[39] H. Uno, B. Claggett, L. Tian, E. Inoue, P. Gallo, T. Miyata, D. Schrag, M. Takeuchi, Y. Uyama, L. Zhao, H. Skali, S. Solomon, S. Jacobus, M. Hughes, M. Packer and L. Wei. Moving beyond the hazard ratio in quantifying the between-group difference in survival analysis. *Journal of clinical Oncology*, 32(22):2380, 2014.

[40] H. Wan. *Issues in group sequential/adaptive designs*. PhD thesis, University of Pennsylvania, 2013.

[41] J.L. Wang and T.P. Hettmansperger. Two-sample inference for survival times based on one-sample procedures for censored survival data. *Journal of the American Statistical Association*, 85:529–536, 1990.

[42] J. Whitehead, A. Whitehead, S. Todd, K. Bolland, and M.S. Sooriyachi. Mid-trial design reviews for sequential clinical trials. *Statistics in Medicine*, 20:165–176, 2001.

[43] J. Wittes. Sample size calculations for randomized controlled trials. *Epidemiology*, 24:39–53, 2002.

[44] D.M. Zucker. Restricted mean life with covariates: modification and extension of a useful survival analysis method. *Journal of the American Statistical Association*, 93:702–709, 1998.

14

Sample Size Estimation and Power Analysis: Longitudinal Data

Sin-Ho Jung

CONTENTS

14.1 Introduction

Often, in RCTs, subjects are randomized to two treatment arms, and evaluated at baseline and intervals across a treatment period. For example, in a study on labor pain, 83 women in labor were randomly assigned to pain medication arm (43 women) or placebo arm (40 women). At 30-minute intervals, the self-reported level of pain was marked on a 100-mm line of visual analog scale, where 0 = no pain and 100 = extreme pain. The maximum number of measurements for each woman was $m = 6$, but there were numerous missing values at later measurement times with a monotone missing pattern. A simple approach to such study objective was to estimate and compare the slopes of pain scores over time in two treatment arms. In this study, the outcome variable is continuous.

GENISOS (Genetics vs. Environment In Scleroderma Outcome Study) is an observational study designed as a collaboration between the University of Texas-Houston Health Science Center and the University of Texas Medical Branch at Galveston and the University of Texas-San Antonio Health Science Center (Reveille et al. 2001). Scleroderma, or systemic sclerosis, is a multi-system disease of unknown etiology characterized by cutaneous and visceral fibrosis, small blood vessel damage, and autoimmune features (Medsger, 1997). The study subjects were regularly followed to check the occurrence of pulmonary fibrosis. In this case, the outcome is a binary variable. We may develop a RCT for an experimental treatment of scleroderma using the results of this observational study as the pilot data of a control arm.

The repeated measurements from each subject are typically dependent, so that data analysis methods should appropriately incorporate the dependency when estimating and testing on the time trajectory of longitudinal data. Random effects model methods have been widely used for longitudinal data analysis, refer to e.g. Laird and Ware (1982) for continuous outcomes and Anderson and Aitkin (1985) for binary outcomes. For estimating regression parameters using the likelihood method, random effects model methods require specification of the dependency structure. If the specified dependency structure is different from the true one, then the variance estimator for the regression estimates may not be consistent, so that the testing on regression parameters will be biased. Usually, however, repeated measurements from each subject are so complicatedly correlated that it is difficult or impossible to model it correctly using a known dependency structure.

By extending quasi-likelihood theory (Wedderburn 1974, McCullagh 1983), Liang and Zeger (1986) propose a longitudinal data analysis method, called generalized estimating equation (GEE) method, that does not require specification of the true dependency structure for the inference on the time trajectory, called marginal model. In this chapter, we consider comparing the change rate of repeated measurements over time between two treatment arms using the GEE method. We propose closed-form sample size formulas for between-arm comparison that can be used when designing RCTs or retrospective case-control studies with either continuous or binary longitudinal outcome data.

Oftentimes, the subjects in a longitudinal study miss some visits so that the resulting data are subject to missingness. If the missingness is completely at random (Rubin, 1976), GEE method provides consistent regression estimator whether the working correlation structure specified for the repeated measurements is the correct one or not. However, missing data attenuate the power of tests on the regression coefficients. Our sample size formulas handle missing probabilities under independent or monotone missing pattern as defined in Section 14.4.1. We assume that the working independent correlation model is used in the final data analysis, but the sample size formula is derived based on the assumed correlation model, such as autoregressive model with order 1, called AR(1), or compound symmetry.

14.2 Generalized Estimating Equations (GEE) Method

Suppose that n_k subjects are randomized to treatment arm $k(=1,2)$, $n_1 + n_2 = n$, and $r_k = n_k/n$ denotes the allocation proportion ($r_1 + r_2 = 1$). For subject i ($i = 1, ..., n_k$) in arm k, let y_{kij} denote the outcome measurement made at time t_{kij} ($j = 1, ..., m_{ki}$) with $\mu_{kij} = \mathrm{E}(y_{kij})$ that is expressed as

$$g(\mu_{kij}) = a_k + b_k t_{kij},$$

for a link function $g(\cdot)$. Let $M_k = \sum_{i=1}^{n_k} m_{ki}$ denote the total number of observations.

To simplify the discussions, we use the identity link $g(\mu) = \mu$ for continuous outcome variables and the logit link $g(\mu) = \log\{\mu/(1 - \mu)\}$ for binary outcome variables, but the generalization to the use of other links is straightforward. Let $\mu(\cdot) = g^{-1}(\cdot)$.

The coefficient b_k represents the change rate per unit time in mean response level if y is continuous and that in log-odds if y is binary. So, the null and alternative hypotheses are

$H_0 : b_1 = b_2$ and $H_1 : b_1 \neq b_2$, respectively, to test if two arms have the same change rate in outcome or not. Since patients are randomized at the baseline with $t_{ki1} = 0$, we expect $a_1 = a_2$.

The measurement times may vary from subject to subject due to missing measurements, patients' visits for measurements at unscheduled times, loss to follow-up, or other causes. In this chapter, we assume missing completely at random by Rubin (1976).

The repeated measurements $(y_{kij}, 1 \leq j \leq m_{ki})$ within each subject tend to be correlated. Let $y_{ki} = (y_{ki1}, ..., y_{ki,m_{ki}})^T$ be the vector of longitudinal observations from subject i in arm k with mean vectors $\mu_{ki}(a_k, b_k) = (\mu(a_k + b_k t_{ki1}), ..., \mu(a_k + b_k t_{ki,m_{ki}}))^T$, and V_{ki} denote a working covariance matrix of y_{ki}. The generalized estimating equation (Liang and Zeger, 1986) for (a_k, b_k) is given as $U_k(a, b) = 0$, where

$$U_k(a, b) = \sum_{i=1}^{n_k} Z_{ki}^T D_{ki} V_{ki}^{-1} \{y_{ki} - \mu_{ki}(a, b)\},$$

$$Z_{ki} = \begin{pmatrix} 1 & \cdots & 1 \\ t_{ki1} & \cdots & t_{ki,m_{ki}} \end{pmatrix}^T$$

is the $m_{ki} \times 2$ matrix of covariate values, $D_{ki} = \text{diag}(\mu'(a + bt_{kij}))_{j=1,...,m_{ki,j}}$ is an $m_{ki} \times m_{ki}$ matrix, and $\mu'(\theta) = \partial\mu(\theta)/\partial\theta$ is the derivative of $\mu(\cdot)$.

The GEE estimator, $(\hat{a}_k, \hat{b}_k)$, is obtained by solving the estimating equation. For large n_k, $\sqrt{n_k}(\hat{a}_k - a_k, \hat{b}_k - b_k)$ is asymptotically normal with mean 0 and variance V_k that can be consistently estimated by the so-called sandwich estimator $\hat{V}_k = n_k \hat{A}_k^{-1} \hat{\Sigma}_k \hat{A}_k^{-1}$, where

$$\hat{A}_k = \sum_{i=1}^{n_k} Z_{ki}^T \hat{D}_{ki} \hat{V}_{ki}^{-1} \hat{D}_{ki} Z_{ki}$$

$$\hat{\Sigma}_k = \sum_{i=1}^{n_k} Z_{ki}^T \hat{D}_{ki} \hat{V}_{ki}^{-1} \hat{\epsilon}_{ki} \hat{\epsilon}_{ki}^T \hat{V}_{ki}^{-1} \hat{D}_{ki}$$

$\hat{\epsilon}_{ki} = y_{ki} - \mu_{ki}(\hat{a}_k, \hat{b}_k)$, $\hat{D}_{ki} = \text{diag}(\mu'(\hat{a}_k + \hat{b}_k t_{kij}))$, and $\hat{V}_{ki}$ is obtained by replacing all parameters (for correlations and regression coefficients) included in V_{ki} with their consistent estimators.

The efficiency of GEE estimators will be high if the working correlation structure is close to the true one. In a longitudinal data analysis, however, the true correlation structure is usually unknown or of secondary interest. Using a complicated working correlation structure involves estimation of nuisance parameters, and it can cause divergent replications in estimating the regression parameters, see Crower (1995). As such, in real data analysis, the working independent correlation structure has been one of the most popular choices, for which $V_{ki} = \text{diag}(\text{var}(y_{kij}))_{j=1,...,m_{ki,j}}$, an $m_{ki} \times m_{ki}$ matrix.

14.2.1 Continuous outcome variable case

Let y_{kij} denote the observation, or its transformation, of a continuous outcome collected at time t_{kij} from patient i in treatment arm k. For arm k, $k = 1, 2$, we consider a linear model $E(y_{kij}) = a_k + b_k t_{kij}$ using the identity lnk. For a continuous outcome, we use the working independent and identically distributed (IID) correlation structure for which $V_{ki} = \sigma^2 I_{m_{ki} \times m_{ki}}$. In this case, we have $D_{ki} = I_{m_{ki} \times m_{ki}}$, so that the resulting estimating function is expressed as

$$U_k(a, b) = \sum_{i=1}^{n_k} \sum_{j=1}^{m_{ki}} (y_{kij} - a - bt_{kij}) \begin{pmatrix} 1 \\ t_{kij} \end{pmatrix}.$$

Using the working IID correlation structure, the estimating equation $U_k(a, b) = 0$ has a closed form solution, $(\hat{a}_k, \hat{b}_k)$:

$$\hat{b}_k = \frac{\sum_{i=1}^{n_k} \sum_{j=1}^{m_{ki}} (t_{kij} - \bar{t}_k) y_{kij}}{\sum_{i=1}^{n_k} \sum_{j=1}^{m_{ki}} (t_{kij} - \bar{t}_k)^2}$$

and $\hat{a}_k = \bar{y}_k - \hat{b}_k \bar{t}_k$, where $\bar{t}_k = M_k^{-1} \sum_{i=1}^{n_k} \sum_{j=1}^{m_{ki}} t_{kij}$ and $\bar{y}_k = M_k^{-1} \sum_{i=1}^{n_k} \sum_{j=1}^{m_{ki}} y_{kij}$. By using the independent working IID correlation structure, $\hat{b}_k$ is identical to the least square estimator of b_k obtained when all $(y_{kij}, j = 1, ..., m_{ki}, i = 1, ..., n_k)$ are independent.

For large n_k, $\sqrt{n_k}(\hat{b}_k - b_k)$ is approximately $N(0, v_k)$, and v_k can be consistently estimated by

$$\hat{v}_k = \frac{n_k \sum_{i=1}^{n_k} \{\sum_{j=1}^{m_{ki}} (t_{kij} - \bar{t}_k) \hat{\epsilon}_{kij}\}^2}{\{\sum_{i=1}^{n_k} \sum_{j=1}^{m_{ki}} (t_{kij} - \bar{t}_k)^2\}^2}$$

where $\hat{\epsilon}_{kij} = y_{kij} - \hat{a}_k - \hat{b}_k t_{kij}$. Note that $\hat{v}_k$ is the $(2, 2)$-component of $\hat{V}_k$ when using the working IID correlation structure. Hence, we reject $H_0 : b_1 = b_2$ if

$$\frac{|\hat{b}_1 - \hat{b}_2|}{\sqrt{\hat{v}_1/n_1 + \hat{v}_2/n_2}} > z_{1-\alpha/2}$$

where $z_{1-\alpha/2}$ is the $100(1 - \alpha/2)$ percentile of the standard normal distribution.

While we fit a linear model for each arm in this chapter, Jung and Ahn (2003) investigate the same statistical problem based on a model combining two arms. For patient i, $i = 1, ..., n_1 + n_2$, the outcome measurement y_{ij} is made at time t_{ij} with

$$E(y_{ij}) = \beta_0 + \beta_1 R_i + \beta_2 t_{ij} + \beta_3 R_i t_{ij},$$

where $R_i = 0$ if patient i is randomized to arm 1 and $R_i = 1$ otherwise. Note that $\beta_0 = a_1$, $\beta_1 = a_2 - a_1$, $\beta_2 = b_1$, and $\beta_3 = b_2 - b_1$. Based on this model, the statistical hypothesis will be $H_0 : \beta_3 = 0$ and $H_1 : \beta_3 \neq 0$. Whichever modelling approach we use, we have the same results.

14.2.2   Binary outcome variable case

Let y_{kij} be a binary outcome variable measured at time t_{kij} taking 1 for response (such as tumor response) and 0 for no response with response rate $E(y_{kij}) = \mu_{kij}$. Using the logit link, $g(\mu_{kij}) = \log\{\mu_{kij}/(1 - \mu_{kij})\} = a_k + b_k t_{kij}$, we have

$$\mu_{kij} = \frac{\exp(a_k + b_k t_{kij})}{1 + \exp(a_k + b_k t_{kij})}.$$

We have $V_{ki} = \text{diag}(\exp(a + bt_{kij})/\{1 + \exp(a + bt_{kij})\}^2)_{j=1,...,m_{ki}, m_{ki}}$ for the independence correlation structure, and $D_{ki} = \text{diag}(\exp(a + bt_{kij})/\{1 + \exp(a + bt_{kij})\}^2)_{j=1,...,m_{ki}}$ for the logit link. Note that, when using working independence correlation structure and logit link, we have $V_{ki} = D_{ki}$ and the estimating function is given as

$$U_k(a, b) = \sum_{i=1}^{n_k} \sum_{j=1}^{m_{ki}} \left\{ y_{kij} - \frac{\exp(a + bt_{kij})}{1 + \exp(a + bt_{kij})} \right\} \begin{pmatrix} 1 \\ t_{kij} \end{pmatrix}. \tag{1}$$

For estimating function (1), we solve the equation $U_k(a, b) = 0$ using a numerical method, such as the Newton-Raphson algorithm: at the l-th iteration,

$$\begin{pmatrix} \hat{a}_k^{(l)} \\ \hat{b}_k^{(l)} \end{pmatrix} = \begin{pmatrix} \hat{a}_k^{(l-1)} \\ \hat{b}_k^{(l-1)} \end{pmatrix} + A_k^{-1}(\hat{a}_k^{(l-1)}, \hat{b}_k^{(l-1)}) U_k(\hat{a}_k^{(l-1)}, \hat{b}_k^{(l-1)}),$$

where

$$A_k(a,b) = -\frac{\partial U_k(a,b)}{\partial(a,b)} = \sum_{i=1}^{n_k}\sum_{j=1}^{m_{ki}} \mu_{kij}(a,b)\bar{\mu}_{kij}(a,b) \begin{pmatrix} 1 & t_{kij} \\ t_{kij} & t_{kij}^2 \end{pmatrix}$$

$$\mu_{kij}(a,b) = g^{-1}(a + bt_{kij}) = \frac{e^{a+bt_{kij}}}{1 + e^{a+bt_{kij}}} \qquad (2)$$

and $\bar{\mu}_{kij}(a,b) = 1 - \mu_{kij}(a,b)$.

By Liang and Zeger (1986), for large n_k, $\sqrt{n_k}(\hat{a}_k - a_k, \hat{b}_k - b_k)^T$ is asymptotically normal with mean 0 and variance V_k that can be consistently estimated by

$$\hat{V}_k = n_k A_k^{-1}(\hat{a}_k, \hat{b}_k)\hat{\Sigma}_k A_k^{-1}(\hat{a}_k, \hat{b}_k)$$

where

$$\hat{\Sigma}_k = \frac{1}{n_k}\sum_{i=1}^{n_k}\{\sum_{j=1}^{m_{ki}} \hat{\epsilon}_{kij} \begin{pmatrix} 1 \\ t_{kij} \end{pmatrix}\}^{\otimes 2},$$

$\hat{\epsilon}_{kij} = y_{kij} - \mu_{kij}(\hat{a}_k, \hat{b}_k)$ and $c^{\otimes 2} = cc^T$ for a vector c. Note that $\hat{\Sigma}_k$ is a consistent estimator of the variance-covariance matrix of $U_k(a_k, b_k)$. Let $\hat{v}_k$ be the (2,2)-component of $\hat{V}_k$. We reject $H_0 : b_1 = b_2$ if

$$\left|\frac{\hat{b}_1 - \hat{b}_2}{\sqrt{\hat{v}_1/n_1 + \hat{v}_2/n_2}}\right| > z_{1-\alpha/2}.$$

14.3 Power Analysis and Sample Size Estimation

We want to test on the change rate in the response variable between two arms, i.e. $H_0 : b_1 = b_2$. Based on the asymptotic results from the previous section, we can reject $H_0 : b_1 = b_2$, in favor of $H_1 : b_1 \neq b_2$ when

$$\left|\frac{\hat{b}_1 - \hat{b}_2}{\sqrt{\hat{v}_1/n_1 + \hat{v}_2/n_2}}\right| > z_{1-\alpha/2}. \qquad (3)$$

In this section, we derive a sample size formula for the two-sided level α test (3) to detect $H_1 : |b_2 - b_1| = d(> 0)$ with power $1 - \beta$. When designing a study, we usually schedule fixed visit times $t_1 < \cdots < t_m$ for m longitudinal measurements from each individual. We often set $t_1 = 0$ for the baseline measurement time. When the study is conducted, however, the subjects may miss some visit times due to various causes, which results in missing values, or may not follow the visit schedule correctly so that the observed visit times may be variable. Jung and Ahn (2003) show through simulations in continuous outcome variable case that the sample size formula based on fixed measurement times is very accurate even when the observed measurement times are variable around the scheduled ones.

By simple algebra, we can derive a power function for a given sample size n,

$$1 - \beta = \bar{\Phi}\left(z_{1-\alpha/2} - \frac{d\sqrt{n}}{\sqrt{v_1/r_1 + v_2/r_2}}\right) \qquad (4)$$

and a sample size formula to detect the specified difference $|b_2 - b_1| = d$ with power $1 - \beta$,

$$n = \frac{(z_{1-\alpha/2} + z_{1-\beta})^2(v_1/r_1 + v_2/r_2)}{d^2}, \qquad (5)$$

where $\bar{\Phi}(\cdot) = 1 - \Phi(\cdot)$, $\Phi(\cdot)$ is the cumulative distribution function of the standard normal distribution, and $v_k = \lim_{n \to \infty} \hat{v}_k$. The expression of v_k is slightly different between continuous and binary outcome variable cases as shown in the following subsections.

In order to allow for missing data, let p_j denote the probability that an individual has a measurement at t_j and $p_{jj'}$ the probability that an individual has measurements at both t_j and $t_{j'}$. Also, let $\rho_{jj'} = \text{corr}(y_{kij}, y_{kij'})$ the correlation coefficient between the measurement at t_j and that at $t_{j'}$ for each individual. We assume that the missing pattern and correlation structure are common in two treatment arms. We will discuss specific models for missing pattern and correlation structure in Section 14.4.

14.3.1 Continuous outcome variable case

When designing a study, we assume that the continuous outcome variable has a constant variance $\text{var}(y_{kij}) = \sigma^2$ over time which is common between two treatment arms.

By Appendix A.1, in continuous outcome variable case, the power function (4) and required sample size (5) are expressed as

$$1 - \beta = \bar{\Phi}\left\{ z_{1-\alpha/2} - \frac{d\sqrt{n}}{\sqrt{v(r_1^{-1} + r_2^{-1})}} \right\}$$

and

$$n = \frac{v(z_{1-\alpha/2} + z_{1-\beta})^2}{d^2 r_1 r_2}. \tag{6}$$

respectively, where

$$v = \frac{\sigma^2(s^2 + c)}{s^4},$$

$$s^2 = \sum_{j=1}^{m} p_j(t_j - \tau)^2$$

$$c = 2\sum\sum_{1 \le j < j' \le m} p_{jj'} \rho_{jj'}(t_j - \tau)(t_{j'} - \tau)$$

and

$$\tau = \frac{\sum_{j=1}^{m} p_j t_j}{\sum_{j=1}^{m} p_j}.$$

Note that the required sample size does not depend on the true values for $\{(a_k, b_k), k = 1, 2\}$, but only on their difference $|b_2 - b_1| = d$. The sample size is proportional to the variance of measurement error σ^2, and decreases as the allocation proportions r_1 and r_2 approach $1/2$.

Chow et al. (2003) propose a sample size formula for analysis of variance with repeated measures. Their formula is similar to the formula for K-sample by Jung and Ahn (2004) under a compound symmetry (CS) correlation structure and no missing data.

14.3.2 Binary outcome variable case

In this section, we discuss Jung and Ahn's (2005) sample size estimation method for comparing the change rates of a binary response outcome over time using logistic regression models. Let $\mu_{kj} = g^{-1}(a_k + b_k t_j)$ denote the probabilities of the binary event under H_1 and $\bar{\mu}_{kj} = 1 - \mu_{kj}$. By Appendix A.2, under H_1, $\hat{v}_k$ converges to

$$v_k = \frac{s_k^2 + c_k}{s_k^4},$$

where

$$s_k^2 = \sum_{j=1}^m p_j \mu_{kj} \bar{\mu}_{kj} (t_j - \tau_k)^2$$

$$c_k = \sum \sum_{j \neq j'} p_{jj'} \rho_{jj'} \sqrt{\mu_{kj} \bar{\mu}_{kj} \mu_{kj'} \bar{\mu}_{kj'}} (t_j - \tau_k)(t_{j'} - \tau_k),$$

and

$$\tau_k = \frac{\sum_{j=1}^m p_j \mu_{kj} \bar{\mu}_{kj} t_j}{\sum_{j=1}^m p_j \mu_{kj} \bar{\mu}_{kj}}.$$

So, under $H_1 : |b_1 - b_2| = d$, the power function (4) and required sample size (5) in binary outcome case are given as

$$1 - \beta = \bar{\Phi}\left(z_{1-\alpha/2} - \frac{d\sqrt{n}}{\sqrt{v_1/r_1 + v_2/r_2}}\right)$$

and

$$n = \frac{(z_{1-\alpha/2} + z_{1-\beta})^2 (v_1/r_1 + v_2/r_2)}{d^2}, \qquad (7)$$

respectively.

We observe that sample size is inversely proportional to d^2. With other parameters fixed, the sample size is minimized at $r_1 = (1 + \sqrt{v_2/v_1})^{-1}$, which is equal to $1/2$ if $v_1 = v_2$, larger than $1/2$ if $v_1 > v_2$ and smaller than $1/2$ if $v_1 < v_2$. Under H_1, v_1, and v_2 are unequal, so that the total sample size n is minimized when a slightly larger portion of the sample is allocated to the group with larger variance v_k. If the specified effect size d is small, then v_1 and v_2 will be similar and the total sample size will be minimized at $r_1 \approx 1/2$. Also, the sample size increases as c_k^2 increases or s_k^2 decreases. Note that s_k^2 is a weighted variance of measurement times.

The sample size formula under binary outcome variable case depends on the probabilities $(\mu_{kj}, j = 1, ..., m)$ under H_1, so that we need to specify all regression parameters $\{(a_k, b_k), k = 1, 2\}$. Let $d = b_2 - b_1$ denote the difference in slope between two arms under H_1. If there exist pilot data for the control arm (arm 1), then we may use the estimates as the parameter values, a_1 and b_1. If $t_1 = 0$ is the time of randomization, i.e. baseline, the intercepts in the two arms are set the same, i.e. $a_1 = a_2$. By setting $b_2 = b_1 + d$, we can specify all regression coefficients under H_1.

If there are no pilot data available, we may specify the binary probabilities at the baseline, μ_{11}, and at the end of follow-up, μ_{1m}, for the control arm. Then we obtain (a_1, b_1) by

$$b_1 = \frac{g(\mu_{1m}) - g(\mu_{11})}{t_m - t_1}, \qquad (8)$$

$$a_1 = g(\mu_{11}) - b_1 t_1 = g(\mu_{11}).$$

And we set $a_2 = a_1$ (since $\mu_{11} = \mu_{21}$ at the baseline) and $b_2 = b_1 + d$.

14.4 Modelling Missing Pattern and Correlation Structure

Calculation of n (or v_k) requires specification of the missing probabilities and the true correlation structure.

14.4.1   Missing pattern

In order to specify $p_{jj'}$ with $j \neq j'$, we need to estimate the missing pattern. If missing at time t_j is independent of missing at time $t_{j'}$ for each individual, then we have $p_{jj'} = p_j p_{j'}$ and call this type independent missing.

In some studies, subjects missing at a measurement time may be missing at all following measurement times as in the labor pain study discussed in Section 14.1. This type of missing pattern is called monotone missingness. In this case, we have $p_{jj'} = p_{j'}$ for $j < j'$ ($p_1 \geq \cdots \geq p_m$). In a monotone missing case, one may want to specify the proportion of patients who will contribute exactly the first j observations, say η_j. Then, noting that $\eta_j = p_j - p_{j+1}$ for $j = 1, ..., m-1$ and $\eta_m = p_m$, we can obtain p_j recursively starting from p_m. Note also that $\sum_{j=1}^{m} \eta_j = p_1$, which equals 1 if all patients have measurements at the first measurement time t_1.

In summary, we consider two missing patterns as candidates to the true one: (a) independent, where $p_{jj'} = p_j p_{j'}$ for $j \neq j'$; (b) monotone, where $p_{jj'} = p_{j'}$ for $j < j'$ ($p_1 \geq \cdots \geq p_m$).

14.4.2   Correlation structure

Now, we specify the true correlation structure $\rho_{jj'}$. A reasonable model for $\epsilon_{ij} = y_{ij} - g^{-1}(a + bt_{ij})$ may be

$$\epsilon_{ij} = u_i + e_{ij},$$

where u_i are subject-specific error terms with variance σ_u^2 and e_{ij} are serially correlated within-subject error terms with variance σ_e^2 and correlation coefficients $\tilde{\rho}_{jj'}$. Assuming that u_i and e_{ij} are independent, we have

$$\text{cov}(\epsilon_{ij}, \epsilon_{ij'}) = \sigma_a^2 + \sigma_e^2 \tilde{\rho}_{jj'}.$$

Often, the variation between subjects (σ_u^2) is much larger than that within subject (σ_e^2). In this case, we have

$$\text{cov}(\epsilon_{ij}, \epsilon_{ij'}) \approx \sigma_u^2$$

and a compound symmetry (CS) correlation structure, $\rho_{jj'} = \rho$ for $j \neq j'$, may be a reasonable approximation to the true one.

On the other hand, if the variation within subject (σ_e^2) dominates over the variation between subjects (σ_u^2), then we will have

$$\text{cov}(\epsilon_{ij}, \epsilon_{ij'}) \approx \sigma_e^2 \tilde{\rho}_{jj'},$$

so that a serial correlation structure may be a reasonable approximation to the true one. One of the most popular serial correlation structure, especially when measurement times are not equidistant, is a continuous autocorrelation model with order 1, AR(1), for which $\rho_{jj'} = \rho^{|t_j - t_{j'}|}$.

We consider two correlation structures as candidate approximations to the true one: (i) CS, where $\rho_{jj'} = \rho$ for $j \neq j'$; (ii) AR(1), where $\rho_{jj'} = \rho^{|t_j - t_{j'}|}$.

14.5   Examples

For sample size estimation, following parameters need to be specified commonly in both continuous and discrete outcome variable cases:

- Type I error rate α and power $1 - \beta$.

- Clinically significant difference, $|b_2 - b_1| = d$.

- Allocation proportion for arm k, r_k.

- Correlation structure and the associated correlation parameter ρ, i.e. (i) $\rho_{jj'} = \rho$ for CS or (ii) $\rho_{jj'} = \rho^{|t_j - t_{j'}|}$ for AR(1).

- Proportion of individuals with an observation at t_j, p_j.

- Missing pattern: (a) independent $(p_{jj'} = p_j p_{j'})$ or (b) monotone $(p_{jj'} = p_{j'}$ for $j < j')$.

In addition, we need to specify $\sigma^2 = \text{var}(\epsilon_{ij})$ in the continuous outcome variable case, and the regression coefficients $\{(a_k, b_k), k = 1, 2\}$ under H_1 in the binary outcome variable case. We demonstrate our sample size formula with real longitudinal studies.

14.5.1   Labor pain study (Continuous outcome case)

Suppose that we want to design a new study on labor pain based on the data reported by Davis (1991). The original study was briefly discussed in Section 14.1. As in the original study, we assume monotone missing. In this study, the measurement times were equispaced, so that we set $t_j = j - 1$ $(j = 1, ..., 6)$ for convenience. From the data, we obtained $\sigma^2 = 815.84$. Suppose we want to detect a difference of σ in mean pain score between two arms at t_6. So, we project $d = \sigma/(t_6 - t_1) = 5.71$ in a new study. We consider a balanced randomization, i.e. $r_1 = r_2 = 1/2$. Also, from the data, the proportion of observed measurements are

$$(p_1, p_2, p_3, p_4, p_5, p_6) = (1, .90, .78, .67, .54, .41).$$

From these results, we obtain $\tau = 2.02$ and $s^2 = 11.42$.

Suppose that we want to detect a difference of $d = 5.71$ with 80% of power $(z_{1-\beta} = 0.84)$ using a two-sided $\alpha = .05$ $(z_{1-\alpha/2} = 1.96)$ test. Under CS, we obtain $\rho = .64$ and $c = -3.13$ from the data, so that we have $v = 815.84 \times (11.42 - 3.13)/11.42^2 = 0.0635$. Hence, from (6), the required sample size is calculated as

$$n = \left[\frac{0.0635 \times (1.96 + 0.84)^2}{5.71^2} \right] + 1 = 50.$$

where $[x]$ is the largest integer not exceeding x.

Under AR(1), from the data, we obtain $\rho = .80$ and $c = 2.31$, so that the required sample size for detecting $d = 5.71$ with $1 - \beta = 80\%$ of power using a two-sided $\alpha = .05$ test is given as $n = 83$. The sample size under AR(1) is larger than that under CS, by about 66%, in this example.

14.5.2   Design of an RCT based on GENISOS (binary outcome case)

As stated in Section 14.1, GENISOS is an observational study. Suppose that we want to develop a new clinical trial to examine the effect of a new drug in preventing the occurrence of pulmonary fibrosis in individuals with scleroderma based on the observations from the observational study. The parameter values specified here for sample size calculation are approximated by the estimates from the current data set of GENISOS.

We want to estimate the sample size for a new clinical trial using $\alpha = .05$ and $1 - \beta = .8$. As in GENISOS, presence or absence of pulmonary fibrosis will be assessed at baseline, and at months 6, 12, 18, 24 and 30. Since the measurement times are equidistant, we set $t_j = j-1$ $(j = 1, ..., 6)$ for $m = 6$ time points. The within-arm correlation structure of the repeated measurements conforms to AR(1) with the adjacent correlation equal to $\rho = .8$, that is, $\rho_{jj'} = .8^{|j-j'|}$. We consider assigning equal number of scleroderma patients in each of two arms, i.e. $r_1 = r_2 = 1/2$.

Approximately 75% of scleroderma patients do not have pulmonary fibrosis at the baseline in the ongoing GENISOS. We project that the proportion of subjects without pulmonary fibrosis is 75% at baseline, i.e. $\mu_{1,1} = .75$, and 50% at 30 months, i.e. $\mu_{1,6} = .50$, in a placebo arm. We assume that a new therapy will prevent or delay further occurrence of pulmonary fibrosis. That is, the proportion of subjects without pulmonary fibrosis will remain 75% during 30-month study in a new therapy arm, i.e. $\mu_{2,1} = \mu_{2,6} = .75$. From these values and (8), we obtain

$$b_1 = \frac{g(.5) - g(.75)}{5 - 0} = -0.220$$

and $a_1 = g(.75) = 1.099$ for logit link $g(\cdot)$. Similarly, we obtain $(a_2, b_2)=(1.099, 0)$, so that $d = 0 - (-0.220) = 0.220$. By (2), the probabilities of no pulmonary fibrosis at the 6 time points are obtained as $(.750, .707, .659, .608, .555, .500)$ for the placebo arm, and $(.750, .750, .750, .750, .750, .750)$ for the treatment arm.

The proportions of observed measurements are expected to be

$$(p_1, p_2, p_3, p_4, p_5, p_6) = (1, .95, .90, .85, .80, .75).$$

Suppose that we expect independent missing. Then, we obtain $p_{jj'}$ from the specified p_j values using $p_{jj'} = p_j p_{j'}$. Now we have all the parameter values required, and obtain $v_1 = 0.305$ and $v_2 = 0.353$. Finally, from (7), we obtain

$$n = \left\lceil \frac{(1.96 + 0.84)^2(0.305/0.5 + 0.353/0.5)}{0.220^2} \right\rceil + 1 = 215.$$

If we assume monotone missing $(p_{jj'} = p_{j \lor j'})$, we obtain $v_1 = 0.324$, $v_2 = 0.308$, and $n = 229$.

14.6 Discussions

We have discussed sample size and power analysis methods for comparing the time trajectory of longitudinal continuous or binary outcome. Due to its robustness and easy computation, we use the GEE with working independent correlation structure for inference. The statistical power and sample size depend on the variance of measurement errors, true correlation structure and missing probability distribution of longitudinal data as well as the difference on the slope of time trajectory, but they are free of the intercepts of the regression models. The missing probability distribution requires specification of marginal and bivariate missing probabilities. In order to model the bivariate probabilities, we considered the popular random missing and monotone missing patterns.

Although we assume a fixed measurement time schedule at the design of a randomized clinical trial, the real visit time for measurements may vary for each patient. Through simulations, Jung and Ahn (2003) show that variable measurement times around the scheduled ones does not change the statistical power much.

Although we have focused on RCT, the formulas can be used for retrospective studies too.

Bibliography

Anderson, D. A. and Aitkin, M. 1985. Variance component models with binary response: Interviewer variability. *Journal of the Royal Statistical Society, Series B*, 47:203–210.

Chow, S. C., Shao, J., and Wang, H. 2003. *Sample size calculations in clinical research.* Taylor & Francis: New York, NY.

Crowder, M. 1995 On the use of a working correlation matrix in using generalised linear models for repeated measures. *Biometrika*, 82:407–410.

Davis, C. S. 1991. Semi-parametric and non-parametric methods for the analysis of repeated measurements with applications to clinical trials. *Statistics in Medicine*, 10:1959–1980.

Jung, S. H. and Ahn, C. 2003. Sample size estimation for GEE method for comparing slopes in repeated measurements data. *Statistics in Medicine*, 22:1305–1315.

Jung, S. H. and Ahn, C. 2004. K-sample test and sample size calculation for comparing slopes in data with repeated measurements. *Biometrical Journal*, 46:554–564.

Jung, S. H. and Ahn, C. 2005. Sample size for a two-group comparison of repeated binary measurements using GEE. *Statistics in Medicine*, 224:2583–2596.

Laird, N. M. and Ware, J. H. 1982. Random-effects models for longitudinal data. *Biometrics*, 38:963–974.

Liang, K. Y. and Zeger, S. L. 1986. Longitudinal data analysis using generalized linear models. *Biometrika*, 73:13–22.

McCullagh, P. 1983. *Generalized linear models.* Chapman and Hall, London.

Medsger, T. A., Jr. 1997. Systemic sclerosis (scleroderma): clinical aspects. In Koopman, W. J. (ed), *Arthritis and allied conditions* (13th edn), pp 1433–1464. Williams and Wilkins, Baltimore, MD.

Reveille, J., Fischbach, M., McNearney, T., et al. 2001. Systemic sclerosis in 3 US ethnic groups: A comparison of clinical, sociodemographic, serologic and immunologic determinants. *Seminars in Arthritis and Rheumatism*, 30:332–346.

Rubin, D. B. Inference and missing data. 1976. *Biometrika*, 63:581–592.

Wedderburn, R. W. M. 1974. Quasi-likelihood functions, generalized linear models, and the Gauss-Newton method, *Biometrika*, 61, 439–447.

APPENDIX: Derivation of Power Functions and Sample Size Formulas

Suppose that m measurement is scheduled for $t_1 < t_2 < \cdots < t_m$, and $\delta_{kij} = 1$ if patient i in arm k has an outcome measurement at time t_j and $= 0$ otherwise. Let $p_j = P(\delta_{kij} = 1)$ for $1 \leq j \leq m$ and $p_{jj'} = P(\delta_{kij}\delta_{kij'} = 1)$ for $j \neq j'$.

A.1. Continuous Outcome Case

For sample size calculation, we assume that the continuous outcome variable has a constant variance $\text{var}(y_{kij}) = \sigma^2$. Then, we have

$$\hat{v}_k = n_k \frac{\sum_{i=1}^{n_k}\{\sum_{j=1}^m \delta_{kij}(t_j - \bar{t}_k)\hat{\epsilon}_{kij}\}^2}{\{\sum_{i=1}^{n_k}\sum_{j=1}^m \delta_{kij}(t_j - \bar{t}_k)^2\}^2},$$

where $\bar{t}_k = \sum_{i=1}^{n_k}\sum_{j=1}^m \delta_{kij}t_j / \sum_{i=1}^{n_k}\sum_{j=1}^m \delta_{kij}$. As $n_k \to \infty$, $\bar{t}_k$ converges to

$$\tau = \frac{\sum_{j=1}^m p_j t_j}{\sum_{j=1}^m p_j},$$

$n_k^{-1}\sum_{i=1}^{n_k}\{\sum_{j=1}^m \delta_{kij}(t_j - \bar{t}_k)\hat{\epsilon}_{kij}\}^2$ converges to $\sigma^2(s^2 + c)$, and

$$n_k^{-1}\sum_{i=1}^{n_k}\sum_{j=1}^m \delta_{kij}(t_j - \bar{t}_k)^2$$

converges to $s^2 = \sum_{j=1}^m p_j(t_j - \tau)^2$, where $c = 2\sum\sum_{1 \leq j < j' \leq m} p_{jj'}\rho_{jj'}(t_j - \tau)(t_{j'} - \tau)$. Hence, $\hat{v}_k$ converges to

$$v = \frac{\sigma^2(s^2 + c)}{s^4}.$$

Without loss of generality, we assume that $b_1 > b_2$ under H_1. Under $H_1 : |b_2 - b_1| = d$, the power is given as

$$1 - \beta = P\left(\frac{|\hat{b}_1 - \hat{b}_2|}{\sqrt{\hat{v}_1/n_1 + \hat{v}_2/n_2}} > z_{1-\alpha/2} \Big| b_1 - b_2 = d\right)$$

$$= P\left(\frac{\hat{b}_1 - \hat{b}_2 - d}{\sqrt{\hat{v}_1/n_1 + \hat{v}_2/n_2}} > z_{1-\alpha/2} - \frac{d}{\sqrt{\hat{v}_1/n_1 + \hat{v}_2/n_2}} \Big| b_1 - b_2 = d\right)$$

$$\approx \bar{\Phi}\left\{z_{1-\alpha/2} - \frac{d\sqrt{n}}{\sqrt{v(r_1^{-1} + r_2^{-1})}}\right\}. \tag{A1}$$

Hence, for a power of $1 - \beta$ under $H_1 : |b_2 - b_1| = d$, the required sample size is obtained by solving equation (A1) with respect to n,

$$n = \frac{v(z_{1-\alpha/2} + z_{1-\beta})^2}{d^2 r_1 r_2}.$$

A.2. Binary Outcome Case

Under the fixed measurement times assumption, we have

$$\hat{A}_k(\hat{a}_k, \hat{b}_k) = \frac{1}{n_k}\sum_{i=1}^{n_k}\sum_{j=1}^m \delta_{kij}\mu_{kj}\bar{\mu}_{kj}\begin{pmatrix} 1 & t_j \\ t_j & t_j^2 \end{pmatrix} + o_p(1)$$

and

$$\hat{\Sigma}_k = \frac{1}{n_k} \sum_{i=1}^{n_k} \{ \sum_{j=1}^{m} \delta_{kij}(y_{kij} - \mu_{kj}) \begin{pmatrix} 1 \\ t_j \end{pmatrix} \}^{\otimes 2} + o_p(1)$$

where $a^{\otimes 2} = aa^T$ for a vector a. So, as $n_k \to \infty$, $\hat{A}_k(\hat{a}_k, \hat{b}_k)$ and $\hat{\Sigma}_k$ converge to

$$A_k = \sum_{j=1}^{m} \delta_j \mu_{kj} \bar{\mu}_{kj} \begin{pmatrix} 1 & t_j \\ t_j & t_j^2 \end{pmatrix}$$

and

$$\Sigma_k = \sum_{j=1}^{m} \sum_{j'=1}^{m} p_{jj'} \rho_{jj'} \sqrt{\mu_{kj} \bar{\mu}_{kj} \mu_{kj'} \bar{\mu}_{kj'}} \begin{pmatrix} 1 & t_j \\ t_{j'} & t_j t_{j'} \end{pmatrix},$$

respectively. Hence, $\hat{V}_k$ converges to $V_k = A_k^{-1} \Sigma_k A_k^{-1}$, and it is easy to show that the 2×2 component of V_k is given as

$$v_k = \frac{s_k^2 + c_k}{s_k^4},$$

where

$$s_k^2 = \sum_{j=1}^{m} p_j \mu_{kj} \bar{\mu}_{kj} (t_j - \tau_k)^2$$

$$c_k = \sum \sum_{j \neq j'} p_{jj'} \rho_{jj'} \sqrt{\mu_{kj} \bar{\mu}_{kj} \mu_{kj'} \bar{\mu}_{kj'}} (t_j - \tau_k)(t_{j'} - \tau_k),$$

and

$$\tau_k = \frac{\sum_{j=1}^{m} p_j \mu_{kj} \bar{\mu}_{kj} t_j}{\sum_{j=1}^{m} p_j \mu_{kj} \bar{\mu}_{kj}}.$$

Without loss of generality, we assume that $b_1 > b_2$ under H_1. Under $H_1 : |b_2 - b_1| = d$, the power under $H_1 : |b_1 - b_2| = d$ is given as

$$1 - \beta = P\left(\frac{|\hat{b}_1 - \hat{b}_2|}{\sqrt{\hat{v}_1/n_1 + \hat{v}_2/n_2}} > z_{1-\alpha/2} \Big| b_1 - b_2 = d \right)$$

$$= P\left(\frac{\hat{b}_1 - \hat{b}_2 - d}{\sqrt{\hat{v}_1/n_1 + \hat{v}_2/n_2}} > z_{1-\alpha/2} - \frac{d}{\sqrt{\hat{v}_1/n_1 + \hat{v}_2/n_2}} \Big| b_1 - b_2 = d \right)$$

$$\approx \bar{\Phi}\left(z_{1-\alpha/2} - \frac{d\sqrt{n}}{\sqrt{v_1/r_1 + v_2/r_2}} \right). \tag{A2}$$

By solving equation (A2) with respect to n, we derive the required sample size for a power of $1 - \beta$

$$n = \frac{(z_{1-\alpha/2} + z_{1-\beta})^2 (v_1/r_1 + v_2/r_2)}{d^2}.$$

Part IV

Monitoring of Randomized Controlled Trials

15

Group Sequential Methods

Michael Proschan

CONTENTS

Clinical trials are monitored over time for quality, safety, efficacy, and futility. This chapter focuses on the statistical aspects of efficacy and futility monitoring using group-sequential boundaries and other tools such as conditional power and stochastic curtailment. Further details may be found in Proschan, Lan, and Wittes (2006) or Jennison and Turnbull (2000). More applied aspects of data monitoring committees may be found in DeMets, Friedman, and Furberg (2006); Ellenberg, Fleming, and DeMets (2003); or Herson (2009).

15.1 Group Sequential Methods

Imagine waiting until the end of a clinical trial to look at results, only to find overwhelming evidence of a treatment effect that would have been readily apparent halfway through. Because that would be unacceptable for a disease with serious consequences, there is an ethical imperative to monitor clinical trials. If a treatment is clearly superior, we might stop the trial, announce the results, and offer the treatment to patients in the control arm, as happened in the Pamoja Tulinde Maishe (PALM) trial of deadly Ebola virus disease in the Democratic Republic of the Congo. At an interim analysis, Regeneron EB3 and monoclonal antibody 114 were declared superior to standard care plus the monoclonal antibody cocktail ZMapp (Mulangu et al., 2019). Alternatively, the true treatment effect in a clinical trial may be much smaller than anticipated, nonexistent, or even harmful. Waiting until the end of the trial to make this discovery may also be unethical. The Cardiac Arrhythmia Suppression Trial (CAST) in patients with a prior heart attack and cardiac arrhythmias was modified and subsequently terminated because some antiarrhythmic drugs were found to be harmful (Cardiac Arrhythmia Suppression Trial Investigators, 1989).

TABLE 15.1
Type I error rate when the null hypothesis is rejected if the absolute value of the Z-score exceeds 1.96 at any of k analyses.

2	3	4	5	10	20
.083	.107	.126	.142	.193	.248

Interim analyses allow the possibility of early stopping for benefit, harm, or futility, but at the cost of some loss in power. After all, the most powerful design analyzes results only at the end of the trial. In fact, the more often we look, the earlier we might stop, but the greater the penalty we must pay for multiple comparisons. The need to pay a penalty for multiple comparisons has been recognized for many years. For example, Armitage, McPherson, and Rowe (1969) showed that the chance of at least one false declaration of treatment benefit can be greatly inflated if one simply performs the usual, unadjusted test statistic multiple times throughout a trial. Even a totally ineffective treatment will eventually reach a p-value of 0.05 or less if enough tests are performed. Table 15.1 shows the type I error rate for k equally-spaced tests. With 5 looks, the type I error rate is approximately 0.142. These numbers apply to a wide variety of statistics commonly used in clinical trials, including paired or unpaired t-statistics, Z-statistics for a single proportion or difference of two proportions, and the logrank statistic comparing two survival curves. They also apply when the treatment effect is adjusted for baseline differences, such as with a linear model like analysis of covariance, or the Cox proportional hazards model for survival.

The solution to the problem of inflated error rate is to increase the level of evidence required at each analysis. This increased level of evidence is often expressed in terms of higher boundaries for the Z-score, but it can also be helpful to think in terms of lower levels required to declare a p-value statistically significant. For example, consider the *Haybittle-Peto* boundary (Haybittle, 1971), one of the earliest boundaries ever proposed. The original proposal was to require an interim Z-score to have magnitude 3 or more to be declared significant using a two-tailed test at $\alpha = 0.05$; If the study continues to its final look, the conventional Z-score boundary of 1.96 is used. A Z-score boundary of 3 corresponds to a two-tailed p-value of 0.0027 or less. We give a more detailed description of the Haybittle-Peto procedure in a subsequent section. Our purpose here is to show that boundaries can be constructed for Z-scores or p-values. They can also be constructed for another curious creature described later, the B-value.

15.1.1   A unified framework

We next show how to put different test statistics on a common footing to allow a unified approach to monitoring. Key references in this regard are Lan and Zucker (1993) and Jennison and Turnbull (1997). We begin with a one-sample t-test with a sample size large enough to treat the variance as if it were known.

Consider a clinical trial with paired differences D on a continuous outcome. Examples might include a crossover trial of diets to lower blood pressure, or a trial in which a topical treatment is applied to one eye, and a placebo treatment to the other eye, etc. The treatment effect is $\delta = \mathrm{E}(D)$. Without loss of generality, assume that large values of δ indicate a beneficial treatment, so the null and alternative hypotheses are

$$H_0 : \delta = 0$$
$$H_1 : \delta > 0. \tag{15.1}$$

The estimate of treatment effect is the sample mean difference,

$$\hat{\delta} = S_n/n, \quad S_n = \sum_{i=1}^{n} D_i.$$

Assume that the variance $\sigma^2 = \text{var}(D)$ is known. Suppose that the final sample size is N, but we will examine the data at an interim analysis after n observations. The Z-scores at the interim and final analyses are:

$$Z_n = \frac{S_n}{\sqrt{n\sigma^2}} \quad \text{and} \quad Z_N = \frac{S_N}{\sqrt{N\sigma^2}},$$

respectively.

The key to being able to control the probability of making a type I error at either the interim or final analyses is to determine the joint distribution of (Z_n, Z_N). The first step is to find the correlation between these Z-scores. Because Z-scores are standardized to have unit variance, the correlation is just the covariance. Write S_N as $S_n + S_N - S_n$, and note that $S_N - S_n = \sum_{i=n+1}^{N} D_i$ is independent of $S_n = \sum_{i=1}^{n} D_i$ because the components of the two sums are non-overlapping.

$$\begin{aligned}
\text{cov}(Z_n, Z_N) &= \frac{1}{\sqrt{n\sigma^2}} \frac{1}{\sqrt{N\sigma^2}} \text{cov}(S_n, S_n + S_N - S_n) \\
&= \frac{1}{\sigma^2\sqrt{nN}} \{\text{cov}(S_n, S_n) + \text{cov}(S_n, S_N - S_n)\} \\
&= \frac{1}{\sigma^2\sqrt{nN}} \{\text{var}(S_n) + 0\} \\
&= \frac{n\sigma^2}{\sigma^2\sqrt{nN}} \\
&= \sqrt{n/N}. \quad (15.2)
\end{aligned}$$

The expression n/N represents the proportion of the trial that has been completed by the time of the interim analysis; N is the total number of patients that will be evaluated by the end of the trial, and n is the number evaluated by the interim analysis. We call t the *information fraction* or *information time*. We will define the concept of information more precisely later; for now, think in terms of sample size. The information fractions are 0 and 1 before any patients have been evaluated and after all have been evaluated, respectively. The result just derived says that the correlation between the interim and final Z-scores is the square root of t, where $t = n/N$.

We can easily extend this result to multiple interim analyses after $n_1, n_2, \ldots, n_k = N$ observations. Define the Z-score stochastic process $Z(t)$ by

$$Z(t_i) = Z_{n_i}, \quad \text{where } t_i = n_i/N.$$

The above argument can be used to show that

$$\text{cov}\{Z(t_i), Z(t_j)\} = \sqrt{t_i/t_j}, \quad t_i \leq t_j.$$

Having determined the correlation structure of $Z(t)$, we now determine its mean structure. The mean of the Z-score at the end of the trial is $\theta = \text{E}\{Z(1)\} = \text{E}(Z_N) = N^{1/2}\delta/\sigma$. The mean at an interim analysis at information time $t = n/N$ is $\text{E}(Z_n) = n^{1/2}\delta/\sigma = (n/N)^{1/2}\text{E}(Z_N)$. Writing this in terms of the Z-score stochastic process, we find that

$$\text{E}\{Z(t)\} = \theta\sqrt{t}, \quad \theta = \text{E}\{Z(1)\}.$$

Under a normality assumption on the Y_i, the joint distribution of the $Z(t_i)$ is multivariate normal. This also holds approximately for large sample sizes even if the observations themselves are not normally distributed. To see this, note that $(S_{n_1}, S_{n_2} - S_{n_1}, \ldots, S_{n_k} - S_{n_{k-1}})$ are independent, and by the central limit theorem, each is asymptotically normal. Therefore, $(Z(t_1), \ldots, Z(t_k))$, being linear combinations of $(S_{n_1}, S_{n_2} - S_{n_1}, \ldots, S_{n_k} - S_{n_{k-1}})$, are asymptotically multivariate normal. Although we assumed known variance, that is also not necessary for large sample sizes because the sample variance estimate is consistent for σ^2. We can collect the above results succinctly and conclude that, at least asymptotically,

- Z1: $Z(t_1), \ldots, Z(t_k)$ are multivariate normal.

- Z2: $\mathrm{E}\{Z(t_i)\} = \theta t_i^{1/2}$, $\theta = \mathrm{E}\{Z(1)\}$ (which is 0 under H_0).

- Z3: $\mathrm{cov}\{Z(t_i), Z(t_j)\} = \sqrt{t_i/t_j}$, $t_i \leq t_j$.

We can use these properties to calculate the probability of false rejection of the null hypothesis of no treatment effect. Notice that under the null hypothesis, the joint distribution of Z-scores at equally spaced values of t is the same regardless of the total time of observation. Whether we take 2 interim looks at respective information fractions $(1/8, 1/4)$ or $(1/4, 1/2)$ or $(1/2, 1)$, the joint distribution of the two Z-scores is the same. An interesting consequence is the following. Imagine continually doubling the number of equally-spaced interim analyses and performing a z-test at nominal level 0.05 at each analysis time. The type I error rate on $[0, 1/2]$ approaches the same limit as the type I error rate on $[0, 1]$ because the joint distribution of the test statistics is the same. But clearly the error rate on $[0, 1]$ must be strictly larger than that on $[0, 1/2]$, unless of course the latter is already 1. This argument shows that the type I error rate from infinitely many equally-spaced interim analyses on $[0, 1/2]$ is 1. But of course the same argument can be used to show that the error rate for infinitely many equally-spaced looks on $[0, 1/4]$ is 1, etc. Indeed, the following is true.

Remark 15.1 *Much of the inflated type I error rate from monitoring without adjustment for multiple comparisons is caused by early looks. The above argument shows that, no matter how small ϵ is, the type I error rate from monitoring at equally-spaced looks on $[0, \epsilon]$ tends to 1 (albeit at a slow rate) as the number of interim analyses tends to infinity. This is not true if we defer the first analysis to some fixed information fraction $t_0 > 0$ (see DeLong, 1981).*

An equivalent way to monitor trials uses a transformation of the Z-score called the B-value (Lan and Wittes, 1988; Lan and Zucker, 1993):

$$B(t) = \sqrt{t}Z(t). \tag{15.3}$$

In the paired data setting, with $t = n/N$, the B-value is

$$
\begin{aligned}
B(t) &= \sqrt{t}Z(t) = \sqrt{\frac{n}{N}} \frac{S_n}{\sqrt{n}\sigma} \\
&= \frac{S_n}{\sqrt{N}\sigma}. \tag{15.4}
\end{aligned}
$$

The B-value and Z-score are the same at the end of the trial (when $t = 1$): $B(1) = Z(1) = S_N/(N\sigma^2)^{1/2}$. If we denote the mean of this final Z-score by θ, then the mean of an interim B-value is

$$\mathrm{E}\{B(t)\} = \mathrm{E}\left(\frac{S_n}{\sqrt{N}\sigma}\right) = \frac{n\delta}{\sqrt{N}\sigma}$$

$$= \frac{\sqrt{N}\delta}{\sigma}\frac{n}{N} = \theta t. \tag{15.5}$$

That is, the mean of the B-value is linear in t, with slope equal to the expected value of the final Z-score. This is one of the advantages of the B-value formulation.

Another advantage of the B-value process is that it has independent increments, meaning that $B(t_1), B(t_2) - B(t_1), \ldots, B(t_k) - B(t_{k-1})$ are independent for $t_1 \le t_2 \le \ldots \le t_k$. This follows immediately from the fact that these increments are sums with non-overlapping components. We will see that the independent increments property greatly facilitates calculation of certain probabilities. The 3 properties of Z-scores translate into the following 3 properties for B-values.

- B1: $B(t_1), \ldots, B(t_k)$ are multivariate normal.

- B2: $\mathrm{E}\{B(t_i)\} = \theta t_i$, $\theta = \mathrm{E}\{Z(1)\}$ (which is 0 under H_0).

- B3: $\mathrm{cov}\{B(t_i), B(t_j)\} = t_i$, $t_i \le t_j$.

Remark 15.2 *It is an amazing fact that the joint distribution of Z-scores and B-values follow properties Z1–Z3 and B1–B3, at least asymptotically, under many different testing settings, including (1) the one-sample or paired t-statistic, (2) the two-sample t-statistic, (3) the one-sample Z-statistic for a proportion, (4) the two-sample Z-statistic for proportions, (5) the logrank Z-statistic comparing two survival distributions, (6) the t-statistic for the treatment coefficient in a linear model adjusting for covariates (i.e., analysis of covariance), (7) the Z-statistic for the treatment coefficient in a Cox proportional hazards model when the proportional hazards assumption holds, and (8) the Z-statistic for a maximum likelihood estimator under the usual regularity conditions.*

The key to understanding how to put all of these settings into the $Z(t)$ or $B(t)$ framework is to think about the concept of information. Consider the simplest case of estimation of treatment effect δ in a paired setting. The sample size is a measure of the information content of our estimator, $\hat{\delta} = \bar{Y}$. The larger the sample size, the greater the information content in $\hat{\delta}$. But we surely get more information from a sample of given size m if $\sigma^2 = 1$ than if $\sigma^2 = 10$. When $\sigma^2 \ne 1$, we ask: how many independent and identically distributed observations $X_1, X_2 \ldots$ from a distribution with variance 1 produce a sample mean with the same variance as our sample mean $\bar{Y}$ of m observations with variance σ^2? Denote the answer by I for "information". Equating the variance of the sample mean of m observations with variance σ^2 to the variance of a sample mean of I observations with variance 1 yields $\sigma^2/m = 1/I$, so $I = m/\sigma^2 = 1/\mathrm{var}(\bar{Y})$. Note that I need not be an integer. Similar reasoning can be used for any estimator $\hat{\delta}$. We ask: what sample size of iid observations with variance 1 yields an estimator with the same variance as $\hat{\delta}$? The answer is $I = 1/\mathrm{var}(\hat{\delta})$. This motivates the following definition.

Definition 15.1 *The information I in an estimator $\hat{\delta}$ with finite variance is defined to be $1/\mathrm{var}(\hat{\delta})$. If I_{int} and I_{end} denote the information at an interim analysis and final analysis, respectively, the information fraction is defined by $t = I_{\mathrm{int}}/I_{\mathrm{end}}$.*

This definition matches the one we gave for the paired t-statistic with equal sample sizes because in that case, the variances of the sample means after n interim and N final observations are σ^2/n and σ^2/N, respectively, so the ratio of inverse variances is n/N.

Example 15.1 Consider a t-test comparing a treatment and control group with respect to change in diastolic blood pressure from baseline to end of study. The null hypothesis is $\mathrm{H}_0 : \delta = 0$, where where $\delta = \mu_T - \mu_C$ is the treatment minus control difference in means.

The final analysis will have 200 participants per arm, and an interim analysis occurs with 40 in the treatment group and 36 in the control group. The variances of the difference between treatment and control sample means at the interim and final analyses are $\text{var}(\bar{Y}_T - \bar{Y}_C) = \sigma^2(1/40 + 1/36)$ and $2\sigma^2/200$, respectively. The information fraction at the interim analysis is

$$t = \frac{\{\sigma^2(1/40 + 1/36)\}^{-1}}{(2\sigma^2/200)^{-1}} = 0.189.$$

We could also have approximated t by dividing the current average per-arm sample size, $(40 + 36)/2 = 38$, by the final per-arm sample size 200: $38/200 = 0.190$. ∎

Example 15.2 Consider a test comparing the proportions of gravely ill patients dying within 30 days of randomization. The null hypothesis is $H_0 : p_T - p_C = 0$, where p_T and p_C are the 30 day mortality probabilities in the treatment and control arms. The final sample size is 100 per arm, and the interim analysis occurs after $n_T = 44$ and $n_C = 47$ patients have been evaluated. The variances of the difference between treatment and control proportions at the interim and final analyses are $\text{var}(\hat{p}_T - \hat{p}_C) = p_T(1 - p_T)/44 + p_C(1 - p_C)/47$ and $p_T(1 - p_T)/100 + p_C(1 - p_C)/100$, respectively. The information fraction at the interim analysis is

$$t = \frac{\{p_T(1 - p_T)/44 + p_C(1 - p_C)/47\}^{-1}}{\{p_T(1 - p_T)/100 + p_C(1 - p_C)/100\}^{-1}},$$

which depends on p_T and p_C. One option is to substitute the observed proportions $\hat{p}_T$ and $\hat{p}_C$ for p_T and p_C. But with this method, our estimates of p_T and p_C change from one interim analysis to the next. Another option is to substitute for $\hat{p}_T$ and $\hat{p}_C$ the best estimate, $\hat{p} = (44\hat{p}_T + 47\hat{p}_C)/(44 + 47)$, of the overall mortality probability. Then the estimated death probability cancels out in the numerator and denominator, and the information time estimate is

$$t = \frac{\{1/44 + 1/47\}^{-1}}{(2/100)^{-1}} = 0.455.$$

Note again that we could have approximated the information fraction using the average per-arm sample sizes $(44 + 47)/2 = 45.5$ and 100 at the interim and final sample sizes: $t \approx 45.5/100 = 0.455$. ∎

We next consider the comparison of survival using the logrank statistic. Assume that the treatment to control log hazard ratio δ is constant over time, and consider the logrank test of the null hypothesis $H_0 : \delta = 0$. Intuitively, it is clear that the information content from such a trial is dictated not by the number of patients, but rather, the number of deaths (or whatever event is being studied). Even a trial with 5,000 patients contains very little information if only 2 people die. We will soon see this intuition corroborated.

At the ith death time, construct a 2×2 table with the numbers n_{Ti} and n_{Ci} of treatment and control patients "at risk" just prior to the death, as shown in Table 15.2. X_i is the indicator that the ith death occurred in the treatment arm, and $E_i = n_{Ti}/(n_{Ti} + n_{Ci})$ and $V_i = n_{Ti}n_{Ci}/(n_{Ti} + n_{Ci})^2$ are the conditional mean and variance, respectively, of X_i given n_{Ti} and n_{Ci}. The logrank statistic with D total deaths at the end is

$$Z = \frac{\sum_{i=1}^{D}(X_i - E_i)}{\sqrt{\sum_{i=1}^{D} V_i}}. \tag{15.6}$$

Note that the numerator is a sum of non-independent terms. Nonetheless, it can be shown that they are uncorrelated under the null hypothesis of no difference between the two survival distributions. Therefore, the denominator is an estimate of the standard deviation

TABLE 15.2
Two-by-two table at the *ith* death time with n_{Ti} and n_{Ci} patients at risk. X_i is the indicator that the *ith* death came from the treatment arm.

	Dead		
Treatment	X_i		n_T
Control			n_C
	1		n

of the numerator. Even though the denominator is itself a random variable, for large sample sizes we can treat the V_i as if they were fixed constants.

A good estimator of the log hazard ratio is identical to Z except that its denominator is not square rooted:

$$\hat{\delta} = \frac{\sum_{i=1}^{D}(X_i - E_i)}{\sum_{i=1}^{D} V_i}. \tag{15.7}$$

To get its estimated variance, again treat the denominator as a constant: $\text{var}(\hat{\delta}) \approx (1/\sum_{i=1}^{D} V_i)^2 \sum_{i=1}^{D} V_i = 1/\sum_{i=1}^{D} V_i$. Therefore, the information fraction at an interim analysis with d deaths is $(\sum_{i=1}^{d} V_i)/(\sum_{i=1}^{D} V_i)$. There is only one problem: at an interim analysis we will not know $V_{d+1}, \ldots, V_D$ because they depend on the numbers at risk in the two arms at future death times. Nonetheless, in a trial with 1-1 allocation, approximately half of the at-risk patients will be in the treatment arm, so each V_i will be close to $(1/2)(1/2) = 1/4$. Accordingly, under the null hypothesis, $\sum_{i=1}^{D} V_i \approx D/4$. Similarly, at an interim analysis with d deaths, $\sum_{i=1}^{d} V_i \approx d/4$. Therefore, the information fraction is approximately

$$t = d/D. \tag{15.8}$$

Under the alternative hypothesis, the numbers of patients at risk in the two arms will not necessarily be equal. If the treatment is effective, fewer treated than control patients will die, leaving more treated patients at risk as the trial continues. Nonetheless, it can be shown that the variance of the numerator of the logrank statistic at the end of the trial is still close to $D/4$ under reasonable assumptions about the size of the treatment-to-control hazard ratio if D is large. This holds under a "local" alternative hypothesis such that power converges to a constant other than 0 or 1 as $D \to \infty$ (see, for example, Schoenfeld, 1981). Therefore, the information fraction at an interim analysis is closely approximated by expression (15.8) whether the null or alternative hypothesis is true. Properties Z1–Z3 and B1–B3 hold for the Z-statistic and B-value using information fraction (15.8) for a wide class of survival methods, including the Cox proportional hazards model (Tsiatis, 1982). See Lan and Zucker (1993), Proschan, Lan, and Wittes (2006) or Jennison and Turnbull (2000) for a review.

Example 15.3 A trial comparing survival using the logrank statistic or the Cox proportional hazards model tests the null hypothesis $H_0 : \delta = 0$, where δ is the log hazard ratio. The trial will continue until 400 events. At an interim analysis with 100 events, the information fraction is $t = 100/400 = 0.25$. Even if the trial has already recruited its planned number of patients, the trial is only 25% complete in terms of its information, so $t = 0.25$.

We summarize the key result:

Result 15.1 *Asymptotically, the joint distribution of Z-statistics $Z(t)$ and B-values $B(t)$ over information time t are given by Z1–Z3 and B1–B3, respectively for all of the settings mentioned in Remark 15.2.*

In fact, Lan and Zucker (1993); Proschan, Lan, and Wittes (2006); Jennison and Turnbull (1997); and Jennison and Turnbull (2000) show that the same result holds for other testing settings as well.

15.1.2   Boundaries

We have seen that, provided that we define information time t correctly, the joint distribution of Z-scores $Z(t)$ is the same for a wide variety of different statistical tests. Boundaries developed to control the type I error rate for repeated t-tests apply equally well to tests of proportions or survival, for example.

Remark 15.3 *Throughout this section, we consider one-sided boundaries. Clinical trials often use two-tailed tests because treatment could be harmful. Nonetheless, benefit is very different from harm, and it is unclear whether the same level of evidence should be used for each. We can always compute two one-tailed p-values.*

The Pocock boundary (Pocock, 1977) is based on a very simple idea: use the same Z-score threshold at equally-spaced analyses over information time. This essentially "raises the bar" from ± 1.96 in a non-monitoring scenario to a larger threshold that depends on the total number of analyses, which are assumed to be after equal increments of information. The Pocock boundary as a function of the number of interim analyses is shown in Table 15.3 for one-tailed $\alpha = 0.005, 0.025$, or 0.05 (two-tailed $\alpha = 0.01, 0.05$, or 0.10). For example, consider 3 interim analyses at information times $t = 1/3, 2/3$, and $3/3 = 1$. For a one-tailed error rate of 0.025, the Z-score boundary is 2.289 at each analysis. The one-tailed *nominal p-value*, meaning the p-value computed as if there were only one look, is 0.011. Therefore, an equivalent way to view the boundary is that the one-tailed p-value must be 0.011 or less to declare statistical significance at a given analysis. This includes the final analysis, and that is why Pocock himself now recommends against his own boundary. It is not palatable to require such strong evidence even at the end of a trial.

O'Brien and Fleming (1979) devised a boundary that is higher than Pocock's early in the trial and lower at the end. In fact, the Z-score boundary at the end of the trial is close to what it would have been with no monitoring. The idea is to use a constant boundary for the B-value instead of the Z-score. At an interim analysis, we reject the null hypothesis and declare a treatment difference if the B-value exceeds a constant a depending on the number of equally-spaced looks. Table 15.4 shows the value of a for different numbers of looks. For example, Table 15.4 shows that for a trial with 4 equally-spaced looks and one-tailed $\alpha = 0.025$ (two-tailed $\alpha = 0.05$), $a = 2.024$. The Z-score boundaries at the 4 looks are $2.024/(1/4)^{1/2} = 4.048$, $2.024/(2/4)^{1/2} = 2.862$, $2.024/(3/4)^{1/2} = 2.337$, and $2.024/(4/4)^{1/2} = 2.024$. These Z-score boundaries are shown in the fourth row of Table 15.5. Table 15.5 also gives the one-tailed p-value associated with the boundaries. For instance, for a two-look trial at one-tailed $\alpha = 0.025$, the Z-score boundary at the first look halfway through the trial is 2.796, which corresponds to a one-tailed p-value of .003. If the trial is not stopped at the halfway point, the final look uses boundary 1.977, with associated one-tailed p-value of 0.024. Unlike Pocock's boundary, O'Brien and Fleming's Z-score boundary at the end of the trial is close to what it would be with no monitoring. For instance, the final boundary for the two-look trial is 1.977 versus 1.960 for a trial with no interim monitoring. Another feature of the O'Brien-Fleming boundary is that it is quite high early in the trial.

TABLE 15.3
One-tailed Pocock boundaries at $\alpha = 0.005$, 0.025, or 0.05 (or two tailed $\alpha = 0.01$, 0.05, or 0.10) . Shown in parenthesis is the one-tailed p-value associated with the boundary.

# Looks (k)	$1\alpha = 0.005$ ($2\alpha = 0.01$)	$1\alpha = 0.025$ ($2\alpha = 0.05$)	$1\alpha = 0.05$ ($2\alpha = 0.10$)
1	2.576 (1p=0.005)	1.960 (1p=0.025)	1.645 (1p=0.050)
2	2.772 (1p=0.003)	2.178 (1p=0.015)	1.875 (1p=0.030)
3	2.873 (1p=0.002)	2.289 (1p=0.011)	1.992 (1p=0.023)
4	2.939 (1p=0.002)	2.361 (1p=0.009)	2.067 (1p=0.019)
5	2.986 (1p=0.001)	2.413 (1p=0.008)	2.122 (1p=0.017)
6	3.023 (1p=0.001)	2.453 (1p=0.007)	2.164 (1p=0.015)
7	3.053 (1p=0.001)	2.485 (1p=0.006)	2.197 (1p=0.014)
8	3.078 (1p=0.001)	2.512 (1p=0.006)	2.225 (1p=0.013)
9	3.099 (1p=0.001)	2.535 (1p=0.006)	2.249 (1p=0.012)
10	3.117 (1p=0.001)	2.555 (1p=0.005)	2.270 (1p=0.012)
∞	∞ (1p=0.000)	∞ (1p=0.000)	∞ (1p=0.000)

For instance, Table 15.5 shows that the first of 5 interim analyses uses Z-score boundary 4.562. It is extremely difficult to cross the boundary early. This is considered an advantage because early in a trial, there can be anomalies caused by an incomplete understanding of the protocol or incomplete training. As staff acquire more experience, errors tend to diminish.

TABLE 15.4
One-tailed O'Brien-Fleming B-value boundaries at $\alpha = 0.005$, 0.025, or 0.05 (or two tailed $\alpha = 0.01, 0.05$, or 0.10). To compute the Z-score boundary at information fraction t_i, divide the tabled value by $t_i^{1/2}$.

# Looks (k)	$1\alpha = 0.005$ ($2\alpha = 0.01$)	$1\alpha = 0.025$ ($2\alpha = 0.05$)	$1\alpha = 0.05$ ($2\alpha = 0.10$)
1	2.576	1.960	1.645
2	2.580	1.977	1.678
3	2.595	2.004	1.710
4	2.609	2.024	1.733
5	2.621	2.040	1.751
6	2.631	2.053	1.765
7	2.640	2.063	1.776
8	2.648	2.072	1.786
9	2.654	2.080	1.794
10	2.660	2.087	1.801
∞	2.807	2.241	1.960

An unfortunate feature of both the O'Brien-Fleming and Pocock boundaries is that they assume the interim analyses occur after equal increments of information. I.e., the information fractions at k planned analyses are at $t = 1/k, 2/k, \ldots, k/k = 1$. A data and safety monitoring board (DSMB) might meet every year, but that does not necessarily mean there will be equal amounts of information between successive looks. For instance, in an event-driven trial using the logrank statistic, information is defined in terms of number of

TABLE 15.5

O'Brien-Fleming Z-score boundaries for k analyses and one-tailed $\alpha = 0.025$ (two-tailed $\alpha = 0.05$). One-tailed p-values corresponding to the boundaries are shown in parentheses.

k	1	2	3	4	5
1	1.960 (.025)				
2	2.796 (.003)	1.977 (.024)			
3	3.471 (.000)	2.454 (.007)	2.004 (.023)		
4	4.048 (.000)	2.862 (.002)	2.337 (.010)	2.024 (.021)	
5	4.562 (.000)	3.226 (.001)	2.634 (.004)	2.281 (.011)	2.040 (.021)

events, and the event rate may be higher at the end of the trial when patients are older. If the information fractions are close to $1/k, 2/k, \ldots, k/k = 1$, the O'Brien-Fleming and Pocock boundaries should still be accurate. Otherwise, there may be a problem.

One method that does not require equal spacing of looks is the Haybittle-Peto boundary mentioned earlier (Haybittle, 1971). The modified verion of this boundary is as follows. Reject the null hypothesis at an interim analysis if the nominal two-tailed p-value (i.e., the ordinary p-value that does not adjust for multiple comparisons) is 0.001 or less. If there are $k-1$ interim analyses before the final one, then the cumulative type I error rate used by the penultimate analysis is at most $(k-1)(0.001)$ by the Bonferroni inequality. Therefore, the p-value required to declare statistical significance at the final analysis is $\alpha - (k-1)(0.001)$. For instance, if there are 3 interim analyses before the final analysis and the two-tailed α is 0.05, the final two-tailed p-value must be $0.05 - 3(0.001) = 0.047$. This oldest of the group-sequential boundaries has important advantages. First, it is valid regardless of the joint distribution of test statistics because it is based on the Bonferroni inequality, which always holds. Second, the early boundaries are high, and the final boundary is close to what would be used if there were no monitoring. Third, the analyses do not have to be equally-spaced in terms of information. The only drawback is that the final boundary drops precipitously, so a Z-score of 2.80 is not statistically significant even near the end of the trial, but is well above the boundary at the end. This can cause logical inconsistencies. For instance, the number of new deaths between the penultimate and final analyses could be higher in the treatment arm, yet the final Z-score is well above its boundary. How could the evidence be convincing only at the end, when in fact it was even stronger earlier?

Figure 15.1 graphically summarizes the classic upper monitoring boundaries of O'Brien and Fleming (1979), Haybittle (1971), and Pocock (1977) for a trial with 5 equally-spaced looks. The O'Brien-Fleming and Haybittle-Peto boundaries are desirable because they are high early in the trial, and near 1.96 (the dashed line) at the end of the trial. The Pocock final boundary is not close to 1.96, and is therefore eschewed in clinical trials. The maximum sample size to achieve a given power level is also lower with O-Brien Fleming than with Pocock boundaries, although the expected sample size if treatment is quite effective is lower for Pocock. For example, suppose we want 90% power for a given treatment effect in a trial with 5 equally-spaced looks. Relative to a trial with no interim monitoring, the increase in maximum sample size required is 20.8% for Pocock and only 2.7% for O'Brien-Fleming. On the other hand, if the true effect is as anticipated, the expected sample size multiplier relative to a trial with no monitoring is 0.648 for O'Brien-Fleming and 0.565 for Pocock.

Lan and DeMets (1983) devised a method of generating boundaries that does not require equally-spaced information fractions. In fact, the *alpha spending function* method does not even require pre-specification of the number of interim analyses. Instead, one specifies a function $\alpha^*(t)$ that specifies the amount of alpha to spend by different information fractions.

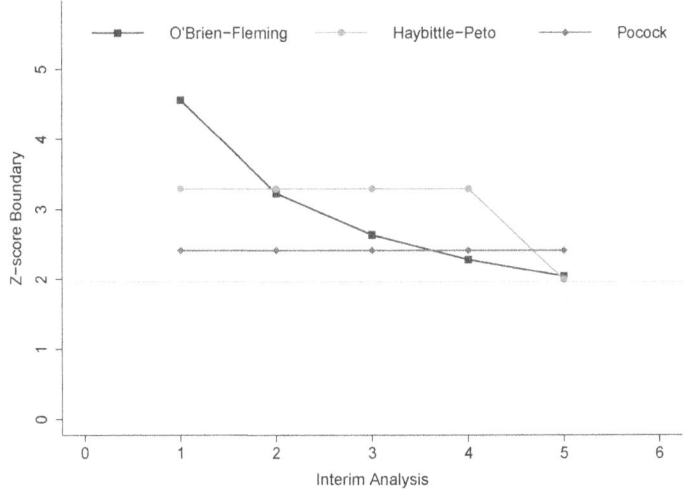

FIGURE 15.1
O'Brien-Fleming, Haybittle-Peto, and Pocock upper boundaries for 5 equally-spaced looks.

TABLE 15.6
Sample size multiplier to achieve 90% power for Pocock, O'Brien-Fleming, and Haybittle-Peto with 5 equally-spaced looks relative to the sample size for a trial with no monitoring that achieves 90% power.

	Pocock	**O'Brien-Fleming**	**Haybittle-Peto**
Maximum sample size ratio	1.208	1.027	1.021
Expected sample size ratio	0.565	0.648	0.755

The only requirements are that $\alpha^*(0) = 0$ (i.e, no alpha is spent before the trial starts), $\alpha^*(1) = \alpha$ (i.e., all of the alpha is spent by the end of the trial), and $\alpha^*(t)$ is an increasing function of t. Using this pre-specified spending function, we generate boundaries as they are needed.

We illustrate the method by constructing a one-tailed 0.025 level boundary using the linear spending function $\alpha^*(t) = 0.025t$. Suppose that the first interim analysis is at information fraction $t = t_1 = 0.30$. The amount of alpha we are allowed to spend is $\alpha^*(0.30) = (0.025)(0.30) = 0.0075$. Thus, the Z-score boundary c_1 at the first analysis must be such that $P(Z(0.30) > c_1) = 0.0075$. Using a standard normal table, we find that $c_1 = 2.432$. Suppose that the second look is at $t = 0.70$. The cumulative amount of α to be spent by $t = t_2 = 0.70$ is $\alpha^*(0.70) = (0.025)(0.70) = 0.0175$. The cumulative crossing probability by t_2 is the probability of crossing at t_1 plus the probability of not crossing at t_1, but then crossing at t_2. Thus, we want

$$P\{Z(t_1) > c_1\} + P\{Z(t_1) \leq c_1, Z(t_2) > c_2\} = \alpha^*(t_2), \text{ so}$$
$$\alpha^*(t_1) + P\{Z(t_1) \leq c_1, Z(t_2) > c_2\} = \alpha^*(t_2);$$
$$\text{i.e., } P\{Z(t_1) \leq c_1, Z(t_2) > c_2\} = \alpha^*(t_2) - \alpha^*(t_1). \quad (15.9)$$

We must determine c_2 to satisfy $P(Z(0.30) \leq 2.432, Z(0.70) > c_2) = 0.0175 - 0.0075 = 0.010$. Using the joint distribution of Z-scores derived earlier, we find from either numerical integration or simulation that $c_2 = 2.253$.

Suppose that neither the first nor second interim boundary is crossed. At the third analysis at t_3, the cumulative type I error rate is $\alpha^*(t_3) = 0.025t_3$. The cumulative crossing probability by t_3 is the probability of crossing by t_2 plus the probability of not crossing by t_2, but crossing at t_3. The former probability is $\alpha^*(t_2)$, while the latter probability is $P(Z(t_1) \leq c_1, Z(t_2) \leq c_2, Z(t_3) > c_3)$. Equating the sum of these probabilities to $\alpha^*(t_3)$, we get

$$\alpha^*(t_2) + P(Z(t_1) \leq c_1, Z(t_2) \leq c_2, Z(t_3) > c_3) = \alpha^*(t_3); \text{ i.e.,}$$
$$P(Z(t_1) \leq c_1, Z(t_2) \leq c_2, Z(t_3) > c_3) = \alpha^*(t_3) - \alpha^*(t_2).$$

Thus, $P(Z(0.30) \leq 2.432, Z(0.70) \leq 2.253, Z(1) > c_3) = .025 - .0175 = .0075$. Using numerical integration, we find that $c_3 = 2.209$.

More generally, having failed to cross the boundary at looks $1, \ldots, i-1$, we determine the boundary c_i such that

$$P(Z(t_1) \leq c_1, \ldots, Z(t_{i-1}) \leq c_{i-1}, Z(t_i) > c_i) = \alpha^*(t_i) - \alpha^*(t_{i-1}).$$

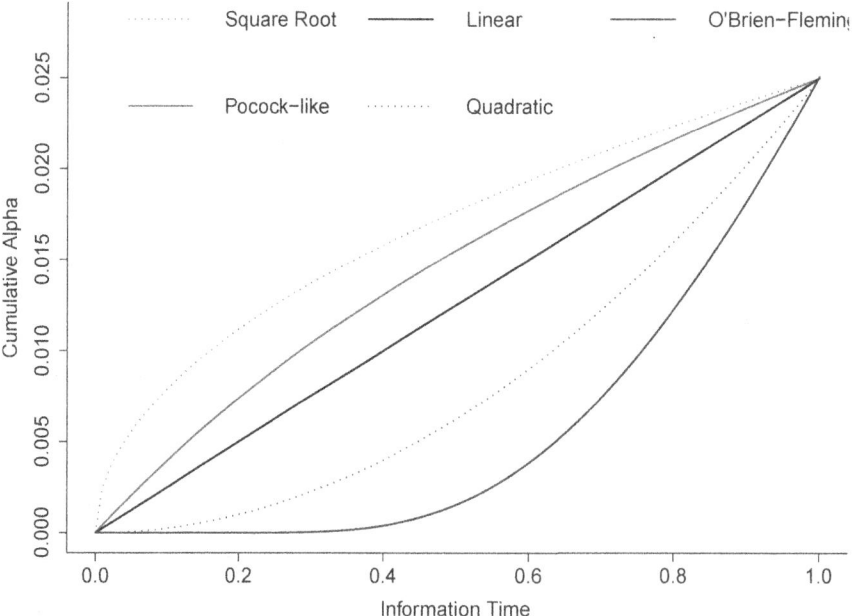

FIGURE 15.2
Different spending functions. The blue functions spend little α early and rise rapidly toward the end of the trial. They are much more desirable than the red spending functions, which should be avoided.

The properties of alpha spending function boundaries depend on the selected spending function, among which there are infinitely many. Therefore, what makes a good alpha spending function? Consider the power family (Kim and DeMets, 1987)

$$\alpha^*(t) = \alpha t^\rho, \tag{15.10}$$

where $\rho \geq 0$ (see Figure 15.2). We have already seen one member of this family, the linear spending function corresponding to $\rho = 1$. Its Z-score boundary at the end of the trial was larger than we would like in the example above: 2.209. The quadratic spending function corresponding to $\rho = 2$ results in a lower spending rate early and a higher spending rate late in the trial. This is because when t is small, $\alpha^*(t) = \alpha t^2$ is tiny. The quadratic spending function rises steeply at the end. The result is that the final boundary is closer to 1.96 than it would be with the linear spending function. For example, in the example above with looks at $t = 0.30, 0.70$, and 1, the Z-score boundaries for the quadratic spending function are 2.841, 2.296, and 2.069. The final boundary of 2.069 is much closer to 1.96 than is the final linear boundary of 2.209.

One of the most commonly used spending functions for a one-tailed test at level α is

$$\alpha^*(t) = 2\{1 - \Phi(z_{\alpha/2}/t^{1/2})\}, \tag{15.11}$$

where $z_{\alpha/2}$ is the upper $\alpha/2$ point of a standard normal distribution function. For a one-tailed test at level 0.025, $z_{\alpha/2} = z_{.0125} = 2.241$, so (15.11) is $2\{1 - \Phi(2.241/t^{1/2})\}$ (see Figure 15.2). This function spends very little alpha early. That is, $\alpha^*(t)$ is tiny until t reaches approximately $1/2$, at which point $\alpha^*(t)$ ascends more steeply. As a result, early boundaries are quite high, whereas late boundaries are close to what would be used if there were no monitoring.

Spending functions that spend alpha at a rapid pace early in the trial have the opposite effect: early boundaries are not as high as they are with, say, the O'Brien-Fleming-like spending function, whereas late boundaries are high. For instance, the following function generates boundaries similar to Pocock's when the interim analyses are equally-spaced with respect to information time.

$$\alpha^*(t) = \alpha \ln\{1 + (e - 1)t\} \tag{15.12}$$

(see Figure 15.2). Because it is undesirable to have relatively low early boundaries and relatively high late boundaries, spending functions like (15.12) should be avoided.

15.2 The Effect of Monitoring on Power

Standard conservative monitoring methods such as Haybittle-Peto, O'Brien-Fleming, and O'Brien-Fleming-like spending functions, require a very small increase in sample size to achieve the same power as a trial with no monitoring. We saw this in Table 15.6 for the O'Brien-Fleming boundary: we needed to increase the sample size by only about 3%, relative to a trial with no monitoring, to achieve the same 90% power.

Why does this happen? In a large trial with no monitoring, the approximate sample size to achieve power $1 - \beta$ is found by solving

$$E(Z) = (z_{\alpha/2} + z_\beta) \tag{15.13}$$

for sample size in a t-test or test of proportions setting, or number of events in the survival setting. Here $E(Z)$ is the mean of the Z-score under the alternative hypothesis of interest. Now suppose we use the following conservative approximation for sample size in a trial with monitoring. Let c be the boundary at the planned end of the trial. Power in the monitored trial is at least as large as if we solve

$$E(Z) = (c + z_\beta). \tag{15.14}$$

But this sample size will only be slightly larger than the solution to (15.13) because c is only slightly larger than $z_{\alpha/2}$ for boundaries like O'Brien-Fleming, Haybittle-Peto, or conservative spending functions. We see in the next section that conservative futility analyses also entail minimum loss in power. These statements do not hold for very aggressive efficacy or futility boundaries.

15.3   Futility/Stochastic Curtailment

Another aspect of monitoring clinical trials is futility analysis. We focus mainly on conditional power and stochastic curtailment. At the end, we discuss beta spending functions. It may become apparent at an interim analysis that continuation will not lead to a statistically significant result. Two questions arise for futility analysis:

1. Given the data we have observed at an interim analysis, how likely is it that we will have a null result at the end of the trial?

2. If we do have a null result at the end, will it answer the original question?

Two different tools are used for these two questions: conditional power is used for the first, and revised unconditional power is used for the second. We will see how important it is to compute both conditional and revised unconditional power when making futility decisions.

Before detailing the two powers mentioned above, we consider some practical considerations. Many things can go wrong in a clinical trial, including

1. Poorer than expected recruitment.

2. Higher than expected rates of noncompliance or missing data.

3. A lower than expected event rate.

4. A smaller than expected treatment effect.

The first three items jeopardize the ability of the trial to answer the original question, whereas the fourth item answers the original question. Therefore, we must think about the first three items very differently from the last one. It is useful to imagine going back in time and asking the question: if we had known at the trial planning stage the actual recruitment, event rate, and rates of noncompliance and missing data that we know now, would we have undertaken the trial? To answer that question, we compute *revised unconditional power.* We recompute the power we had at the beginning of the trial, using our original treatment effect and revised, within-trial estimates of the required nuisance parameters. If revised unconditional power is low, our trial may not be able to answer its original question.

A low value of revised unconditional power is not necessarily the death knell for a trial. For instance, suppose that the overall or control event rate is much lower than expected, resulting in reduced power for detecting a 25% relative reduction due to treatment. That is not a problem if the observed treatment effect is much larger than expected. It would make no sense to stop for futility on the basis of low revised unconditional power without considering the observed treatment effect. A complementary tool to revised unconditional power is *conditional power,* namely the conditional probability of a statistically significant result at the end of the trial given the data observed by the time of our interim analysis.

Conditional power is very different from unconditional power in at least two ways. Conditional power can be computed only by someone with access to data on the treatment effect. Revised unconditional power does not require looking at the treatment effect. In fact, one could compute revised unconditional power on the basis of the overall data without knowledge of treatment assignments. The other big difference is in how they are used. Conditional power tells us how likely a null result is, given the data thus far. Revised unconditional power tells us whether a null result would still answer the original question. If conditional and unconditional power are both low, that tells us that a null result is likely and will not be informative. That is, if revised unconditional power is low, a null result will not rule out the originally hypothesized treatment effect. Therefore, if conditional and revised unconditional power are both low, one may want to stop the trial. On the other hand, we definitely would not want to stop if conditional power is high. The more problematic case is when conditional power is low and revised unconditional power is still high. In some settings, we may want to continue the trial to demonstrate that the treatment does not work. This might apply if the trial is (1) in a condition whose consequences are not life-threatening, e.g., a short-term diet trial or a trial of a treatment to reduce arthritis pain, and (2)treatment is in widespread use despite the lack of prior randomized trial evidence. Under those circumstances, it might be worthwhile to continue the trial to demonstrate that the putative treatment actually does not work. If, on the other hand, the disease has serious consequences, it would not be ethical to continue a trial just to prove that a treatment does not work, especially if that treatment had substantial side effects.

We next consider the actual calculation of unconditional and conditional power. Unconditional power is intimately connected to the expected Z-score at the end of the trial, $\theta = E\{Z(1)\}$. Many standardized test statistics are asymptotically normally distributed with unit variance. The following equation can be used to derive approximate power, sample size, or detectable effect

$$\theta = z_\alpha + z_\beta, \tag{15.15}$$

where z_β denotes the upper β point of a standard normal curve and power is $1 - \beta$. If we want 90% power for a one-tailed test at level $\alpha = 0.025$, then $z_\alpha = 1.96$, $1 - \beta = 0.90$, and $z_\beta = z_{0.10} = 1.28$. Therefore, θ must be $1.96 + 1.28 = 3.24$. For 85%, or 80% power, $\theta = 1.96 + 1.04 = 3.00$, or $1.96 + 0.84 = 2.80$, respectively. For example, the expected t-statistic with n patients in each arm is approximately $\theta = \delta/(2\sigma^2/n)^{1/2}$ under the hypothesis that the between arm difference in means is δ. Equating this to $z_\alpha + z_\beta$ and solving for n gives us the approximate per-arm sample size formula

$$n = 2\sigma^2(z_\alpha + z_\beta)^2/\delta^2. \tag{15.16}$$

For the logrank statistic (15.6), we pointed out that (15.7) is a good estimate of δ, the log hazard ratio, while the denominator of (15.6) is approximately $(D/4)^{1/2}$. Accordingly, the expected value of the logrank Z-statistic (15.6) is $\theta \approx \delta \sqrt{D/4}$. Equating this to $z_\alpha + z_\beta$ yields the number of events required for power $1 - \beta$ when the true log hazard ratio is δ:

$$D = 4(z_\alpha + z_\beta)^2/\delta^2. \tag{15.17}$$

Example 15.4 In a clinical trial with change in diastolic blood pressure as primary outcome, we expect the standard deviation to be $\sigma = 5$, and we want 85% power to detect a between-arm difference of 3. The per-arm sample size using (15.16) is $n = 2(5^2)(1.96 + 1.04)^2/3^2 = 50$. Suppose that at an interim analysis with $n = 30$ participants per arm, the pooled variance is $s^2 = 64$. In other words, our best estimate of the common standard deviation is 8, which exceeds the pre-trial estimate of 5. What is the actual power of our trial using the revised estimate of σ? The revised expected Z-score at the end of the trial is only $\theta = \delta/(2\sigma^2/n)^{1/2} = 3/(2 \cdot 64/50)^{1/2} = 1.875$ instead of $\theta = 3$ required for 85% power. Substitute into equation (15.15) to find that $1.875 = 1.96 + z_\beta$, so $z_\beta = -0.085$. Take Φ of both sides to find that $1 - \beta = \Phi(z_\beta) = \Phi(-0.085) = 0.47$. In other words, our revised estimate of power is 47%. We would never have planned a trial with 47% power. Consequently, if we get a null result at the end, it will not be very meaningful because it will not rule out the originally hypothesized treatment effect of $\delta = 3$. ∎

Example 15.5 In a clinical trial with coronary heart disease events as primary outcome, we expect treatment to reduce the event rate by 25%. That is, the treatment-to- control hazard ratio is 3/4. We re-parameterize so that large values of the Z-statistic indicate that the treatment is beneficial. Thus, we use the control-to-treatment hazard ratio of $1/(3/4) = 4/3$. The required number of events for 90% power is $D = 4(1.96 + 1.28)^2/\{\ln(4/3)\}^2 \approx 508$. Using the anticipated control event rate, we determine the number of patients and duration of followup to reach 508 events. At an interim analysis, we realize that we initially overestimated the event rate. Unfortunately, this is common. We now realize that our trial will only have about 380 events instead of 508. What is the power of the trial with 380 events instead of 508? The expected Z-score with 380 events is approximately $\theta = \delta(D/4)^{1/2} = \ln(4/3)(380/4)^{1/2} = 2.804$. To determine the revised power, return to equation (15.15): $2.804 = 1.96 + z_\beta$, so $z_\beta = 0.844$ and $1 - \beta = \Phi(0.844) = 0.80$. That is, our revised estimate of power is 80% instead of 90%. Although power is diminished, it is still reasonably high. A null result will be somewhat meaningful, unlike in Example 15.4. ∎

In Examples 15.4 and 15.5, we evaluated one key component of futility analysis, namely revised unconditional power. We found that a null result would not be meaningful in Example 15.4, but would be meaningful in Example 15.5 because power is still reasonably high. Does this mean that we should abandon trial 1 and continue trial 2? We are missing a key piece of information, namely whether, considering the current results, we are likely to get a null result at the end. If not, then it is immaterial whether a null result would be meaningful. Therefore, we want to compute conditional power, the probability of a statistically significant result at the end of the trial, given the observed results by arm so far. To compute conditional power, we use the B-value formulation and exploit the fact that the B-value has independent increments. At an interim analysis at information fraction t, we have observed the standardized statistic $Z(t)$. We convert this to a B-value by multiplying by $t^{1/2}$. Conditional power is the conditional probability that $Z(1)$ exceeds its critical value c, given that $B(t) = b$. Note that $Z(1)$ is the same as $B(1)$. Thus, we need to find the conditional distribution of $B(1)$ given $B(t) = b$. We know that $B(t)$ and $B(1) - B(t)$ are independent. Also, $B(1) - B(t)$ is normally distributed with mean $\theta \cdot 1 - \theta \cdot t = \theta(1 - t)$ and variance $\mathrm{var}\{B(1)\} + \mathrm{var}\{B(t)\} - 2\mathrm{cov}\{B(1), B(t)\} = 1 + t - 2t = 1 - t$. Therefore, conditional power can be computed as follows.

$$
\begin{aligned}
P\{B(1) > c \mid B(t) = b\} &= P\{B(1) - B(t) > c - b \mid B(t) = b\} \\
&= P\{B(1) - B(t) > c - b\} \\
&= 1 - \Phi\left\{\frac{c - b - \theta(1 - t)}{\sqrt{1 - t}}\right\} \\
&= \Phi\left\{\frac{\theta(1 - t) + t^{1/2}z(t) - c}{\sqrt{1 - t}}\right\}.
\end{aligned}
\tag{15.18}
$$

For commonly used boundaries, the final critical value c will be close to the nominal level with no monitoring, 1.96.

It is common to compute conditional power under various assumptions about the treatment effect. For instance, we could use the originally hypothesized effect or the current observed treatment effect, or something in between. We prefer to use the originally hypothesized effect for futility decisions. The idea is that if conditional power is low even assuming the original effect, which might now look optimistic, that is a good sign that the trial may be futile. This raises another question: how low must conditional power be to stop a trial? This is clearly subjective, but it is common to set a threshold of somewhere between 10% and 20%. Stopping when conditional power drops below a fixed threshold like 15% is called *stochastic curtailment,* although some people erroneously use the terms stochastic curtailment and conditional power interchangeably.

Example 15.6 Return to Example 15.4. Suppose that at the interim analysis with 30 per arm, the treatment effect and standard deviation are $\hat{\delta} = \bar{Y}_C - \bar{Y}_T = 3.5$ and $s = 8$, respectively. The information fraction is $t = 30/50 = 0.60$. The expected Z-score at the end of the trial using standard deviation 8 is $\theta = 3/\{2(8^2)/50\}^{1/2} = 1.875$. The current Z-score is $Z(0.60) = 3.5/\{2(8^2)/30\}^{1/2} = 1.694$. Therefore, conditional power using formula (15.18) is $\Phi[\{1.875(1 - 0.60) + (0.60)^{1/2}(1.694) - 1.96\}/(1 - 0.60)^{1/2}] = \Phi(0.162) = 0.56$. This conditional power is not close to values typically considered futile, namely 0.10 – 0.20. We still have a greater than 50% chance of getting a statistically significant result. Therefore, we probably would not seriously consider futility stopping, even though the revised unconditional power is low. ∎

Example 15.7 Return to Example 15.5. At an interim analysis with 190 events, we realize the folly of our original event rate estimate. We change the target number of events to 380. Therefore, the current information fraction is $190/380 = 0.50$. The expected Z-score at trial's end was computed in Example 15.5 as $\theta = 2.804$. Suppose that the current logrank Z-score is actually in the wrong direction, $Z(0.50) = -0.75$. Conditional power is $\Phi[\{2.804(1-0.50) + (0.50)^{1/2}(-0.75) - 1.96\}/(1-0.50)^{1/2}] = \Phi(-1.539) = 0.06$. Note that this conditional power assumes the original treatment effect estimate of a 25% reduction in coronary heart disease events. Given the current results, that estimate seems suspect. Nonetheless, even under that optimistic assumption, there is only about a 6% chance of observing a statistically significant result by the end of the trial. This strongly suggests that continuation is futile.

If this example had involved a less serious disease with non life-threatening consequences and the treatment had been in widespread use despite the lack of any clinical trial data to support it, it may have been worthwhile to continue the trial to prove that the treatment does not work. This is not an ethical option for a serious disease like CHD. It would not be ethical to continue the trial just to prove that treatment was harmful. ∎

One potential concern about stochastic curtailment is that the type II error rate increases. After all, we might stop for futility when, if we had continued, we might have obtained a statistically significant test statistic at the end of the trial. The following result of Lan, Simon, and Halperin should alleviate those concerns.

Result 15.2 *(Lan, Simon, and Halperin, 1982) Suppose that stochastic curtailment is used with conditional power threshold ϵ; i.e., the trial is stopped if conditional power ever drops below ϵ. If the type II error rate without monitoring is β, then the type II error rate using stochastic curtailment is at most $\beta/(1 - \epsilon)$ irrespective of the number of interim analyses. The type II error rate is exactly $\beta/(1 - \epsilon)$ if the trial is monitored continuously.*

For example, suppose that the stochastic curtailment rule stops whenever conditional power (under the original hypotheses) drops below 0.10 (i.e., $\epsilon = 0.10$). If the trial with no monitoring had 85% power (i.e., $\beta = 0.15$), power with monitoring is at least $1 - 0.15/0.90 = 0.83$.

Other tools for futility monitoring can also be used, such as the triangular test and beta spending functions. We give a very brief description of these methods. The *triangular test* (Whitehead and Stratton, 1983) is similar in spirit to the sequential probability ratio test, except that it involves two groups. Linear upper and lower boundaries for the score statistic meet at the end of the trial. At an interim analysis, we declare efficacy (respectively, futility) if the score statistic crosses the upper (resp., lower) boundary. The "continuation" region between the lower and upper boundaries forms a triangle, hence the name. It has been proposed in the context of emerging infectious diseases to test multiple new treatments with unknown efficacy because the expected sample size can be substantially lower than with more traditional, conservative boundaries. One reason for smaller expected sample sizes is that the lower boundary is binding and is used to decrease the upper boundary and maintain the correct type I error rate. Regulatory agencies tend not to favor binding lower boundaries. DSMBs often regard boundaries as guidelines because the decision to stop a trial involves other considerations as well. Futility boundaries are considered especially advisory. If a DSMB does not stop the trial when the lower boundary is crossed, the type I error rate of the triangular test is inflated.

A *beta spending function* $\beta^*(t)$ specifies the cumulative type II error rate by time t (Pampallona, Tsiatis, and Kim 2001); $\beta^*(1)$ is β, the targeted type II error rate (e.g., $\beta = 0.10$ for 90% power). At analysis i with information fraction t_i, a lower boundary L_i is selected such that the cumulative type II error rate by time t_i, computed under the alternative hypothesis, is $\beta^*(t_i)$. Equivalently, the probability of first exiting the lower boundary at time t_i,

$$P\left(L_1 \leq Z(t_1) \leq c_1, \ldots, L_{i-1} \leq Z(t_{i-1}) \leq c_{i-1}, Z(t_i) < L_i\right),$$

is set to $\beta^*(t_i) - \beta^*(t_{i-1})$. The final boundary L_k can be chosen to be c_k, the final efficacy boundary, so a unique conclusion is guaranteed at the end of the trial. The beta spending function can be as aggressive or conservative as desired. Aggressive means spending beta quickly to allow earlier stopping if treatment is ineffective. This can be attractive in a multi-armed trial when no effective treatments are known and there is an imperative to rapidly find a treatment that works. The penalty to pay for aggressive monitoring is an increase in maximum sample size. Still, the expected sample size can be favorable because unfavorable treatments are weeded out early. As with the triangular test, we could use the fact that there is a lower boundary to reduce the upper boundary and still maintain the type I error rate. We recommend against this practice because, as mentioned in the above discussion of the triangular test, DSMBs often regard futility boundaries as advisory rather than binding.

15.4 Problems with Post-Trial Inference

An entire book could be written on the topic of inference following a trial that uses group-sequential monitoring. Instead, we give a very brief introduction to the difficulties caused by monitoring. The main problem is that methods that fail to account for monitoring tend to overstate the level of evidence of a treatment effect. To see this, forget about monitoring and consider two independent, unbiased estimators, $\hat{\delta}_1$ and $\hat{\delta}_2$, of a parameter δ. Suppose that if $\hat{\delta}_1$ is large enough, we report $\hat{\delta}_1$, but if $\hat{\delta}_1$ is not large enough, we report $\hat{\delta}_2$ as our estimate of δ. Even though $\hat{\delta}_1$ and $\hat{\delta}_2$ are both unbiased, our procedure tends to overestimate δ. Suppose we modify our procedure by reporting $\hat{\delta}_1$ if it is large enough, and reporting the average, $(\hat{\delta}_1 + \hat{\delta}_2)/2$, if $\hat{\delta}_1$ is not large enough. This still "cherry picks" and tends to overestimate δ. This modified procedure is equivalent to monitoring a trial at two equally-spaced looks; if the interim result is large enough (i.e., above the boundary), it will be reported, but if it is not large enough, the trial continues and the final estimate is the average of the independent estimates from the first and second halves. Methods of reducing bias caused by monitoring are discussed in Chapter 8 of Jennison and Turnbull (2000) and Chapter 7 of Proschan, Lan, and Wittes (2006).

A p-value calculated without accounting for monitoring also tends to overstate the level of evidence for a treatment effect. We can account for monitoring by extending the usual definition of a p-value as the null probability of a result at least as extreme, in the direction of the alternative hypothesis, as the observed result. But the sufficient statistic following a monitored trial is the pair $(\tau, Z(\tau))$, where τ is the information fraction when the trial was stopped (either early or at the planned end) and $Z(\tau)$ is the Z-statistic at that time (see page 172 of Jennison and Turnbull, 2000 or page 114 of Proschan, Lan, and Wittes, 2006). There is not a unique way to define "at least as extreme" when the sufficient statistic is a pair. We might order the sample space in terms of the Z-score, ignoring the time the trial was stopped. Alternatively, we might treat earlier stopping as more extreme than later stopping, and for two outcomes that stop at the same time, we order by the Z-score. This is known as the *stagewise ordering*. There are several other orderings, including by the value of the maximum likelihood estimate of the treatment effect.

To illustrate one advantage of the stagewise ordering over competitors, consider a trial with 3 equally spaced looks, and suppose the observed Z-score at the first look is 3.5, crossing its boundary. The p-value using the Z-score ordering is

$$P\{Z(\tau) \geq 3.5\} = P\{\tau = 1/3 \text{ and } Z(1/3) \geq 3.5\} + P\{\tau = 2/3 \text{ and } Z(2/3) \geq 3.5\}$$
$$+ P\{\tau = 1 \text{ and } Z(1) \geq 3.5\}.$$

It seems peculiar that to determine how much evidence there is at the first analysis, we need to compute probabilities of future events! For all we know, if the trial had continued, the second analysis might have taken place at information fraction 0.77 instead of 0.67. Knowing that would change the p-value. The stagewise p-value p_{sw} can be computed without knowing the future:

$$
\begin{aligned}
p_{sw} &= P(\text{stopping earlier}) + P\{\text{not stopping earlier but } Z(1/3) \geq 3.5\} \\
&= 0 + P\{Z(1/3) \geq 3.5\} = 0.0002.
\end{aligned}
\tag{15.19}
$$

Incidentally, the stagewise p-value in this example is the same as in a single-look trial with the same Z-score. This is always true when the trial stops at the first analysis.

P-values can also be computed for testing nonzero null hypotheses, irrespective of how the boundaries were constructed. In other words, even if boundaries were constructed for testing the zero null hypothesis, we can still compute a p-value for a nonzero null. For example, with the stagewise ordering, we use the nonzero null to compute the probability of either crossing the boundary earlier, or not crossing earlier, but achieving a more extreme current Z-score. A p-value less than α provides evidence that the true effect exceeds the null value. Moreover, we can invert such tests to construct confidence intervals for the treatment effect. See Chapter 8 of Jennison and Turnbull (2000) or Chapter 7 of Proschan, Lan, and Wittes (2006) for more details.

15.5 Conclusions

Group Sequential monitoring of clinical trials allows early stopping for efficacy, harm, or futility. Many different testing scenarios, such as comparisons of means, proportions, or survival, can be unified into the same Brownian motion framework. Therefore, the same boundaries can be used in these diverse settings. Popular efficacy boundaries are quite high early in the trial, allowing the final boundary to be close to what it would be with no interim monitoring. Consequently, such boundaries necessitate only a minimal increase in sample size relative to a trial with no monitoring. Alpha spending functions provide flexible monitoring that does not require the number or timing of interim analyses to be pre-specified, although it is good practice to do so.

Conditional and unconditional power are useful methods of monitoring for futility. The conditional probability of a statistically significant result at trial's end, given the current data, is a very natural metric. Conditional and revised unconditional power can be used together; conditional power tells us whether a null result is likely, while revised unconditional power tells whether such a null result would still be meaningful. Beta spending functions can also be used for futility. One can choose conservative futility monitoring that stops only if results are quite bleak, or use more aggressive futility monitoring. Like aggressive efficacy monitoring, aggressive futility monitoring requires a substantially increased maximum sample size. Nonetheless, the expected sample size is favorable when the true treatment effect is negligible. This can be appealing in an emerging disease setting with no proven, but multiple potential, treatments because ineffective treatments are quickly weeded out.

Inference following a trial that uses group Sequential monitoring is complicated by two factors: (2) the trial may have stopped on a "random high" and (2) the sufficient statistic consists of both the information fraction when the trial was stopped and the Z-score at that time. Factor 1 means that ignoring the monitoring leads to a biased treatment effect estimate and a potentially misleading p-value. Factor 2 means that if we do account for interim monitoring when computing a p-value, there are different ways to order outcomes in terms of strength of evidence.

Bibliography

[1] Armitage, P., McPherson, C.K., and Rowe, B.C. (1969). Repeated significance tests on accumulating data. *Journal of The Royal Statistical Society A* **132**, 235–244.

[2] Cardiac Arrythmia Suppression Trial Investigators (1989). Preliminary report: Effect ofencainide and flecainide on mortality in a randomized trial of arrhythmia suppression after myocardial infarction. *The New England Journal of Medicine* **321**, 406–412.

[3] DeLong, D.M. (1981). Crossing probabilities for a square root boundary by a Bessel process. *Communications in Statistics–Theory and Methods A* **10**, 2197–2213.

[4] DeMets, D.L., Friedman, L.M., Furberg, C.D. (2006). *Data Monitoring in Clinical Trials*. Springer, New York.

[5] Ellenberg, S.S., Fleming, T.R., and DeMets, D.L. (2003). *Data Monitoring Committees in Clinical Trials: A Practical Perspective*. John Wiley and Sons, New York.

[6] Haybittle, J.L. (1971). Repeated assessment of results in clinical trials of cancer treatment. *British Journal of Radiology* **44**, 793–797.

[7] Herson, J. (2009). *Data and Safety Monitoring Committees in Clinical Trials*. Chapman and Hall/CRC, Boca Raton, Fl.

[8] Jennison, C. and Turnbull, B.W. (1997). Group-sequential analysis incorporating covariate information. Journal of The American Statistical Association **92**, 1330–1341.

[9] Jennison, C. and Turnbull, B.W. (2000). *Group Sequential Methods with Applications to Clinical Trials*. Chapman and Hall/CRC, Boca Raton, Fl.

[10] Kim, K. and DeMets, D.L. (1987). Design and analysis of group sequential tests based on the type I error spending rate function. *Biometrika* **74**, 149–154.

[11] Lan, K.K.G. and DeMets, D. (1983). Discrete sequential boundaries for clinical trials. *Biometrika* **70**, 659–663.

[12] Lan, K.K.G. and Wittes, J. (1988). The B-value: A tool for monitoring data. *Biometrics* **44**, 579–585.

[13] Lan, K.K.G. and Zucker, D.M. (1993). Sequential monitoring of clinical trials: The role of information and Brownian motion. *Statistics in Medicine* **12**, 753–765.

[14] Lan, K.K.G., Simon, R., and Halperin, M. (1982). Stochastically curtailed tests in long-term clinical trials. *Communications in Statistics-Sequential Analysis* **1**, 207–219.

[15] Mulangu, S., Dodd, L.E., Davey, R.T. et al. (2019). A randomized controlled trial of Ebola virus disease therapeutics. *New England Journal of Medicine* **381**, 2293–2303

[16] O'Brien, P.C. and Fleming, T.R. (1979). A multiple testing procedure for clinical trials. *Biometrics* **35**, 549–556.

[17] Pampallona, S., Tsiaitis, A.A., and Kim, K. (2001). Interim monitoring of group sequential trials using spending functions for the type I and type II error probabilities. *Drug Information Journal* **35**, 1113–1121.

[18] Pocock, S.J. (1977). Group sequential methods in the design and analysis of clinical trials. *Biometrika* **64**, 191–199.

[19] Proschan, M.A., Lan, K.K. Wittes, J.T. (2006). *Statistical Monitoring of Clinical Trials: A Unified Approach.* Springer, New York.

[20] Schoenfeld (1981). The asymptotic properties of nonparametric tests for comparing survival distributions. *Biometrika* **68**, 316-319.

[21] Whitehead, J. and Stratton, I. (1983). Group sequential clinical trials with triangular continuation regions. *Biometrics* **39**, 227–236.

16

Sample Size Re-Estimation

Tobias Mütze and Tim Friede

CONTENTS

16.1 Introduction

Sample size calculation is an essential part of planning a clinical trial since a sufficient number of subjects ensures that the trial's objectives can be achieved with high statistical confidence [14]. Clinical trials with a low statistical confidence in the results, i.e. underpowered trials, are considered to be unethical and might fail to detect an actually effective treatment [45]. Moreover, an inappropriately sized trial might increase the costs unnecessarily. The sample size of a clinical trial depends not only on the targeted rates for the type I and type II errors and the effect size but also on nuisance parameters such as the outcome variability or overall event rates. However, in practice, the nuisance parameters often cannot be specified precisely resulting in unreliable sample sizes. While the treatment effect size is also unknown, usually the clinically meaningful minimal effect size, also known as the clinically relevant effect size, is either known or can be guesstimated with a much higher precision than the actual treatment effect size, and can be considered when planning the sample size. Thus, adjusting the sample size mid-trial based on a nuisance parameter estimate can guarantee a reliable sample size for the clinically meaningful minimal effect size independent of the nuisance parameter. When the initially planned sample size is based on an unreliable guess of the treatment effect size, adjusting the sample size mid-trial based on a treatment effect estimate makes sense to assure an adequately powered clinical trial. In the first part of this chapter the focus is on adjusting the sample size based on a nuisance parameter estimate and in the second part the focus is on adjusting the sample size based on treatment effect estimates.

The lack of information about the nuisance parameter can arise from the change of (primary) endpoints between development phases or a high between-trial variability [81]. To avoid the misspecification of nuisance parameters resulting in inappropriately sized clinical trials, Wittes and Brittain (1990) [98] proposed to adjust the sample size based on nuisance parameter estimates from an internal pilot study. An internal pilot study is an integrated part of the main clinical trial, with the data from the internal pilot study being included in the final statistical analysis of the main trial. In more detail, let $n = n(\Delta, \boldsymbol{\nu})$ be the sample size as a function of the effect Δ and the nuisance parameter vector $\boldsymbol{\nu}$. Moreover, let n_P be the sample size for the internal pilot study. Then, a nuisance parameter based sample size re-estimation is conducted after n_P patients are recruited by estimating the nuisance parameter vector $\boldsymbol{\nu}$ based on the available results of the already recruited patients and plugging in the resulting estimate $\hat{\boldsymbol{\nu}}_P$ into the sample size formula. Thus, the re-estimated sample size is given by $\hat{n}_{reest} = n(\Delta, \hat{\boldsymbol{\nu}}_P)$. Sample size re-estimation in which only the nuisance parameter is estimated is referred to as nuisance parameter based sample size re-estimation. However, the re-estimated sample size $\hat{n}_{reest}$ is in general not the final sample size $\hat{n}_{final}$ because the final sample size has to be at least as large as the internal pilot study. Moreover, Wittes and Brittain (1990) [98] suggest to not reduce the initially planned sample size $\hat{n}$, i.e.

$$\hat{n}_{final} = \max\left\{\hat{n}_{reest}, \hat{n}\right\}.$$

In contrast, Birkett and Day (1994) [8] proposed to allow for reducing the initially planned sample size and, thus, the final sample size is given by

$$\hat{n}_{final} = \max\left\{\hat{n}_{reest}, n_P\right\}.$$

In practice, the number of patients in clinical studies is often limited due to financial and time constrains. Therefore, Gould (1992) [41] introduced an upper limit n_{max} for the final sample size $\hat{n}_{final}$ resulting in the final sample size

$$\hat{n}_{final} = \min\left\{\max\left\{n_P, \hat{n}_{reest}\right\}, n_{max}\right\}.$$

Independently of how the sample size is re-estimated, the number of patients to be recruited after the internal pilot study is

$$\hat{n}_{recruit} = \hat{n}_{final} - n_P.$$

If $\hat{n}_{recruit} = 0$, the recruitment is stopped. Procedures for estimating the nuisance parameter are categorized by whether or not the blindness of the results from the internal pilot study is maintained [26]. From a trial logistics perspective, unblinding the internal pilot study to re-estimate the sample size has the disadvantage that it requires an independent Data Monitoring Committee (DMC) to keep the results confidential and to ensure the integrity of the trial. In contrast, a blinded sample size re-estimation can be performed without an independent DMC and still maintain the integrity of the trial [70].

In a nuisance parameter based sample size adjustment, the treatment effect is held fixed for the initial sample size planning and the mid-trial sample size adjustment and, in particular, not to be based on data from the ongoing trial. However, information about the actual treatment effect itself is often uncertain, too. This might result in choices of treatment effects which are over-optimistic or identical to the clinically meaningful minimal effect. An over-optimistic choice of the treatment effect can result in an underpowered trial not detecting the actual treatment effect. If a clinical trial is planned with the clinically meaningful minimal effect and the treatment effect is actually larger, the clinical trial can be overpowered and take up resources unnecessarily. Therefore, it has been proposed to adjust the sample size mid-trial based on not only nuisance parameter estimates but also treatment effect estimates or to use a group sequential design. Treatment effect based sample size adjustment is always unblinded. It is worth emphasizing that re-estimating the sample size by calculating the effect estimate $\hat{\Delta}$ from the unblinded internal pilot study and simply substituting the effect Δ by an estimate $\hat{\Delta}$ in the sample size formula $n(\Delta, \nu)$ inflates the type I error rate, see for instance [51, 86].

From a regulatory perspective, sample size adjustment of an ongoing clinical trial is discussed in Section 4.4 of ICH guideline E9 and emphasis is put on how the blindness and the type I error rate are affected: "The steps taken to preserve blindness and consequences, if any, for the type I error and the width of confidence intervals should be explained." The draft guidance *Adaptive Design Clinical Trials for Drugs and Biologics* [22] lists blinded sample size re-estimation as an "generally well-understood adaptive design" [22, Section V] and recommends their application: "Sample size adjustment using blinded methods to maintain desired study power should generally be considered for most studies" [22, Section V.B]. The PhRMA working group on adaptive designs in clinical drug development [36] recommends that "[...] sample size re-estimation based on nuisance parameters should be routinely considered [...]" but also emphasizes that "sample size re-estimation should never be a substitute for adequate up-front planning [...]." In summary, blinded sample size adjustment is well-accepted from a regulatory perspective as long as the type I error is not inflated.

This chapter is structured as follows. In Section 16.2, we present nuisance parameter based sample size re-estimation with emphasis on continuous and count data. In Section 16.3, we discuss effect-based sample size re-estimation. This chapter concludes with a discussion in Section 16.4.

16.2 Nuisance Parameter Based Sample Size Re-Estimation

In the first two parts of this section we discuss examples of nuisance parameter based sample size re-estimation for normal data and for count data. An overview of other issues and recent developments in nuisance parameter based sample size re-estimation is provided in the third part.

16.2.1 Sample size re-estimation for normal data

16.2.1.1 Motivating example

We illustrate the issues of uncertainty concerning nuisance parameters in the sample size planning with the multi-center, double-blind, randomised, controlled trial reported by Mac-Donald et al. (2008) [62]. The objective of the study was the comparison of the effects of lumiracoxib and ibuprofen on blood pressure. The primary endpoint was the change from baseline to 4 weeks in blood pressure determined by ambulatory blood pressure monitoring. During the planning phase of the study, information on the outcome standard deviation was limited and uncertain with observed standard deviations ranging from 6.5 mmHg to 15 mmHg [7]. Motivated by published and sponsor internal studies, a primary endpoint standard deviation of around 11 mmHg and an assumed treatment effect of 2 mmHg were considered to calculate the initial total sample size of 1020 patients. Due to the uncertainty concerning the standard deviation, a blinded sample size adjustment was performed after 600 patients had already completed the study. During the sample size review, the standard deviation was estimated to be 8.33 mmHg which is clearly smaller than the initially considered standard deviation. Using a standard deviation of 8.33 mmHg, a total sample size of 548 patients is required to detect a difference of 2 mmHg with a power of 80% using a two-sided significance level of 5%. Since the re-estimated sample size is smaller than the number of patients which completed the study at the time of the sample size review and clearly smaller than the 787 patients who were already randomized, the recruitment was stopped after the interim sample size re-calculation. The trial reported by MacDonald et al. (2008) [62] found that the lumiracoxib significantly decreased the blood pressure compared to baseline while ibuprofen significantly increased the blood pressure compared to baseline. The reported mean difference was -5.0 mmHg (95% CI: $[-6.1, -3.8]$) which is much larger than the assumed effect of 2 mmHg.

16.2.1.2 Statistical model and sample size re-estimation

Let X_{ij} be a continuous random variable modeling the outcome of patient $j = 1, \ldots, n_i$ receiving a control ($i = 1$) or a treatment ($i = 2$). We assume that the random variable X_{ij} has mean μ_i and variance σ^2. Smaller means μ_i represent a favorable result. The hypothesis of interest is whether the difference in means is smaller than or equal to a margin δ, that is

$$H_0 : \mu_2 - \mu_1 \geq \delta \quad \text{vs.} \quad H_1 : \mu_2 - \mu_1 < \delta. \tag{16.1}$$

The margin δ is chosen to be larger than zero, $\delta > 0$, for testing non-inferiority of the treatment compared to the control and equal to or smaller than zero, $\delta \leq 0$, for testing superiority. The hypothesis is commonly tested using a one-sided Student's t-test. Let $(1 - r) : r$ with $r \in (0, 1)$ be the allocation ratio and δ^* the mean difference under the alternative H_1. Then, the sample size for a one-sided Student's t-test of the hypothesis H_0 with one-sided significance level α and power $1 - \beta$ under the mean difference $\delta^* < \delta$ is given by

$$n = \frac{1}{r(1-r)} \frac{(q_\alpha + q_\beta)^2}{(\delta^* - \delta)^2} \sigma^2. \tag{16.2}$$

Here, q_p denotes the p-quantile of a standard normal distribution. The variance σ^2 is the nuisance parameter since the randomization ratio, the margin δ, and the error rates α, β are fixed prior to planning and conducting the trial. Therefore, in this setting, nuisance parameter based sample size re-estimation aims to estimate the variance σ^2 based on results from an internal pilot study. We denote an arbitrary nuisance parameter estimator based on the results of the internal pilot study by $\hat{\sigma}_P^2$. Then, the re-estimated sample size $\hat{n}_{reest}$ is obtained by substituting the variance σ^2 in (16.2) by the estimator $\hat{\sigma}_P^2$, that is

$$\hat{n}_{reest} = \frac{1}{r(1-r)} \frac{(q_\alpha + q_\beta)^2}{(\delta^* - \delta)^2} \hat{\sigma}_P^2. \tag{16.3}$$

Depending on the estimator $\hat{\sigma}_P^2$, in some settings, such as when the internal pilot study is small, re-estimating the sample size using (16.3) results in a power some percentage points smaller than the target power. For these settings, Zucker et al. (1999) [102] proposed the inflation factor

$$\gamma = \frac{(t_{n_P-2,\alpha} + t_{n_P-2,\beta})^2}{(q_\alpha + q_\beta)^2}$$

to be applied to the re-estimated sample size $\hat{n}_{reest}$. Here, $t_{n,p}$ is the p-quantile of a t-distribution with n degrees of freedom. The adjusted re-estimated sample size is then calculated by

$$\hat{n}_{reest}^{(adj)} = \gamma \, \hat{n}_{reest}.$$

In the following, we discuss various unblinded and blinded nuisance parameter estimators and compare the resulting sample size re-estimation procedures by means of a Monte Carlo simulation study.

16.2.1.3 Unblinded nuisance parameter estimation

The most intuitive unblinded estimator is the pooled sample variance $\hat{\sigma}_{Pool}^2$ from the internal pilot study, i.e.

$$\hat{\sigma}_{Pool}^2 = \frac{1}{n_P - 2} \sum_{i,j} \left(X_{ij} - \bar{X}_{i\cdot} \right)^2,$$

where $\bar{X}_{i\cdot}$ denotes the sample mean in group $i = 1, 2$. This estimator was proposed in the original paper by Wittes and Brittain (1990) [98] which introduced the internal pilot study and four years later was studied in more detail by Birkett and Day (1994) [8]. The sample size re-estimation procedure based on the unblinded estimator $\hat{\sigma}_{Pool}^2$ inflates the type I error rate [26, 56] and is not recommended to be applied in confirmatory trials if no steps are taken to control the type I error rate.

Stein (1945) [88] proposed an unblinded sample size re-estimation method in which the variance estimator from the internal pilot study is employed in the test statistic of the final analysis. While this method is exact, it does not consider all information available at the end of the study and does require a larger sample size to obtain the same power compared to an approach which estimates the variance for the final test statistic using all data. When the sample size is re-estimated unblinded with the pooled variance estimator, the sample variance estimator at the end of the study is biased. Miller (2005) [66] calculated a bias correction for the final sample variance which reduces the type I error rate inflation in designs with unblinded sample size re-estimation.

16.2.1.4 Blinded nuisance parameter estimation

In the following, we discuss various approaches for estimating the nuisance parameter σ^2 blinded. Several approaches for estimating the nuisance parameter blinded are based on the sample variance of the pooled data, that is the data set which ignores the treatment indicator. We denote the random variables modeling the blinded results of the internal pilot study by $X_1, \ldots, X_{n_P}$ and the corresponding sample mean by $\bar{X}.$. Then, the sample variance of the pooled data is given by

$$S_P^2 = \frac{1}{n_P - 1} \sum_{i=1}^{n_P} \left(X_i - \bar{X}. \right)^2.$$

The sample variance is often referred to as "lumped variance" [102] or "one-sample variance" [57]. The one-sample variance estimator has the expected value

$$\mathbb{E}\left[S_P^2\right] = \sigma^2 + \frac{r(1-r)n_P}{n_P - 1}(\mu_1 - \mu_2)^2.$$

Thus, the one-sample variance is an unbiased estimator of the nuisance parameter σ^2 if the mean difference is zero. Otherwise, the one-sample variance is biased. Gould and Shih (1992) [92] and later Zucker et al. (1999) [102] suggested to estimate the nuisance parameter σ^2 with the bias-corrected one-sample variance where the correction is calculated under the assumed alternative, that is

$$\hat{\sigma}_{OSU}^2 = S_P^2 - \frac{r(1-r)n_P}{n_P - 1}(\delta^*)^2.$$

Thus, the estimator $\hat{\sigma}_{OSU}^2$ is unbiased on the considered alternative. However, Kieser and Friede [29, 57] showed that estimating the nuisance parameter σ^2 with the one-sample variance itself, that is $\hat{\sigma}_{OS}^2 = S_P^2$, is recommended. Xing and Ganju (2005) [99] proposed a blinded variance estimator for randomized block designs. Let k be the number of blocks, m the block size, T_j the sum of all observations for block $j = 1, \ldots, k$, and $\bar{T}.$ the mean of the k block sums. Then, the Xing-Ganju variance estimator is given by

$$\hat{\sigma}_{XG}^2 = \frac{1}{m(k-1)} \sum_{j=1}^{k} \left(T_j - \bar{T}. \right)^2.$$

The Xing-Ganju variance estimator $\hat{\sigma}_{XG}^2$ does not require unblinding since it only utilizes the block sums and not the individual observations. Moreover, the Xing-Ganju variance estimator $\hat{\sigma}_{XG}^2$ is an unbiased estimator for the nuisance parameter σ^2.

Further methods for an blinded nuisance parameter estimation include an Expectation Maximization (EM) algorithm proposed by Gould and Shih (1992) [42]. However, Friede and Kieser (2002) [23] and Waksman (2007) [92] later demonstrated the inappropriateness of said EM algorithm for sample size re-estimation.

Next, we compare the variance of the nuisance parameter estimators as presented in [26]. A small variability of the nuisance parameter estimators is desirable since it corresponds to a small variability of the re-estimated sample size. The variances of the blinded variance estimators are given by

$$\text{Var}\left[\hat{\sigma}_{OS}^2\right] = \text{Var}\left[\hat{\sigma}_{OSU}^2\right] = \frac{2\sigma^4}{n_P - 1}\left(1 + 2\frac{r(1-r)n_P}{n_P - 1}\frac{(\mu_1 - \mu_2)^2}{\sigma^2}\right),$$

$$\text{Var}\left[\hat{\sigma}_{XG}^2\right] = \frac{2\sigma^4}{n_P - m}m.$$

The variance of the Xing-Ganju variance estimator is the smallest for block size $m = 2$.

16.2.1.5   Comparison of sample size re-estimation procedures

In the following, we study the operational characteristics of the blinded and unblinded sample size re-estimation procedures and compare their performance to the fixed design's performance. We restrict ourselves to the sample size re-estimation procedures based on the most common estimators $\hat{\sigma}_{OS}^2$, $\hat{\sigma}_{XG}^2$, and $\hat{\sigma}_{Pool}^2$. The operating characteristics are the power, the re-estimated sample size distribution, and the type I error rate.

To compare the power of the sample size re-estimation designs and the fixed design, the power is simulated for scenarios in which the standard deviation σ_{Plan} in the planning phase of the trial is misspecified, i.e. it differs from the true standard deviation σ. The internal pilot study sample size n_P is 50% of the sample size n_{Plan} obtained with the standard deviation σ_{Plan} in the planning phase. The effect size, that is the mean difference $\delta^* = \mu_2 - \mu_1$ in the alternative, is motivated by the clinical trial published by MacDonald et al. (2008) [62]. In the simulations, the true mean difference is assumed to be the mean difference δ^* under the alternative. The margin is set to $\delta = 0$ such that we obtain the hypothesis testing problem $H_0 : \mu_2 \geq \mu_1$ versus $H_1 : \mu_2 < \mu_1$. A complete list of parameters can be found in Table 16.1. The total sample sizes of the fixed design for the standard deviations

TABLE 16.1
Scenarios for the power-related Monte Carlo simulation study.

Parameter	Value
One-sided significance level α	0.025
Power $1 - \beta$	0.8
Allocation ratio	1:1 ($r = 0.5$)
Effect $\delta^* = \mu_2 - \mu_1$ under H_1	$-0.24, -0.6$
Superiority margin δ	0
True standard deviation σ	1
Standard deviation σ_{Plan} in planning phase	$0.6, 0.7, \ldots, 1.3, 1.4$
Internal pilot study ratio n_P/n_{Plan}	0.5

$\sigma_{Plan} = 0.6, 1, 1.4$ and the effects -0.24, -0.6 are shown in the Table 16.2. The power of

TABLE 16.2
Total sample size n and total internal pilot study size n_P in the fixed design for standard deviation σ_P.

Effect δ^*	Standard deviation σ_{Plan}	n	n_P
-0.24	0.6	200	100
	1	548	274
	1.4	1072	536
-0.6	0.6	34	17
	1	90	45
	1.4	174	87

the designs with sample size re-estimation is compared to the power of the fixed design in Figure 16.1. The sample size is re-estimated as stated in (16.3), i.e. the re-estimated sample size is not multiplied with an inflation factor. The results for the designs with sample size re-estimation are based on 250,000 Monte Carlo replications. Figure 16.1 clearly highlights the

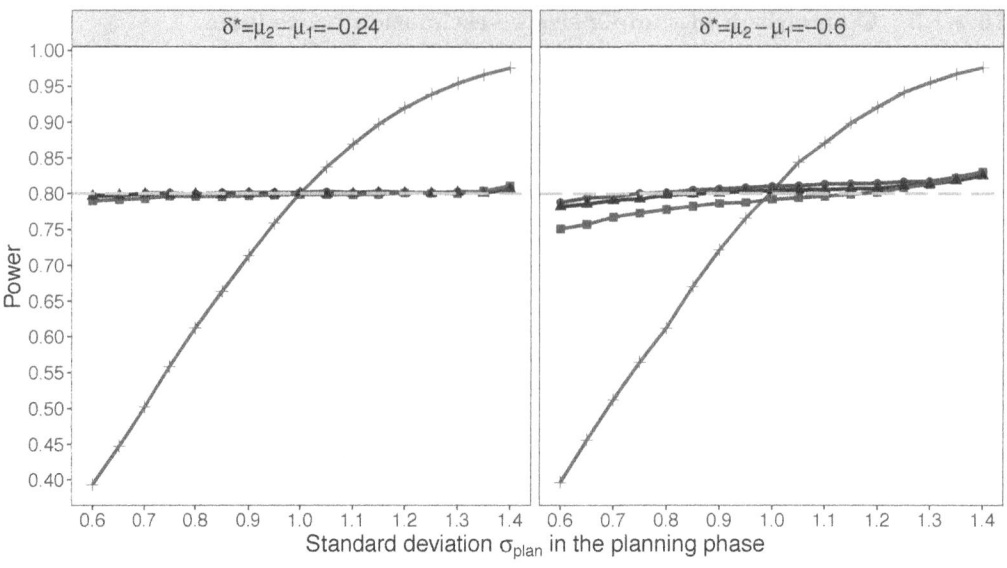

FIGURE 16.1
Power of the fixed design and the designs with blinded and unblinded sample size re-estimation procedures. The true standard deviation is $\sigma = 1$ and the true mean difference is equal to δ^*.

robustness of sample size re-estimation designs concerning misspecification of the nuisance parameter in comparison with the fixed design. Moreover, the design with blinded sample size re-estimation based on the one-sample variance estimator and the design with unblinded sample size re-estimation meet the target power exactly for the effect size of $\delta^* = -0.24$. For the same effect size, the design with blinded sample size re-estimation procedure based on the Xing-Ganju variance estimator is slightly underpowered by about one to two percentage points when the standard deviation σ_{Plan} in the planning phase is only half of the true standard deviation. For the scenario with the larger effect for which the trial size itself and consequently the internal pilot study size are smaller, the design with blinded sample size re-estimation based on the Xing-Ganju variance estimator is clearly underpowered with by five percentage points for a standard deviation of $\sigma_{Plan} = 0.5$ in the planning phase. For the same scenarios, the power of the design with a blinded sample size re-estimation based on the one-sample variance estimator is only about one to two percentage point below the target power. The power of all designs with sample size re-estimation increases in σ_{Plan}. An increasing σ_{Plan} corresponds to an increase of the internal pilot study sample size which in turn is almost as large as the required sample size of $n = 90$ resulting in overpowered designs with sample size re-estimation. The distribution of the re-estimated sample size is shown in Figure 16.2. Figure 16.2 shows that the Xing-Ganju variance estimator results in re-estimated sample size with a higher variability than the one-sample variance estimator and the pooled variance estimator. The re-estimated sample size based on the Xing-Ganju and the pooled variance estimator is unbiased, however, as Figures 16.1 and 16.2 highlight, the sample size needs to be overestimated for the design with sample size re-estimation to reach the desired power. The variability of the re-estimated sample size decreases in σ_{Plan}, i.e. the variability decreases when the size of the internal pilot study increases. Furthermore, the

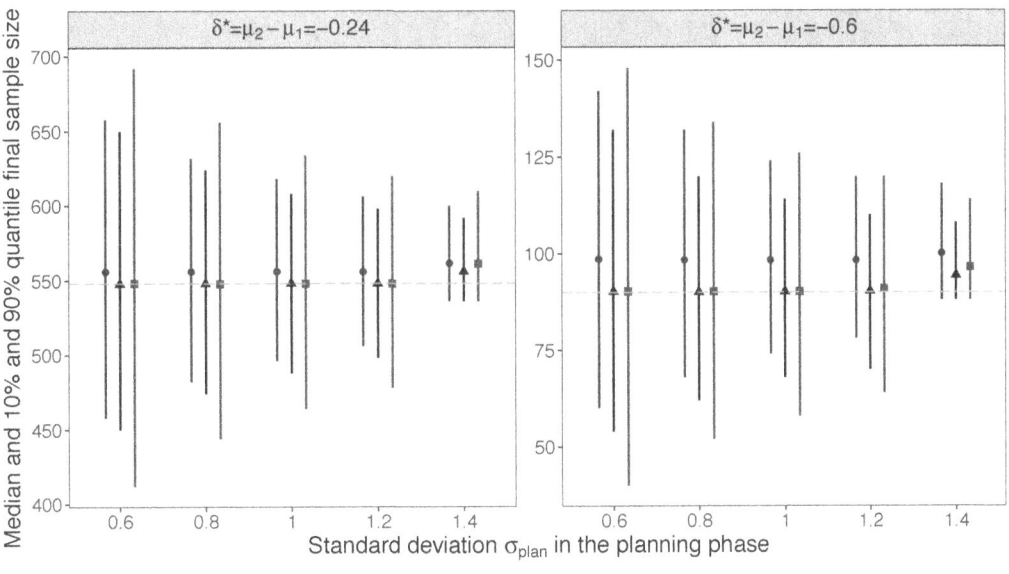

FIGURE 16.2
Median as well as 10% and 90% quantiles of the final sample size for designs with blinded and unblinded sample size re-estimation procedures. The true standard deviation is $\sigma = 1$. The fixed design total sample sizes are $n = 548$ and $n = 90$ for $\delta^* = -0.24$ and $\delta^* = -0.6$, respectively.

re-estimated sample sizes are biased upwards when $\sigma_{Plan} \gg \sigma$. For $\sigma_{Plan} \gg \sigma$, the internal pilot study sample sizes are of similar size as the required sample size for the true standard deviation $\sigma = 1$. Thus, the sample size re-estimation results in a sample size increase for $\hat{n}_{reest} > n_P$ but not in a sample size decrease if $\hat{n}_{reest} < n_P$ which induces an upwards bias in the re-estimated sample size. Figures 16.1 and 16.2 show the results of sample size re-estimation without inflated sample size. The underpowering of clinical trials with sample size re-estimation can be mitigated through inflating the re-estimated sample size. However, inflating the re-estimated sample size also increases its variability.

Controlling the type I error rate is one major regulatory requirement concerning designs of confirmatory trials with sample size re-estimation. For the scenarios listed in Table 16.1, we simulate the type I error rate of the sample size re-estimation procedures under the null hypothesis H_0, that is $\mu_1 = \mu_2 = 0$. Each simulated type I error rate is based on $1\,000\,000$ Monte Carlo replications corresponding to a simulation error of 0.00015 for a type I error rate of 0.025. The results are shown in Figure 16.3. Figure 16.3 shows that the type I error rate of the blinded sample size re-estimation based on the one-sample variance estimator is not inflated. However, unblinded sample size re-estimation and blinded sample size re-estimation based on the Xing-Ganju variance estimator inflates the type I error rate for internal pilot studies with less than 100 subjects, that is for the effect $\delta^* = -0.6$ and a small standard deviation σ_{Plan} in the planning phase.

Studying the power, final sample size distribution, and type I error rate of designs with sample size re-estimation is crucial for understanding their general performance. For comparing various designs with sample size re-estimation, a single metric of the efficacy of a sample size re-estimation procedure is desirable. In what follows, we adapt the metric

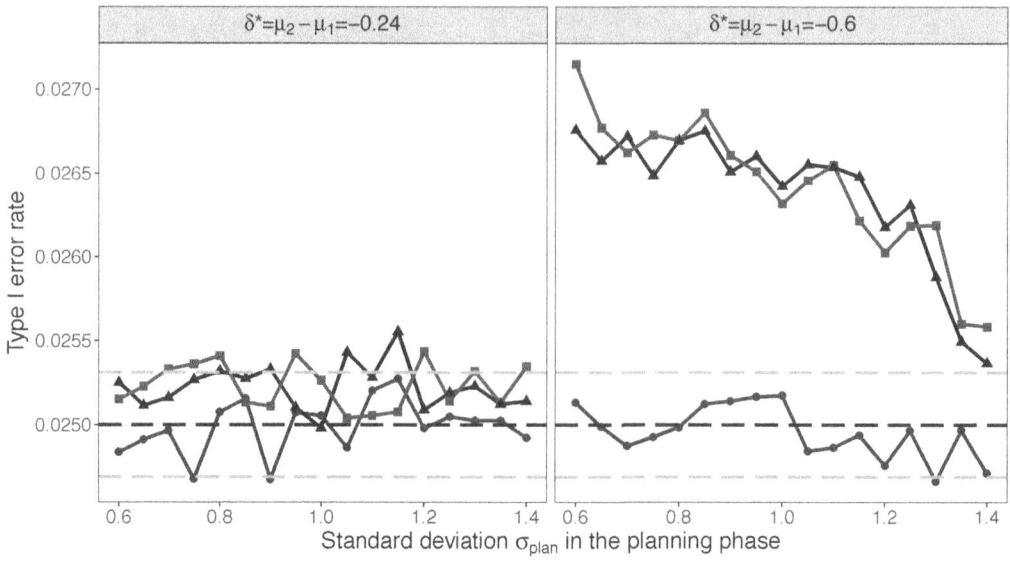

FIGURE 16.3
Type I error rate of the designs with sample size re-estimation procedures. The black line marks $\alpha = 0.025$ and the grey lines depict $\alpha = 0.025$ plus/minus two times the simulation error for an error rate of 0.025. The true standard deviation is equal to $\sigma = 1$.

proposed by Jennison and Turnbull (2006) [53] for comparing adaptive and nonadaptive designs based on the expected information. We define the efficiency metric of a design A as the ratio of the sample size in a fixed design, with type I error rate and power identical to design A, and the median final sample size of design A. More precisely, let $med(\hat{n}_{f,A})$ the median final sample size of design A. Furthermore, let α_A be the type I error rate of design A at $\mu_2 = \mu_1$, and denote β_A the type II error rate under the alternative, i.e., at $\mu_2 - \mu_1 = \delta^*$. Then, we define the efficiency of design A to be

$$MN_A = \frac{\sigma^2}{r(1-r)} \frac{(q_{\alpha_A} + q_{\beta_A})^2}{(\delta^* - \delta)^2} \frac{1}{med(\hat{n}_{f,A})}.$$

The efficiency measure MN_A is positive and the larger it is, the more efficient is design A. To compare the designs A and B, we define the relative efficiency

$$MR_{A,B} = \frac{MN_A}{MN_B} = \frac{(q_{\alpha_A} + q_{\beta_A})^2\, med(\hat{n}_{f,B})}{(q_{\alpha_B} + q_{\beta_B})^2\, med(\hat{n}_{f,A})}.$$

To compare the various sample size re-estimation procedures, we calculate the relative efficiency of each of the procedures in comparison to the fixed sample designs in dependency of the standard deviation σ_{Plan} in the planning phase. The results are shown in Figure 16.4. It can be seen in Figure 16.4 that the relative efficiencies of the designs with a sample size re-estimation are around one and that they are in general very similar for the scenarios with a mean difference of $\delta^* = -0.24$. In other words, the flexibility of re-estimating the final sample size and correcting incorrectly planned sample size comes with no relevant cost in efficiency. For the scenarios with a mean difference in the alternative of $\delta^* = -0.6$, the relative efficiency of the design with unblinded sample size re-estimation is again around one.

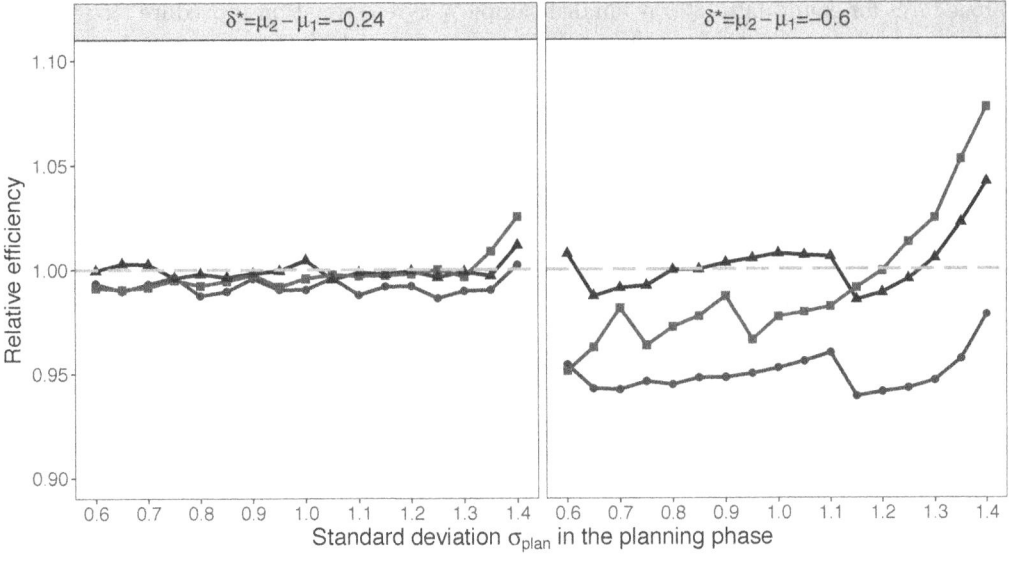

FIGURE 16.4
Relative efficiency of the designs with sample size re-estimation in comparison for the fixed sample designs against the standard deviation σ_{Plan} in the planning phase. The true standard deviation is $\sigma = 1$.

However, the relative efficiencies of the designs with blinded sample size re-estimation are small with the relative efficiency of the design based on the one-sample variance estimator being the smallest around 0.95. While there is a small loss of efficiency when designing a trial with a sample size re-estimation based on the one-sample variance estimator, for the scenarios with $\delta^* = -0.6$, it is the only re-estimation procedure controlling the type I error rate.

Concluding, two-arm parallel group designs with blinded sample size re-estimation based on the one-sample variance estimator are robust concerning the misspecification of the standard deviation in the planning phase. Moreover, they satisfy regulatory requirements by maintaining the blindness and controlling the type I error rate in superiority trials. They also meet the target power. The designs with unblinded sample size re-estimation or blinded sample size re-estimation based on the Xing-Ganju variance estimator do not control the type I error rate and are not achieving the target power.

16.2.2 Sample size re-estimation for count data

Nuisance parameter based sample size re-estimation for count data received great attention in the last years. In this section we provide an overview about existing literature and then discuss sample size re-estimation for the negative binomial distribution in detail. Friede and Schmidli (2010) [33] developed a nuisance parameter based blinded sample size re-estimation procedure for Poisson and quasi-Poisson models. Friede and Schmidli (2010) [34] proposed a blinded sample size re-estimation approach for negative binomial outcomes. Schneider et al. (2013) [84] studied the robustness of various blinded sample size re-estimation procedures for overdispersed count data and compared unblinded and blinded sample size re-estimation procedures for count data [82]. A blinded sample size re-estimation procedure for recurrent events, modeled as a non-homogeneous Poisson process with time trends, was developed by Schneider et al. (2013) [83]. A sample size re-estimation procedure for count data based on an EM algorithm was proposed by Cook et al. (2009) [18]. Some controversy arose regarding the applicability of such algorithms in settings with count data [17, 84]. Asendorf et al. (2017) [2] proposed a model for longitudinal count data based on negative binomial integrated value autoregressive processes and developed a blinded sample size re-estimation procedure for this model. Moreover, Asendorf et al. (2018) [3] extended their previous work to also incorporate within-patient time trends in the event rates. We briefly summarize the basic idea of blinded sample size re-estimation procedures for count data and their performance. In a two-arm trial setting, let λ_i and n_i, $i = 1, 2$, denote the group specific event rates and sample sizes, respectively. The total sample size is denoted by $n = n_1 + n_2$. For Poisson data, the nuisance parameter is the overall event rate $\bar{\lambda} = (\lambda_1 n_1 + \lambda_2 n_2)/n$. Unbiased estimates of the overall event rate can be obtained from the blinded data. For overdispersed count data, the nuisance parameters are the overall event rate and an overdispersion parameter. Both parameters are estimated from the blinded data by fitting the overdispersed count data model to the pooled treatment groups. The resulting nuisance parameter based blinded sample size re-estimation procedures do not inflate the type I error and meet the desired power. In contrast, the unblinded procedures for sample size re-estimation inflate the type I error and, thus, are not recommended unless steps are taken to control the type I error rate.

We present a motivating example in clinical trials in heart failure next before discussing blinded sample size re-estimation with negative binomial outcomes in detail.

16.2.2.1 Motivating example

Heart failure with preserved ejection fraction refers to a medical condition where patients show the symptoms of heart failure but have a normal left-ventricular ejection fraction [63]. In clinical trails in heart failure with preserved ejection fraction an important indication of treatment efficacy is the number of heart failure hospitalizations. A common model for heart failure hospitalizations in both heart failure with preserved ejection fraction and heart failure with reduced ejection fraction is the negative binomial distribution [76–78]. The CHARM-Preserved trial is a double-blind, multi-center, placebo-controlled clinical trial in heart failure with preserved ejection fraction [100]. The object of the trial was assessing the efficacy of candesartan in patients with chronic heart failure and preserved left-ventricular ejection fraction. A total of 3023 patients were randomly assigned to either candesartan (1514 patients) or placebo (1509 patients). The median follow-up time was 36.6 months. The heart failure hospitalizations observed during the CHARM-Preserved trial are listed in Table 16.3. The estimated rate ratio for recurrent heart failure hospitalizations is $\theta = 0.71$

TABLE 16.3
Heart failure hospitalizations in CHARM-Preserved.

	Placebo	Candesartan
Number of patients	1509	1514
Total follow-up years	4374.03	4424.62
Patients with ≥ 1 admission	278	230
Total admissions	547	392

when a negative binomial model is applied [78].

16.2.2.2 Negative binomial outcomes

Let X_{ij} denote the number of events for patient $j = 1, \ldots, n_i$ in treatment group $i = 1, 2$ at follow-up time t_{ij}. The control group is indexed with $i = 1$ and the treatment group with $i = 2$. We assume the randomization ratio $k : 1$, i.e. $n_2 = kn_1$. The event rate per time unit in treatment group i is denoted by $\lambda_i > 0$. Smaller rates are associated with a more effective treatment. We model the number of events with a negative binomial distribution, that is

$$X_{ij} \sim \text{NegBin}(t_{ij}\lambda_i, \phi).$$

Here, ϕ denotes the shape parameter which is assumed to be identical among the groups. The negative binomial distribution is parametrized such that the expected value and the variance of the random variable X_{ij} are given by

$$\mathbb{E}[X_{ij}] = t_{ij}\lambda_i,$$
$$\text{Var}[X_{ij}] = t_{ij}\lambda_i(1 + \phi t_{ij}\lambda_i).$$

We are interested in testing whether treatment $i = 2$ is superior to control $i = 1$, that is whether $\theta = \lambda_2/\lambda_1 < 1$. This corresponds to the statistical testing problem

$$H_0 : \theta \geq 1 \quad \text{vs.} \quad H_1 : \theta < 1.$$

Let $\hat{\lambda}_i$ denote the maximum likelihood estimator of the rate λ_i, $\beta_i = \log(\lambda_i)$ denotes the log-rate, and $\hat{\beta}_i = \log(\hat{\lambda}_i)$ the maximum likelihood estimator of the log-rate. Then, the Wald test statistic for testing the null hypothesis H_0 is given by

$$T = \frac{\hat{\beta}_2 - \hat{\beta}_1}{\sqrt{\frac{1}{\hat{I}_{\beta_2}} + \frac{1}{\hat{I}_{\beta_1}}}} \overset{H_0}{\underset{\text{asymp.}}{\sim}} \mathcal{N}(0, 1).$$

Here, $\hat{I}_{\beta_i}$ is the plug-in estimator of the Fisher information of log-rates β_j [59],

$$I_{\beta_i} = \sum_{j=1}^{n_i} \frac{t_{ij}\exp(\beta_j)}{1 + \phi t_{ij}\exp(\beta_i)} = \sum_{j=1}^{n_i} \frac{t_{ij}\lambda_i}{1 + \phi t_{ij}\lambda_i}.$$

An asymptotic level α Wald test for the null hypothesis H_0 is obtained when the null hypothesis is rejected for T smaller than the α-quantile of a standard normal distribution. In the following, for illustrative purposes we focus on equal follow-up times $t_{ij} \equiv t$ at the final analysis. However, the follow-up times might differ at the end of the internal pilot when the sample size is re-estimated. The presented methodology holds for unequal follow-up times t_{ij}, too, however, no closed form expression for the sample size n_1 exists. For a given rate ratio θ^* in the alternative, the sample size n_1 in the control group required to obtain a power of $1 - \beta$ is given by

$$n_1 = \left(\frac{q_\alpha + q_\beta}{\log(\theta^*)}\right)^2 \left(\frac{1 + k\theta^*}{t\lambda_1\theta^* k} + \phi\frac{1 + k}{k}\right)$$
$$= \left(\frac{q_\alpha + q_\beta}{\log(\theta^*)}\right)^2 \left(\frac{(1 + k\theta^*)^2}{t\bar{\lambda}\theta^* k(1 + k)} + \phi\frac{1 + k}{k}\right).$$

Here, $\bar{\lambda}$ is the overall event rate per time unit. The sample size depends on two nuisance parameters: the overall event rate $\bar{\lambda}$ and the shape parameter ϕ. We denote the blinded sample from the internal pilot study and their follow-up times at the end of the internal pilot study by $X_1, \ldots, X_{n_P}$ and $t_1, \ldots, t_{n_P}$, respectively. The nuisance parameter estimates $\hat{\bar{\lambda}}_P$ and $\hat{\phi}_P$ are the maximum likelihood estimates obtained from the blinded sample by maximizing the log-likelihood function $\log \mathcal{L}(\cdot, \cdot)$ of $i = 1, \ldots, n_P$ independent negative binomial distributed random variables with rate $t_i \bar{\lambda}$ and shape parameter ϕ, i.e.

$$
\log \mathcal{L}(\bar{\mu}, \phi) = \sum_{i=1}^{n_P} \left[\log \Gamma \left(X_i + \frac{1}{\phi} \right) - \left(\frac{1}{\phi} + X_i \right) \log \left(1 + \phi t_i \bar{\lambda} \right) + X_i \log \left(\phi t_i \bar{\lambda} \right) \right.
$$
$$
\left. - \log(X_i!) - \log \Gamma \left(\frac{1}{\phi} \right) \right].
$$

The performance of the blinded sample size re-estimation procedure is assessed by means of a Monte Carlo simulation study for scenarios motivated by the results of the CHARM-Preserved trial. For the simulation study, we assume that each patient has a fixed follow-up time of three years at the end of the trial and that the initial accrual period is set to 1.5 years with a uniform recruitment. The accrual rate is determined such that the number of patients required for the fixed design planned with shape parameter $\phi = 5$ and control group rate $\lambda_1 = 0.125$ would be recruited after the accrual period of 1.5 years. For a one-sided significance level $\alpha = 0.025$, a power of $1 - \beta = 0.8$, and a randomization ratio of $k = 1$, the group specific sample sizes in the fixed design are $n_1 = n_2 = 1017$ for rate ratio $\theta^* = 0.7$ and $n_1 = n_2 = 2523$ for rate ratio $\theta^* = 0.8$. The sample size re-estimation is performed after 66.67% of the recruitment period, that is one year into the study. The power and the distribution of the re-estimated sample size for nuisance parameter based sample size re-estimation with negative binomial outcomes are listed in Table 16.4. Table 16.4 shows that the power of the blinded sample size re-estimation procedure is for all cases between 79% and 80.2% which means that the design with blinded sample size re-estimation is robust against the misspecification of the control rate λ_1 and the shape parameter ϕ in the planning stage of the trial. The mean of the re-estimated sample size is larger than the sample size required by the fixed design. As mentioned before, the mean sample size must in general be overestimated for the sample size re-estimation procedure to achieve the target power. In addition to the power related operating characteristics listed in Table 16.4, we study the type I error rate of the design with blinded samples size re-estimation. For the respective simulation study, the same scenarios as before are considered with the difference that now the rates are identical, i.e., $\lambda_1 = \lambda_2$. The results are presented in Table 16.5. For each scenarios, 500 000 Monte Carlo replications are performed which corresponds to a simulation error of 0.00022 for a simulated type I error rate of 0.025. Table 16.5 shows that no practically relevant type I error inflation can be detected for the design with blinded sample size re-estimation. Moreover, the mean sample size under the null hypothesis is always smaller than the fixed design sample size.

16.2.3 Further issues and recent developments

So far, we have discussed nuisance parameter based sample size re-estimation with particular emphasis on methods for two-arm trials with normal and count data which maintain blindness. In the following we provide an overview about further issues and recent developments in the area of nuisance parameter based sample size re-estimation.

16.2.3.1 Non-inferiority trials

In Section 16.2.1.5 we compared various blinded and unblinded sample size re-estimation procedures for testing superiority with normal data. We concluded that sample size re-

TABLE 16.4
Power and the distribution of the re-estimated sample size. The simulations are performed under the alternative $\lambda_1/\lambda_2 = \theta^*$. Q25 and Q75 refer to the 25% and 75% quantiles of the final sample size. The results are based on 25 000 Monte Carlo replications.

Scenario			Fixed design	Power	Distribution $\hat{n}_1$			
θ^*	ϕ	λ_1	n_1		Mean	SD	Q25	Q75
0.7	4	0.1	994	0.79372	1030.99	273.77	825	1185
		0.125	894	0.79824	931.80	216.95	756	1057
		0.15	827	0.80188	864.24	174.89	712	965
	5	0.1	1117	0.79184	1155.73	314.29	927	1336
		0.125	1017	0.7934	1049.59	252.53	866	1196
		0.15	950	0.79688	983.44	217.93	820	1111
	6	0.1	1240	0.79092	1281.65	355.35	1030	1476
		0.125	1140	0.7906	1178.26	291.66	969	1349
		0.15	1074	0.79756	1107.20	247.40	930	1252
0.8	4	0.1	2444	0.79372	2474.34	411.84	2184	2727
		0.125	2207	0.79384	2232.92	333.04	1990	2443
		0.15	2050	0.79696	2081.39	280.18	1867	2258
	5	0.1	2759	0.79476	2793.13	465.95	2462	3081
		0.125	2523	0.79572	2552.79	388.07	2280	2793
		0.15	2365	0.79912	2394.35	334.39	2158	2602
	6	0.1	3074	0.80012	3112.82	519.32	2749	3432
		0.125	2838	0.79372	2873.92	434.84	2569	3145
		0.15	2680	0.7964	2709.74	374.67	2446	2947

estimation based on the blinded one-sample variance estimator performs best concerning power and control of the type I error rate. For non-inferiority trials it has been shown that the one-sample variance estimator and the Xing-Ganju variance estimator, too, inflate the type I error rate [24, 28, 35]. Sample size re-estimation can still be employed even when the type I error rate is inflated as long as measures for controlling the actual type I error are in place. We discuss such measures later on. However, blinded sample size re-estimation does not inflate the type I error for non-inferiority trials in general. For instance, for negative binomial outcomes [34] and binary outcomes [31] blinded sample size re-estimation procedures did not inflate the type I error rate in non-inferiority trials.

16.2.3.2 Controlling the type I error rate

If a sample size re-estimation procedure does not control the type I error rate, procedures for dealing with the inflation must be in place to comply with regulatory requirements. One approach for controlling the type I error rate of a blinded sample size re-estimation procedure is to perform the final test as well as the sample size adjustment with an adjusted type I error rate α_{adj} and not with the target type I error rate α [28, 35]. The adjusted type I error rate α_{adj} is the largest rate for which the inflation is not exceeding the target α.

16.2.3.3 Size of the internal pilot study

The size and the timing, of the internal pilot study affects not only the logistics of the clinical trial but also the performance characteristic of the sample size re-estimation procedure and

TABLE 16.5
Type I error rate and the distribution of the re-estimated sample size under the null hypothesis $\lambda_1 = \lambda_2$. Q25 and Q75 refer to the 25% and 75% quantiles of the final sample size. The results are based on 500 000 Monte Carlo replications.

Scenario			Fixed design	Power	Distribution $\hat{n}_1$			
θ^*	ϕ	λ_1	n_1		Mean	SD	Q25	Q75
0.7	4	0.10	994	0.02546	939.99	224.49	761	1066
		0.12	894	0.02529	858.21	174.65	704	957
		0.15	827	0.02533	805.85	140.07	678	883
	5	0.10	1117	0.02514	1059.01	264.25	864	1211
		0.12	1017	0.02544	971.44	214.46	809	1098
		0.15	950	0.02522	914.37	182.09	773	1023
	6	0.10	1240	0.02531	1181.29	298.41	968	1353
		0.12	1140	0.02523	1092.84	247.48	915	1239
		0.15	1074	0.02516	1033.92	213.64	880	1161
0.8	4	0.10	2444	0.02496	2339.84	371.16	2073	2571
		0.12	2207	0.02500	2128.13	297.89	1904	2315
		0.15	2050	0.02522	1991.54	244.92	1792	2146
	5	0.10	2759	0.02503	2653.40	424.00	2355	2915
		0.12	2523	0.02511	2437.93	350.76	2190	2659
		0.15	2365	0.02486	2294.56	302.62	2080	2486
	6	0.10	3074	0.02446	2970.77	473.56	2638	3264
		0.12	2838	0.02535	2754.99	395.75	2477	3004
		0.15	2680	0.02513	2610.63	344.02	2370	2827

the minimal sample size of the clinical trial. A small internal pilot study results in imprecise estimates of the nuisance parameter which then produce a variable re-estimated sample size. A large internal pilot study increases the risk that the internal pilot study is already larger than the required sample size which results in an overpowered clinical trial. Sandvik et al. (1996) [80] suggested a method for determining the size of the internal pilot study based on a confidence interval for the standard deviation and Singer (1999) [87] proposed an improvement. The optimal choice of the size of the internal pilot study was studied for unblinded sample size re-estimation in a t-test scenario by Denne and Jennison (1999) [20]. A more general discussion of the size of the internal pilot study was given by Friede and Schmidli (2010) [33, Section 6].

16.2.3.4　Covariates

Both the one-sample variance estimator and the Xing-Ganju variance estimator have been extended to incorporate covariates. Ganju and Xing (2009) [37] proposed a blinded variance estimator for the linear model based on a randomized block design. Friede and Kieser (2011) [27, 28] studied a one-sample variance estimator for blinded sample size re-estimation in a linear model with covariates. Unblinded sample size re-estimation in designs with covariates, or more precisely in fixed-effect linear models, was studied by Coeffey and Muller (1999, 2000) [15, 16].

16.2.3.5 Other endpoints and more complex designs

Nuisance parameter based sample size re-estimation procedures, in particular procedures that maintain the blindness, have been studied for a variety of endpoints and designs. We provide an overview of the available literature and discuss the main ideas.

For binary and categorical data, the overall event rate is the nuisance parameter. Gould (1992) [41] and Friede and Kieser (2004) [25] proposed a blinded sample size re-estimation procedure for testing equality of event rates for binary data. The idea was extended to categorical data [9] and to studies with relative risks [72]. Friede et al. (2007) [31] studied blinded sample size re-estimation for testing non-inferiority with binary outcomes. A blinded sample size re-estimation procedure for binary composite endpoints was proposed by Sander et al. (2016) [79]. Hees and Kieser (2017) [48] introduced blinded sample size re-estimation for a binary endpoint in two-arm superiority trials with historical control data.

A sample size re-estimation procedure for binary data which at least requires partial unblinding was proposed by Herson and Wittes (2007) [49]. Blinded sample size re-calculation for survival data based on the overall survival rates has been studied by Whitehead et al. [96, 97].

In linear models, the nuisance parameter is in general the covariance matrix of the error term. Besides the sample size re-estimation procedures mentioned above in the context of covariates [15, 16], the majority of research on sample size re-estimation in linear models focused on procedures which do not maintain the blindness. Zucker and Denne (2002) [101] compared various unblinded sample size re-estimation approaches for repeated measures in the linear mixed model framework. Gurka et al. (2007) [43] studied unblinded sample size adjustment for mixed models with a compound symmetry covariance matrix. A blinded sample size re-estimation for longitudinal data using generalized estimating equations was proposed by Wachtlin and Kieser (2013) [91].

Blinded sample size re-estimation procedures based on a one-sample variance estimator have been studied for further designs, for instance by Golkowski et al. (2014) [40] for crossover bioequivalence trials and by Placzek and Friede (2016) [71] for clinical trials with nested subgroups. Blinded sample size re-estimation for normal endpoints was compared with blinded continuous monitoring based on the one-sample variance estimator by Friede and Miller (2012) [30]. In designs with continuous monitoring, the sample size is re-estimated continuously, that is after each observation. While a continuous monitoring is associated with more complex logistics, it reduces the variability of the re-estimated sample size compared to a single blinded sample size re-estimation.

Posch et al. (2016) [73] studied the effect of blinded sample size re-estimation for the comparison of two normal means on the coverage probability of confidence intervals and calculated the bias of the effect and variance estimators. Depending on the rule for selecting the re-estimated sample size, the effect and variance estimators can be biased. Therefore, Posch et al. (2016) [73] recommended prespecifying binding sample size selection rules for the re-estimation in confirmatory trials and to study on a case-to-case basis the effect of the respective rule for sample size re-estimation on characteristics of the effect and variance estimators.

16.2.3.6 Multi-arm trials

Literature with explicit focus on nuisance parameter based sample size re-estimation in multi-arm trials is rare even though various methods can be extended from two-arm trials to multi-arm trials. Kieser and Friede (2000) [55] studied the one-sample variance estimator in the context of F-tests in multi-arm trials. Mütze and Friede (2016) [68] extended the one-sample variance estimator and the Xing-Ganju variance estimator [99] to three-arm trials in the 'gold standard' design, i.e., three-arm trials with an experimental treatment, an active control, and a placebo control, and compared their performance. In contrast to two arm-trials, sample size re-estimation based on the one-sample variance estimator results in overpowering in three-arm trials. However, Mütze and Friede (2016) [68] proposed an adjusted version of the Xing-Ganju variance estimator which meets the target power for a wide range of scenarios.

16.2.3.7 Incorporating historical data into the sample size re-estimation

Schmidli et al. (2016) [81] predicted the variance of a new clinical trial from previous trials using the meta-analytic-predictive (MAP) approach. The predicted variance is then considered for planning the new trial. Following the work of Schmidli et al. (2016) [81], Mütze et al. (2016) [69] studied the MAP approach in the context of nuisance parameter based sample size re-estimation and showed that prior information can reduce the variability of the re-estimated sample size when no data conflict between the ongoing trial and historical trials is present. In the case of data conflicts the target power is not met and the sample size is clearly under- or overestimated. Thus, due to the risks of not meeting the target power and the limited benefits with respect to reducing the variability of the re-estimated sample size, it can generally not be recommended to incorporate prior data into the nuisance parameter based sample size re-estimation.

Hartley (2012) [46] proposed nuisance parameter based blinded sample size re-estimation for comparison of two normal means which incorporates historical information not only on the nuisance parameter but also on the treatment effect. Compared with the sample size re-estimation of Gould and Shih (1992) [42], which does not implicitly account for uncertainty in the treatment effect, the method by Hartley has a smaller absolute deviation from the target conditional power. Other methods for incorporating historical data include a Bayesian blinded sample size re-estimation procedure for risk differences of binomial data [47].

16.3 Effect-Based Sample Size Re-Estimation

The focus in this part of the chapter is on effect-based sample size re-estimation. When planning a clinical trial, the actual treatment effect is obviously not known and its quantification is in general one of the goals of the trial. Historically, one of the first trial designs which could deal with the uncertainty in the effect size are (group) sequential designs. In group sequential designs interim analyses are performed in addition to the final analysis to test whether the trial can be stopped early for efficacy or futility. Thus, a clinical trial with a group sequential design can be planned with the smallest clinical relevant effect and if the actual treatment effect is larger, the trial is likely to stop early for efficacy after an interim analysis and an overpowering of the trial can be avoided. For more information about group sequential designs we refer to the previous chapter and to Jennison and Turnbull (2000) [50]. In classical group sequential designs, the group sizes and the maximum number of groups are determined prior to the trial. The sample size is flexible in the sense that the trial might be stopped early for efficacy or futility. If the group sizes are chosen or adjusted based on effect estimates from the ongoing trial, the type I error rate can be inflated [19, 51]. Type I error inflation after an effect based sample size adjustment is not limited to group sequential designs but also occurs in basic designs such as a two-sample mean comparison [86]. Thus, one of the challenges of effect-based sample size re-estimation is determining the statistical test which controls the type I error rate. Cui et al. (1999) [19] proposed a weighted test statistic in group sequential designs to guarantee type I error control when the sample size is adjusted based on the treatment effect estimator. For the same model, Lehmacher and Wassmer (1999) [60] suggested the inverse normal method which controls the type I error rate after sample size re-estimation by using an inverse normal distribution function to combine p-values from various stages. The two different approaches result in the same solution. Müller and Schäfer (2001) [67] showed that every group sequential design implicitly defines a rule for p-value combination and, thus, defines a testing procedure which controls the type I error rate for any adaptation. The latter method is a special case of the p-value combination approach for adaptive designs proposed by Bauer and Köhne (1994) [6]. A more general approach to adaptive designs with multiple interim analyses based on the recursive use of the p-value combination test was proposed by Brannath et al. (2002) [11]. For a classical study comparing two normal means, Proschan and Hunsberger (1995) [75] proposed a procedure for extending a study based on conditional power. Shen and Fisher (1999) [85] also suggested an approach for effect-based sample size adjustment using a weighted test statistic. The sample size adjustment suggested by Shen and Fisher (1999) [85] is generally referred to as the variance spending approach. Gao et al. (2008) [38] showed the weighted test by Cui at al. (1999) [19] is special cases of the p-value combination method from Müller and Schäfer (2001) [67].

In the following we provide a more detailed explanation of the approaches from Cui et al. (1999) [19], Proschan and Hunsberger (1995) [75] (conditional power), and Bauer and Köhne (1994) [6] (combination tests) which all control the type I error rate in adaptive designs. Then, we discuss adjusting the sample size based on the treatment effect estimate for the various testing procedures. We conclude with a discussion.

16.3.1 Controlling the type I error rate

16.3.1.0.1 Cui, Hung, and Wang method

Let $X_1, \ldots, X_n$ and $Y_1, \ldots, Y_n$ be two sets of independent random variables with means μ_x and μ_Y, respectively, and variance 1, that is

$$X_1, \ldots, X_n \sim \mathcal{N}(\mu_X, 1),$$
$$Y_1, \ldots, Y_n \sim \mathcal{N}(\mu_Y, 1).$$

With $\delta = \mu_Y - \mu_X$ the hypothesis testing problem of interest is

$$H_0 : \delta \leq 0 \quad \text{vs.} \quad H_1 : \delta > 0.$$

The hypothesis H_0 is tested using a Z-test with test statistic $Z = \sqrt{n/2}(\bar{Y}. - \bar{X}.)$. Here, $\bar{X}.$ and $\bar{Y}.$ denote the group means. We consider a group sequential design with a maximum of K data looks, i.e., if the trial is not stopped early, the design includes $K - 1$ interim analyses and a final analysis. Moreover, the design includes a sample size re-estimation at data look l. Let n denote the initially planned maximum sample size and n_k the cumulative sample size at data look $k = 1, \ldots, K$. In a classical group sequential design, at look k the null hypothesis H_0 is rejected if the test statistic $Z_k = \sqrt{1/(2n_k)} \sum_{i=1}^{n_k} Y_i - X_i$ is larger than a pre-specified critical value c_k. Let the re-estimated sample size $\hat{n}$ be a function of the effect estimate $\hat{\delta}_l$ at look l, that is $\hat{n} = \hat{n}(\hat{\delta}_l)$. Then, $b = (\hat{n} - n_l)/(n - n_l)$ indicates by which factor the remaining sample size after the sample size re-estimation is changed and $\hat{n}_{l+j} = n_i + b(n_{l+j} - n_i)$ $(j = 1, \ldots, K - l)$ denotes the new sample size at look $l + j$. In a group sequential design with effect based sample size re-estimation, the type I error rate is inflated if the null hypothesis is tested as described above. Instead, Cui, Hung, and Wang

(1999) [19] proposed the test statistics $\tilde{Z}_k = Z_k$ for $k = 1, \ldots, l$ and

$$\tilde{Z}_{l+j} = Z_l \sqrt{w_j} + \frac{\sum_{i=n_l+1}^{\hat{n}_{l+j}} Y_i - X_i}{\sqrt{b(n_{l+j} - n_l)}} \sqrt{1 - w_j}$$

with weight $w_j = n_l/n_{j+l}$ and $j = 1, \ldots, K - l$. The weights w_j do not depend on the re-estimated sample size. The decision of whether the null hypothesis H_0 can be rejected is made based on the critical values c_k from the classical group sequential design. The method for sample size re-estimation is arbitrary. We discuss the originally suggested approach for sample size adjustment in the next section.

16.3.1.0.2 *Conditional error function*

In the context of comparing two normal means, Proschan and Hunsberger (1995) [75] proposed a procedure for extending a trial based on conditional power which controls the type I error rate. We consider a parallel two group design with two stages. The random variables and the null hypothesis are defined as above. We assume that, after n_1 data points, the sample size is re-estimated based on the value z_1 of the Z-score $Z_1 = \sqrt{n_1/2}(\bar{Y}_1. - \bar{X}_1.)$ resulting in the decision to recruit additional $n_2 = n_2(z_1)$ subjects. Let $Z_2 = \sqrt{n_2/2}(\bar{Y}_2. - \bar{X}_2.)$ denote the standardized effect based solely on the second stage data. Here, $\bar{X}_k.$ and $\bar{Y}_k.$ are the group means based for stage $k = 1, 2$. With $\bar{X}.$ and $\bar{Y}.$ the group means at the end of the trial and the total sample size $n = n_1 + n_2(z_1)$, we define the final test statistic $Z = \sqrt{n/2}(\bar{Y}. - \bar{X}.)$. We briefly illustrate the type I error inflation when the null hypothesis H_0 is rejected if the test statistic Z is larger than the $(1 - \alpha)$-quantile $q_{1-\alpha}$ of the standard normal distribution and then we define the conditional error function. We define the conditional probability that the null hypothesis H_0 is rejected conditioned on the effect estimate at the interim analysis by

$$CP_\delta(n_2, q_{1-\alpha}|z_1) = \mathbb{P}_\delta\left(Z \geq q_{1-\alpha}|Z_1 = z_1\right).$$

The conditional probability CP_δ is evaluated under a fixed value of δ given $Z_1 = z_1$. With $Z = (\sqrt{n_1}Z_1 + \sqrt{n_2}Z_2)/\sqrt{n_1 + n_2}$, the conditional probability $CP_\delta(n_2, q_{1-\alpha}|z_1)$ can be written as

$$CP_\delta(n_2, q_{1-\alpha}|z_1) = 1 - \Phi\left(q_{1-\alpha}\sqrt{\frac{n_1 + n_2}{n_2}} - z_1\sqrt{\frac{n_1}{n_2}} - \sqrt{\frac{n_2}{2}}\delta\right).$$

Here, $\Phi(\cdot)$ is the cumulative distribution function of the standard normal distribution. Then, since the Z-score Z_1 is standard normally distributed, the actual type I error rate of a effect-based sample size re-estimation procedure can be calculated by

$$\int_{\mathbb{R}} CP_{\delta=0}(n_2(z_1), q_{1-\alpha}|z_1)\phi(z_1)\mathrm{d}z_1.$$

Here, $\phi(\cdot)$ is the density of the standard normal distribution. Proschan and Hunsberger (1995) [75] showed that, depending on the choice of the function $n_2(\cdot)$, the maximum type I error rate can be up to $\alpha_{\max} = \alpha + \exp(-q_{1-\alpha}^2/2)/4$. Shun et al. (2001) [86] quantified the type I error inflation when $n_2(\cdot)$ is obtained by replacing the effect δ in the sample size formula by the estimate $\hat{\delta}$. Therefore, if the sample size is re-estimated based on effect estimates from the ongoing trial, the type I error rate is inflated.

A conditional error function $A(z_1)$ is an increasing function with values in $[0, 1]$ which fulfills

$$\int_{\mathbb{R}} A(z_1)\phi(z_1)\mathrm{d}z_1 = \alpha.$$

Let $c = c(n_2, z_1)$ be the solution to $CP_{\delta=0}(n_2, c|z_1) = A(z_1)$, that is

$$c = \frac{\sqrt{n_2}q_{1-A(z_1)} + \sqrt{n_1}z_1}{\sqrt{n_1 + n_2}}.$$

Then, the test with the critical value $c = c(n_2, z_1)$ instead of $q_{1-\alpha}$ controls the type I error rate. Proschan and Hunsberger (1995) [75] proposed the circular conditional error function

$$A_{circ}(z_1) = \begin{cases} 0 & \text{if } z_1 < q_{1-\alpha_0} \\ 1 - \Phi(\sqrt{k^2 - z_1^2}) & \text{if } q_{1-\alpha_0} < z_1 < k \\ 1 & \text{if } z_1 \geq k \end{cases}.$$

The conditional error function is called circular because the term $y = \pm\sqrt{k^2 - x^2}$ defines a circle with radius k around the origin. The circular conditional error function is defined such that the study stops without being able to reject H_0, that is stopping for futility, if the p-value obtained from the first n_1 observations is equal to or greater than α_0. The study stops for efficacy if $z_1 \geq k$ and continues with an increased sample size for $z_1 \in (q_{1-\alpha_0}, k)$. The value k is calculated such that the mandatory conditions of a conditional error function are fulfilled. Another class of conditional error functions are linear conditional error functions

$$A_{lin}(z_1) = \begin{cases} 0 & \text{if } z_1 < q_{1-\alpha_0} \\ 1 - \Phi(a + bz_1) & \text{if } q_{1-\alpha_0} < z_1 < k \\ 1 & \text{if } z_1 \geq k \end{cases}.$$

Wassmer (2000) [94] outlined how the conditional error function approach can be defined based on p-values. Moreover, Wassmer (2000) [94] listed the conditional error functions for which the conditional error approach in two-stage designs is equal to the inverse normal approach proposed by Lehmacher and Wassmer (1999) [60] and to the procedure proposed by Bauer and Köhne (1994) [6]. For example, in terms of the stage 1 p-values, the circular conditional error function is given by

$$A_{circ}(p_1) = \begin{cases} 1 & \text{if } p_1 \leq \alpha_1 \\ 1 - \Phi\left(\sqrt{\Phi^{-1}(1-\alpha_1)^2 - \Phi^{-1}(1-p_1)^2}\right) & \text{if } \alpha_1 < p_1 \leq \alpha_0 \\ 0 & \text{if } p_1 > \alpha_0 \end{cases}.$$

With the circular conditional error function, the null hypothesis H_0 would be rejected if the p-value based on the stage 2 data is smaller than or equal to the conditional error function evaluated at the stage 1 p-value, that is $p_2 \leq A_{circ}(p_1)$.

16.3.1.0.3 Combination tests

In the following we outline the approach from Bauer and Köhne (1994) [6] for combining two p-values through Fisher's combination test. We focus on two-stage designs with an arbitrary adaptation since the theory is not specific to any particular adaptation. Let p_1 be the p-value obtained just from the first stage data and p_2 the p-value from just the second stage data. Then, Fisher's combination test rejects the null hypothesis H_0 at a significance level α if

$$p_1 p_2 \leq c_\alpha = \exp\left(-\chi^2_{4,1-\alpha}/2\right),$$

with $\chi^2_{m,p}$ the p-quantile of a χ^2-distribution with m degrees of freedom. Based on Fisher's combination test, Bauer and Köhne (1994) [6] proposed a procedure for analyzing two-stage adaptive trials which stops the trial after the first stage without rejecting H_0, that is stopping for futility, if first stage p-value p_1 is larger than or equal to threshold α_0, $p_1 \geq \alpha_0$. Stopping for futility is excluded when $\alpha_0 = 1$. Moreover, the trial is stopped for efficacy after the first stage, that is H_0 is rejected, if the first stage p-value p_1 is smaller than or equal to threshold α_1, $p_1 \leq \alpha_1$. If the first stage p-value is between the efficacy and futility bounds, the trial continues to the second stage, the design specifications of which can be chosen based on the first stage data. Eventually, the null hypothesis H_0 is rejected after the second stage, if the condition of Fisher's combination test, $p_1 p_2 \leq c_\alpha$, is fulfilled. The futility and efficacy bounds α_0 and α_1 must be chosen such that the global type I error is equal to α which, under the assumption that the p-values p_1 and p_2 are independent standard uniformly distributed, corresponds to the condition

$$\alpha_1 + \int_{\alpha_1}^{\alpha_0} \int_0^1 \mathbf{1}_{\{p_1 p_2 \leq c_\alpha\}} \mathrm{d}p_2 \mathrm{d}p_1$$

$$= \alpha_1 + \int_{\alpha_1}^{\alpha_0} \int_0^{c_\alpha/p_1} \mathrm{d}p_2 \mathrm{d}p_1 = \alpha_1 + c_\alpha(\ln(\alpha_0) - \ln(\alpha_1)) = \alpha.$$

Bauer and Köhne (1994) [6] provided an extension of the procedure to more than two stages. Furthermore, Brannath et al. (2002) [11] proved the assumption of independent standard uniformly distributed p-values p_1 and p_2 can be relaxed to the assumption that p_1 and $p_2|p_1$ are stochastically larger than or equal to the standard uniform distribution. The test based on p-value combination does in general loose some power compared to a uniformly most powerful test in a non-adaptive setting. For the case of comparing two normal means the power loss has been quantified by Bauer and Köhne (1994) [6] and by Banik et al. (1996) [4].

The combination test proposed by Bauer and Köhne (1994) [6] and the conditional error function approach with a circular conditional error function have been compared by Wassmer (1998) [93] and very similar decision rules were observed. In general, for every combination test with combination function $C(p_1, p_2)$ and a critical value c, a conditional error function $A(p_1)$ can be defined such that the rejecting areas are identical, that is $C(p_1, p_2) \leq c$ is identical to $p_2 \leq A(p_1)$. The conditional error function is given by

$$A(p_1) = \begin{cases} 1 & \text{if } p_1 \leq \alpha_1 \\ \max\{y \in [0,1] : C(p_1, y) \leq c\} & \text{if } \alpha_1 < p_1 \leq \alpha_0 \\ 0 & \text{if } p_1 > \alpha_0 \end{cases} .$$

Here, we introduced the combination test approach using the p-value combination function $C(p_1, p_2) = p_1 p_2$. However, p-value combination is not limited to Fisher's product of p-values. The general requirement is that the combination function $C(p_1, p_2)$ is nondecreasing in both arguments and continuous in the second argument. Other choices include the combination functions

$$C(p_1, p_2) = p_1 p_2^w,$$

which results in the weighted Fisher's product test [11]. The inverse normal combination test proposed by Lehmacher and Wassmer (1999) [60] is obtained with the combination function

$$C(p_1, p_2) = 1 - \Phi\left(w_1 \Phi^{-1}(1 - p_1) + w_2 \Phi^{-1}(1 - p_2)\right).$$

The conditional error function approach as proposed by Proschan and Hunsberger (1995) [75] is obtained with the conditional error function [11]

$$C(p_1, p_2) = 1 - \Phi\left((\Phi^{-1}(1 - p_1))^2 + (\max\{0, \Phi^{-1}(1 - p_2)\})^2\right).$$

Optimal conditional error functions are discussed by Brannath and Bauer (2004) [10]. An extension of the combination test to multiple interim analyses was developed by Brannath et al. (2002) [11].

16.3.2   Sample size adaptation

In the following methods for adapting the sample size for the previously introduced testing procedures are discussed. We start with sample size re-estimation for the conditional error function approach and use the comparison of two normal means as an illustration. The same approach holds for combination tests since each combination test defines a conditional error function. The simplest approach for re-calculating the sample size is substituting the effect δ in the unconditional sample size formula $2(q_\beta + q_\alpha)^2/\delta^2$ by an interim estimate $\hat{\delta}$. However, this approach does not incorporate all available information on the success probability of the ongoing trial into the sample size determination. The success probability of the ongoing trial for a given stage 2 sample size n_2 and effect δ conditioned on the stage 1 results is the conditional power $CP_\delta(n_2, c|z_1)$. Thus, to incorporate all available information from stage 1 into the stage 2 sample size n_2, the sample size n_2 should be determined based on the conditional power. There to, the conditional power at δ is equated with $1 - \beta$, that is $1 - \beta = CP_\delta(n_2, c|z_1)$ which corresponds to a group specific stage 2 sample size

$$n_2 = 2\frac{(q_\beta + q_{A(z_1)})^2}{\delta^2}.$$

It remains to discuss how to select the effect δ in the sample size formula. If the minimal clinically relevant effect size δ_{min} can be specified, it can be chosen for the sample size calculation. This guarantees that the target conditional power $1 - \beta$ is obtained for every relevant effect size $\delta^* \geq \delta_{min}$. The effect δ can also be estimated by the interim estimate $\hat{\delta}$. However, the disadvantages are that the sample size can become very large when $\hat{\delta}$ is close to 0 and that such a designs can be ineffective [51, 53, 90]. Another set of methods for adjusting the sample size are based on Bayesian statistics. For a discussion of those methods, we refer to Chapter 7.4 in the book of Wassmer and Brannath (2016) [95].

Cui, Hung, and Wang (1999) [19] suggested to increase the sample size at look l only when the conditional power at the effect estimate $\hat{\theta}_l$ is smaller than $\gamma_I CP_\delta(n - n_l, c_K|z_l)$ for a given constant $\gamma_I \leq 1$, that is

$$CP_{\hat{\delta}_l}(n - n_l, c_K|z_l) < \gamma_I CP_\delta(n - n_l, c_K|z_l).$$

The re-estimated sample could be determined to be $\hat{n} = (\delta/\hat{\delta}_l)^2 n$ with δ the effect in the initial sample size planning.

16.3.3   Further issues and recent developments

Up to now, we discussed adapting a clinical trial based on a single variable or a Z-score. However, in practice such adaptions are not always possible. For example, when comparing two normal means, the outcome variance is often not known either. Posch et al. (2004) [74] studied the conditional rejection probability for Student's t-test and Timmesfeld et al. (2007) [89] proposed a method for increasing the sample size for Student's t-test without inflating the type I error rate. A general approach for applying the conditional error function approach in the presence of nuisance parameters has been proposed by Gutjahr et al. (2011) [44]. A general framework for adapting clinical trials which does not only depend on the Z-score but considers further information from the interim study is provided by Liu et al. (2012) [61].

Adjusting the sample size based on treatment effect estimates can inflate but also deflate the type I error rate of the standard test statistics and to control the type I error rates adjusted tests are commonly applied. Chen et al. (2004) [13] quantified the inflation and deflation of the type I error rate depending on the conditional power and proposed increasing the sample size only when the conditional power is in a range in which the type I error rate would not be inflated. By construction such an effect based sample size adjustment would not inflate the type I error rate and, therefore, not require any adjustment of the test statistic or the critical value. The range of the conditional power in which the type I error of the standard test statistic is not inflated is referred to as the promising zone. The idea was extended to group sequential designs by Gao et al. (2008) [38]. Mehta and Pocock (2011) [64] further studied the promising zone idea with a focus on its application in clinical trials. In a comment on the latter publication, Glimm (2012) [39] showed that the promising zone approach is a special case of the conditional error function approach. Emerson et al. (2011) [21] also commented on the promising zone approach questioning the benefit of adaptive designs based on promising zones compared to group sequential designs and criticizing that the promising zone is defined based on control of the type I error rate. Jennison and Turnbull (2015) [52] highlighted that adapting the sample size such that a certain conditional power is obtained under the estimated effect $\hat{\delta}$ is not efficient. Moreover, they derived efficient designs based on the inverse normal combination test and sample size rules that optimize the trade-off between a conditional power gain and the additional sample size under a fixed alternative δ^*.

16.4 Discussion

In this chapter we presented nuisance parameter based and effect-based sample size re-estimation. As their names imply, the methods adjust the sample size of an ongoing trial based on nuisance parameter and effect estimates, respectively. While we presented both sample size re-estimation approaches separately, they can be combined. For instance, in a setting with uncertainty in both the nuisance parameter and the effect size, the sample size could be adjusted in a blinded manner based on a nuisance parameter estimate early on in the trial and at a later point unblinded based on an effect estimate [65]. Both sample size re-estimation approaches are able to adjust for misspecification of trial specifications in the planning phase. However, it is important to keep in mind that adjusting the sample size mid-trial cannot replace a careful evaluation of trial specifications during the planning phase of a trial. Moreover, the more complex the trial design, the more effort must be spent on assessing the validity, efficacy, and robustness using performance characteristics such as power, type I error rate, expected sample size, probability of stopping early, etc. for various potential analysis methods. For a more detailed discussion and multiple examples of assessing a trial design in the context of clinical development programs we refer to Benda et al. (2010) [7] and Friede et al. (2010) [32].

An important aspect of whether a statistical methodology is applied in practice is often the availability of software. In the following we give an overview of available software for sample size re-estimation. When the sample size is re-estimated based on effect estimates, the final statistical analysis is performed using methods for adaptive designs. A variety of software is available for analyzing adaptive trials and we refer to Bauer et al. (2016) [5] for a comprehensive list. The software available for the effect-based sample size re-estimation is rare. The R package *gsDesign* [54] includes a function for effect-based sample size re-estimation using conditional power in various two-stage adaptive designs. The commercial software *East* from Cytel offers with their extensions *East ADAPT* and *East SURVADAPT* methods for effect-based sample size re-estimation based on, for instance, the promising zone and conditional power. A similar functionality is offered through the commercial software *ADDPLAN* from ICON. The R package *spass* [1] implements methods for blinded sample size re-estimation in a variety of models, e.g. designs with subgroups, longitudinal data, and count data. Nuisance parameter based sample size re-estimation for bioequivalence studies are implemented in the R package *Power2Stage* [58]. Both R and SAS code for effect and nuisance parameter based sample size re-estimation can also be found in a book by Chang (2014) [12].

Acknowledgements

Tobias Mütze is supported by the DZHK (German Centre for Cardiovascular Research) under grant GOE SI 2 UMG Information and Data Management.

Bibliography

[1] T. Asendorf, R. Gera, S. Islam, M. Harden, and M. Placzek. R package spass: Study Planning and Adaptation of Sample Size, 2016.

[2] T. Asendorf, R. Henderson, H. Schmidli, and T. Friede. Modelling and sample size reestimation for longitudinal count data with incomplete follow up. *Statistical Methods in Medical Research*, 28(1):117–133, 2019.

[3] T. Asendorf, R. Henderson, H. Schmidli, and T. Friede. Sample size re-estimation for clinical trials with longitudinal negative binomial counts including time trends. *Statistics in Medicine*, 38(9):1503–1528, 2019.

[4] N. Banik, K. Köhne, and P. Bauer. On the power of Fisher's combination test for two stage sampling in the presence of nuisance parameters. *Biometrical Journal*, 38(1):25–37, 1996.

[5] P. Bauer, F. Bretz, V. Dragalin, F. König, and G. Wassmer. Twenty-five years of confirmatory adaptive designs: opportunities and pitfalls. *Statistics in Medicine*, 35(3):325–347, 2016.

[6] P. Bauer and K. Köhne. Evaluation of experiments with adaptive interim analyses. *Biometrics*, 50(4):1029–1041, 1994.

[7] N. Benda, M. Branson, W. Maurer, and T. Friede. Aspects of modernizing drug development using clinical scenario planning and evaluation. *Drug Information Journal*, 44(3):299–315, 2010.

[8] MA. Birkett and SJ. Day. Internal pilot studies for estimating sample size. *Statistics in Medicine*, 13(23-24):2455–2463, 1994.

[9] K. Bolland, MR. Sooriyarachchi, and J. Whitehead. Sample size review in a head injury trial with ordered categorical responses. *Statistics in Medicine*, 17(24):2835–2847, 1998.

[10] W. Brannath and P. Bauer. Optimal conditional error functions for the control of conditional power. *Biometrics*, 60(3):715–723, 2004.

[11] W. Brannath, M. Posch, and P. Bauer. Recursive combination tests. *Journal of the American Statistical Association*, 97(457):236–244, 2002.

[12] M. Chang. *Adaptive design theory and implementation using SAS and R*. CRC Press, 2014.

[13] YH. Chen, DL. DeMets, and KKG. Lan. Increasing the sample size when the unblinded interim result is promising. *Statistics in Medicine*, 23(7):1023–1038, 2004.

[14] SC. Chow, H. Wang, and J. Shao. *Sample size calculations in clinical research*. CRC press, 2007.

[15] CS. Coffey and KE. Muller. Exact test size and power of a Gaussian error linear model for an internal pilot study. *Statistics in Medicine*, 18(10):1199–1214, 1999.

[16] CS. Coffey and KE. Muller. Some distributions and their implications for an internal pilot study with a univariate linear model. *Communications in Statistics - Theory and Methods*, 29(12):2677–2691, 2000.

[17] RJ. Cook. Authors' redress on 'robustness of methods for blinded sample size re-estimation with overdispersed count data'. *Statistics in Medicine*, 32(22):3955–3957, 2013.

[18] RJ. Cook, PJ Bergeron, JM. Boher, and Y. Liu. Two-stage design of clinical trials involving recurrent events. *Statistics in Medicine*, 28(21):2617–2638, 2009.

[19] L. Cui, HM. Hung, and SJ. Wang. Modification of sample size in group sequential clinical trials. *Biometrics*, 55(3):853–857, 1999.

[20] JS. Denne and C. Jennison. Estimating the sample size for a t-test using an internal pilot. *Statistics in Medicine*, 18(13):1575–1585, 1999.

[21] SS. Emerson, GP. Levin, and SC. Emerson. Comments on 'adaptive increase in sample size when interim results are promising: A practical guide with examples'. *Statistics in Medicine*, 30(28):3285–3301, 2011.

[22] US Food and Drug Administration. Adaptive design clinical trials for drugs and biologics, 2010.

[23] T. Friede and M. Kieser. On the inappropriateness of an EM algorithm based procedure for blinded sample size re-estimation. *Statistics in Medicine*, 21(2):165–176, 2002.

[24] T. Friede and M. Kieser. Blinded sample size reassessment in non-inferiority and equivalence trials. *Statistics in Medicine*, 22(6):995–1007, 2003.

[25] T. Friede and M. Kieser. Sample size recalculation for binary data in internal pilot study designs. *Pharmaceutical Statistics*, 3(4):269–279, 2004.

[26] T. Friede and M. Kieser. Sample size recalculation in internal pilot study designs: a review. *Biometrical Journal*, 48(4):537–555, 2006.

[27] T. Friede and M. Kieser. Blinded sample size recalculation for clinical trials with normal data and baseline adjusted analysis. *Pharmaceutical Statistics*, 10(1):8–13, 2011.

[28] T. Friede and M. Kieser. Sample size reassessment in non-inferiority trials. *Methods of Information in Medicine*, 50(3):237–243, 2011.

[29] T. Friede and M. Kieser. Blinded sample size re-estimation in superiority and noninferiority trials: Bias versus variance in variance estimation. *Pharmaceutical Statistics*, 12(3):141–146, 2013.

[30] T. Friede and F. Miller. Blinded continuous monitoring of nuisance parameters in clinical trials. *Journal of the Royal Statistical Society: Series C (Applied Statistics)*, 61(4):601–618, 2012.

[31] T. Friede, C. Mitchell, and G. Müller-Velten. Blinded sample size reestimation in non-inferiority trials with binary endpoints. *Biometrical Journal*, 49(6):903–916, 2007.

[32] T. Friede, R. Nicholas, N. Stallard, S. Todd, N. Parsons, E. Valdés-Márquez, and J. Chataway. Refinement of the clinical scenario evaluation framework for assessment of competing development strategies with an application to multiple sclerosis. *Drug Information Journal*, 44(6):713–718, 2010.

[33] T. Friede and H. Schmidli. Blinded sample size reestimation with count data: methods and applications in multiple sclerosis. *Statistics in Medicine*, 29(10):1145–1156, 2010.

[34] T. Friede and H. Schmidli. Blinded sample size reestimation with negative binomial counts in superiority and non-inferiority trials. *Methods of Information in Medicine*, 49(6):618, 2010.

[35] T. Friede and H. Stammer. Blinded sample size recalculation in noninferiority trials: A case study in dermatology. *Drug Information Journal*, 44(5):599–607, 2010.

[36] P. Gallo, C. Chuang-Stein, V. Dragalin, B. Gaydos, M. Krams, and J. Pinheiro. Adaptive designs in clinical drug development – an executive summary of the PhRMA working group. *Journal of Biopharmaceutical Statistics*, 16(3):275–283, 2006.

[37] J. Ganju and B. Xing. Re-estimating the sample size of an on-going blinded trial based on the method of randomization block sums. *Statistics in Medicine*, 28(1):24–38, 2009.

[38] P. Gao, JH. Ware, and C. Mehta. Sample size re-estimation for adaptive sequential design in clinical trials. *Journal of Biopharmaceutical Statistics*, 18(6):1184–1196, 2008.

[39] E. Glimm. Comments on 'adaptive increase in sample size when interim results are promising: A practical guide with examples' by C.R. Mehta and S.J. Pocock. *Statistics in Medicine*, 31(1):98–99, 2012.

[40] D. Golkowski, T. Friede, and M. Kieser. Blinded sample size re-estimation in crossover bioequivalence trials. *Pharmaceutical Statistics*, 13(3):157–162, 2014.

[41] AL. Gould. Interim analyses for monitoring clinical trials that do not materially affect the type I error rate. *Statistics in Medicine*, 11(1):55–66, 1992.

[42] AL. Gould and WJ. Shih. Sample size re-estimation without unblinding for normally distributed outcomes with unknown variance. *Communications in Statistics – Theory and Methods*, 21(10):2833–2853, 1992.

[43] MJ. Gurka, CS. Coffey, and KE. Muller. Internal pilots for a class of linear mixed models with Gaussian and compound symmetric data. *Statistics in Medicine*, 26(22):4083–4099, 2007.

[44] G. Gutjahr, W. Brannath, and P. Bauer. An approach to the conditional error rate principle with nuisance parameters. *Biometrics*, 67(3):1039–1046, 2011.

[45] SD. Halpern, JHT. Karlawish, and JA. Berlin. The continuing unethical conduct of underpowered clinical trials. *The Journal of the American Medical Association*, 288(3):358–362, 2002.

[46] AM. Hartley. Adaptive blinded sample size adjustment for comparing two normal means – a mostly Bayesian approach. *Pharmaceutical Statistics*, 11(3):230–240, 2012.

[47] AM. Hartley. A bayesian adaptive blinded sample size adjustment method for risk differences. *Pharmaceutical Statistics*, 14(6):488–514, 2015.

[48] K. Hees and M. Kieser. Blinded sample size recalculation in clinical trials incorporating historical data. *Contemporary Clinical Trials*, 2017.

[49] J. Herson and J. Wittes. The use of interim analysis for sample size adjustment. *Drug Information Journal*, 27(3):753–760, 1993.

[50] C. Jennison and BW. Turnbull. *Group sequential designs with applications to clinical trials.* Boca Raton, Chapman & Hall/CRC, 2000.

[51] C. Jennison and BW. Turnbull. Mid-course sample size modification in clinical trials based on the observed treatment effect. *Statistics in Medicine*, 22(6):971–993, 2003.

[52] C. Jennison and BW. Turnbull. Adaptive sample size modification in clinical trials: start small then ask for more? *Statistics in Medicine*, 34(29):3793–3810, 2015.

[53] C. Jennison and CW. Turnbull. Adaptive and nonadaptive group sequential tests. *Biometrika*, 93(1):1–21, 2006.

[54] A. Keaven. R package gsDesign: Group sequential design, 2016.

[55] M. Kieser and T. Friede. Blinded sample size reestimation in multiarmed clinical trials. *Drug Information Journal*, 34(2):455–460, 2000.

[56] M. Kieser and T. Friede. Re-calculating the sample size in internal pilot study designs with control of the type I error rate. *Statistics in Medicine*, 19(7):901–911, 2000.

[57] M. Kieser and T. Friede. Simple procedures for blinded sample size adjustment that do not affect the type I error rate. *Statistics in Medicine*, 22(23):3571–3581, 2003.

[58] D. Labes and H. Schuetz. R package power2stage: Power and sample-size distribution of 2-stage bioequivalence studies, 2015.

[59] JF. Lawless. Negative binomial and mixed Poisson regression. *Canadian Journal of Statistics*, 15(3):209–225, 1987.

[60] W. Lehmacher and G. Wassmer. Adaptive sample size calculations in group sequential trials. *Biometrics*, 55(4):1286–1290, 1999.

[61] Q. Liu, MA. Proschan, and GW. Pledger. A unified theory of two-stage adaptive designs. *Journal of the American Statistical Association*, 97(460):1034–1041, 2012.

[62] TM. MacDonald, JY. Reginster, TW. Littlejohn, D. Richard, K. Lheritier, G. Krammer, and R. Rebuli. Effect on blood pressure of lumiracoxib versus ibuprofen in patients with osteoarthritis and controlled hypertension: a randomized trial. *Journal of Hypertension*, 26(8):1695–1702, 2008.

[63] MT. Maeder and DM. Kaye. Heart failure with normal left ventricular ejection fraction. *Journal of the American College of Cardiology*, 53(11):905–918, 2009.

[64] CR. Mehta and SJ. Pocock. Adaptive increase in sample size when interim results are promising: a practical guide with examples. *Statistics in Medicine*, 30(28):3267–3284, 2011.

[65] CR. Mehta and AA. Tsiatis. Flexible sample size considerations using information-based interim monitoring. *Drug Information Journal*, 35(4):1095–1112, 2001.

[66] F. Miller. Variance estimation in clinical studies with interim sample size reestimation. *Biometrics*, 61(2):355–361, 2005.

[67] HH. Müller and H. Schäfer. Adaptive group sequential designs for clinical trials: combining the advantages of adaptive and of classical group sequential approaches. *Biometrics*, 57(3):886–891, 2001.

[68] T. Mütze and T. Friede. Blinded sample size re-estimation in three-arm trials with 'gold standard' design. *Statistics in Medicine*, 36(23):3636–3653, 2017.

[69] T. Mütze, H. Schmidli, and T. Friede. Sample size re-estimation incorporating prior information on a nuisance parameter. *Pharmaceutical Statistics*, 17(2):126–143, 2018.

[70] AJ. Phillips and ON. Keene. Adaptive designs for pivotal trials: discussion points from the PSI adaptive design expert group. *Pharmaceutical Statistics*, 5(1):61–66, 2006.

[71] M. Placzek and T. Friede. Clinical trials with nested subgroups: Analysis, sample size determination and internal pilot studies. *Statistical Methods in Medical Research*, 27(11):3286–3303, 2018.

[72] M. Pobiruchin and M. Kieser. Sample size calculation and blinded sample size recalculation in clinical trials where the treatment effect is measured by the relative risk. *Communications in Statistics - Simulation and Computation*, 42(7):1643–1653, 2013.

[73] M. Posch, F. Klinglmueller, F. König, and F. Miller. Estimation after blinded sample size reassessment. *Statistical Methods in Medical Research*, 27(6):1830–1846.

[74] M. Posch, N. Timmesfeld, F. König, and HH. Müller. Conditional rejection probabilities of Student's t-test and design adaptations. *Biometrical Journal*, 46(4):389–403, 2004.

[75] MA. Proschan and SA. Hunsberger. Designed extension of studies based on conditional power. *Biometrics*, 51(4):1315–1324, 1995.

[76] JK. Rogers, PS. Jhund, AC. Perez, M. Böhm, JG. Cleland, L. Gullestad, J. Kjekshus, DJ. van Veldhuisen, J. Wikstrand, and H. Wedel. Effect of rosuvastatin on repeat heart failure hospitalizations: the corona trial (controlled rosuvastatin multinational trial in heart failure). *JACC: Heart Failure*, 2(3):289–297, 2014.

[77] JK. Rogers, JJV. McMurray, SJ. Pocock, F. Zannad, H. Krum, DJ. van Veldhuisen, K. Swedberg, H. Shi, J. Vincent, and B. Pitt. Eplerenone in patients with systolic heart failure and mild symptoms analysis of repeat hospitalizations. *Circulation*, 126(19):2317–2323, 2012.

[78] JK. Rogers, SJ. Pocock, JJV. McMurray, CB. Granger, EL. Michelson, J. Östergren, MA. Pfeffer, SD. Solomon, K. Swedberg, and S. Yusuf. Analysing recurrent hospitalizations in heart failure: a review of statistical methodology, with application to charm-preserved. *European Journal of Heart Failure*, 16(1):33–40, 2014.

[79] A. Sander, G. Rauch, and M. Kieser. Blinded sample size recalculation in clinical trials with binary composite endpoints. *Journal of Biopharmaceutical Statistics*, 27(4):705–715, 2017.

[80] L. Sandvik, J. Erikssen, P. Mowinckel, and EA. Rødland. A method for determining the size of internal pilot studies. *Statistics in Medicine*, 15(14):1587–1590, 1996.

[81] H. Schmidli, B. Neuenschwander, and T. Friede. Meta-analytic-predictive use of historical variance data for the design and analysis of clinical trials. *Computational Statistics & Data Analysis*, 113:100–110, 2017.

[82] S. Schneider, H. Schmidli, and T. Friede. Blinded and unblinded internal pilot study designs for clinical trials with count data. *Biometrical Journal*, 55(4):617–633, 2013.

[83] S. Schneider, H. Schmidli, and T. Friede. Blinded sample size re-estimation for recurrent event data with time trends. *Statistics in Medicine*, 32(30):5448–5457, 2013.

[84] S. Schneider, H. Schmidli, and T. Friede. Robustness of methods for blinded sample size re-estimation with overdispersed count data. *Statistics in Medicine*, 32(21):3623–3635, 2013.

[85] Y. Shen and L. Fisher. Statistical inference for self-designing clinical trials with a one-sided hypothesis. *Biometrics*, 55(1):190–197, 1999.

[86] Z. Shun, W. Yuan, WE. Brady, and H. Hsu. Type I error in sample size re-estimations based on observed treatment difference. *Statistics in Medicine*, 20(4):497–513, 2001.

[87] J. Singer. A method for determining the size of internal pilot studies by L. Sandvik, J. Erikssen, P. Mowinckel and E. A. Rødland, Statistics in Medicine, 15, 1587–1590 (1996). *Statistics in Medicine*, 18(9):1151–1153, 1999.

[88] C. Stein. A two-sample test for a linear hypothesis whose power is independent of the variance. *The Annals of Mathematical Statistics*, 16(3):243–258, 1945.

[89] N. Timmesfeld, H. Schäfer, and HH. Müller. Increasing the sample size during clinical trials with t-distributed test statistics without inflating the type I error rate. *Statistics in Medicine*, 26(12):2449–2464, 2007.

[90] AA. Tsiatis and C. Mehta. On the inefficiency of the adaptive design for monitoring clinical trials. *Biometrika*, 90(2):367–378, 2003.

[91] D. Wachtlin and M. Kieser. Blinded sample size recalculation in longitudinal clinical trials using generalized estimating equations. *Therapeutic Innovation & Regulatory Science*, 47(4):460–467, 2013.

[92] JA. Waksman. Assessment of the Gould-Shih procedure for sample size re-estimation. *Pharmaceutical Statistics*, 6(1):53–65, 2007.

[93] G. Wassmer. A comparison of two methods for adaptive interim analyses in clinical trials. *Biometrics*, 54(2):696–705, 1998.

[94] G. Wassmer. Basic concepts of group sequential and adaptive group sequential test procedures. *Statistical Papers*, 41(3):253–279, 2000.

[95] G. Wassmer and W. Brannath. *Group sequential and confirmatory adaptive designs in clinical trials.* Springer, 2016.

[96] J. Whitehead. Predicting the duration of sequential survival studies. *Drug Information Journal*, 35(4):1387–1400, 2001.

[97] J. Whitehead, A. Whitehead, S. Todd, K. Bolland, and MR. Sooriyarachchi. Mid-trial design reviews for sequential clinical trials. *Statistics in Medicine*, 20(2):165–176, 2001.

[98] J. Wittes and E. Brittain. The role of internal pilot studies in increasing the efficiency of clinical trials. *Statistics in Medicine*, 9(1):65–72, 1990.

[99] B. Xing and J. Ganju. A method to estimate the variance of an endpoint from an on-going blinded trial. *Statistics in Medicine*, 24(12):1807–1814, 2005.

[100] S. Yusuf, MA. Pfeffer, K. Swedberg, CB. Granger, P. Held, JJV. McMurray, EL. Michelson, B. Olofsson, J. Östergren, CHARM Investigators, and Committees. Effects of candesartan in patients with chronic heart failure and preserved left-ventricular ejection fraction: the CHARM-Preserved trial. *The Lancet*, 362(9386):777–781, 2003.

[101] DM. Zucker and J. Denne. Sample size redetermination for repeated measures studies. *Biometrics*, 58(3):548–559, 2002.

[102] DM. Zucker, JT. Wittes, O. Schabenberger, and E. Brittain. Internal pilot studies II.: comparison of various procedures. *Statistics in Medicine*, 18(24):3493–3509, 1999.

17

Adaptive Designs

Gernot Wassmer, Franz Koenig and Martin Posch

CONTENTS

17.1 Introduction

This chapter is devoted to confirmatory adaptive designs as suggested by the pioneering papers of Peter Bauer and colleagues in the last decade of the past century [2, 4, 5]. This and other related proposals [67, 76] were the starting point of the development of a broad class of statistical methods that enable the redesign of a study at interim analyses while not compromising the Type I error rate of hypotheses tests. Bauer et al. [3] summarize the developments of this kind of adaptive design from different perspectives including methodological issues, regulatory statements, as well as practical feasibility. Wassmer and Brannath [98] provide a recent review of the general principles of adaptive designs that can be regarded as a generalization of the classical group sequential methodology (see Chapter 15). For a more general classification of adaptive designs see Dragalin [24].

An important application of adaptive designs is the data-driven reassessment of the sample size. This and related topics were described in Chapter 16. We emphasize that from early on more general kinds of adaptations have been considered. Most importantly, dropping or adding treatment arms in a multi-armed trial or selecting a sub-population in population enrichment designs were proposed as typical applications of adaptive design. In general, this involves testing more than one hypothesis and thus multiple testing procedures that control the experimentwise error rate in the strong sense need to be derived. This means that the probability of rejecting at least one true null hypothesis is bounded by the specified level, irrespective which and how many of the null hypotheses are true. The focus of this chapter is therefore on adaptive designs for multiple hypotheses (see also Chapter 19).

We note that many other developments to enhance the flexibility and efficiency of clinical trial designs exist, such as blinded sample size reassessment [31], Bayesian adaptive designs [8], response-adaptive designs [42], and Type I error control ensured by critical values which are found either by exact calculation [34, 83] or via simulation [74] for a prespecified set of adaptation rules. In this chapter we explicitly do not describe these approaches but will focus on confirmatory adaptive designs in the sense of [2, 5].

A general method to derive multiple testing procedures that control the experimentwise error rate in the strong sense is the closed testing principle due to Marcus, Peritz, and Gabriel [62]. In adaptive designs, Type I error rate control was achieved by making use of the combination testing principle as suggested in [2, 5]. Proschan and Hunsberger [76], on the other hand, proposed the conditional error function approach, and demonstrated its use for the data-driven sample size recalculation. Many of the later approaches follow or extend these approaches. Although different in their appearance combination tests and the conditional error function approach are closely related [71] and are both based on a common principle which has been called *conditional invariance principle* [13]. We will show how the general principles of adaptive and multiple testing designs can be combined such that the resulting procedures fulfill the regulatory demand of strong Type I error rate control in the presence of adaptations [27, 29, 30]. For recent reviews of adaptive designs for multiple hypotheses, see, e.g., [16, 72, 96].

The structure of this chapter is as follows: We first describe the basic methodology for adaptive designs with multiple hypotheses. We restrict our description to two-stage designs, i.e., a design with one interim analysis. This is done, on the one hand, for simplicity because the procedures can easily be generalized to designs with more than two stages. On the other hand, if a clinical trial should be considered as confirmatory by regulators, a recent review by Elsässer et al. [26] showed the majority of adaptive design proposals submitted to the European Medicines Agency suggested indeed just one adaptive interim analysis and that in most cases the adaptations go beyond early stopping for efficacy and/or futility. The methodology will be applied to the practically most relevant situations that are adaptive treatment arm selection and population enrichment designs. In the last section, we describe some extensions and applications of the designs and briefly discuss the regulatory acceptance.

17.2　General Principles

We consider adaptive two-stage designs with a single interim analysis. Design characteristics of the second stage can be chosen based on the data from the first stage as well as external information. To achieve strict Type I error control despite potential design adaptations, we first introduce the combination testing principle for a single hypothesis, H_0, to be tested in two stages. We then show how this principle can be extended to multiple hypotheses in the framework of the closed testing principle.

17.2.1 The combination testing principle

The main idea of the combination testing principle for adaptive two-stage designs is – instead of pooling across stages – to combine the information via stage-wise calculated test statistics in a predefined way. Let p and q denote the stage-wise (one-tailed) p-values for H_0, such that p is based only on the first stage and q only on the second stage data. A two-stage combination test is defined by a combination function $C(p, q)$ which is left continuous and monotonically increasing in both arguments, boundaries for early stopping α_1, α_0, and a critical value c for the final analysis. The trial is stopped at the interim analysis and H_0 is rejected if $p \leq \alpha_1$ and the trial is stopped for futility if $p \geq \alpha_0$. If the trial proceeds to the second stage, H_0 is rejected if $C(p, q) \leq c$, where c solves

$$\alpha_1 + \int_{\alpha_1}^{\alpha_0} \int_0^1 \mathbf{1}_{[C(x,y) \leq c]} dy \, dx = \alpha \,. \tag{17.1}$$

Here, the indicator function $\mathbf{1}_{[\cdot]}$ equals 1 if $C(p, q) \leq c$ and 0 otherwise. By the definition of c, the combination test is a level α test. This still holds if the design of the second stage (e.g., the choice sample size and/or test statistic) is based on the interim data. The only requirement is that under H_0 the distribution of the second stage p-value q conditioned on p is larger than or equal to the uniform distribution [14]. For example, this holds true if new patients are recruited in the second stage and valid level-α tests are used at each stage. This means that at least conservative tests should be used at each stage. For an exact probabilistic foundation see Brannath, Gutjahr, and Bauer [12].

As seen from (17.1), the critical value c depends on the boundaries α_1 and α_0. For example, for a fixed value of α_1, decreasing values of α_0 result in larger values of c and hence a larger rejection region for the second stage. As a consequence, if the observed first stage p-value p exceeds the futility threshold α_0, the null hypothesis H_0 must be retained, irrespective of the second stage results. Thus, in practice, $p \geq \alpha_0$ implies a futility stop because, if this rule is not followed, the Type I error rate might be inflated. Thus, choosing $\alpha_0 < 1$ implies a *binding* futility rule. We note that even for $\alpha_0 = 1$ the study can be stopped for futility without compromising the Type I error rate. This implies a *non-binding* futility rule because one is not forced to stop the study for futility if $p \geq \alpha_0^\star$, where $\alpha_0^\star$ is not used in the determination of α_1 and c. This comes at the cost of power but with a higher degree of flexibility. Nowadays regulatory agencies like the US Food and Drug Administration (FDA) suggest the use of non-binding futility bounds [58].

The workflow of an adaptive two-stage design can be characterized by three important phases: the planning phase before the trial starts, the adaptive interim analysis after stage 1, and the final analysis after stage 2. At the planning stage one has to specify the hypothesis of interest, the combination test (including the function C and the first and second-stage boundaries α_1, α_0 and c), and the design of the first stage (including, e.g., sample size, allocation ratio and test statistic of the first stage). In the adaptive interim analysis it is first assessed whether the trial can be terminated due to crossing one of the early stopping boundaries. Otherwise the second stage design can be fixed using all information available at the adaptive interim analysis. This allows data dependent changes of the sample size, allocation ratio, test statistic, etc.

Most prominent examples of combination functions are Fisher's combination test, i.e., $C(p, q) = pq$ [2, 5] and the weighted inverse normal combination function [5, 22, 57] (see also Chapter 16). The latter combination function is defined as $C(p, q) = 1 -$

$\Phi\left[v\,\Phi^{-1}(1-p) + w\,\Phi^{-1}(1-q)\right]$, where v, w denote predefined weights such that $v^2 + w^2 = 1$, Φ is the cumulative distribution function (cdf) of the standard normal distribution and Φ^{-1} its inverse. For the one-tailed test of the mean of normally distributed observations with known variance the inverse normal combination test with stage-wise prefixed sample sizes n_1, n_2 and weights $v^2 = n_1/(n_1 + n_2)$, $w^2 = n_2/(n_1 + n_2)$ is equivalent to a classical two stage group sequential test. This equivalence also applies for adaptive designs with more than two stages and can be regarded as a decisive advantage of the inverse normal method [98]. As a consequence, the critical values for the inverse normal method, α_1, α_0, and c, can be computed with standard software for group sequential trials (for a recent review, see [3]). Note, however, that this equivalence typically does not apply to testing problems with multiple hypotheses.

Besides formal testing, for reporting study results it is common to provide confidence intervals and overall p-values. There are several possible definitions for defining these quantities. One prominent example is the concept of repeated confidence intervals (RCIs) and related p-values proposed by Jennison and Turnbull [46, 47] for group sequential designs. For the definition of RCIs for adaptive designs we consider the shifted hypotheses

$$H_0(\delta) : \theta = \delta \,,$$

where θ denotes the parameter of interest. At stage k, $k = 1, 2$, the RCI contains those δ values for which $H_0(\delta)$ cannot be rejected (at stage k), using the specified boundaries for the two-stage combination test. This interval can be calculated repeatedly at both stages irrespective of whether the hypothesis H_0 could be rejected and/or the trial was stopped for efficacy or for futility. For adaptive designs with a single hypothesis, RCIs were proposed by [57]. There is also the possibility to define confidence intervals that are based on the stage-wise ordering [14]. Whereas RCIs can even be calculated (for example) if we could have rejected H_0 at the interim but we choose to continue, the latter method relies on the stage 1 stopping rule being strictly enforced. So in contrast to RCIs they can only be calculated once, at the end of the trial. We note that the calculation of confidence intervals in adaptive designs that are based on the stage-wise ordering is currently restricted to the two-stage case.

Overall (repeated) p-values for H_0 are defined as the smallest significance level for which the study results, given a specified family of boundaries, yield rejection of H_0. These quantities are generally found by a numerical root finding algorithm and can be calculated at both stages stage of the trial. By definition, at stage k, $k = 1, 2$, an overall p-value falls below the overall significance level α if and only if H_0 can be rejected at stage k. Hence, these p-values are consistent with the test decision. A corresponding overall p-value that is based on the stage-wise ordering (so-called monotone p-values) was proposed by Brannath, Posch, and Bauer [14].

In principle, not the same hypotheses need to be tested at each stage. It is also possible that, at the interim analysis, the hypothesis H_0 is replaced by a different hypothesis H_0'. The resulting two-stage procedure based on the combination test is then a level-α test for the intersection hypothesis $H_0 \cap H_0'$. Rejection of this hypothesis means that H_0 or H_0' or both are false. In order to make inferences about the individual hypotheses, H_0 and H_0', a test procedure that controls the experimentwise error rate in the strong sense needs to be applied. Strong experimentwise error rate control is achieved if the probability of at least one erroneous rejection is bounded by α, irrespective of which and how many of the hypotheses are true. A general principle that guarantees strong control is the closed testing principle due to Marcus, Peritz, and Gabriel [62].

17.2.2 The closed testing principle

To describe the closed testing principle and its application for two-stage adaptive designs, assume that we are interested in testing G elementary hypotheses $H_0^1, \ldots, H_0^G$. That is, we will extend the adaptive two-stage design to testing a set of hypotheses. We then derive the corresponding closed system of hypotheses. This consists of all possible intersection hypotheses $H_0^{\mathcal{J}} = \bigcap_{g \in \mathcal{J}} H_0^g, \mathcal{J} \subseteq \{1, \ldots, G\}, \mathcal{J} \neq \emptyset$, including the global null hypothesis $H_0 = H_0^1 \cap \ldots \cap H_0^G$. For each $H_0^{\mathcal{J}}$ a suitable level-α test needs to be defined (the so-called "local" tests). The closed test rejects an elementary hypothesis H_0^g (controlling the experimentwise error rate) if all hypotheses $H_0^{\mathcal{J}}$ with $\mathcal{J} \subseteq \{1, \ldots, G\}$ such that $g \in \mathcal{J}$ can be rejected with their local level-α tests. We note that in the closed test procedure any level-α tests can be chosen as local tests. A simply example, that can be generally applied, is the use of the Bonferroni test but many other tests are available too. If the Bonferroni test is applied within the closed test procedure, this leads to the well known step-down Bonferroni-Holm procedure [40].

To reject an elementary hypothesis H_0^g with the closed test procedure all hypotheses along paths that lead to H_0^g have to be rejected with the local level-α test. Thus, for a set of three hypotheses H_0^1, H_0^2, H_0^3 the closed system of hypotheses and how a rejection of an elementary hypothesis can be reached is shown in Figure 17.1: only if the global hypothesis can be rejected do we test hypotheses consisting of two elementary hypotheses, and only if an intersection hypothesis is rejected do we test the corresponding elementary null hypotheses.

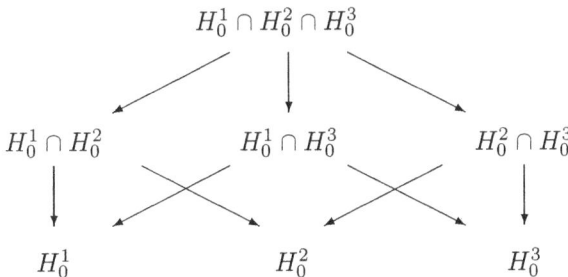

FIGURE 17.1
Closed test procedure for three null hypotheses H_0^1, H_0^2, and H_0^3. Arrows point in the direction of the next hypothesis that can be tested if we reject the current null.

17.2.3 Adaptive designs for multiple hypotheses

Adaptive closed tests combine the closed testing principle with the combination test principle when testing more than one elementary null hypothesis. The general rule for the adaptive closed two-stage test is that we test each of the intersection hypotheses with a suitable combination test [4, 41, 49, 73].

According to the closed testing principle, for each (intersection) hypothesis a level α test has to be prespecified. In the planning phase a level α combination test $C(p, q)$ has to be prespecified for each (intersection) hypothesis. This might permit early stopping to reject or accept an (intersection) null hypothesis. Formally, already rejected intersection hypotheses need not (but can) be re-tested, but hypotheses which have been discarded at the interim for futility are not allowed to be re-tested at stage 2. Note that in principle different combination functions as well as different stopping boundaries could be used for each of the intersection hypotheses. One key characteristic of the adaptive closed test is that for each stage separate test statistics for the intersection hypotheses are calculated and afterwards the stage-wise information is combined via the pre-specified combination function. To address multiplicity within a stage, standard multiplicity adjustments from designs with a fixed sample can be applied resulting in multiplicity adjusted p-values. This means that in addition to prespecifying a combination function for each intersection hypothesis also an appropriate multiplicity adjusted p-value has to be pre-specified for each intersection hypothesis and stage. The adjustment for the first stage p-value has to be laid down in the planning phase, whereas the multiplicity adjustment for the second stage has only to be fixed at the adaptive interim analysis. This means that for a specific intersection hypothesis different multiplicity adjustment strategies could be applied for the two stages. So each intersection hypothesis is tested by the specified combination test using the multiplicity adjusted p-values for the corresponding intersection hypothesis, see Figure 17.2 for testing three elementary null hypotheses. In the figure, the letter for the p-values (p and q) denotes stage and the subscript indexes hypotheses, thus p_g and q_g are referring to the unadjusted p-values for testing H_0^g.

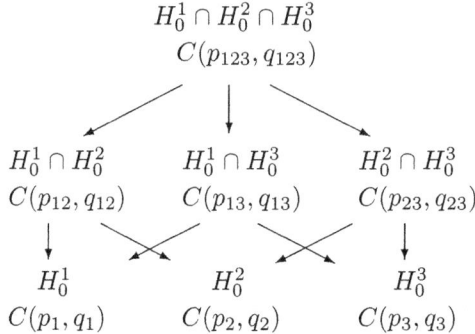

FIGURE 17.2
Combination tests to be performed for the closed system of hypotheses for $G = 3$.

Consequently, a combination test of an intersection hypothesis is well defined even if some of the corresponding elementary hypotheses are deselected at the interim analysis. For example, a hypothesis relating to a sub-population might be de-selected at interim (see Section 17.4) such that no data for calculating a p-value for the deselected hypothesis would be available. It is possible to define second stage tests for intersection hypotheses containing a deselected hypotheses as tests of intersection hypotheses comprising the remaining non-deselected hypotheses. This holds because any test for an intersection hypothesis is a valid test for those hypotheses that are contained in the intersection hypothesis. For example, when testing three elementary null hypotheses, a p-value of a test for H_0^3 can be defined as a p-value for the test of the global hypothesis $H_0 = H_0^1 \cap H_0^2 \cap H_0^3$. Therefore, in the event that H_0^1 and H_0^2 are dropped from stage 2, a valid test for H_0 is the one that combines the first stage p-value for H_0 with the second stage p-value for H_0^3. That is, if p_{123} denotes the p-value for H_0 from the first stage data and q_3 denotes the second stage p-value for H_0^3, the global hypothesis is rejected at the final analysis if $C(p_{123}, q_3) \leq c$. This way of testing the global hypothesis is also applied to the intersection hypotheses. If the combination tests for testing H_0, $H_0^1 \cap H_0^3$, $H_0^2 \cap H_0^3$, and H_0^3 yield significance, H_0^3 can be rejected. Figure

17.3 illustrates the principle for this example. For a detailed numerical example we refer to Bretz et al. [16].

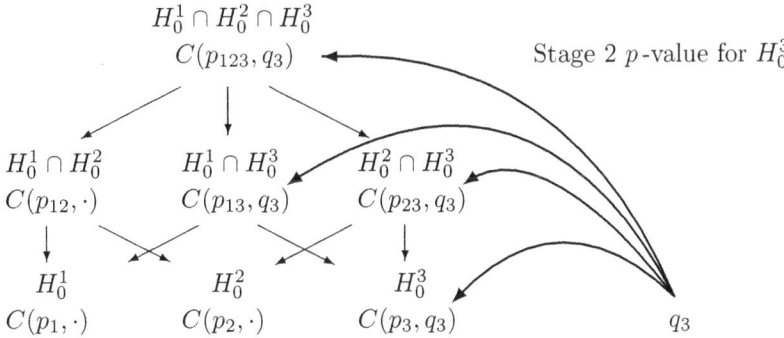

FIGURE 17.3
Combination tests to be performed for the closed system of hypotheses for $G = 3$ if hypothesis H_0^3 is selected for the second stage.

Generally, for the second stage, let $\mathcal{E} \subset \{1, \ldots, G\}$ denote the index set of all excluded H_0^g. As described above, for testing $H_0^{\mathcal{J}}$ the test for $H_0^{\mathcal{J} \setminus \mathcal{E}}$ with $\mathcal{J} \cap \mathcal{E} \neq \emptyset$ is used. Therefore, given a combination function C, at the second stage a selected hypothesis, H_0^s, is rejected at experimentwise level α if

$$\max_{\mathcal{J} \ni s} C(p_{\mathcal{J}}, q_{\mathcal{J} \setminus \mathcal{E}}) \leq c, \tag{17.2}$$

where c denotes the critical value for the combination test. If one hypothesis, H_0^s, is selected, condition (17.2) reduces to

$$\max_{\mathcal{J} \ni s} C(p_{\mathcal{J}}, q_s) \leq c.$$

For this case, the advantage of this kind of adaptive two-stage design becomes obvious: As the multiplicity adjustment is performed within each stage, no adjusted p-values are necessary for the second stage. For any intersection hypotheses containing H_0^s the second stage p-value for H_0^s is plugged into the respective combination function. This comes along with an increased power compared to the procedure where no selection took place.

17.2.4 Assessing the performance of an adaptive design

Assessing the power and other operational characteristics (e.g., bias of effect estimates) of an adaptive trial for multiple hypotheses is a complex task and is usually performed by simulation. Essentially, amongst others, the performance of an adaptive design depends on:

- the considered combination test procedure (type of rejection boundaries and weights),

- the binding or non-binding futility rules,

- the multiple testing adjustment strategy for testing the intersection hypotheses within each stage,

- the timing of the interim analysis and first stage sample sizes,

- the recruitment and dropout rate,

- the handling of missing data and drop outs (e.g., last observation carried forward, baseline observation carried forward, responder analysis, see Chapter 18)

- the parameter and effect sizes of interest,

- the time until the primary endpoint is available,

- the selection procedure (based on endpoint, surrogate, or external information),

- the sample size adaptation rules to determine the second stage sample sizes in each treatment group.

We note that, for Type I error rate control, specifically the selection procedure and the sample size adaptation rules need not be fixed for the real trial at hand, however, for assessing the performance of the adaptive procedure a specific rule needs to be fixed for the simulations. Furthermore, it has to be noted that in the context of closed tests there is no unique power definition and one has to think of an appropriate success criterion when planning the study, e.g., the power to reject at least one, all, or a specific subset of hypotheses [79]. Bias adjusted estimates were proposed too (e.g., [10, 11, 20, 51]), but no clear recommendation can be given and thus the impact of potential adaptations on naïve estimates should evaluated by simulations in the planning phase [6, 35].

In the following two sections we consider two examples of adaptively modifying multiple hypotheses after an interim analysis. The first example illustrates the selection of a treatment arm at interim, and the second example considers the selection of a pre-specified sub-population. We briefly describe the methodology, discuss some relevant issues that need to be considered and specified, and provide case studies.

17.3   Treatment Arm Selection Designs

Multi-arm trial designs with an adaptive interim analysis in order to select treatment arms were mainly developed for the many-to-one comparison setting [4, 41, 73]. They have also been referred to as adaptive seamless designs [19, 37, 48, 59, 78, 97].

17.3.1   The procedure

In the many-to-one comparison setting, G experimental treatment groups are tested against a control group. Assuming a normally distributed endpoint with a common variance σ^2, we consider testing

$$H_0^g : \mu_0 = \mu_g, \; g = 1, \ldots, G,$$

where μ_0 is the mean of the control group and μ_g, $g = 1, \ldots, G$, refer to the means in the active treatment groups. The global null hypothesis is

$$H_0 = \bigcap_{g=1}^{G} H_0^g : \mu_0 = \cdots = \mu_G .$$

When performing the closed test, we also need to consider the intersection null hypotheses $H_0^{\mathcal{J}}$, $\mathcal{J} \subseteq \{1, \ldots, G\}$, stating that all μ_g, $g \in \mathcal{J}$ are equal and equal to μ_0.

To apply the adaptive closed test, the p-values of the global and intersection tests from the two stages will be used. Thus, tests for the global and intersection hypotheses in the closed system of hypotheses need to be specified. Several tests were proposed in the literature, the most prominent test for many-to-one comparisons being the Dunnett test. Note that this test is valid for normally distributed endpoints with a (common) unknown variance. To compute the stage-wise p-values, the multivariate t distribution (with correlation matrix having product correlation structure, known as the Dunnett distribution [33]) is used. Alternatively, the simple but conservative Bonferroni test can be used to compute the stage-wise p-values. Then, the adjusted p-value to test a hypothesis $H_0^{\mathcal{J}}$, $\mathcal{J} \subseteq \{1,\ldots,G\}$, is given by

$$p_{\mathcal{J}} = \min\{|\mathcal{J}| \min_{g\in\mathcal{J}}\{p_g\}, 1\},$$

where p_g denotes the p-value for testing H_0^g (usually calculated from the univariate t-distribution). We note that this adjusted p-value may become equal to 1. Hence, when using it with the inverse normal method we may obtain combination test statistics equal to 1 if the first stage adjusted p-value is equal to 1. This implies an implicit futility stopping criterion. Further tests for the global and intersection hypotheses are the Simes and the Šidák tests. Note that the p-values from both tests yield valid level α test procedures because the correlation between the test statistics is always positive because all comparisons are made against a common control [43]. One can also use a test that is based on a fixed ordering of the hypotheses under consideration which is called the a priori hierarchical intersection test. Then each intersection hypothesis $H_0^{\mathcal{J}}$ is tested with the test for the hypothesis with highest rank in $\mathcal{J}$. For example, if the treatment arms refer to different doses of a drug one might fix in advance that the hierarchy goes from 1 up to G. This means that a hypothesis H_0^g can be rejected if and only if the hypothesis H_0^{g+1}, $g = 1,\ldots,G-1$, can be rejected.

Applying the adaptive two-stage testing procedure, at the first interim analysis, it is possible to stop the trial while showing a significant treatment effect in one (or more) treatment arms. This is the case if all hypotheses which are a subset of the hypothesis referring to this (or these) treatment arms are rejected at significance level α_1. It is also possible to stop the trial due to crossing a futility boundary or conditional power calculations (see Chapter 16). In most cases, however, the first stage is used to select the treatment arm(s) to be considered in the subsequent stage of the trial and/or to reassess the sample size for the subsequent stage.

For $G = 3$, the case of selecting two active treatment arms is illustrated in Figure 17.4. For example, applying the adaptive closed test, in order to show significance for H_0^3 and dropping treatment 1, the combination test statistics $C(p_{123}, q_{23})$, $C(p_{23}, q_{23})$, $C(p_{13}, q_3)$, and $C(p_3, q_3)$ must fall short of the critical bound c where the index in p and q refers to the considered intersection of hypotheses.

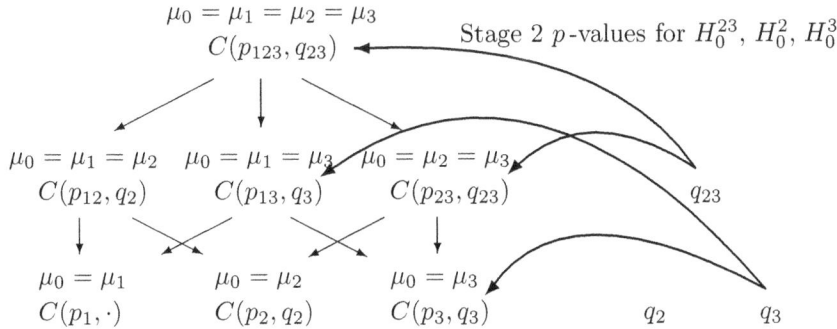

FIGURE 17.4
Combination tests to be performed for the closed system of hypotheses ($G = 3$) for testing hypothesis H_0^3 if treatment arms 2 and 3 are selected for the second stage.

The concept of RCIs and related overall p-values can be adopted to the multi-armed setting. In this testing situation, given a specified family of boundaries (e.g., an O'Brien & Fleming family of boundaries for its use in the inverse normal combination test), overall (repeated) p-values for a hypothesis H_0^g are defined as smallest significance level for which the test results yield rejection of the considered (single) hypothesis H_0^g when applying the closed test procedure. By definition – as for the single hypothesis case – at stage k, an overall p-value falls below the overall significance level α if and only if the corresponding hypothesis H_0^g can be rejected at stage k. Hence, these p-values account for the sequential adaptive and step-down nature of the closed testing principle and are completely consistent with the test decision [98]. As an illustration, the overall p-value is equal to α if the test statistic is equal to the critical value at the given stage. At an early stage, this usually corresponds to a very small p-value. In some sense, this might seem counterintuitive but is the price to be paid for the possible repeated calculation.

RCIs for the multi-armed setting are more difficult to construct. In general, confidence intervals that are based on stepwise closed test procedures are problematic and even for single-stage designs there are no straightforward solutions. A solution for adaptive closed tests proposed by Posch et al. [73] is to replace the stepwise procedure by the corresponding single-step procedure. That is, consider the shifted hypotheses

$$H_0^g(\delta_g) : \mu_g - \mu_0 = \delta_g, \; g = 1, \dots, G,$$

with corresponding p-values $p_g(\delta_g)$ and $q_g(\delta_g)$ at each stage. As for the overall p-value the confidence intervals are computed separately for each treatment arm g that was selected. For example, when using cut-off tests like the Bonferroni, the idea is to replace the adjustment for the intersection hypotheses with the adjustment for the global hypothesis (at the first stage) or the adjustment, if necessary, for the hypothesis that refers to the selected arms at the second stage. For example, if one active treatment arm $s \in \{1, \dots, G\}$ is selected for the second stage, using the inverse normal method and Bonferroni intersection tests, the lower bound of the confidence interval for treatment arm s is obtained by

$$\max\{\delta_s : 1 - \Phi\big(\Phi^{-1}(1 - \min\{1, G\,p_s(\delta_s)\}) + \Phi^{-1}(1 - q_s(\delta_s))/\sqrt{2}\big) \geq c\}.$$

The upper bound is determined analogously. Note that for the a priori hierarchical intersection test, these RCIs are not defined.

If only the treatment arm with the smallest p-value was selected in the first stage of the trial, the resulting confidence intervals are completely consistent with the test decision. In general, however, they are not. For example, it might happen that a hypothesis is rejected but the lower bound of the CI is below 0. Simultaneous confidence intervals that are generally consistent with adaptive closed tests are provided in [61]. These confidence intervals, however, can be "non-informative", i.e., there are outcomes where the confidence interval does not give more information than the test.

For assessing the performance of an multi-armed design, if the many-to-one comparison relates to different dose groups, often as a starting point a linear relationship between treatment arms and effect size can be assumed. However, also other relationships, for example, defined through an exponentially shaped parameter curve, an umbrella like shape, a logistic shape, a sigmoid Emax shape, or a parameter shape with constant effect sizes for all treatment arms, should be assessed. For details see, for example, [16, 18, 25, 97].

17.3.2 Binary and survival endpoints

For binary endpoints, the adaptive closed test can be easily adapted for testing

$$H_0^g : \pi_0 = \pi_g, \; g = 1, \dots, G,$$

where π_g denotes the unknown proportion in treatment group g, $g = 0, \ldots, G$. At each stage k, $k = 1, 2$, we consider the test statistic

$$Z_{g,k} = \frac{\hat{\pi}_{g,k} - \hat{\pi}_{0,k}}{\sqrt{\hat{\bar{\pi}}_k(1 - \hat{\bar{\pi}}_k)}} \left(\frac{1}{n_{0,k}} + \frac{1}{n_{g,k}} \right)^{-1/2},$$

where $\hat{\pi}_{0,k}$ and $\hat{\pi}_{g,k}$ are the observed rates at stage k in the two treatment groups and $\hat{\bar{\pi}}_k = (n_{0,k}\hat{\pi}_{0,k} + n_{g,k}\hat{\pi}_{g,k})/(n_{0,k} + n_{g,k})$ is the observed overall rate at stage k. These test statistics are approximately normal and so the stage-wise p-values are calculated with the use of the normal cdf. So it is straightforward to combine the p-values with a suitable combination test. For the many-to-one treatment arm comparison scenario, this can be done for each pairwise comparison such that this procedure can be used within the adaptive closed test procedure as described above. As intersection tests, the Bonferroni, the Simes and the Šidák tests can be applied. The Dunnett test can be used for an approximate strong control of the experimentwise Type I error rate as well. We further note that other tests, such as Fisher's exact test, can also be used to compute the p-values per stage and to combine them with the specified combination test.

The calculation of confidence intervals is possible with the use of a combination test and an appropriately defined test statistics or p-values for testing the hypotheses $H_0^g(\delta_g)$: $\pi_g - \pi_0 = \delta_g$, $g = 1, \ldots, G$. For calculating the p-value for testing $H_0^g(\delta_g)$, a number of proposals have been made, for example, the method proposed by Farrington and Manning [28]. The application of this technique in adaptive designs with a single hypothesis was described in [94].

For survival trials, consider testing the pairwise comparisons

$$H_0^g : \lambda_g = 1, \ g = 1, \ldots, G,$$

where λ_g denotes the hazard ratio comparing the hazards of treatment arm g compared with the control treatment arm. Denoting $LR_{g,1}^*$ and $LR_{g,2}^*$ the log-rank statistics and $d_{g,1}$ and $d_{g,2}$, $g = 1, \ldots, G$, the number of events in the comparison of treatment group g with the control group up to the interim and the final stage, respectively,

$$Z_{g,2} = \frac{\sqrt{d_{g,2}}\, LR_{g,2}^* - \sqrt{d_{g,1}}\, LR_{g,1}^*}{\sqrt{d_{g,2} - d_{g,1}}}, \ g = 1, \ldots, G,$$

is the approximately independent increment of the logrank statistic for analysis set g in stage k. The approximately normally distributed and independent test statistics $Z_{g,1} = LR_{g,1}^*$ and $Z_{g,2}$ (more precisely, the corresponding p-values) can be used within the combination test yielding approximate control of the Type I error rate in the adaptive case. Therefore, the methods described above can be applied to survival trials. As intersection hypotheses tests, all methods (including Dunnett's method) can be applied yielding approximately strong control of the experimentwise Type I error rate. The calculation of overall p-values and RCIs is possible too [95].

A problem arises if not only the log rank statistic based on the primary endpoint itself but another (correlated) endpoint is used for the design adaptation at interim. This is typical for survival trials because the primary endpoint is often a long-term endpoint and thus a correlated endpoint (usually a surrogate) is used for the design adaptation. Especially, if information on patients who are still under risk is used, the adaptive design that is based on the combination testing principle may fail to control the Type I error rate. This is true even for two-armed survival trials incorporating an adaptive sample size re-estimation procedure and was first noted by Bauer and Posch [7]. Several proposals have been made to overcome this problem. Essentially, the proposals are either based on a modification of the combination testing principle [21, 44, 45, 60, 65, 82] or by requiring additional assumptions regarding the joint distribution of the primary and the short-term endpoints [23, 81].

17.3.3 Case studies

From a recent summary in [3] we report two case studies where an adaptive seamless design with treatment selection at interim was used.

The Hemangeol Trial

This is a trial in a pediatric population. Hemangeol (a pediatric-specific oral propranolol solution) was developed as the treatment for proliferating infantile hemangioma requiring systemic therapy. A Phase II/III trial was performed to assess the efficacy and safety of the medicine in infants 1 to 5 months of age. The primary endpoint was complete recovery at week 16 and a two-stage confirmatory adaptive trial with regimen selection at interim was used to identify the appropriate dose and duration for further study in the second stage. Infants were randomly assigned to receive placebo or one of four propranolol regimens (1 or 3 mg of propranolol base per kilogram of body weight per day for 3 or 6 months). Prespecified possible adaptations to be made after the interim analysis, as outlined in the protocol and statistical analysis plan, were selection of one or two regimens, sample-size reassessment, and stopping for futility. As combination test the inverse normal method was specified and for testing the intersection tests, Simes test was used. At the interim analysis, the independent data and safety monitoring committee selected one of the drug regimens and did not recommend adjusting the planned sample size [56]. At the final analysis the selected regimen was shown to be superior and marketing authorization of Hemangeol was granted by both FDA and European Medicines Agency (EMA) in 2014. Heritier, Lô, and Morgan [39] provide statistical details of the adaptive design.

The ESCAMI Trial

The first major clinical trial using the adaptive design methodology for many-to-one comparisons with dose-selection at an adaptive interim analysis was the ESCAMI trial which is described in Zeymer et al. [100]. It was an international, prospective, randomized, double-blind, placebo-controlled Phase II trial in patients undergoing thrombolytic therapy or primary angioplasty for acute ST-elevation myocardial infarction. The primary endpoint was the immediate response variable infarct size measured by the cumulative release of α-HDBH within 72 hours after administration of the drug. The trial was conducted as a two-stage design with Fisher's combination test and trend tests for the stage-wise p-values. In the interim analysis, two out of four doses of the study drug were selected. At the end no hypotheses could be rejected, and so the trial did not succeed in showing that the drug was superior to placebo at any of the investigated dose levels. For further details on the study results, see Bauer et al. [3].

Other examples of adaptive treatment selection designs have been implemented since then [66]. A recent review of adaptive designs undertaken in clinical research is also provided by Hatfield et al. [38].

17.4 Population Enrichment Designs

Adaptive enrichment designs are interesting for situations where studies of unselected patients might be unable to detect a drug effect and it seems necessary to "enrich" the study with potential responders, which are defined as a subpopulation of the unselected patient population where the specification of the subgroup is through a baseline variable. If the selection of the subpopulation is done in an adaptive way, we might use "adaptive population enrichment designs" [91–93]. We note that Bob Temple [85] was the first who used the term "enrichment" for this and similar kinds of patient selection, see also [86, 87]. Brannath et al. [15] proposed the adaptive closed test for enrichment designs as described here and used Bayesian decision tools for the selection rule, see also [36]. A systematic review of (also exploratory) procedures for enrichment designs is provided in [70], see also [77].

17.4.1   The procedure

Assume that there is a full population F with $G - 1$ pre-specified subpopulations of interest denoted as $S_1, S_2, \ldots, S_{G-1}$ such that $S_g \subset F$. Let S_G denote the full population F. Assuming normally distributed outcomes consider a set of G elementary hypotheses

$$H_0^{S_g} : \mu_0^g = \mu_1^g, \ g = 1, \ldots, G,$$

where $H_0^{S_g}$ tests the effect of the experimental treatment μ_1^g versus control μ_0^g in subpopulation S_g. As before, the closed system of hypotheses consists of all possible intersection hypotheses

$$H_0^{\mathcal{J}} = \bigcap_{g \in \mathcal{J}} H_0^{S_g}, \ \mathcal{J} \subseteq \{1, \ldots, G\}.$$

The global test decision follows from testing the global hypothesis

$$H_0 = \bigcap_{g=1}^G H_0^{S_g}$$

with a suitable global intersection test. If the global null hypothesis H_0 can be rejected, all other intersection hypotheses are then tested. By performing the closed test procedure an elementary hypothesis $H_0^{S_g}$ can be rejected if the combination test fulfills the rejection criterion for all $H_0^{\mathcal{J}}$ with $\mathcal{J} \ni g$.

   As above, given a combination function C, at the second stage the hypothesis belonging to a selected subpopulation s is rejected if

$$\max_{\mathcal{J} \ni s} C(p_{\mathcal{J}}, q_{\mathcal{J} \setminus \varepsilon}) \leq c,$$

where $\varepsilon \subset \{1, \ldots, G\}$ denotes the index set of all excluded H_0^g, $\mathcal{J} \cap \varepsilon \neq \emptyset$, and c denotes the critical value for the combination test.

   As for the adaptive treatment arm selection case, for the closed test procedure in enrichment designs several choices of the tests for the global and intersection hypotheses are available. The choice of an intersection test that is based on the maximum statistic similar to Dunnett's test is possible for $G = 2$, i.e, for one subpopulation $S \subset F$. This was proposed by [32, 80] for the known variance case. More general, for normal responses with a common unknown variance the adjusted p-value, p^{adj}, for testing the global null hypothesis H_0 is given by

$$p^{\text{adj}} = 1 - F_{\Sigma, df}\left(\max\{Z^F, Z^S\} \right),$$

where $F_{\Sigma, df}(\cdot)$ is the cdf of the bivariate t distribution with correlation matrix Σ having element

$$\sigma_{SF} = \sqrt{\dfrac{n_0^S + n_1^S}{n_0^F + n_1^F}}$$

and $df = n_0^F + n_1^F - 4$ degrees of freedom,

$$Z^g = \frac{\bar{x}_1^g - \bar{x}_0^g}{\hat{\sigma}\sqrt{1/n_0^g + 1/n_1^g}}$$

is the directional test statistic for analysis set $g \in \{S, F\}$, and $\hat{\sigma}^2$ is the residual variance estimate corresponding to a two-factorial model with factors treatment and subgroup [98]. This test provides exact Type I error rate control and can even be generalized to more than one subpopulation accounting for the specific structure and computability of the multivariate t-distribution [33]. The Bonferroni, Šidák, Simes, and the hierarchical test can be applied as well. Again, both Simes and Šidák derived p-values for intersection hypotheses yield valid level α test procedures since in it can be assumed that the elements of Σ are always non-negative. Therefore, Type I error rate control is guaranteed.

For $G = 2$, using the adaptive closed test procedure, at an interim stage one can decide to continue to stage 2 to test H_0^F and H_0^S, H_0^F only, or H_0^S only. For testing H_0^S only, the procedure is illustrated in Figure 17.5, the curved arrows indicating which combination tests have to be carried out in order to show significance of the selected subgroup S.

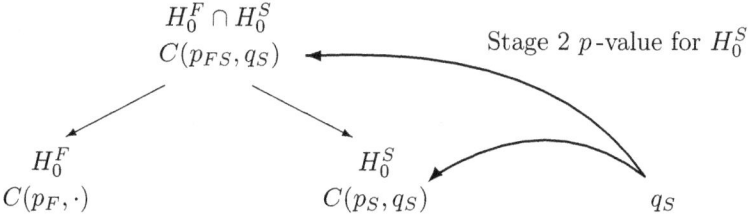

FIGURE 17.5
Combination tests to be performed for the closed system of hypotheses $(G = 2)$ if subpopulation S referring to hypothesis H_0^S is selected for the second stage.

Note that for a two-stage trial where no subpopulation is selected at interim, the complete set of intersection hypotheses is tested at each stage, yielding p-values $p_{\mathcal{J}}$ and $q_{\mathcal{J}}$ for each intersection hypothesis $H_0^{\mathcal{J}}$. These p-values are combined according to the specified combination test, for example, the inverse normal method or Fisher's combination test. This combination test might have a power disadvantage as compared to the single-stage non-adaptive test test where the p-values would be obtained from the pooled data. However, we have the advantage that data driven adaptations including subgroup selection are possible, thereby improving power.

Overall (repeated) p-values and simultaneous confidence intervals can be constructed as for the treatment arm selection designs. For the simultaneous confidence intervals, we consider the shifted hypothesis

$$H_0^{S_g}(\delta_g) : \mu_1^g - \mu_0^g = \delta_g, \ g = 1, \dots, G,$$

with corresponding stage-wise p-values $p_g(\delta_g)$ and $q_g(\delta_g)$ and derive the multiplicity adjusted shifted p-values with the use of the selected intersection hypothesis test. Then the method described for the treatment arm selection case is used.

17.4.2 Effect specification

Designing a population enrichment trial is a complex task and the statistical performance of an enrichment design is usually assessed by simulation. Specifically, the operating characteristics depend on the effect sizes and prevalences of the considered subpopulations S_g. If a simple subpopulation S is considered, this is to specify the prevalence of S and the assumed effect size in S and $F\backslash S$. The effect size in F is then a weighted average of the effect sizes in the disjunct subgroups S and $F\backslash S$, respectively. If we want to consider a range of possible effect sizes there is typically a large number of possible configurations because each effect size in S is combined with the effect sizes in $F\backslash S$.

For $G = 3$, things become even more difficult. Assume there are two dichotomous indicators I_1 and I_2 that classify the patients in the full population as I_1^- or I_1^+ and I_2^- or I_2^+. I_1 may be a baseline characteristic such as gender, race, performance status, or disease stage. I_2 may be a genomic biomarker (positive versus negative) or a genomic signature (good versus poor). We may be interested in showing the treatment effect of a test drug versus placebo in the full population and in I_1^+ and I_2^+. In this case, we have three analysis sets: $S_1 = I_1^+$, $S_2 = I_2^+$, and $S_3 = F$.

In general, $S_1 \cap S_2 \neq \emptyset$ and therefore the operating characteristics of the test procedure depend on the effect sizes in $S_1\backslash S_2$, $S_2\backslash S_1$, $S_1 \cap S_2$ and $F\backslash(S_1 \cup S_2)$ and their assumed prevalences. $S_1 \cap S_2 = \emptyset$ can be assumed if an investigation of effect sizes is considered in different patient populations (for example, countries) besides the effect in a full population. This reduces the number of necessary specifications for the prevalences and the effect sizes a bit. If we consider the nested case where $S_1 \subset S_2 \subset F$, such as in [92], $S_1 \cap S_2 = S_1$ and $S_1\backslash S_2 = \emptyset$. Note that if we are also interested in showing an effect in $S_1 \cap S_2$ (i.e., interaction effects), then we specify $G = 4$ subpopulations $S_1 = I_1^+$, $S_2 = I_2^+$, $S_3 = I_1^+ \cap I_2^+$, and $S_4 = F$.

17.4.3 Binary and survival endpoints

For other than continuous and normally distributed endpoints the specification of tests for hypotheses referring to specific subgroups (including the full population) may become a bit more difficult. We report the summary of these still unresolved issues from [98].

In the binary case, the derivation of closed adaptive tests for the set of G elementary null hypotheses

$$H_0^{S_g} : \pi_0^g = \pi_1^g, \ g = 1, \ldots, G,$$

is formally straightforward when using stage-wise adjusted p-values like the Bonferroni, Šidák, or Simes test which do not account for the correlation structure of the test statistics. If we wish to account for the correlation between the G test statistics by using the asymptotic multivariate normal distribution of the effect estimates for the adjusted stage-wise p-values additional problems may arise in the estimation of the correlation matrix which depends on the different null proportions in the different subpopulations. As far as we know, no research has been done on these issues yet.

In the survival case, when testing

$$H_0^{S_g} : \lambda^g = 1, \ g = 1, \ldots, G,$$

one needs to use the stratified log-rank test with the disjoint subgroups as strata in order to achieve asymptotic Type I error rate control where the disjoint subgroups result from the specific configuration of the subpopulations [15]. Utilization of the approximate multivariate normal distribution for improving the stage-wise p-values is more straightforward here but, as far as we know, has not been published yet.

Furthermore, for the survival case it was described what problems can arise if the population selection is based on a short-term rather than on the primary endpoint. A solution specifically referring to the population enrichment case was given in [45].

17.4.4 Case studies

The statistical methodology for adaptive enrichment designs as it is described here is relatively new, at least it is one of the newest emerging fields of current research in adaptive designs. Therefore, although many therapeutic areas for the potential application of adaptive enrichment designs have been published recently [9, 63–65, 89, 90, 99] there is still a lack of concrete trial examples that were planned and conducted with this statistical methodology. Wassmer and Dragalin [99] presented hypothetical case studies from I-SPY 2 which is an ongoing clinical trial in patients with high-risk primary breast cancer [1]. They showed how the described confirmatory adaptive enrichment designs might be suitable alternatives to the traditional trial designs and demonstrated how to assess the operational characteristics of such designs.

17.5 Discussion and Further Developments

In this chapter we have shown how two general tools – combination tests and the closed testing principle – can be used to derive test procedures that enable data dependent selection of hypotheses and other design modifications at an interim analysis. A major characteristic and advantage of these methods is that no specific adaptation rules need to be pre-specified for Type I error rate control. This permits adaptations that are not only based on the interim data, but also – for example – on emerging new external information and expert knowledge. This is particularly valuable in long term clinical trials and therefore these methods are specifically attractive for such trials. Additionally, the sample size can be adjusted also for other reasons than to control the power of the primary hypothesis test. For instance, one can increase the sample size to obtain more information on important secondary efficacy or safety endpoints.

An even more general approach than the combination testing approach is given by the conditional rejection probability (CRP) principle due to Müller and Schäfer [67, 68]. The idea of this approach is to consider, at the adaptive interim analysis, the conditional Type I error rate $\alpha(x_k)$ which is defined as the conditional probability, under H_0, of rejecting H_0 in one of the subsequent stages, given the data observed up to this stage, x_k. The remainder of the trial can be performed as a test at level $\alpha(x_k)$ where the design of this test is arbitrary. That is, at an interim analysis one can choose to re-design the remainder of the study according to a one-stage or a multistage design with appropriately chosen decision regions, new weights for combining the test statistic, or even a change in the hypotheses tested in further analyses. Particularly, this involves a free choice of the number of stages considered in a group sequential test design. Furthermore, the extension to multiple hypotheses is straightforward.

The CRP principle essentially coincides with the "recursive" application of combination tests as proposed by Brannath, Posch, and Bauer [14] when testing means with normally distributed data, the variance is assumed to be known, and the inverse normal combination test was chosen as the adaptive test. Generally, however, the calculation of the conditional Type I error rate is based on the original test statistic, for example, on the t test statistic or the test statistic of the Dunnett, Bonferroni, or some other test. In the t test situation (i.e., if the variance is not assumed to be known), the calculation of the conditional Type I error rate may become complicated [75, 88].

An interesting application of the CRP principle is for multi-armed designs as proposed by König et al. [53]. In the population enrichment case, for $G = 2$, it is also possible to derive a test procedure that is based on this principle. This was proposed by Friede, Parsons, and Stallard [32] who reported in the correction that in realistic situations the difference in power as compared to the combination test approach is only small and there is actually no power advantage of the CRP methodology. For most other testing situations, the (exact) implementation of the CRP approach, i.e., the exact determination of the conditional Type I error rate based on the originally defined test statistics is not possible, as it may depend on nuisance parameters. Nevertheless, in some settings it can be replaced by the conditional Type I error rate of the inverse normal combination test. The latter even yields an exact level α testing procedure if an exact way for calculating the p-values for the separate stages is available.

Another way to derive multiple tests that control the experimentwise error in the strong sense is based on directed, weighted graphs. This was introduced by Bretz et al. [17] and adopted for the adaptive case by [52, 84].

We also note that another option provided by adaptive multiple tests is to adapt the second stage test statistic or the structure of the underlying test procedure in a multiple testing situation. This application was investigated for trend tests [54] and non-parametric procedures [69]. Kieser, Schneider, and Friede [50] proposed a bootstrap procedure for the adaptive selection of the test statistic. Lawrence [55] considered the change of the test statistic in survival trials. Finally, in studies with multiple endpoints, it might be appropriate to either deselect an endpoint that will, for example, likely fail to show a significant outcome at the end of the trial. Or it might be desirable to select a specific test based on interim outcomes. A trial with adaptive endpoint selection that involved an adaptive change of hypotheses is described in [98].

Adaptive designs have raised much interest in the last two decades among academia, industry, and regulators. There are three regulatory guidance documents available for the use of adaptive clinical trials [27, 29, 30]. All guidances emphasize the need for Type I error rate control as the primary statistical concern when using such designs. Two of the documents were summarized in [26]. The latter paper describes a survey of scientific advice letters from the European Medicines Agencies and showed that mostly there was a positive opinion about these designs. Most frequent concerns raised by the regulators were insufficient justification of the adaptive strategy in general. In general, however, adaptive designs have been well accepted when properly planned.

Acknowledgment

This project has received funding from the European Union's Seventh Framework Programme for research, technological development, and demonstration under grant agreement number FP7 HEALTH 2013-602144 InSPiRe (GW, MP) and grant agreement number 602552 IDeAl (FK). We also thank the referee for making very constructive comments on an earlier version of the manuscript.

Bibliography

[1] A. D. Barker, C. C. Sigman, G. J. Kelloff, N. M. Hylton, D. A. Berry, and L. J. Esserman. I–SPY 2: An adaptive breast cancer trial design in the setting ofneoadjuvant chemotherapy. *Clinical Pharmacology and Therapeutics*, 86:97–100, 2009.

[2] P. Bauer. Multistage testing with adaptive designs. *Biometrie und Informatik in Medizin und Biologie*, 20:130–148, 1989.

[3] P. Bauer, F. Bretz, V. Dragalin, F. König, and G. Wassmer. 25 years of confirmatory adaptive designs: opportunities and pitfalls. *Statistics in Medicine*, 35:325–347, 2016.

[4] P. Bauer and M. Kieser. Combining different phases in the development of medical treatments within a single trial. *Statistics in Medicine*, 34:1833–1848, 1999.

[5] P. Bauer and K. Köhne. Evaluation of experiments with adaptive interim analyses. *Biometrics*, 50:1029–1041, correction in *Biometrics* 1996, 52: 380, 1994.

[6] P. Bauer, F. König, W. Brannath, and M. Posch. Selection and bias – two hostile brothers. *Statistics in Medicine*, 29(1):1–13, 2010.

[7] P. Bauer and M. Posch. Letter to the Editor: Modification of the sample size and the schedule of interim analyses in survival trials based on data inspections. *Statistics in Medicine*, 23:1333–1335, 2004.

[8] S. M. Berry, B. P. Carlin, J. J. Lee, and P. Muller. *Bayesian Adaptive Methods for Clinical Trials*. CRC press, 2010.

[9] D. L. Bhatt and C. Mehta. Adaptive designs for clinical trials. *The New England Journal of Medicine*, 375(1):65–74, 2016.

[10] J. Bowden and E. Glimm. Unbiased estimation of selected treatment means in two-stage trials. *Biometrical Journal*, 50(4):515–527, 2008.

[11] J. Bowden and E. Glimm. Conditionally unbiased and near unbiased estimation of the selected treatment mean for multistage drop-the-losers trials. *Biometrical Journal*, 56(2):332–349, 2014.

[12] W. Brannath, G. Gutjahr, and P. Bauer. Probabilistic foundation of confirmatory adaptive designs. *Journal of the American Statistical Association*, 107:824–832, 2012.

[13] W. Brannath, F. König, and P. Bauer. Multiplicity and flexibility in clinical trials. *Pharmaceutical Statistics*, 6(3):205–216, 2007.

[14] W. Brannath, M. Posch, and P. Bauer. Recursive combination tests. *Journal of the American Statistical Association*, 97:236–244, 2002.

[15] W. Brannath, E. Zuber, M. Branson, F. Bretz, P. Gallo, M. Posch, and A. Racine-Poon. Confirmatory adaptive designs with Bayesian decision tools for a targeted therapy on oncology. *Statistics in Medicine*, 28:1445–1463, 2009.

[16] F. Bretz, F. König, W. Brannath, E. Glimm, and M. Posch. Tutorial in biostatistics: Adaptive designs for confirmatory clinical trials. *Statistics in Medicine*, 28:1181–1217, 2009.

[17] F. Bretz, W. Maurer, W. Brannath, and M. Posch. A graphical approach to sequentially rejective multiple test procedures. *Statistics in Medicine*, 28(4):586–604, 2009.

[18] F. Bretz, J. C. Pinheiro, and M. Branson. Combining multiple comparison and modeling techniques in dose–response studies. *Biometrics*, 61:738–748, 2005.

[19] F. Bretz, H. Schmidli, F. König, A. Racine, and W. Maurer. Confirmatory seamless phase II/III clinical trials with hypotheses selection at interim: General concepts. *Biometrical Journal*, 48(4):623–634, 2006.

[20] M. Carreras and W. Brannath. Shrinkage estimation in two-stage adaptive designs with midtrial treatment selection. *Statistics in Medicine*, 32(10):1677–1690, 2013.

[21] M. Carreras, G. Gutjahr, and W. Brannath. Adaptive seamless designs with interim treatment selection: a case study in oncology. *Statistics in Medicine*, 34:1261–1440, 2015.

[22] L. Cui, H. M. J. Hung, and S. J. Wang. Modification of sample size in group sequential clinical trials. *Biometrics*, 55:853–857, 1999.

[23] L. Di Scala and E. Glimm. Time-to-event analysis with treatment arm selection at interim. *Statistics in Medicine*, 30(26):3067–3081, correction in *Statistics in Medicine* 2013, 32: 1974, 2011.

[24] V. Dragalin. Adaptive designs: Terminology and classification. *Drug Information Journal*, 40:425–435, 2006.

[25] V. Dragalin, F. Hsuan, and S. K. Padmanabhan. Adaptive designs for dose–finding studies based on sigmoid E–max model. *Journal of Biopharmaceutical Statistics*, 17:1051–1070, 2007.

[26] A. Elsäßer, J. Regnstrom, T. Vetter, F. König, R. J. Hemmings, M. Greco, M. Papaluca-Amati, and M. Posch. Adaptive clinical trial designs for European marketing authorization: a survey of scientific advice letters from the European Medicines Agency. *Trials*, 15(1):383, 2014.

[27] EMA. *Reflection Paper on Methodological Issues in Confirmatory Clinical Trials Planned with an Adaptive Design*. European Medicines Agency, London, UK, 2007.

[28] C. P. Farrington and G. Manning. Test statistics and sample size formulae for comparative binomial trials with null hypothesis of non–zero risk difference or non–unity relative risk. *Statistics in Medicine*, 9:1447–1454, 1990.

[29] FDA. *Draft Guidance for Industry Adaptive Design Clinical Trials for Drugs and Biologics*. Food and Drug Administration. Center for Drug Evaluation and Research (CDER) and Center for Biologics Evaluation and Research (CBER), Rockville MD, USA, 2010.

[30] FDA. *Guidance for Industry and Food and Drug Administration Staff. Adaptive Designs for Medical Device Clinical Studies*. Food and Drug Administration. Center for Devices and Radiological Health (CDRH) and Center for Biologics Evaluation and Research (CBER), Rockville MD, USA, 2016.

[31] T. Friede and M. Kieser. Sample size recalculation in internal pilot study designs: a review. *Biometrical Journal*, 48(4):537–555, 2006.

[32] T. Friede, N. Parsons, and N. Stallard. A conditional error function approach for subgroup selection in adaptive clinical trials. *Statistics in Medicine*, 31(30):4309–4320 correction in *Statistics in Medicine* 2014, 32: 2513–2514., 2012.

[33] A. Genz and F. Bretz. *Computation of multivariate normal and t probabilities*, volume 45. Springer, Dordrecht, Heidelberg, London, New York, 2009.

[34] A. C. Graf, P. Bauer, E. Glimm, and F. König. Maximum type 1 error rate inflation in multiarmed clinical trials with adaptive interim sample size modifications. *Biometrical Journal*, 56:614–630, 2014.

[35] A. C. Graf, G. Gutjahr, and W. Brannath. Precision of maximum likelihood estimation in adaptive designs. *Statistics in medicine*, 35(6):922–941, 2016.

[36] A. C. Graf, M. Posch, and F. König. Adaptive designs for subpopulation analysis optimizing utility functions. *Biometrical Journal*, 57:76–89, 2015.

[37] L. V. Hampson and C. Jennison. Optimizing the data combination rule for seamless phase II/III clinical trials. *Statistics in Medicine*, 34(1):39–58, 2015.

[38] I. Hatfield, A. Allison, L. Flight, S. A. Julious, and M. Dimairo. Adaptive designs undertaken in clinical research: a review of registered clinical trials. *Trials*, 17(1):1, 2016.

[39] S. Heritier, S. N. Lô, and C. C. Morgan. An adaptive confirmatory trial with treatment selection: practical experiences and unbalanced randomization. *Statistics in Medicine*, 30:1541–1554, 2011.

[40] S. Holm. A simple sequentially rejective multiple test procedure. *Scandinavian Journal of Statistics*, 6:65–70, 1979.

[41] G. Hommel. Adaptive modifications of hypotheses after an interim analysis. *Biometrical Journal*, 43:581–589, 2001.

[42] F. Hu and W. F. Rosenberger. *The Theory of Response-Adaptive Randomization in Clinical Trials*. John Wiley & Sons, Chichester, 2006.

[43] M. F. Huque. Validity of the Hochberg procedure revisited for clinical trial applications. *Statistics in Medicine*, 35:5–20, 2016.

[44] S. Irle and H. Schäfer. Interim design modifications in time-to-event studies. *Journal of the American Statistical Association*, 107:341–348, 2014.

[45] M. Jenkins, A. Stone, and C. Jennison. An adaptive seamless phase II/III design for oncology trials with subpopulation selection using correlated survival endpoints. *Pharmaceutical Statistics*, 10:347–356, 2011.

[46] C. Jennison and B. W. Turnbull. Interim analysis: the repeated confidence interval approach. *Journal of the Royal Statistical Society B*, 51:305–361, 1989.

[47] C. Jennison and B. W. Turnbull. *Group Sequential Methods with Applications to Clinical Trials*. Chapman & Hall/CRC, Boca Raton, London, New York, Washington, D.C., 2000.

[48] C. Jennison and B. W. Turnbull. Adaptive seamless designs: selection and prospective testing of hypotheses. *Journal of Biopharmaceutical Statistics*, 17(6):1135–1161, 2007.

[49] M. Kieser, P. Bauer, and W. Lehmacher. Inference on multiple endpoints in clinical trials with adaptive interim analyses. *Biometrical Journal*, 41:261–277, 1999.

[50] M. Kieser, B. Schneider, and T. Friede. A bootstrap procedure for adaptive selection of the test statistic in flexible two–stage designs. *Biometrical Journal*, 44:641–652, 2002.

[51] P. K. Kimani, S. Todd, and N. Stallard. Estimation after subpopulation selection in adaptive seamless trials. *Statistics in Medicine*, 34(18):2581–2601, 2015.

[52] F. Klinglmüller, M. Posch, and F. König. Adaptive graph-based multiple testing procedures. *Pharmaceutical Statistics*, 13:345–346, 2014.

[53] F. König, W. Brannath, F. Bretz, and M. Posch. Adaptive Dunnett tests for treatment selection. *Statistics in Medicine*, 27:1612–1625, 2008.

[54] T. Lang, A. Auterith, and P. Bauer. Trend tests with adaptive scoring. *Biometrical Journal*, 42:1007–1020, 2000.

[55] J. Lawrence. Strategies for changing the test statistic during a clinical trial. *Journal of Biopharmaceutical Statistics*, 12:193–205, 2002.

[56] C. Léauté-Labrèze, P. Hoeger, J. Mazereeuw-Hautier, L. Guibaud, E. Baselga, G. Posiunas, R. J. Phillips, H. Caceres, J. C. Lopez Gutierrez, R. Ballona, et al. A randomized, controlled trial of oral propranolol in infantile hemangioma. *New England Journal of Medicine*, 372(8):735–746, 2015.

[57] W. Lehmacher and G. Wassmer. Adaptive sample size calculations in group sequential trials. *Biometrics*, 55(4):1286–1290, 1999.

[58] M. Lin, S. Lee, B. Zhen, J. Scott, A. Horne, G. Solomon, and E. Russek-Cohen. CBER's experience with adaptive design clinical trials. *Therapeutic Innovation & Regulatory Science*, 50(2):195–203, 2016.

[59] J. Maca, S. Bhattacharya, V. Dragalin, P. Gallo, and M. Krams. Adaptive seamless phase II/III designs — background, operational aspects, and examples. *Drug Information Journal*, 40:463–473, 2006.

[60] D. Magirr, T. Jaki, F. König, and M. Posch. Adaptive survival trials. *arXiv preprint arXiv:1405.1569*, 2014.

[61] D. Magirr, T. Jaki, M. Posch, and F. Klinglmüller. Simultaneous confidence intervals that are compatible with closed testing in adaptive designs. *Biometrika*, 100(4):985–996, 2013.

[62] R. Marcus, E. Peritz, and K. R. Gabriel. On closed testing procedures with special reference to ordered analysis of variance. *Biometrika*, 63:655–660, 1976.

[63] C. R. Mehta and P. Gao. Population enrichment designs: case study of a large multinational trial. *Journal of Biopharmaceutical Statistics*, 21(4):831–845, 2011.

[64] C. R. Mehta, P. Gao, D. L. Bhatt, R. A. Harrington, S. Skerjanec, and J. H. Ware. Optimizing trial design sequential, adaptive, and enrichment strategies. *Circulation*, 119(4):597–605, 2009.

[65] C. R. Mehta, H. Schäfer, H. Daniel, and S. Irle. Biomarker driven population enrichment for adaptive oncology trials with time to event endpoints. *Statistics in Medicine*, 33(26):4515–4531, 2014.

[66] C. C. Morgan, S. Huyck, M. Jenkins, L. Chen, A. Bedding, C. S. Coffey, B. Gaydos, and J. K. Wathen. Adaptive design: Results of 2012 survey on perception and use. *Therapeutic Innovation & Regulatory Science*, 48:473–481, 2014.

[67] H.-H. Müller and H. Schäfer. Adaptive group sequential designs for clinical trials: Combining the advantages of adaptive and of classical group sequential approaches. *Biometrics*, 57:886–891, 2001.

[68] H.-H. Müller and H. Schäfer. A general statistical principle for changing a design any time during the course of a trial. *Statistics in Medicine*, 23:2497–2508, 2004.

[69] M. Neuhäuser. An adaptive location–scale test. *Biometrical Journal*, 43:809–819, 2001.

[70] T. Ondra, A. Dmitrienko, T. Friede, A. Graf, F. Miller, N. Stallard, and M. Posch. Methods for identification and confirmation of targeted subgroups in clinical trials: a systematic review. *Journal of Biopharmaceutical Statistics*, 26(1):99–119, 2016.

[71] M. Posch and P. Bauer. Adaptive two stage designs and the conditional error function. *Biometrical Journal*, 41:689–696, 1999.

[72] M. Posch, P. Bauer, and W. Brannath. Flexible designs. In R. B. D'Agostino, L. M. Sullivan, and J. M. Massaro, editors, *Encyclopedia of Clinical Trials*. Wiley, New York, 2009.

[73] M. Posch, F. König, M. Branson, W. Brannath, C. Dunger-Baldauf, and P. Bauer. Testing and estimating in flexible group sequential designs with adaptive treatment selection. *Statistics in Medicine*, 24:3697–3714, 2005.

[74] M. Posch, W. Maurer, and F. Bretz. Type I error rate control in adaptive designs for confirmatory clinical trials with treatment selection at interim. *Pharmaceutical Statistics*, 10(2):96–104, 2011.

[75] M. Posch, N. Timmesfeld, F. König, and H.-H. Müller. Conditional rejection probabilities of Student's t–test and design adaptation. *Biometrical Journal*, 46:389–403, 2004.

[76] M. A. Proschan and S. A. Hunsberger. Designed extension of studies based on conditional power. *Biometrics*, 51:1315–1324, 1995.

[77] K. Rufibach, M. Chen, and H. Nguyen. Comparison of different clinical development plans for confirmatory subpopulation selection. *Contemporary Clinical Trials*, 47:78–84, 2016.

[78] H. Schmidli, F. Bretz, A. Racine, and W. Maurer. Confirmatory seamless phase II/III clinical trials with hypotheses selection at interim: applications and practical considerations. *Biometrical Journal*, 48(4):635–643, 2006.

[79] S. Senn and F. Bretz. Power and sample size when multiple endpoints are considered. *Pharmaceutical Statistics*, 6:161–170, 2007.

[80] B. Spiessens and M. Debois. Adjusted significance levels for subgroup analysis in clinical trials. *Contemporary Clinical Trials*, 31:647–656, 2010.

[81] N. Stallard. A confirmatory seamless phase II/III clinical trial design incorporating short-term endpoint information. *Statistics in Medicine*, 29(9):959–971, 2010.

[82] N. Stallard, T. Hamborg, N. Parsons, and T. Friede. Adaptive designs for confirmatory clinical trials with subgroup selection. *Journal of Biopharmaceutical Statistics*, 24(1):168–187, 2014.

[83] N. Stallard and S. Todd. Sequential designs for phase III clinical trials incorporating treatment selection. *Statistics in Medicine*, 22(5):689–703, 2003.

[84] T. Sugitani, F. Bretz, and W. Maurer. A simple and flexible graphical approach for adaptive group-sequential clinical trials. *Journal of Biopharmaceutical Statistics*, 55(3):341–359, 2014.

[85] R. Temple. Special study designs: Early escape, enrichment, studies in non–responders. *Communications in Statistics – Theory and Methods*, 23:499–531, 1994.

[86] R. Temple. Enrichment designs: Efficiency in development of cancer treatments. *Journal of Clinical Oncology*, 23:4838–4839, 2005.

[87] R. Temple. Enrichment of clinical study populations. *Clinical Pharmacology & Therapeutics*, 88(6):774–778, 2010.

[88] N. Timmesfeld, H. Schäfer, and H.-H. Müller. Increasing the sample size during clinical trials with t-distributed test statistics without inflating the Type I error rate. *Statistics in Medicine*, 26:2449–2464, 2007.

[89] C. Tournoux-Facon, Y. de Ryckee, and P. Tubert-Bitter. How a new stratified adaptive phase II design could improve targeting population. *Statistics in Medicine*, 30:1555–1562, 2011.

[90] C. Tournoux-Facon, Y. de Ryckee, and P. Tubert-Bitter. Targeting population entering phase III trials: A new stratified adaptive phase II design. *Statistics in Medicine*, 30:801–811, 2011.

[91] S. J. Wang and J. H. M. Hung. Adaptive enrichment with subpopulation selection at interim: Methodologies, applications and design considerations. *Contemporary Clinical Trials*, 36(2):673–681, 2013.

[92] S. J. Wang, J. H. M. Hung, and R. T. O'Neill. Adaptive patient enrichment designs in therapeutic trials. *Biometrical Journal*, 51:358–374, 2009.

[93] S. J. Wang, R. T. O'Neill, and J. H. M. Hung. Approaches to evaluation of treatment effect in randomized clinical trials with genomic subset. *Pharmaceutical Statistics*, 6(3):227–244, 2007.

[94] G. Wassmer. Data–driven analysis strategies for proportion studies in adaptive group sequential test designs. *Journal of Biopharmaceutical Statistics*, 13(4):585–603, 2003.

[95] G. Wassmer. Planning and analyzing adaptive group sequential survival trials. *Biometrical Journal*, 48(4):714–729, 2006.

[96] G. Wassmer. Adaptive interim analyses in clinical trials. In A. Pong and S.-C. Pong, editors, *Handbook of Adaptive Designs in Pharmaceutical and Clinical Development*. CRC Press, Boca Raton, London, New York, 2010.

[97] G. Wassmer. On sample size determination in multi–armed confirmatory adaptive designs. *Journal of Biopharmaceutical Statistics*, 21(4):802–817, 2011.

[98] G. Wassmer and W. Brannath. *Group Sequential and Confirmatory Adaptive Designs in Clinical Trials*. Springer, Science and Business Media, 2016.

[99] G. Wassmer and V. Dragalin. Designing issues in confirmatory adaptive population enrichment trials. *Journal of Biopharmaceutical Statistics*, 25:651–669, 2015.

[100] U. Zeymer, H. Suryapranata, J. P. Monassier, G. Opolski, J. Davies, G. Rasmanis, G. Linssen, U. Tebbe, R. Schröder, R. Tiemann, T. Machnig, and K. L. Neuhaus. The Na+/H+ exchange inhibitor eniporide as an adjunct to early reperfusion therapy for acute myocardial infarction1results of the evaluation of the safety and cardioprotective effects of eniporide in acute myocardial infarction (ESCAMI) trial. *Journal of the American College of Cardiology*, 38(6):E1644–E1650, 2001.

Part V

Practical Issues in Analysis of Randomized Controlled Trials

18

Multiple Testing

Yi Liu, Jason Hsu and Szu-Yu Tang

CONTENTS

One view of multiple comparisons is there is an extensive menu of methods to choose from, depending on the problem and perhaps on taste. The perspective we give in this chapter instead is, there are basic principles of multiple testing underlying the methods, and these principle can be applied to construct new methods as decision-making in clinical trials evolves.

Error rates in multiple testing are defined in Section 18.1, and the Partitioning Principle and Closed Testing principle are stated in Section 18.2.

Holm's test and Hochberg's test are presented in Section 18.4, not as menus items to choose from, but as examples of methods that can be constructed from the Partitioning Principle.

In Section 18.5, we use the example of decision-making in clinical trials with multiple doses and multiple endpoints to demonstrate how to construct new methods from basic principles.

Section 18.6 uses the same example to illustrate the choices that may be available in implementing multiple tests constructed from basic principles on the computer. An exciting recent development that allows users to customize their own multiple tests is briefly referenced in Section 18.6.1.

Finally, Section 18.8 anticipates some future multiple comparison developments in clinical trials.

18.1   Error Rates in Multiple Comparisons

Tukey (1953), which has been re-printed as Tukey (1994), laid the foundation of multiple comparisons. Page 4 of Tukey (1994) lists the following three error rates[1]:

error rate per comparison: proportion of incorrect statements, among infinitely many statements made;

error rate per family: number of incorrect statements in a family, averaged over infinitely many families;

error rate familywise: proportion of families with at least one incorrect statement, among infinitely many families.

To illustrate these concepts, consider 4 studies each with a family of 100 comparisons using 100 confidence intervals. Table 18.1 shows the value of the three error rates for each of the study example.

TABLE 18.1
Illustration of three Error Rates concepts in four examples.

	Study 1	Study 2	Study 3	Study 4
No. of CIs not covering true value	**0**	**5**	**7**	**3**
Per comparison errors	0	0.05	0.07	0.03
Per family error	0	5	7	3
Familywise error	0	1	1	1

When any error rate is discussed in Tukey (1994), all three error rates appear together. So it was not the case that Tukey thought error rate familywise – the probability of which is now popularly known as Familywise Error Rate (FWER), is especially more meaningful to control than any other error rate.

In the 1990s, the concept of FWER control was very useful in pointing out the inadequacy of controlling experimentwise error rate, defined as the probability of making at least one incorrect statement when *all* the null hypotheses are true. The distinction is FWER is the *supremum* of the probability of making at least one incorrect statement taken over all possible null/alternative parameter configurations, which can occur when not all the null hypotheses are true. Examples of multiple comparison methods for which the supremum

---

[1]Instead of testing hypotheses, Tukey (1953) provides confidence intervals. So, technically, these are non-coverage rates of confidence intervals.

occurs when some, but not all, of the null hypotheses are true include Fisher's Protected Least Significant Difference (LSD) (Fisher, 1935), and methods based on joint-ranking, see Section 3.3.2 and 5.1.8 of Hsu (1996) for more details. Since experimentwise error rate is often referred to as week control of FWER in the literature, we refer to this definition of FWER by taking the supremum as the the strong control of FWER.

Regardless of which multiple testing error rate is to control, there is further subtlety in the calculation of Type I error rate (i.e. FWER when testing a single null hypothesis), in order to guarantee that error rate control translates to controlling incorrect (regulatory) decision rate. Since clinical trials involve studies of different treatments for different diseases, this translation relies on Kolmogorov's version of the Law of Large Numbers that allows for not identically distributed random variables (Serfling 1980). As discussed in the classic statistical decision theory book by James O. Berger (1985), calculating the supremum of the probability of making at least one incorrect statement should be cognizant of unknown parameters whose values are not explicitly specified in the null hypotheses. For testing a single null hypothesis involving multiple parameters, this recognition can be seen in the definition of Type I error rate in Casella and Berger (1990) as the supremum of the probability of false rejection taken over all possible parameter values, but is sometimes overlooked. For example, Huang et al (2006) and Kaizar, Li, and Hsu (2011) showed that validity of permutation multiple tests for equality of means requires the assumption that variances and co-variances between endpoints are identical between the groups compared.

18.2 Principles of Multiple Testing

Consider testing a family of scientific null hypotheses H_{0i}, $i = 1, \ldots, k$. Let $\theta_i, i = 1, \ldots, k$, be the parameters of interest and let $\boldsymbol{\theta} = (\theta_1, \ldots, \theta_k)$. Formulate the k null hypotheses as

$$H_{0i} : \theta_i \in \Theta_{0i}, \text{ vs } H_{1i} : \theta_i \in \Theta_{1i} \ i = 1, \ldots, k. \tag{18.1}$$

where $\Theta_i = \Theta_{0i} \cup \Theta_{1i}$

The following three settings will be used for illustration throughout this chapter assuming larger parameter value indicating better treatment effect.

Dose × Endpoints It is quite common to test efficacy of the investigational drug with multiple doses in phase II or III clinical trials. For example, a sponsor is interested to see whether high and low doses of a new compound has efficacy in primary and secondary endpoints comparing to a placebo or standard care arm. The scientific questions are then which of the four dose × endpoint combinations have efficacy. Let P and S denote primary and secondary endpoints and subscripts $_1$ and $_2$ denote low and high doses. If $\theta_i^j, i = 1, 2, j = P, S$ denote efficacy in Low/High doses and Primary/Secondary endpoints compared to the placebo/standard care arm, then the statistical null hypotheses can be expressed as following

$$H_{0ij} : \theta_i^j \leq \delta_j, i = 1, 2, j = P, S. \tag{18.2}$$

where δ_j denotes a clinically meaningful difference for endpoint j.

Pairwise comparisons In earlier phase of clinical development, such as phase II studies, if multiple treatment arms are included (e.g. multiple single agent drug candidates, or multiple combinations with an investigational drug), the scientific questions are typically which treatment is better than which other treatment. The statistical formulation in the case of three treatment arms is the following

$$H_{0ij} : \theta_{ij} = \mu_i - \mu_j = 0, 1 \le i < j \le 3. \tag{18.3}$$

where μ_1, μ_2, μ_3 denotes the treatment effect for each arm.

Subgroups Personalized medicine (or equivalently, tailored therapeutics) has been the direction for the entire public heath sector. Relating to drug development, we are typically concerned with finding whether there are subgroups of an overall patient population that exhibit a differential response to treatment. Suppose there exists a biomarker that separates the population into two groups, denoted as marker positive (g^+) and marker negative (g^-), then the statistical null hypotheses may be

$$H_{0g^+} : \theta_{g^+} \le 0, H_{0g^-} : \theta_{g^-} \le 0, \text{ and } \overline{H}_0 : \overline{\theta} \le 0 \tag{18.4}$$

where $\theta_{g^+}, \theta_{g^-}, \overline{\theta}$ denotes efficacy in g^+, g^- and overall population.

18.2.1 Partitioning principle

The Partitioning Principle (Takeuchi 1973, Takeuchi 2010, Stefansson et al., 1988, Finner and Strassburger, 2002) is a general principle for constructing multiple tests. We demonstrate the construction of multiple tests that control FWER using this principle.

For testing (1), partition testing proceeds as follows.

P1: Partition $\Theta_1 \times \cdots \times \Theta_k$ as follows. For each $I \subseteq \{1, \ldots, k\}$, form

$$\Theta_I = \{\boldsymbol{\theta} | \theta_i \in \Theta_{0i} \text{ for all } i \in I \text{ and } \theta_j \notin \Theta_{0j} \text{ for all } j \notin I\}.$$

There are up to 2^k such parameter subspaces.

P2: For each $I \ne \emptyset$, test the partition null hypothesis

$$H_{0I}^* : \boldsymbol{\theta} \in \Theta_I$$

at level-α. There are at most $2^k - 1$ hypotheses to be tested.

P3: For each i, infer $\theta_i \notin \Theta_{0i}$ if and only if all H_{0I}^* with $i \in I$ are rejected. This is a logical conclusion because H_{0i} is the union of H_{0I}^* with $i \in I$.

Since the partition null hypotheses H_{0I}^*'s are disjoint, at most one H_{0I}^* can be true. Therefore, by the definition of strong FWER control, as long as each H_{0I}^* is tested as level α, the supremum is also no more than α. In other words, even though there is no multiplicity adjustment in testing the up to $2^k - 1$ hypotheses H_{0I}^*, partition testing controls FWER strongly.

Interestingly, Tukey's (1953) method for pairwise comparisons, which predates the Partitioning Principle, can be thought of as derived from *extreme* partitioning. The parameter space is partitioned into the infinitely many possible (vector) parameter values of ($\theta_{ij} = \mu_i - \mu_j, 1 \le i < j \le 3$). For each possible parameter value, the null hypothesis that it is the true parameter value is rejected if

$$\max_{1 \le i < j \le 3} (\hat{\theta}_{ij} - \theta_{ij}) > q \tag{18.5}$$

where $\hat{\theta}_{ij}$ is the sample estimate of θ_{ij} and q is the critical value so that the test is of level-α

(without multiplicity adjustment, in accordance with the Partitioning Principle). Then, the set of parameters that fail to be rejected is exactly the Tukey's $1-\alpha$ level confidence set.

So, why is there interest in less extreme partitioning also? The reason is, different scientific inference may be of interest in different parts of the partition Θ_I. Within Θ_I, it is known that the null hypotheses $H_{0j}, j \notin I$, are false, so rejecting those hypotheses cause no Type I error. Thus, instead of full multiplicity adjustment for all dimensions of Θ_I, one can tailor local level α test of each rejection region for Θ_I, as illustrated in the remainder of this chapter.

In principle, the Partitioning Principle can be used to construct multiple tests that control error rates other than FWER. See Xu and Hsu (2007) for example. However, this chapter will focus on FWER, since that is the error rate currently controlled in clinical trials.

18.2.2 Closed testing principle

The closed testing principle Marcus, Peritz, and Gabriel (1976) is similar to the Partitioning Principle but tests less restrictive hypotheses.

C1: For each $I \subseteq \{1, \ldots, k\}$, form the closure by testing the following intersection hypothesis

$$H_{0I}^{\cap} : \boldsymbol{\theta} \in \{\boldsymbol{\theta} | \theta_i \in \Theta_{0i} \text{ for all } i \in I\}.$$

C2: Test each $H_{0I}^{\cap}$ at level-α.

C3: For each i, infer $\theta_i \notin \Theta_{0i}$ if and only if all $H_{0I}^{\cap}$ with $i \in I$ are rejected.

Closed Testing guarantees strong control of FWER because a level-α test for $H_{0I}^{\cap}$ is also a level-α test for H_{0I}^{*}. However, closed testing can be less flexible in forming the partition null hypotheses H_{0I}^{*} in the presence of decision path as will be shown in next few sections. Gatekeeping methods such as Dmitrienko et al (2008) and Dmitrienko and Tamhane (2011) are based on closed testing. Bretz et al. (2009) invented a graphical human-computer interface that lets users flexibly design multiple tests for a variety of scientific problems, which the computer can then execute using the closed testing principle.

18.3 A Simple Example

In the Dose $\times$ Endpoints example, consider making inference on the primary endpoint only. The scientific questions are which dose or doses have efficacy, and the initial statistical null hypotheses to be tested are $H_{0i}^{P} : \theta_i^{P} \leq 0, i = L, H$.

Default application of the Partitioning Principle would partition the parameter space into

$$H_{0HL}^{*} : \text{Neither High dose nor Low dose is efficacious}$$
$$H_{0H}^{*} : \text{High dose is not efficacious but Low dose is efficacious}$$
$$H_{0L}^{*} : \text{High dose is efficacious but Low dose is not efficacious}$$
$$H_{aHL}^{*} : \text{Both High dose and Low dose are efficacious}$$

test each of H_{0HL}^{*}, H_{0H}^{*}, and H_{0L}^{*} at level-α, and then collate the results as described in

P3. That is, high dose will be rejected if and only if both $H_{0HL}^{\star}$ and $H_{0H}^{\star}$ are rejected and low dose will be rejected if and only if both $H_{0HL}^{\star}$ and $H_{0L}^{\star}$ are rejected.

Default application of the Closed Testing principle would form the intersection null hypotheses

$$H_{0HL}^{\cap} : \text{Neither High dose nor Low dose is efficacious}$$
$$H_{0H}^{\cap} : \text{High dose is not efficacious}$$
$$H_{0L}^{\cap} : \text{Low dose is not efficacious}$$

test each of $H_{0HL}^{\cap}$, $H_{0H}^{\cap}$, and $H_{0L}^{\cap}$ at level-α, and then collate the results as described in C3. That is, high dose will be rejected if and only if both $H_{0HL}^{\cap}$ and $H_{0H}^{\cap}$ are rejected and low dose will be rejected if and only if both $H_{0HL}^{\cap}$ and $H_{0L}^{\cap}$ are rejected.

Testing $H_{0HL}^{\star}$, which is the same as $H_{0HL}^{\cap}$, requires multiplicity adjustment. That can be done in a variety of ways, as we will show next.

In this particular setting of low and high dose efficacy testing, there are sometimes expectation or prior knowledge indicating that higher dose is more efficacious than lower dose. Then it may be of interest to require decision to follow the path that only when high dose is claimed efficacious, low dose then can be tested. This reflects a pre-defined *path* in decision making to first test H_{0H}^{P} then test H_{0L}^{P}, as depicted in Figure 18.1.

Low dose   High dose

FIGURE 18.1
Decision path between low and high doses.

With such a path, Hsu and Berger (1999) proposed to partition the entire parameter space into three mutually exclusive and exhaustive subspaces:

$$H_{0H}^{\downarrow} : \text{High dose is not efficacious}$$
$$H_{0L}^{\downarrow} : \text{High dose is efficacious but Low dose is not efficacious}$$
$$H_{aHL}^{\downarrow} : \text{Both High dose and Low dose are efficacious.}$$

False efficacy claims correspond to rejections of true $H_{0H}^{\downarrow}$ or $H_{0L}^{\downarrow}$, so controlling FWER at level-α in testing them guarantees the probability of incorrectly inferring efficacy in High or Low dose is no more than α. This guarantee requires no assumption on monotonicity of true efficacy in doses, because the union of $H_{0H}^{\downarrow}$, $H_{0L}^{\downarrow}$, and $H_{aHL}^{\downarrow}$ exhausts the entire parameter space. Interestingly, testing $H_{0H}^{\downarrow}$ and $H_{0L}^{\downarrow}$ leads naturally to testing $H_{0H}^{\downarrow}$ first followed by testing H_{0L}^{P}, without multiplicity adjustment, as follows.

Suppose $H_{0H}^{\downarrow}$ is rejected. Then obviously one can infer High dose is efficacious. Suppose $H_{0H}^{\downarrow}$ and $H_{0L}^{\downarrow}$ both are rejected, then since the union of $H_{0H}^{\downarrow}$ and $H_{0L}^{\downarrow}$ is "either High dose or Low dose is not efficacious", the inference is "both High dose and Low dose are efficacious". On the other hand, if $H_{0H}^{\downarrow}$ fails to be rejected, only $H_{0L}^{\downarrow}$ is rejected, then no efficacy inference can be logically given. Thus, the decision-making path can be executed as testing with predetermined *steps*:

$$\boxed{\text{Step 1}}$$

If $H_{0H}^{\downarrow}$ is rejected

then infer High dose is efficacious and go to Step 2;

else stop.

Step 2

If $H_{0L}^{\downarrow}$ is rejected

then infer Low dose is efficacious and stop;

else stop.

Note that simultaneously testing $H_{0H}^{\downarrow}$ and $H_{0L}^{\downarrow}$ needs no multiplicity adjustment, because at most one of them can be true. Thus, each of $H_{0H}^{\downarrow}$ and $H_{0L}^{\downarrow}$ is tested at level-α. Further, a test which rejects $H_{0L}^{\downarrow}$ no more than 5% of the time when Low dose is ineffective, regardless of whether High dose is effective, will reject no more than 5% of the time in particular when Low dose is ineffective and High dose is effective. That is, a level-α test for H_{0L}^{P} is also a level-α test for $H_{0L}^{\downarrow}$. So, $H_{0H}^{\downarrow}$ and $H_{0L}^{\downarrow}$ can be tested by level-α tests for H_{0H}^{P} and H_{0L}^{P} respectively.

This particular decision-making process is sometimes called *fixed sequence* testing, using a *serial gate-keeping* approach (Westfall and Krishen 2001). Not bother testing $H_{0L}^{\downarrow}$ unless $H_{0H}^{\downarrow}$ is rejected gives the impression that there is a *gate* that one goes through. No harm in testing $H_{0L}^{\downarrow}$ though, it is just not particularly useful unless $H_{0H}^{\downarrow}$ is rejected. So, a *path* metaphor seems more evocative of how path-partitioning in formulation naturally ensures that decision-making follows a proper path.

The simple Partitioning proof above shows that for testing in a pre-determined steps fashion (i.e. stopping when one fails to reject), no multiplicity adjustment is needed to control FWER. No Closed Testing involved, only intuitive Partitioning, the proof applies to testing for *non-inferiority* to be followed by testing for *superiority*, for example. It also applies to testing for efficacy in the Primary endpoint, to be followed by testing for efficacy in the Secondary endpoint.

18.4   Shortcutting

By default, with k hypotheses to be tested, partition testing partitions the entire parameter space (including both null and alternative space) into 2^k disjoint subsets, and tests $2^k - 1$ partition null hypotheses.

Step-wise multiple tests appear to test a smaller number of null hypotheses, seemingly a sequence (or a single path) of *randomly indexed* null hypotheses. They are very popular in practice due to this convenient feature. However, that is an illusion. In reality, all partition hypotheses are tested.

Illusion is made possible by recognizing, under some conditions (to be discussed here), there are shortcuts to partition testing which reduce the number of tests that actually needs to be executed.

Such a test typically chooses a set of test statistics $T_i, i = 1, \ldots, k$, for the hypotheses $H_{0i}, i = 1, \ldots, k$. It then combines T_i to test each H_{0I}^{*} in such a way that, if H_{0I}^{*} is rejected, then many of H_{0J}^{*} with $J \subset I$ can be rejected without actual testing. Specifically, suppose the test statistic for H_{0I}^* is of *max T* form:

$$\text{Reject } H_{0I}^* \text{ if } \max_{i \in I} T_i > c_I, \tag{18.6}$$

If a T_{i^*}, $i^* \in I$, is large enough for testing H_{0I}^* ($T_{i^*} > c_I$) and causes H_{0I}^* to be rejected, then T_{i^*} will also be large enough to cause H_{0J}^* for $i^* \in J, J \subset I$, to be rejected with ($T_{i^*} > c_J$), *provided the critical value for a test involving a bigger set of parameters is bigger than the critical value for a test involving a smaller set of parameters.*

A more precise set of sufficient conditions for such shortcutting to be valid is as follows.

SD_1: The test for H_{0I}^* is of the form of rejecting H_{0I}^* if $max_{i \in I} T_i > c_I$;

SD_2: $sup_{H_{0I}^*} P\{\max_{i \in I} T_i > c_I\} \le \alpha$;

SD_3: The values of the test statistics T_i, $i = 1, \ldots, k$, are not re-computed for different H_{0I}^*;

SD_4: Critical values c_I have the property that if $J \subset I$ then $c_J \le c_I$.

Let T_{iI}, $i \in I \subseteq \{1, 2, \ldots, k\}$, be the subset of the T_i with indices in I. Let $[1]^I, \ldots, [|I|]^I$, where $|I|$ is the number of elements in I, denote the random indices such that

$$T_{[1]^I} \le \cdots \le T_{[|I|]^I}. \tag{18.7}$$

Then short-cutting partition testing is possible because if $T_{[|I|]^I} > c_I$, then $T_{[|J|]^J} > c_J$ for all $J \ni [|I|^I], J \subset I$.

For example, suppose $k = 3$ and $T_1 < T_3 < T_2$. If $T_2 > c_{\{1,2,3\}}$ so that $H_{\{1,2,3\}}$ is rejected, then one does not need to test $H_{\{1,2\}}$, $H_{\{2,3\}}$, or $H_{\{2\}}$ because by conditions SD_1, SD_3, SD_4, it is known that the results would be rejections.

More formally, letting $[1], \ldots, [k]$ be the random indices such that $T_{[1]} < \cdots < T_{[k]}$, if $T_{[k]} \le c_{\{1,\ldots,k\}}$, then $H_{0\{1,\ldots,k\}}$ cannot be rejected so no individual H_{0i} will be rejected. One should thus accept $H_{0[k]}^*$ and stop.

If the largest test statistic value $T_{[k]}$ is larger than the critical value $c_{\{1,\ldots,k\}}$, then every H_{0J}^*, $J \ni [k]$, will be rejected and thus $H_{0[k]}$ can be rejected.

We therefore move on to compare the second-largest test statistic value $T_{[k-1]}$ with the critical value $c_{\{[1],\ldots,[k-1]\}}$. If $T_{[k-1]} > c_{\{[1],\ldots,[k-1]\}}$, then all H_{0J}^*, $J \ni [k-1]$, will be rejected since $T_{[|J|]^J} \ge T_{[k-1]}, J \ni [k-1]$, so $H_{0[k-1]}$ can be rejected, and so on.

Thus, if conditions SD_1 to SD_4 are satisfied, then partition testing has the following single path shortcut (with the sequence determined by the sample):

Step 1: If $T_{[k]} > c_{\{[1],\ldots,[k]\}}$, then infer $\theta_{[k]} \notin \Theta_{0[k]}$ and go to step 2; else stop.

Step 2: If $T_{[k-1]} > c_{\{[1],\ldots,[k-1]\}}$, then infer $\theta_{[k-1]} \notin \Theta_{0[k-1]}$ and go to step 3; else stop.

$\ldots$

Step k: If $T_{[1]} > c_{\{[1]\}}$, then infer $\theta_{[1]} \notin \Theta_{0[1]}$ and stop; else stop.

A multiple test of this form is called a step-down test because it steps from the *most* significant to the *least* significant. Shortcutting does not have to take this single path form. For example, suppose $k = 3$ and conditions SD_1 to SD_4 are satisfied. For any index i, half of the $2^3 = 8$ partition of the parameter space contains that index, half do not. Even if $T_1 < T_3 < T_2$, one can start by checking whether $T_3 > c_{\{1,2,3\}}$. If it is, then $H_{\{1,3\}}$, $H_{\{2,3\}}$, and $H_{\{3\}}$ need not be tested because by conditions SD_1, SD_3, SD_4, it is known that the results would be rejections. But if not, then (instead of stopping) one still has to check whether $T_2 > c_{\{1,2,3\}}$. Stepping down from the most significant reduces the potential number of tests to be executed.

18.4.1   Holm's method is a shortcut

Shortcutting of closed or partition testing is necessary if the number of null hypotheses k to be tested is so large as to make the execution of $2^k - 1$ intersection or partition null hypotheses computationally infeasible. This is the case for Genome Wide Association Studies (GWAS), in which the number of Single Nucleotide Polymorphisms (SNPs) tested may be in the millions. However, in typical clinical trial situations, computers can execute closed or partition testing in real time, so the advantage of shortcutting is an appearance of simplicity, as exemplified by Holm's methodn (Holm, 1979).

Let p_i, $i = 1, \ldots, k$, denote the sample p-values corresponding to the test statistics T_i for H_{0i}, $i = 1, \ldots, k$, computed without multiplicity adjustment. Let $[1], \ldots, [k]$ denote the random indices such that

$$p_{[1]} \leq \cdots \leq p_{[k]}.$$

That is, $[i]$ is the anti-rank of p_i among $p_1, \ldots, p_k$.

Holm's method uses the rejection region

$$\min_{i \in I} p_i < \alpha/|I| \qquad (18.8)$$

to test H^*_{0I}, which is of (18.6) form and satisfies conditions SD_1 to SD_4. The resulting method has a step-down shortcut, and it proceeds as follows.

Step 1. If $p_{[1]} < \alpha/k$, reject $H_{0[1]}$ and go to Step 2; otherwise stop.
Step 2. If $p_{[2]} < \alpha/(k-1)$, reject $H_{0[2]}$ and go to Step 3; otherwise stop.
$\cdots$
Step k. If $p_{[k]} < \alpha$, reject $H_{0[k]}$ and stop.

In reality, Step 1 of Holm's method tests *half* of the 2^k parameter subspaces mentioned in P1 of section 18.2.1, the half that involves the index [1]. Step 2 of Holm's method tests *half* of the remaining 2^{k-1} parameter subspaces, the half that involves the index [2], and so forth. In the end, only one parameter subspace will not be tested, i.e. $I = \emptyset$ (corresponding to the subspace where none of the null hypotheses are true).

Holm's method tests each partition null hypothesis using a single test statistic value, $\min_{i \in I} p_i$. That need not be the case, as we illustrate using Hochberg's method.

18.4.2   Hochberg's method is also a shortcut

Hochberg's (1988) test, a step-up method whose seeming simplicity cloaks its complexity in construction, proceeds as follows.

Step 1. If $p_{[k]} < \alpha$, reject $H_{0[i]}$, $i = 1, \ldots, k$, and stop; otherwise go to Step 2.
Step 2. If $p_{[k-1]} < \alpha/2$, reject $H_{0[i]}$, $i = 1, \ldots, k-1$, and stop; otherwise go to Step 3.
$\cdots$
Step k. If $p_{[1]} < \alpha/k$, reject $H_{0[i]}$, $i = 1$, and stop.

So Hochberg's method looks just like Holm's method, except it is a step-up method, as it steps from the *least* significant to the *most* significant. The mechanism by which Hochberg's method shortcuts is more complex than that of Holm's method, however.

Let p_{i^I}, $i \in I \subseteq \{1, 2, \ldots, k\}$, be the subset of the p-values with indices in I. Let $[1]^I, \ldots, [|I|]^I$, where $|I|$ is the number of elements in I, denote the random indices such that

$$p_{[1]^I} \leq \cdots \leq p_{[|I|]^I}. \tag{18.9}$$

Simes (1986) proved that, if the these p-values are *independent* under Θ_I, then

$$P_{\Theta_I}\{p_{[i]^I} < \frac{i\alpha}{|I|} \text{ for some } i, i \in I\} = \alpha. \tag{18.10}$$

This cute equality can be thought of an improvement of the Bonferroni inequality using the knowledge that the p-values are independent, as we illustrate below.

For example, when $|I| = 2$, each of the two hypotheses would be tested at $\alpha/2$ according to the Bonferroni inequality used by Holm's method. If the p-values are independent, then it is conservative by the amount of $(\alpha/2)^2$, because under Θ_I

$$P(\min_{i \in I} p_i < \alpha/2) = 1 - P(\min_{i \in I} p_i \geq \alpha/2) = 1 - (1 - \alpha/2)^2 = \alpha - (\frac{\alpha}{2})^2.$$

To remove this conservatism when the p-values are independent, one can naturally test each of the two hypotheses at level $1 - \sqrt{1 - \alpha}$. What Simes' test does instead is to start with the Bonferroni acceptance region

$$\left\{ p_{[1]^I} \geq \frac{1\alpha}{|I|} \right\} = \{\text{smaller p-value} \geq \alpha/2\}, \tag{18.11}$$

and remove from it a portion with probability $(\alpha/2)^2$. The portion

$$\left\{ p_{[2]^I} < \frac{2\alpha}{|I|} \right\} = \{\text{larger p-value} < \alpha\},$$

which is a corner of the acceptance region (18.11), has probability $(\alpha/2)^2$.

With this recognition, Huang and Hsu (2007) showed that Hochberg's method can be thought of as starting with testing each partition hypothesis H_{0I}^* by Simes' (1986) test:

$$\text{reject } H_{0I}^* \text{ if } p_{[i]^I} < \frac{i\alpha}{|I|} \text{ for some } i. \tag{18.12}$$

resulting in a multiple test, known as Hommel's (1988) test, which does *not* allow a shortcut. Hochberg's method then uses the more conservative rejection region

$$\text{reject } H_{0I}^* \text{ if } p_{[i]^I} < \frac{\alpha}{|I| - i + 1} \text{ for some } i \tag{18.13}$$

to test H_{0I}^* instead. Through a somewhat intricate analysis not reproduced here, Huang and Hsu (2007) showed the resulting partition test has a step-up shortcut which is Hochberg's method.

Comparing (18.13) with (18.8), one sees whereas Holm's method uses only the minimum p-values with indices in I to test H_{0I}^*, Hochberg's method uses *all* the ordered p-values with indices in I to test H_{0I}^*. When the p-values are dependent, distribution of the minimum p-value is analytically tractable and step-down conditions can be checked, so there are step-down methods that compute critical values taking joint distribution into account (see, for example, the step-down version of Dunnett's method described in Section 3.1.1.2 of Hsu 1996). However, joint distribution of all the ordered p-values is complicated when the p-values are dependent, causing difficulty in checking step-up conditions. For this reason, when p-values are dependent, step-up methods still use critical values computed under the independence assumption, relying on probabilistic inequalities to prove their conservatism under certain dependence structure (a distribution with the multivariate total positivity of order two property or a scale mixture thereof, see Sarkar 1998 for more details).

If a method constructed from the Partition Principle can be executed fast enough for the result be displayed to the user in real time without shortcutting, then shortcutting is not essential. Our purpose of showing how Holm's method and Hochberg's method can be constructed from the Partition Principle is not to particularly recommend them or not recommend them, or to demonstrate how to shortcut. Rather, demonstrating that they are results of using *different* tests for each partition hypothesis shows we can construct multiple tests from the Partitioning Principle, to appropriately solve new, more complex clinical trial problems.

18.5 Paths in Decision-Making

Modern clinical trials often have decision paths. For example, in clinical trials with multiple endpoints, the endpoints may be *alternative primary, co-primary,* or *primary and secondary* endpoints. See Offen, Chuang-Stein, Dmitrienko and Maca (2007) for definitions. When there are ordered endpoints (e.g. primary vs secondary endpoint), efficacy in a lower-ordered endpoint is only relevant if efficacy in all higher-ordered endpoints has been shown. There are thus defined paths to decision-making.

When multiple doses are coupled with multiple endpoints in a clinical trial, the question of interest is for which dose and endpoint combinations is the compound effective. For a given dose, lower-ordered endpoint is only relevant if efficacy in all higher-ordered endpoints has been shown.

Let μ_{ij} denote the mean response of group i for endpoint j, $i = 0, 1, ..., k$, $j = 1, ..., m$, where $i = 0$ denote the placebo group and $j = 1$ denote the primary endpoint. Define $\theta_{ij} = \mu_{ij} - \mu_{0j}$ to be the difference in mean efficacy measurement between dose group i and the placebo group for endpoint j. To be general, we allow the clinically meaningful differences for endpoints to differ, denoted as δ_j for endpoint j. Assuming a larger measurement indicates a better treatment, the statistical inference of interest is to test, for each dose endpoint combination,

$$H_{0ij} : \theta_{ij} \leq \delta_j, \quad i = 1, ..., k, \ j = 1, ..., m. \tag{18.14}$$

Marketing approval of a new drug is based on establishing efficacy in the primary endpoint alone. For a given dose, stating efficacy of a lowered ordered endpoints only if efficacy in the primary endpoint has shown at that dose makes for logical statement on a drug label:

Efficacy in [the primary endpoint] has been established for all doses. In addition, efficacy in [the secondary endpoint] has been established for medium and high doses. In addition, efficacy in [the tiertiary endpoint] has been established for the medium dose.

Decision paths are thus within, but not between, doses. Figure 18.2 illustrates such paths with k doses and m ordered endpoints where it states:

For a given dose i, the inference $\theta_{ij} > \delta_j$ in (18.14) is not given unless the inference $\theta_{ij^\star} > \delta_{j^\star}$ is established for all $j^\star < j$.

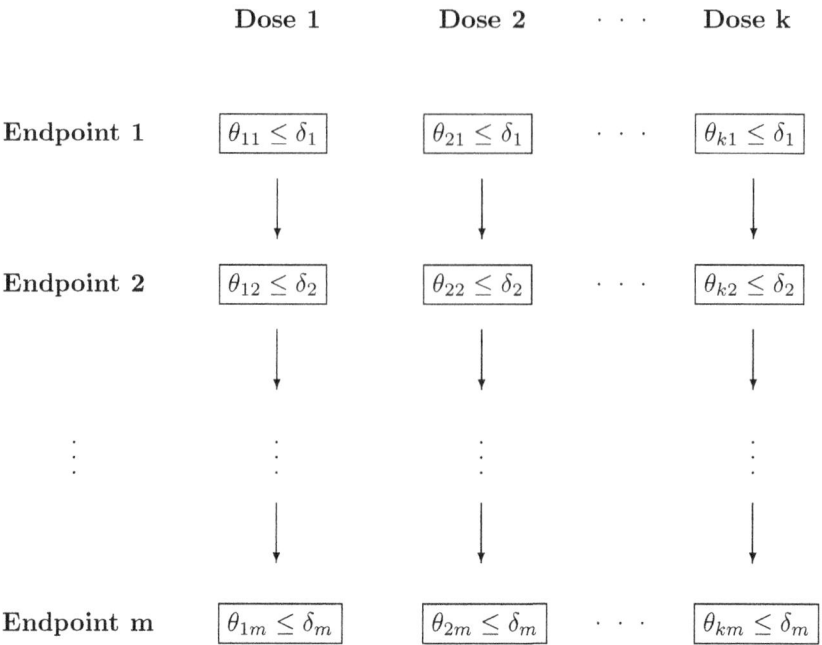

FIGURE 18.2
Decision paths for k doses m endpoints, with one path for each dose going from endpoint 1 to endpoint m.

18.5.1 Decision path respecting principle

When there are decision paths, Liu and Hsu (2009) proposed the following:

> **Decision Path Respecting Principle:** *Null hypotheses should be formulated so that decision-making naturally follows decision paths.*

Default closed testing forms all possible intersections of the null hypotheses, oblivious to paths in the decision-making process. Without placing restriction on how each intersection hypothesis is tested, Closed Testing may lead to efficacy claim in the secondary endpoint even if efficacy in the primary endpoint has not been shown. Careful restrictions on the choice of test statistics and critical values for testing the intersection hypotheses must be placed to avoid this.

A more direct approach is to use the Partitioning Principle to implement the Decision Path principle. In the ordered endpoints setting, instead of testing all intersections of null hypotheses, (a smaller number of) partitioning null hypotheses can be formulated so that, for a given dose i, the inference $\theta_{ij} > \delta_j$ in (18.14) is not given unless the inference $\theta_{ij^\star} > \delta_{j^\star}$ is established for all $j^\star < j$, with no restriction on how each partitioning null hypothesis is tested, so long as it is a level-α test. In fact, in the case of a single dose ($k = 1$), Partition testing led to the predetermined steps method of Hsu and Berger (1999).

18.5.2 A specific dose $\times$ endpoint example

If only primary and secondary endpoints, and low dose and high dose are involved, the decision path in Figure 18.2 becomes Figure 18.3 as shown here. For notation conveniences, instead of using the second subscript to index the endpoint, we use superscripts ^{P} and ^{S} to denote primary and secondary endpoints.

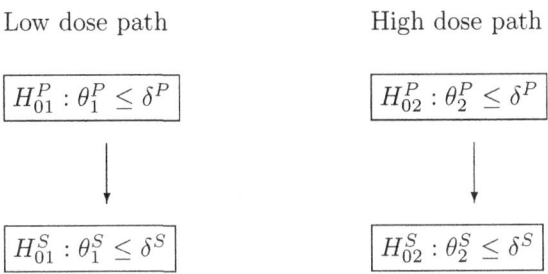

Low dose path   High dose path

$$H_{01}^P : \theta_1^P \leq \delta^P \qquad H_{02}^P : \theta_2^P \leq \delta^P$$

$$H_{01}^S : \theta_1^S \leq \delta^S \qquad H_{02}^S : \theta_2^S \leq \delta^S$$

FIGURE 18.3
Decision paths for low and high doses.

Given decision paths in Figure 18.3, the parameter space is partitioned in two stages:

- Path partition: Partition within each path.

- Disjointness partition: Further partition by taking intersections to make hypotheses between paths disjoint.

Path partition is within each dose i. Starting with the primary endpoint, we test H_{0i}^P : $\theta_i^P \leq \delta^P$. If it is rejected, then efficacy in the primary endpoints has been established.

Then follow the path to the secondary endpoint. But, instead of testing $H_{0i}^S : \theta_i^S \leq \delta^S$, we make it disjoint with H_{0i}^P and test $H_{0i}^{\star S} : \theta_i^S \leq \delta^S$ and $\theta_i^P > \delta^P$. If both H_{0i}^P and $H_{0i}^{\star S}$ are rejected, then we logically conclude efficacy in both the primary and the secondary endpoints.

Whereas the reason for path partitioning is inference in the secondary endpoint is irrelevant unless efficacy in the primary endpoint is established, disjointness partitioning is for proper multiplicity adjustment. If two or more hypotheses can be true simultaneously, then multiplicity adjustment is needed in testing them. If the hypotheses tested are made disjoint, then since (at most) one of them is true, FWER is controlled strongly if each is tested at level α.

Between paths, connect an *edge* between subspaces of the parameter space that are not disjoint, as illustrated in Figure 18.4. Edges represent hypotheses that can be true simultaneously. (There is no edge between $\{\theta_1^S > \delta^S$ and $\theta_1^P > \delta^P\}$ and $\{\theta_2^S > \delta^S$ and $\theta_2^P > \delta^P\}$, since the intersection of these two hypotheses, the ideal situation of efficacy in all doses and endpoints, need not be tested.) Take intersections of connected subspaces to form new hypotheses. The resulting set of hypotheses, as presented in Table 18.2, partitions the parameter space. Therefore, so long as each partition hypothesis is tested at level α, the FWER is controlled strongly at level α.

In general, each partition hypothesis is represented by a non-empty intersection of *at most one hypothesis from each column* of Figure 18.2, and the number of partitioning hypotheses to be tested reduces to $(m+1)^k - 1$. Inferences on which of the $m \times k$ combinations of dose and endpoint are efficacious are then obtained by collating the results from the $(m+1)^k - 1$ tests.

Inference $\theta_{ij} > \delta_j$ is made if all hypotheses implying that $\theta_{ij} \le \delta_j$ could be true are rejected. This includes partitioning hypotheses which do not explicitly state an inequality for θ_{ij}.

For example, inference $\theta_1^P > \delta^P$ is made if hypotheses PH1–PH3 in Table 18.2 are rejected. On the other hand, inference $\theta_1^S > \delta^S$ is made only if hypotheses PH1–PH3, PH4, PH6–PH7 are rejected, because hypotheses PH1–PH3 include the possibility that $\theta_1^S \le \delta^S$ by allowing θ_1^S to be any value. This is how partition testing stays on decision-paths.

Note that path-respecting partition testing places no restriction on how each of the partition null hypothesis is tested (so long as each is a level-α test).

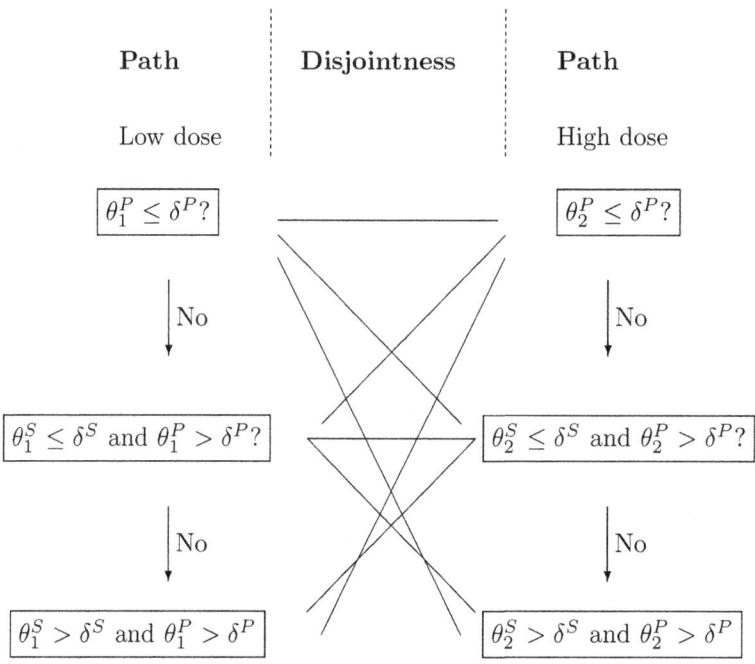

FIGURE 18.4
Within path and between paths partitioning in the setting of Figure 18.3.

TABLE 18.2
Path-respecting partition hypotheses for Figure 18.3.

Index	Partition hypothesis		
PH1	$\theta_1^P \le \delta^P$ and $\theta_2^P \le \delta^P$		
PH2	$\theta_1^P \le \delta^P$ and $\theta_2^P > \delta^P$		and $\theta_2^S \le \delta^S$
PH3	$\theta_1^P \le \delta^P$ and $\theta_2^P > \delta^P$		and $\theta_2^S > \delta^S$
PH4	$\theta_1^P > \delta^P$ and $\theta_2^P \le \delta^P$	and $\theta_1^S \le \delta^S$	
PH5	$\theta_1^P > \delta^P$ and $\theta_2^P \le \delta^P$	and $\theta_1^S > \delta^S$	
PH6	$\theta_1^P > \delta^P$ and $\theta_2^P > \delta^P$	and $\theta_1^S \le \delta^S$	and $\theta_2^S \le \delta^S$
PH7	$\theta_1^P > \delta^P$ and $\theta_2^P > \delta^P$	and $\theta_1^S \le \delta^S$	and $\theta_2^S > \delta^S$
PH8	$\theta_1^P > \delta^P$ and $\theta_2^P > \delta^P$	and $\theta_1^S > \delta^S$	and $\theta_2^S \le \delta^S$

In contrast to Partition Testing, by testing all intersection null hypotheses, Closed Testing requires care to stay on paths. Take the partition hypothesis PH1 in Table 18.2, for instance. Closed testing separates it into four intersection hypotheses CT1–CT4, as shown in Table 18.3. To stay on paths, Xu et al (2009) showed that one way is to test each set of closed testing intersection hypotheses corresponding to the same partition hypothesis using the same rejection region, as exemplified in Table 18.3.

TABLE 18.3
Typical path-respecting closed testing where $t_i^j, i = 1, 2; j = P, S$ denotes the test statistic for each dose endpoint combination.

Index	Intersection hypothesis	Rejection rule
CT1	$\theta_1^P \leq \delta^P$ and $\theta_2^P \leq \delta^P$ and $\theta_1^S \leq \delta^S$ and $\theta_2^S \leq \delta^S$	$t_1^P > c_1$ or $t_2^P > c_1$
CT2	$\theta_1^P \leq \delta^P$ and $\theta_2^P \leq \delta^P$ and $\theta_1^S \leq \delta^S$	$t_1^P > c_1$ or $t_2^P > c_1$
CT3	$\theta_1^P \leq \delta^P$ and $\theta_2^P \leq \delta^P$ and $\theta_2^S \leq \delta^S$	$t_1^P > c_1$ or $t_2^P > c_1$
CT4	$\theta_1^P \leq \delta^P$ and $\theta_2^P \leq \delta^P$	$t_1^P > c_1$ or $t_2^P > c_1$

18.6 Setting Priorities in Multiple Testing for Each Study

Each clinical study is different. We use the dose $\times$ endpoint example to illustrate exactly where there is flexibility for users to customize a multiple test, according to priorities between different doses and different endpoints, for each study.

For each partition hypothesis in Table 18.2, we test what needs to be tested. If N denotes the (i, j) pairs for which a partition hypothesis states $\theta_{ij} \leq \delta_j$, then we test

$$\theta_{ij} \leq \delta_j \text{ for all } (i, j) \in N.$$

For example, in testing the partition hypothesis which states

$$\theta_1^P \leq \delta^P \text{ and } \theta_2^P > \delta^P \text{ and } \theta_2^S \leq \delta^S \tag{18.15}$$

one thinks in terms of testing the hypothesis which states

$$\theta_1^P \leq \delta^P \text{ and } \theta_2^S \leq \delta^S \tag{18.16}$$

and then form the rejection region using the test statistics corresponding to these two parameters. Strong control of FWER is maintained, because a level-α test for (18.16) is also a level-α test for (18.15).

Suppose a large test statistic t_{ij} value is indicative of $\theta_{ij} > \delta_j$. Perhaps the most common test of a 1-sided hypothesis is the so-called maxT test, which rejects if

$$\max_{(i,j) \in N} (t_{ij} - c_{ij}) > 0, \tag{18.17}$$

Note this is a more general form of (18.6).

Let $t_i^j, i = 1, 2; j = P, S$ denotes the test statistic for each dose endpoint combination in the two doses, two endpoints case. Our illustration of possible multiple tests assumes high and low doses should be treated the same, so we use the same critical value for both doses within each endpoint. However, endpoints may be treated differently, so different critical values between endpoints are allowed, as shown in Table 18.4.

TABLE 18.4
MaxT rejection regions for the path-respecting partition null hypotheses in Table 18.2.

Index	Rejection rule		
PH1	$t_1^P > c_1$ or $t_2^P > c_1$		
PH2	$t_1^P > c_1$		or $t_2^S > c_2$
PH3	$t_1^P > c_3$		
PH4		$t_2^P > c_1$ or $t_1^S > c_2$	
PH5		$t_2^P > c_3$	
PH6			$t_1^S > c_1$ or $t_2^S > c_1$
PH7			$t_1^S > c_3$
PH8			$t_2^S > c_3$

Low dose   **High dose**

PH3. $\theta_1^P \le \delta^P$ PH1. $\theta_1^P \le \delta^P$ and $\theta_2^P \le \delta^P$ PH5. $\theta_2^P \le \delta^P$

$p_1^P < \alpha$ $p_1^P < \alpha_1$ or $p_2^P < \alpha_1$ $(\alpha_1 \ge \alpha/2)$ $p_2^P < \alpha$

PH4. $\theta_1^S \le \delta^S$ and $\theta_2^P \le \delta^P$ PH2. $\theta_1^P \le \delta^P$ and $\theta_2^S \le \delta^S$

$p_1^S < \alpha_s$ or $p_2^P < \alpha_p$ $(\alpha_p + \alpha_s = \alpha)$ $p_1^P < \alpha_p$ or $p_2^S < \alpha_s$ $(\alpha_p + \alpha_s = \alpha)$

PH7. $\theta_1^S \le \delta^S$ PH6. $\theta_1^S \le \delta^S$ and $\theta_2^S \le \delta^S$ PH8. $\theta_2^S \le \delta^S$

$p_1^S < \alpha$ $p_1^S < \alpha_1$ or $p_2^S < \alpha_1$ $(\alpha_1 \ge \alpha/2)$ $p_2^S < \alpha$

FIGURE 18.5
Path-respecting and FWER-controlling rejection regions, with p_i^P (p_i^S) being the p-value for t_i^P (t_i^S). Index in each box shows which hypothesis in Table 18.2 and Table 18.4 it corresponds to.

Strong FWER control requires each null hypothesis be tested at level-α. Testing any hypothesis at a size smaller than α risks losing power.

Hypotheses PH3, PH5, PH7–PH8 each involves a single test statistic, so c_3 is set to reject each such hypotheses if its p-value is less than α, as illustrated in Table 18.4 and Figure 18.5.

Hypotheses PH1 and PH6 in Table 18.4 involve pairs of test statistics, (t_1^P, t_2^P) and (t_1^S, t_2^S). Within each endpoint, the test statistics for high and low doses are correlated if both doses are compared to a common Control. Assuming these pairs have the same joint distribution, treating the two doses equally would assign the same critical value c_1 for both doses, for both endpoints, as illustrated in Table 18.4 and Figure 18.5.

If each endpoint is normally distributed following a GLM, then (t_1^P, t_2^P) and (t_1^S, t_2^S) have the same multivariate t distribution. For a balanced design, if we calculate c_1 taking correlation into account (with infinite error degrees of freedom) so that the maxT tests for hypotheses PH1 and PH6 are exactly size 0.025, then the tests would rejecting hypothesis PH1 if either of the p-values for t_1^P, t_2^P is less than 0.0135, and reject hypothesis PH6 if either of the p-values for t_1^S, t_2^S is less than 0.0135, as illustrated in Figure 18.6.

Low dose   **High dose**

| PH3. $\theta_1^P \leq \delta^P$ |
$p_1^P < 0.025$

| PH1. $\theta_1^P \leq \delta^P$ and $\theta_2^P \leq \delta^P$ |
$p_1^P < 0.0135$ or $p_2^P < 0.0135$

| PH5. $\theta_2^P \leq \delta^P$ |
$p_2^P < 0.025$

| PH4. $\theta_1^S \leq \delta^S$ and $\theta_2^P \leq \delta^P$ |
$p_1^S < 0.0125$ or $p_2^P < 0.0125$

| PH2. $\theta_1^P \leq \delta^P$ and $\theta_2^S \leq \delta^S$ |
$p_1^P < 0.0125$ or $p_2^S < 0.0125$

| PH7. $\theta_1^S \leq \delta^S$ |
$p_1^S < 0.025$

| PH6. $\theta_1^S \leq \delta^S$ and $\theta_2^S \leq \delta^S$ |
$p_1^S < 0.0135$ or $p_2^S < 0.0135$

| PH8. $\theta_2^S \leq \delta^S$ |
$p_2^S < 0.025$

FIGURE 18.6
Rejection regions that take joint distribution into account, with strong FWER = 0.025 and without endpoint prioritization.

Hypotheses PH2 and PH4 in Table 18.4 involve different endpoints under different doses. As pointed out by Xu et al. (2009), test statistics (t_1^P, t_2^S) and (t_2^P, t_1^S) not only do *not* have a multivariate t distribution, their distributions depend on correlation between endpoints which is unknown. Liu, Hsu and Ruberg (2007) showed that, while estimating the variance allows the unknown variance be pivoted out in univariate testing, estimating the variance-covariance matrix does *not* allow the unknown correlation be pivoted out in maxT testing. Thus, as proposed in Xu et al. (2009), current practice is to use the Bonferroni inequality in calculating critical values for testing hypotheses PH2, PH4, guaranteeing conservatism. That is, hypotheses PH2 and PH4 are tested by allocation α to the two endpoints additively, as shown in Figure 18.5 where α_s and α_p can be any potential allocation with $\alpha_s + \alpha_p = \alpha$.

There are infinitely many possible α_s and α_p in testing hypotheses PH2 and PH4. One possibility is to treat the two endpoints equally, allocating $\alpha/2$ to each, as shown in Figure 18.6.

Another possibility is to give the primary endpoint higher priority, keeping the α_1 in testing hypothesis PH1 for the primary endpoint (i.e. let $\alpha_p = \alpha_1$ in Figure 18.5) and allocate $\alpha_s = \alpha - \alpha_p$ to the secondary endpoint, in testing hypotheses PH2 and PH4. That is, test the hypothesis involving both endpoints (hypothesis PH2 and PH4 in Table 18.4) by assigning the critical value c_1 to the primary endpoint first, and then calculate the critical value for the secondary endpoint c_2 so that the level of the test for such hypothesis is level-α. To illustrate the latter possibility, in the balanced GLM setting, with $\alpha = 0.025$ and α_1 already set at 0.0135, α_p should be set the same as $\alpha_1 = 0.0135$ and $\alpha_s = 0.025 - 0.0135 = 0.0115$ as shown in Figure 18.7.

The Graphical Approach, described below, provides a graphical user interface that lets the user expresses his/her priorities not only in terms of endpoints but doses as well, via the c_{ij}'s in (18.17).

18.6.1   The graphical approach

Bretz et al. (2009) invented a graphical representation for closed testing. The gMCP R package implements a Graphical User Interface (GUI) for it (Rohmeyer and Klinglmueller2015).

Low dose   **High dose**

PH3. $\theta_1^P \leq \delta^P$		PH1. $\theta_1^P \leq \delta^P$ and $\theta_2^P \leq \delta^P$		PH5. $\theta_2^P \leq \delta^P$
$p_1^P < 0.025$		$p_1^P < 0.0135$ or $p_2^P < 0.0135$		$p_2^P < 0.025$

PH4. $\theta_1^S \leq \delta^S$ and $\theta_2^P \leq \delta^P$		PH2. $\theta_1^P \leq \delta^P$ and $\theta_2^S \leq \delta^S$
$p_1^S < 0.0115$ or $p_2^P < 0.0135$		$p_1^P < 0.0135$ or $p_2^S < 0.0115$

PH7. $\theta_1^S \leq \delta^S$		PH6. $\theta_1^S \leq \delta^S$ and $\theta_2^S \leq \delta^S$		PH8. $\theta_2^S \leq \delta^S$
$p_1^S < 0.025$		$p_1^S < 0.0135$ or $p_2^S < 0.0135$		$p_2^S < 0.025$

FIGURE 18.7
Rejection regions that take join distribution into account (for balanced GLM), with strong
FWER $= 0.025$ and endpoint prioritization. Index in each box shows which hypothesis in
Table 18.2 and Table 18.4 it corresponds to.

In the multiple doses primary-secondary endpoints setting, tests are represented by directed, weighted graphs, with each node corresponding to a hypothesis that a dose-endpoint combination lacks efficacy. Our specific example is represented by Figure 18.8, which corresponds to Figure 18.3, with H1, H2, H3, H4 corresponding to $H_{01}^P : \theta_1^P \leq \delta^P$, $H_{02}^P : \theta_2^P \leq \delta^P$, $H_{01}^S : \theta_1^S \leq \delta^S$, $H_{02}^S : \theta_2^S \leq \delta^S$ respectively.

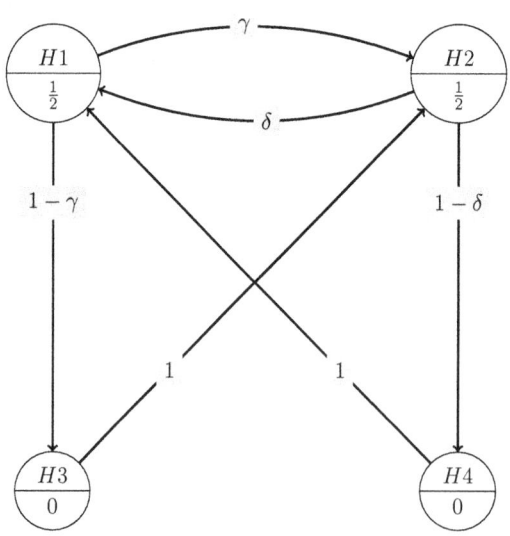

FIGURE 18.8
Graphical Approach representation of the example in Figure 18.3.

Numbers in the node circles (e.g., $\frac{1}{2}$ and 0 in Figure 18.8) represent initial allocation of fractions of FWER α to the nodes. Weights attached to directed edges (e.g., δ and $1 - \delta$ in Figure 18.8) emanating from a node represent additive distribution of its fraction of α to other nodes, in testing intersection hypotheses that do not include this node. For example, there are two edges emanating from H2, one with an edge of δ to H1 and one with an edge of $1 - \delta$ to H4. This means if H2 is rejected at $\alpha/2$ level, then $\delta * \alpha/2$ is re-allocated to H1 and $(1 - \delta) * \alpha/2$ is re-allocated to H4.

To respect decision paths, one can allocate zero α to secondary endpoints nodes, and giving them positive distribution only from their primary endpoint nodes. This means for example, H3 does not have any alpha to be tested initially and only when H1 is rejected then a portion of its alpha can be allocated to H3 so that H3 can be tested.

Not having an edge between nodes is the same as having an edge with zero weight. If $\delta = 0$ in Figure 18.8, and partition hypotheses PH1 and PH6 in Table 18.4 are tested without taking joint distribution into account, then partition hypothesis PH1 would be rejected if either $p_1^P < \alpha/2$ or $p_2^P < \alpha/2$. Partition hypothesis PH2 would be rejected if either $p_1^P < \alpha/2$ or $p_2^S < \alpha/2$, because (without H1 in partition hypothesis PH2) H3 gets the $\alpha/2$ initially allocated to H1. Similarly, partition hypothesis PH4 would be rejected if either $p_1^S < \alpha/2$ or $p_2^P < \alpha/2$, because (without H2 in partition hypothesis PH4) H4 gets the $\alpha/2$ initially allocated to H2. Partition hypothesis PH6 would be rejected if either $p_1^S < \alpha/2$ or $p_2^S < \alpha/2$, because (without H1 and H2 in partition hypothesis PH6) H3 gets the $\alpha/2$ initially allocated to H1 while H4 gets the $\alpha/2$ initially allocated to H2. Finally, one can work out that each of partition hypotheses PH3, PH5, PH7, PH8 is rejected if its p-value is less than α. Thus, the multiple test is as shown in Figure 18.9, which is not as powerful as can be, because the tests for partition hypotheses PH1 and PH6 do not take joint distribution into account.

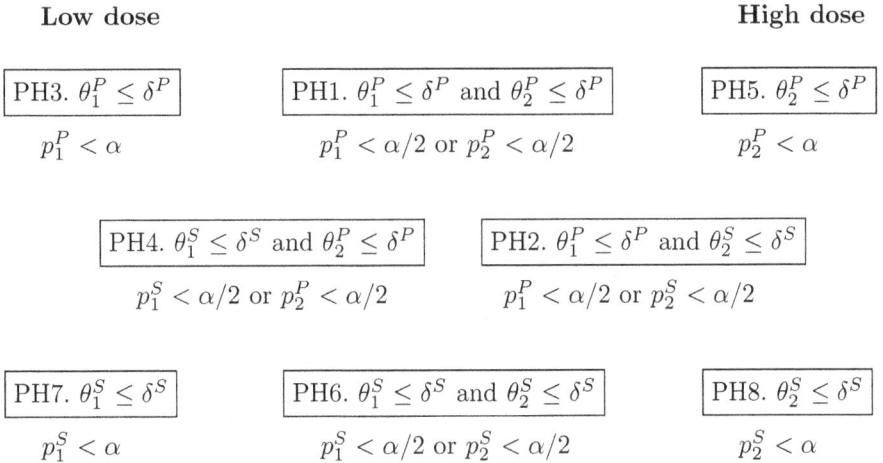

Low dose   **High dose**

PH3. $\theta_1^P \leq \delta^P$
$p_1^P < \alpha$

PH1. $\theta_1^P \leq \delta^P$ and $\theta_2^P \leq \delta^P$
$p_1^P < \alpha/2$ or $p_2^P < \alpha/2$

PH5. $\theta_2^P \leq \delta^P$
$p_2^P < \alpha$

PH4. $\theta_1^S \leq \delta^S$ and $\theta_2^P \leq \delta^P$
$p_1^S < \alpha/2$ or $p_2^P < \alpha/2$

PH2. $\theta_1^P \leq \delta^P$ and $\theta_2^S \leq \delta^S$
$p_1^P < \alpha/2$ or $p_2^S < \alpha/2$

PH7. $\theta_1^S \leq \delta^S$
$p_1^S < \alpha$

PH6. $\theta_1^S \leq \delta^S$ and $\theta_2^S \leq \delta^S$
$p_1^S < \alpha/2$ or $p_2^S < \alpha/2$

PH8. $\theta_2^S \leq \delta^S$
$p_2^S < \alpha$

FIGURE 18.9
Rejection regions that test each hypothesis in Table 18.2 and Table 18.4 by unweighted Bonferroni, without endpoint prioritization ($\delta = 0$).

If $\delta = 0$ in Figure 18.8, and partition hypotheses PH1 and PH6 in Table 18.4 are tested with joint distribution taken into account, then partition hypothesis PH1 would be rejected if either $p_1^P < 0.0135$ or $p_2^P < 0.0135$. Partition hypothesis PH2 would be rejected if either $p_1^P < \alpha/2$ or $p_2^S < \alpha/2$, because (without H1 in partition hypothesis 2) H3 gets the $\alpha/2$ initially allocated to H1. Similarly, partition hypothesis PH4 would be rejected if either $p_1^S < \alpha/2$ or $p_2^P < \alpha/2$, because (without H2 in partition hypothesis PH4) H4 gets the $\alpha/2$ initially allocated to H2. And one can work out that each of partition hypotheses PH3,

PH5, PH7, PH8 is rejected if its p-value is less than α. Thus, the multiple test is as shown in Figure 18.6.

A positive δ in Figure 18.8 gives higher priority to the primary endpoint in testing partition hypotheses PH1 and PH6. If $\delta = 0.0783$ then partition hypothesis PH2 would be rejected if either $p_1^P < 0.0135$ or $p_2^S < 0.0115$, because $0.0125 \times (1. + 0.0783) = 0.0135$ and $0.0125 \times (1 - 0.0783) = 0.0115$. Similarly, partition hypothesis PH4 would be rejected if either $p_1^S < 0.0115$ or $p_2^P < 0.0135$. Each of partition hypotheses PH3, PH5, PH7, PH8 is rejected if its p-value is less than α, so the multiple test is the one in Figure 18.7. Specifying this Graph, with $\delta = 0.0783$, is referred to as the "consonant parametric graphical approach" in Bretz et al. (2011). The concept of *Consonance* is discussed here, together with the concept of *Coherence*.

18.7 Logical Relationships Among Parameters Tested

To avoid illogical statistical inference, logical relationship among the parameters tested in null hypotheses should be respected. Ding, Lin, Hsu (2014) proposed essentially the following.

> ***Parameter Logic Respecting Principle:*** *Statistical methods should be constructed so that decision-making respects logics among the scientific parameters.*

It is useful to distinguish logical relationships inherent in the scientific parameters of interest, and those that arise from intersecting scientific null hypotheses in the construction of multiple tests.

18.7.1 Logic induced in multiple test construction

In the Pairwise comparisons example of Section 18.2, intersection of all the scientific null hypotheses in (18.3) is

$$H_0 : \mu_1 = \cdots = \mu_k. \tag{18.18}$$

If (18.18) is false, then at least one of (18.3) is false. If thus seems logical to require that if (18.18) is rejected, then at least one of (18.3) is rejected as well. This is the *Consonance* requirement of Gabriel (1969): "A null hypothesis can be rejected only if at least one of the hypotheses implied by it is rejected". The violation of consonance is termed as dissonance.

Similarly, if any one of (18.3) is false, then (18.18) is false. If thus seems logical to require that if any one of (18.3) is rejected, then (18.18) is rejected as well. This is the *Coherence* requirement of Gabriel (1969): "No hypothesis should be 'accepted' if any hypothesis implied by it is rejected". The violation of coherence is termed as incoherence.

Perhaps a motivation for Gabriel's proposals was, at the time, an unfortunate common practice was to pursue multiple comparisons only if the F-test for (18.18) rejects. Quite aside from the fact that this practice fails to control FWER (see Section 6.3 of Hsu 1996), it leads to dissonant and incoherent inferences, due to the fact that the F-test has an ellipsoidal acceptance region while Tukey's method has a polygonal acceptance region, as illustrated in Figure 18.10. Using Tukey's method to test (18.18) and 2-sample t-tests to test (18.3) avoids such paradoxes.

In our multiple dose × endpoint example, if one chooses the first option of allocating α equally to primary and secondary endpoints in testing hypotheses PH2 and PH4 and executes closed testing (instead of partition testing), then it may appear there is the possibility

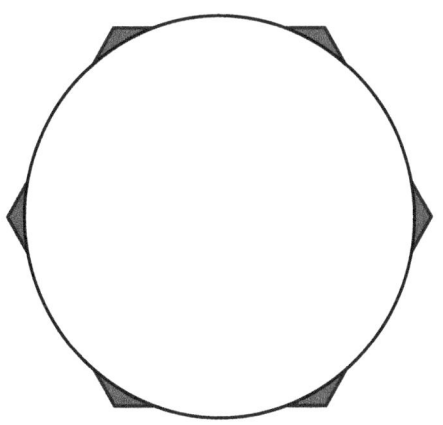 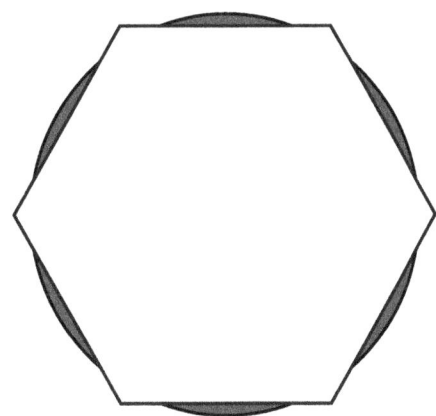

FIGURE 18.10
Dissonance (left panel): F-test rejects but Tukey's method fails to find significant difference in shaded region. Incoherence (right panel): F-test fails to reject but Tukey's method finds significant difference in shaded region.

of dissonance with hypothesis CT1 in Table 18.3. As noted in Bretz et al (2011), closed testing intersection hypothesis CT1

$$\theta_1^P \leq \delta^P \text{ and } \theta_2^P \leq \delta^P \text{ and } \theta_1^S \leq \delta^S \text{ and } \theta_2^S \leq \delta^S$$

in Table 18.3 uses the following rejection region with $\alpha = 0.025$

$$p_1^P < 0.0135 \text{ or } p_2^P < 0.0135 \qquad (18.19)$$

and closed testing intersection hypotheses such as the ones below

$$\theta_1^P \leq \delta^P \text{ and } \theta_2^S \leq \delta^S \qquad (18.20)$$
$$\theta_2^P \leq \delta^P \text{ and } \theta_1^S \leq \delta \qquad (18.21)$$

use the following rejection region

$$p_1^P < 0.0125 \text{ or } p_2^S < 0.0125.$$

$$p_2^P < 0.0125 \text{ or } p_1^S < 0.0125.$$

If we observe the following p-values: $p_1^P = 0.013$, $p_2^P = 0.06$, $p_1^S = 0.1$, $p_2^S = 0.2$, then intersection hypothesis CT1 is rejected but none of the individual null hypotheses H_{0i}^j, $i = 1, 2; j = P, S$ implied by it can be rejected. This violates the consonance requirement and shortcut condition SD_4 in Section 18.4, making single-path shortcut execution not possible.

There is also a possibility of incoherence because intersection hypothesis CT1 may be accepted while (18.20) which is implied by it is rejected e.g. $p_1^P = 0.06$, $p_2^P = 0.1$, $p_1^S = 0.1$, $p_2^S = 0.002$. But once it is recognized that rejection region (18.19) for testing hypothesis CT1 is actually testing partition hypothesis PH1

$$\theta_1^P \leq \delta^P \text{ and } \theta_2^P \leq \delta^P$$

in Table 18.2 which does not imply closed testing hypothesis (18.20), then there is no concern about illogical inference.

## 18.7.2	Logic inherent in scientific parameters

In personalized medicine development, the patient population is thought of as a mixture of two or more subgroups that might derive differential efficacy. An important decision to make is whether to target the entire patient population (so-called all-comers), or just a subgroup of the patients, or a combination of subgroups of the patients.

Consider the Subgroups example of Section 18.2, the case of a binary marker g that separates the entire population into two subgroups, g^+ and g^-. Abbreviate "treatment" and "control" by Rx and C, respectively. Denote by μ^{Rx} and μ^C the true mean responses over the entire patient population if the entire population had received treatment or control, respectively. Denote by $\mu^{Rx}_{g^+}$, $\mu^{Rx}_{g^-}$, $\mu^C_{g^+}$, $\mu^C_{g^-}$ the corresponding mean responses in the g^+ and g^- subgroups.

If efficacy is measured by the *difference* of expectation of treatment and control outcomes, then

$$\mu_{g^+} = \mu^{Rx}_{g^+} - \mu^C_{g^+}, \quad \mu_{g^-} = \mu^{Rx}_{g^-} - \mu^C_{g^-}.$$

represent efficacy of the drug in the g^+ and g^- subgroups. If population prevalence of the g^+ subgroup is γ^+ (independent of the random assignment of patients to treatment and control), then efficacy in the combined population is

$$\bar{\mu} = \mu^{Rx} - \mu^C = \gamma^+ \times \mu_{g^+} + (1 - \gamma^+) \times \mu_{g^-} \tag{18.22}$$

Ding, Lin, and Hsu (2014) noted that exact relationship such as (18.22) holds only if efficacy is measured as a difference of means. Note also that relationship (18.22) does not guarantee the relationship in the null parameter spaces, because

$$\{\mu_{g^+} \leq 0\} \bigcap \{\mu_{g^-} \leq 0\} \neq \{\bar{\mu} \leq 0\}.$$

Due to the intrinsic relationship between efficacy in the combined population and $g+$ and $g-$ subgroups, there are a few desirable features that any hypothesis testing method proposed should satisfy. For example, if a higher mean response is better, then it is obviously desirable for a statistical method to guarantee that, if both $\mu_{g^+} > 0$ and $\mu_{g^-} > 0$ are inferred, then the inference $\bar{\mu} < 0$ does not occur, to avoid Simpson's Paradox. Call this PL_1 (PL stands for Parameter Logic).

In fact, since $\mu_{g^+} > 0$ and $\mu_{g^-} > 0$ implies $\bar{\mu} > 0$, it might be desirable for a statistical method to guarantee that, if both $\mu_{g^+} > 0$ and $\mu_{g^-} > 0$ are inferred, then $\bar{\mu} > 0$ is inferred as well. Call this PL_2.

Further, since $\bar{\mu} > 0$, implies either $\mu_{g^+} > 0$ or $\mu_{g^-} > 0$ or both, it might be desirable for a statistical method to guarantee that, if $\bar{\mu} > 0$ is inferred, then at least one of $\mu_{g^+} > 0$ and $\mu_{g^-} > 0$ is inferred as well. Call this PL_3.

To summarize:

PL_1: If both $\mu_{g^+} > 0$ and $\mu_{g^-} > 0$ are inferred, then the inference $\bar{\mu} < 0$ does not occur;

PL_2: If both $\mu_{g^+} > 0$ and $\mu_{g^-} > 0$ are inferred, then $\bar{\mu} > 0$ is also inferred;

PL_3: If $\bar{\mu} > 0$ is inferred, then at least one of $\mu_{g^+} > 0$ and $\mu_{g^-} > 0$ is inferred as well.

Appropriate choice of measure of efficacy and estimation technique will ensure PL_1. However, as Ding, Lin, and Hsu (2014) noted, no exact relationship analogous to (18.22) holds if efficacy is measured as odds ratio of response rates for binary outcome, or if efficacy is measured as ratio of hazard rates for time-to-event outcomes. Thus, whether PL_2 and PL_3 should be required in general is a topic for future discussion.

18.8 Going Forward

In the 1990s, the concept of FWER control was very useful in pointing out the inadequacy of controlling the *experimentwise* error rate, defined as the probability of making at least one incorrect statement *when all the null hypotheses are true*. The distinction is FWER is the supremum of the probability of making at least one incorrect statement, which can occur when not all the null hypotheses are true. Examples of multiple comparison methods for which the supremum occurs when some, but not all, of the null hypotheses are true include Fisher's Protected LSD, and methods based on joint-ranking. So the concept of FWER control served an important purpose. (Controlling the experimentwise error rate is now called weak control of FWER.) But as statistical knowledge advances and clinical trials become more complex, it is perhaps time to go beyond FWER control.

Traditional error rates, including those cited in Section 18.1 and the False Discovery Rate (FDR) (Benjamini and Hochberg 1995) popular in discovery studies, are defined on the number or proportion of incorrect statements. They treat all incorrect statements as equally consequential. When an analysis involves primary and secondary endpoints, and the decision path involves accelerated and final approvals, assigning different losses to different incorrect statements and controlling risk (expected loss) is a possibility worth considering.

Bibliography

[1] Y. Benjamini and Y. Hochberg. Controlling the false discovery rate: A practical and powerful approach to multiple testing. *Journal of the Royal Statistical Society, Series B*, 57(2):289–300, 1995.

[2] J.O. Berger. *Statistical Decision Theory and Bayesian Analysis*. Springer-Verlag, New York, USA, second edition, 1985.

[3] F. Bretz, W. Maurer, W. Brannath, and M. Posch. A graphical approach to sequentially rejective multiple test procedures. *Statistics in Medicine*, 28:586–604, 2009.

[4] F. Bretz, M. Posch, E. Glimm, F. Klinglmueller, W. Maurer, and K. Rohmeye. Graphical approaches for multiple comparison procedures using weighted bonferroni, simes, or parametric tests. *Biometrical Journal*, 53:894–913, 2011.

[5] G. Casella and R.L. Berger. *Statistical Inference*. Duxbury Press, Belmont, CA, USA, 1990.

[6] Y. Ding, H.M. Lin, and J.C. Hsu. Subgroup mixable inference in personalized medicine, with an application to time-to-event outcomes. *Statistics in Medicine*, 35:1580–1594, 2016.

[7] A. Dmitrienko and Tamhane A. Mixtures of multiple testing procedures for gatekeeping applications in clinical trials. *Statistics in Medicine*, 30:1473–1488, 2011.

[8] A. Dmitrienko, Tamhane A., and Wiens B.L. General multistage gatekeeping procedures. Biometrical Journal, 50:667–677, 2008.

[9] A. Dmitrienko, W. Offen, O. Wang, and D. Xiao. Gatekeeping procedures in dose-response clinical trials based on the Dunnett test. *Pharmaceutical Statistics*, 5:19–28, 2006.

[10] H. Finner and K. Strassburger. The partitioning principle: a powerful tool in multiple decision theory. *Annals of Statistics*, 30:1194 –1213, 2002.

[11] R.A. Fisher. *The Design of Experiments*. Oliver and Boyd, Edinburgh and London, UK, 1935.

[12] A. Genz and F. Bretz. Numerical computation of multivariate t-probabilities with application to power calculation of multiple contrasts. *Journal of Statistical Computation and Simulation*, 63:361–378, 1999.

[13] Y. Hochberg. A sharper Bonferroni procedure for multiple tests of significance. *Biometrika*, 75:800–802, 1988.

[14] S. Holm. A Simple Sequentially Rejective Multiple Test Procedure. *Scandanavian Journal of Statistics*, 6:65–70, 1979.

[15] G. Hommel. A stagewise rejective multiple test procedure based on a modified Bonferroni test. *Biometrika*, 75:383–386, 1988.

[16] J.C. Hsu. The factor analytic approach to simultaneous inference in the general linear model. *Journal of Computational and Graphical Statistics*, 1:151–168, 1992.

[17] J.C. Hsu. *Multiple Comparisons: Theory and Methods*. Chapman and Hall/CRC, Boca Raton, London, New York, Washington D.C., USA, 1st edition, 1996.

[18] J.C. Hsu and R.L. Berger. Stepwise confidence intervals without multiplicity adjustment for dose response and toxicity studies. *Journal of the American Statistical Association*, 94:468–482, 1999.

[19] Y. Huang and J.C. Hsu. Hochberg's step-up method: Cutting corners off Holm's step-down method. *Biometrika*, 94:965–975, 2007.

[20] Y. Huang, H. Xu, V. Calian, and J.C. Hsu. To permute or not to permute. *Bioinformatics*, 22(18):2244–2248, 2006.

[21] Rohmeyer K. and Klinglmueller F. gMCP - an R package for a graphical approach to weighted Multiple Test Procedures. *R package version 0.8.10*, 1:1–60, 2015.

[22] E.E. Kaizar, Y. Li, and J.C. Hsu. Permutation multiple tests of binary features do not uniformly control error rates. *Journal of the American Statistical Association*, 106(18):1067–1074, 2011.

[23] Y. Liu and J.C. Hsu. Testing for efficacy in primary and secondary endpoints by partitioning decision paths. *Journal of the American Statistical Association*, 104(18):1661–1670, 2009.

[24] Y. Liu, J.C. Hsu, and S. Ruberg. Partition testing in dose-response studies with multiple endpoints. *Pharmaceutical Statistics*, 6:181–192, 2007.

[25] R. Marcus, E. Peritz, and K.R. Gabriel. On closed testing procedures with special reference to ordered analysis of variance. *Biometrika*, 63:655–660, 1976.

[26] W. Offen, C. Chuang-Stein, A. Dmitrienko, J. Maca, L. Meyerson, R. Muirhead, P. Stryszak, A. Boddy, K. Chen, K. Copley-Merriman, W. Dere, S. Givens, D. Hall, D. Henry, J.D. Jackson, A. Krishen, T. Liu, S. Ryder, A.J. Sankoh, J. Wang, and C.H. Yeh. Multiple co-primary endpoints: Medical and statistical solutions a report from the multiple endpoints expert team of the pharmaceutical research and manufacturers of america. *Drug Information Journal*, 41:31–46, 2007.

[27] S. Sarkar. Some probability inequalities for ordered MTP2 random variables: A proof of the Simes conjecture. *The Annals of Statistics*, 26:494–504, 1998.

[28] R.J. Serfling. Some probability inequalities for ordered MTP2 random variables: A proof of the Simes conjecture. *The Annals of Statistics*, 26:494–504, 1998.

[29] R. J. Simes. An improved Bonferroni procedure for multiple tests of significance. *Biometrika*, 73:751–754, 1986.

[30] G. Stefansson, W.C. Kim, and J.C. Hsu. An improved Bonferroni procedure for multiple tests of significance. *Biometrika*, 73:751–754, 1986.

[31] K. Takeuchi. *Studies in Some Aspects of Theoretical Foundations of Statistical Data Analysis (in Japanese)*. Toyo Keizai Shinposha, Tokyo, Japan, 1973.

[32] K. Takeuchi. Basic ideas and concepts for multiple comparison procedures. *Biometrical Journal*, 52:722–734, 2010.

[33] J.W. Tukey. *The Problem of Multiple Comparisons. Dittoed manuscript of 396 pages*. Department of Statistics, Princeton University, USA, 1953.

[34] J.W. Tukey. *The problem of multiple comparisons*, volume VIII: The Collected Works of John W. Tukey. Chapman & Hall, New York and London, 1994.

[35] P.H. Westfall and A. Krishen. Optimally weighted, fixed sequence and gatekeeper multiple testing procedures. *Journal of Statistical Planning and Inference*, 99:25–40, 2001.

[36] H. Xu and J.C. Hsu. Using the partitioning principle to control the generalized family error rate. *Biometrical Journal*, 49:52–67, 2007.

[37] H. Xu, I. Nuamah, J. Liu, P. Lim, and A. Sampson. A Dunnett-Bonferroni-based parallel gatekeeping procedure for dose-response clinical trials with multiple endpoints. *Pharmaceutical Statistics*, 8(4):301–316, 2009.

19

Subgroup Analysis

Rui Wang

CONTENTS

19.1 Introduction

The primary objective of a randomized clinical trial (RCT) is to investigate whether one treatment is better than its alternatives on average. However, treatment effects may vary across different patient subpopulations. When translating trial results to guide patient management, one is often most concerned with the question of which treatment strategy is most appropriate for achieving a desired outcome, for a particular patient, or a group of patients with similar characteristics. For example, while there is clear evidence suggesting beneficial effects of beta blockers for Caucasians with heart failures, the evidence for benefit in African-Americans has been ambiguous. In fact, scientists have discovered a race-specific gene variant, GRK5, which resembles the effect of beta blockers and may account for the heterogeneity of the Beta Blocker effect across race [36]. Sacks et al [48] reported results from a RCT which demonstrated that pravastatin had protect effect for the risk of a fatal coronary event or a nonfatal myocardial infarction (MI) among patients whose low-density lipoprotein (LDL) cholesterol levels of 115 to 174 mg per deciliter. However, additional subgroup analysis revealed an absence of benefits in the subgroup of patients whose LDL levels were below 125 mg/dL. The US Food and Drug Administration (FDA) uses subgroup analyses in evaluating the safety and efficacy to justify approval of treatment in subgroups of patients [44]. For example, while the administration of drotrecogin alfa activated was found to reduce the rate of death in patients with a clinical diagnosis of severe sepsis [7, 44], its associated risk of intracrinial hemorrhage justified its approval in a subgroup of more severely ill patients where the benefit outweighed the risk.

Subgroup analyses are often undertaken to assess the heterogeneity of treatment effects among different subpopulations defined by a baseline characteristic. This type of subgroup analyses can be regarded as hypothesis-generating and exploratory. In contrast, sometimes investigators are interested in learning how treatment effects vary according to a specific baseline factor, motived by a previous study. This type of subgroup analyses is often listed as a primary or secondary study objective and may be regarded as confirmatory.

Although subgroup analyses are useful for guiding patient treatment selection, they introduce analytic challenges and can lead to overstated and misleading results. Poorly planned and conducted analyses may lead to false positive findings due to multiplicity and testing post-hoc data driven hypotheses [41, 46, 59] and are also prone to false negative findings due to insufficient statistical power [61]. The Second International Study of Infarct Survival investigators [5] provided an interesting example to illustrate the difficulty in reliable identification of subgroup differences even with a large trial of more than 17,000 patients: Compared to placebo, the mortality rate for patients on aspirin was slightly higher for those born under Zodiac signs Gemini and Libra but substantially lower for those born under other Zodiac signs. This example underscores the necessity of considering biological plausibility when interpreting a statistically significant subgroup analysis result to avoid false positive findings. This chapter describes analytic methods for conducting subgroup analyses, considerations for subgroup analyses in the design stage, and reporting and interpretation of subgroup analyses in randomized clinical trials.

19.2 Methods for Conducting Subgroup Analyses

19.2.1 Commonly used methods

The most commonly used method for detecting treatment effects heterogeneity is through the use of a statistical model incorporating interaction terms between the treatment group and the baseline factor. We will use the following three examples to illustrate how subgroup analyses based on testing for interaction terms can be conducted when the outcome variable is continuous, binary, or a time-to-event.

In the Childhood Adenotonsillectomy Trial (CHAT) [38], children with the obstructive sleep apnea syndrome were randomized to early adenotonsillectomy or watchful waiting with supportive care to assess whether early adenotonsillectomy would result in improvement in polysomnographic, cognitive, behavioral, and health outcomes at 7 months. The investigators were interested in whether the treatment effect varied according to the obesity status. Let Y denote the primary endpoint, the attention and executive function score on the Developmental Neuropsychological Assessment, $X = 1$ or 0 denote the early adenotonsillectomy or watchful waiting respectively, and $Z = 1$ or 0 denote being obese or not. A standard linear regression model to assess treatment effect modification by obesity status is:

$$Y = \alpha + \beta X + \gamma Z + \delta X \cdot Z + \epsilon, \qquad (1)$$

where α represents the intercept, β represents the treatment effect in the non-obese group, and $\beta + \delta$ represents the treatment effect in the obese group. ϵ represents the random error in the outcomes. A formal test of the treatment effect modification by the obesity status can be conducted through testing the null hypothesis of $\delta = 0$.

When the outcome is binary, a logistic regression model is often used. Kirpalani et al. [33] conducted a randomized clinical trial to examine whether the rate of survival to 36 weeks of post-menstrual age without bronchopulmonary dysplasia differed after non-invasive respiratory support with either nasal intermittent positive-pressure ventilation (IPPV) or nasal continuous positive airway pressure (CPAP). The primary outcome was a

TABLE 19.1
Heterogeneity of treatment effects depends on metric.

Baseline risk status	Mortality Risk			
	Treatment	Control	RR	$R_c - R_t$
Scenario 1				
low	0.05	0.1	0.5	0.05
high	0.25	0.5	0.5	0.25
Scenario 2				
low	0.05	0.1	0.5	0.05
high	0.45	0.5	0.9	0.05

composite endpoint including death before 36 weeks of post-menstrual age and survival with bronchopulmonary dysplasia. Let $Y = 1$ or 0 denote the occurrence or non-occurence of the primary outcome, respectively, and $X = 1$ or 0 denote IPPV or CPAP, respectively. One of the pre-planned subgroup analyses was to examine the treatment effect modification by the birth-weight stratum. Let $Z = 0$ or 1 denote whether or not the infant birth weight was less or greater than 750 grams. A logistic model can be fit to test the treatment by birth-weight stratum interaction as follows:

$$\text{logit}(p) = \alpha + \beta X + \gamma Z + \delta X \cdot Z, \tag{2}$$

where $p = P(Y = 1|X, Z)$, and $\text{logit}(p) = \log[p/(1-p)]$. Here, the treatment effect is quantified by the odds ratio of death or survival with bronchopulmonary dysplasia in the IPPV group as compared to the CPAP group, $\exp(\beta)$ represents the treatment effect among those with birth weight less than 750 grams, and $\exp(\beta + \delta)$ represents the treatment effect among those with birth weight greater than 750 grams. It was estimated that the treatment effect among those with birth weight less than or greater than 750 grams were 1.30 or 0.90, respectively, after adjusting for baseline covariates and stratification factors. The birth-weight stratum-specific odds ratio estimates seemed to indicate a heterogeneity in treatment effects, that is, compared to infants in the CPAP group, the estimated odds of death or survival with bronchopulmonary was higher in the IPPV group among those with birth weight less than 750 grams, but lower among those with birth weight greater than 750 grams. However, we should avoid making inferences based on point estimates only, ignoring the uncertainty around these estimates. In fact, the 95% confidence interval for the odds ratio was $(0.83 - 2.04)$ or $(0.64 - 1.26)$ among those with birth weight less than 750 grams or greater than 750 grams, respectively; both intervals included the null value of 1, representing an absence of treatment effect. Furthermore, the presence of a treatment effect heterogeneity should be determined by a formal interaction test for the null hypothesis of $\delta = 0$.

The interaction term in the logistic regression model compares log-odds ratios across different subgroups. Let R_t or R_c denote the risk in the treatment or control group, respectively. Other commonly used treatment effect measures for binary outcomes include risk ratio RR, defined as R_t/R_c, and risk difference RD, defined as $R_c - R_t$. The detection of an interaction effect depends on the metric used to quantify treatment effects. As illustrated in Table 19.1, under Scenario 1, the estimated treatment effects as measured by RR are the same for the low and high baseline risk groups, but different when measured by risk difference. Scenario 2 illustrates the opposite situation.

Choice of treatment effect metrics would ideally depend on the question of interest. For a treatment with serious side effects, the gained benefit from a reduction in absolute risk from 1% to 0.5% might not outweigh the side effects while a reduction from 50% to

25% would be much more substantial, although both correspond to a relative reduction of 50%. Risk difference can be a more relevant metric when evaluating the treatment benefits relative to other factors such as side effects or costs. One way to test interaction on the risk difference scale is through the use of the relative excess risk due to interaction (RERI) [45]. Let p_{00}, p_{01}, p_{10}, or p_{11} denote the probability of event in each of the four subgroups defined by treatment arm and the subgroup factor, say gender, such that the first index represents treatment (1) or control (0), and the second index represents males (1) or females (0). Let $\mathrm{RR}_{11} = p_{11}/p_{00}$ represent the relative risk comparing the males in the treatment group to the females in the control group, $\mathrm{RR}_{01} = p_{01}/p_{00}$ represent the relative risk comparing the males in the control group to the females in the control group, and $\mathrm{RR}_{10} = p_{10}/p_{00}$ represent the relative risk comparing the females in the treatment group to the females in the control group. The RERI is defined as

$$\mathrm{RERI} = \mathrm{RR}_{11} - \mathrm{RR}_{01} - \mathrm{RR}_{10} + 1.$$

Since

$$\mathrm{RERI} = \frac{p_{11}}{p_{00}} - \frac{p_{01}}{p_{00}} - \frac{p_{10}}{p_{00}} + 1 = \frac{p_{11} - p_{01} - (p_{10} - p_{00})}{p_{00}},$$

testing whether $\mathrm{RERI} = 0$ is equivalent to testing whether the risk difference is the same for males and females. Assuming that the odds ratio approximates relative risk (e.g., in the rare disease settings), we can obtain a point estimate for RERI using the estimated logistic regression coefficients from model (2) [29]:

$$\widehat{\mathrm{RERI}} = \exp(\hat{\beta} + \hat{\gamma} + \hat{\delta}) - \exp(\hat{\beta}) - \exp(\hat{\gamma}) + 1,$$

where we use $\hat{\beta}, \hat{\gamma}$, or $\hat{\delta}$ to denote an estimate for β, γ, or δ, respectively. A standard error estimate for $\widehat{\mathrm{RERI}}$ can be obtained using the standard delta method based on a Taylor expansion or through the bootstrap methods. Assmann et al. [3] found that confidence intervals constructed based on a bootstrap percentile method yielded better performance through simulations. The bootstrap percentile confidence interval is formed by using the 2.5th and 97.5th percentiles of the bootstrap sampling distribution for $\widehat{\mathrm{RERI}}$.

For time-to-event endpoints, a Cox proportional hazards model is often employed. In a randomized trial to compare strategies for multivessel revascularization in patients with diabetes, subjects were randomized to either percutaneous coronary intervention (PCI) with drug-eluting stents ($X = 1$) or coronary-artery bypass grafting (CABG) ($X = 0$) [19]. The primary endpoint was a composite of death from any cause, nonfatal MI, or nonfatal stroke. Several pre-specified subgroup analyses were conduct to assess whether the greater benefit of CABG over PCI was consistent across subgroups defined by gender, race, disease type, and other factors. Let T denote the time to the composite endpoint and $Z = 1$ or 0 denote males and females, respectively. To test for treatment effect modification by gender, we can fit the following Cox proportional-hazards model:

$$\mathrm{h}(t) = \mathrm{h}_0(t)\exp(\alpha + \beta X + \gamma Z + \delta X \cdot Z),$$

where $\mathrm{h}(t)$ is the hazard function, representing the hazard of the composite endpoint at time t, and $\mathrm{h}_0(t)$ is the baseline hazard function, representing the hazard function corresponding to the groups of individuals with all covariates equal to 0. We can then test the null hypothesis of $\delta = 0$ as before. Here, the treatment effect is measured by the hazard ratio. The treatment effect is represented by $\exp(\beta)$ and $\exp(\beta + \delta)$ in the females and males, respectively. The interaction test yielded a p-value of 0.46, suggesting that treatment effect was consistent across genders.

Worthy of mention also is that in this example, the 95% confidence interval for the hazard ratio in the male group excluded the null value of 1 but that for the female group included the null value. It would be erroneous to conclude that the treatment effect is different between males and females based on separate testing of no treatment effect within each subgroup. Such stratified analysis approach increases the risk of obtaining a false positive result (i.e., claiming there is a treatment effect within one subgroup when there is not) due to multiplicity and also a false negative result (i.e., failing to detect the subgroup-sepcific treatment effect when there is one) as such analyses are associated with reduced power to detect a similar treatment effect due to the fact that data are divided into smaller data sets. In fact, there were 1356 males and 544 females in the trial. It is possible that lack of statistical significance in the female group is due to smaller sample size. Similar phenomenon was also present in Bombardier et al. [8], which reported results from a randomized clinical trial comparing toxicities of rofecoxib and naproxen in patients with rheumatoid arthristis. Four percent of the study subjects were aspirin indicated, that is, they met the criteria of the FDA for the use of aspirin for secondary cardiovascular prophylaxis but were not taking low-dose aspirin therapy. These subjects appeared to have an elevated risk for MIs; they accounted for 38% of the subjects who had MIs in the study. Therefore, the investigators assessed the relative risk of MIs comparing naproxen vs. rofecoxib in the subgroups defined by whether or not aspirin was indicated. There was a significantly higher rate of MIs in patients receiving rofecoxib than in those receiving naproxen overall and among patients for whom aspirin was indicated. However, the difference in the rates of MIs did not reach statistical significance in the subgroup of patients for whom aspirin was not indicated; even though the relative risk was as high as 3, the 95% confidence interval was very wide ([0.91–12.78]) as the event rates were very small [17]. The misleading conclusion that rofecoxib in safe in this subgroup would have a large public health impact as most patients belong to this group.

In the examples discussed above, we consider binary covariates. The same methods apply to ordinal or continuous covariates such as education level or age. In practice, investigators often categorize a continuous covariate to assess treatment effect modification for the ease of interpretation. The cutoff points for categorization need to be pre-specified. Usually the choices of cutoff points are either based on clinical relevance and biological rationale or using quantiles to ensure a more balanced allocation of subjects into different subgroups. Searching a cutoff point to divide subjects into subgroups corresponding to the biggest differences in treatment effect would lead to inflated type I error for the null hypothesis of no treatment effect differences across the subgroups defined by the cutoff point and should be avoided.

For cluster randomized trials where a group of subjects, rather than individuals, are randomized to different treatment groups, outcomes from subjects within a cluster are correlated. Mixed effect models based on individual-level regression can incorporate the interaction terms to examine the evidence for effect modification by both individual-level and cluster-level covariates [24].

One common mistake is to claim statistical significance of a treatment effect heterogeneity ignoring the total number of subgroup analyses conducted. As it has been widely recognized, the probability of a false positive finding increases drastically as the number of subgroup analyses increases (see for example [34]). Wang and Ware [61] described several ways for addressing the issue of multiplicity relating to subgroup analyses. They discussed two general types of formal statistical methods. One type controls the family-wise false positive rate — the probability of at least one Type I error in the presence of multiple tests, such as the Bonferroni method [49], the Holm method [27], the Simes-Hochberg method [26, 51], and the Hommel's method [28]. The other type controls the false discovery rate — the expected proportion of Type I errors among the rejected hypotheses[6]. In summary, they recommended the use of methods that control for family-wise false positive rate for pre-specified subgroup analyses of confirmatory nature; methods that control for false discovery rate may be employed for exploratory subgroup analyses. A less formal approach to address the issue of multiplicity in reporting subgroup analyses is to note the total number of subgroup analyses performed and the expected number of tests that would be significant by chance.

19.2.2 Qualitative interaction

The heterogeneity in treatment effects has been classified as being either qualitative or quantitative, differed by whether or not there is a directional change of treatment effects in difference subgroups of patients [41]. Qualitative interactions are especially important since they often have important implications for patient management. Several tests have been proposed for assessing whether there are qualitative interactions across I disjoint patient subgroups in the setting where two treatments are compared. There are two types of null hypotheses considered. Let δ_i denote the true difference in treatment efficacy within patient subset i for $i = 1, \ldots, I$. One null hypothesis is

$$H_0 : \delta_i > 0 \text{ for all } i \text{ or } \delta_i < 0 \text{ for all } i, \tag{19.1}$$

that is, one treatment is more beneficial than the other treatment across all subgroups. Gail and Simon [23] proposed a likelihood ratio test of H_0. Assume that estimates D_i of δ_i are independent and normally distributed with mean δ_i and known variance σ_i^2. In practice, consistent estimates of σ_i^2 will be used. This likelihood ratio test is constructed by noting that testing the null hypothesis of no qualitative interaction is equivalent to testing the hypothesis that $\delta_i \geq 0$ for all i or $\delta_i \leq 0$ for all i. The null hypothesis of no qualitative interaction is rejected if both

$$Q^- = \sum_{i=1}^{I} (D_i^2/\sigma_i^2) I(D_i > 0) > c,$$

and

$$Q^+ = \sum_{i=1}^{I} (D_i^2/\sigma_i^2) I(D_i < 0) > c,$$

where $I(D_i > 0) = 1$ if $D_i > 0$ and 0 otherwise, and $I(D_i < 0) = 1$ if $D_i < 0$ and 0 otherwise. The critical value c for varying numbers of subgroups from 2 to 30 and varying significance level from 0.001 to 0.20 are provided in Table 1 of [23]. Suppose that there is a minimal treatment difference of clinical relevance and it is of interest to test the hypothesis that a clinical meaningful treatment difference exists uniformly across subgroups. This test can be used to test the hypothesis that all treatment differences are greater (or less) than a pre-specified δ_0. This is accomplished by applying the test to the translated estimates $(D_i - \delta_0)$. Also of note is that this test is constructed assuming independent estimates of treatment effects D_i in each subgroup. Therefore, in practice, it is suggested that D_i be estimated from separate models for each subset rather than fitting one model to the entire dataset even though the latter can sometimes yield asymptotically uncorrelated D_i.

When sample sizes within subgroups are small, it may be advantageous to consider using the exact critical values provided in [50] for the Gail and Simon test. These critical values are derived from the exact null distribution and can have better performance than asymptotic critical values in finite samples. In addition, if the underlying distribution of the outcomes substantially deviates from normality with long tails, even though the least-squares estimates (such as the difference in means in each subgroup) are still asymptotically normal based on the central limit theorem, the test using a robust estimator with smaller variance will be more efficient. Silvapulle [50] evaluated the Pitman asymptotic relative efficiency [43] of two tests sequences as the Type I and Type II error rates tend to positive limits $\alpha > 0$ and $\beta > 0$ under a sequence of local alternatives. They found that the asymptotic efficiency of the test using a robust estimator relative to that using the least squares estimator was the same as the asymptotic efficiency of the robust estimator relative to the least squares estimator, that is, the asymptotic efficiency properties of the robust estimator would translate to asymptotic power of the test based on the same estimator.

The null hypothesis of no qualitative interaction can also be tested using the range test described in [42]. This test rejects the null hypothesis at level α if both

$$\max_{1 \leq i \leq I}\{D_i/\sigma_i\} > c_r, \quad \text{and} \quad \min_{1 \leq i \leq I}\{D_i/\sigma_i\} < -c_r,$$

where the corresponding critical value c_r is provided in Table 1 of [42]. The authors also performed simulation studies to compare the power of the proposed range test and that of the likelihood ratio test proposed in [23] for the detection of qualitative interactions. They found that the range test tends to be more powerful when there is a preponderance of effects in one direction across all subgroups while the likelihood ratio test is more powerful otherwise. Li and Chan [35] extended the range test by performing the usual range test on the extreme values of all the subgroups first as in the original range test and subsequently on all subgroups of the subsets in a stepwise manner. The overall Type I error control is accomplished by using critical values at each step that correspond to the pre-specified Type I error minus a small positive value, which is determined through extensive simulations. By using all possible subgroups, the extended range test can have greater power than the original range test in some settings.

Ciminera et al [15] described a 'pushback' procedure to assess whether there is substantial evidence of qualitative interaction. Although developed for the setting of evaluating the treatment by center interaction, this method can be used more generally for assessing qualitative interactions across other types of pre-specified subgroups. This procedure involves 3 steps. In Step 1, they calculate center-specific standardized deviates from a reference, for example, the median of subgroup-specific treatment effects. In Step 2, they re-order and 'push-back' these deviations by amounts determined by the corresponding order statistics from a reference distribution. In Step 3, they restore the standardized values to the original scale and claim substantial evidence of a qualitative interaction based on the appearance of opposite signs among the de-standardized values across subgroups. This procedure can be implemented with several variations using any combination of reference distributions, 'pushback' amounts, and methods for standardization in Step 1. This procedure can be liberal and has been shown to have inflated Type I errors in some settings such as where some subgroups have large positive effects and there are many other subgroups with effects that are close to zero [11].

Another type of null hypothesis was considered in Pan and Wolfe [40]. They considered the problem of detecting a qualitative interaction which represented the situation where the treatment effects were not all greater than a value d, the minimal treatment difference of clinical relevance. That is, the authors were interested in testing the null hypothesis vs. the alternative hypothesis

$$H_0(d) : \delta_i > -d \text{ for all } i = 1, \ldots, I \text{ or } \delta_i < d \text{ for all } i = 1, \ldots, I \tag{19.2}$$

$$H_1(d) : \text{there is at least one } \delta_i > d \text{ and at least one } \delta_j < -d, \text{ where } i \neq j = 1, \ldots, I.$$

This problem differs from testing the null hypothesis that a substantial treatment difference

TABLE 19.2
Treatment effect (Fluoxetine vs. CBT) across 5 subgroups based on CDRS-R score at baseline.

| | **CDRS-R score at baseline** | | | | |
	<50	50–55	55–61	61–69	≥ 69
n_i	29	45	36	37	40
D_i	−1.14	−6.35	6.38	4.11	17.50
σ_i	4.69	3.12	4.39	4.19	4.40

exists uniformly across subgroups. Assuming that $D_1, \ldots, D_I$ are independently normally distributed with means $\delta_1, \ldots, \delta_I$ and known variances σ_i^2, the method consists of constructing a z-confidence interval (L_i, U_i) for within subgroup treatment effect δ_i and rejecting the null hypothesis $H_0(d)$ if at least one subgroup confidence interval is entirely above d and at least one subgroup confidence interval is entirely below $-d$. The z-confidence interval is constructed as $(D_i - \Phi^{-1}(1 - \frac{1-\text{PE}}{2}) * \sigma_i, D_i + \Phi^{-1}(1 - \frac{1-\text{PE}}{2}) * \sigma_i)$, where $\text{PE} = 2(1 - \alpha)^{\frac{1}{I-1}} - 1$ and $\Phi(\cdot)$ is the cumulative distribution function of a standard normal distribution. The range test can be viewed as a special case when $d = 0$.

We illustrate the various methods for detecting qualitative interactions using data from the Treatment for Adolescents with Depression Study (TADS) [37]. In TADS, patients with diagnosis of major depressive disorder were randomized to Fluoxetine alone, cognitive-behavioral therapy (CBT) alone, their combination, or placebo. Here we focus on the comparison of the effect of Fluoxetine and CBT on the value of Children's Depression Rating Scale-Revised (CDRS-R) total score at 12 weeks and whether this effect depends on CDRS-R score at baseline. A total of 187 subjects were randomized to receiving Fluoxetine alone (n=97) or to CBT alone (n=90). We create 5 subgroups with equal number of subjects based on their CDRS-R score at baseline using the quintiles. The sample size n_i, estimated treatment effect D_i (difference in means) and its associated standard error estimate σ_i within each subgroup are summarized in Table 19.2.

To apply the likelihood ratio test for the null hypothesis of no qualitative interactions specified in (1.1) in [23], we obtain $Q^- = 18.89$, $Q^+ = 4.19$ and compare $\min(Q^+, Q^-) = 4.19$ to cutoff values. According to the Table 19.1 of [23], the cutoff value for a significance level of 0.1 or 0.2 is 3.39 or 4.96, respectively. Therefore, the p-value for the likelihood ratio test would be between 0.1 and 0.2. Comparing the test statistic $\min(Q^+, Q^-) = 4.19$ to the exact cutoff values in [50] requires an additional parameter $\nu = \sum_{i=1}^{I}(n_i - 1)$, which is 182. The corresponding cutoff value for a significance level of 0.05 provided in Table 2 of [50] is about 6.6 which is close to 6.5 in [23]. Regardless of asymptotic or exact cutoff values are used, the test will lead to the same conclusion in this example. For the range test, we obtain $\max_{1 \leq i \leq I}\{D_i/\sigma_i\} = 3.98$ and $\min_{1 \leq i \leq I}\{D_i/\sigma_i\} = -2.03$. The critical values provided in Table 1 of [42] for a significance level of 0.1 and 0.05 are 1.94 and 2.23, respectively. Using the testing rule specified in (1.2) the range test would lead to a p-value between 0.05 and 0.1. Table 19.3 summarizes the results from each steps of the 'pushback' procedure [15] where the pushback values are determined using medians of order statistics from the Gaussian distribution, where the columns of the table are arranged in the ascending order of the treatment effect estimates D_i. The standardized deviations are given by $[D_i - \text{median}(D_i)]/\sigma_i$. The pushback values are $\Phi^{-1}(\frac{3i-1}{16})$, $i = 1, \ldots, 5$, where Φ is the cumulative distribution function of the standard normal distribution. The pushed-back standardized deviations are obtained by subtracting the pushback values from the

TABLE 19.3
Pushback procedure for TADS based on the median of order statistics from the Gaussian distribution.

	CDRS-R score at baseline				
	50–55	<50	61–69	55–61	≥69
D_i (ordered)	−6.35	−1.14	4.11	6.38	17.50
σ_i	3.12	4.69	4.19	4.39	4.40
Standardized deviations	−3.35	−1.12	0	0.52	3.04
Pushback values	−1.15	−0.49	0	0.49	1.15
Pushed-back standardized deviations	−2.2	−0.63	0	0.03	1.89
Restore to the original scale	−2.76	1.16	4.11	4.23	12.44

standardized deviations. The last step involves restoring the pushed-back standardized deviations to the original scale by multiplying σ_i and then adding median(D_i). We observe opposite signs among the restored values, indicating substantial evidence of qualitative interaction.

To construct the confidence interval within each subgroup as in Pan and Wolfe [40], we first calculate the confidence coefficient PE which is 0.975 for a significance level of 0.05. The resulting confidence intervals for each of the 5 groups as in Table 19.2 are $(-11.62, 9.34)$, $(-13.33, 0.63)$, $(-3.42, 16.17)$, $(-5.25, 13.47)$, and $(7.67, 27.33)$. Although the confidence interval for the treatment effect in the fifth subgroup (CDRS-R score at baseline above 69) is entirely above 0, all other intervals contain 0. Therefore the p-value for testing the null hypothesis in (1.2) for any $d \geq 0$ is greater than 0.05.

19.2.3 Graphical methods

Methods for detecting quantitative or qualitative interaction described above consider how treatment effects vary across disjoint subject subgroups. It is also of interest to assess how treatment responses differ according to a continuous marker or overlapping subgroups. This section discusses several graphical methods developed to address this problem.

Song and Pepe [53] proposed the selection impact (SI) curve to facilitate treatment selection based on whether or not the value of a single biomarker, say Y, exceeds a threshold c. Let D denote a dichotomous response and $D = 1$, or 0 denote success or failure, respectively. Consider the setting where there are two candidate treatments ($T = 1$ or 0 denote a surgery procedure A and conventional treatment B, respectively) and larger values of Y are potentially associated with better performance of surgery vs. conventional treatment. Consider the treatment policy where surgery will be selected if $Y > c$, otherwise, conventional treatment will be selected. The threshold c determines the proportion of patients assigned to each treatment. Let $v(c) = P[Y < c]$ denote the proportion of subjects assigned to conventional treatment. As c increases to $-\infty$ to ∞, v increases from 0 to 1. Let y_v denote the vth quantile of Y in the population so that $v = P[Y < y_v]$. The population response rate corresponding to this policy is

$$\begin{aligned}
\theta(v) &= P[D = 1|(Y > y_v, T = 1) \text{ or } (Y < y_v, T = 0)] \\
&= (1 - v)P[D = 1 \mid Y > y_v, T = 1] + vP[D = 1 \mid Y < y_v, T = 0].
\end{aligned}$$

The SI curve is constructed by plotting $\theta(v)$, the population response rate, against v. Figure 19.1 provides a schematic illustration of a SI curve. The upper and lower dotted line

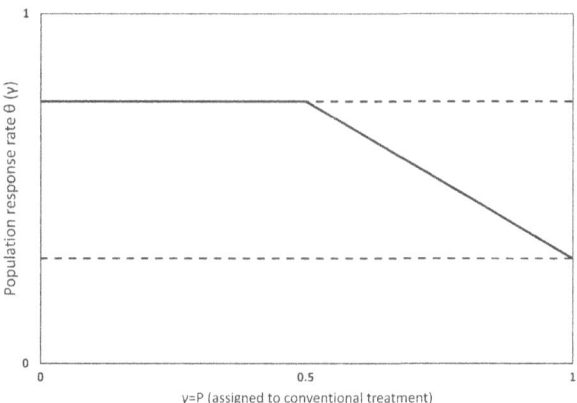

FIGURE 19.1
Schematic illustration of a SI curve.

represent the population response rate when surgery is always assigned ($\theta(0)$) and that when conventional treatment is always assigned ($\theta(1)$). The solid SI curve represents a case where the population response rate remains the same until up to 50% of population with the highest Y values are assigned to surgery. That is, assigning about 50% with the lowest values of the biomarker Y to conventional treatment can achieve the same response rate in the population as assigning all subjects to surgery. The distance between the upper (lower) dotted curve and the SI curve represents the treatment difference among the subgroup of subjects with biomarker values greater (less) than the corresponding threshold. $\theta(v)$ can be estimated nonparametrically by substituting the empirical estimates for the probabilities $P[D = 1, Y > y_v, T = t]$ and $P[Y > y_v, T = t]$, where $t = 0, 1$, or parametrically based on regression modeling

$$\text{logit}\{P[D = 1 \mid Y > y_v, T]\} = \alpha^T R(v, T),$$

where $R(v, T)$ is a q-dimensional function of v and T. The standard errors and confidence band for $\theta(v)$ can be obtained using the bootstrap method. The SI curve can also be constructed similarly for continuous outcomes by considering suitable definitions for $\theta(v)$ such as

$$\theta(v) = E(D \mid (Y > y_v, T = 1) \ or \ (Y < y_v, T = 0)).$$

One advantage of the SI curve over the standard approach by adding an interaction term $T * Y$ in a regression model is that the SI curve is metric-free while the definition of the interaction depends on the parametric assumptions of the model.

Bonetti and Gelber [9, 10] proposed the subpopulation treatment effect pattern plot (STEPP) method. The STEPP illustrates how treatment effects change across different but potentially overlapping patient subpopulations. They considered both settings where the treatment effect is the coefficient from Cox proportional hazards models or the difference in survival probabilities at a fixed time point between two treatment. The STEPP approach has been extended to continuous, binary and count outcomes[63]. This approach first estimating treatment effects in each subject subgroup, $\hat{\theta}_1, \ldots, \hat{\theta}_K$, where K denote the total number of subgroups, and then testing the null hypothesis of equal treatment effects across

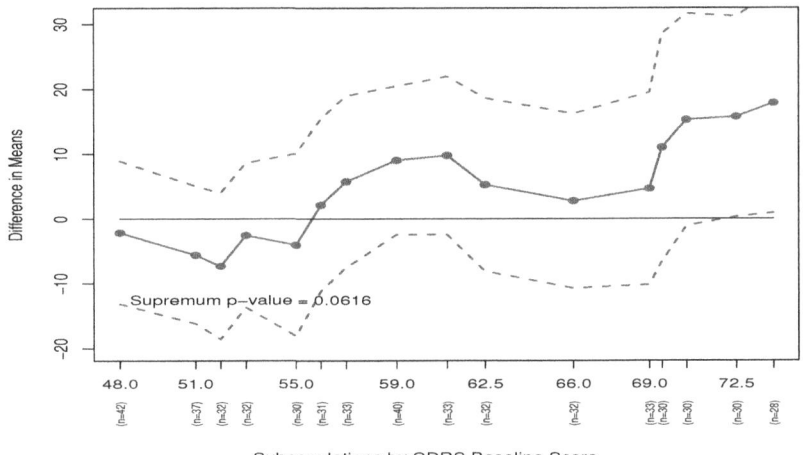

FIGURE 19.2
Subpopulation treatment effect pattern plot for TADS data according to CDRS summary score at baseline.

all subgroups, i.e., $H_0 : \theta_1 = \theta_2 = \cdots = \theta_K$. The test can be constructed by considering a supremum test statistic $T = max_{j=1,...,K}\{\frac{|\hat{\theta}_j - \hat{\theta}_{ALL}|}{\hat{\sigma}_j}\}$, where $\hat{\theta}_{ALL}$ is the treatment effect estimate using data from all subjects and $\hat{\sigma}_j = \sqrt{\widehat{var}(\hat{\theta}_j - \hat{\theta}_{ALL})}$. A p-value can be obtained by comparing the observed test statistic to a permutation distribution under the null hypothesis. The STEPP method takes into account both multiplicity and correlation among estimates due to overlapping. The subgroups can be formed by dividing the range of a continuous covariate into sub-intervals or by multiple covariates. For example, this method allows the comparison of the treatment effect in subjects younger than 60 years old and that in males. Assessing patterns of treatment effects along a continuous variable this way avoids the dependence on the parametric assumptions of the regression models and also avoids having to arbitrarily dividing subjects into disjoint subgroups. In addition, overlapping subgroups, compared to disjoint ones, necessarily have larger sample sizes within subgroups, which leads to more precise estimates of subgroup-specific treatment effects. This method is implemented in an add-on R package "STEPP"[32]. Figure 19.2 presents the STEPP for the TADS data using the same outcome and covariates as before. The "sliding window" approach was used to define subpopulations so that approximately 30 subjects were included in each subpopulation and a minimum of 5 subjects were replaced between any two subsequent subgroups, resulting in 16 subpopulations. The treatment effect estimates (CBT vs. Fluoxetine) are shown together with the corresponding 95% confidence band. The numbers in parentheses below the x-axis are the number of subjects in each subgroups. The label on the x-axis for each subgroup is the median of CDRS baseline summary score in that group. Consistent with the findings before, data suggested that when compared with CBT, Fluoxetine was associated with a lower (better) CDRS-R total score for those with lower CDRS baseline summary score and a higher (worse) CDRS-R total score for those with higher CDRS baseline summary score. The p-value for the supremum statistic was 0.0616.

19.2.4 Multivariate tests of interaction

The ultimate goal of a subgroup analysis is to guide treatment choices with a view to optimizing the outcome for each person. Subgroup analyses based on individual covariates one at a time, while useful, have serious limitations. For example, the results of a study may suggest that a given treatment works better in women and not so well in Asians. While this interpretation is relatively straightforward, it presents problems in providing treatment recommendations for Asian women. Therefore, approaches for a systematic exploration of multiple markers are needed. In what follows, we first discuss approaches for an overall test for the null hypothesis of no treatment-covariate interaction for any covariate, then describe approaches to identify baseline factors that can categorize patients into subpopulations with varying magnitudes or directions of treatment effects to guide the treatment selection.

Let Y denote the study endpoint, Z denote the vector of baseline covariates of length p, and $X = 1$, or 0 denote the intervention or control arms, respectively. One standard approach to test the null hypothesis of no treatment-covariate interactions for any covariate is a likelihood ratio test comparing the model with all treatment-covariate interaction terms and the model without. More specifically, this approach involves fitting the full model (M_f):

$$\phi[E(Y)] = \alpha + \beta X + \gamma Z + \delta X * Z,$$

and the reduced model (M_r):

$$\phi[E(Y)] = \alpha^* + \beta^* X + \gamma^* Z,$$

calculating the difference in -2Log-likelihood of the two models M_f and M_r, and comparing the resulting test statistic to a χ^2 distribution with degree of freedom p. Here we use $\phi(\cdot)$ to denote the link function. When Y is a continuous outcome, taking $\phi(\cdot)$ as an identity function yields linear regression models. When Y is a dichotomous outcome, $\phi(\cdot)$ can be the logit function.

The likelihood ratio test is an asymptotically optimal test under mild regularity conditions. However, in many settings, especially in clinical trials with modest sample sizes, when the number of candidate covariate is large, the number of parameters in M_f (or even M_r) quickly increases and can be greater than the number of observations. The likelihood ratio test will no longer be applicable because the parameters in the models may not be identifiable.

Wang et al. [60] propose a simple permutation test for such settings for a continuous outcome Y. Let $\mathbf{D} = (\mathbf{Y}, \mathbf{X}, \mathbf{Z})$ denote the observed ($n \times (p + 2)$) data matrix, where n denotes the number of observations. The permutation test involves three steps. In Step 1, the original data matrix $\mathbf{D}$ is transformed to $\tilde{\mathbf{D}}$ that can be partitioned into two parts ($\tilde{\mathbf{D}}_p, \tilde{\mathbf{D}}_f$) where the variables comprising each part are independent under the null hypothesis. The transformation is defined by subtracting the overall treatment effect from the outcome. In practice, a consistent estimate for the overall treatment effect is used. In Step 2, the rows of $\tilde{\mathbf{D}}_p$ are permuted to obtained datasets $\tilde{\mathbf{D}}^\ell$, $\ell = 1, \ldots, m$, that are equally likely as $\tilde{\mathbf{D}}$ under the null hypothesis of no treatment-covariate interactions for any covariate, where m denotes the number of permutations. In Step 3, $\tilde{\mathbf{D}}^\ell$'s are reverse-transformed to obtained datasets that are equally likely as the observed dataset. The permutation test is constructed by comparing a test statistic calculated from the observed dataset $T(\mathbf{D})$ to the permutation distribution formed by test statistics calculated from the reverse-transformed permuted datasets. One natural test statistic is the difference in prediction errors calculated from the full model and those from the reduced model. Note that in calculating the prediction errors, we can incorporate variable selection procedure such as the LASSO [57] to obtain the prediction error estimates to handle the case with a larger number of covariates and a modest sample size.

The proposed permutation test can be used when the number of covariates is large so that the number of parameters in the full model with all the 2-way treatment-covariate interaction terms is larger than the sample size and hence the likelihood ratio test cannot be used. It also maintains approximately correct Type I error regardless of the error distribution while the likelihood ratio may lead to inflated Type I error for skewed error distributions. The anticipated power loss associated with the permutation test is modest compared to the likelihood ratio test in settings with normally distributed error distributions and small numbers of covariates.

In the TADS study, in addition to the CDRS-R baseline score considered in previous sections, an abundance of baseline information was available. Over 4000 variables were collected at baseline, reflecting depression symptoms, psychopathology and cognitive functions. Because these variables were collected from various questionnaires that contained overlapping information, a subset of 116 candidate variables that were either demographical or having the words "total", "scale", or "score" in their descriptions were selected for further consideration. The full model including the main effect of treatment, all 116 covariates, and all possible two-way interactions between treatment and covariates contained 234 parameters which was greater than the total sample size of 187. The observed data matrix $\mathbf{D}$ was a 187×118 matrix where each row represented the observed data for each individual and the columns consisted of the outcome $\mathbf{Y}$, treatment indicator $\mathbf{X}$, and the covariate matrix $\mathbf{Z}$. We first transformed the observed data matrix $\mathbf{D}$ to $\tilde{\mathbf{D}}$ by subtracting from the outcome the treatment effect estimated under the null hypothesis of no treatment-covariate interactions. That is, $\tilde{\mathbf{D}} = (\tilde{\mathbf{Y}}, \mathbf{X}, \mathbf{Z})$, where $\tilde{Y} = Y - \alpha X$. Under the null hypothesis of no treatment-covariate interaction combined with the randomization design, the treatment indicator was independent of the random vector $(\tilde{Y}, \mathbf{Z}^T)$. Therefore we could permute $\mathbf{X}$ and obtained permuted data matrices $\tilde{\mathbf{D}}^\ell = (\tilde{\mathbf{Y}}, \mathbf{X}^\ell, \mathbf{Z})$. The last step involved reverse-transform $\tilde{\mathbf{D}}$ by adding back $\alpha \mathbf{X}$ to the column representing the transformed outcomes. For observed data matrix, this led to the original data matrix; for each permuted data matrix $\tilde{\mathbf{D}}^\ell$, this led to reverse-transformed permuted data matrices that were equally likely as the original data matrix under the null hypothesis. The test statistic we considered was difference in predictor errors obtained through two ways. The first approach related the outcome Y with the treatment indicator X and all the covariates incorporating variable selection. The second approach involved fitting a model relating Y and all the covariates for each treatment group, then obtaining fitted values using treatment-specific models. We used the LASSO [57] implemented in an R package 'glmnet' [22] to obtain fitted models for prediction. A p-value was obtained by comparing the test statistic evaluated on the observed dataset to those obtained based on the permuted datasets. Because the p-value is dependent on the random seeds, we repeated the tests 100 times. The median (Q1, Q3) for p-values from these 100 tests were 0.026 (0.020, 0.031), suggesting the presence of treatment-covariate interactions.

In the presence of multiple candidate baseline factors, an overall test for the null hypothesis of no treatment-covariate interaction effects for any covariate serves as an initial screening step in identifying subgroup effects. If this overall test is statistically significant, it provides evidence that there exists treatment-covariate interactions and warrants further search for these covariates; otherwise, it suggests that the evidence for treatment-covariate interactions is low and it may not be necessary to continue searching for specific covariates that interact with the treatment. In calculating the test statistic for the permutation test, a set of selected covariates that predicts outcome for each treatment group are identified and treatment-specific models relating the outcome and the selected covariates are fit. Note that in a linear model, the null hypothesis of no interaction effects for any covariates can also be expressed as the null hypothesis that the coefficients for the covariates across these treatment-specific models are the same. Therefore, comparing the coefficients of covariates in these treatment-specific models can provide some indications for covariates that warrant further investigation. In the TADS example, based on their associations with the outcome, the LASSO procedure did not select any variable for the Fluoxetine group; while CDRS-R baseline summary score was selected for the CBT group. This suggested that CDRS-R baseline summary score may be a potential effect modifier.

Sometimes treatment effects may depend on the severity of disease: the sickest patients may benefit the most from an effective treatment. Suppose that the outcome can be accurately predicted with multiple baseline covariates $\mathbf{X}$ using some regression model so that the disease severity can be expressed by a linear combination $S = \mathbf{X}'\beta$. The interest is to test for an interaction between the treatment and the disease severity. Follmann and Proschan [20] developed methods to both estimate β and test for an interaction along S. This is accomplished by testing $H_0 : \beta_0 = \beta_1$ vs. $H_A : a\beta_0 = \beta_1$, $a \neq 1$, where β_k is a vector of regression coefficients for group $k = 0, 1$. A likelihood ratio test is derived based on standardized estimates of β_0 and β_1. The resulting test statistic has an asymptotic chi-squared distribution with one degree of freedom. In practice, the adequacy of the chi-squared approximation can be checked by estimating a common β for both groups under the null hypothesis and checking if the quadratic form $\hat{\beta}'\widehat{\text{cov}}(\hat{\beta})^{-1}\hat{\beta}$ is at least $(2p)^2$. If not, a permutation reference distribution may be used by permuting the treatment indicators and calculating the same test statistic on the resulting permuted datasets. However, when β is close to 0, $X'\beta$ may not be a useful index for disease severity.

Kent and Hayward [31] promoted the use of risk-stratified subgroup analyses where several patient attributes were combined into a score that described a single dimension of risk along which treatment effects were likely to vary. Some methodological advances have been made. Cai et al. [14] first used estimated individual-level treatment differences between two treatment strategies to create an index for clustering subjects, then calibrated the average treatment differences for each cluster of subjects using a non-parametric approach. Zhao et al. [64] generalized this approach to include model selection so that their approach could be applied to settings where the number of covariates is large. Their approach can be used to identify a subgroup of future subjects who would have a desired treatment effect.

Data mining techniques have also been considered. Classification and regression tree (CART) analysis is another useful tool for investigating interactions among baseline factors without imposing parametric assumptions on the relationship between the outcome and candidate variables [12]. Ruberg et al. [47] illustrated, as exploratory analyses, how classification trees built through recursive partitioning could be used to identify a small set of variables and their cut-off values to define a subgroup of subjects who may have an enhanced treatment response. However, the authors cautioned that such approaches could be prone to over-fitting and validation using independent data would be necessary. Subsequently, Foster et al. [21] proposed the "Virtual Twins" method, which involved predicting response probabilities for each subject under either treated or control conditions, and then used the differences in these probabilities as the outcome in a classification or regression true, to identify a subgroup of patients who may have an enhanced treatment effect. They also investigated the use of re-sampling strategies to obtain an honest estimate of the magnitude of the treatment effect in the identified subgroup and find that a bias-corrected bootstrap procedure could reduce, but not eliminate, the bias resulting from the over-fitting. Su et al. [54] applied the CART methodology to censored survival data to explore the interactions between the treatment and patient characteristics and defined subgroups where treatment effects vary.

19.3 Power Consideration of Subgroup Analysis

It is recommended to consider plans for subgroup analyses during the design of a study. If investigators are interested in testing confirmatory hypotheses about specific variables, these hypotheses should be included in the primary or secondary objectives. It is well recognized that tests for interactions are associated with limited statistical power. In a simple setting of clinical trials with two treatment arms and a continuous outcome, it is well-known that, for interactions of the same magnitude as the overall effect to be detected with the same power, sample sizes needs to be inflated by a factor of four even with equal number of subjects in each of the mutually exclusive subgroups (see for example, [61]). Furthermore, the effect sizes for interactions are often smaller than those for main effects [16]. Figure 4 of [13] provides the inflation factors by which the sample size would have to be inflated for the interaction test to have the same power as that provided by the original sample size for the overall treatment effect. As the Figure illustrates, if the magnitude of interactions is half of the size of the overall effect, the inflation factor is about 16. This factor can increase dramatically to 100 or greater for interactions smaller than 20% of the overall effect. Trials with 80% power for the overall effect had 29% power to detect an interaction effect of the same magnitude [13] for subgroups of equal sample sizes. The power decreases to around 26% for a 1:2 ratio, 23% for a 1:3 ratio, and 18% for 1:5. To improve power for detecting important interactions, stratified randomization of treatment assignments to ensure sufficient representation in the subgroups of interest may be considered.

19.4 Subgroup Analysis Reporting and Interpretation

Assessment of the quality of subgroup analysis reporting has revealed a lack of completeness and clarity (see for example, [4, 25, 59]). Key information such as the total number of subgroup analyses conducted and whether the analysis is pre-specified or post-hoc are often not reported. Multiple comparison issues are often ignored. Subgroup analysis reporting guidelines have been developed to encourage better practice (see for example, [2, 59]).

Su et al. [55] reviewed 3662 journal articles reporting results from randomized controlled trials published in 118 core clinical journals in 2007. They concluded that the credibility of subgroup effects was usually low. Criteria to evaluate the credibility of subgroup analyses have been proposed (see for example [39, 55, 56, 62]). These criteria call for the evaluation of results of subgroup analyses in light of the total number of subgroup analyses conducted and how likely the apparent subgroup effect can be explained by chance, whether the hypothesis was stated a priori, the magnitude of the subgroup effect, whether the subgroup effect is similar across different outcomes within the study and whether it has been seen in other studies, and whether there are pre-existing biological support for the hypothesized subgroup effect.

19.5 Final Remarks

Subgroup analyses can be very informative and are needed to refine guidance for patient management. Properly planned, conducted, and reported subgroup analyses allow us to target interventions to specific subpopulations. Subgroup analyses based on one variable at a time are subject to several limitations. Integrated statistical methods for defining patient subpopulations based on multiple covariate simultaneously are much needed to overcome these limitations.

We focus on methods in the frequentist framework. Bayesian methods based on shrinkage estimation for subset-specific treatment effects may also be considered. Such methods may be particularly suited in settings where the usual frequentist hypothesis testing framework meets serious challenges such as the issue of multiplicity and the tendency for inflated Type I errors and Type II errors. Dixon and Simon [18] and Simon [52] developed Bayesian methods to produce subgroup-specific treatment effects based on a combination of observed within-subgroup differences and overall treatment differences, with weights according to a priori view of the likelihood of qualitative interactions, for clinical trials with continuous, binary or right-censored endpoints, and multiple binary covariates. Jones et al. [30] further extended this to more general class of models with fixed and/or random effects and interactions.

This chapter discusses subgroup analyses based on baseline covariates measured in randomized clinical trials. Subgroup analyses based on post-baseline factors are subject to all the considerations we have discussed for analyses based on baseline factors, but are more susceptible to bias because treatment can affect subgroup membership and the advantage of randomization at baseline is lost. As a result, differences in outcome can be confounded by measured or unmeasured differences in subject characteristics. Furthermore, group membership may not be predicted at baseline, which limits the translation of such findings to guide patient care in practice. For example, Van den Berghe et al. [58] investigated whether intensive insulin therapy reduced mortality in patients in medical intensive care units (ICUs). They found that for patients who stayed in the ICU for less than three days, the mortality was greater in those who receiving the intensive insulin therapy compared to those receiving the conventional therapy; whereas for those who stayed for longer than three days, the mortality was reduced for those who receiving the intensive insulin therapy. However, as Aberegg [1] pointed out, this finding has "limited usefulness for clinicians" as ICU duration cannot be predicted accurately at admission.

Bibliography

[1] K. S. Aberegg. Intensive insulin therapy in the medical ICU [correspondence].

[2] D. G. Altman, K. F. Schulz, D. Moher, M. Egger, F. Davidoff, D. Elbourne, P. C. Gøtzsche, and T. Lang. The revised CONSORT statement for reporting randomized trials: explanation and elaboration. *Annals of Internal Medicine*, 134(8):663–694, 2001.

[3] S. F. Assmann, D. W. Hosmer, S. Lemeshow, and K. A. Mundt. Confidence intervals for measures of interaction. *Epidemiology*, 7(3):286–290, 1996.

[4] S. F. Assmann, S. J. Pocock, L. E. Enos, and L. E. Kasten Subgroup analysis and other (mis) uses of baseline data in clinical trials. *The Lancet*, 355(9209):1064–1069, 2000.

[5] C. Baigent, R. Collins, P. Appleby, S. Parish, P. Sleight, and R. Sleight. Randomized trial of intravenous streptokinase, oral aspirin, both, or neither among 17,187 cases of suspected acute myocardial infarction: ISIS-2. *The Lancet*, 2:349–60, 1988.

[6] Y. Benjamini and Y. Hochberg. Controlling the false discovery rate: a practical and

powerful approach to multiple testing. *Journal of the Royal Statistical Society. Series B (Methodological)*, pages 289–300, 1995.

[7] G. R. Bernard, J. Vincent, P. Laterre, S. P. LaRosa, J. Dhainaut, A. Lopez-Rodriguez, J. S. Steingrub, G. E. Garber, J. D. Helterbrand, E. W. Ely, and C. J. Fisher Jr. Efficacy and safety of recombinant human activated protein C for severe sepsis. *New England Journal of Medicine*, 344(10):699–709, 2001.

[8] C. Bombardier, L. Laine, A. Reicin, D. Shapiro, R. Burgos-Vargas, B. Davis, R. Day, M. Bosi Ferraz, C. J. Hawkey, M. C. Hochberg, T.K. Kvien, and T.J. Scnitzer. Comparison of upper gastrointestinal toxicity of rofecoxib and naproxen in patients with rheumatoid arthritis. *New England Journal of Medicine*, 343(21):1520–1528, 2000.

[9] M. Bonetti and R. D. Gelber. A graphical method to assess treatment–covariate interactions using the Cox model on subsets of the data. *Statistics in Medicine*, 19(19):2595–2609, 2000.

[10] M. Bonetti and R. D. Gelber. Patterns of treatment effects in subsets of patients in clinical trials. *Biostatistics*, 5(3):465–481, 2004.

[11] D. D. Boos, C. Brownie, and J. Zhang. Estimating the magnitude of interaction. *Institute of Statistics Mimeo Series*, 2285, 1996.

[12] L. Breiman, J. Friedman, C. J. Stone, and R. A. Olshen. *Classification and Regression Trees*. CRC press, 1984.

[13] S. T. Brookes, E. Whitely, M. Egger, G. D. Smith, P. A. Mulheran, and T. J. Peters. Subgroup analyses in randomized trials: risks of subgroup-specific analyses; power and sample size for the interaction test. *Journal of Clinical Epidemiology*, 57(3):229–236, 2004.

[14] T. Cai, L. Tian, P. H. Wong, and L. J. Wei. Analysis of randomized comparative clinical trial data for personalized treatment selections. *Biostatistics*, 12(2):270–282, 2010.

[15] J. L. Ciminera, J. F. Heyse, H. H. Nguyen, and J. W. Tukey. Tests for qualitative treatment-by-centre interaction using a pushback-procedure. *Statistics in Medicine*, 12(11):1033–1045, 1993.

[16] L. M. Collins, J. J. Dziak, and R. Li. Design of experiments with multiple independent variables: a resource management perspective on complete and reduced factorial designs. *Psychological Methods*, 14(3):202, 2009.

[17] G.D. Curfman, S. Morrissey, and J.M. Drazen. Expression of Concern: Bombardier et al., "Comparison of Upper Gastrointestinal Toxicity of Rofecoxib and Naproxen in Patients with Rheumatoid Arthritis," N Engl J Med 2000;343:1520–8. *New England Journal of Medicine*, 353:2813–2814, 2005.

[18] D. O. Dixon and R. Simon. Bayesian subset analysis. *Biometrics*, pages 871–881, 1991.

[19] M. E. Farkouh, M. Domanski, L. A. Sleeper, F. S. Siami, G. Dangas, M. Mack, M. Yang, D. J. Cohen, Y. Rosenberg, S. D. Solomon, A.S. Desai, and B.J. Gersh. Strategies for multivessel revascularization in patients with diabetes. *New England Journal of Medicine*, 367(25):2375–2384, 2012.

[20] D. A. Follmann and M. A. Proschan. A multivariate test of interaction for use in clinical trials. *Biometrics*, 55(4):1151–1155, 1999.

[21] J. C. Foster, J. M. G. Taylor, and S. J. Ruberg. Subgroup identification from randomized clinical trial data. *Statistics in Medicine*, 30(24):2867–2880, 2011.

[22] J. Friedman, T. Hastie, and R. Tibshirani. Regularization paths for generalized linear models via coordinate descent. *Journal of Statistical Software*, 33(1):1, 2010.

[23] M. Gail and R. Simon. Testing for qualitative interactions between treatment effects and patient subsets. *Biometrics*, 41(2):361–372, 1985.

[24] R. J. Hayes and L. H. Moulton. *Cluster Randomised Trials*. CRC press, 2017.

[25] A. V. Hernández, E. Boersma, G. D. Murray, J. D. F. Habbema, and E. W. Steyerberg. Subgroup analyses in therapeutic cardiovascular clinical trials: are most of them misleading? *American Heart Journal*, 151(2):257–264, 2006.

[26] Y. Hochberg. A sharper bonferroni procedure for multiple tests of significance. *Biometrika*, 75(4):800–802, 1988.

[27] S. Holm. A simple sequentially rejective multiple test procedure. *Scandinavian Journal of Statistics*, 6(2):65–70, 1979.

[28] G. Hommel. A stagewise rejective multiple test procedure based on a modified bonferroni test. *Biometrika*, 75(2):383–386, 1988.

[29] D. W. Hosmer and S. Lemeshow. Confidence interval estimation of interaction. *Epidemiology*, 3(5):452–456, 1992.

[30] H. E. Jones, D. I. Ohlssen, B. Neuenschwander, A. Racine, and M. Branson. Bayesian models for subgroup analysis in clinical trials. *Clinical Trials*, 8(2):129–143, 2011.

[31] D. M. Kent and R. A. Hayward. Limitations of applying summary results of clinical trials to individual patients: the need for risk stratification. *Journal of the American Medical Association*, 298(10):1209–1212, 2007.

[32] W. K. Yip, with contributions from A. Lazar, D. Zahrieh, C. Cole, A. Lazar, M. Bonetti, V. Wang, W. Barcella, and R. Gelber. *stepp: Subpopulation Treatment Effect Pattern Plot (STEPP)*, 2014. R package version 3.0-11.

[33] H. Kirpalani, D. Millar, B. Lemyre, B. A. Yoder, A. Chiu, and R. S. Roberts. A trial comparing noninvasive ventilation strategies in preterm infants. *New England Journal of Medicine*, 369(7):611–620, 2013.

[34] S. W. Lagakos. The challenge of subgroup analyses-reporting without distorting. *New England Journal of Medicine*, 354(16):1667, 2006.

[35] J. Li and I. S. F. Chan. Detecting qualitative interactions in clinical trials: an extension of range test. *Journal of Biopharmaceutical Statistics*, 16(6):831–841, 2006.

[36] S. B. Liggett, S. Cresci, R. J. Kelly, F. M. Syed, S. J. Matkovich, H. S. Hahn, A. Diwan, J. S. Martini, L. Sparks, R. R. Parekh, J. A. Spertus, W. J. Koch, S. L. R. Kardia, and G. W. Dorn II. A GRK5 polymorphism that inhibits β-adrenergic receptor signaling is protective in heart failure. *Nature Medicine*, 14(5):510–517, 2008.

[37] J. March, S. Silva, S. Petrycki, J. Curry, K. Wells, J. Fairbank, B. Burns, M. Domino, S. McNulty, B. Vitiello, and J. Severe. Fluoxetine, cognitive-behavioral therapy, and their combination for adolescents with depression: Treatment for adolescents with depression study (TADS) randomized controlled trial. *Journal of the American Medical Association*, 292(7):807–820, 2004.

[38] C. L. Marcus, R. H. Moore, C. L. Rosen, B. Giordani, S. L. Garetz, H. G. Taylor, R. B. Mitchell, R. Amin, E. S. Katz, R. Arens, S. Paruthi, H. Muzumdar, D. Gozal, N. H. Thomas, J. Ware, D. Beebe, K. Snyder, L. Elden, R. C. Sprecher, P. Willging, D. Jones, J. P. Bent, T. Hoban, R. D. Chervin, S. S. Ellenberg, and S. Redline. A randomized trial of adenotonsillectomy for childhood sleep apnea. *New England Journal of Medicine*, 368(25):2366–2376, 2013.

[39] A. D. Oxman and G. H. Guyatt. A consumers guide to subgroup analyses. *Annals of Internal Medicine*, 116(1):78–84, 1992.

[40] G. Pan and D. A. Wolfe. Test for qualitative interaction of clinical significance. *Statistics in Medicine*, 16(14):1645–1652, 1997.

[41] R. Peto. Statistical aspects of cancer trials. In: The treatment of cancer (Halnan K, ed). London: Chapman & Hall, 1982, 867–871.

[42] S. Piantadosi and M. H. Gail. A comparison of the power of two tests for qualitative interactions. *Statistics in Medicine*, 12(13):1239–1248, 1993.

[43] E. J. G. Pitman *Lecture Notes on Nonparametric Statistical Inference: Lectures Given for the University of North Carolina,[Chapel Hill], 1948*. University of North Carolina, 1948.

[44] J. H. Powers, D. Lin and D. Ross. FDA evaluation of antimicrobials: subgroup analysis. *CHEST Journal*, 127(6):2298–2301, 2005.

[45] K. J. Rothman, S. Greenland and T. L. Lash. *Modern Epidemiology*. Lippincott Williams & Wilkins, 2008.

[46] P. M. Rothwell, Z. Mehta, S. C. Howard, S. A. Gutnikov and C. P. Warlow. From subgroups to individuals: general principles and the example of carotid endarterectomy. *The Lancet*, 365(9455):256–265, 2005.

[47] S. J. Ruberg, L. Chen and Y. Wang. The mean does not mean as much anymore: finding sub-groups for tailored therapeutics. *Clinical Trials*, 7(5):574–583, 2010.

[48] F. M. Sacks, M. A. Pfeffer, L. A. Moye, J. L. Rouleau, J. D. Rutherford, T. G. Cole, L. Brown, J. W. Warnica, J. M. Arnold, C. C. Wun, B. R. Davis and E. Braunwald. The effect of pravastatin on coronary events after myocardial infarction in patients with average cholesterol levels. *New England Journal of Medicine*, 335(14):1001–1009, 1996.

[49] J. P. Shaffer. Multiple hypothesis testing. *Annual review of psychology*, 46(1):561–584, 1995.

[50] M. J. Silvapulle. Tests against qualitative interaction: Exact critical values and robust tests. *Biometrics*, 57(4):1157–1165, 2001.

[51] R. J. Simes. An improved Bonferroni procedure for multiple tests of significance. *Biometrika*, 73(3):751–754, 1986.

[52] R. Simon. Bayesian subset analysis: application to studying treatment-by-gender interactions. *Statistics in Medicine*, 21(19):2909–2916, 2002.

[53] Xiao Song and Margaret Sullivan Pepe. Evaluating markers for selecting a patient's treatment. *Biometrics*, 60(4):874–883, 2004.

[54] X. Su, T. Zhou, X. Yan, J. Fan and S. Yang. Interaction trees with censored survival data. *The International Journal of Biostatistics*, 4(1):2, 2008.

[55] X. Sun, M. Briel, J. W. Busse, J. J. You, E. A. Akl, F. Mejza, M. M. Bala, D. Bassler, D. Mertz, N. Diaz-Granados, P. O. Vandvik, G. Malaga, S. K. Srinathan, P. Dahm, B. C. Johnston, P. Alonso-Coello, B. Hassouneh, S. D. Walter, D. Heels-Ansdell, N. Bhatnagar, D. G. Altman and G. H. Guyatt. Credibility of claims of subgroup effects in randomised controlled trials: systematic review. *British Medical Journal*, 344:e1553, 2012.

[56] X. Sun, J. P. A. Ioannidis, T. Agoritsas, A. C. Alba and G. Guyatt. How to use a subgroup analysis: users guide to the medical literature. *Journal of the American Medical Association*, 311(4):405–411, 2014.

[57] R. Tibshirani. Regression shrinkage and selection via the Lasso. *Journal of the Royal Statistical Society. Series B (Methodological)*, 58(1): 267–288, 1996.

[58] G. Van den Berghe, A. Wilmer, G. Hermans, W. Meersseman, P. J. Wouters, I. Milants, E. Van Wijngaerden, H. Bobbaers and R. Bouillon. Intensive insulin therapy in the medical ICU. *New England Journal of Medicine*, 354(5):449–461, 2006.

[59] R. Wang, S. W. Lagakos, J. H. Ware, D. J. Hunter and J. M. Drazen. Statistics in medicine – reporting of subgroup analyses in clinical trials. *New England Journal of Medicine*, 357(21):2189–2194, 2007.

[60] R. Wang, D. A. Schoenfeld, B. Hoeppner and A. E. Evins. Detecting treatment-covariate interactions using permutation methods. *Statistics in Medicine*, 34(12):2035–2047, 2015.

[61] R. Wang and J. H. Ware. Detecting moderator effects using subgroup analyses. *Prevention Science*, 14(2):111–120, 2013.

[62] H. Wedel, D. Demets, P. Deedwania, B. Fagerberg, S. Goldstein, S. Gottlieb, A. Hjalmarson, J. Kjekshus, F. Waagstein, J. Wikstrand and MERIT-HF Study Group. Challenges of subgroup analyses in multinational clinical trials: experiences from the MERIT-HF trial. *America Heart Journal*, 142(3):502–511, 2001.

[63] W. Yip, M. Bonetti, B. F. Cole, W. Barcella, X. V. Wang, A. Lazar and R. D. Gelber. Subpopulation treatment effect pattern plot (STEPP) analysis for continuous, binary, and count outcomes. *Clinical Trials*, 13(4):382–390, 2016.

[64] L. Zhao, L. Tian, T. Cai, B. Claggett and L. Wei. Effectively selecting a target population for a future comparative study. *Journal of the American Statistical Association*, 108(502):527–539, 2013.

20

Competing Risks

Haesook Kim

CONTENTS

20.1 Introduction

Competing risks occur when one type of event prevents the occurrence of other types of events or alters the probability of occurrence of the other events. In other words, the probability of many types of failure compete to be observed. Competing risks were initially observed during the smallpox outbreak in the 18th century in Europe. When the case fatality rate from the smallpox outbreak peaked in the 18th century, whether to mandate inoculation against smallpox generated heated debates about the pros and cons of this procedure, and various attempts were made to evaluate its possible effects. One of these attempts was a mathematical approach taken by Bernoulli. Bernoulli proposed a compartmental model to calculate the gain in life expectancy by separating the risk of dying due to smallpox from the risk of dying from other causes. In his memoir published in 1776, Daniel Bernoulli gave this question: "Available life tables reflect the mortality of the population for which they were calculated, taking into account all causes of death including smallpox. How would these life tables change if, because of mandatory vaccination, deaths from smallpox were entirely eliminated?" [14]. Although competing risks data were initially observed in the 18th century, the statistical inference and applications of competing risks data did not receive much attention until recent years.

Competing risks data are inherent to cancer clinical research in which failure can be classified by its types, and the information on each type of failure is as important as the overall survival probability. For instance, cause of death for women with early breast cancer who receive post-operative radiotherapy can be either treatment-related cardiac death due to the radiotherapy (*treatment-related mortality, TRM*) or recurrence of breast cancer (*disease recurrence or disease relapse*). In studies of allogeneic hematopoietic stem cell transplantation (HSCT), patients die from either recurrence of disease (*disease relapse*) or complications related to transplantation (*transplant-related mortality, treatment-related mortality or TRM*) if the transplant does not cure the underlying disease. In both studies, disease recurrence is an important event of interest as is treatment-related mortality for evaluation of efficacy and toxicity of a treatment. If disease recurrence is the event of interest and if an individual dies from TRM, this competing risk removes the individual from being at risk for disease recurrence. Therefore, applying methods of standard survival analysis to an event of interest when a competing risk is present would lead to biased results since standard survival analysis assumes independence of events and does not take types of failure into account.

Competing risks can be categorized as *classic* or *semi-competing*. The *classic* competing risks problem arises when there are mutually exclusive events and the occurrence of one type of event prevents the occurrence of the other types of event. An example of *classic* competing risks can be found in many cancer studies where the outcome of interest is a specific cause of death (e.g., death due to disease recurrence) in which case all other causes of death act as a competing event (e.g., treatment-related mortality). The *semi-competing* situation refers to studies where the occurrence of the outcome of interest may be prevented by the occurrence of the competing event but not vice versa. An example of this situation in HSCT studies is a study of acute graft-versus-host disease (aGVHD). Acute GVHD is an immune reaction caused by transplanted donor immune cells attacking host tissues. After allogeneic transplantation, subjects may develop aGVHD and then die or die early without developing aGVHD due to graft failure. And thus, subjects who die without developing aGVHD are no longer *at risk* of developing aGVHD. In this case, death is a competing risk of aGVHD whereas aGVHD does not obscure observation of death. This is distinct from censoring because regardless of the duration of follow-up time, the outcome of interest, aGVHD, will never occur for those subjects who die without aGVHD.

This chapter is organized as follows. Section 2 presents estimation of cumulative incidence function in the presence of competing risks; Section 3 presents methods for testing differences between cumulative incidence curves in the presence of competing risks; Section 4 presents competing risks regression analysis, Section 5 presents conclusion, and Section 6 presents computing tools. Throughout the presentation, HSCT studies are used to illustrate competing risks data analysis although competing risks are not limited to HSCT studies and occur commonly in many clinical trials.

20.2   Cumulative Incidence Function in the Presence of Competing Risks

Much of early theoretical development in competing risks data focused on a set of latent failure times and cause-specific hazard, $\lambda_i(t)$, that can be formulated as the marginal distribution of latent failure times [6–8, 12, 24]. However, because these latent failure times are

hypothetical and unobervsable, the marginal distribution is unidentifiable unless independence is assumed between competing risks [45]. For example, in the aforementioned HSCT
study, if an individual dies of disease recurrence, then time to TRM (competing event),
is unobservable. The independence assumption is untestable and unjustifiable in the competing risks setting in which the biologic mechanisms among risks of events may be either
unknown or likely interdependent. Cumulative incidence function, on the other hand, is
often more appealing and circumvents the un-identifiability issue [45].

20.2.1   Cumulative incidence function

If there is only one type of event, the hazard function of failure time T is

$$\lambda(t) = \lim_{u \to 0} \frac{\text{Prob}(t \leq T < t + u | T \geq t)}{u}. \tag{20.1}$$

and the cumulative incidence function for the failure is

$$F(t) = \text{Prob}(T \leq t) = \int_0^t f(u)du = \int_0^t \lambda(u)S(u)du \tag{20.2}$$

where $F(t) = 1 - S(t)$ for all t and $S()$ is the survival function. If there are k distinct types
of failure, then the hazard of failing from cause i is

$$\lambda_i(t) = \lim_{u \to 0} \frac{\text{Prob}(t \leq T < t + u, I = i | T \geq t)}{u} \tag{20.3}$$

where i is any one of k distinct types of failure denoted by $I \in 1, \cdots, k$ and the cumulative
incidence function (CIF) for failure i, $F_i(t)$, is

$$F_i(t) = \text{Prob}(T \leq t, I = i) = \int_0^t f_i(u)du = \int_0^t \lambda_i(u)S(u)du \tag{20.4}$$

[15] [24], where $f_i(t)$ is a subdensity function for failure type i in the presence of other failure
types and $S(t)$ is the overall survival probability. (20.4) is also known as *subdistribution
function*. Since there are k failures, $F_i(t) < 1$ and $\sum_{i=1}^k F_i(t) = 1$ as $t \to \infty$.

If other types of failure are present but ignored, then the CIF for failure i is

$$F_i^*(t) = \int_0^t \lambda_i(u)S_i^*(u)du \tag{20.5}$$

where $S_i^*(t)$ is a cause-specific survival function for cause i that censors competing risks.
Since competing events are censored, $F_i^*(t) + S_i^*(t) = 1$ for all t and $F_i^*(t) = 1$ as $t \to \infty$.
Because events from causes other than i are treated as censored observations in $S_i^*(t)$, $S(t) \leq$
$S_i^*(t)$ and $\int_0^t \lambda_i(u)S(u)du \leq \int_0^t \lambda_i(u)S_i^*(u)du$ since $\lambda_i(t)$ is same between two equations at
each time point. Therefore, $F_i(t) \leq F_i^*(t)$.

20.2.2   Estimation of CIF in the presence of competing risks

Let $0 < t_1 < \cdots < t_l$ represent the l ordered distinct failure times for any cause of failure.
If t is discrete, the hazard of failing from cause i is

$$\lambda_i(t_j) = \frac{\text{Prob}(T = t_j, I = i)}{\text{Prob}(T > t_{j-1})}, \quad j = 1, \cdots, l$$

and its estimate is

$$\hat{\lambda}_i(t_j) = \frac{d_{ij}}{n_j}$$

where d_{ij} is the number of failure i at time t_j and n_j is the number of subjects at risk just prior to t_j. Let $d_j = \sum_{i=1}^{k} d_{ij}$ and $\hat{\lambda}(t_j) = \sum_{i=1}^{k} \hat{\lambda}_i(t_j)$. Then the Kaplan-Meier (KM) estimate of the overall survival function in (20.4) is

$$\hat{S}(t) = \prod_{j:t_j<t} (1 - \hat{\lambda}(t_j)) = \prod_{j:t_j<t} (1 - \frac{d_j}{n_j}). \tag{20.6}$$

Thus, the estimate of the CIF (20.4) [24] is

$$\hat{F}_i(t) = \sum_{j:t_j<t} \frac{d_{ij}}{n_j} \hat{S}(t_{j-1}). \tag{20.7}$$

If we ignore competing risks and use the naive KM method instead, the estimate of CIF for failure i (20.5) is

$$\hat{F}_i^*(t) = \sum_{j:t_j<t} \frac{d_{ij}}{n_j} \hat{S}_i^*(t_{j-1}), \tag{20.8}$$

where

$$\hat{S}_i^*(t) = \prod_{j:t_j<t} (1 - \frac{d_{ij}}{n_j}). \tag{20.9}$$

Because $\hat{S}(t) \leq \hat{S}_i^*(t)$, $\hat{F}_i(t) \leq \hat{F}_i^*(t)$. Therefore, when there are competing risks, the KM method in standard survival analysis overestimates the cumulative incidence function, and the magnitude of overestimation in the KM method depends on the level of incidence rates of competing events.

We illustrate the computation of CIF of an event of interest in the presence of competing event using the competing risks (CR) method (20.7) and the KM method (20.8) using an hypothetical example. Suppose that there are ten patients with the ordered failed or censored times as shown here

10(R), 20+, 35(R), 40 (T), 50+, 55 (R), 70 (T), 71 (T), 80(R), 90+

where + denotes a censored time, (R) denotes death due to disease recurrence, and (T) denotes death due to treatment-related complications. Detailed calculation for CIF using both methods is presented in Tables 20.1 and 20.2. At $t = 40$, a patient dies of TRM and the KM method treats this as a censored observation. Therefore the KM survival estimate $\hat{S}_R^*(40) = 0.79$, and the corresponding CIF, $\hat{F}_R^*(40) = 0.21$. At $t = 55$, $\hat{S}_R^*(55)$ is dropped to 0.63, and $\hat{F}_R^*(55)$ is increased to 0.37. Of note, $\hat{S}_R^*(t) + \hat{F}_R^*(t) = 1$ at each time point. If the CR method is used instead, at $t = 40$, $\hat{S}_R(t) = 0.68$ which is smaller than 0.79. However, this difference does not immediately affect the cumulative incidence rate of disease recurrence because there is no incidence of disease recurrence at $t = 40$. At $t = 55$, $\hat{F}_R(55) = 0.35$, which is lower than the KM estimate of disease recurrence, 0.37. This is because the KM method treats one death due to TRM at $t = 40$ as censored, and this results in overestimation of the survival rate at $t = 50$ (0.79). At $t = 80$, the difference in the cumulative incidence rate of disease recurrence between the KM (0.69) and CR method (0.48) is more apparent, 0.21. This is because $\hat{S}_R^*(71)$ is substantially overestimated (0.63), compared to $\hat{S}_R(71)$ (0.27). Because of this, $\hat{F}_R^*(80)$ is subsequently overestimated. Of note, using the CR method, $\hat{S}(t) + \hat{F}_R(t) + \hat{F}_T(t) = 1$ for all t. Figure 20.1 shows the cumulative incidence curves of disease recurrence using the KM and CR methods in the presence of TRM as a competing risk. Two cumulative incidence curves jump at the same time whenever a disease recurrence occurs, but the cumulative incidence rates are different.

TABLE 20.1
KM method applied to hypothetical data example.

Time	COD	No. at Risk	R	C	$\hat{S}_R^*(t)$	$\hat{F}_R^*(t)$
10	R	10	1	0	1*(9/10)=0.9	0+1*(1/10)=0.1
20+	-	9	0	1	0.9*(9/9)=0.9	0.1+0.9*(0/9)=0.1
35	R	8	1	0	0.9*(7/8)=0.787	0.1+0.9*(1/8)=0.212
40	T	7	0	1	0.787*(7/7)=0.787	0.212+0.787*(0/7)=0.212
50+	-	6	0	1	0.787*(6/6)=0.787	0.212+0.787*(0/6)=0.212
55	R	5	1	0	0.7875*(4/5)=0.63	0.212+0.787*(1/5)=0.37
70	T	4	0	1	0.63*(4/4)=0.63	0.37+0.63*(0/4)=0.37
71	T	3	0	1	0.63*(3/3)=0.63	0.37+0.63*(0/3)=0.37
80	R	2	1	0	0.63*(1/2)=0.315	0.37+0.63*(1/2)=0.685
90+	-	1	0	1	0.315*(1/1)=0.315	0.685+0.315*(0/1)=0.685

COD: cause of death. R: indicator of relapse. C: indicator of censored time.

TABLE 20.2
CR method applied to hypothetical data example.

Time	COD	No. at Risk	Any event	C	$\hat{S}(t)$	$\hat{F}_R(t)$
10	R	10	1	0	1*(9/10)=0.9	0+1*(1/10)=0.1
20+	-	9	0	1	0.9*(9/9)=0.9	0.1+0.9*(0/9)=0.1
35	R	8	1	0	0.9*(7/8)=0.787	0.1+0.9*(1/8)=0.212
40	T	7	1	0	0.787*(6/7)=0.675	0.212+0.787*(0/7)=0.212
50+	-	6	0	1	0.675*(6/6)=0.675	0.212+0.675*(0/6)=0.212
55	R	5	1	0	0.675*(4/5)=0.54	0.212+0.675*(1/5)=0.347
70	T	4	1	0	0.54*(3/4)=0.405	0.347+0.54*(0/4)=0.347
71	T	3	1	0	0.405*(2/3)=0.27	0.347+0.405*(0/3)=0.347
80	R	2	1	0	0.27*(1/2)=0.135	0.347+0.27*(1/2)=0.482
90+	-	1	0	1	0.135*(0/1)=0.135	0.482+0.135*(0/1)=0.482

COD: cause of death. R: indicator of relapse. T: indicator of TRM.

Example 1: Effect of conditioning intensity on the outcome of allogeneic hematopoietic stem cell transplantation (HSCT) for patients older than 50 years of age.

To assess the impact of conditioning intensity on disease relapse and transplant-related mortality (TRM), we analyzed 152 patients age older than 50 years who underwent HSCT with myeloablative conditioning regimen (MAC)(N = 81) or reduced intensity conditioning regimen (RIC)(N = 71) [2]. Figure 20.2 shows the cumulative incidences of disease relapse and TRM among 81 patients who received MAC using the KM method (20.8) that ignores competing risks and the CR method (20.7), respectively. If we use the KM method, the

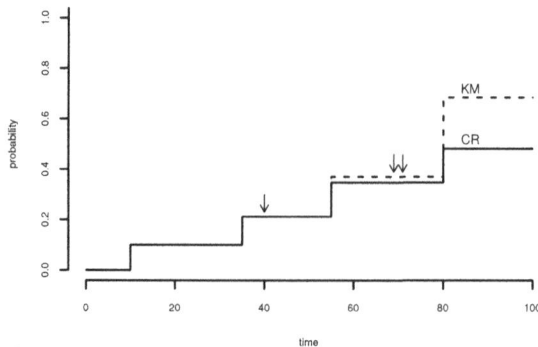

FIGURE 20.1
Cumulative incidence of disease recurrence using the CR method (solid line) and the KM method (dotted line).

3-year cumulative incidence rate is 50% for relapse and 58% for TRM, and thus the combined rate of 108%. If we use the CR method, the 3-year cumulative incidence rate is 30% for relapse and 50% for TRM, so the combined rate of 80%, which corresponds to one minus the event-free survival at 3 year post transplantation (i.e., $1 - \hat{S}(3)$). It is noticed that even though these two events are mutually exclusive, and some patients were still alive and relapse-free at 3 years post transplantation, the combined rate of relapse and TRM is over 100% if we use the naive KM method, indicating the amount of bias introduced by the KM method.

20.3 Testing for Differences between Cumulative Incidence Curves in the Presence of Competing Risks

20.3.1 Gray test

When there is only one type of event, there is a one-to-one relationship between the hazard function (20.1) and the cumulative incidence function (20.2) and survivor function. However, this relationship doe not hold in the presence of competing risks. [45] In other words, the same cause-specific hazard can correspond to multiple different survivor functions. Since there exists no one-to-one relationship between the cause-specific hazard (20.3) and the cumulative incidence function (20.4) in the presence of competing risks, the comparison of cause-specific hazards of failure type i between different groups can be quite different from the comparison of cumulative incidence functions of failure type i (i.e., subdistribution of failure type i). To be able to directly compare subdistributions (20.4) in the presence of competing risks, Gray [19] further defined a hazard function that corresponds to the subdistribution of failure type i.

$$
\begin{aligned}
\gamma_i(t) &= \lim_{u \to 0} \frac{1}{u} \Pr\{t \le T < t+u, I = i | T \ge t \cup (T \le t \cap I \ne i)\} \\
&= \frac{f_i(t)}{1 - F_i(t)}.
\end{aligned} \tag{20.10}
$$

$\gamma_i(t)$ is the hazard of failing from failure type i at time t in the presence of competing risks, given that a subject has survived or has already failed from a competing cause. This hazard,

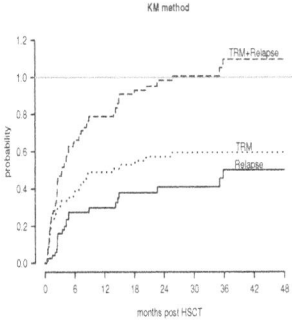 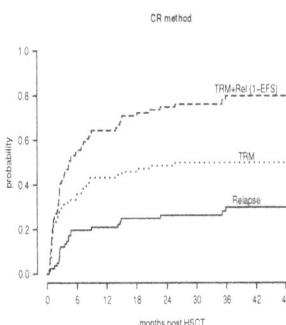

FIGURE 20.2
Cumulative incidence of disease relapse (solid line), TRM (dotted line), and TRM and relapse combined (dashed line) using the KM method (left panel) and CR method (right panel). The gray line in the left panel represents probability of 1.

known as *subdistribution hazard*, corresponds to the cumulative incidence function (20.4) and is the key concept of Gray test [19] and the Fine and Gray model [16].

For an illustration, let us consider a clinical trial conducted to compare cumulative incidence functions between two treatment groups (k=A,B) where patients can experience one of two types of failure (i=1, 2). Testing the group difference in cumulative incidence function in failure type 1 is

$$H_0 : F_{1A} = F_{1B} = F_1^0 \tag{20.11}$$

where F_{1A} is a subdistribution function of failure type 1 in the treatment group A. Since testing $F_{1A} = F_{1B}$ is not the same as testing $\lambda_{1A} = \lambda_{1B}$, to test $F_{1A} = F_{1B}$, the Gray test compares weighted averages of the subdistribution hazards, $\gamma_{1k}(t) = f_{1k}(t)/(1 - F_{1k}(t))$, $k = A, B$, where $f_{1k}(t)$ is the subdensity function of $F_{1k}(t)$. That is,

$$\int_0^\tau w(t) \left(\frac{d\hat{F}_{1A}(t)}{1 - \hat{F}_{1A}(t-)} - \frac{d\hat{F}_{1B}(t)}{1 - \hat{F}_{1B}(t-)} \right) = \int_0^\tau w(t) \left(\hat{\gamma}_{1A}(t) - \hat{\gamma}_{1B}(t) \right), \tag{20.12}$$

where $\hat{F}_{1k}(t)$ is an estimate of F_{1k} for failure type 1 in group k ($k = A, B$) at t as defined in (20.4), and $w(t)$ is a weight function. (20.12) is basically comparing weighted averages of the subdistribution hazards $\gamma_i(t)$. Under H_0, the subdistribution hazard ratio of the two treatments equals to 1 and constant over time (i.e., proportional hazards).

20.3.2 Estimation of Gray statistic

If event times are discrete and complete without censoring, then the subdistribution hazard for failure type i (20.10) at time $t_j (j = 1, \cdots, l)$ can be estimated as

$$\hat{\lambda}_i(t_j) = \frac{d_{ij}}{R_{ij}}$$

where d_{ij} denotes the number of failure type i at time t_j and R_{ij} is the modified risk set that includes all subjects who did not experience any event until time t_j and all subjects who failed from competing events before t_j. Since the number of subjects at risk of failing for event i at t_j is expected to be greater than n_j as we need the number of subjects free from event i, Gray modified the number at risk for each type of failure and proposed a 'correction factor', $\frac{1-\hat{F}_i(t-)}{\hat{S}(t-)}$ which is greater or equal to 1. Because this modified risk set equals to n_j until the first competing event occurs and larger than n_j after the occurrence of the first competing event, the estimated subdistribution hazard always equals to or smaller than the estimated cause specific hazard.

To estimate cumulative incidence functions in the presence of competing risks (20.11) and to compare these estimates, we first need to estimate the modified risk set, R_{ij}. Let n_{jA} and n_{jB} be the number of subjects at risk at t_j who are free from all types of failure in treatment A and B, respectively. The estimates of the modified risk sets for failure type 1 at t_j for the treatment groups A and B are

$$R_{1A}(t_j) = n_{jA}\frac{1-\hat{F}_{1A}(t-)}{\hat{S}_A(t-)} \quad R_{1B}(t_j) = n_{jB}\frac{1-\hat{F}_{1B}(t-)}{\hat{S}_B(t-)}, \text{ and}$$

$$R_1(t_j) = R_{1A}(t_j) + R_{1B}(t_j).$$

$R_1(t_j)$ is the total number of subjects from two treatment groups combined at risk at t_j for failure type 1. Of note, the calculation for failure type 2 is identical and thus should be straightforward. Then, the score (numerator of the Gray statistic) for failure type 1 in group A is

$$
\begin{aligned}
z_A(t) &= \sum_{t \in (0, t_l)} w(t)(\frac{d_{1A}}{R_{1A}} - \frac{d_1}{R_1}) \quad \text{if } w(t) = R_{1A}(t) \\
&= \sum_{t \in (0, t_l)} (d_{1A} - \frac{R_{1A}}{R_1}d_1)
\end{aligned}
$$

where $R_1(t) = R_{1A}(t) + R_{1B}(t)$. The quadratic term of this score divided by its variance, V, is

$$z'_A V^{-1} z_A \sim \chi_1^2.$$

(note: χ_{K-1}^2 for K groups where $K > 2$). By symmetry, this quantity is identical for group B. For the formula of the variance, we refer to Gray [19].

Example 3: Returning to Example 1.

Table 20.3 presents a summary of the 3-year cumulative incidences of relapse and TRM and a comparison between MAC and RIC using the log-rank test and Gray test. The 3-year cumulative incidence of relapse in RIC was 61% using the KM method, but 46% using the CR method; the 3-year cumulative incidence of relapse in MAC was 50% using the KM method, but 30% using the CR method suggesting that the naive KM method overestimates the cumulative incidences of these events. Also, the results reveal that although the cumulative incidence of combined events (relapse and TRM) in a single endpoint analysis is similar between two types of transplantation (80% vs. 78%), RIC is associated with a higher incidence of relapse but a lower incidence of TRM whereas MAC is associated with

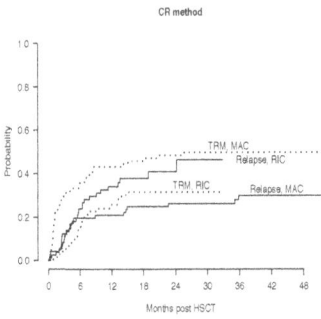

FIGURE 20.3
Cumulative incidence of relapse and TRM for MAC and RIC using the CR method.

a higher incidence of TRM and a lower incidence of relapse. i.e., the result reflects the different immunologic effects of these two types of transplantation on relapse and TRM (Figure 20.3). Therefore, to be able to evaluate and compare transplant strategies comprehensively, both standard survival analysis (i.e., a single endpoint analysis) and competing risks data analysis need to be performed as each of these analyses addresses different study questions. Detailed discussions of these analyses are provided in Kim, Kim and Armand [26, 27].

TABLE 20.3
Summary of 3-year cumulative incidence of relapse and TRM.

	KM method			CR method		
	RIC	**MAC**	**p-value (log-rank)**	**RIC**	**MAC**	**p-value (Gray test)**
3-yr CI of relapse	61%	50%	0.35	46%	30%	0.05
3-yr CI of TRM	38%	58%	0.008	32%	50%	0.01
3-yr CI of relapse and TRM combined	99%	108%		78%	80%	

20.4 Competing Risks Regression Analysis

The most commonly used regression methods for competing risks data in the medical research are the cause-specific hazard regression using Cox model [39] and the subdistribution hazard regression proposed by Fine and Gray [16]. This is due largely to the readily available statistical software packages with various well developed functions for model checking and assumption checking. In the cause-specific hazard regression using Cox model, failures from other causes are treated as censored observations and the effect of covariates is on the instantaneous probability of failing from cause i given a subject experienced no event until time t. This analysis is identical to performing a standard Cox regression analysis for a single type of failure.

Since the simple relationship between a single endpoint and a single cause-specific hazard does not hold in the presence of competing risks, Fine and Gray [16] proposed a direct regression modeling of the effect of covariates on the cumulative incidence function for competing risks data. Later, Klein and Andersen [28] proposed another direct modeling method that tests the effects of covariates on the cumulative incidence function using pseudo values on the generalized estimating equation approach [34].

Since these two approaches (i.e., cause-specific and direct modeling) are from different stochastic quantities, results from these models might be substantially different and thus they should be interpreted carefully. We herein present cause-specific Cox model, the Fine and Gray model [16] and the Klein and Andersen model [28]. We will illustrate these approaches using real data examples.

20.4.1 Cause-specific hazard regression model

Cause-specific hazard of failure type i $(i = 1, 2, \cdots, k)$ in the presence of other events is given in (20.3). As in the standard Cox model, the effect of covariates on the cause-specific hazard for failure type i can be constructed as

$$\lambda_i(t|X) = \lambda_{i,0}(t) \exp\{X\beta\}$$

where $\lambda_{i,0}(t)$ is an unspecified baseline hazard rate for failure type i.

This model can be used to test the effects of the covariates on the cause-specific hazard for failure type i and not on the cumulative incidence function of failure type i, and the effects of the covariates on the competing event(s) are ignored. Although the initial model was written for time fixed covariates under the proportional hazards assumption, the model can be easily modified to incorporate internal (i.e., $\beta X(t)$) and external (i.e., $\beta(t)X$) time dependent covariates [24].

20.4.2 Fine and Gray model

The Fine and Gray model [19] is a Cox proportional hazards like model for the subdistribution hazard (20.10). The model uses the partial likelihood principle and weighted estimated equations to obtain consistent estimators of the covariate effects on the subdistribution hazard. As in the Cox model, the Fine and Gray model for failure type i, (20.13), is a relative risk model that decomposes into the baseline hazard and regression effect of covariates.

$$\begin{aligned} \gamma_i(t; X) &= \lim_{u \to 0} \frac{1}{u} \Pr\{t \leq T < t + u, I = i | T \geq t \cup (T \leq t \cap I \neq i), X\} \\ &= \frac{f_i(t; X)}{1 - F_i(t; X)} = \gamma_{i,0}(t) e^{X'\beta}, \end{aligned} \tag{20.13}$$

where $\gamma_{i,0}(t)$ is the subdistribution baseline hazard for failure type i, X is a vector of covariates, and β is a vector of coefficients. The risk set is

$$R_j = \{k : (min(C_k, T_k) \geq T_j) \cup (T_k \leq T_j \cap I \neq i \cap C_k \geq T_j)\}, \tag{20.14}$$

where $j \neq k$ and C and T denote the censoring and failure time, respectively. The first part of the risk set represents subjects who have not failed from any cause, and the second part represents subjects who have failed from a competing risk and censoring failure type i has not occurred. In other words, subjects who failed from competing causes remain in the risk set indefinitely. The subdistribution hazard is linked directly to the cumulative incidence function of the event of interest.

$$F_i(t|X) = 1 - \exp(-\Gamma_i(t; X)) = 1 - \exp(-\int_0^t \gamma_i(t; X)),$$

where $\gamma_i(t; X))$ is the subdistribution hazard for failure type i given covariates X. As in the standard Cox model, subdistrubition hazards are assumed to be proportional over time. This model handles time by covariate interactions, but not internal time-dependent covariates (i.e., $\beta X(t)$) as described in Prentice et al [39]. As internal time-dependent covariates such as a simple one time jump (e.g., occurrence of acute GVHD after HSCT) are common in competing risks data, this is a limitation of the model [33].

20.4.3   Klein and Andersen model

Klein and Andersen [28] proposed a competing risks regression model using pseudo-values. The key idea of the method is to replace all observations, including censored observations, with pseudo values so that standard statistical methods for multivariate data can be used. More specifically, the estimated pseudo-value for h individual at a grid point τ_j $(j = 1, \cdots, l)$ is

$$\hat{C}_h(\tau_j) = n\hat{C}(\tau_j) - (n-1)\hat{C}^{-h}(\tau_j)$$

where $\hat{C}(\tau_j)$ is the estimated cumulative incidence function at a grid point τ_j using all subjects (n), and $\hat{C}^{-h}(\tau_j)$ is the estimated cumulative incidence function after deleting subject h at τ_j (i.e. jackknife approach, [36]). Note that grid points need to be specified a priori based on the failure times and the recommended number of grid points (l) is five. Then the effect of covariates on the cumulative incidence function for subject h at τ_j is modeled through a link function $g(x) = log[-log(1 - z)]$.

$$g(C_h(\tau_j|X)) = \alpha\tau_j + \beta X_h\tau_j,$$

where X is a covariate vector. The regression coefficients and standard errors can be estimated by using the generalized estimating equation approach [34].

Example 4. The data presented in Table 20.4 is based on our allogeneic transplantation data comparing TRM (Event 1) and relapse (Event 2) between peripheral blood stem cell transplantation (Group A) and double unit cord blood transplantation (Group B) [11]. However, the data are not identical to the one used in the original publication. To illustrate the difference of two models and to make a similar sample size between two cohorts, some data have been excluded or altered. In the table, the variable *event* has 3 categories: 0 for censored observations, 1 for TRM (Event 1) and 2 for relapse (Event 2). Table 20.5 is a summary of results from the Fine and Gray (F&G) model, the Klein and Andersen model (K&A) and cause-specific Cox model. The subdistribution hazard ratio (sHR) from F&G model for Group A over Group B in Event 1 is 3.58 (p = 0.0005) and the hazard ratio (HR) from the cause-specific Cox model is 3.47 (p = 0.001). The sHR for Group A over Group B in Event 2 from the F&G model is 0.47 (p = 0.046) and the HR from the cause-specific Cox model is 0.64 (p = 0.25). Although results from both competing risks regression analyses are consistent, the results from the K&A model are more significant than those from the F&G model. These results suggest that there is a significant difference in cumulative incidence of Event 1 between two groups and this is consistent in cause-specific hazard. However, the significant difference in cumulative incidence of Event 2 between two groups does not lead to a significant difference in cause-specific hazard of Event 2. This is illustrated in Figure 20.4. In Figure 20.4, the cumulative incidence curves of Event 2 for Group A and Group B appear to be significantly wide (left panel), but the difference of the cumulative cause-specific hazard between two groups is narrower compared to the difference in cumulative incidence curves between two groups (right panel).

TABLE 20.4

CR method.

Group A	
event	0 0 0 0 0 0 0 0 0 0 0 0 0 0 0 0 0 0 0 0 0 0 0 0 0
	1 1 1 1 1 1 1 1 1 1 1 1 1 1 1 1 1 1 1 1 1 1 1 1
	2 2 2 2 2 2 2 2 2 2 2 2 2 2 2 2 2 2 2 2 2 2 2 2 2 2 2 2 2
time	5.7 10.7 11.2 11.7 11.9 12.2 12.2 12.5 15.2 17.4 17.9 18.2 18.7 18.9 20.4 20.8
	21.7 23.9 30.8 32.4 32.7 32.9
	0.9 2.1 2.6 3.3 3.8 4 4.7 5.1 5.5 5.6 6 6.1 6.4 6.6 6.8 7.9 9 11.9 13 13.2 14.6
	0.1 0.2 0.3 2 2.6 2.7 3 3 3.2 3.4 4 4.1 4.5 5.3 5.5 5.7 5.7 6.4 6.5 6.8 7.8 9.1
	9.9 11.3 13.2 13.5 18.9 24.2
Group B	
event	0 0 0 0 0 0 0 0 0 0 0 0 0 0 0 0 0 0 0 0
	1 1 1 1 1 1 1 1 1 1 1 1 1 1 1 1 1 1 1 1 1 1 1 1 1 1 1 1 1 1 1 1 1 1 1 1 1 1 1 1 1 1 1
	2 2 2 2 2 2 2 2 2 2 2 2 2 2 2 2 2 2 2 2 2 2 2
time	13.6 13.7 29.6 30 30.5 30.5 30.9 37.4 43.2 48 50.9 59.5 63.4 64.3 69.5 69.6 70.3 72.9
	0.36 0.39 0.46 0.49 0.53 0.62 0.72 0.82 0.82 0.85 0.89 0.89 0.99 0.99 1.1 1.1 1.2 1.2 1.6
	2.1 2.1 2.3 2.4 2.9 3.3 3.4 4.4 5.7 5.8 6.8 7.1 7.3 7.9 8.4 8.6 13.8 15 18.1 20.4 25.7
	0.4 0.6 1.4 2 2.3 2.4 2.5 2.5 2.5 2.6 3.5 3.9 4.3 4.3 4.7 4.7 8.9 14.2 14.7 14.9 22.7 35.1 35.8

Example 5: HSCT for patients with T cell lymphoma

Allogeneic HSCT is an accepted therapy for B cell lymphoma but the indications for allogeneic HSCT and the outcomes of the procedure are less clear in T cell lymphoma. To investigate clinical outcomes after HSCT for T cell lymphoma, we identified 52 patients who underwent transplant between April 4, 1997 and February 17, 2009 at the Dana-Farber/Brigham and Women's Cancer Center [17]. Table 20.6 shows the results of the standard Cox model for the combined events of TRM and relapse as a single endpoint and the Fine and Gray model for TRM and relapse as competing risks.

As in any regression analysis, competing risks regression analysis is also useful in identifying prognostic factors, other than the factor of interest, for each type of failure. For example, in Table 20.6, good prognosis is associated with the event-free survival (i.e., relapse or TRM as a single endpoint) in Cox model ($\hat{\beta}$=-1.34, HR=0.26, p=0.002), but it is

TABLE 20.5

Summary of competing risks regression model and cause-specific Cox model.

	Event 1				Event 2			
	$\hat{\beta}$	$\exp(\hat{\beta})$	s.e. $(\hat{\beta})$	p-value	$\hat{\beta}$	$\exp(\hat{\beta})$	s.e. $(\hat{\beta})$	p-value
F & G model	1.28	3.58	0.376	0.0005	−0.75	0.47	0.367	0.046
K & A model	1.86	6.01	0.47	<0.0001	−1.02	0.36	0.46	0.026
CS Cox model	1.24	3.47	0.38	0.001	−0.44	0.64	0.38	0.25

F & G: Fine and Gray. CS: cause-specific. K & A model: Klein and Andersen model.

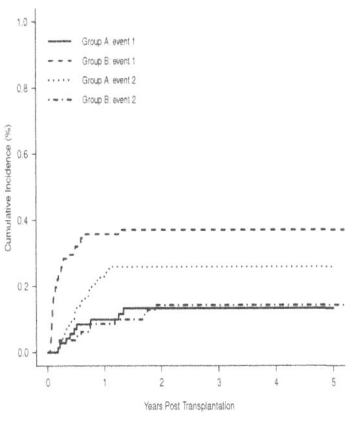 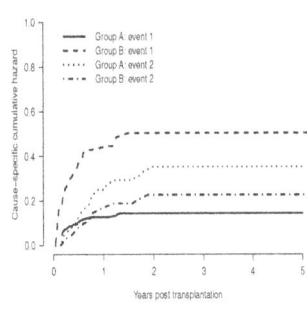

FIGURE 20.4

Cumulative incidence of curves for Event 1 and Event 2 for Group A and Group B (left panel), and cumulative cause-specific hazard for Event 1 and Event 2 for Group A and Group B (right panel).

TABLE 20.6

Summary of competing risks regression model and Cox model.

	Relapse: CRR			TRM: CRR			Relapse & TRM: Cox		
	$\hat{\beta}$	HR	p-value	$\hat{\beta}$	HR	p-value	$\hat{\beta}$	HR	p-value
Age (≥50 vs. <50)	−0.38	0.68	0.52	0.71	2.02	0.33	0.51	1.66	0.26
Patient-donor sex									
mismatch (MF vs. Other)	0.92	2.51	0.08	0.14	1.15	0.81	0.86	2.36	0.02
Conditioning (RIC vs. MAC)	1.64	5.13	0.01	−1.19	0.31	0.05	0.37	1.45	0.43
Good vs. Poor prognosis	−1.66	0.19	0.005	−1.18	0.31	0.07	−1.34	0.26	0.002
Year (≥2002 vs. <2002)	−1.98	0.14	0.014	−1.65	0.19	0.009	−1.83	0.16	0.002

MF: male patient and female donor. RIC: reduced intensity conditioning. MAC: myeloablative conditioning. Year: year of transplantation. Good prognosis includes unclassified peripheral T-cell, angioimmunoblastic T-cell, and anaplastic large-cell and T/null cell histology [17].

unknown from this model whether this is for relapse or TRM prevention. The competing risks model (Fine and Gray model) reveals that good prognosis is significantly associated with relapse.

Over-fitting can be an issue in regression analysis. Analogous to other regression analysis, a backward or forward model selection can be performed in the Fine and Gray model. In the above example, we consider a backward approach for relapse using the Akaike information criteria(AIC) [1] and the Bayesian information criteria (BIC)[44]. We first fit the full model, then remove one variable each time on the basis of significance in the full model. Table 20.7 shows results of BIC and AIC scores for 5 models. The first 5 columns were generated by the add-on R program **modsel.crr** (see Section 6 for computing Tools), and corrected AIC (AICc) was calculated based on the formula, $-2 * L + 2 * p * [1 + \frac{p+1}{n-p-1}][22]$, where L is the maximized log likelihood value and p is the number of parameters.

Applying the general rule-of-thumb [23], BIC suggests model 5, which includes prognosis only, whereas AICc suggests model 2, which excludes age only.

TABLE 20.7
Summary of BIC and AIC.

Model	Variables included	n	L	Df.fit	BIC	BIC.diff	AICc
null		52	−77.61	0	155.22	0	155.22
1	age, sex mismatch, conditioning, prognosis, year	52	−71.28	5	162.32	7.10	153.87
2	sex mismatch, conditioning, prognosis, year	52	−71.52	4	158.84	3.61	151.88
3	conditioning, prognosis, year	52	−72.82	3	157.5	2.28	152.15
4	conditioning, prognosis	52	−74.99	2	157.88	2.66	154.23
5	prognosis	52	−76.59	1	157.13	1.91	155.26

Df.fit: the number of parameters. BIC.diff or AIC.diff: difference with the respect to the minimum value observed from the set of candidate models.

20.4.4 Remarks

Competing risks regression analysis has been, indeed, very useful for determining clinical utility of treatment. For regression analysis of competing risks data, cause-specific hazard regression and subdistribution hazard regression have been the most commonly used in medical research due largely to the readily available statistical software packages, such as R, Stata, and SAS that offer a variety of built-in functions. Since these two models were constructed from different stochastic quantities, the results of both modeling methods should be interpreted carefully.

As to the Fine and Gray model, the model does not handle time-dependent covariates. To address this issue, Cortese and Andersen [9] recently proposed a few approaches to deal with time-dependent covariates. One approach was a multi-state model on cause-specific hazard, not on cumulative incidence function. The other approach was to use landmark analysis which also comes with shortcomings [4]. As time dependent covariates are common in competing risks data, this can be a limiting factor of this model. Also, Andersen and Keiding [3] recently questioned the interpretability and usefulness of the Fine and Gray model. In their opinion, risk set should not be conditioned on having reached on absorbing state. i.e., the risk set in the Fine and Gray model is improper because subjects who failed from competing risks are included in the risk set.

The Klein and Andersen model based on pseudo-values using a GEE-model with complementary log-log link gives results similar to the Fine and Gray model and can be interpreted similarly. Recent simulation studies [5] showed that standard errors of the pseudo-value approach in the Klein and Andersen model are larger than in the Fine and Gray model and thus pseudo-value approach was recommended not to be used for estimation when standard methods are available, but it might be useful where standard methods do not exist [5].

There are other methodologies developed for regression analysis in the presence of competing risks in recent years. We refer to Haller et al. [21] for an overview over current methodologies for the regression analysis of competing risks data.

20.5 Conclusion

Competing risks occur frequently in medical research although the presence is not always recognized. As a summary, we would like to highlight four important aspects of competing risks data analysis.

- **Recognition** Recognizing competing risks data is important since it may have a significant impact on the interpretation of the results. Recently Koller et al. [29] reviewed medical and statistical articles that dealt with competing risks data and reported that despite a rapid development in competing risks methodology in the past decades, recognition of competing risks in the medical community is still marginal. HSCT study is a classical example of competing risks data as most of clinical endpoints have immunologically intertwined competing events. In a recent survey of HSCT studies conducted by Kim and Armand [27], engraftment was rarely considered in a competing risks framework.

- **Presentation** In the analysis of competing risks data, it is important to present both the results of the event of interest and the results of competing risks. It is particularly important if the cumulative incidence rate of the competing event is high. In the example of chronic graft-versus-host disease (cGVHD) after umbilical cord blood transplantation (UCBT) compared to the peripheral blood transplantation (PBSCT) [11], the cumulative incidence of cGVHD was high after PBSCT compared to the one after UCBT. However, this is largely due to the fact that many patients died of transplant related complications early after UCBT, such as infection. Therefore, presenting the cumulative incidence of cGVHD without presenting the competing event may mislead the true effect of UCBT.

- **Choice of competing event** In the survey of 116 HSCT papers presented in Kim and Armand [27], the choice of competing events for an endpoint of interest varies greatly and is far from clear. In order to compare results across studies, competing events should be clearly defined for an event of interest in each disease since this might have significant implications on the estimate of the event of interest itself.

- **Interpretation** Results of competing risks data analysis should be interpreted carefully. In the example of the aforementioned cGVHD [11], one might conclude that UCBT is preventive of cGVHD. However, since time to cGVHD for those subjects who died of transplant related mortality (TRM) without developing cGVHD is latent, it is unknown if patients who died of TRM would have developed cGVHD if they didn't die without cGVHD. In Example 5, the cause-specific hazard of Event 2 was not significantly different between Group A and Group B, whereas the cumulative incidence of Event 2 was significantly different. Thus, a higher cumulative incidence of an event should not be automatically translated into a higher cause-specific hazard and vice versa.

20.6 Computing Tools

Various computing tools are available for competing risks data analysis.

- **Gray Test**:
  cuminc in the R package **cmprsk**, developed by Robert Gray. The package can be down-loaded from the website (http://www.r-project.org).

- **Fine and Gray Model**:
  crr in the R package **cmprsk**, developed by Robert Gray. The package can be down-loaded from the website (http://www.r-project.org). **crr** fits time-varying non-proportional covariates, $\beta(t)X$, via time by covariate interaction, but does not handle

time-dependent covariates, $\beta X(t)$. **crr** outputs a matrix of Schoenfeld residuals, and plots of these residuals against failure times can be used for checking the proportional hazards assumption.

- **Stratified Fine and Gray Model**: CRRS is a documented R function for the stratified Fine and Gray model [46]. It is available at the *Biometrics* website http://www.biometrics.tibs.org.

- **Adds-on functions to Fine and Gray Model**:
  Two adds-on functions to **crr** are available on http://www.stat.unipg.it/ luca/R/. **CumIncidence** [42] is a R program to calculate the confidence intervals of cumulative incidence functions and plot them(20.3).
  modsel.crr [43] is a model selection tool among candidate competing risks models.

- **Klein and Anderson Model**:
  The SAS and R programs for the Klein and Andersen model is available on the Center for International Blood and Marrow Transplant Research (CIBMTR) website. The URL is
  http://www.cibmtr.org/ReferenceCenter/Statistical/Education or
  http://www.mcw.edu/biostatistics/statisticalresources/CollaborativeSoftware.htm
  The analysis can also be performed using the R function *geese* in the R package *geepack* [20] or PROC GENMOD in SAS.

- **Power calculation:** For power calculation, Pintilie [37] modified the existing sample size formula and extended to the competing risks setting. **power** is a R program available on http://www.uhnres.utoronto.ca/labs/hill/People_Pintilie.htm.

Acknowledgements

The research was supported by PO1CA142106-10 and RO1CA183559 from the National Cancer Institute.

Bibliography

[1] A. Hirotugu (1974). A new look at the statistical model identification. IEEE Transactions on Automatic Control 19 (6): 716–723.

[2] E. P. Alyea, H. T. Kim, V. Ho, C. Cutler, J. Gribben, D. J. DeAngelo, S. J. Lee, S. Windawi, J. Ritz, R. M. Stone, J. H. Antin, and R. J. Soiffer. (2005) Comparative outcome of nonmyeloablative and myeloablative allogeneic hematopoietic cell transplantation for patients older than 50 years of age. *Blood* 105:1810–1814.

[3] P. K. Andersen, and N. Keiding. Interpretability and importance of functionals in competing risks and multistate models. *Statistics in Medicine.* 2012 May 20;31(11–12):1074–88. doi: 10.1002/sim.4385. Epub 2011 Nov 14.

[4] J. R. Anderson, K. C. Cain, and R.D. Gelber. (1983). Analysis of survival by tumor response and other comparisons of time-to-event by outcome variables. *Journal of Clinical Oncology.* 1983 Nov;1(11):710–9.

[5] P. K. Andersen, and M. P. Perme. (2010) Pseudo-observations in survival analysis. *Statistical Methods in Medical Research,* 19(1), 71–99.

[6] A. P. Basu, and J. K. Ghosh. (1978). Identifiability of the multinormal distribution under competing risks model (with J.K. Ghosh). *Journal of Multivariate Analysis,* 8:413–429.

[7] A. P. Basu, and J. K. Ghosh. (1980). Identifiability of distributions under competing risks and complementary risks model. *Communications in Statistics - Theory and Methods,* 9, 1515–1525.

[8] A. P. Basu, and J. P. Klein. (1982). Some recent results in competing risks theory. Survival Analysis, 216–229 Crowley, John (ed.) and Johnson, Richard A. (ed.) *Institute of Mathematical Statistics* (Hayward).

[9] G. Cortese, and P. K. Andersen. (2009). Competing Risks and Time-Dependent Covariates. *Biometrical Journal,* 51:138–158.

[10] D.R. Cox, and D. Oakes. (1984). Analysis of Survival Data. *Chapman and Hall* 91–110.

[11] Y. B. Chen, J. Aldridge, H. T. Kim, K. K. Ballen, C. Cutler, G. Kao, D. Liney, G. Bourdeau, E. P. Alyea, P. Armand, J. Koreth, J. Ritz, T. R. Spitzer, R. J. Soiffer, J. H. Antin, and V. T. Ho. Reduced-Intensity Conditioning Stem Cell Transplantation: Comparison of Double Umbilical Cord Blood and Unrelated Donor Grafts. *Biology of Blood and Marrow Transplantation.* 2012 May;18(5):805–12.

[12] M. Crowder. (1994). Identifiability Crises in Competing Risks. *International Statistical Review.* 62, 379–391.

[13] M. Crowder. (2001). *Classical Competing Risks.* Chapman & Hall/CRC.

[14] K. Dietz, and J. A. Heesterbeek. (2002). Daniel Bernoulli's epidemiological model revisited. Mathematical Biosciences. Nov-Dec;180:1–21.

[15] H. A. David, and M. L. Moeschberger. (1978). *The theory of competing risks.* London: Griffin.

[16] J. P. Fine, and R. J. Gray. (1999). A proportional hazards model for the subdistribution of a competing risk. *Journal of the American Statistical Association.* 94:496–509.

[17] E. D. Jacobsen, H. T. Kim, V. T. Ho, C. S. Cutler, J. Koreth, D. C. Fisher, P. Armand, E. P. Alyea, A. S. Freedman, R. J. Soiffer, and J. H. Antin. A large single-center experience with allogeneic stem-cell transplantation for peripheral T-cell non-Hodgkin lymphoma and advanced mycosis fungoides/Sezary syndrome. *Annals of Oncology.* 2011 Jul;22(7):1608-13. Epub 2011 Jan 20. PMCID:PMC3121969

[18] J. J. Gaynor, E. J. Feuer, C. C. Tan, D. H. Wu, C. R. Little, D. J. Straus, B. D. Clarkson, and M. F. Brennan. On the Use of Cause-Specific Failure and Conditional Failure Probabilities: Examples From Clinical Oncology Data. *Journal of the American Statistical Association.* Vol. 88, No. 422 (Jun., 1993), pp. 400–409

[19] R. J. Gray. A class of K-sample tests for comparing the cumulative incidence of a competing risk(1988). Ann. Statist. 16:1140–1154.

[20] S. Hjsgaard, U. Halekoh, and J. Yan. (2006) The R Package geepack for Generalized Estimating Equations. *Journal of Statistical Software*, 15, 2, pp1–11.

[21] B. Haller, G. Schmidt, and K. Ulm. Applying competing risks regression models: an overview. *Lifetime Data Analysis.* 2013 Jan;19(1):33–58.

[22] C. M. Hurvich, C. Tsai. (1995) Model selection for extended quasi-likelihood models in small samples Biometrics, 51, 1077–1084.

[23] H. Jeffreys. (1961). Theory of probability, 3rd ed., Oxford University Press.

[24] J. D. Kalbfleisch, and R. L. Prentice. (2002). The statistical analysis of failure time data John Wiley & Sons (New York; Chichester).

[25] E. L. Kaplan, P. Meier. (1958). Nonparametric estimation from incomplete observations. *Journal of the American Statistical Association.* 53:457–81.

[26] H.T. Kim. (2007). Cumulative incidence in a competing risks setting and competing risks regression analysis. *Clinical Cancer Research.* 13(2):559–65.

[27] H. T. Kim, and P. Armand. (2013). Clinical Endpoints in Allogeneic Hematopoietic Stem Cell Transplantation Studies: The Cost of Freedom. *Biology of Blood and Marrow Transplantation* 2013;19(6):860–6.

[28] J. P. Klein, P. K. Andersen. (2005). Regression modeling of competing risks data based on pseudovalues of the cumulative incidence function. *Biometrics.* 61:223–229.

[29] M. T. Koller, H. Raatz, E. W. Steyerberg, M. Wolbers. (2012) Competing risks and the clinical community:irrelevance or ignorance. *Statistics in Medicine* 31:1089–1097

[30] A. Latouche, R. Porcher. (2007). Sample size calculations in the presence of competing risks. *Statistics in Medicine.* 30;26(30):5370–80.

[31] A. Latouche, V. Boisson, S. Chevret, R. Porcher. (2007) Misspecified regression model for the subdistribution hazard of a competing risk. *Statistics in Medicine* 26(5):965–974. doi:10.1002/sim.2600

[32] A. Latouche, R. Porcher, S. Chevret. A note on including time-dependent covariate in regression model for competing risks data. *Biometrical Journal* 2005 Dec;47(6):807–14.

[33] A. Latouche, R. Porcher, and S. Chevret. (2004). Sample size formula for proportional hazards modelling of competing risks. *Statistics in Medicine* 23(21):3263–74.

[34] K. Y. Liang, and S. L. Zeger. (1986) Longitudinal data analysis using generalized linear models. Biometrika 73:13–22.

[35] E. Maki. (2006). Power and sample size considerations in clinical trials with competing risk endpoints. *Pharmaceutical Statistics.* 5(3):159–71.

[36] R. G. Miller. (1974). The jackknife - a review. Biometricak 6(1):1-15. doi:-.1093/-biomet/61.1.1

[37] M. Pintilie. (2002). Dealing with competing risks: Testing covariates and calculating sample size Statistics in Medicine, 21:3317–3324.

[38] M. Pintilie. (2006). *Competing Risks: A Practical Perspective.* Wiley, New York.

[39] R. Prentice, J. Kalbfleisch, A. Peterson, N. Flournoy, V. Farewell, N. Breslow. (1978) The analysis of failure times in the presence of competing risks. *Biometrics* 34:541–554.

[40] T. H. Scheike, and M. J. Zhang. Flexible competing risks regression modeling and goodness-of-fit. *Lifetime Data Analysis.* 2008 Dec;14(4):464–83.

[41] T. H. Scheike, and M. J. Zhang. Analyzing Competing Risk Data Using the R timereg Package. *Journal of Statistical Software* 2011 Jan;38(2). pii: i02.

[42] L. Scrucca, A. Santucci, and F. Aversa. (2007). Competing risk analysis using R: an easy guide for clinicians. Bone Marrow Transplantation. Aug;40(4):381–7. Epub 2007 Jun 11.

[43] L. Scrucca, A. Santucci, and F. Aversa (2010). Regression Modeling of Competing Risk Using R: An In Depth Guide for Clinicians. Bone Marrow Transplantation. Jan 11. [Epub ahead of print].

[44] G. Schwarz. (1978).Estimating the dimension of a model. The Annals of Statistics, 6, 461–464.

[45] A. Tsiatis. (1975). A nonidentifiability aspect of the problem of competing risks. *Proceedings of the National Academy of Sciences of the United States of America.* 72(1):20–22.

[46] B. Zhou, A. Latouche, V. Rocha, and J. Fine. (2011). Competing risks regression for stratified data. Biometrics. Jun;67(2):661–70.

21

Joint Models for Longitudinal and Time to Event Data

Hélène Jacqmin-Gadda, Loïc Ferrer and Cécile Proust-Lima

CONTENTS

21.1 Introduction

Clinical trials often involve repeated measurements of quantitative variables as long as the collection of times of onset of clinical events. The quantitative variables are often biological markers. Typical examples are the prostate-specific antigen in prostate cancer trials and the CD4+ T cell count or HIV viral load in HIV patients. But many other quantitative non biological markers may also be collected repeatedly during trials, such as cognitive scores in dementia or quality of life measures and pain scales in many pathologies. In this chapter, we will use marker to designate these quantitative variables measured repeatedly. In most clinical trials the primary endpoint is the clinical event such as death, onset of the disease or of a clinical relapse while the repeated measures of the markers are collected for secondary analyses. However, change over time of a quantitative marker is increasingly used as the primary outcome for comparing treatment when the clinical events are rare or delayed. As repeated measures of a quantitative marker bring more information than time to (a possibly rare) event, using the marker as primary outcome makes possible to reduce the sample size or the duration of the trial. Nevertheless, a gain in efficiency is possible only if the quantitative marker is a good surrogate marker of the clinical event (see Chapter 29), which implies that most of the treatment effect on the clinical event is mediated by effect of the treatment on the trajectory of the marker.

When investigating direct and indirect effects of a treatment on the clinical event, the association between the treatment and the clinical event may be estimated while adjusting for the repeated measures of the marker. This may be done by including the marker as a time-dependent covariate in a time-to-event model, e.g., a semi-parametric proportional hazard model. However, this approach yields biased estimates because the marker is measured with error and only at discrete measurement times such as the marker value is not known at all event times [27]. As estimation of a Cox model with a time-dependent covariate requires the knowledge of the marker value at all event times for all the subjects at risk, the imputation of the last observation is often used. Flaws of this approach, that assumed constancy of the marker between measurements, have been demonstrated, especially when time between visits is large. At last, estimation of a Cox model with a time-dependent covariate relies on the assumption that the distribution of the time-dependent variable is not affected by the occurrence of the event, i.e., that the time-dependent covariate is *exogenous* or *external* [19]. This is almost never true for markers collected in clinical trials, especially for surrogate markers. Formally, a time-dependent variable $Y(t)$ is *exogeneous* if:

$$f_{Y(t)}(y \mid T \geq s, \mathcal{H}(s)) = f_{Y(t)}(y \mid T = s, \mathcal{H}(s)) \text{ for } t \geq s$$

or

$$\begin{aligned} \alpha(s \mid \mathcal{H}(s)) &= \mathrm{P}(s \leq T < s + ds \mid T \geq s, \mathcal{H}(s)) & (21.1) \\ &= \mathrm{P}(s \leq T < s + ds \mid T \geq s, \mathcal{H}(t), t \geq s). & (21.2) \end{aligned}$$

where $\mathcal{H}(s)$ is the set of measures of Y until time s and T is the time-to-event. For *endogenous* or *internal* time-dependent variable, the joint modelling of the change over time of the marker and of the risk of the clinical event is required for estimating the effect of the treatment on the risk of event adjusting for the change of the marker. This has been one of the main motivations for the development of joint models for longitudinal data and time to event over the last two decades.

Joint models are also useful in clinical trials to assess the treatment effect on the trajectory of the marker when the follow-up is truncated by the onset of an event. For instance, the follow-up may be stopped by death. It is also frequent that the collection of repeated measures of the marker stops when the patient experiences the main clinical event of interest or any event that could dramatically change the course of the marker (e.g., change of treatment). When the risk of event depends on unobserved values of the marker, the change over time of the marker cannot be estimated using only observed marker measures and neglecting the event. One way to tackle this issue is to model jointly the marker change over time and the risk of event.

Joint models are built by combining a mixed model describing the marker course over time and a time-to-event model. A latent structure accounts for the association between the two outcomes. Two approaches have been developed depending on the assumed relationship between the outcomes. In the joint shared random effects model (JSREM), functions of the random effects from the mixed model are included as covariate in the time-to-event model. This is the earliest and the most common approach in joint modelling [11, 42]. In clinical trials, JSREM are used for evaluating surrogacy or investigating the causal effect of the treatment [32]. In the framework of incomplete longitudinal data with informative dropout, JSREM are used to model jointly the trajectory of the marker and the time-to-dropout. In this context, they are named selection models because they model the selection process. The joint latent class mixed model (JLCMM) has been proposed as an alternative approach when the population is expected to be heterogeneous [22, 31]. It assumes that the population is divided in several latent classes with class-specific trajectories of the marker and class-specific risks of events. More formally, this is a mixture model where some parameters in the two sub-models, the mixed model and the time-to-event model, are class-specific. In clinical trials, JLCMM are an alternative to explore the causal pathway of the treatment and may be useful to explore heterogeneity of the treatment effect in the population.

21.2 Illustrative Example

We will illustrate the use of joint models in clinical trials using the Radiation Therapy Oncology Group (RTOG) Protocol 9202 clinical trial that compares two treatment arms in patients with locally advanced prostate cancer and prostate-specific antigen (PSA) level below 150 ng/mL [14]. The two treatment groups received the same external-beam radiotherapy (RT) and 4 months of initial adjuvant androgen deprivation (AD) and were then randomly assigned to receive either no additional hormone therapy (Arm 1) or 24 additional months of AD (Arm 2). Patients follow-up visits were scheduled every 3 months for the first year, every 4 months for year 2, every 6 months from years 3 to 5 and then annually. Each visit included a clinical exam and a measurement of PSA level which is the main biological marker used for monitoring patients with prostate cancer. Several combined end points were studied in this trial but for the purpose of illustration in this chapter, we will use only the combined end point defined by clinical progression (local or distant recurrence or death) or administration of a new hormonal treatment. This event will be thereafter referred to as clinical progression. Repeated PSA measurements were collected after the end of RT until the occurrence of the event or the censoring. In the remainder of the chapter, we used the following Box-Cox transformation on the PSA level for normalizing the response, denoted $tPSA$:

$$tPSA = \frac{(\text{PSA} + 0.1)^{-0.2} - 1}{-0.2}.$$

Table 21.1 gives a brief description of the study sample and Figure 21.1 displays spaghetti plots by treatment group (i.e, individual curves of observed tPSA measures according to time) with the mean curve estimated by a loss function [6]. Figure 21.2 displays the same individual trajectories according to the clinical progression. It highlights that clinical progression is preceded by an increase of PSA level while no increase is observed in subjects without clinical progression. This suggests that PSA dynamics might constitute a candidate surrogate marker of clinical progression. Figure 21.3, which depicts the Kaplan-Meier estimate by treatment arm, shows that the risk of event is lower in Arm 2; the log-rank test is significant, with p-value < 0.001.

TABLE 21.1
Description of the data, RTOG Protocol 9202: median (5th and 95th percentiles) for continuous variables and n (%) for categorical variables.

Treatment arms	Arm 1	Arm 2	Pooled
Number of patients	554	562	1116
Number of PSA measures per patient	11.0 (3.7, 19.0)	12.0 (4.0, 20.0)	11.0 (4.0, 19.0)
iPSA*	20.4 (4.8, 95.0)	19.3 (4.4, 87.5)	19.7 (4.5, 92.0)
Clinical T-stage			
2	262 (47.3%)	257 (45.7%)	519 (46.5%)
3	279 (50.4%)	279 (49.6%)	558 (50.0%)
4	13 (2.3%)	26 (4.6%)	39 (3.5%)
Gleason score			
2–6	236 (42.6%)	221 (39.3%)	457 (40.9%)
7	184 (33.2%)	202 (35.9%)	386 (34.6%)
8–10	134 (24.2%)	139 (24.7%)	273 (24.5%)
Number of events[†]	359 (64.8%)	318 (56.6%)	677 (60.7%)
Time of event[†]	3.9 (1.4, 9.3)	5.1 (1.6, 10.3)	4.5 (1.4, 10.1)
Time of censoring	9.2 (4.3, 11.5)	9.8 (4.2, 11.6)	9.6 (4.3, 11.5)

Times are in years since the end of radiotherapy.
* Pre-therapy PSA value (ng/ml).
[†] Clinical recurrence, hormonal therapy or death.

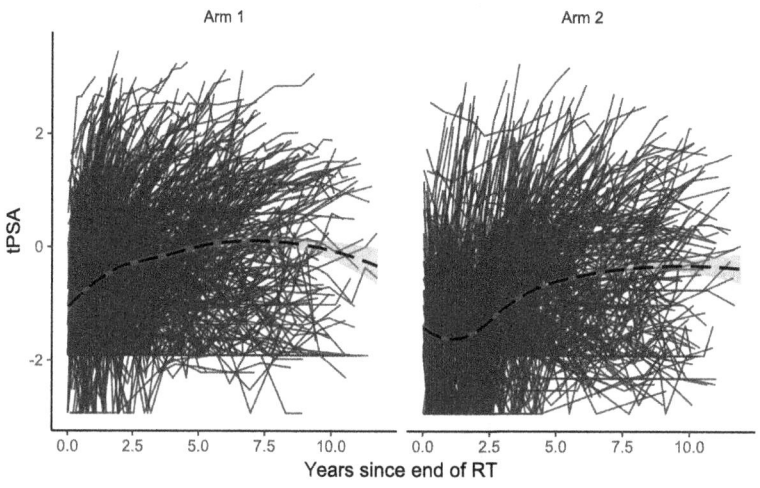

FIGURE 21.1
Individual trajectories of transformed PSA (tPSA) according to treatment arm. The black dashed lines are the mean trajectories estimated by local polynomial regressions, with 95% confidence intervals in grey. RTOG Protocol 9202.

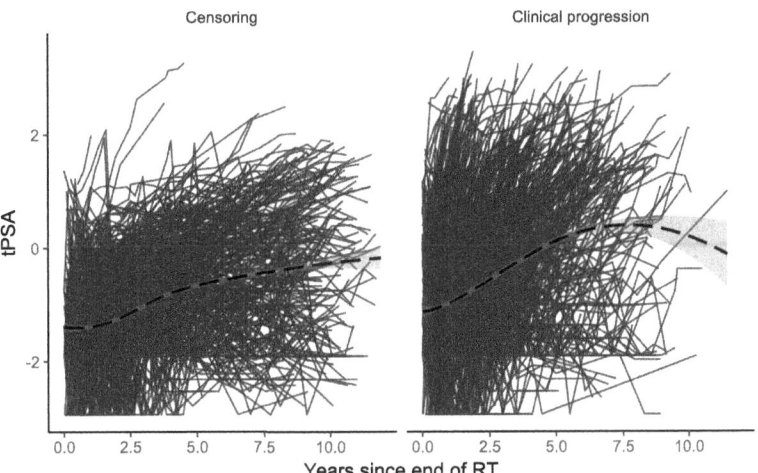

FIGURE 21.2
Individual trajectories of transformed PSA in pooled dataset according to clinical progression or censoring. The black dashed lines are the mean trajectories estimated by local polynomial regressions, with 95% confidence intervals in grey. RTOG Protocol 9202.

21.3 Joint Shared Random-Effect Models

21.3.1 Model definition for Gaussian markers

The longitudinal marker, that may take either continuous or discrete values, is assumed to be defined in continuous time and observed at discrete times. Thus, its change over time may be considered as a continuous-time stochastic process $\{Y(t), t > 0\}$. The observation of this process for subject i, $i = 1, ..., N$, and occasion j, $j = 1, ..., n_i$ at time t_{ij} is $Y_i(t_{ij})$ sometimes denoted Y_{ij}. T_i is the time of event for subject i that may be right censored by the censoring time C_i such that we only observe $\tilde{T}_i = T_i \wedge C_i$ and the failure indicator $\delta_i = I_{\{T_i < C_i\}}$.

Joint models have been first developed for continuous Gaussian longitudinal markers by combining a linear mixed model for describing the change over time of the marker and a time-to-event model, most often, a proportional hazard model. We first describe joint models in this context and detail in Section 21.3.2 how they can be extended to discrete outcomes.

In joint shared random-effect models, the link between the two outcomes is handled by including one or several functions of the random effects from the mixed model as explanatory variables in the time-to-event model. The model is thus defined by the two following sub-models [39]:

$$Y_{ij} = \boldsymbol{X}_{1i}(t_{ij})^\top \boldsymbol{\beta} + \boldsymbol{Z}_i(t_{ij})^\top \boldsymbol{b}_i + \epsilon_{ij} = \tilde{Y}_i(t_{ij}) + \epsilon_{ij} \tag{21.3}$$

$$\lambda_i(t) = \lambda_0(t) \exp(\boldsymbol{X}_{2i}(t)^\top \boldsymbol{\gamma} + \boldsymbol{h}_i(\boldsymbol{b}_i, t)^\top \boldsymbol{\eta}), \tag{21.4}$$

where ϵ_{ij} are Gaussian random errors, $\epsilon_{ij} \sim \mathcal{N}(0, \sigma^2)$; $\lambda_i(t)$ is the hazard function and $\lambda_0(t)$ the baseline hazard function (mostly parametric, see Section 21.3.3); $\boldsymbol{\beta}$, $\boldsymbol{\gamma}$ and $\boldsymbol{\eta}$ are vectors of regression parameters; $\boldsymbol{b}_i$ is a vector of random effects which prior distribution is assumed to be Gaussian, $\boldsymbol{b}_i \sim \mathcal{N}(0, B)$; $\boldsymbol{X}_{1i}(t)$ and $\boldsymbol{X}_{2i}(t)$ are vectors of explanatory variables possibly time-dependent and $\boldsymbol{Z}_i(t)$ is a sub-vector of $\boldsymbol{X}_{1i}(t)$ (to insure $E(\boldsymbol{b}_i) = 0$). Most often,

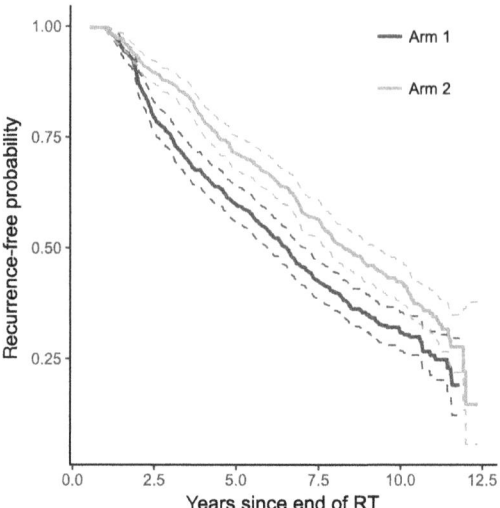

FIGURE 21.3
Kaplan-Meier curves according to treatment arm with 95% confidence interval. RTOG Protocol 9202.

the vector $\boldsymbol{Z}_i(t)$ includes an intercept and functions of time to account for inter-individual variations of Y over time through the random-effects. The random intercept accounts for inter-individual variations of the mean level of the marker. If time t is included in the vector $\boldsymbol{Z}_i(t)$, the associated random effect accounts for inter-individual variations of the slope of the linear time-trend. If the time trend is not linear, other functions of time (such as polynomial or spline functions) may be included in $\boldsymbol{X}_{2i}(t)$ to model the averaged time trend of the population and in $\boldsymbol{Z}_i(t)$ to account for the inter-individual variations of the time trend. The choice of the best mixed model among models with different structures for the random effects may be based on the Akaike criterion [1] or test for variance component [17]. When $\boldsymbol{X}_{2i}(t)$ includes time-dependent covariates, they are assumed to be exogenous (see definition in Section 21.1) with values known at every time (with negligible measurement error).

The specification of the function $\boldsymbol{h}_i(\boldsymbol{b}_i, t)$ and the associated vector of parameters $\boldsymbol{\eta}$ define the hypothesized dependence structure between the time of event T and the marker Y. Testing the hypothesis $H_0: \boldsymbol{\eta} = 0$ makes possible to test if the two outcomes are independent.

When it is assumed that the hazard function depends on the current expected value of the marker [40], $\boldsymbol{h}_i(\boldsymbol{b}_i, t) = \tilde{Y}_i(t) = \boldsymbol{X}_{1i}^\top(t)\boldsymbol{\beta} + \boldsymbol{Z}_i^\top(t)\boldsymbol{b}_i$ and

$$\lambda_i(t) = \lambda_0(t)\exp(\boldsymbol{X}_{2i}(t)^\top\boldsymbol{\gamma} + \tilde{Y}_i(t)\eta_1). \tag{21.5}$$

When the risk of event also depends on the current trend of the marker that may be measured by the slope of its trajectory,

$$\lambda_i(t) = \lambda_0(t)\exp(\boldsymbol{X}_{2i}(t)^\top\boldsymbol{\gamma} + \tilde{Y}_i(t)\eta_1 + \tilde{Y}_i(t)'\eta_2), \tag{21.6}$$

where $\tilde{Y}_i(t)'$ is the derivative of the function $\tilde{Y}_i(t)$ with respect to t at time t.

Another formulation relies on the hypothesis that the event risk depends only on the individual deviations, i.e., the random effects:

$$\lambda_i(t) = \lambda_0(t)\exp(\boldsymbol{X}_{2i}(t)^\top\boldsymbol{\gamma} + \boldsymbol{b}_i^\top\boldsymbol{\eta}). \tag{21.7}$$

We point out that in this last formulation, the hazard function does not change with the course of the marker. Subjects with positive deviations to the mean have a higher (lower, respectively) risk of event from time 0 when $\eta > 0$ ($\eta < 0$, respectively).

Henderson et al [15] proposed a different formulation of the joint model for continuous Gaussian data by introducing two correlated Gaussian processes with zero expectation, $W_{1i}(t)$ and $W_{2i}(t)$:

$$Y_{ij} = \boldsymbol{X}_{1i}(t_{ij})^\top \boldsymbol{\beta} + W_{1i}(t_{ij}) + \epsilon_{ij} \text{ with } \epsilon_{ij} \sim \mathcal{N}(0, \sigma^2) \tag{21.8}$$

and

$$\lambda_i(t) = \lambda_0(t) \exp(\boldsymbol{X}_{2i}(t)^\top \boldsymbol{\gamma} + W_{2i}(t)). \tag{21.9}$$

The process $W_{1i}(t)$ may for instance be defined by:

$$W_{1i}(t) = \boldsymbol{Z}_i^\top(t)\boldsymbol{b}_i + w_i(t)$$

where $w_i(t)$ is an autoregressive process or a Brownian motion. This model accounts more flexibly for the intra-individual variations of the marker change over time but the estimation may be computationally intensive. Note that when $W_{1i}(t) = \boldsymbol{Z}_i^\top(t)\boldsymbol{b}_i$ and $W_{2i}(t) = \eta W_{1i}(t)$, the model is identical to the one defined by (21.3) and (21.4) with $\boldsymbol{h}_i(\boldsymbol{b}_i, t) = \boldsymbol{Z}_i^\top(t)\boldsymbol{b}_i$.

21.3.2 Model definition for discrete markers

When the marker takes discrete values and its distribution may be assumed to belong to the exponential family, its change over time can be described by a generalized linear mixed model:

$$g(\mathrm{E}(Y_{ij}|\boldsymbol{b}_i)) = \boldsymbol{X}_{1i}(t_{ij})^\top \boldsymbol{\beta} + \boldsymbol{Z}_i(t_{ij})^\top \boldsymbol{b}_i = \tilde{Y}_i(t_{ij}) \tag{21.10}$$

where $g(.)$ is the appropriate link function. Note that in the above formula, $\tilde{Y}_i(t_{ij})$ is the value of the linear predictor for subject i at time t_{ij} and not the expected value of Y_{ij}. The time-to-event submodel has the same form as previously (21.4) with dependence through the function $\boldsymbol{h}_i(\boldsymbol{b}_i, t)$ that may include the current value of the linear predictor $\tilde{Y}_i(t)$, its derivative or the random effects $\boldsymbol{b}_i$.

21.3.3 Estimation

21.3.3.1 Likelihood

Parametric JSREM may be estimated by maximising the joint likelihood of observed data for the two outcomes i.e $(Y_i, \tilde{T}_i, \delta_i)$. Let us denote θ, the vector including the parameters from the 2 sub-models (21.3) or (21.10) and (21.4), i.e. the regression parameters $\boldsymbol{\beta}$, $\boldsymbol{\gamma}$ and η, possibly the variance σ^2, the parameters from the variance matrix B and those of the baseline risk function $\lambda_0(t)$. Thanks to the conditional independence assumption between the marker Y and the event time T given the random effects, the log-likelihood may be written as:

$$L(\theta) = \sum_{i=1}^{N} \log \mathcal{L}(Y_i, \tilde{T}_i, \delta_i)$$

$$= \sum_{i=1}^{N} \log \int p_{Y_i|\boldsymbol{b}_i}(Y_i \mid \boldsymbol{b}) S_i(\tilde{T}_i \mid \boldsymbol{b}) \lambda_i(\tilde{T}_i \mid \boldsymbol{b})^{\delta_i} f_{\boldsymbol{b}_i}(\boldsymbol{b}) d\boldsymbol{b}, \tag{21.11}$$

where $f_{\boldsymbol{b}_i}(\boldsymbol{b})$ is the multivariate Gaussian density function with mean zero and variance $\boldsymbol{B}$ while the probability density function for Y_i is given by

$$p_{Y_i|\boldsymbol{b}_i}(Y_i \mid \boldsymbol{b}) = \prod_{j=1}^{n_i} p_{Y_{ij}|\boldsymbol{b}_i}(Y_{ij} \mid \boldsymbol{b})$$

with $p_{Y_{ij}|\boldsymbol{b}_i}(Y_{ij} \mid \boldsymbol{b})$ defined according to (21.3) or (21.10), and the survival function is

$$S_i(\tilde{T}_i \mid \boldsymbol{b}) = \exp\left(-\int_0^{\tilde{T}_i} \lambda_i(s \mid \boldsymbol{b})ds\right).$$

Tsiatis and Davidian [38] gave a rigorous derivation of this likelihood showing that it requires the assumption that the observation process is not informative. Specifically, the measurement times for the marker before the event and the censoring time for the event must depend neither on the time of occurrence of the event nor on the unobserved values of the marker. If there are any missing marker measures not due to the event occurrence, they must be missing at random (MAR), that is, not dependent on the current or future value of the marker.

The likelihood includes two integrals: one multivariate integral over the random effects and one univariate over t for computing $S_i(\tilde{T}_i \mid \boldsymbol{b})$. The integral over the random effects in (21.11) has no closed form and is computed using either Gaussian quadrature, Monte Carlo or Laplace approximation. The integral in the survival function has a closed form in some rare cases where the hazard function includes no time-dependent covariate (and thus $\boldsymbol{h}_i(\boldsymbol{b}_i, t) = \boldsymbol{b}_i$) or when the baseline risk function is exponential. In most other cases, it is computed by Gaussian quadrature. Then the likelihood can be maximized by Newton-like algorithms (Newton-Raphson or Quasi-Newton) or by the EM algorithm [10] in which random effects are considered as missing data.

Software development is an intensive area of current research. At the time of writing this chapter, the majority of software handle primarily Gaussian longitudinal markers. In the R package JM [33, 35], the optimisation procedure combines the EM algorithm for the first iterations and then a Quasi-Newton algorithm. The integrals over the random effects are computed by *pseudo-adaptive* Gauss-Hermite quadrature [34] where the quadrature points are centered at the Best Linear Unbiased Predictor of the random effects $\boldsymbol{b}_i$ from the linear mixed model and rescaled using the Hessian matrix of $\log \mathrm{P}(Y_i \mid \boldsymbol{b}_i)f_{\boldsymbol{b}_i}(\boldsymbol{b}_i)$. The R package joineR [26] makes possible to estimate semi-parametric JSREM defined according to the formulation (21.8) and (21.9) (with a non parametric baseline risk function) using Gauss-Hermite quadrature and EM algorithm [42]. The R package frailtypack also estimates joint models by relying on Gaussian quadrature for the integral over random-effects and optimization via a robust marquardt-levenberg algorithm (https://cran.r-project.org/web/packages/frailtypack). The stjm module in STATA [8] combines the Newton-Raphson algorithm and adaptive quadrature while JMFit macro in SAS [45] uses the optimization and integration techniques of PROC NLMIXED.

21.3.3.2 Bayesian estimation

Instead of maximising directly the joint likelihood (21.11), some authors take a Bayesian approach and estimate the posterior distribution $p(\theta, \mathbf{b})$, with $\mathbf{b} = (\boldsymbol{b}_1, ..., \boldsymbol{b}_N)$:

$$p(\theta, \boldsymbol{b}) \propto \prod_{i=1}^N \prod_{j=1}^{n_i} p_{Y_{ij}|\boldsymbol{b}_i}(Y_{ij} \mid \boldsymbol{b})S_i(\tilde{T}_i \mid \boldsymbol{b})\lambda_i(\tilde{T}_i \mid \boldsymbol{b})^{\delta_i} f_{\boldsymbol{b}_i}(\boldsymbol{b}).$$

Prior distributions must be chosen for all the parameters of the models and the posterior distribution is then obtained using Monte Carlo Markov Chain [3, 4]. This approach has been implemented in the R package JMbayes that makes possible to estimate JSREM for both continuous and categorical longitudinal marker [36].

21.3.3.3 Model diagnostic

Different aspects of the joint shared random effect model can be assessed. The fit of the mixed model may be evaluated graphically by comparing the means over the subjects of the observed values $Y_i(t)$ and the means of the conditional prediction for a series of time intervals. The latter are obtained by $\hat{E}(Y_i(t) \mid \boldsymbol{X}_{1i}(t), \hat{\boldsymbol{b}}_i, \hat{\theta})$ where $\hat{\boldsymbol{b}}_i$ is the empirical Bayes estimator (or BLUP) of the random effect $\boldsymbol{b}_i$. These conditional predictions are computed for all the individuals observed within the considered time interval. As, by design, in JSREM the dropout due to the event occurrence depends on the random effects, it is important to use conditional predictions given the random effects rather than marginal predictions.

The fit of the time-to-event sub-model may be checked by comparing the mean of the predicted individual survival curves $S_i\left(t \mid \boldsymbol{X}_{2i}, \hat{\boldsymbol{b}}_i, \hat{\theta}\right)$ with the Kaplan-Meier estimate of the survival curve. Finally, the dependence structure between T and Y may be checked by comparing models with different functions $\boldsymbol{h}_i(\boldsymbol{b}_i, t)$ using selection criteria such as AIC or likelihood ratio test when the models are nested. This test allows to assess the conditional independence assumption between T and Y versus specific alternatives.

21.3.4 Joint shared random-effect models for clinical trials

As explained in introduction, two main applications of JSREM in clinical trials regard the assessment of direct and indirect effects on the clinical event, an important step in the evaluation of surrogacy of a marker and the analysis of longitudinal data with informative dropouts.

21.3.4.1 Distinguishing direct and indirect treatment effects

Joint models can be used to evaluate if the treatment effect is mediated by the longitudinal marker. With this goal in mind, it is of great importance to define a flexible mixed model for describing the marker change over time and a flexible latent structure for linking the risk of event and the marker trajectory.

By denoting τ_i the treatment variable ($\tau_i = 1$ for one treatment and 0 for the other), the objective is to test the hypothesis $H_0 : \boldsymbol{\gamma} = 0$ in the following JSREM:

$$E(Y_{ij}|\boldsymbol{b}_i) = \boldsymbol{X}_{1i}(t_{ij})^\top \boldsymbol{\beta} + \boldsymbol{Z}_i(t_{ij})^\top \boldsymbol{b}_i = \tilde{Y}_i(t_{ij}) \tag{21.12}$$

$$\lambda_i(t) = \lambda_0(t) \exp(\tau_i \gamma + \boldsymbol{h}_i(\boldsymbol{b}_i, t)^\top \boldsymbol{\eta}). \tag{21.13}$$

Flexible mixed models can be obtained by including several functions of time in $\boldsymbol{X}_{1i}(t_{ij})$ and $\boldsymbol{Z}_i(t_{ij})$ such as splines functions that make possible to fit most shapes of trajectory. For flexibly modelling the dependence between T and Y, several functions of $\tilde{Y}_i(t)$, $\tilde{Y}_i(t)'$ or the random effects may be included in $\boldsymbol{h}_i(\boldsymbol{b}_i, t)$ and the various models may be compared by the likelihood ratio test or selection criteria when not nested.

The estimated value of $\boldsymbol{\gamma}$ in the above model has to be compared with the estimated value of $\boldsymbol{\gamma}^*$ in the unadjusted model:

$$\lambda_i(t) = \lambda_0(t) \exp(\tau_i \gamma^*). \tag{21.14}$$

If the treatment is efficient and its effect is mediated by the longitudinal marker Y, γ^* should typically be much larger than γ (in absolute value) and ideally γ should be non-significantly different from 0 whereas γ^* is significant. We point out that this test does not validate the surrogacy of marker Y. Indeed, the validation of a surrogate marker should involve a meta-analysis and the computation of coefficients of determination at the individual and trial levels (see Chapter 29). The next paragraph illustrates the use of a joint model to assess the direct effect of a hormonal therapy on clinical progression of prostate cancer after adjustment for the prostate-specific antigen.

Example:
In the RTOG clinical trial, the Kaplan-Meier curves for the two arms of treatment and the log-rank test highlighted the superiority of Arm 2. This result was confirmed in the standard Cox model for the risk of event in which only the treatment τ_i was included as covariate ($\tau_i = 0$ if subject i is in Arm 1, $\tau_i = 1$ otherwise):

$$\lambda_i(t) = \lambda_0(t) \exp(\tau_i \gamma^*).$$

Indeed, the estimate of γ^* was $\hat{\gamma}^* = -0.342$ (standard-error: 0.077, $p < 0.001$).

To evaluate if the treatment effect was mediated by the change in PSA level, we estimated a joint shared random effect model. Due to the length of the estimation procedure of joint models, the basic specification of the longitudinal and survival sub-models is usually done by considering separate models, i.e. assuming no association between the two sub-models ($\boldsymbol{\eta} = 0$). Here this strategy was considered. The submodel for tPSA trajectory (after the Box-Cox transformation introduced in 21.2) was a mixed model including the pre-treatment PSA, the stage of the tumor at diagnosis (Tstage), a measure of aggressiveness of the tumor (Gleason score), the treatment arm and its interaction with functions of time as covariates. Several polynomial and spline functions of time were considered to reflect at best the longitudinal trajectory of tPSA. Based on the Akaike criterion, natural cubic splines with 2 interior nodes at 6 and 24 months were finally chosen. For the survival submodel, a proportional hazard model depending on the treatment arm was used. To make the model flexible enough while avoiding computational burden of a semi-parametric model, the logarithm of the baseline risk function was modeled by cubic B-splines with 5 internal knots positioned at the quantiles of the event times.

Then, the dependence structure between the two outcomes was chosen by comparing joint models where the function $\boldsymbol{h}_i(\boldsymbol{b}_i, t)$ was either the true current marker level, the current slope or a bivariate function combining both. Based on the Akaike criteria, we selected the model depending on both the current predicted tPSA level and the current derivative of tPSA trajectory:

$$\lambda_i(t) = \lambda_0(t) \exp(\tau_i \gamma + \eta_1 \tilde{Y}_i(t) + \eta_2 \tilde{Y}_i'(t)).$$

The models were estimated with the R package JM which is extensively described with code examples in [33] and [35]. The estimates displayed in Table 21.2 show that the risk of event increases significantly with the current PSA level ($p < 0.001$) and the current slope of PSA ($p < 0.001$). Thus the risk of clinical progression increases with the growth of PSA. Adjusted for these characteristics of PSA trajectory, the risk of clinical progression was not associated with the treatment (p = 0.742). This shows that the treatment effect is completely mediated by the time course of PSA. This suggests that the PSA time course is a good candidate as a surrogate marker.

TABLE 21.2
Parameters estimates in the survival submodel of the joint shared random effect model, RTOG Protocol 9202.

	Value	StdErr	p-value
γ	0.031	0.093	0.742
η_1	0.715	0.044	< 0.001
η_2	1.186	0.114	< 0.001

The fit of the model is assessed in Figure 21.4. The left panel compares the mean of observed values and subject-specific predictions of tPSA as a function of time. The right panel displays the Kaplan-Meier estimate with 95% confidence band of the survival function without clinical progression and the mean predicted survival curve from the joint model (averaging over the subject-specific predictions). These plots show the good fit of the two submodels.

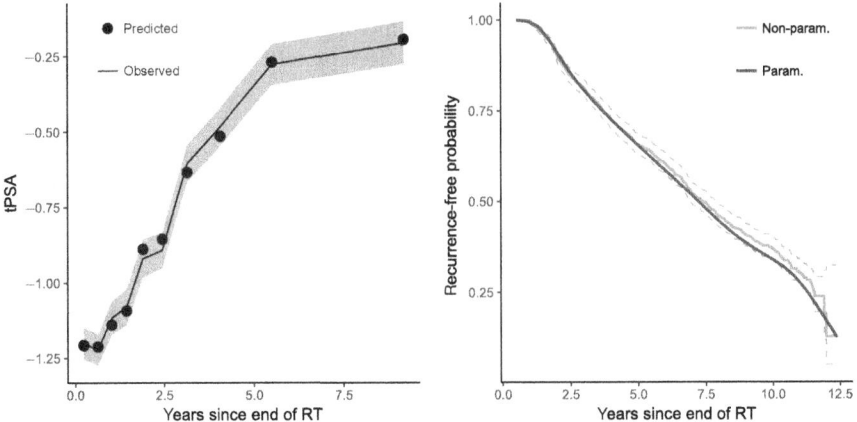

FIGURE 21.4

Left: comparison between observed mean values and means of subject-specific predictions of the transformed longitudinal biomarker. The grey zone characterizes the 95% confidence interval of the observed trajectory; right: predicted survival curves from the joint model and the Kaplan-Meier estimator. The dotted grey lines depict the 95% confidence interval of the non-parametric estimate. RTOG Protocol 9202.

21.3.4.2 Incomplete data

Dropouts often arise in clinical trials either because of loss to follow-up or by design when the follow-up is stopped after some pre-specified event occurrence. This yields to missing data that must be accounted for in the longitudinal analysis of the marker trajectory. According to Little and Rubin's terminology [23], maximum likelihood estimates of the mixed models obtained using only observed marker measurements are unbiased if the data are missing at random. In practice, this means that the probability of dropout can depend on covariates included in the mixed model or on previous observed values of the marker. When the dropout probability may depend on unobserved values of the marker (e.g., on the change in the marker value between the last measures and the event) or on unobserved characteristics of the marker trajectory (e.g., on the random effects), the missing data are informative. In the latter case, maximum likelihood estimates are biased while joint modelling of the marker trajectory and the risk of dropout (or the risk of the event that may induce dropout) leads to unbiased estimates if it is well specified. Note that it is impossible in practice to be sure that the missing data are not informative. It is thus often useful to perform a sensitivity analysis with joint models and different specifications of the dependence structure to evaluate the robustness of the mixed model analysis under the MAR assumption. The next paragraph gives an illustration of the use of JSREM for dealing with informative dropouts.

Example:

In this example, the main objective was to assess the impact of treatment on the PSA change over time. Thus the outcomes were the repeated measures of tPSA and the main model was a linear mixed model with the change over time modeled by natural cubic splines with 2 interior nodes at 6 and 24 months and including only the treatment arm and its interaction with functions of time as explanatory variables. It was compared with the model assuming instead a biphasic linear trajectory with a change of slope at 24 months. The AIC was much better for the model using splines (AIC=22494.04) than for the model assuming two linear slopes (AIC=23668.04). In the splines model, the global test for the treatment effect was highly significant ($p < 0.001$) and the estimated mean curve of PSA displayed in Figure 21.5 demonstrated that mean PSA levels in Arm 1 (left panel) were always higher that mean PSA levels in Arm 2 (right panel).

To account for informative dropout following clinical progression, we estimated a joint shared-random effect model with the above linear mixed sub-model for tPSA and a proportional hazard model adjusted for the pre-treatment PSA, the T-stage, the Gleason score, the treatment arm, the current level and the current slope of PSA for the time-to-clinical progression. Figure 21.5 displays the estimated mean curve of PSA change over time for each treatment arm in the two models. The global test for the treatment effect on PSA trajectories remained highly significant when accounting for the informative dropout ($p < 0.001$ in the joint model) but the estimated mean curves of tPSA are clearly different. The increase over time appears slighter when dropout is not accounted for because subjects who dropped out from the study are mostly those who experienced the most dramatic increase of PSA.

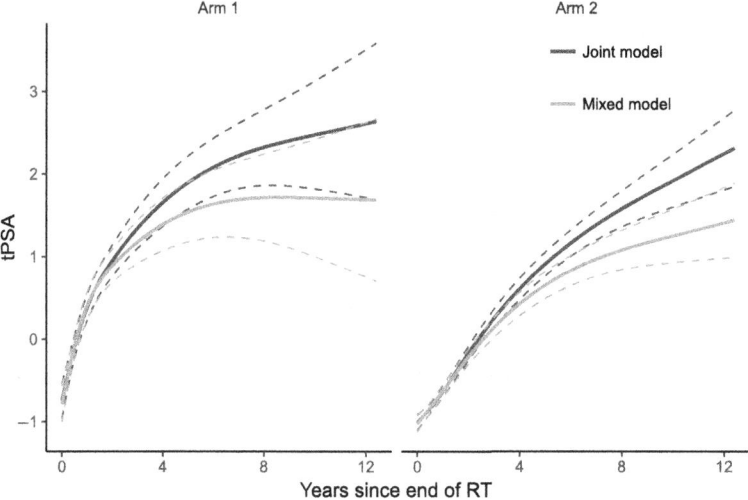

FIGURE 21.5
Fitted marginal trajectories (with 95% confidence intervals) using linear mixed model or joint model in the two treatment arms. RTOG Protocol 9202.

As previously, the goodness-of-fit of the two sub-models was checked by comparing observed and predicted values of the longitudinal marker on one hand, and non-parametric and predicted survival curves on the other hand. The results were similar to those obtained in Figure 21.4.

21.4 Joint Latent Class Models

21.4.1 Model definition

Similarly to JSREM, joint latent-class mixed models (JLCMM) combine a mixed model for the evolution of the marker and a time-to-event model, but the latent structure linking the two outcomes is different. In JLCMM the association is captured by categorical random effects, the latent classes, that are different from the continuous random effects that account for the correlation between repeated measures of the marker in the mixed model [22, 31]. This is an asset of JLCMM to model separately the correlation between repeated measures of the marker and the association between the marker and the time-to-event. JLCMM rely on the idea that the population is heterogeneous and made of several sub-populations, unobserved (or latent) classes, with class-specific trajectories for the marker and class-specific risks for the event.

Using the same notation as previously for the marker and the event, i.e., Y_{ij}, $\tilde{T}_i$ and δ_i, and defining the discrete latent variable c_i which equals g if subject i belongs to latent class g ($g = 1, ..., G$ and G is the number of latent classes), a JLCMM is defined by three submodels. A logistic regression model for nominal categorical data is used for the class-membership probability:

$$\pi_{ig} = \mathrm{P}(c_i = g \mid \boldsymbol{X}_{3i}) = \frac{e^{\xi_{0g} + \boldsymbol{X}_{3i}^\top \boldsymbol{\xi}_{1g}}}{\sum_{l=1}^G e^{\xi_{0l} + \boldsymbol{X}_{3i}^\top \boldsymbol{\xi}_{1l}}} \tag{21.15}$$

with constraints $\xi_{0G} = 0$ and $\boldsymbol{\xi}_{1G} = 0$ to ensure identifiability. For continuous marker, the second submodel is most often a linear mixed model with class-specific parameters that describes the conditional distribution of Y_i given $c_i = g$:

$$Y_{ij} = \boldsymbol{Z}_i(t_{ij})^\top \boldsymbol{b}_i + \boldsymbol{X}_{1i}(t_{ij})^\top \boldsymbol{\beta}_g + \epsilon_{ij} \tag{21.16}$$

with $\epsilon_i = (\epsilon_{i1}, ..., \epsilon_{in_i})^\top \sim \mathcal{N}(0, \boldsymbol{\sigma}_i)$. Given the latent class, the random effects $\boldsymbol{b}_i$ are assumed to be Gaussian with class-specific expectations $\boldsymbol{\mu}_g$ and possibly class-specific variance matrix $\boldsymbol{B}_g$, e.g. $\boldsymbol{B}_g = \omega_g^2 \boldsymbol{B}$ with $\boldsymbol{B}$ an unstructured variance matrix and $\omega_G = 1$. At last, a proportional hazard model with class-specific parameters may be used to define the distribution of the time-to-event given $c_i = g$:

$$\lambda_i(t \mid c_i = g) = \lambda_0(t; \boldsymbol{\zeta}_g) \exp(\boldsymbol{X}_{2i}(t)^\top \boldsymbol{\gamma}_g) \tag{21.17}$$

The four vectors $\boldsymbol{X}_{1i}(t)$, $\boldsymbol{X}_{2i}$, $\boldsymbol{X}_{3i}(t)$ and $\boldsymbol{Z}_i(t)$ include explanatory variables. In contrast with (21.3), $\boldsymbol{X}_{1i}(t)$ and $\boldsymbol{Z}_i(t)$ must not overlap to ensure identifiability of parameters $\boldsymbol{\mu}_g$ for $g = 1, ...G$; $\boldsymbol{Z}_i(t)$ usually includes functions of time to model both the mean change of the marker over time and inter-individual variability. In the model formulation (21.16) and (21.17), all the parameters were indexed by g suggesting they are all class-specific, but, in practice, some of them may be common over the classes, for instance to represent the marginal effect of an explanatory variable over all the classes. The choice of the parameters that are class-specific and especially, the choice of the covariates that can have a class-specific effect, depends of the objective of the analysis, as illustrated below. Note that when the longitudinal outcome is not continuous with Gaussian errors, equation (21.16) may be replaced by any generalization of the linear mixed model theory that woud be appropriate for the outcome, in which parameters would be assumed to be class-specific.

Formulas (21.16) and (21.17) highlight that JLCMMs assume conditional independence between the marker Y and the time-to-event T given the classes. Thus the association between the two outcomes comes entirely from the shared latent classes. A test for evaluating this assumption will be discussed in Section 21.4.2.2.

21.4.2 Estimation

21.4.2.1 Likelihood

JLCMM are estimated by maximising the log-likelihood for a given number of latent classes G. Then models with different values of G are compared by some criteria, mainly the BIC [2]. We denote $\boldsymbol{\theta}_G$ the vector including all the parameters involved in the three sub-models defined in (21.15), (21.16) and (21.17) with G latent classes. Thanks to the conditional independence assumption between the repeated measures of Y and the event time T given the latent classes, the log-likelihood may be written:

$$
\begin{aligned}
L(\boldsymbol{\theta}_G) &= \sum_{i=1}^{N} \log \mathcal{L}(Y_i, \tilde{T}_i, \delta_i) \qquad (21.18) \\
&= \sum_{i=1}^{N} \log \left(\sum_{g=1}^{G} \pi_{ig} f_{Y_i|c_i}(Y_i \mid c_i = g) \lambda_i(\tilde{T}_i \mid c_i = g)^{\delta_i} S_i(\tilde{T}_i \mid c_i = g) \right),
\end{aligned}
$$

where all the functions in the second line depend on $\boldsymbol{\theta}_G$. The class-membership probability π_{ig} is defined by (21.15), the hazard function $\lambda_i(t \mid c_i = g)$ is defined by (21.17) and $S_i(t \mid c_i = g)$ is the corresponding survival function in class g. The density $f_{Y_i|c_i}(Y_i \mid c_i = g)$ of the vector of marker measures for subject i given class g is a multivariate Gaussian density with mean $\boldsymbol{Z}_i \boldsymbol{\mu}_g + \boldsymbol{X}_{1i} \boldsymbol{\beta}_g$ and variance matrix $\boldsymbol{Z}_i \boldsymbol{B}_g \boldsymbol{Z}_i^\top + \boldsymbol{\sigma}_i$ where $\boldsymbol{Z}_i$ and $\boldsymbol{X}_{1i}$ are the matrices with row j consisting of $\boldsymbol{Z}_i(t_{ij})^\top$ and $\boldsymbol{X}_{1i}(t_{ij})^\top$, respectively.

This log-likelihood is maximised iteratively using for instance Newton type algorithms. JLCMM estimation is implemented in the R package `lcmm` (function `Jointlcmm`) [30] in which the optimization algorithm is a Marquardt algorithm [25]. As for all mixture models, the log-likelihood can be multimodal, and iterative algorithms might converge to local maxima. It is thus recommended to run the estimation procedure several times with different starting values to ensure convergence toward the global log-likelihood maximum.

21.4.2.2 Model diagnostic

This section describes methods to assess the quality of the discrimination between classes, the conditional independence assumption and the goodness of fit of the two sub-models [29, 31]. First, using $\hat{\boldsymbol{\theta}}_G$, the posterior class-membership probabilities may be computed for each subject by:

$$
\begin{aligned}
\hat{\pi}_{ig}^{Y,T} &= P(c_i = g \mid Y_i, (\tilde{T}_i, \delta_i); \hat{\boldsymbol{\theta}}_G) \\
&= \frac{\hat{\pi}_{ig} f_{Y_i|c_i}(Y_i \mid c_i = g; \hat{\boldsymbol{\theta}}_G) \lambda_i(\tilde{T}_i \mid c_i = g; \hat{\boldsymbol{\theta}}_G)^{\delta_i} S_i(\tilde{T}_i \mid c_i = g; \hat{\boldsymbol{\theta}}_G)}{\sum_{l=1}^{G} \hat{\pi}_{il} f_{Y_i|c_i}(Y_i \mid c_i = l; \hat{\boldsymbol{\theta}}_G) \lambda_i(\tilde{T}_i \mid c_i = l; \hat{\boldsymbol{\theta}}_G)^{\delta_i} S_i(\tilde{T}_i \mid c_i = l; \hat{\boldsymbol{\theta}}_G)} \qquad (21.19)
\end{aligned}
$$

A posterior classification may be built by classifying each subject in the class corresponding to his/her highest posterior probability. The quality of the discrimination between the classes may then be assessed by a table displaying the mean posterior probability to belong to each class computed for subjects classified in each posterior class. The classification is good when the mean probabilities on the diagonal are close to 1 and those outside from the diagonal are close to 0 [31]. This evaluation is required before interpreting the classes. In our experience, the quality of discrimination is variable according to the clinical event and the marker under study. When predicting recurrence of prostate cancer from repeated measures of PSA, the diagonal probabilities are typically higher than 0.9 [31] while joint modelling time-to-dementia and cognitive decline led to diagonal probabilities between 0.6 and 0.8 [37].

Then, the conditional independence assumption between Y and T given the latent classes can be evaluated by a score test [18]. The alternative hypothesis is defined by a joint model with shared random-effects and latent classes defined by (21.15), (21.16) and the submodel (21.17) replaced by:

$$\lambda_i(t \mid c_i = g) = \lambda_{0g}(t; \zeta_g) \exp(\boldsymbol{X}_{2i}(t)\boldsymbol{\gamma}_g + \boldsymbol{b}_{ig}^*\boldsymbol{\eta}) \tag{21.20}$$

where $\boldsymbol{b}_{ig}^*$ is the vector of centered random effects from the mixed model (21.16) ($\boldsymbol{b}_{ig}^* = \boldsymbol{b}_{ig} - \boldsymbol{\mu}_g$). The null hypothesis is defined by $\mathcal{H}_0$: $\boldsymbol{\eta} = 0$. The score statistic is a weighted estimate of the covariance between the class-specific residuals of the survival model and the class-specific empirical Bayes estimators of the random effects:

$$U(0, \boldsymbol{\theta}_G) = \sum_{i=1}^{N} \sum_{g=1}^{G} \pi_{ig}^{Y,T} \left(\delta_i - \Lambda_{ig}(\tilde{T}_i; \boldsymbol{\theta}_G) \right) \mathrm{E}(\boldsymbol{b}_{ig}^* \mid c_i = g, Y_i, \boldsymbol{\theta}_G) \tag{21.21}$$

where $\Lambda_{ig}(\tilde{T}_i; \boldsymbol{\theta}_G)$ is the cumulative risk function at $\tilde{T}_i$ for subject i in class g. Parameters $\boldsymbol{\theta}_G$ are estimated under $\mathcal{H}_0$ and $U(0, \hat{\boldsymbol{\theta}}_G)^\top \mathrm{var}(U)^{-1} U(0, \hat{\boldsymbol{\theta}}_G)$ has asymptotically a χ^2 distribution with the number of degrees of freedom equals to the size of vector $\boldsymbol{\eta}$. Two estimators of $\mathrm{var}(U)$ can be used [18].

Finally, the fit of each submodel to the data may be assessed graphically as for JSREM. However, it is recommended to assess the fit for each class. The weighted means of the class-specific predictions must be compared to the weighted means of the observations. The weights are the posterior class membership probabilities $\hat{\pi}_{ig}^{Y,T}$. For the mixed model, class-specific predictions given the random effects are $\hat{Y}_{ijg} = \boldsymbol{Z}_{ij}^\top(\hat{\boldsymbol{\mu}}_g + \hat{\boldsymbol{b}}_{ig}^*) + \boldsymbol{X}_{lij}^\top \hat{\boldsymbol{\beta}}_g$ where $\hat{\boldsymbol{b}}_{ig}^*$ are the empirical Bayes estimates of $\boldsymbol{b}_{ig}^*$. These predictions may then be averaged over the subjects or over the classes for goodness-of-fit analyses.

For the time-to-event sub-model, the estimated individual survival functions conditional to the classes $\hat{S}_{ig}(t)$ can also be averaged by subject or by class. The predicted survival curves averaged per class may be compared with survival curves estimated by Kaplan-Meier weighted by the posterior class membership probabilities.

21.4.3 Joint latent class models for clinical trials

The main advantage of JLCMM over JSREM is that it accounts for the heterogeneity of the population. The description of the population in terms of typical profiles of change for the marker associated with typical profiles of risk for the event allows simple graphical representations that help to communicate results to clinicians. Sometimes the classes have meaningful interpretation and identifying sub-populations with very different profiles may be helpful for a better care of the patients.

In clinical trials, JLCMM might be an alternative to JSREM for exploring causal pathway of treatment effect, i.e., evaluating if the treatment effect is mediated by the evolution of the marker. However, this is achieved at the price of many parameters as illustrated in the example below which reduces the interest of these models in this context. We do not recommend the use of JLCMM to estimate the effect of the treatment on the marker change over time accounting for informative dropout. Indeed, it was previously highlighted that latent classes are mainly driven by the heterogeneity in the marker and not by the association with dropout which is often weak [9]. Nevertheless, these models remain useful for exploring the dependence structure between the event and the marker, or the heterogeneity of the treatment effect between sub-populations, e.g. in the presence of responders and not responders [44].

Example:

A JLCMM was estimated on the data from the RTOG clinical trial with the aim to assess if the treatment effect was mediated by the heterogeneity of the PSA course over time. As in the JRSEM example, the time-trend for tPSA was modelled by natural splines with 2 interior nodes at 6 and 24 months and the mixed model was adjusted for the pre-treatment PSA, the Tstage, the Gleason score, the treatment arm and its interaction with functions of time. All the fixed effects in the mixed model were class-specific. That way, the latent classes captured the overall heterogeneity of PSA dynamics. The time-to-event model was a proportional hazard model with class-specific weibull baseline risks and the treatment (τ_i) as the unique explanatory variable (with a common effect over classes):

$$\lambda_i(t \mid c_i = g; \boldsymbol{\zeta}_g, \gamma) = \lambda_0(t; \boldsymbol{\zeta}_g) \exp(\tau_i \gamma)$$

The models were estimated with the R package lcmm which is extensively described with code examples in [30] and [7]. Table 21.3 displays the BIC, number of parameters and sizes of the posterior classes for models with G = 1 to 6 classes. According to BIC, the best model was the model with 5 classes. The posterior classes were defined using the posterior probabilities (21.19). Figure 21.6 depicts the estimated mean trajectories of PSA and the recurrence-free survival according to treatment group and latent classes for a subject with iPSA = 20 ng/mL, Tstage = 2, Gleason = 6. Adjusted for the latent classes, the risk of clinical progression was no more associated with the treatment ($\hat{\gamma} = -0.015$, p = 0.886) and the class-specific recurrence-free survival curves are the same for the two treatment groups. Figure 21.7 shows that the JLCMM model fitted well both the observed weighted means of PSA measures and the weighted Kaplan-Meier estimates of the survival curve. Thus the JLCMM also concludes to the mediation of the treatment effect by the PSA dynamics. However, this is achieved by including 91 parameters, among which 76 parameters describe the longitudinal sub-model (mostly due to the large number of class-specific parameters) compared to the 36 parameters in the JSREM. This is why this model is not as appropriate as the JSREM in this context although it may be of interest when the type of dependency between the longitudinal marker and the risk of event is complicated and not known *a priori*. Note that in terms of goodness-of-fit, the JLCMM provides a much better fit with 833 less points in BIC compared to the JSREM.

TABLE 21.3

Summary of the JLCMM estimation, RTOG Protocol 9202.

G	LogLikelihood	p*	BIC	Frequency of the latent classes (%)					
				1	2	3	4	5	6
1	−13389.96	27	26969.39	100.0					
2	−12881.87	43	26065.50	24.6	75.4				
3	−12629.80	59	25673.64	7.3	21.2	71.5			
4	−12471.29	75	25468.89	7.1	10.2	63.7	19.0		
5	−12363.53	91	25365.65	6.6	11.5	56.7	16.0	9.2	
6	−12331.71	107	25414.30	6.4	10.8	0.8	56.1	16.1	9.8

* Total number of parameters of the models (including variance parameters)

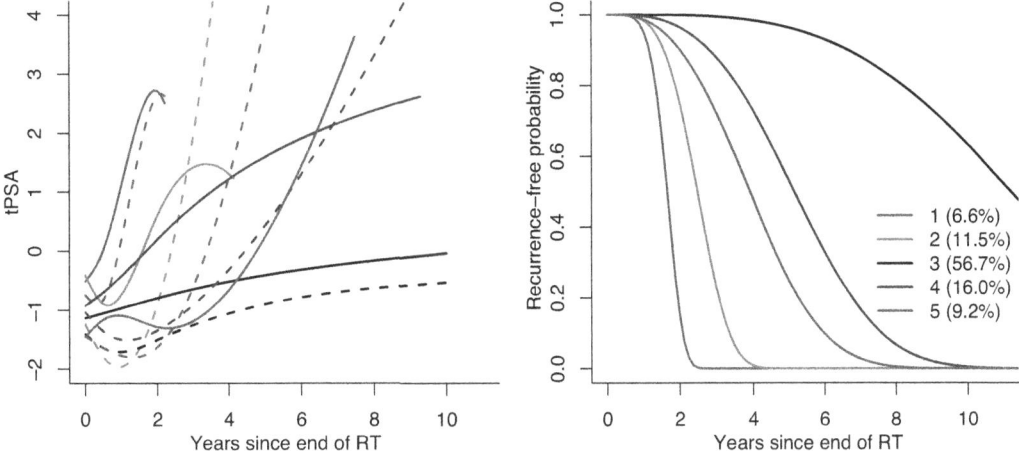

FIGURE 21.6
Left: class-specific predicted PSA profiles according to treatment arm (arm 1 in plain lines and arm 2 in dashed lines) for a man with iPSA=20 ng/mL, Tstage = 2,Gleason = 6; right: class-specific recurrence-free predicted probability (curves in both arms are indistinguishable). 5-class JLCMM, RTOG Protocol 9202.

21.5   Conclusion and Recent Developments

In this chapter, the principles of the two main types of joint models for one longitudinal marker and one time-to-event have been presented as well as examples. In clinical trial context, joint models are mainly used for exploring the causal pathway for the treatment effect, by distinguishing direct and indirect treatment effects, evaluating surrogacy and accounting for informative dropout. Although JLCMM may be more flexible, JSREM are more appropriate especially since they involve fewer parameters and are easier to interpret. In other contexts, joint models make possible to build dynamic prediction tools for the event based on repeated measures of the marker and, for this purpose, JLCMM may exhibit better predictive ability [31].

During the last 20 years, many extensions of joint models for multiple markers or multiple times of events have been proposed that can be of interest in the analysis of clinical trials. For instance, combining information from several longitudinal markers may increase the power of the trial or may help build a better combined surrogate marker. Such analysis can be carried out with joint models for multiple longitudinal markers [4, 5, 21, 43]. In these approaches, each marker is modeled by a mixed model and random effects are correlated between markers. However these models may be untractable when the number of markers is large because the number of random effects (and fixed parameters) increases proportionally to the number of markers. A more parsimonious approach assumes that the markers are noisy measures of transformations of a common latent process. Using non-linear marker-specific transformations, this approach allows the handling of non gaussian markers, possibly with ceiling and floor effects. This is especially useful when studying cognitive aging because cognition is not directly measured; it is assessed by a battery of highly correlated cognitive tests with various metrologic properties leading to non-standard asymmetric distributions with ceiling and/or floor effects [28, 29].

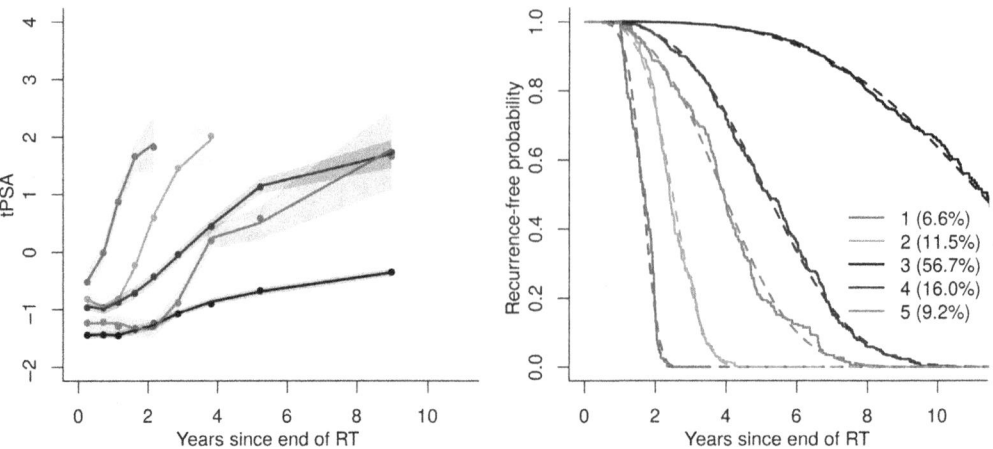

FIGURE 21.7
Left: mean observed (plain lines) and mean subject-specific predicted values (•) in each latent class weighted by the posterior class-membership probabilities along with the 95% point-wise confidence bands of the weighted mean observed trajectories (shades) up to the last available data in this class; right: Kaplan-Meier estimate (plain lines) and mean predicted (dashed lines) recurrence-free probability in each latent class weighted by the posterior class-membership probabilities. 5-class JLCMM, RTOG Protocol 9202.

On the other hand, multiple events are often collected in clinical trials (e.g. successive recurrences of cancer, competing events or different clinical events). In the latent class framework, joint models have been described for recurrent events [13] and multiple non Gaussian markers [29] or for interval-censored semi-competing risks [37]. In the shared random effects framework, joint models for competing risks have been proposed [16, 20, 41] as well as joint models for recurrent and terminal events [24]. Recently, Ferrer et al [12] made available a R function using the JM and mstate packages to estimate joint shared random effects models combining mixed models and multi-state models. This chapter was written in 2015 and that it thus neglects developments done after 2015.

Acknowledgements:

The authors sincerely thank the Radiation Therapy Oncology Group for providing the RTOG 9202 dataset (supported by U10 CA21661, and CCOP U10 CA37422 from the U.S. National Cancer Institute), and especially James Dignam, associate professor at the University of Chicago.

Bibliography

[1] H. Akaike. A new look at the statistical model identification. *IEEE transactions on automatic control*, 19(6):716–723, 1974.

[2] D. J. Bauer and P. J. Curran. Distributional assumptions of growth mixture models: implications for overextraction of latent trajectory classes. *Psychological Methods*, 8(3):338–63, 2003.

[3] E. Brown and J. Ibrahim. A Bayesian semiparametric joint hierarchical model for longitudinal and survival data. *Biometrics*, 59(2):221–228, 2003.

[4] E. Brown, J. Ibrahim, and V. DeGruttola. A flexible B-spline model for multiple longitudinal biomarkers and survival. *Biometrics*, 61(1):64–73, 2005.

[5] Y. Chi and J. Ibrahim. Joint models for multivariate longitudinal and multivariate survival data. *Biometrics*, 62(2):432–445, 2006.

[6] W. Cleveland. Robust locally weighted regression and smoothing scatterplots. *Journal of the American Statistical Association*, 74(368):829–836, 1979.

[7] D. Commenges and H. Jacqmin-Gadda. *Dynamical biostatistical models*, volume 86. CRC Press, 2015.

[8] M. Crowther, K. Abrams, and P. Lambert. Joint modeling of longitudinal and survival data. *Stata Journal*, 13(1):165–184, 2013.

[9] E. Dantan, C. Proust-Lima, L. Letenneur, and H. Jacqmin-Gadda. Pattern mixture models and latent class models for the analysis of multivariate longitudinal data with informative dropouts. *The International Journal of Biostatistics*, 4(1):1–26, 2008.

[10] A. Dempster, N. Laird, and D. Rubin. Maximum likelihood from incomplete data via the EM algorithm. *Journal of the Royal Statistical Society. Series B (methodological)*, pages 1–38, 1977.

[11] C. Faucett and D. Thomas. Simultaneously modelling censored survival data and repeatedly measured covariates: a Gibbs sampling approach. *Statistics in Medicine*, 15(15):1663–1685, 1996.

[12] L. Ferrer, J. Dignam, V. Rondeau, T. Pickles, and C. Proust-Lima. Joint modelling of longitudinal and multi-state processes: application to clinical progressions in prostate cancer. *Statistics in Medicine*, 35(22):3933–3948, 2016.

[13] J. Han, E. Slate, and E. Peña. Parametric latent class joint model for a longitudinal biomarker and recurrent events. *Statistics in Medicine*, 26(29):5285–5302, 2007.

[14] G. Hanks, T. Pajak, A. Porter, D. Grignon, H. Brereton, V. Venkatesan, E. Horwitz, C. Lawton, S. Rosenthal, H. Sandler, et al. Phase III trial of long-term adjuvant androgen deprivation after neoadjuvant hormonal cytoreduction and radiotherapy in locally advanced carcinoma of the prostate: the radiation therapy oncology group protocol 92–02. *Journal of Clinical Oncology*, 21(21):3972–3978, 2003.

[15] R. Henderson, P. Diggle, and A. Dobson. Joint modelling of longitudinal measurements and event time data. *Biostatistics*, 1(4):465–480, 2000.

[16] W. Hu, G. Li, and N. Li. A Bayesian approach to joint analysis of longitudinal measurements and competing risks fialure time data. *Statistics in Medicine*, 28:1601–1619, 2009.

[17] H. Jacqmin-Gadda and D. Commenges. Tests of homogeneity for generalized linear models. *Journal of the American Statistical Association*, 90(432):1237–1246, 1995.

[18] H. Jacqmin-Gadda, C. Proust-Lima, J. M. G. Taylor, and D. Commenges. Score test for conditional independence between longitudinal outcome and time to event given the classes in the joint latent class model. *Biometrics*, 66(1):11–19, 2010.

[19] J. Kalbfleisch and R. Prentice. *The statistical analysis of failure time data*, volume 360. 2011.

[20] N. Li, R. Elashoff, G. Li, and C. Tseng. Joint analysis of bivariate longitudinal ordinal outcomes and competing risks survival times with nonparametric distributions for random effects. *Statistics in Medicine*, 31(16):1707–1721, 2012.

[21] H. Lin, C. McCulloch, and S. Mayne. Maximum likelihood estimation in the joint analysis of time-to-event and multiple longitudinal variables. *Statistics in Medicine*, 21(16):2369–2382, 2002.

[22] H. Lin, B. W. Turnbull, C. E. McCulloch, and E. H. Slate. Latent class models for joint analysis of longitudinal biomarker and event process data: application to longitudinal prostate-specific antigen readings and prostate cancer. *Journal of the American Statistical Association*, 97:53–65, 2002.

[23] R.J.A. Little and D.B. Rubin. *Statistical Analysis with Missing Data*. Wiley, New-York, 1987.

[24] L. Liu and X. Huang. Joint analysis of correlated repeated measures and recurrent events processes in the presence of death, with application to a study on acquired immune deficiency syndrome. *Journal of the Royal Statistical Society: Series C (Applied Statistics)*, 58(1):65–81, 2009.

[25] D. Marquardt. An algorithm for least-squares estimation of nonlinear parameters. *Journal of the society for Industrial and Applied Mathematics*, 11(2):431–441, 1963.

[26] P. Philipson, I. Sousa, P. Diggle, P. Williamson, R. Kolamunnage-Dona, R. Henderson, and G. Hickey. joiner: Joint modelling of repeated measurements and time-to-event data. *R package version 1.2.2,https://github.com/graemeleehickey/joineR/*, 2017.

[27] R. Prentice. Covariate measurement errors and parameter estimation in a failure time regression model. *Biometrika*, 69:331–42, 1982.

[28] C. Proust-Lima, H. Amieva, and H. Jacqmin-Gadda. Analysis of multivariate mixed longitudinal data: a flexible latent process approach. *British Journal of Mathematical and Statistical Psychology*, 66(3):470–487, 2013.

[29] C. Proust-Lima, J.F. Dartigues, and H. Jacqmin-Gadda. Joint modeling of repeated multivariate cognitive measures and competing risks of dementia and death: a latent process and latent class approach. *Statistics in Medicine*, 35(3):382–398, 2016.

[30] C. Proust-Lima, V. Philipps, and B. Liquet. Estimation of extended mixed models using latent classes and latent processes: The R package lcmm. *Journal of Statistical Software*, 78(2), 2017.

[31] C. Proust-Lima, M. Sene, J.M.G. Taylor, and H. Jacqmin-Gadda. Joint latent class models for longitudinal and time-to-event data: A review. *Statistical Methods in Medical Research*, 23(1):74–90, 2014.

[32] D. Renard, H. Geys, G. Molenberghs, T. Burzykowski, M. Buyse, T. Vangeneugden, and L. Bijnens. Validation of a longitudinally measured surrogate marker for a time-to-event endpoint. *Journal of Applied Statistics*, 30(2):235–247, 2003.

[33] D. Rizopoulos. JM: An R package for the joint modelling of longitudinal and time-to-event data. *Journal of Statistical Software*, 35(9):1–33, 2010.

[34] D. Rizopoulos. Fast fitting of joint models for longitudinal and event time data using a pseudo-adaptive gaussian quadrature rule. *Computational Statistics & Data Analysis*, 56(3):491–501, 2012.

[35] D. Rizopoulos. *Joint models for longitudinal and time-to-event data: With applications in R*. CRC Press, 2012.

[36] D. Rizopoulos. The R package JMbayes for fitting joint models for longitudinal and time-to-event data using MCMC. *Journal of Statistical Software*, 72(7):1–45, 2016.

[37] A. Rouanet, P. Joly, J.F. Dartigues, C. Proust-Lima, and H. Jacqmin-Gadda. Joint latent class model for longitudinal data and interval-censored semi-competing events: Application to dementia. *Biometrics*, 72(4):1123–1135, 2016.

[38] A.A. Tsiatis and M. Davidian. Joint modeling of longitudinal and time-to-event data: an overview. *Statistica Sinica*, 14(3):809–834, 2004.

[39] A.A. Tsiatis, V. Degruttola, and M.S. Wulfsohn. Modeling the relationship of survival to longitudinal data measured with error. Applications to survival and CD4 counts in patients with AIDS. *Journal of the American Statistical Association*, 90(429):27–37, 1995.

[40] Y. Wang and J.M.G Taylor. Jointly modeling longitudinal and event time data with application to acquired immunodeficiency syndrome. *Journal of the American Statistical Association*, 96(455):895–905, 2001.

[41] P.R. Williamson, R. Kolamunnage-Dona, P. Philipson, and A.G. Marson. Joint modelling of longitudinal and competing risks data. *Statistics in Medicine*, 27(30):6426–6438, 2008.

[42] M. Wulfsohn and A.A. Tsiatis. A joint model for survival and longitudinal data measured with error. *Biometrics*, 53(1):330–339, 1997.

[43] J. Xu and S.L. Zeger. The evaluation of multiple surrogate endpoints. *Biometrics*, 57(1):81–87, 2001.

[44] W. Xu and D. Hedeker. A Random-Effects Mixture Model for Classifying Treatment Response in Longitudinal Clinical Trials. *Journal of Biopharmaceutical Statistics*, 11(4):253–273, 2001.

[45] D. Zhang, M.H. Chen, J. Ibrahim, M. Boye, and W. Shen. Jmfit: a SAS macro for joint models of longitudinal and survival data. *Journal of Statistical Software*, 71(3), 2016.

Part VI

Miscellaneous Topics in Randomized Controlled Trials

22

Design and Analysis Methods for Developing Personalized Treatment Rules

Emily Butler and Michael Kosorok

CONTENTS

22.1   Introduction

Precision or personalized medicine is the practice of tailoring healthcare plans to patients as individualized as possible. In an ideal setting caregivers are able to account for patient history, genetic information, response to previous treatments, and other relevant factors when making treatment decisions throughout the course of the disease. These treatments would be assigned based on how each patient is progressing at each stage and which sequence of treatments is best earns the optimal expected long term response. These treatment plans have been coined dynamic treatment regimes (DTRs) and estimating DTRs is the main goal of precision medicine for patients with a chronic disease. These treatment plans are unique in the sense that they are not predetermined at the initial doctor's visit as a general treatment plan for all patients but are determined at each follow up visit, which makes the treatment plan dynamic through time. The evidence-based treatment decisions are determined by the patient's response to the course of treatment and uses prognostic and treatment information to define their history (5). DTRs have four components: decision points (time points at which the decisions are made), tailoring variables (patient's prognostic information used to make treatment decisions), intervention components (the type or dose/intensity/duration of the treatment), and decision rules (a function that links the tailoring variables to the intervention options at decision points) (6).

In the statistical framework, DTRs are developed to create mathematically dependent treatment plans that mimic how clinicians treat patients who need chronic medical treatment. Initial interest in this method of treating patients began in the mid-1980s when researchers were evaluating two stage dynamic treatment strategies for cancer patients; in the 1990s, the idea of a two stage customized treatment plan expanded when psychiatrists were interested in expanding to k-stage treatment plans; and by the early 2000s, clinical trials and treatment strategies were implemented by doctors studying DTRs in substance abuse and mental health research (13). Once the investigator has chosen a selection of potential DTRs, they need to be evaluated in a special clinical trial called a Sequential Multiple Assignment Randomization Trials (SMART) (19). The development of SMARTs began when traditional clinical trial designers were looking to identify an intermediate outcome between randomization and the primary outcome and make an appropriate reaction to this outcome. The treatment plan can be adjusted based on the patient's progress or just carry-over the patient's to a new set of treatments at the following stage.

It is important to highlight the distinction between DTRs and SMARTs. A DTR is a treatment strategy that tells the caregiver which adaptive, sequential treatment plan will most likely lead to the highest probability of success. A SMART design is an experimental trial with the purpose of determining which DTRs are optimal for which patients. A SMART is able to evaluate the various DTRs because it compares them by randomizing patients to a set preselected treatments at each phase. A SMART can have two goals: compare a small number of pre-specified DTRs embedded in the SMART design or construct new DTRs which may not be naturally embedded in the design.

The material presented in this literature review is meant to be a survey of current statistical methods used to estimate DTRS in the context of using data collected in a SMART trial. Kosorok and Moodie (10) published a book titled Adaptive Treatment Strategies in Practice: Planning Trials and Analyzing Data for Personalized Medicine which covers this and similar topics in more detail. In this book, the authors discuss select topics such as SMART designs, observational studies, reinforcement learning and various methods of estimating DTRs. Please refer to this books for an in depth description of a number of useful methods for DTRs.

This review covers a brief introduction on how to design a SMART and what statistical methodologies exist to make inference from this data, with particular emphasis on recent developments in the area. Section 2 will highlight how a SMART design fits into the overall development of a treatment strategy, the importance of pilot studies, practical considerations such as sample size, calculations, and handling missing data, as well as the future of data collection. Section 3 introduces methodological results for the single stage paradigm and Section 4 provides methodological results for the multiple stage paradigm. Section 5 highlights related topics, such as variable selection and developing DTRs for when there are multiple outcomes and observational data.

22.2   Study Design

A SMART is a clinical trial design that helps determine which sequence of treatments is best for which type of patients. They investigate the best sequencing of intervention components, what tailoring variables should be used, when and how frequently should these tailoring variables be assessed, and should one treatment be assigned or should patients have the ability to choose from a list of options. This design is novel in that the treatments are assessed as an entire treatment sequence and not isolated by each individual phase of the trial. This involves multiple randomizations over time where each randomization corresponds to a treatment decision. Additionally, this design can distinguish between responders and non-responders by assigning rescue or maintenance therapies (5; 6). Pragmatically, another benefit of this design is that SMARTs require a smaller sample size than their randomized clinical trial counterparts because of their efficient use of subjects throughout stages. Allowing for a rescue therapy for patients experiencing extreme adverse events will deter dropout as well as better capture their response profile overtime.

Consider a generic example for illustration. A 2 stage SMART enrolls 200 participants with equal randomization. It assigns 100 patients to treatment A and 100 patients to treatment B at stage 1. At the end of stage 1, the patient's status is assessed and the they are classified as either responsive or non-responsive, where what defines responsive or non-responsive should be determined a priori by the investigator. Each of the patients is then re-randomized according to their response classification. For instance, patients who responded to treatment A are assigned to stay on treatment A at stage 2, while the patients who did not respond to treatment A are randomized to treatment C or D. A similar situation could arise for treatment B where responders stay on treatment B and non-responders are randomized to treatments E or F. In this scenario, there are 6 DTRs: {A, A}, {A, C}, {A, D} {B, B}, {B, E}, {B, F}. Alternatively, responsive patients could also be re-randomized. There could be more than 2 treatment options at each stage, or the non-responsive patients could switch from A to B or vice versa. The randomization also does not need to be balanced, and the stages could be extended beyond only 2. In this example data would be collected at 3 time points: baseline, time 1 (after the first stage of the study but before the patient is re-randomized to the second treatment), and time 2 (after the second stage or at the end of the study). Generally speaking, randomization does not need to depend on responder status, although this is the case in this example.

There are numerous reasons it is advantageous to use a SMART. For example, it has the desirable quality of making better use of the pre-determined sample size and can answer more clinical questions than a RCT. In a two phase SMART, similar to RCT, the first phase of the trial can be assessed by comparing the mean outcomes between the first two lines of treatment and the second phase can compare the effect of the treatment options for responders and non-responders, regardless of their first line of treatment. This increases the power of the hypothesis tests, since they recycle the patients in the second phase of the study. In addition to hypothesis testing as in RCT, SMARTs can also be used to generate hypotheses. Most importantly, the design then has the ability to compare the embedded treatment regimes that pool information across multiple experimental conditions by, again, recycling patients through the trial (6). While the main goal of a SMART is to mine data used to develop DTRs, they have other uses, such as discovering which treatments work best sequentially to obtain an improved outcome, investigating the interplay between trajectories of the patients disease progression and treatment sequences, comparing different treatment sequences, and investigating the benefit of both prognostic information and clinical data in determining individualized treatments (1).

As with most experimental designs, it is important to begin with an information gathering pilot study. A good pilot study is a great practice for implementing a larger design, gives critical value estimates, is important for sample size estimation, and provides a preliminary look at the utility of the proposed treatments. The novelty of SMARTs raises feasibility concerns which makes a pilot study even more crucial for designing an effective and efficient SMART. Almirall et al. (1) highlights the following important topics that researchers must be aware of when designing a SMART and how solutions can be elicited through a pilot study: (1) determining primary outcome. This will be used to assess response status (if applicable), when treatment decisions should be made, what criterion is used to make the decisions, frequency of assessment, how sensitive this measure is, justification of its use, and feasible application in clinical treatment; (2) deciding which tailoring variables should be collected, such as baseline characteristics or time varying measures; (3) deciding how to control for missing data. This should be guided by how it would be handled in clinical practice; (4) deciding between up-front randomization (randomization at the beginning of the trial) or real-time randomization (randomization at each decision point, which allows for clinical information to be used in randomization); (5) highlighting the difference between research assessments for data analysis to develop adaptive treatment strategies and assessments of the adaptive treatment strategies used to inform the sequential treatment assessment; (6) identifying concerns clinicians have regarding sequences of treatments offered and assessment of what determines response versus nonresponse; (7) assessing for patient acceptability; (8) testing the language of consent forms; and (9) illuminating unanticipated tailoring variables that would be useful in the subsequent SMART.

The pilot study provides crucial information to increase the probability of success of a SMART. As previously stated, one goal of a SMART is to identify embedded DTRs not embedded in the original design. A great example of this is the analysis of SMART data collected from a study for the treatment of advanced prostate cancer. Wang et al. (26) created a new method to compare dynamic treatment regimes and along the way, changed the definition of the DTRs after the trial ended. This analysis is different than previous analyses using the same data because the authors changed the definition of viable DTRs based on what had been predetermined as a "missing observation." The protocol required that patients were re-randomized to a new treatment and if they did not complete it, they were classified as missing. However, the treatment plan determined by the protocol was not feasible for patients with toxicity or disease progression. They altered one of the DTRs to include those patients who had to leave the study because of toxicity and specified that the second treatment was the recovery treatment they were given after leaving the trial. The viable DTRs were now defined by efficacy, toxicity and disease progression. The authors also redefined this endpoint because it needed to quantify the health experience of the patient over a pre-specified fixed period, not just their final tumor size or toxicity.

While there is still methodological work to be done or expanded upon, which we will see later, there is still research to be done at the design phase. One clear gap in the literature is a universal sample size formula. While there are sample size calculation publications for SMART designs, but there is no universally accepted calculation for general use. One of these developments is the development of upper bound sample size estimates for censored data. Unfortunately, this sample size formula does not have great generalizability properties because the upper bounds are based on a Kaplan-Meier estimate and the log-rank statistic. Li and Murphy (14) developed a sample size formula for a two stage SMART for failure time outcomes. The difficulty in such a calculation stems from the variances of the common test statistics. These test statistics depend on the joint distribution of the time, early response determination, and the primary failure time, which are likely to be dependent. The sample size is derived using upper bounds on the variances in place of the usual variances and, hence, the resulting formula only requires the same assumptions of a traditional single stage randomized clinical trial. Using the upper bounds of the variances, the proposed samples size formulas for the Kaplan-Meier estimate (n_K) and the log rank statistic (n_L) are

$$n_K \leq \frac{(Z_{1-\alpha/2} + Z_{1-\beta})^2 \sigma_B^2}{\left\{\bar{F}_1(\tau) - \bar{F}_2(\tau)\right\}^2}$$

$$n_L \leq \left(\frac{1}{pq} + \frac{1}{(1-p)q}\right) \frac{(Z_{1-\alpha/2} + Z_{1-\beta})^2}{\xi^2 \int_0^t \bar{F}_c(t) \mathrm{d}F_1(t)}$$

where Z is the z-score for a standard normal distribution, α is the type I error, β is the type II error, $\bar{F}$ is the survival function, τ is the time at the end of the study, $p = p(A_1 = 1)$ and $q = p(A_2 = 1 | R = 1)$ are the randomization probabilities, R is the indicator of randomization, and ξ is the log hazard ratio. We define

$$\sigma_B^2 = \frac{\bar{F}_1^2(\tau)}{pq} \int_0^\tau \frac{d\Lambda_1(t)}{\bar{F}_1(t)\bar{F}_c(t)} + \frac{\bar{F}_2^2(\tau)}{(1-p)q} \int_0^\tau \frac{d\Lambda_2(t)}{\bar{F}_2(t)\bar{F}_c(t)}$$

where Λ is the cumulative hazard function. This sample size calculation has been proven to provide the desired power if the hazards of the alternative are proportional. This sample size calculation is most notable because the nature of chronic diseases allows the outcome, or surrogate outcome, to be thought of in terms of failure, even if death is not the primary endpoint. This means that this sample size formula will be applicable in many settings.

Another important research topic is developing strategies that prevent, and methodology that controls, for missing data. The construct of a SMART requires numerous randomizations and multiple treatment prescriptions which presents unique challenges when analyzing incomplete data. Imputation strategies for handling missing data collected from SMARTs is an understudied area at this time. Shortreed et al. (24) presented the following five missing data issues: (1) transition between treatment stages does not always occur at pre-specified times, but instead can be determined by a patient outcome; (2) some outcome variables are irregularly spaced while some tailoring variables are collected at regularly scheduled study visits; (3) observing some tailoring variables is dependent on a patient's history, which results in structural missingness for the data-dependent portion of the observed data. (4) individuals are simply lost to follow up leaving the treatment stage; and (5) some individuals are lost to follow up entering the treatment stage. Their proposed solution is a flexible imputation strategy that is a time ordered, nested, conditional imputation method which exploits the nearly monotone pattern of missing data found in this type of longitudinal study. It ensures that a complete multivariate prediction distribution exists while obtaining desirable traits for inference across longitudinal outcomes. Assuming missingness at random, this method works best when the data is imputed with a pseudo-Gibbs sampler, which applies repeated iterations through the model. Multiple imputation is one of many strategies used when working with missing data, and the type of strategy often depends on the structure of the data and the nature of the missingness. This method was not compared to other imputation strategies such as inverse probability weighting or likelihood methods, and there is no contingency plan when the missingness is not monotone. It is clear the presented work is exciting and promising progress, but more headway is still needed.

An exciting area of expansion is data collection and treatment allocation using mobile technology. There is strong interest is the ability to increase access to fast, accurate care through mobile technologies that include cell phones, sensors and monitors. The goal is development of evidence based Just in Time Adaptive Interventions (JITAIs) that collects real time data from patients and uses that data to inform the real time delivery of intervention options, such as treatment, dose, and timing of care (20). For example, trying to intervene in heavy drinking and smoking, a mobile phone would be administered and participants would be prompted 3 times a day to assess their smoking urge, affect, and drinking behaviors. Urge management interventions would be delivered only if the individual reports the urge to smoke at a specific time. Anytime during the day the user can text either lapse or crave, and a series of encouraging text messages will be sent back to their cell phone. Another example is managing eating disorders. When treating college women with eating disorders, the subject would be provided with a cell phone which receives 5 prompts regarding mood, eating behaviors, exposure, etc. When she reports what is considered a negative mood she is recommended to use one of the treatments provided via a CD. In all instances, the interventions are adapted and delivered through a mobile medium such that patient information can be obtained at any time and responses can be administered at any time. The variety of potential interventions includes reach out interventions, behavioral strategies, cognitive strategies, and goal setting. Tailoring variables can be collected actively (self-reported via prompting or user initiated) or passively (activity level, location, social media activities, number of ignored recommended interventions, etc.). The decision points vary depending on the goal of the treatment plan and can include a random prompt, user requested help, or indication of specific experiences. The decision rules can be deterministic (if the patient reports more than X then give them this, otherwise give them that) or stochastic (determining the probability of an intervention). The corresponding thresholds can be determined and optimized using reinforcement learning (which will be covered in later sections). Even though this concept is in the early stages of development, there are obvious feasibility issues for this kind of treatment implementation, such as cost and monitoring adherence. Innovative methods like this have the ability to change the way patients are treated and can serve as a guide for future treatment and collection methods.

A SMART is an innovative approach to efficiently collect data which accurately estimates DTRs for specific types of patients. The basic design structure has been created, trials are ongoing in the clinical setting and new advances are being developed day by day, but more work needs to be done. The flexibility of the design makes developing broad techniques difficult, but the need and the talent is there to continue to make advancements in what has become an extremely timely, interesting and practical trial design.

22.3 Analysis Techniques: Single Stage

The list of methodology presented here and in the subsequent section is neither complete nor representative of all available methods, but a simple summary of recent methods employed across a broad range. The purpose is to introduce popular techniques, highlight advancements and display a plethora of methodological options applicable in multiple areas of interest.

Important notation must be introduced so that an individualized treatment rule (ITR) for the single stage paradigm can be properly defined. An ITR differs from a DTR in that it is the personalized rule for a single treatment setting while a DTR is the sequence of decision rules for a multiple treatment setting. Assuming the data is collected from a single stage two arm trial, the treatments will be annotated as $A \in \{-1, 1\}$. These are independent of the patient prognostic variables denoted as $\boldsymbol{X} = (X_1, \ldots, X_p)^T$, where $\boldsymbol{X}$ is a p-dimensional matrix. The observed clinical outcome, Y, can be considered a reward function where larger values are desired. The ITR is a map from the prognostic variable space, $\boldsymbol{X}$, to the treatment space, A, and the optimal ITR is the A which maximizes the expected reward. The distribution of $(\boldsymbol{X}, A, Y)$ is denoted by P with the respective expectation denoted as E. The distribution of $(\boldsymbol{X}, A, Y)$ given the ITR, D (i.e. that $A = D(\boldsymbol{X})$), is denoted as P^D and the corresponding expectation as E^D. The expected reward under D is

$$V(D) = E^D(Y) = E\left[\frac{I\{A = D(\boldsymbol{X})\}}{A\pi + \frac{1-A}{2}}Y\right]$$

where $\pi = P(A = 1)$. This $V(D)$ is referred to as the value function for a given D. The optimal ITR, denoted D^*, is estimated as:

$$D^* \in argmax_D V(D) = argmax_D E\left[\frac{I\{A = D(\boldsymbol{X})\}}{A\pi + \frac{1-A}{2}}Y\right]$$

and is considered the treatment rule, D, which maximizes the value function $V(D)$. The optimal treatment regime is defined as the one that maximizes the average expected outcome (32).

One way to estimate ITRs is to restructure the estimation procedure into a classification problem where the optimal classifier corresponds to the optimal treatment decision. The optimal classifier can be found by estimating the Bayes classifier, which is the one that minimizes the expected weighted misclassification error. This framework allows for estimation of mean outcomes under existing methods such as regression estimation, inverse probability weighted estimation (IPWE) or augmented inverse probability weighted estimation (AIPWE) (29). The class of treatment decisions is data driven because it is chosen by minimizing the L1 the expected weighted misclassification error and does not need to be pre-specified.

Define the contrast function as

$$C(\boldsymbol{X}) = \mu(1, \boldsymbol{X}) - \mu(-1, \boldsymbol{X})$$

which can be thought of as the mean difference between treatment options for a given set of prognostic variables. The optimal ITR estimation problem can be transformed into a weighted classification problem such that the optimal treatment rule D^* is found by

$$D^* = argmax_D E\left[D(\boldsymbol{X})C(\boldsymbol{X})\right] = argmax_D E\left(|C(\boldsymbol{X})|\left[I\left\{C(\boldsymbol{X}) > 0\right\} - D(\boldsymbol{X})\right]^2\right)$$

This means that the optimal treatment rule, D^*, is found to be the one that maximizes $E(|C(\boldsymbol{X})|[I\{C(\boldsymbol{X}) > 0\} - D(\boldsymbol{X})]^2)$, which is a weighted classification problem. Each subject belongs to two classes such that class $Z = 1$ contains those subjects who would benefit more from treatment $A = 1$ as opposed to treatment $A = -1$, e.g. $\mu(1, \boldsymbol{X}) > \mu(-1, \boldsymbol{X})$, and $Z = 0$ the opposite. Each observation is also given a weight, $W = |C(\boldsymbol{X})|$, which is the loss that would incur from misclassification. Hence, the optimal ITR is the expected weighted misclassification error under the classification rule $D(\boldsymbol{X})$. Within this classification construct, the problem then decomposes into two critical steps. First, one must construct a suitable estimator of the contrast function using regression, and then invert this to find the estimated optimal treatment rules with an interpretable form using classification methods. This can be extended to the multiple stage scenario as well. This classification prospective falls under the machine learning umbrella. Machine learning, most specifically reinforcement learning, has recently been implemented since it sidesteps the problem of completely modeling the underlying generation model as is necessary in some estimation techniques. Reinforcement learning is a dynamic programming system that dictates which actions are recommended to optimize the expectation of a given reward.

Qian and Murphy (21) propose a modification of this which first estimates the conditional mean response using L_1 penalized least squares (L_1-PLS) with a rich linear model and then uses that to derive the estimated treatment rule. If the conditional mean is modeled correctly, this method consistently estimates the optimal treatment rule. The finite sample upper bounds of the difference between the mean response from the optimal treatment rule and the mean response from the estimated treatment rule holds even if the linear model for the conditional mean response is incorrect. If the part of the conditional mean model involving the treatment effect is correct then the upper bounds imply that the estimated treatment rule is consistent. These upper bounds can also inform how to choose the tuning parameters involved in the L_1-penalty to create the best rate of convergence. To obtain the ITR the estimated prediction error is minimized then the conditional mean model is maximized over the treatment A. To control for overfitting, L_1-PLS is implemented since the L_1 penalty innately does variable selection. The resulting treatment rules are cheaper to implement and easier to interpret.

The forgoing methods are considered indirect methods of estimation. Indirect estimation refers to techniques that first estimate a quantity reflecting the conditional distribution of the outcome, such as conditional mean, and then uses the resulting model to deduce the optimal ITR (12). Indirect estimation can be desirable because the initial estimation regarding the outcome can be built using traditional statistical modeling techniques. Unfortunately, optimal ITR estimation requires that the conditional outcome be modeled correctly. Indirect estimation methods often and easily experience model misspecification because of the difficulty of modeling high dimensional, time dependent factors. In high dimensional situations the two-step procedure of estimation and maximization equations can be poor fits. In contrast, direct methods of estimation are solutions to the problems proposed by these other techniques which directly achieves this maximization without requiring the initial estimation step be done with indirect approaches. This direct class of methods immediately estimates the value function for all pre-specified treatment rules and then obtains the optimal treatment rule by maximizing the estimator. Direct estimation methods tend to produce treatment regime estimates that are more precise than indirect methods in the single stage setting due to the associated reduced bias (32).

Zhang et al. (30) approaches estimating dynamic treatment rules by assuming a posited regression model. This defines the class of treatment rules while recognizing that it is possible for the model to be misspecified. The optimal treatment regime is estimated by directly maximizing the estimator for the overall population mean outcome under all possible specified treatment plans using a suitable inverse probability weighted estimator. When using observational data, this estimator has the ability to control for possible confounders by estimating propensity scores and exploiting the predicted outcome, which ensures precision of the estimate. Let D^* be the optimal treatment decision, which is the one that corresponds to the largest value of $E[Y^*(D)]$, where

$$Y^*(D) = Y^*(1)D(\boldsymbol{X}) + Y^*(-1)\left\{1 - D(\boldsymbol{X})\right\}$$

is the potential outcome. The potential outcome is the outcome that would be observed if a randomly chosen patient were to receive treatment regime D. Consider treatment rules of the form $D_\eta(\boldsymbol{X}) = D(\boldsymbol{X}, \eta)$ in the class of all possible treatment rules which is indexed by η and will contain D^* if $\mu(A, \boldsymbol{X}; \beta)$ the posited regression model is correctly specified. Therefore, estimating $\eta^* = argmax_\eta E[Y^*(D_\eta)]$ and defining $D_\eta^* = D(\boldsymbol{X}, \eta^*)$ will provide an estimator for D^*. To estimate $E[Y^*(D_\eta)]$, an IPWE or a doubly robust AIPWE can be employed. This estimator is directly maximized in η to obtain an η^* and hence $\widehat{D}_\eta^*(\boldsymbol{X}) = D(\boldsymbol{X}, \widehat{\eta}^*)$. This can easily be extended to the multiple decision situation by estimating $Q(\eta) = E[Y^*(D_\eta)]$ as a function of η.

Direct methods of estimation can also be restructured into a classification problem which can utilize computer science techniques. This looks at the data by comparing the difference between subjects with observed high and low rewards so that the determination of the actual treatment decisions is associated with the actual treatments received for the different groups. This method is referred to as outcome weighted learning (OWL or O-learning). Developed by Zhao et al. (32), O-learning is a nonparametric approach which directly optimizes the value function, $V(D)$, where each subject is weighted proportional to their clinical outcome divided by the propensity score, which is the probability of receiving the assigned treatment given the covariates. In the case of a clinical trial, the propensity simplifies to the constant probably of receiving the assigned treatment. Finding the D^* that maximizes $V(D) = E[\frac{I\{A=D(\boldsymbol{X})\}}{A\pi + \frac{1-A}{2}}Y]$ is equivalent to finding the D^* that minimizes $\bar{V}(D) = E\left[\frac{I\{A \neq D(\boldsymbol{X})\}}{A\pi + \frac{1-A}{2}}Y\right]$ which sets the stage to view this as a weighted classification error. Minimizing the previous expected value can be approximated by minimizing

$$n^{-1} \sum_{i=1}^{n} \frac{Y_i}{A_i \pi + \frac{1-A_i}{2}} I\left[A_i \neq sign\{f(\boldsymbol{X}_i)\}\right]$$

to find the optimal f^* and then setting

$$D^*(\boldsymbol{X}) = sign\{f^*(\boldsymbol{X})\}$$

since $D^*(\boldsymbol{X})$ can always be represented as $sign\{f^*(\boldsymbol{X})\}$. This implies the goal is to find a decision rule which chooses treatments based on their specific prognostic variables. On average, patients with large rewards will be recommended the same treatment that they actually received while patients with small rewards will receive the opposite. This is considered 0-1 loss in the machine learning scenario and is difficult to minimize due to non-convexity and discontinuity. This problem is alleviated by transforming the problem, using a surrogate for the 0-1 loss, so that the goal becomes minimizing

$$n^{-1} \sum_{i=1}^{n} \frac{Y_i}{A_i \pi + \frac{1-A_i}{2}} \{1 - A_i f(\boldsymbol{X})\}^+ + \lambda ||f||^2$$

where $\boldsymbol{X}^+ = max(\boldsymbol{X}, 0)$, and $||f||$ is the norm of f. Therefore, this problem is now a weighted classification problem that can be solved using support vector machine methods.

O-learning has many applications but there are also many ways to extend this line of thinking. Chen et al. (4) presented a one stage clinical trial design for penalized dose finding using a robust analysis method based on the O-learning framework. The method converts the individualized dose selection problem into a penalized weighted regression with truncated L_1 loss. The dose level is assumed to be found on a continuum and a non-trivial extension of O-learning for binary treatments is proposed. The dose finding problem becomes a weighted regression with random outcome where the individual responses are the weights. In the linear case, this framework has the goal of minimizing the loss plus penalty of the form

$$\min_{f} \left\{ \frac{1}{n} \sum_{i=1}^{n} \frac{R_i l_\phi \{A_i - f(\boldsymbol{X}_i)\}}{2\phi_n p(A_i | \boldsymbol{X}_i)} + \lambda_n ||f||^2 \right\}$$

where $\phi = \phi_n$ is non-random parameter in real space, λ_n controls the severity of the penalty on f, $l_\phi \{A_i - f(\boldsymbol{X}_i)\} = \min(\frac{|A - f(\boldsymbol{X})|}{\phi}, 1)$, $R^*(a)$ is the potential outcome and $R = \int I(A = a) R^*(a) p(a|x) da$. The complexity of $f(x)$ is penalized to prevent overfitting. This function is nonconvex and hence difficult to optimize, so an adaptive difference convex (DC) algorithm is implemented (25). Considering a linear loss function, the objective function is

$$S = \frac{\lambda_n}{2} ||w||_2^2 + \frac{1}{\phi_n} \sum_{i=1}^{n} R_i \min\left(\frac{|A_i = D(\boldsymbol{X}_i)|}{\phi_n}, 1 \right)$$

where λ_n is now the tuning parameter. This algorithm minimizes the sequence of convex sub-problems with the intent of solving the original non-convex minimization problem. Therefore, the convex sub-problem becomes a weighted penalized median regression problem. Ultimately, the algorithm concludes when $||w^{t+1} - w^t||$ is smaller than some prespecified constant, where $w = \sum_{i \in T} (\alpha_i - \bar{\alpha}_i) x_i$. Expanding to the nonlinear framework, the decision function then becomes a function of w and some unknown transformation on $\boldsymbol{X}$. A Gaussian kernel is used to construct a dual problem for nonlinear learning that is solved using quadratic programming. To practically implement this procedure, a nonconvex loss function and a DC algorithm for optimization is employed.

Another common and extremely relevant application of O-learning is estimating ITRs for censored data. Realistically many chronic diseases measure short term success of a treatment as a failure or success. It is desirable to develop methods of estimating treatment regimes that are applicable to survival analysis because it has clear relevance to personalized medicine. When considering censored data, notation is slightly altered. The value function is redefined as

$$V(D) = E^D(T) = E[T|\boldsymbol{X}, A = D(\boldsymbol{X})] = E\left[\frac{I\{A = D(\boldsymbol{X})\}}{A\pi + \frac{1-A}{2}}T\right]$$

where $T = min(\tau, \tilde{T})$ where $\tilde{T}$ is the survival time and τ is the end of the study (34). Even though the outcome is redefined, the optimal treatment rule is still the treatment rule which maximizes the value function. The goal is to estimate D^* using censored data following the OWL framework. There are two approaches to estimation. First, one can maximize the estimator of the average survival time. To account for right censoring, the estimated mean survival time is reassigned as the weighted misclassification rate. These weights are comprised of both the observed outcome and the inverse probability of censored weights. To offset bias from a misspecified censoring model, a second method, a doubly robust variation of outcome weighted learning, is formulated. In both instances, the treatment rule is consistent for the optimal rule when the model for either the survival times or censoring times is correctly specified. Note: it is not required that both models be correctly specified. A convex relaxation idea from support vector machines is invoked for construction of the necessary estimation algorithm.

The methodological techniques available for estimating ITRs in the single stage scenario encompass a broad spectrum. Some of these techniques have been extended to apply to the estimation of DTRs but not all have, making this an important area of future work. Because of the nature of sequential decision making, some of the associated techniques cannot easily be extended beyond the single stage setting, so it is important to continue making progress in both areas.

22.4   Analysis Techniques: Multiple Stages

There is a lot of interest and value in creating techniques which accurately estimate optimal DTRs because the multiple stage scenario most similarly mimics the natural course of a chronic disease. Considering patients often need multiple treatments, individuals respond differently to different treatments at different points in their progression and the longevity of the disease can be unknown, these techniques are important for an adequate treatment plan. It is important to note that each method requires assumptions on the design structure, which may or may not be flexible for each method. Design considerations should be made in line with the chosen analysis technique to ensure all assumptions are met.

In order to properly define DTRs in this setting, notation will be presented that expands on that which is used in the single stage paradigm. Consider a trial with T decision points. For $t = 1, \ldots, T$, let $A_t \in \{-1, 1\}$ be the dichotomous treatment assignment at the tth stage and $\boldsymbol{X}_t$ be the patients' prognostic variables before the tth decision point but after the A_{t-1} treatment assignment. The outcome, or reward, at the tth stage is Y_t where larger values are assumed more desirable. Y_t is assumed to depend on all previous prognostic information $(\boldsymbol{X}_1, \ldots, \boldsymbol{X}_t)$, all treatment history $(A_1, \ldots, A_t)$ and previous outcomes $(Y_1, \ldots, Y_{t-1})$. The overall outcome of interest is the total reward $\sum_{t=0}^{T} Y_t$. The DTR is a set of sequential decision rules $D = (D_1, \ldots, D_t)$ which is a map from total patient history, $\boldsymbol{H}_t = (\boldsymbol{X}_1, A_1, \ldots, A_{t-1}, \boldsymbol{X}_t)$, to the treatment space. The value function is then defined as

$$V(D) = E^D \left[\sum_{t=1}^{T} Y_t \right]$$

where E^D is the expectation under the measure P^D which is the probiility distribution for distribution of $(\boldsymbol{X}_1, A_1, Y_1, \ldots, \boldsymbol{X}_T, A_T, Y_T, \boldsymbol{X}_{T+1})$. The value function is the expected long term benefit if the population were to follow regimen D and can also be defined as

$$V(D) = E^D \left[\frac{\left(\sum_{t=1}^{T} Y_t \right) \prod_{t=1}^{T} I\{A_t = D_t(\boldsymbol{H}_t)\}}{\prod_{t=1}^{T} \pi_t(A_t, \boldsymbol{H}_t)} \right].$$

Similar to the single stage situation, the value that maximizes the value function $V(D)$

$$D^* \in argmax_D V(D)$$

is the optimal DTR D^* (33).

One of the most popular indirect methods of estimation is a computer science method called Q-learning (27). Q-learning is a form of reinforcement learning and is a dynamic programming procedure that uses backwards recursion to solve the complex Bellman equation more efficiently using regression models. In the two stage setting, Q-functions are defined as

$$Q_2(\boldsymbol{h}_2, a_2) = E\left[Y | \boldsymbol{H}_2 = \boldsymbol{h}_2, A_2 = a_2\right]$$

$$Q_1(\boldsymbol{h}_1, a_1) = E\left[\max_{a_2} Q_2(\boldsymbol{h}_2, a_2) | \boldsymbol{H}_1 = \boldsymbol{h}_1, A_1 = a_1\right].$$

The Q-functions are conditional expectations where $Q_2(\boldsymbol{h}_2, a_2)$ evaluates the quality of choosing treatment a_2 for patients with history $\boldsymbol{h}_2$ and $Q_1(\boldsymbol{h}_1, a_1)$ evaluates the quality of choosing treatment a_1 for patients with history $\boldsymbol{h}_1$ assuming that the best second stage intervention has chosen. This can be extended to more than 2 phases such that

$$Q_t(\boldsymbol{h}_t, a_t) = E\left[\max_{a_t} Q_{t+1}(\boldsymbol{h}_{t+1}, a_{t+1}) | \boldsymbol{H}_t = \boldsymbol{h}_t, A_t = a_t\right]$$

would evaluate the quality of choosing a_t for patients with history $\boldsymbol{h}_t$ assuming the best intervention is chosen at all future stages. In practice, these Q-functions are not known but must be estimated. For illustration, consider the linear form as

$$Q_t(\boldsymbol{h}_t, a_t) = \boldsymbol{h}_{t,1}^T \beta_{t,1} + a_t \boldsymbol{h}_{t,2}^T \beta_{t,2}.$$

In the two stage scenario, estimating the Q-functions, $\widehat{Q}_t(\boldsymbol{h}_t, a_t)$, is a three step procedure. First, using ordinary least squares regression, the estimates $\widehat{\beta}_{2,1}$ and $\widehat{\beta}_{2,2}$ are obtained by regressing the patient history on Y_2. Those estimates are used to estimate the fitted Q function at the second stage $\widehat{Q}_2(\boldsymbol{h}_2, a_2) = \boldsymbol{h}_{2,1}^T \widehat{\beta}_{2,1} + a_2 \boldsymbol{h}_{2,2}^T \widehat{\beta}_{2,2}$. The stage 1 pseudo outcome is $\tilde{Y}_1 = Y_1 + \max_{a_2} \widehat{Q}_2(\boldsymbol{h}_2, a_2)$. Note that if the outcome is only collected at the final stage (in other words, there is only one Y so there is no Y_2, Y_1), the stage 2 outcome is Y and the stage 1 psuedo outcome is $\tilde{Y} = \max_{a_2} \widehat{Q}_2(\boldsymbol{h}_2, a_2)$. The first stage patient history is regressed on $\tilde{Y}_1$ to obtain the estimates $\widehat{\beta}_{1,1}$ and $\widehat{\beta}_{1,2}$. The first stage fitted Q function is $\widehat{Q}_1(\boldsymbol{h}_1, a_1) = \boldsymbol{h}_{1,1}^T \widehat{\beta}_{1,1} + a_1 \boldsymbol{h}_{1,2}^T \widehat{\beta}_{1,2}$. Finally, the estimated optimal treatment decision is given by

$$D_t^*(\boldsymbol{h}_t) = argmax_{a_t} \widehat{Q}_t(\boldsymbol{h}_t, a_t).$$

Estimating the Q-functions is similar for three or more stage implementation where the predicted future outcome is used to create the estimates for the previous stage estimated Q-function.

While Q-learning is a very popular estimation technique, it is not without its limitations and suffers from some undesirable properties such as irregularity, non-smoothness, and asymptotic bias (22). The irregularity problem occurs in Q-learning when the last stage treatment is non-unique for some subjects in the population, which causes bias and inaccurate inference. To remedy this, Goldberg et al. (8) uses special adaptive weights within the penalization. This corrects for the non-regularity condition by concentrating on the indifference hyperplane of patient covariates where two treatment have the same effect. The indifference hyperplane is the covariate region where there is no difference between treatments. Solving this irregularity condition involves correctly identifying the covariate values which lie on this hyperplane. This adaptive penalized Q-learning procedure can handle continuous covariates and performs better than regular penalized Q-learning method.

Instead of the typical first stage of Q-learning (which involves solving the minimization problem) the adaptive minimization problem involves solving

$$\Phi_{2n}(\theta_2) = \sum_{i=1}^{n} \{Y_{2i} - Q_2(\boldsymbol{h}_{2i}, a_{2i}; \beta_{2,1}, \beta_{2,2})\}^2 - \frac{\lambda_n}{n} \sum_{i=1}^{n} \widehat{\omega}_{ni} |\beta_{2,2}' \boldsymbol{H}_{2i(2)}|$$

where $\widehat{\omega}_{ni}$ are the data driven weights and λ_n is the tuning parameter. Then, $\tilde{\theta}_2$ (in traditional Q learning this is the set of parameters which minimizes the ordinary least squares regression function at the second treatment decision time point) is the minimizer of Φ_{2n} and the remaining steps of Q learning are the same after substituting in $\tilde{\theta}_2$ for the normal estimator. In order to obtain the oracle property (which means the estimator behaves asymptotically as if the indifference plane is already known) the selection of weights is critical. The goal is to find weights that penalize the observations that are close to or are on the indifference hyperplane and that provide weights that go asymptotically to zero for observations far from the hyperplane. This will help define where the indifference hyperplane is and resolve the irregularity problem.

As in most statistical research areas, after developing an estimator the next step is to assess its properties, oftentimes with the use of inference techniques such as confidence intervals. When estimating optimal DTR, common approaches such as Q-learning involve estimation and interference of parameters that are non-smooth functions of the underlying generative distribution. As was mentioned before, these estimates are irregular and asymptotically biased. Standard asymptotic approximations to the sampling distributions cannot be used to directly form reliable confidence intervals or carry out hypothesis testing (12). One method to construct confidence intervals is an m-out-of-n bootstrap procedure to correct the nonsmoothness. The confidence sets are constructed in a way to adapt to the irregularity present in the underlying generative model. The data driven adaptive choice of m produces asymptotically correct confidence sets under fixed alternatives. This method has the added benefit of conceptual and computational simplicity with a corresponding R package (3).

The proposed adaptive scheme to select m is a class of resample sizes given by $m = n^{f(p)}$. The suggested simple form is proposed to be

$$\widehat{m} = n^{\frac{1+\alpha(1-\widehat{p})}{1+\alpha}}$$

where $\alpha > 0$ is a tuning parameter. α controls the smallest acceptable sample size and may be dictated by practical considerations or tuned using the data. A bootstrap algorithm is used for choosing α using data which appears to reduce conservatism. When the parameter of interest is a linear function of the parameters, $(c'\theta_{1,n})$, the algorithm first draws B_1 m-out-of-n first stage bootstrap samples and estimates $c^T\widehat{\theta}_{1,n}^{b_1}$. α is fixed at the smallest value in the grid and $\widehat{m}^{b_1}$ is then calculated using the equation above. This is repeated by drawing B_2 $\widehat{m}^{b_1}$-out-of-n second stage bootstrap samples and calculating $c^T\widehat{\theta}^{(b_1,b_2)}$, which is a double bootstrapped version of the estimate. For all b_1, compute $\left(\frac{\eta}{2}\right)$x100 and $\left(1 - \frac{\eta}{2}\right)$x100 percentiles which are the lower bounds and upper bounds defined as $\widehat{l}_{DB}^{b_1}$ and $\widehat{u}_{DB}^{b_1}$ respectively. The coverage rate of the double bootstrap confidence interval from all first stage bootstrap data sets is

$$\frac{1}{B_1} \sum_{b_1=1}^{B_1} I\left(c^T\widehat{\theta}_{1,n}^{b_1} - \frac{\widehat{u}_{DB}^{b_1}}{\sqrt{\widehat{m}^{b_1}}} \leq c^T\theta_{1,n} \leq c^T\widehat{\theta}_{1,n}^{b_1} - \frac{\widehat{l}_{DB}^{b_1}}{\sqrt{\widehat{m}^{b_1}}}\right).$$

Increase α to the next highest value on the grid until the coverage rate is at or exceeds the nominal value and in that case pick the current value of α as the final value. The process is repeated until the coverage rate of the double bootstrap confidence interval attains a nominal coverage rate or all the options on the grid are exhausted.

Another methodology that can accommodate the irregularity from using Q-learning to estimate parameters is the locally consistent Adaptive Confidence Interval (ACI) (12). When construction of DTRs using Q-learning, there is particular interest in reducing bias of first stage coefficients. If the Q-function is near 0 with high probability there will be issues approximating the distribution of $\sqrt{n}(\widehat{\beta}_1 - \beta_1^*)$. Once the asymptotically biased parameters are identified, given the correct amount of shrinkage, a shrinkage estimator can reduce the bias. However, shrinking too aggressively leads to bad performance in finite samples. Constructing valid confidence intervals for irregular estimators is a difficult task because estimating the sampling distribution of the estimator cannot be done uniformly. The proposed solution is a locally consistent confidence interval for linear combinations of the first stage coefficients. The interest is not in construction of second stage confidence intervals because they can be estimated using standard methods for least square estimators. Since it is not possible to construct a uniformly convergent estimator of the limiting distribution of $\sqrt{n}(\widehat{\beta}_1 - \beta_1^*)$, for a given constant c the proposed method bounds $c^T\sqrt{n}(\widehat{\beta}_1 - \beta_1^*)$ between two regular uniformly convergent upper and lower bounds. These smooth bounds can be bootstrapped to form a confidence set for $c^T\beta_1^*$. The extension to more than two stages is straightforward as the last stage uses standard methods for least squares estimation, so the ACI would be used on all previous stages.

Similar to previously discussed analysis techniques, for medical research it is important to develop these techniques to accommodate censored data. Goldberg and Kosorok (7) developed a Q-learning algorithm that allows for censored data when the outcome of interest is survival time and allows for a flexible number of stages in a randomized trial. Q-learning is expanded upon by using inverse probability censoring weighting to account for censored observations.

For each $t = 1, \ldots, T$, let the state S_t be the pair $S_t = (\boldsymbol{X}_t, Y_{t-1})$ where $\boldsymbol{X}_t$ is either a vector of covariates describing the condition of the patient before time t or it is null. If $\boldsymbol{X}_t$ is null then a failure happened during the tth stage. Let Y_{t-1} be the length of time between decision points t and $t-1$. Hence, $\sum_{j=1}^{t} Y_j$ is the total survival time, or reward, up to and including stage t. Let $C \in [0, \tau]$ be the censoring variable. The goal is to find a policy that maximizes the expected rewards. Then, the optimal policy, D^*, is the one that approximately maximizes over all policies of $E_{0,\pi}[(\sum_{t=1}^{T} Y_t) \wedge \tau]$ where $\bar{T}$ is the random number of stage for the subject. This optimal policy is found using a three step algorithm. First the problem is mapped to an auxiliary problem. The auxiliary problem creates modified trajectories of a fixed length T and the modified sum of the rewards is less than or equal to τ to account for censoring. Next, the Q-functions are approximated $\{\widehat{Q}_1, \ldots, \widehat{Q}_T\}$ using the original Q-function framework. Last, the optimal treatment rule, D^*, is found by maximizing $\widehat{Q}_t$ over all possible a_t.

Recall the methodology introduced in the previous section for estimation of ITRs in a single stage. Two of those methods will be expanded on when estimating DTRs for multiple stages of treatment. In the single decision scenario presented by Zhang et al. (30) the estimation procedure was restructured into a classification problem. In this case the optimal classifier corresponds to the optimal treatment decision. The optimal classifier was found by estimating the Bayes classifier which is the one that minimizes the expected weighted misclassification error. This can be expanded upon for the two decision point scenario based on reassessing the problem as a monotone coarsening problem using an AIPWE to estimate the mean outcome (31). Assign Y^* to be the often unobserved potential outcome and Y_D^* to be the potential outcome associated with treatment regime D. The optimal treatment regime D^* is that which satisfies $E[Y^*(D^*)] \geq E[Y^*(D)]$, meaning it is that which maximizes the expected potential outcome. The problem is cast into a monotone coarsening problem where the coarsening happens at random if, for each t, the probability that the data are coarsened at level t given the full data depends only on the data observed at level t. Then, from Robins et al. (23), under these coarsening assumptions if the coarsening mechanism is correctly defined then asymptotically linear consistent estimators for $E[Y^*(D_\eta)]$ for a fixed η have the form

$$\frac{\sum_{i=1}^n I(C_{\eta,i} = \infty)}{K_{\eta,k} \bar{\boldsymbol{X}}_{k,i}} Y_i + \frac{\sum_{i=1}^n \left\{ I(C_{\eta,i} = k) - \lambda_{\eta,k}(\bar{\boldsymbol{X}}_{k,i}) I(C_{\eta,i} > k) \right\}}{K_{\eta,k}(\bar{\boldsymbol{X}}_{k,i})} L_k(\bar{\boldsymbol{X}}_{k,i})$$

where $L_k(\bar{\boldsymbol{X}}_{k,i})$ are arbitrary functions, $C_{\eta,i}$ is the discrete coarsening variable, $K_{\eta,K}(\bar{\boldsymbol{X}}_K) = \prod_{k'=1}^k \{1 - \lambda_{\eta,k'}(\bar{\boldsymbol{X}}_{k'})\}$, and $\lambda_{\eta,k'}(\boldsymbol{X}'_k)$ is the hazard function. The left side of the above estimator is on its own a consistent estimator if $\lambda_{\eta,k}(\bar{\boldsymbol{X}}_k)$ is correctly specified. Then the entire estimator is a doubly consistent robust estimator for $E[Y^*(D\eta)]$ if either $\lambda_{\eta,k}(\bar{\boldsymbol{X}}_k)$ are correctly specified or if $L_k(\bar{\boldsymbol{X}}_{k,i}) = Y^*(D_\eta) | \{ \bar{\boldsymbol{X}}_k^*(\bar{D}_{\eta_{k-1}}) = \bar{x}_k \}$.

O-learning was presented in Section 3 as a machine learning approach which directly optimizes the value function $V(D)$ where each subject's weight is proportional to their clinical outcome. This is a weighted classification error problem since finding the D^* that maximizes $V(D)$ is equivalent to finding the D^* that minimizes $\bar{V}(D)$. O-learning can also be expanded to the two stage paradigm using a few strategies. One such way is backwards outcome weighted learning (BOWL) which modifies existing algorithms to solve a sequence of weighted classification problems (33). The algorithm is backwards fitting and at each time point T, the algorithm is as follows. The goal in the first stage is to minimize

$$\frac{n^{-1} \sum_{i=1}^n [Y_{iT} \phi \{A_{iT} f_T(\boldsymbol{H}_{iT})\}]}{\pi_T(A_{it}, \boldsymbol{H}_{it})} + \lambda_{T,n} ||f_T||^2$$

with respect to f_T where $\widehat{f}_T$ is the minimizer. The optimal decision rule is

$$\widehat{D}_T(\boldsymbol{h}_T) = sign \left\{ \widehat{f}_T(\boldsymbol{h}_T) \right\},$$

and this stage is essentially equivalent to the single stage outcome weighted learning found in Zhao et al. (32) and has a similar dual objective function as found in support vector machines. The second stage is, for $t = T-1, T-2, \ldots, 1$, to backward sequentially minimize

$$n^{-1} \sum_{i=1}^{n} \frac{(\sum_{j=1}^{T} Y_{ij}) \prod_{j=t+1}^{T} I\left\{A_{ij} = \widehat{D}_j(\boldsymbol{H}_{ij})\right\}}{\prod_{j=1}^{T} \pi_j(A_{ij}, \boldsymbol{H}_{ij})} x\phi\left\{A_{ij}, f_t(\boldsymbol{H}_{it})\right\} + \lambda_{t,n} \|f_t\|^2$$

where $\widehat{D}_{t+1}, \ldots, \widehat{D}_T$ are previously obtained.

A disadvantage of BOWL is the number of observations utilized by the algorithm decreases geometrically as t decreases. The authors explain this can be solved using iterative outcome weighted learning (IOWL) which involves re-estimating the optimal treatment rule at stage 2 after the stage 1 rule is estimated. This estimate is based on the subset of patients whose stage 1 treatment assignments are consistent with the optimal rule. The procedure would continue with a re-estimation of the stage 1 treatment rule based on the new optimal stage 2 rule. IOWL allows the exploration of different subjects through iterative re-estimation.

Zhao et al. (33) also present simultaneous outcome weighted learning (SOWL) which frames estimation of DTRs as a single classification problem. This is an effective way of looking at the problem because a multiple stage treatment plan has not previously been estimated simultaneously using a single algorithm. The method directly optimizes the empirical counterpart of the value function in one step. Since this problem is computationally difficult (mostly because of the discontinuity of the indicator functions) a continuous and concave surrogate function is used in lieu of the product of indicators that would usually be required. In the two decision point scenario, the surrogate reward function is chosen to mimic hinge loss: $\psi(Z_1, Z_2) = min(Z_1 - 1, Z_2 - 1, 0) + 1$ where $Z_1 = A_1 f_1(H_1)$ and $Z_2 = A_2 f_2(H_2)$. Hence the SOWL estimator maximizes

$$n^{-1} \sum_{i=1}^{n} \frac{(\sum_{j=1}^{2} Y_{ij}) \psi\left\{A_{i1} f_1(\boldsymbol{H}_{i1}), A_{i2} f_2(\boldsymbol{H}_{i2})\right\}}{\prod_{j=1}^{2} \pi_j(A_{ij}, \boldsymbol{H}_{ij})} - \lambda_n \left(\|f_1\|^2 + \|f_2\|^2\right),$$

where the tuning parameter λ_n controls the amount of penalization. This can easily be extended to more than 2 stages.

Much exciting and significant work is being done in developing treatment rules for dynamic sequential decision making. To this effect, there have been promising advancements, but these methods often times need to be expanded upon or adapted to various specific settings. Science will forever be changing and research will be forever trying to keep up. While existing methodology can always be improved and generalized, there will always be a need for new and innovative mechanisms for estimating optimal DTRs.

22.5   Related Topics

22.5.1   Variable selection

Variable selection is an important component of estimating optimal DTRs because tailoring variables are used to adapt the treatment plan to the individual. The goal is to avoid a priori hand picking tailoring variables, but instead use the data to select a subset of the tailoring variables that estimates a decision rule similar to the optimal rule chosen when using all variables. Including all possible variables as tailoring variables is inefficient and will often lead to over fitting. Once the tailoring variables are selected, they can be used when optimizing DTRs.

Biernot and Moodie (2) discuss two computer science techniques that can be used for variable selection: the S-score criterion and the use of reducts. The S-score of a variable shows the expected increase in response that is observed by choosing the treatment based on the value of that variable. It combines the interaction of the covariate with the treatment and the proportion of the population exhibiting variability in that covariate. Higher values indicate stronger relationships between the variable and the treatment, and shows that a large proportion of patients would experience change in the optimal action if the variable was taken into consideration. This scoring is used to rank potential variables but each variable is evaluated separately, meaning correlation between variables is not taken into consideration. The S-score could also be used sequentiality such that the variable with the highest score is first selected, then the variable with the second highest score given the first variable is selected and so on.

The reducts approach was developed from rough set theory in computer science. The positive region is a set of all observations that can be uniquely classified into one equivalence class based on the non-decision variables. The reduct is the minimal set of tailoring variables that classifies individuals into unique decision equivalence classes as well as the complete set of variables does. Reducts help eliminate redundant variables while preserving information regarding the similarity of individuals in the sample. In the scenario with multiple reducts, one can select the variables most frequently seen in the reducts or can select amongst reducts by choosing the set of covariates with the highest S-score. This last hybrid method is believed to combine the strengths of these two methods. Unfortunately, reducts cannot be used on a continuous outcome.

Another way to approach variable selection is to simultaneously estimate optimal treatment regimes and identify significant variables. This is done with a penalized regression model that finds which variables interact with the treatment using a new loss based framework. Lu et al. (16) introduces a method which does not require estimating the baseline mean function for the outcome of interest and is easily adaptable to shrinkage methods for variable selection based on their loss structure making it quickly implementable with current software. The authors suggest the loss function

$$L_{n,\phi}(\beta, \gamma) = \frac{1}{n} \sum_{i=1}^{n} \left[Y_i - \phi(\boldsymbol{X}_i; \gamma) - \beta^T \tilde{\boldsymbol{X}}_i \{A_i - \alpha(\boldsymbol{X}_i)\} \right]^2$$

where n is the number of observations, Y_i is the ith patient's outcome, $\boldsymbol{X}_i$ is the ith patient's prognostic variables, $\tilde{\boldsymbol{X}} = (1, \boldsymbol{X}^T)^T$, $A_i \in \{-1, 1\}$ represents the dichotomous treatment choice, $\alpha(x)$ denotes the propensity score, and ϕ is an arbitrary function with a constant model for $\phi : \phi(x; \gamma) = \gamma$ and a linear model for $\phi : \phi(x; \gamma) = \gamma^T \tilde{x}$. This characterization of the loss function increases simplicity in adopting shrinkage penalties for variable selection. Employing the adaptive lasso penalty (or alternatively, the SCAD or minimax concavity penalty) the solution is the β which satisfies

$$\min_{\beta} L_{n,\phi}(\beta, \tilde{\gamma}) + \lambda_n \sum_{j=1}^{p+1} w_j |\beta_j|$$

where λ_n is a tuning parameter and w_j are the weights such that $w_j^{-1} = |\tilde{\beta}_j|$ is used. Aside from estimating the optimal DTR, these β values are used to determine which variables are important in selecting the optimal DTR such that the important variables are those with nonzero coefficients.

Variable selection is an important part of estimating optimal DTRs because a parsimonious selection of tailoring variables will make the estimation faster and more reliable. Three methods have been presented here for these purposes, but more methodology has been published. It is imperative to use a selection technique that is relevant for the data set and can be effectively integrated into the analysis plan.

22.5.2   Multiple outcomes

While development of these advanced estimation techniques is necessary to effectively personalize medical care, treatment of the entire person, not just one disease, should be considered as well. In the clinical setting, the patient or caregiver will likely be interested in balancing competing outcomes such as survival, quality of life, and financial burden. While survival may be the ultimate goal for a cancer patient, a single mother of two may prefer a treatment that allows her to work (higher quality of life) which could lead to a longer treatment course, or patients may need to balance the financial burden with their treatment plan. This is a very new area of study inside the precision medicine umbrella, but it is important and quickly developing.

A crucial step in balancing competing outcomes for personalized patient care is eliciting the patient's or physician's preferences regarding the ideal tradeoff between the outcomes. Lizotte et al. (15) produced an inverse preference elicitation approach which first considers all of the actions available at any given state. Then, for each action, asks what range of preferences makes that action a good choice. This provides a large amount of information about the potential actions at each state. The patients also have the ability to see if their preference is near the boundary or see if a small change in preference results in a change of recommended treatment. In this situation, the patient can feel confident that both treatments perform well and make the decision based on other potentially minor preferences. This method provides an efficient algorithm that computes the optimal policy for varying reward functions and provides insight into how the choice of reward influences the optimal treatment decision.

Alternatively, Laber et al. (11) developed a way to construct DTRs that does not require tradeoffs between outcomes by eliciting a clinically significant difference for each respective outcome. When the algorithm concludes that no single treatment is best, the patient's or doctor's preferences are able to be incorporated. They are free to choose the treatment arbitrarily based on other qualities that matter to them, such as cost. This method involves set-valued dynamic treatment regimes that take as input the current patient history and provide as output a set of recommended treatments.

Considering just the static set valued decision rules for the single decision time point, let Y and Z be the competing outcomes and Δ_Y, Δ_Z represent the predetermined clinically meaningful difference in the respective outcomes. In the ideal situation, the algorithm will produce one recommended treatment if that treatment provides significant benefit to one outcome without producing significant detriment to the other. However, in all other cases, the algorithm will produce a set of recommended treatments and the decision is left up to the clinician or patient. With

$$\tau_Y(h) = E[Y|H = h, A = 1] - E[Y|H = h, A = -1]$$
$$\tau_Z(h) = E[Z|H = h, A = 1] - E[Z|H = h, A = -1]$$

then the ideal decision rule $\pi_\Delta^{ideal}(h)$ is either

1. $sign\{\tau_Y(h)\}$ if $|\tau_Y(h)| > \Delta_Y$ and $sign\{\tau_Y(h)\}\tau_Z(h) > -\Delta_Z$
2. $sign\{\tau_Z(h)\}$ if $|\tau_Z(h)| > \Delta_Z$ and $sign\{\tau_Z(h)\}\tau_Y(h) > -\Delta_Y$
3. $\{-1, 1\}$ otherwise

Generalizing this procedure to dynamic set valued decision rules for two or more decision points, the algorithm is backwards regressive and Q-learning with linear working models is used to estimate $r_Y(h)$ and $r_Z(h)$. Using ordinary least squares, the optimal treatment set can be estimated from patient history. At the second stage, estimating $\pi_{2\Delta}^{ideal}$ is essentially the same as described for the single decision point situation. To find $\pi_{2\Delta}^{ideal}$, it is assumed that the best single treatment decision (not a set-valued decision) was made, τ_2. Then, $\pi_{1\Delta}^{ideal}(h_1, \tau_2)$, at the first stage is

1. $sign\{\tau_Y(h_1, \tau_2)\}$ if $|\tau_Y(h_1, \tau_2)| > \Delta_Y$ and $sign\{\tau_Y(h_1, \tau_2)\} \tau_Z(h_1, \tau_2) > -\Delta_Z$
2. $sign\{\tau_Z(h_1, \tau_2)\}$ if $|\tau_Z(h_1, \tau_2)| > \Delta_Z$ and $sign\{\tau_Z(h_1, \tau_2)\} \tau_Y(h_1, \tau_2) > -\Delta_Y$
3. $\{-1, 1\}$ otherwise

Hence, the optimal decision rule is the set valued rule

$$\pi_{1\Delta}^{ideal}(h_1) = \bigcup_{\tau_2 \in C\left(\pi_{2\Delta}^{ideal}\right)} \pi_{1\Delta}^{ideal}(h_1, \tau_2)$$

where $C(\pi_{2\Delta}^{ideal})$ is the set of all treatment options compatible with $\pi_{2\Delta}^{ideal}$ and τ_2 is compatible with π_2 if and only if $\tau_2(h_2) \in \pi_2(h_2) \quad \forall \quad h_2$.

Progressing from estimating techniques for one outcome to multiple outcomes is the next step in the natural course of this specialization. Practically, patients and doctors will have more than one goal when developing a treatment plan and these complicated preferences should be taken into consideration if possible. As this is a new area, there is continuing progress.

22.5.3   DTRs for observational data

As clinical researchers were experimenting with new trial designs to improve patient treatment plans, epidemiologists were simultaneously investigating the relationships of time varying continuous exposures to various outcomes in observational data (13). They developed a longitudinal generalization of Rubin's potential outcomes (9) for inference between exposure and outcomes for observational data which naturally led to an interest in developing DTRs for this data. These DTRs assume the exposures are assigned in a way that is conditionally independent of the potential future responses given the history of the patients and treatments up to the current state. This resembles the assumptions made when developing treatment plans using randomized prospective data. Sometimes situations arise where a randomized trial is impossible or impractical, so it is efficacious to perform an observational study instead or, quite simply, observational data may exist from a preexisting study. Using this resource can be more expedient and reduce significant financial burden because new patients are not needed and no treatments are given. The development of DTRs is often exploratory and hence it is potentially important to be able to estimate these treatment plans using large samples of observational data with the intention of validating the DTR in a confirmatory randomized trial. Furthermore, collecting observational data on time-varying outcomes, predictors and confounders can sometimes emulate a randomized trial that lacks baseline randomization.

An effective estimation technique when eliciting DTRs from observation data is the parametric g-formula. The parametric g-formula uses Robins' G-estimation to naturally estimate DTRs and can appropriately adjust for time-dependent cofounding variables (28). This formula is an alternative to inverse probability weighting that provides more efficient estimates but requires more parametric modeling assumptions. It can be computed for the potential outcome by non-parametrically estimating the value of each density function for all of the possible histories of patients. The formula takes the sum over the histories but requires that all possible covariates need to be categorical. For high dimensional data, the g-formula can only be carried out by estimating the density functions using parametric modeling assumptions, then taking the sum over the histories via Monte Carlo simulation. Because of distributional a priori knowledge for certain histories, when estimating the g-formula parametric models are not needed to be imposed over all components of the densities and histories.

G-estimation in the context of estimating DTRs has advantages over traditional parametric approaches for producing consistent estimators. However, these estimators are asymptotically biased under a given structural nested mean model for certain data distributions (coined exceptional laws) and exhibit non-regular behavior. To combat this, Moodie and Richardson (17) presented a new approach called Zeroing Instead of Plugging In (ZIPI). ZIPI provides estimators nearly identical to those provided by g-estimation but with the benefit of reducing bias in those situations when decision rule parameters are not shared across intervals. More specifically in the context of constructing DTRs, the observed longitudinal distribution function is "exceptional" if at some interval there is a positive probability that the true optimal decision rule is not unique. For a distribution to be exceptional, the blip function must include at least one covariate (such as the previous treatment), and the probability that the true blip function has value 0 is positive. The proposed ZIPI method is considered a modification of g-estimation when there is no parameter sharing and detects and reduces bias in the presence of exceptional laws.

Moodie et al. (18) extended one of the more frequently utilized methods of optimal DTR estimation, Q-learning, to accommodate observational data. A soft threshold approach is used which has a good performance in terms of bias and coverage in the non-regular settings. This approach shrinks the problematic term in the potential outcome towards zero. When using Q-learning for observational data, the basic approach requires the construction of a propensity score, $\pi(x) = P(A = 1 | \boldsymbol{X} = \boldsymbol{x})$, or treatment model followed by some form of adjustment. It assumes the treatment received is independent of known covariates given the propensity score. This leads to unbiased estimates of the treatment effect based on the conditional expectation modeling the outcome given the propensity score. In inverse probability weighting analysis, the weights are used to create a pseudo-sample so that the treatment does not depend on the variables in the pseudo-sample. Including covariates into the models for the Q-functions can be implemented in four ways which perform well: including the covariates as linear terms in the Q-function, including the propensity score as a linear term in the Q-function, including quintiles of the interval-specific propensity score (which depends on a time varying confounding variable) as covariates in the jth interval Q-function, and IPTW weighted with $\boldsymbol{H}_1$ and $\boldsymbol{H}_2$ defined as in traditional Q-learning.

Not all data can be collected using a randomized clinical trial, so it is imperative to develop methodologies that can estimate optimal DTRs using this observational data. This data is difficult because there is no randomization, but the assumption of independence between exposure and predicted future outcome simplifies the problem to an extent. While the parametric g-formula and Q-learning are great options for analysis, not all data will benefit from these techniques and more work would provide more resources and flexibility for scientists.

22.6 Conclusion

Data collected from SMARTs is an integral part of effectively developing DTRs. This has been a hot topic in clinical trial design in recent years and appears to be the future of clinical practice. They are flexible, informative studies which utilize all of the participants and can answer more questions than a traditional RCT. Pilot studies are essential for designing a SMART so that resources are optimized and the essential information is properly collected. While there has been progress made in properly designing SMARTs, there is still a lot of work to be done particularly in sample size estimation and handling missing data. A plethora of direct and indirect techniques have been presented here to highlight some of the most current methods available so the reader can make informed decisions when creating their analysis plan for a SMART design or even when working with observational data. One of many future directions in this area is practical implementation in clinical practice which involves estimating DTRs for competing outcomes, an area which is quickly expanding. The recent progress over the last ten years has been very exciting, but there are still many areas that could use further development and many topics that have not been explored yet. The future of implementing SMARTs for developing DTRs to personalize medicine is bright and promising.

Bibliography

Daniel Almirall, Scott N. Compton, Meredith Gunlicks-Stoessel, Naihua Duan, and Susan A. Murphy. Designing a Pilot Sequential Multiple Assignment Randomized Trial for Developing an Adaptive Treatment Strategy. *Statistics in Medicine*, 31(17):188–1902, 2012.

Peter Biernot and Erica E.M. Moodie. A Comparison of Variable Selection Approaches for Dynamic Treatment Regimes. *The International Journal of Biostatistics*, 6(1): 1557–4679, 2010.

Bibhas Chakraborty, Eric B Laber, and Yingqi Zhao. Inference for Optimal Dynamic Treatment Regimes Using an Adaptive m-Out-of-n Bootstrap Scheme. *Biometrics*, 69 (3):714–723, 2013.

Guanhua Chen, Donglin Zeng, and Michael R Kosorok. Personalized Dose Finding using Outcome Weighted Learning. Manuscript submitted for publication.

Linda M. Collins, Susan A. Murphy, and Victor Stretcher. The Multiphase Optimization Strategy (MOST) and the Sequential Multiple Assignment Randomized Trial (SMART): New Methods for More Potent eHealth Interventions. *American Journal of Preventative Medicine*, 32(5):112–118, 2007.

Linda M. Collins, Inbal Nahum-Shani, and Daniel Almirall. Optimization of behavioral dynamic treatment regimens based on the sequential, multiple assigment, randomization trial (SMART). *Clinical Trials*, 11(4):426–434, 2014.

Yair Goldberg and Michael R. Kosorok. Q-learning with Censored Data. *Annals of Statistics*, 40(1):529–560, 2012.

Yair Goldberg, Rui Song, and Michael R Kosorok. Adaptive Q-learning. *Institute of Mathematical Statistics Collections*, 9(1):150–162, 2013.

Paul W. Holland. Statistics and Causal Inference. *Journal of the American Statistical Association*, 81(396):945–960, 1986.

Michael R Kosorok and Erica EM Moodie. *Adaptive Treatment Strategies in Practice: Planning Trials and Analyzing Data for Personalized Medicine*, volume 21. SIAM, 2016.

Eric B. Laber, Daniel J. Lizotte, and Bradley Ferguson. Set-valued Dynamic Treatment Regimes for Competing Outcomes. *Biometrics*, 70(1):53–61, 2014.

Eric B. Laber, Daniel J. Lizotte, Min Qian, William E. Pelham, and Susan A. Murphy. Dynamic Treatment Regimes: Technical Challenges and Applications. *Electronic Journal of Statistics*, 8(1):1225–1272, 2014.

Philip W. Lavori and Ree Dawson. Introduction to Dynamic Treatment Strategies and Sequential Multiple Assignment Randomization. *Clinical Trials*, 11(4):393–399, 2014.

Zhiguo Li and Susan A Murphy. Sample size formulae for two-stage randomized trials with survival outcomes. *Biometrika*, 98(3):503–518, 2011.

Daniel J. Lizotte, Michael Bowling, and Susan A. Murphy. Linear Fitted-Q Iteration with Multiple Reward Functions. *Journal of Machine Learning Research*, 13:3253–3295, 2012.

Wenbin Lu, Hao Helen Zhang, and Donglin Zeng. Variable Selection for Optimal Treatment Decision. *Statistical Methods in Medical Research*, 22(5):493–504, 2013.

Erica E. M. Moodie and Thomas S Richardson. Estimating Optimal Dynamic Regimes: Correcting Bias under the Null. *Scandinavian Journal of Statistics*, 37(1):126–146., 2009.

Erica E.M. Moodie, Bibhas Chakraborty, and Michael S. Kramer. Q-learning for Estimating Optimal Dynamic Treatment Rules from Observational Data. *Candian Journal of Statistics*, 40(4):629–645, 2012.

Susan A Murphy. An experimental design for the development of adaptive treatment strategies. 2005.

Inbal Nahum-Shani. What is a JITAI? In *Proceedings of Workshop on Just In Time Adaptive Interventions (JITAIs)*, 2013.

Min Qian and Susan A. Murphy. Performance Guarantees for Individualized Treatment Rules. *Annals of Statistics*, 39(2):1180–1210, 2011.

James M. Robins. *Proceedings of the Second Seattle Symposium in Biostatistics*, chapter Optimal Structural Nested Models for Optimal Sequential Decisions, pages 189–326. Springer, 2004.

James M. Robins, Andrea Rotnitzky, and Lue Ping Zhao. Estimation of regression coefficients when some regressors are not always observed. *Journal of the American Statistical Association*, 89(427):846–866, 1994.

Susan M. Shortreed, Eric Laber, T. Scott Stroup, Joelle Pineau, and Susan A. Murphy. A multiple imputation strategy for sequential multiple assignment randomized trials. *Statistics in Medicine*, 33(24):4202–4214, 2014.

Pham Dinh Tao and Le Thi Hoai An. Convex analysis approach to d. c. programming: Theory, Algorithms and Applications. *ACTA Mathematica Vietnamica*, 22(1):289–355, 1997.

Lu Wang, Andrea Rotnitzk, Xihong Lin, Randall E. Millikan, and Peter F. Thall. Evaluation of Viable Dynamic Treatment Regimes in a Sequentially Randomized Trial of Advanced Prostate Cancer. *Journal of the American Statistical Association*, 107(498): 493–508, 2012.

Christopher J.C.H Watkins and Peter Dayan. Q-Learning. *Machine Learning*, 8(3-4): 279–292, 1992.

Jessica G. Young, Lauren E. Cain, James M. Robins, Eilis J. O'Reilly, and Miguel A. Hernan. Comparative Effectiveness of Dynamic Treatment Regimes: An Application of the Parametric G-Formula. *Statistics in Biosciences*, 3(1), 2011.

Baqun Zhang, Anastasios A. Tsiatis, Marie Davidian, Min Zhang, and Eric Laber. Estimating Optimal Treatment Regimes from a Classification Perspective. *Stat*, 1(1):103–114, 2012.

Baqun Zhang, Anastasios A Tsiatis, Eric B Laber, and Marie Davidian. A Robust Method for Estimating Optimal Treatment Regimes. *Biometrics*, 68(4):1010–1018, 2012.

Baqun Zhang, Anastasios A. Tsiatis, Eric B. Laber, and Marie Davidian. Robust Estimation of Optimal Dynamic Treatment Regimes for Sequential Treatment Decisions. *Biometrika*, 100(3):681–694, 2013.

Yingqi Zhao, Donglin Zeng, A. John Rush, and Michael R. Kosorok. Estimating Individualized Treatment Rules Using Outcome Weighted Learning. *Journal of the American Statistical Association*, 107(499):1106–1118, 2012.

Yingqi Zhao, Donglin Zeng, Eric B. Laber, and Michael R. Kosorok. New Statistical Learning Methods for Estimating Opitmal Dynamic Treatment Regimes. *Journal of the American Statistical Association*, 2014.

Yingqi Zhao, Donglin Zeng, Eric B. Laber, Rui Song, Ming Yuan, and Michael R. Kosorok. Doubly Robust Learning for Estimating Individualized Treatment with Censored Data. *Biometrika*, 102(1), 2015.

Yufan Zhao, Donglin Zeng, Mark A. Socinski, and Michael R. Kosorok. Reinforcement Learning Strategies for Clinical Trials in Nonsmall Cell Lung Cancer. *Biometrics*, 67(4): 1422–1433, 2011.

23

Safety Evaluation in Clinical Trials

H. Amy Xia, Brenda J. Crowe and Jesse A. Berlin

CONTENTS

23.1   Introduction

Safety evaluation is critically important in clinical trials and is continuous throughout the drug's lifecycle. It may be exploratory in nature in early phase trials, but the randomized controlled trials (RCTs) in later phases of development enable better characterization of the product safety and tolerability profile, benefitting from the larger sample size and study design to control bias (International Conference on Harmonization 1995).

The focus of this chapter is on the evaluation of adverse events (AEs) in clinical trials during drug development, prior to marketing authorization. However, the same principles apply to safety evaluation for clinical trials post authorization, throughout the lifecycle of a product. A well-defined and structured, systematic and proactive process to promptly identify and evaluate potential safety risks is the key to a successful safety program in drug development. The foremost reason for early safety signal detection is to protect study participants so that they are not exposed to unnecessary safety risks that may exceed the potential benefits. Early identification of potential safety concerns is also desirable for improving the quality of safety evidence collected across a development program. After a potential safety concern is identified, adjustments may be made to the existing and future study protocol(s), to limit safety risks to study subjects and enable focused data collection and data analysis to improve the understanding of the targeted safety issue. This effort can improve the overall quality of the safety evidence of the program, which will be reflected in the summary of safety data that is included in the filing package. It should lead to a better understanding of the post marketing commitment needs and contribute to the continued refinement of the benefit-risk profile of the product.

For any researchers and industry or regulatory professionals working in this field, we highly recommend two publications as key references. Both publications emphasize the importance of a proactive and systematic approach.

The first is a report of the Council for International Organizations of Medical Sciences (CIOMS) Working Group VI. CIOMS is an international, non-governmental, non-profit organization established jointly by WHO and UNESCO in 1949. The CIOMS mission is to advance public health through guidance on health research including ethics, medical product development and safety (https://cioms.ch/about/). The CIOMS Working Group VI (2005) report includes a set of detailed recommendations for a systematic approach to safety monitoring and evaluation during drug development. This publication recommended that a Safety Management Team should be formed for each development program. The Safety Management Team is responsible for reviewing available safety information on a regular basis so that safety issues can be identified and evaluated, decisions can be made, and risk minimization actions can be taken in a timely manner. Reviews by the Safety Management Team should start with first-in-human studies and continue throughout the entire lifecycle of the product. Identified risks and potential risks arising during the development program, along with plans to address them, should be documented in the Development Risk Management Plan.

The second is written by the Safety Planning, Evaluation, and Reporting Team (SPERT), which was a team formed in 2006 by the Pharmaceutical Research and Manufacturers of America to recommend a pharmaceutical industry standard for safety planning, data collection, evaluation, and reporting. SPERT's recommendations were based on review of relevant literature and on consensus reached in their discussions. SPERT (Crowe et al., 2009) built on the proactive and systematic approach recommended by CIOMS VI. They recommended that sponsors create a Program Safety Analysis Plan (PSAP), a complementary document to the Development Risk Management Plan that specifies the detailed procedure for safety data collection and data analysis.

Additionally, when planning safety analyses for Phase II–III studies and integrated submission documents, we highly recommend researchers refer to analysis and display white papers created as part of an FDA/PHUSE collaboration (PHUSE 2013, 2015, 2017, 2018). PHUSE is an independent, not-for-profit organization run by volunteers, established in 2004. The organization provides a global platform for the discussion of topics encompassing the work of data managers, biostatisticians, statistical programmers, and eClinical IT professionals. In 2012, PHUSE and FDA created a collaboration to provide an open, transparent and collaborative forum in a non-competitive environment in which academia, regulators, industry, technology providers and others can address computational science needs in support of health product development and regulatory review (Rosario et al., 2012). More information on PHUSE and the collaboration can be found at phuse.eu.

23.2 Elements of a Systematic Approach to Clinical Trial Safety Data Evaluation

In the systematic approach to safety evaluation in clinical trials, one of the most important elements is planning. In this section, we will discuss two aspects of planning—the use of a PSAP and planning prospectively for meta-analyses.

23.2.1 The program safety analysis plan (PSAP)

The PSAP is a program-level safety analysis plan, which may either be stand-alone or embedded in another program-level document. It describes the strategy for data collection and evaluation of clinical trial data within the same development program. Unlike a regular SAP that covers the statistical analyses for a single study, the PSAP covers clinical trials carried out in the entire program across multiple indications being developed either sequentially or in parallel. The PSAP would include details of how to integrate the individual study results into the overall analysis. The PSAP is a recommended planning document but is not yet a required regulatory document.

The PSAP is a living document that may either be updated periodically or be amended as needed in response to emerging safety information. It serves as the basis for forming the integrated statistical analysis plan (iSAP) for the Summary of Clinical Safety. The iSAP is generally prepared before a regulatory filing. The PSAP may continue to be maintained throughout the product lifecycle post market approval.

A PSAP serves as a tool to proactively plan for the program-level data collection, evaluation, and analysis of safety data. The PSAP encompasses two parts: part one provides guidance on structured and standardized data collection across the development program and the first draft is expected to be completed before the start of phase II studies (see Section 4 for more detailed discussion); part two is an analytical section that describes planned analyses that will eventually be included in the Summary of Clinical Safety. This section specifies AEs with a prior hypothesis about an association with drug exposure and AEs of special interest[1] (AESIs). It specifies methods to detect safety signals in studies in the drug development program. It also specifies methods to integrate safety data from multiple studies, and methods to analyze and graph data for reporting. Similar to a single-trial SAP, the PSAP is expected to discuss statistical methods to address issues such as missing data, multiplicity, etc. The PHUSE white papers listed in Section 1 are useful references when populating the analytical section of a PSAP.

The first version of the PSAP including the analytical section should typically be completed by the end of phase II and before pivotal phase III studies start. This will allow the sponsors to discuss the key content of the PSAP with the FDA, or other regulatory agencies, at the end-of-phase II meeting. The PSAP may be discussed at other development

---

[1] An adverse event of special interest (serious or non-serious) is one of scientific and medical concern specific to the sponsor's product or program, for which ongoing monitoring and rapid communication by the investigator to the sponsor may be appropriate. Such events may require further investigation in order to characterize and understand them. Depending on the nature of the event, rapid communication by the trial sponsor to other parties may also be needed (e.g., regulators). (CIOMS Work Group VI 2005, p. 220)

milestone meetings as needed as the program evolves and safety data accumulate (Sutter 2019). For a more detailed description of the PSAP and its implementation considerations, please see Crowe et al., (2015), Xia and Jiang (2014), Chuang-Stein and Beltangady (2011).

23.2.2   Facilitating combining data across studies, including planning meta-analyses (be prepared)

Individual studies typically have insufficient statistical power to assess safety events that occur only infrequently. Meta-analytic methods should be considered in this situation. Meta-analysis can be utilized for safety questions that are addressed retrospectively and for those that are prospectively defined and require additional attention and planning. In this section, we will discuss the planning of meta-analysis, which is another important element of systematic planning of safety evaluation.

The International Conference on Harmonisation (ICH) E9 guideline (1998) clearly stated that meta-analyses in the drug development setting should be prospectively planned "so that the relevant trials are clearly identified and any necessary common features of their designs are specified in advance." The guideline further specified the common features to be included as an essential part of a planned meta-analysis: common definitions of variables across studies in a program; consistent methods for measuring key variables; consistent timing of assessments relative to study entry; and common definition of prognostic factors.

Planning a meta-analysis should go beyond planning the operational logistics of endpoints definitions and measurement specifications. A more critical component of planning is to specify in advance the scientific questions to be addressed and even the hypothesis to be tested so that the meta-analysis is not simply a post-hoc exploratory analysis. Planning a meta-analysis is like designing a "meta-experiment", similar to planning individual experiments (studies), with the considerations to minimize bias that could be introduced by confounding factors such as study design and differences in patient characteristics across studies.

As a simple example, assuming patient sex is being considered as a potential treatment effect modifier and a pair of two confirmative trials are required for the development program. One may choose one of the two options:

(1) Design two studies each including both men and women stratified by sex

(2) Design two separate studies, one in men and the other in women

One drawback of the second design is that it completely confounds sex and "study", and it is impossible to determine whether any differences in results are due to sex or "study" effect, such as study conduct, enrollment sites, time of the study, etc. However, the first design may be impractical and undesirable in some settings, for example, if patients with one sex require a different dose or standard of care, or are hard to find and take much longer time to enroll, then separate studies may be a better choice. This simple case provides an example of what might be considered when designing a "meta-experiment."

23.3   Approaches to Characterizing the Product Safety Profile

As clinical development moves from early to late phase, information on the product safety profile also accumulates. The safety data may initially come from preclinical and toxicology studies, then from the phase I–III clinical trials and integrated analyses of the trials. After drugs are approved and marketed, less common or rare events may be observed and reported. Unlike in clinical trials, where patients are often carefully selected under narrowly defined inclusion and exclusion criteria, during post-marketing, real world patients with comorbidities and complicated medical conditions may provide further data to the safety profile of the product and, in this setting, epidemiologic data play an important role.

In this section, we focus on the characterization of the product safety profile during the development phase of a product, using data from clinical trials and meta-analyses of the RCTs, guided by the PSAP. Safety assessment in this setting should be dynamic and adjustment to the development program may need to be made to reflect the newly emerging safety information. For example, additional safety data may need to be collected; the frequency of data collection may need to be increased; analysis methods may need to be updated in the statistical analysis plan (and the PSAP); all these may take place over the course of drug development as the knowledge on the safety profile of the product accumulates.

23.3.1 Known or pre-specified safety issues

23.3.1.1 Specific safety issues that should always be considered for all products

As discussed in the previous sections, safety issues are identified as data accumulate during drug development. However, some safety questions may be anticipated based on knowledge of the drug mechanism of action or data available from related development programs. In addition, certain drug-induced toxicities are recommended by various regulatory and/or industry guidelines as events that should always be explicitly considered in planning safety data collection and assessment for any new drug development programs. These safety events have led to many new drug development failures in the past. Failure can occur during the drug development or approval process or can result in the withdrawal of a drug from the market where the public health impact could be substantial, e.g., if depriving patients of the benefits of that drug cause more harm than preventing the AE. Such safety events include, but are not limited to the following: QT prolongation (International Conference on Harmonisation 2005), liver toxicity (hepatotoxicity) (U.S. Food and Drug and Administration 2009), nephrotoxicity, immunogenicity, and bone marrow toxicity (CIOMS Working Group VI 2005).

A comprehensive approach is recommended to assess these events. For example, in order to assess hepatotoxicity with a comprehensive approach, liver damage may be evaluated using data from different sources in combination: laboratory abnormalities defined using Hy's law criteria (U.S. Food and Drug and Administration 2009); laboratory data in combination with adverse event (AE) data, various combinations of AEs (such as liver damage requiring transplant, hepatitis, or jaundice), as well as information from medical history and concomitant medication to rule out any underlying disease conditions.

23.3.1.2 Product-specific adverse events of special interest (AESIs)

In addition to the pre-specified safety events that should always be evaluated, early planning to assess potential safety issues anticipated for the product (product-specific) based on knowledge of the drug mechanism and/or data available from related development programs is also important. The early planning should include consideration of how best to define and manage product-specific AESIs. An extensive discussion of AESIs was presented by the CIOMS Working Group VI (2005) report and will not be repeated in this chapter.

23.3.1.3 Adverse events not specified in advance

As previously noted, the known safety profile of a medicine evolves over time. For example, various AEs will become of interest after the phase III studies have been unblinded.

The same analytical principles apply for these AEs, but certain desirable things won't be possible, such as prospective adjudication of events. As noted in Section 4.1, retrospective adjudication, if required, presents additional challenges compared with prospective adjudication. While it will usually be too late to collect detailed clinical information, effort should be put into reducing bias by, for example, keeping the adjudicators blinded to treatment code.

23.3.2　Data sources for safety evaluation including specific safety studies

In this section, we discuss data from other sources, in addition to data accumulated from clinical trials within a development program, which may inform the safety profile and help in subsequent safety planning. Here are some example categories of such data sources:

(1) Toxicology and Other Nonclinical Data

Toxicology and other nonclinical data are intended to inform potential safety concerns in humans. This knowledge can enable sponsors to pre-specify certain events as AESIs and proactively require focused data collection and/or reporting.

(2) Knowledge of the Epidemiology of the Disease and Patient Population

The epidemiology of the disease for which a treatment is indicated, and the characteristics of the patient population are of critical importance in understanding the safety profile of a drug product. These may include, but are not limited to, the prevalence of risk factors for potential or known adverse reactions associated with the product (or products of the same or similar mechanism) and the prevalence or incidence of the common major comorbidities and their associated mortality in the patient population. Events characteristic of the underlying patient population are sometimes called "anticipated events." For example, patients with rheumatoid arthritis (RA) are known to have higher risk of cardiovascular diseases than the general population, even after adjusting for other risk factors such as age and sex. The causes of the cardiovascular comorbidity are multifactorial, potentially including the complications of the disease itself, treatments used for the disease, and other reasons not completely explained by traditional cardiovascular risk factors (Quyyumi 2006). In this case, the knowledge of the related epidemiology is important for any drug development program in RA. It would probably result in more extensive monitoring and (possibly adjudicated) assessment of cardiovascular events. It could also provide background rates of AESIs for assessment of safety risk, which are especially helpful for development or post-marketing studies without a placebo or active control group. Special caution is warranted when comparing safety event rates from a clinical trial to external background rates due to potential confounding factors such as patient population differences (e.g., selected patients in trials and "all comers" in epidemiologic studies), observation/follow-up times, concomitant medications, standard of care, differences in definitions of the AE, and many others.

(3) Post-marketing Spontaneous AE Reporting Systems

Data from spontaneous AE reporting systems (e.g. FDA's Adverse Event Reporting System) can continue to help identify safety signals in the post-marketing phase of products. Since the FDA Sentinel Initiative was started (U.S. Food and Drug Administration 2008b, Platt et al., 2009, U.S. Food and Drug Administration 2010a, b) large databases of electronic health records and administrative claims have been utilized more frequently for the purpose of gaining further understanding and characterizing known risks, and signal detection (United States Food and Drug and Administration 2011).

23.4   Planning for Clinical Data Collection and Standardization

In contrast to efficacy data, for which the data format and collection schedule may be strictly defined and the analyses pre-specified; safety data collection and analysis need to be more flexible due to the unpredictable nature of safety events. Nonetheless, high quality data are crucial for thorough and valid safety analyses. Therefore, comprehensive approaches are desired to combine routine collection of safety data with targeted and dynamic data monitoring and assessment aimed at specific AESIs identified earlier in the development program. In addition, since many safety questions require aggregated assessment across studies, a high degree of standardization and consistency within a development program are necessary to facilitate the integrated assessment.

23.4.1   Definition of safety outcomes and adjudication

To evaluate AEs consistently, standard medical definitions (search terms) should be used whenever available. However, sometimes there may not be a standard definition and multiple variations might exist. In such cases, it is important to proactively reach out to regulatory agencies and obtain agreement on the operational definitions of the AESIs and the adequate measurement of the safety outcomes in question. As a good practice in scientific inquiry, it is always desired to pre-define safety outcomes in a transparent and reproducible manner, independent of disease areas and products, before initiating the investigation. In drug development, the pre-specified definitions may need to be completed at a critical milestone such as prior to the start of pivotal trials. Although the definitions may need to be modified as safety evidence accumulates, and there may be a need to develop a new definition retrospectively, such a practice should be avoided if possible and justifications need to be provided and discussed.

It is preferable to establish a pool of AESI definitions (e.g., using well-established Standardized MedDRA [Medical Dictionary for Regulatory Activities] Queries or SMQs) independent of the therapeutic area or product (Crowe et al., 2013). Using SMQs has many benefits. It ensures consistency in AE searching strategy; allows proper comparisons across trials and even across products; avoids unnecessary revisions and retrospective definitions; and migrates simultaneously with the MedDRA version upgrades.

MedDRA SMQs have broad and narrow search definitions. There are pros and cons for adopting either one as the pre-defined searching strategy. Some may favor broad searches (noting that a "broad" search includes both broad and narrow terms) because they ensure the inclusion of all relevant terms and will be less likely to miss any related safety event. Others may prefer broad searches because narrow searches may result in unintended loss of statistical power, e.g., by missing events that might have the same pathophysiological pathway and by reducing the total number of events. On the other hand, some may argue that defining events too broadly and including terms that are "non-differential" (i.e., equally likely to occur in each treatment arm), can underestimate the true relative risk and potentially mask a safety signal (O'Neill 1988). A broad classification might include events that are either less likely to be related to the drug-induced mechanism of action or more likely to be misclassified in clinical trials by their very nature. The observed relative risk will generally be closer to the null for the broader classification than for the narrower classification when misclassification is non-differential (Proschan, Lan, and Wittes 2006, Xia and Jiang 2014). This is consistent with the theory of signal detection: broad search allows more noise and therefore improves sensitivity but reduces specificity; narrow search on the contrary, may reduce sensitivity but improves specificity.

For general safety signal detection, Section 10.4 of the PHUSE (2017) white paper has the following recommendations:

> The Medical Dictionary for Regulatory Activities provides several levels of granularity – lowest level term (LLT), PT, high-level term (HLT), high-level group term (HLGT), and system organ class (SOC). As noted in Section 6.2 in ICH E9 (International Conference on Harmonization 1995) and Section IV (f) in the CIOMS VI report (CIOMS Working Group VI 2005), the "preferred term" is generally recommended for reporting purposes. Additionally, using the MedDRA PT for reporting is currently predominant industry practice. For now, we support the use of MedDRA PT (nested within SOC) in the standard set of static displays but believe additional steps should be taken to address potential over-granularity, and we believe further research is needed in this area. We currently believe an additional static display incorporating MedDRA PT nested within HLT and SOC could facilitate an assessment that is less granular than the MedDRA PT and is recommended for integrated summaries. However, since the MedDRA HLT level does not address all potential over-granularity issues (e.g., HLTs do not combine PTs that are in different SOCs), a review of AEs for signal detection should include a review of whether excessive splitting exists in the data that would potentially require additional grouping of terms. Such a review could be done manually (by reviewing the entire TEAE table reported using PTs), and/or by utilizing interactive tools that allow for selecting terms to group interactively. The FDA uses an interactive display tool called JReview as well as an internally developed tool called MedDRA-Based Adverse Event Diagnostics (MAED) that allows reviewers to create custom queries with specific PTs to look for safety signals. At the time of submission (when decisions are made with respect to determining ADRs to communicate in labeling), a discussion in the Summary of Clinical Safety on how events were reviewed to address these potential pitfalls and the result of the review is often warranted

Some safety outcomes may require developing an event-specific case-report form (CRF) to collect additional detailed information. For example, the collection of additional and more-detailed data for subjects with a suspected cardiac AE (e.g., venous thrombotic event) may be required by the protocol to facilitate safety assessment. In this case, a specific definition of a venous thrombotic event (usually including deep vein thrombosis and pulmonary embolism) may require diagnostic confirmation (e.g., venography). However, a standard AE CRF is not designed to collect the required confirmatory evidence. A specific CRF may be designed to collect the confirmatory test results. A case identification number will be used to provide link between the AE event and the specific CRF with the additional diagnostic information.

In some situations, adjudication by an expert or a group of experts (an adjudication committee) may be desirable for certain AEs. For example, an excess of hepatic events noted in a phase II study may prompt adjudication of such events in phase III. The adjudication committee provides a medical classification and standardized, unbiased evaluation of an AE of special interest. Generally, adjudication is considered high quality and credible if it is done prospectively with review being blinded to study treatment assignment, and using pre-specified criteria. Retrospective adjudication presents additional challenges, at least in part because it may be too late to collect the detailed clinical information required for the adjudication.

23.4.2   Standardization of safety data collection

Standardization can be applied to the following aspects in safety data collection:

- Design aspects of the clinical studies (for example, a common, or at least compatible, visit structure)

- Coding procedures for AEs, medical history, concomitant medications, and other areas

- Data collection procedures and definitions (using Clinical Data Acquisition Standards Harmonization [CDASH] where possible)

- Case-report form design and instructions, including processes and procedures for eliciting safety outcomes (CIOMS Working Group VI 2005, p 79)

- Definitions of AEs of special interest

- Definitions of subgroups of patients (e.g., risk groups of special interest, such as those with "mild renal impairment," could be defined by using the same threshold values for a laboratory test)

- Adjudication process

Standardization plays a crucial role in the integration of safety evidence across studies within a development program. Standardization across companies, although more challenging and complex, could facilitate data integration and analysis of safety events of interest across products in a disease area and patient population; more importantly, it would allow regulatory reviewers to standardize data collection and analysis methods, and potentially conduct cross-company meta-analyses to investigate potential class effects.

23.5   Safety Data Analysis and Reporting

Safety assessment in drug development should focus on the identification of potential harms; characterization and quantification of known harms; identification of risk factors or subgroups for potential and known harms; and understanding the relationship between exposure (dose and duration) and safety outcomes.

The CIOMS Working Group VI (2005) proposed the conduct of periodic aggregate safety data reviews as a recommended approach. The focus of these reviews will likely be on unblinded data from the completed trials and blinded data from the ongoing trials. Even blinded data can provide valuable and timely information to help identify unusual patterns that may be suggestive of a safety signal. The aggregated safety reviews can be enhanced by using graphical methods to review patient profile data; graphical approaches to complement statistical analyses; and other advanced methods, such as meta-analysis or competing risk analysis during development. Should an aggregate safety data review identify important safety issues, any of the following may occur: prompt communication to relevant parties (such as regulatory authorities or institutional review boards), changes to informed consent, data collection, monitoring procedures, conduct of trials, the development plan or amendments to protocols.

Unblinded review of ongoing trial data by a data monitoring committee (DMC) or a select group of people internal to the sponsor, but independent from the staff members managing the conduct of the trial, may be warranted and necessary if one or more safety events of clinical importance (e.g., a life-threatening AE) were reported. The unblinded review can help determine if a safety signal is present and if steps are needed to ensure the safety of clinical trial subjects. As a good practice in study planning, study teams should consider if unblinded safety assessments are needed based on the study design and the accumulated safety data, and if so, whether they will be carried out by the DMC if available, or by the internal independent review group. The detailed specifications should be documented in the DMC charter or internal independent review team charter.

There are many analytical challenges in analyzing safety data from clinical trials. Safety endpoints frequently lack well-defined, prespecified hypotheses, which can lead (paradoxically) to both multiplicity issues and low statistical power. Multiplicity concerns can arise from multiple looks at safety endpoints, or from the large number and types of safety endpoints that are typically analyzed and reported.

23.5.1 Considerations for individual studies

23.5.1.1 Defining the safety analysis set

The intention-to-treat (ITT) principle implies that the primary statistical analysis should include all randomized subjects. The ICH E9 (1998) guideline recommended using the ITT principle for individual clinical trials because it is important in preventing bias and in providing a secure foundation for statistical tests. In meta-analysis of efficacy, it is always important to consider if the individual studies included in the meta-analysis followed the ITT principle. The CONSORT (Consolidated Standards of Reporting Trials) group (Ioannidis et al., 2004) recommended using the ITT population for safety data analysis, in general, as well. However, the CIOMS VI Working Group (CIOMS Working Group VI 2005) made a counter argument that ITT analyses may not always be most appropriate for the analysis of safety data, since this approach may have a tendency to underestimate the true differences between the groups.

CIOMS VI further suggested that another definition of the safety analysis population, such as subjects who received a prespecified minimum number of doses of the study drug, might be more appropriate. As they noted, this definition may also result in bias, and the direction of the bias is unknown due to lack of knowledge about the relationship between the reasons for early stopping of treatment and the safety outcomes. In fact, ICH E9 (1998) suggests that it may be reasonable to exclude any subjects who did not receive any study medication or who have no data post-randomization, provided that any drop outs before first study treatment are random and unrelated to study treatment assignment. Due to this complexity, teams might consider sensitivity analyses using different population definitions from the primary population and also review safety issues reported from those excluded subjects who received fewer than the required number of doses. Additionally, alternative statistical methods may be useful. For example, Wang et al., (2015) proposed a method to assess safety in randomized trials with noncompliance.

23.5.1.2 Accounting for time on or off treatment

Compliance with the ITT principle also necessitates complete follow-up of all randomized subjects for study outcomes, including safety. In real practice, this is difficult to achieve, and the length of follow-up time may vary from patient to patient and by treatment group. The length of time that each patient is "at risk" of having an AE has to be considered in any assessment of risk. For meta-analysis across studies, differences in follow-up time from study to study will need to be considered as well (Berlin et al., 2013). Follow-up time may not be the same as the duration of treatment (exposure) since some studies may be designed with separate on-treatment and post-treatment follow-up periods, or may allow subjects to remain in the study after treatment is discontinued. The CIOMS Working Group VI (2005, p 150) highlighted the importance of this post-treatment follow-up period and recommended that it should be extended to at least 5 half-lives of the drug after treatment is stopped; or even beyond that if safety events with long latency, e.g., cancer, are being investigated. Each protocol should clearly define the required follow-up period and end of study for each patient that will be used for inferential statistical analysis.

Safety event rates are often calculated per person-time when follow-up time varies, but this approach assumes constant hazard rates over time. In situations where this assumption is violated, it may be necessary to break the observation period into reasonable sub-periods (e.g., month 1, month 2, etc.) and analyze by each period separately (U.S. Food and Drug Administration 2005).

For studies in which patients have different drop-in and drop-out times, survival analysis techniques, such as the Kaplan-Meier, Cox proportional hazards, piece-wise exponential models, and other time-to-event methods, may be helpful. These methods can properly handle differential dropout rates and provide less-biased estimates than methods ignoring person-time. However, none can properly adjust for differential dropout rates when the reasons for dropping out are related to the AE that is being assessed (i.e., there is so-called informative censoring). For example, in a situation where placebo-treated patients tend to discontinue due to lack of efficacy and test drug-treated patients tend to discontinue because of AEs, even though dropout rates appear to be similar between groups, person-time rates or survival methods may still result in biased results in comparing the treatment groups.

Depending on study design and the nature of a safety event, different analysis methods may be considered. For events that tend to occur early in treatment if they are going to occur at all (such as hypersensitivity reactions), ignoring person-time may be most appropriate. For treatment taken on an as-needed basis, analyses based on total doses received (after adjusting for time at risk) may be more helpful since time on study may not be closely correlated with drug exposure.

23.5.2   Meta-analysis of adverse event data

Combining results from multiple randomized studies has many known benefits resulting from the increase in overall sample size. In general, the benefits include increased precision of treatment effect estimates and increased statistical power to test hypotheses relative to individual studies. Furthermore, many important questions unanswered within individual studies (e.g., subgroup effects), may be assessed with the gain in power from the enlarged sample size.

Combined analysis is not the same as crude pooling. Crude pooling, which means that we ignore the fact that the data come from more than one study, should generally be avoided (Lièvre, Cucherat, and Leizorovicz 2002, Chuang-Stein and Beltangady 2011, Crowe, Wang, and Nilsson 2014, Crowe et al., 2016). Crude pooling of AE numbers across different trials to estimate treatment effect can be misleading, especially when the randomization ratio varies among studies and study is a confounding factor to the treatment effect.

Meta-analysis methods can be used to assess product safety using the combined data from multiple studies. Such meta-analyses can be used both to estimate an average treatment effect across studies and to explore reasons for heterogeneity among study-specific effects (Vanhonacker 1996, Thompson, Smith, and Sharp 1997, Schmid 1999, Thompson and Sharp 1999, Greenland and O'Rourke 2001, Higgins et al., 2002, Sterne et al., 2002). Berlin et al., (2013) published a manuscript on answers to frequently asked questions with respect to meta-analysis of clinical trial safety data in a drug development program. They addressed the following topics: choice of studies to pool, effects of the method of ascertainment, use of individual patient-level data compared to trial-level summary data, the need for multiplicity adjustments, heterogeneity of effects and its sources, and choice of fixed effect versus random effects models.

In combined analyses of AE data across trials, it might be desirable to preserve the stratification by the original randomization stratification factors (e.g., stratification by study, or by clinical subgroup and study simultaneously). It is important to examine the stratum-specific results if the study stratification factors are considered potential modifiers of treatment effect. If effect modifier is of interest, one could stratify analyses on the relevant factors, regardless of the stratification used in the original studies.

23.5.3 Multiplicity

As noted at the beginning of this section, multiplicity is a key analytical challenge in analyzing safety data. It can arise from multiple looks at a specific endpoint over time, and from the large number of different types of safety endpoints that are typically analyzed.

The SPERT paper describes a 3-tiered approach to safety endpoints: those for which a prior hypothesis is clearly defined (Tier 1); commonly occurring events for which there is no prior hypothesis (Tier 2); and uncommon or rare events for which there is no prior hypothesis (Tier 3). The paper suggested multiplicity adjustment for multiple looks for Tier 1 events; and multiplicity adjustment for multiple endpoints for Tier 2. For Tier 3 endpoints, there may generally be too few events to make any meaningful statistical comparisons, with or without adjustment for multiplicity.

Cumulative meta-analysis may be conceptualized as a series of meta-analyses when a new meta-analysis is performed as the data from a new clinical trial become available (Antman et al., 1992), or at prespecified decision points. Cumulative meta-analysis can be applied to the repeated integrations of AESIs for which there are prespecified hypotheses (Tier 1 events) to be tested throughout the course of development. An example is the evaluation of cardiovascular risks in Type 2 diabetic drug development (U.S. Food and Drug Administration 2008a). Bayesian philosophy of continuous learning is well reflected in cumulative meta-analysis. We learn as we accumulate data, and today's posterior becomes tomorrow's prior (Stangl and Berry 2000). Special caution on multiplicity may arise when interpreting cumulative meta-analyses. Methods for addressing multiplicity (e.g., in analyses done at prespecified time points) have been developed in the context of cumulative meta-analysis (Berkey et al., 1996, Hu, Cappelleri, and Lan 2007).

Multiplicity resulting from simultaneously examining many safety endpoints (Tier 2 and 3 events) is another challenge. Any multiplicity adjustment must strike a reasonable balance between false positives (Type-I error) and false negatives (Type-II error) and assist in interpreting the safety findings. A number of multiplicity adjustment methods have been proposed in this context, including the "Double False Discovery Rate" multiplicity-adjustment procedure by Mehrotra and Heyse (2004) and the Bayesian hierarchical modeling methods by Berry and Berry (2004). The Double False Discovery Rate method requires a clear distinction between Tier 2 and Tier 3 events because uncommon events are removed from the first pre-processing step; but this distinction is not necessary with the Bayesian approach. One of the advantages of the Bayesian approach is that the entire AE dataset is analyzed and the extremes are modulated through borrowing strength among different AEs under the existing hierarchical MedDRA coding structure. For example, if many AE preferred terms (PTs) under the same System Organ Class (SOC) were reported in multiple subjects in the treatment arm and none were reported in the control arm, the Bayesian approach, by borrowing strength among different AEs within the same SOC, may be able to detect the strengthened safety signal. The same signal could be missed with traditional frequentist methods due to lack of statistical significance at the individual PT level.

Xia, Ma, and Carlin (2011) extended the work of Berry and Berry (2004) by accounting for various exposure or follow up times among different subjects under the Poisson model. They recommended using simulation studies to select a signal detection threshold, and demonstrated that the Bayesian approach outperformed other methods in the scenarios that they simulated.

23.5.4   Signal detection for common events

Screening AEs in clinical trials for safety signal is a complex task as the range of possible adverse effects is very large and new and unforeseeable effects are always possible. It requires thorough review of safety data with statistical guidance, yet should not be dependent entirely on formal statistical criteria for making conclusions. Clinical judgment is always necessary so as to avoid basing decisions on what is likely to be artifact. Southworth and O'Connell (2009) cited an example in which the non-specific event of "pain" appears to be statistically significant. However, a closer look revealed further detail of the events which included "tender inflamed nostrils," "toothache," "tenderness in big toe" and "pain in anterior lower part of leg when going to bed." Considering the lack of a meaningful pattern or similarity among these events, the authors suggested (but could not confirm) that the difference between treatment groups was likely due to chance. They further noted that the p-value for this association was 0.002, possibly statistically significant even after multiplicity adjustment. Conversely, given the low frequency of some severe events, a real signal might miss a statistical threshold yet be clinically important. The bottom line is that clinical perspective, based on the knowledge of the class of drugs or the drug mechanism of action, plays an important role in safety signal detection.

In the spontaneous reports of AEs in the post-approval setting, signal generation is commonly implemented. Similar methods can be applied to the drug development programs while clinical trials are ongoing. Unlike in the spontaneous report setting, the number of exposed patients (denominator for calculation) is known in the development program. Signal generation can be accomplished through a combination of inferential and descriptive statistics with the goal being to screen for safety signals rather than making definitive conclusions. A set of hierarchical criteria may be developed based on p-values (remembering that multiplicity may be a concern), the magnitude of relative risk estimates, and evidence of dose-response. A recent publication described a set of hierarchical criteria that can be used to screen large numbers of AEs (Crowe et al., 2013). Bayesian approaches are promising in this setting as described in the previous section.

Recent technology advancement in data mining techniques and visualization tools can bring safety screening and signal detection to a next level of sophistication. Several methods are presented by Southworth and O'Connell (2009). These methods include a so-called inside-out data mining method that treats AEs as explanatory variables, with treatment allocation as the "outcome" variable; a support method that fits separate regression models to each AE, with and without a term for the treatment effect; and a hierarchical Bayesian model for the analysis of counts of AEs. These methods focus on strength of evidence, rather than p-values. They can be implemented in conjunction with graphical methods and other individual-level safety data (e.g., lab values, concomitant medications) to enable a deeper and broader profiling of patient safety.

23.5.5   Descriptive analysis of infrequent adverse events

For AEs that are infrequent (Tier 3 per SPERT definition), there are unlikely to be any prespecified hypotheses. Nonetheless, clinical evaluation of these events is important for the overall benefit-risk profile of a product. Some of the rare safety events are severe and of great clinical importance and may meet criteria for Serious and Unexpected Suspected Adverse Reactions (SUSAR). Analysis of rare events (particularly those with zero events being observed in 1 or more arms) is challenging. Preferred methods for rare events are addressed in a few papers (Sweeting, Sutton, and Lambert 2006, Sweeting, Sutton, and Lambert 2004, Bradburn et al., 2007).

Relative risks and odds ratios can be misleading in comparing safety risks of rare events, because they tend to magnify the differences between groups. For example, a change in event rate of 1 per 10,000 to 3 per 10,000 triples relative risk, however it may not carry the same implications as tripling (or even doubling) event rates of a common event. In this context, severity of the event is also important to consider. When analyzing and reporting rare events, comparisons in both absolute and relative metrics should be provided, and the results should be assessed with clinical judgment.

23.5.6 Reporting

As discussed in the previous section, appropriate metrics should be used to report safety data. Depending on the objective and questions to be addressed, the choice may be different. In general, absolute risk (e.g., frequency, subject incidence, or incidence rate per person-time of exposure) is considered a good descriptive summary statistic for both common and rare events. Risk difference, relative risk, or odds ratio and their corresponding confidence intervals or p-values are generally reported for comparisons between groups (Chuang-Stein and Beltangady 2011, Deeks 2002, Localio, Margolis, and Berlin 2007). While relative metrics such as risk ratio and odds ratio are very useful for signal detection, risk differences may be required for assessment of benefit-risk. In addition to treatment, patient demographics (e.g., age, sex, etc.) and baseline characteristics (e.g., comorbidities, disease status, etc.) may be risk factors (confounders) for certain safety events. These potential risk factors should be investigated as predictors of the safety event within each treatment group and overall. There may also be treatment by subgroup interactions, that is, particular subgroups of patients may have increased safety risk due to treatment, compared with other subgroups. A proper assessment of the risk factors would call for prospective collection of relevant risk factor information.

23.6 Conclusions

In recent years, both industry and regulatory agencies have placed an increased emphasis on a systematic approach to safety evaluation and early detection of potential harms. There is a desire to identify safety issues for new compounds (or new indications for existing compounds) early in the drug development process. A well-defined, coordinated, program-wide approach to safety evaluation during new product development programs is the key to success. The creation of a prospectively-defined program-level data collection and analysis plan (the PSAP, as recommended by SPERT (Crowe et al., 2009)) and regular reviews of aggregate data by a multidisciplinary safety management team (as recommended by the CIOMS Working Group VI (2005)) are two key elements of such a systematic approach.

Potential harms may be identified at any time in a drug development program. A comprehensive, dynamic and proactive approach allows data collection strategies to be modified in time to collect additional data to help understand the safety issue. This effort can improve the overall quality of the safety evidence of the program, which will be reflected in the summary of safety data that is included in the filing package. It should lead to a better understanding of the post marketing commitment needs and contribute to the continued refinement of the benefit-risk profile of the product. These efforts have the potential to enhance understanding of the safety profile of the product and promote use of the drug so that the benefits will outweigh the risks for the target patient population.

Disclaimer: The views expressed herein represent those of the authors and do not necessarily represent those of their employers or any other parties.

Bibliography

Antman, E M, J Lau, B Kupelnick, F Mosteller, and TC Chalmers. 1992. "A comparison of results of meta-analyses of randomized control trials and recommendations of clinical experts." *JAMA*:240–248.

Berkey, CS, F Mosteller, J Lau, and EM Antman. 1996. "Uncertainty of the time of first significance in random effects cumulative meta-analysis." *Control Clin Trials* 17:357–371.

Berlin, J. A., Brenda Crowe, E. Whalen, H. A. Xia, C. E. Koro, and J. Kuebler. 2013. "Meta-analysis of clinical trial safety data in a drug development program: answers to frequently asked questions." *Clinical Trials* 10 (1):20-31. doi: 10.1177/1740774512465495.

Berry, S M, and D A Berry. 2004. "Accounting for multiplicities in assessing drug safety: A three-level hierarchical mixture model." *Biometrics* 60:418–426.

Bradburn, M. J., J. J. Deeks, J. A. Berlin, and A. Russell Localio. 2007. "Much ado about nothing: A comparison of the performance of meta-analytical methods with rare events." *Statistics in Medicine* 26 (1):53-77. doi: 10.1002/sim.2528.

Chuang-Stein, Christy, and Mohan Beltangady. 2011. "Reporting cumulative proportion of subjects with an adverse event based on data from multiple studies." *Pharmaceutical Statistics* 10 (1):3–7.

CIOMS Working Group VI. 2005. *Management of safety information from clinical trials.* *Geneva*: Council for International Organizations of Medical Sciences

Crowe, Brenda, Andreas Brueckner, Charles Beasley, and Pandurang Kulkarni. 2013. "Current practices, challenges, and statistical issues with product safety labeling." *Stat Biopharm Res* 5 (3):180–193. doi: 10.1080/19466315.2013.791640.

Crowe, Brenda, Christy Chuang-Stein, Sally Lettis, and Andreas Brueckner. 2016. "Reporting adverse drug reactions in product labels." *Therapeutic Innovation & Regulatory Science* 50 (4):455–463. doi: 10.1177/2168479016628574.

Crowe, Brenda, Wei V. Wang, and Mary E. Nilsson. 2014. Advances in techniques for combining data from multiple clinical trials. In *Advances in collating and using trial data*, edited by Matthieu Resche-Rigon and Sylvie Chevret: Future Science.

Crowe, Brenda, H Amy Xia, Mary E. Nilsson, Seta Shahin, Wei V. Wang, and Qi Jiang. 2015. "The program safety analysis plan: An implementation guide." In *Quantitative evaluation of safety in drug development: Design, analysis, and reporting*, edited by Qi Jiang and H Amy Xia, 55–68. London: Chapman & Hall.

Crowe, Brenda, H. Amy Xia, Jesse Berlin, Douglas J. Watson, H. Shi, S. L. Lin, Juergen Kuebler, Robert C. Schriver, Nancy. Santanello, George Rochester, J. B. Porter, Manfred Oster, Devan Mehrotra, Z. Li, E. C. King, E. S. Harpur, and David B. Hall. 2009. "Recommendations for safety planning, data collection, evaluation and reporting during drug, biologic and vaccine development: a report of the safety planning, evaluation, and reporting team." *Clin Trials* 6 (5):430–440. doi: 10.1177/1740774509344101.

Deeks, J. J. 2002. "Issues in the selection of a summary statistic for meta-analysis of clinical trials with binary outcomes." *Stat Med* 21 (11):1575–1600. doi: 10.1002/sim.1188.

Greenland, S, and K O'Rourke. 2001. "On the bias produced by quality scores in meta-analysis, and a hierarchical view of proposed solutions." *Biostatistics* 2:463–471.

Higgins, J., S. Thompson, J. Deeks, and D. Altman. 2002. "Statistical heterogeneity in systematic reviews of clinical trials: a critical appraisal of guidelines and practice." *J Health Serv Res Policy* 7 (1):51–61.

Hu, M, J C Cappelleri, and K KG Lan. 2007. "Applying the law of iterated logarithm to control type I error in cumulative meta-analysis of binary outcomes." *Clin Trials* 4:329–340.

International Conference on Harmonisation. 1998. E9: Statistical Principles for Clinical Trials. *International Conference on Harmonization Guidelines.*

International Conference on Harmonisation. 2005. E14: The clinical evaluation of QT/QTc prolongation and proarrhythmic potential for non-antiarrhythmic drugs. Accessed February 13, 2016.

International Conference on Harmonization. 1995. "ICH E9 statistical principles for clinical trials ICH Harmonised Tripartite Guideline." International Conference on Harmonization, accessed 25 May 2015. http://www.ich.org/products/guidelines/efficacy/efficacy-single/article/statistical-principles-for-clinical-trials.html.

Ioannidis, J. P., S. J. Evans, P. C. Gotzsche, R. T. O'Neill, D. G. Altman, K. Schulz, D. Moher, and Consort Group. 2004. "Better reporting of harms in randomized trials: an extension of the CONSORT statement." *Ann Intern Med* 141 (10):781–788.

Lièvre, Michel, Michel Cucherat, and Alain Leizorovicz. 2002. "Pooling, meta-analysis, and the evaluation of drug safety." *Curr Control Trials Cardiovasc Med* 3 (1):6.

Localio, A R, D J Margolis, and J A Berlin. 2007. "Relative risks and confidence intervals were easily computed indirectly from multivariable logistic regression." *J Clin Epidemiol* 60:874–882.

Mehrotra, D V, and J F Heyse. 2004. "Use of the false discovery rate for evaluating clinical safety data." *Statistical Methods in Medical Research* 13:227–238.

O'Neill, Robert T. 1988. "Assessment of safety." In *Biopharmaceutical Statistics for Drug Development*, edited by Karl E Peace. New York: Marcel Dekker.

PHUSE. 2013. Analyses & Displays Associated with Measures of Central Tendency - Focus on Vital Sign, Electrocardiogram, & Laboratory Analyte Measurements in Phase 2-4 Clinical Trials & Integrated Submission Documents.

PHUSE. 2015. Analyses and Displays Associated with Outliers or Shifts from Normal to Abnormal: Focus on Vital Signs, Electrocardiogram, and Laboratory Analyte Measurements in Phase 2-4 Clinical Trials and Integrated Summary Documents.

PHUSE. 2017. Analysis and Displays Associated with Adverse Events: Focus on Adverse Events in Phase 2-4 Clinical Trials and Integrated Summary Documents.

PHUSE. 2018. Analyses and Displays Associated with Demographics, Disposition, and Medications in Phase 2-4 Clinical Trials and Integrated Summary Documents.

Platt, R, M Wilson, K A Chan, J S Benner, J Marchibroda, and M McClellan. 2009. "The new Sentinel Network–improving the evidence of medical-product safety." *N Engl J Med* 361:645–647.

Proschan, M A, K KG Lan, and J T Wittes. 2006. *Statistical Methods for Monitoring Clinical Trials*. New York: Springer.

Quyyumi, A A. 2006. "Inflamed Joints and Stiff Arteries. Is Rheumatoid Arthritis a Cardiovascular Risk Factor?" *Circulation* 114:1137–1139.

Rosario, Lilliam A, Timothy J Kropp, Stephen E Wilson, and Charles K Cooper. 2012. "Join FDA/PhUSE working groups to help harness the power of computational science." *Drug Information Journal* 46 (5):523–524.

Schmid, C H. 1999. "Exploring heterogeneity in randomized trials via meta-analysis." *Drug Inf* J 33:211–224.

Southworth, H, and M O'Connell. 2009. "Data mining and statistically guided clinical review of adverse event data in clinical trials." *Journal of Biopharmaceutical Statistics* 5:803–817.

Stangl, D K, and D A Berry. 2000. *Meta-analysis in medicine and health policy*: Marcel Dekker.

Sterne, J A, P Juni, K F Schulz, D G Altman, C Bartlett, and M Egger. 2002. "Statistical methods for assessing the influence of study characteristics on treatment effects in 'meta-epidemiological' research." *Stat Med* 21:1513–1524.

Sutter, Sue. 2019. US FDA Looks To Standardize Premarketing Safety Assessments. *Pink Sheet* 08 Sep 2019.

Sweeting, Michael J, Alex J Sutton, and Paul C Lambert. 2006. "Correction." *Stat Med* 25:2700.

Sweeting, Michael J., Alexander J. Sutton, and Paul C. Lambert. 2004. "What to add to nothing? Use and avoidance of continuity corrections in meta-analysis of sparse data." *Stat Med* 23 (9):1351–75. doi: 10.1002/sim.1761.

Thompson, S G, and S J Sharp. 1999. "Explaining heterogeneity in meta-analysis: a comparison of methods." *Stat Med* 18:2693–2708.

Thompson, S G, T C Smith, and S J Sharp. 1997. "Investigating underlying risk as a source of heterogeneity in meta-analysis." *Stat Med* 16:2741–2758.

U.S. Food and Drug Administration. 2005. Reviewer guidance: Conducting a clinical safety review on a new product application and preparing a report on the review.

U.S. Food and Drug Administration. 2008a. Guidance for industry: Diabetes mellitus-evaluating cardiovascular risk in new antidiabetic therapies to treat Type 2 diabetes. Accessed February 13, 2016.

U.S. Food and Drug Administration. 2008b. The Sentinel Initiative: A national strategy for monitoring medical product safety.

U.S. Food and Drug Administration. 2010a. FDA's Sentinel Initiative - Ongoing projects. Accessed February 13, 2016.

U.S. Food and Drug Administration. 2010b. The Sentinel Initiative: An update on FDA's progress in building a national electronic system for monitroing the postmarket safety of FDA-approved drugs and other medical products. Accessed February 13, 2016.

U.S. Food and Drug and Administration. 2009. Guidance for Industry Drug-Induced Liver Injury: Premarketing Clinical Evaluation. Accessed February 13, 2016.

Vanhonacker, W R. 1996. "Meta-analysis and response surface extrapolation: a least squares approach." *Amer Stat* 50:294–299.

Wang, Y., J. A. Berlin, J. Pinheiro, and M. A. Wilcox. 2015. "Causal inference methods to assess safety upper bounds in randomized trials with noncompliance." *Clin Trials* 12 (3):265–275. doi: 10.1177/1740774515572352.

Xia, H. A., and Q. Jiang. 2014. "Statistical evaluation of drug safety data." *Therapeutic Innovation and Regulatory Science* 48 (1):109–120.

Xia, H. Amy, H Ma, and B P Carlin. 2011. "Bayesian Hierarchical Modeling for Detecting Safety Signals in Clinical Trials." *Journal of Biopharmaceutical Statistics* 21:1006–1029.

24

Non-Inferiority Trials

Brian L. Wiens

CONTENTS

24.1 Background and History

Active control non-inferiority trials compare an investigational treatment to an active control. The intention of the analysis is to demonstrate that the investigational treatment is not much worse than the active control. By extension, this will also imply that the investigational treatment is superior to placebo through an indirect comparison, if the active control has been previously shown superior to placebo in a historical comparison.

What are now called non-inferiority trials were proposed in the 1980s to counter a tendency to conclude similarity between any treatments that did not demonstrate statistically significant differences. A conclusion of similarity is not appropriate when two treatments are not significantly different in an underpowered study. Similarly and not as widely understood, a study that is overpowered to find an important difference might produce a conclusion that two treatments are different, but the magnitude might be too small to be of interest. For these reasons, Blackwelder (1982) proposed testing the hypotheses as follows:

$$H_0: \mu_C - \mu_T \geq \delta$$

$$H_1: \mu_C - \mu_T < \delta$$

In this notation, the subscript C denotes the control group, the subscript T denotes the test group, μ is the population mean, $\delta > 0$ is the non-inferiority margin, and larger values are considered better without loss of generality. If the non-inferiority margin is chosen appropriately, rejecting the null hypothesis is tantamount to concluding that the test treatment is not importantly worse than the control treatment.

The mathematics of analysis of non-inferiority trials is generally straightforward. However, the non-mathematical aspects can be complex and nuanced, requiring great care. Additionally, it is not always possible to prove that the non-mathematical aspects were handled appropriately, which can lead to concern about whether the study conclusions are appropriate.

Finally, we note that one-sample non-inferiority analyses are also possible, comparing a single treatment to a standard. This paper will only consider two-sample analyses, and we refer a reader interested in one-sample non-inferiority studies to, for example, Wellek (2010).

24.2   Basics

In this section, we outline some of the basic issues, assumptions and applications of non-inferiority clinical trials. Since the seminal article in 1982, many issues have been found with design, analysis and interpretation of non-inferiority trials. Some of the issues have been fully resolved (or, at least, are currently believed to be fully resolved) while other issues are less well understood or have solutions that are not universally agreed. In this section, we start with issues that we treat as well-understood and with widely accepted solutions.

24.2.1   Historical studies

An essential element of designing a non-inferiority trial is the consideration of the historical comparison of the active control to placebo. An objective in many, if not most, non-inferiority trials are to demonstrate that an investigational treatment is superior to placebo. Because this involves an indirect comparison to placebo, it is vital to accurately estimate the magnitude of treatment effect of the active control versus placebo. These historical comparisons will be used to set the non-inferiority margin and will also influence other details of the study design.

It is tempting to find one relevant historical comparison of the active control to placebo and use that comparison only. That temptation must be avoided. All comparisons of the active control to placebo should be considered, even if some carry more influence than do others.

The non-inferiority trial must match aspects of study design of the historical comparisons of active control to placebo to the extent possible. Aspects that should be matched include inclusion and exclusion criteria, treatment duration, dosage and posology and endpoints. Additionally, sponsors should choose study sites that have similar standards of treatment as in the historical comparisons—same geographic regions, same countries, even the same investigators can be used. Although not all aspects of study design can be perfectly matched, sponsors should attempt to match as many as possible, as closely as possible, with the understanding that each difference will reduce the credibility at the conclusion of the trial in ways that are not easily quantified. Notably, if concomitant care, diagnosis or other aspects to clinical practice have changed since the historical comparison of the active control to placebo, this will affect the design, and in an extreme situation may make a non-inferiority trial impossible.

24.2.2   Parameters and margins

Non-inferiority trials are analyzed on the same parameter as the historical trials that compared the active control to placebo. Arithmetic means and proportions are therefore common, with geometric means, medians and time-to-event parameters also used at times. Analyses of ordinal data can also be considered, but such data have limitations that will be discussed in this section. At times, the primary endpoint for the non-inferiority trial may be different than the primary endpoint from the historical comparisons, but it will be a parameter that can be calculated from data recorded in the historical comparisons.

Once a parameter is chosen, the non-inferiority margin must be chosen. (As an aside, this might be an iterative process: if a margin cannot be chosen for a given parameter, it is possible that a different parameter will be needed to support decision-making.) The margin must be chosen prospectively, with agreement obtained from all interested parties (notably, regulatory agencies) before the trial begin. Because of the subjectivity involved, no argument after the data are collected can be used to convincingly support a non-inferiority margin.

There are multiple criteria for a non-inferiority margin, and the chosen margin must satisfy all criteria. The two most important are the putative placebo margin and the clinical importance margin, dubbed M_1 and M_2, respectively, in guidance from the United States Food and Drug Administration. (FDA, 2010) The putative placebo margin is a number smaller than the treatment difference between the active control and placebo. Because it is smaller than the estimated treatment difference from prior studies, it is very unlikely that placebo will be shown to differ from the active control by less than this quantity; thus, rejecting the null hypothesis implies that the investigational treatment is superior to placebo. The clinical importance margin is a value that, if true, represents a treatment difference that is of little relevance to a clinician or a patient. If the investigational treatment is worse than the active control by less than such a quantity, it can be substituted for the active control without concern about worse outcomes. While both M_1 and M_2 must be considered, it is expected that $M_1 > M_2$, and so M_2 will eventually be selected as the margin in the hypothesis test.

To obtain the putative placebo margin, the quantitative difference between the active control and placebo must be estimated. The purpose is to estimate a putative placebo effect—the difference that would be observed between the active control and placebo in the upcoming trial, if placebo was used. Again, this cannot just be the difference from one comparison of the active control to placebo, but must consider all comparisons. Additionally, an average treatment effect is not sufficient: as an important consideration is the difference that would have been observed between the investigational treatment and placebo, had placebo been used, and that might be different from the average effect. Rather, several subjective issues must be considered.

- Consistency of quantitative treatment effects is important. When most or all of the historical trials produced similar treatment effects, not only is it easier to calculate a putative placebo effect, it is easier to support this effect as well.

- Outlying trials must be carefully considered. It is tempting to simply discard a historical trial that has a much larger or much smaller treatment effect than other comparisons of the active control to placebo, but (especially when the outlying treatment effect value is much smaller) it is dangerous to do so. In the event that some aspect of the outlying trial design or conduct affected the observed effect, the same detail could do so in the non-inferiority trial. When that aspect is not obvious, it must be assumed that it could also occur in the non-inferiority trial.

- Historical trials that most closely match the design and conduct of the non-inferiority trial will carry more weight. This is a subjective assessment. Again, it is tempting to find a reason to omit trials that have smaller observed treatment effects, and it is always possible to reverse engineer an explanation for why a particular trial must be downweighted. Because of the incentive to exclude trials that have smaller treatment effects, the standard for such trials should be higher, not lower, than for other trials.

- A point estimate of the treatment effect is not sufficient. The lower bound of a confidence interval instead will be more appropriate. Because it gives a reasonable limit on the smallest true difference between the active control and placebo, the lower bound of a confidence interval will be less than the difference observed between placebo and the active control, if the two were compared directly. Thus, an investigational treatment that is within this margin is almost certainly superior to placebo, based on the indirect comparison. Note that the confidence interval need not be a 95% confidence interval, but a lower confidence level will reduce the assurance that a successful non-inferiority trial effectively excludes placebo.

The putative placebo margin can be based in part on calculations and in part on subjective assessment of the various aspects of the historical comparisons of the active control to placebo, but the clinical importance margin is even more subjective. Additionally, many interested parties must be considered and all may have a different value that is clinically relevant. These interested parties may include not just physicians and patients, but also third-party payers (governmental or non-governmental), regulators, pharmacists, hospitals, other caregivers and advocacy groups. When the putative placebo margin is chosen, it will be smaller than the true difference. The ratio between the margin and the true difference will come into play when discussing the analysis, and will be designated as $\phi = \delta/(\mu_C - \mu_P)$. However the denominator of this ratio will often be the lower bound of a confidence interval, not the true difference, so ϕ plays the role of a retention factor, relating how much of the minimum true difference must be retained for a conclusion of non-inferiority.

The clinically important difference can be different for different investigational treatments even in the same indication, depending on other aspects of the treatments. A treatment with toxic side effects, or that is more difficult to administer, may require a smaller clinically important difference because patients are more willing to give up efficacy for a treatment that has fewer such difficulties than for a treatment that involves side effects or complicated posology. Formal assessment of benefits versus risks are being considered as part of a patient-weighted benefit-risk analysis process (Mühlbacher et al., 2016), and this conceivably could contribute to justification of M2 in certain situations.

At times, quantitative arguments may be used to support a clinically important margin. When the parameter is binary, the number needed to treat (NNT) can be used to understand how a difference in event rates impacts physicians' treatment decisions. A non-inferiority margin of 20 percentage points implies that a physician must treat up to 5 patients (0.20^{-1}) with the inferior treatment to observe an additional event, so the NNT is 5. A non-inferiority margin of 4 percentage points results in an NNT of 25, so a physician must treat up to 25 patients with the inferior treatment to observe one additional event. For an event of moderate but reversible morbidity such as a condition requiring hospitalization, the difference in NNT between 5 and 25 could be dramatic; for an event of irreversible morbidity or mortality, even NNT of 25 could be too large. Toxicity can also be evaluated with NNT, so differences in rates of adverse events can be converted into an NNT needed to observe an additional adverse event, the margin on an efficacy endpoint can be converted into an NNT needed to observe an additional treatment failure, and the relative importance of the adverse event and efficacy endpoint can be compared to determine whether the difference in NNTs is appropriate.

When the putative placebo margin M_1 and clinically important margin M_2 are both determined, the two can be compared. In general it is expected that the putative placebo margin will be larger than the clinically important margin. (If this relationship does not hold, it may be that the active control has an advantage compared to placebo that is not clinically important.) The margin chosen for hypothesis testing will generally be the smaller of the putative placebo margin and the clinically important margin, and sometimes may be rounded down or otherwise decreased to achieve the final margin.

Non-inferiority trials that have multiple endpoints, either co-primary or secondary, can have margins chosen for each of the endpoints. The process for each endpoint will follow the process described above. With multiple co-primary endpoints, all of which must be met for a successful study, the margins for the various endpoints must be considered in unison so that a specified difference in all endpoints is sufficient for an investigational treatment to be considered both superior to placebo and also clinically unimportantly different from the active control.

The section on choosing a non-inferiority margin is lengthy. This attention given to this topic is not illogical. If the non-inferiority margin is not well chosen no post hoc analysis can salvage the trial.

Example: Ranibizumab is a biologic drug that is approved for treatment of wet acute macular degeneration (AMD), requiring monthly intravitreal injections. To assess the efficacy of ranibizumab with less frequent dosing, an active control non-inferiority trial was undertaken. (Schmidt-Erfurth et al., 2011) The primary endpoint for the non-inferiority assessment after one year of treatment was number of letters read on a standard visual acuity chart, which has 5 letters per line in decreasing size. The number of letters read is used to assess visual acuity, with reading more letters indicating better vision.

The historical comparisons of ranibizumab dosed monthly compared to placebo used a different endpoint for primary assessment of efficacy, the number of subjects who improved by at least 15 letters. Multiple such comparisons were made (monotherapy: Rosenfield et al., 2006; combination therapy with verteporfin: Heier et al., 2006). In both studies, the difference in mean number of letters read was reported, and the effect of ranibizumab on the number of letters was large. In the monotherapy study, subjects randomized to receive ranibizumab had a mean improvement of 7 letters while subjects randomized to receive placebo had a mean loss of 10 letters for an estimated treatment effect of 17 letters; in the combination therapy study, subjects randomized to receive ranibizumab had a mean improvement of 5 letters while subjects randomized to receive placebo had a mean loss of 8 letters for an estimated treatment effect of 13 letters. Standard deviations were not provided in the primary publication of the monotherapy study, while in the combination therapy study the standard deviation was 16.3 and 14.7 in placebo and ranibizumab arms, respectively.

The non-inferiority margin in the active control trial assessing less frequent dosing was set to 6.8 letters (Schmidt-Erfurth et al., 2011). This was justified as being half of the lower bound of the confidence interval from the monotherapy study (implying that the designers of the non-inferiority study apparently had access to the standard deviation from the monotherapy study). A lower bound on the CI from the combination therapy study, using normal theory on the simple summary statistics, was approximately 8 letters. While the margin of 6.8 is still below this lower bound, it is not half of this lower bound. If subjects enrolled in the non-inferiority study will often receive verteporfin, a smaller margin might be necessary, as study designers must be comfortable that showing a difference that is only slightly less than 6.8 letters implies that the less frequent dosing of ranibizumab is indeed superior to placebo.

Additionally, justification for the margin for the non-inferiority analysis only considered M_1, the putative placebo margin. No consideration to M_2, the clinically irrelevant margin, was given as justification of the 6.8 letters. Consideration of the impact of losing up to 6.8 letters can be assessed informally but most people with sufficient vision: to understand the impact of losing 5 letters, one can look at an eye test chart and ask whether losing the ability to read the last legible line would interfere with quality of life. However, the authors also mention that the "currently accepted margin" was 5.0 letters, without giving justification, and M_2 might have played a role in determining that value.

24.2.3 Study design and conduct

Design and conduct are important in any clinical trial, but they have special importance in non-inferiority trials. Study design must reflect the design of historical studies comparing the active control to placebo. Conduct must rigorously follow the written study design.

Matching the study design of historical comparisons of the active control to placebo is required to justify the choice of non-inferiority margin. Some of these aspects that must be matched are fairly obvious: same dosage, same indication, etc. Other aspects are more nuanced and perhaps more difficult. When patients receive different concomitant medications in the non-inferiority trial than patients received in the historical comparison of the active control to placebo, the non-inferiority margin chosen from the historical comparison (M_1) may no longer be valid. Advances in diagnosis may result in patients who are in earlier stages of the disease process, and change is assessment of disease may result in differing prognoses, with unknown effects on the non-inferiority margin. Examples of therapeutic areas in which these could be of issue include infectious diseases (with pathogens evolving resistance to existing therapies), oncology (with RECIST and WHO definitions of disease progression being used; see Choi et al., 2005) and heart failure (with new anti-hypertensive drug classes approved). Some of these cannot be controlled because it is unethical to prevent patients from receiving the best available therapy, while others might be addressed by using assessments that were used previously, even though more advanced methods are available.

Note that attempting to match study designs by limiting study participants to a subset of all available patients may not be a solution. Limiting enrollment to subjects who are not receiving the newest class of concomitant medications may change the patient population in unpredictable (and unmeasured) ways, resulting in again a lack of applicability of the non-inferiority margin.

Once the study design is finalized, it is vital that the study be conducted rigorously in accordance with the written protocol. Deviations from the protocol may result in obfuscation of any differences between the investigational treatment and the active control, making it more likely that similarity is observed. Examples of study conduct issues that can mask differences include enrolling study subjects who do not have the disease being studied, subjects who do not receive the assigned treatment in sufficient quantity for an effect to be demonstrated, and subjects who do not return for follow-up visits to assess the impact of the treatment. These are issues in superiority trials, but have special concern in non-inferiority trials. Some of these issues can make two treatments look more similar to each other, which may make it more likely to demonstrate non-inferiority (although if variance is increased, this may not be the case). In a superiority trial, such problems can increase the confidence that a new treatment has advantages, if those advantages are observed even with conduct that is not conducive to demonstrating a difference. In a non-inferiority trial, such problems will not increase the confidence in a conclusion of non-inferiority.

24.2.4 Test statistics, confidence intervals and decision rules

The null and alternative hypotheses will take the form shown in the introductory section above. For this section, we will consider the data to be continuous, but the ideas extend to dichotomous and time-to-even endpoints in most cases. The null hypothesis $H_0 : \mu_C - \mu_T \geq \delta$ must be rejected in favor of $H_1 : \mu_C - \mu_T < \delta$ to conclude non-inferiority. A test statistic can be calculated as

$$Z_1 = \frac{(\overline{x_C} - \overline{x_T}) - \delta}{\sigma}$$

This test statistic Z_1 has a familiar form: the numerator is the difference between the estimated treatment effect $(\overline{x_C} - \overline{x_T})$ and the hypothesized treatment effect (δ). The denominator is the standard error of the numerator (σ). However, there are multiple ways to calculate the standard error, leading to various test statistics. With strong assumptions about the distribution of the underlying data, or mild assumptions and a large enough sample size to invoke the central limit theorem, Z_1 is distributed as a standard normal random variable under the null hypothesis. The null hypothesis is rejected if $Z_1 < z_\alpha$, where z_α is the deviate of the standard normal distribution. In practice, $\alpha = 0.025$, one-sided, before adjustments for multiple comparisons, half of the typical two-sided α in superiority studies.

As noted in the introduction of this chapter, the mathematical aspects of non-inferiority trial analysis are typically straightforward. One exception is the calculation of variance, or the standard error of the test statistic. While some calculations are simple, there may be advantages to more complex calculations in some situations.

If the standard error ignores any variability in the value of δ, the resulting test statistic is used in what is called the fixed margin approach. By assuming no variability in the margin, the denominator becomes $\sigma = (\sigma^2_C + \sigma^2_T)^{1/2}$, where σ^2_i is the variance of group i.

By including some quantity in the denominator to account for uncertainty in the margin, the denominator will become larger, the ratio will become smaller, and rejecting the null hypothesis will become less likely. How much less likely (how much less power) depends on the change to the denominator.

If one assumes simply that uncertainty in the choice of margin will be independent of the non-inferiority trial, a simple quantity added to $\sigma = (\sigma^2_C + \sigma^2_T)^{1/2}$ will be used. One obvious choice is to include the entire value of ϕ $(\sigma^2_C + \sigma^2_P)^{1/2}$, where ϕ is the ratio of the margin δ to a conservative estimate of the true difference between the active control and placebo. The denominator then becomes $[(\sigma^2_C + \sigma^2_T) + \phi^2(\sigma^2_C + \sigma^2_P)]^{1/2}$. This denominator is used in what is called the synthesis approach. Such an analysis starts with a non-inferiority margin determined primarily by the retention fraction required to demonstrate non-inferiority (where the retention fraction serves both as a putative placebo margin and a clinically important margin), and the analysis synthesizes data from the historical comparison of active control to placebo with data from the non-inferiority trial in a single test statistic (Holmgren, 1999).

Even more extreme, the denominator can be chosen to incorporate some uncertainty in a different way than is done with the synthesis method. Observing that the upper bound of the confidence interval on $\mu_C - \mu_T$ must be less than δ, a fraction of the lower bound of the confidence interval on $\mu_P - \mu_C$, the test is equivalent to the test statistic above with the denominator being $(\sigma^2{}_C + \sigma^2{}_T)^{1/2} + \phi(\sigma^2{}_C + \sigma^2{}_P)^{1/2}$ (Snapinn and Jiang, 2007). Alternatively, a weighted average of different terms can be used as an *ad hoc* approach to choosing a denominator, although this might be more for purposes of understanding properties than for evaluating actual non-inferiority trials. (Snapinn and Jiang, 2007).

Calculation of the standard error is more complicated yet for binary outcomes. Because the variance is a function of the mean, the estimated variance under the null hypothesis is different for a non-inferiority assessment than for a superiority assessment. Recall that for a standard test of equality of two proportions, a pooled estimate of the common success rate is used to calculate the variance: $(x_1 + x_2)/(n_1 + n_2)$, where x_i and n_i are the number of success and total number of observations, respectively, in group i. This is the maximum likelihood estimate of the variance under the null hypothesis that the two proportions are identical. However, in a non-inferiority analysis, the maximum likelihood under the null hypothesis that the two proportions differ by exactly δ is desired. For that value, the restricted maximum likelihood estimate of Farrington and Manning (1990) is more appropriate, and is available in standard software packages. The restricted MLE under the non-inferiority null hypothesis and the usual MLE under the superiority null hypothesis will differ more when the margin is large and when at least one of the observed proportions is near 0 or 1.

When non-inferiority is defined in terms of ratios of means or proportions, rather than differences of means or proportions, the test statistic also must change accordingly. A log transformation can often be used to produce a test statistic that has a similar form. Retaining ratios in the test statistic is also possible, and various methods can be used to estimate the asymptotic variances (Fieller's Theorem, Satterthwaite's method, Delta Method among them). Again, the difference among the test statistics is in how the variance is estimated. (For a comprehensive discussion of the various test statistics, see Rothmann, Wiens and Chan, 2012, pp 335–338.)

In practice, it is common to produce not test statistics but confidence intervals for purposes of decision making. If the upper bound U of a two-sided $(1 - 2\alpha) \times 100\%$ confidence interval on $\mu_C - \mu_T$ is smaller than the non-inferiority margin of δ, the null hypothesis is rejected and non-inferiority is concluded in a one-sided test of size α. Each of the test statistics can be easily converted into a confidence interval, with choice of the proper critical value. Again, the main difference among the various methods is how the variance is estimated when the confidence interval is calculated.

Confidence intervals are an important part of reporting the results of a non-inferiority trial. The non-inferiority margin is often the subject of debate, and reporting the confidence bounds can help all parties evaluate the difference in treatments according to their own standards. Additionally, for a quantitative (and somewhat subjective) assessment of the benefit versus the risk of the investigational treatment, confidence intervals can help the reviewer understand not just whether the trial met the prestated objective, but also the magnitude of difference in benefit that might exist. This is especially important when the two treatments have different safety profiles, or when there is a known safety issue of the active control versus placebo. Thus, two-sided confidence intervals are preferred over one-sided intervals, despite the one-sided nature of hypothesis testing. This is to report the maximum benefit (or minimum deficit) of the investigational treatment compared to the control, not just the maximum deficit (minimum benefit), to more comprehensively assess the benefit compared to the risk.

24.2.5 Reporting and interpretation

Results of non-inferiority trials should be reported so that the audience understands. The intended audience will include statisticians, physicians, patients, regulators, payers, and others involved in the interpretation of the trial and application of information to clinical practice. Confidence intervals and results of hypothesis testing are important but not sufficient.

Study reports and documents filed with regulators will typically be reviewed by clinical trialists who are well versed in non-inferiority trials. Especially the statistical reviewers for regulators will understand nuances. Preparing for clinicians may require more explanation, and review by those who are not full time medical reviewers (such as FDA advisory committee members) will require even more care to present results.

Reports in medical journals require different attention than reports for regulatory review. An expanded CONSORT statement has been proposed for non-inferiority trials, to ensure that such trials are publicly presented in a manner that facilitates understanding (Piaggio et al., 2006). Important points including ensuring that the title and abstract clearly state the non-inferiority nature of the study design, the margin is explicitly stated and justified (and was prespecified), assessing whether study subjects were similar to those in the historical comparisons of the active control to placebo, and interpreting results appropriately. Again, because many readers are not experts in non-inferiority trials, it is necessary to address some fairly basic concerns in writing results on non-inferiority trials for publication in medical journals.

Finally, we offer an opinion on reporting p-values for non-inferiority trials. Le Hananff et al., (2006), in a systematic review, found that 48.1% of 162 reports of non-inferiority or equivalence clinical trial results included p-values. However, half of the p-values were for null hypothesis of no difference, rather than for the null hypothesis used in non-inferiority or equivalence testing. In general, p-values are difficult to understand, even for highly trained and experienced statisticians. Clinicians can interpret small p-values as evidence of a difference rather than evidence against a null hypothesis. This is an incorrect interpretation, of course, but the challenge for the statistical practitioner is to either educate the audience, or present a result that is not as prone to misinterpretation. Therefore, we recommend reporting a two-sided confidence interval, a margin, and a conclusion of testing (non-inferiority is concluded or is not concluded). If a p-value is reported with the results of a non-inferiority trial, the associated null hypothesis must be clearly defined. [Notably, the ranibizumab papers referenced earlier (Schmidt-Erfurth et al., 2011) reported p-values, and clearly noted they were associated with the non-inferiority null hypothesis.]

24.2.6 Power and sample size assessment

In general, power calculations are straightforward for non-inferiority trials. Power is the probability that the null hypothesis is rejected, when the null is false, so power $= 1 - \beta = P(U < \delta \mid \mu_C - \mu_T = \delta_1 < \delta)$. Although this is not the same as the power for a hypothesis in a superiority trial, calculations are parallel: given the same variance for normally distributed data, the power to reject the superiority trial null hypothesis $H_0: \mu_{Pbo} - \mu_T \geq 0$ when $\mu_{Pbo} - \mu_T = -\varepsilon$ is the same and the power to reject the non-inferiority trial null hypothesis $H_0: \mu_C - \mu_T \geq \delta$ when $\mu_C - \mu_T = \delta - \varepsilon$. The relationship is not quite a simple for binomial data, because a change in the proportion comes with a change in variance. For proportions near 0.5, the difference will often be minor, so standard sample size formulae can be used to obtain an estimate of required sample size.

However, an estimate of sample size is not a substitute for an exact calculation. For such calculations, two options are proposed: sample size software and simulations. Many widely available software packages allow calculations of sample size for non-inferiority trials, including PASS (NCSS, LLC), SAS® (SAS Institute, Inc.) and nQuery Advisor (Statistical Solutions, Ltd). In SAS, for example, the null ratio and null difference statements with the two sample means option specify the margin. Such software packages will be sufficient for calculating power and sample size for many non-inferiority trials. Other packages certainly are available, as is codes in R, but the packages mentioned are familiar to the author, and no claim is made that these options are the best.

However, at times even more precision and transparency are desired. This may be true for binary data, when the software package does not allow the user's preferred option of estimating variance (or, worse, when it is not clear how variance is estimated), or when over- or under-dispersion are desired due to stratification or important covariates. This may also be true for time-to-event data when the user desires to use nonstandard survival curves or nonproportional hazard functions. In such situations, simulations will provide not only more accurate results, but more control over and flexibility of input assumptions. Simulation can also provide accurate power estimates for complex multiple comparison procedures with correlated hypothesis tests, something not available with most software packages. The disadvantages include a requirement that the user be adept at programming simulations using software such as SAS or R, a process that is longer than using commercially available software, and potential delays from reviewers on ethical review committees or at regulatory agencies who may ask for code to evaluate power claims.

Another issue in power calculations is the choice of the point in the alternative space at which to calculate the power. Some authors (e.g., Wellek, 2010, p. 23) state that the obvious choice is the point at which equality is achieved: $\mu_C - \mu_T = 0$. Though this is an obvious point, it may not be the only or even the best point to consider, because a minor deviation $\mu_C - \mu_T > 0$ can have an important impact on power even if such a difference is clinically unimportant. It is easy to find situations in which a true difference that is only 10% of the margin reduces power from 80% to 60%, an important loss in power due to a very small difference in means. In some situations, $\mu_C - \mu_T = 0$ might be the best point to use—when two formulations of a single drug are compared, or there is other reason to believe that the two groups are identical, it makes sense. When two drugs with different mechanisms of action are compared, there may be no reason to believe efficacy outcomes will be identical. Rather than use $\mu_C - \mu_T = 0$ as the default point at which to calculate power, the reader is encouraged to use a small positive value to investigate robustness.

24.2.7 Equivalence and non-inferiority

Equivalence and non-inferiority are similar concepts. Equivalence implies that an investigational treatment is not much better or much worse than a control treatment, while non-inferiority allows for the investigational treatment to be better than, but not much worse than, the control.

Non-inferiority trials assume that higher efficacy is better, but equivalence trials assume that too much effect and too little effect are both undesirable. The most common example of equivalence in clinical trials is in bioequivalence, in which the drug availability of a new formulation and a reference formulation are compared. Higher bioavailability of the new formulation is not desirable, as it may lead to toxicity; lower bioavailability may lead to lack of efficacy. (Schuirmann, 1987) A similar concept is used in evaluation of vaccine products, in which three (or more) lots of a new vaccine candidate are compared for immunogenicity. All lots must have similar effect to demonstrate that the product can be manufactured consistently (Wiens and Iglewicz, 1999; Wiens and Iglewicz, 2000).

Most issues applicable to non-inferiority trials also apply to equivalence trials. A noticeable difference is the formulation of the hypotheses. Commonly called two one-sided tests, equivalence requires both rejecting H_0: $\mu_C - \mu_T \geq \delta_U$ in favor of H_1: $\mu_C - \mu_T < \delta_U$ and rejecting H_0: $\mu_C - \mu_T \leq -\delta_L$ in favor of H_1: $\mu_C - \mu_T > -\delta_L$. The hypotheses can be written more succinctly as H_0: $\mu_C - \mu_T \geq \delta_U$ or $\mu_C - \mu_T \leq -\delta_L$ versus H_1: $\delta_L > \mu_C - \mu_T > -\delta_U$. Often, $\delta_U = \delta_L$ versus so the hypotheses can be further shortened to H_0: $|\mu_C - - \mu_T| \geq \delta$ versus $H_1 : |\bar{\mu}_C - \mu_T j| < \delta$. Even if the hypotheses are written in such shorthand, two test statistics will be calculated, with the smaller being compared to the critical value. A confidence interval approach to testing decisions will require that a two-sided confidence interval be entirely within the interval $(-\delta_L, \delta_U)$ to conclude equivalence. This effectively requires that two one-sided hypotheses be tested, and equivalence be concluded only if both null hypotheses are rejected. If a p-value is presented, it is typically the larger of the two p-values from the two one-sided tests.

The process for choosing the margins for equivalence testing will be similar to the process for non-inferiority testing. An exception is the upper margin, in which no putative placebo margin can be calculated. Rather, only clinical judgment will be of interest. Because of this, the synthesis approach to testing will not generally be used.

24.3   Issues and Evolving Ideas

In this section we discuss some issues for which the solution is not widely agreed. There are several areas of disagreement among statisticians (Huitfeldt and Hummel, 2011). The three issues chosen for further discussion—analysis sets, missing data and adaptive design—are much better understood in the context of superiority trials than non-inferiority trials. That can be an advantage or a disadvantage—an advantage because there may be a road map for a general solution for non-inferiority trials, but a disadvantage is the modifications required to apply a solution that works in the superiority setting are not appreciated.

Recent guidance from the International Conference on Harmonisation (ICH) introduced the concept of estimands in clinical trials (International Conference on Harmonisation, 2017). We note that the understanding of the concepts of estimands is evolving (Akacha et al., 2017). As the understanding becomes more widespread, information in the following two sections, in particular, may become outdated.

24.3.1   Analysis sets

In superiority trials, it is common to analyze results by the intention to treat principle, in which all randomized subjects contribute to the analysis according to randomized treatment. Non-inferiority trials have been more commonly analyzed according to different criteria, in which patients who were noncompliant or had other protocol deviations did not contribute to the analysis. Instead, a per protocol approach was used to exclude subjects who did not, in retrospect, meet inclusion criteria, did not receive a sufficient amount of randomized treatment, or did not have follow-up evaluations.

The reason for using the intention to treat approach for clinical trials is to avoid bias. Issues that arise after randomization, or that are adjudicated after randomization, can be affected by the assigned treatment, even if the study is blinded. Toxicity, difficulties of administration and lack of efficacy can all influence a subject to discontinue the assigned treatment, skip follow-up evaluations or discontinue the study. More egregiously, a subject who received the wrong treatment will, when analyzed as randomized, contribute to the wrong treatment group. Any of these can make two treatments with different efficacy appear to have similar efficacy.

Per protocol approaches, however, do not solve these problems. Removing randomized subjects due to events that occur after randomization can introduce bias as surely as including them can. Instead, an strategy that appropriately includes such subjects should be the goal. This strategy will have three objectives: avoiding missing data and other protocol deviations, properly assessing the trial in the presence of missing data, and planning sensitivity analyses to assess the impact of these events on the interpretation of the trial.

Avoiding missing data and other protocol deviations is related to the topic discussed above, rigorous study conduct. If a non-inferiority study is not conducted rigorously, no amount of post hoc analysis can salvage it. Obtaining complete data on all patients is the goal of any clinical trial (National Institutes report) Even if some data points are not used in the analysis, it is better to have the data and not use it (for the primary analysis) than to need it and not have it. And, if not used in the primary analysis, having data available can be used in sensitivity analyses or to otherwise subjectively assess whether data excluded from the primary analysis has an important impact on study conclusions. Similarly, avoiding other protocol deviations will increase confidence that the study was rigorously conducted and that any differences in treatments, if they exist, would not have been obscured by study conduct.

Missing data will be discussed in the next section. With respect to the topic of analysis datasets, the per protocol analysis essentially handles missing data by ignoring subjects with missing data. This is unacceptable unless data are known to be missing completely at random (MCAR), a very strong assumption that is difficult to justify. In the absence of justification of MCAR as the only mechanism of missingness, a per protocol analysis is not an acceptable option for non-inferiority trials.

Another issue in choosing an analysis set is the justification of the margin (or retention fraction). These values are calculated from historical comparison of the active control to placebo, the analyses of which almost certainly used an intention to treat approach rather than a per protocol approach to analysis datasets. The margin for a non-inferiority analysis that uses a per protocol analysis set cannot be justified from historical comparisons that used intention to treat approaches.

Another concern with using an intention to treat approach is in the inclusion of subjects who did not meet all inclusion and exclusion criteria in a trial. In clinical trials, the criteria for study subjects is rigorously defined to ensure that subjects are able to provide informed consent, have the condition under investigation, can be evaluated for outcomes and will not be placed at undue risk by participating in the trial. It is common that a small proportion of subjects do not, in retrospect, meet all criteria. Reasons for deviations from the prespecified criteria include errors at the investigative site, incomplete patient recall or understanding, or test for which results are not available before treatment must begin. A general rule is that events that occur after randomization should not exclude a subject from the analysis, but events that occur before randomization might exclude a subject from the analysis without inducing bias if handled appropriately. Handling appropriately requires that subjects be evaluated consistently using only data obtained before randomization. In an extreme situation, this can be accomplished by using an independent eligibility committee comprised of physicians and statisticians who review only pre-randomization information and retroactively determine which subjects were eligible. Subjects determined not to have met entry criteria can be excluded from the primary analysis with minimal introduction of bias, again noting that the number of such subjects should be small compared to the total enrollment (Wiens and Zhao, 2007).

Subjects who are not treated with the investigational or active control, or who receive the wrong treatment, also pose a problem. Subjects who do not receive treatment and who do not have follow-up information can be handled with missing data, discussed here. Subjects, who have follow-up information after receiving no study treatment, or the wrong study treatment, are typically handled in superiority trials by analyzing according to the randomized treatment. In a superiority trial, this will be conservative, as two treatments that are different will appear more similar, and two treatments that are identical will appear identical. In a non-inferiority trial, this may not be conservative, as an inferior investigational treatment will appear better, or a superior active control will appear worse, with this analysis convention. Rather, non-inferiority trials may be analyzed as treated, rather than as randomized, if the treatment errors can convincingly be ascribed to reasons not related to assigned treatment. (FDA, 2010) As an alternative, if treatment errors cannot be convincingly ascribed to reasons not related to assigned treatment, the best approach might be to treat such subjects as having missing data, and apply sensitivity analyses described in the next section.

Again, the discussion of study subjects who do not rigorous adhere to protocol specified activities has addressed ways to include such subjects in analyses. A per protocol analysis ignores such subjects and omits them from analysis. Because this excludes subjects based on post randomization events, it can introduce bias. A better strategy is to include such subjects in the analysis in an appropriate manner, and follow with sensitivity analyses to assess the impact, while noting that a study in which a high percentage of randomized subjects have important protocol deviations will have little external validity.

Rather than considering an analysis set in the manner of intention to treat or per protocol, one may instead consider in terms of estimands, as noted earlier. At present, there is not a consensus on a single definition of an estimand to be used in analysis of non-inferiority studies. Intercurrent events may increase or decrease the type I error rate or power of such analyses, and should be critically considered when planning an analysis of a non-inferiority trial. Intercurrent events that include or cause discontinuation of protocol-assigned therapy are especially important for non-inferiority analyses, as subjects in either treatment group will be treated the same when treatment is discontinued, which may result in an inflated type I error rate.

24.3.2 Missing data

All clinical trials are expected to have some amount of missing data, even if heroic measures are taken to gather as much complete data as possible. Analyses must therefore be valid even if some data are missing, so proper planning is required. Elements of an effective analysis strategy will include choosing primary inferential methods that are valid with some amount of missing data. Missing data mechanisms are often grouped into three classes: MCAR, missing at random (MAR), and missing not at random (MNAR). This listing puts the mechanisms in order of decreasing difficulty in proving the relationship but increasing difficulty in analyzing the resulting data. Briefly, MCAR assumes that the missingness is unrelated to any data, observed or unobserved. MAR assumes that the missingness is unrelated to unobserved data, but may be related to observed data (at other timepoints or other endpoints). MNAR assumes that missingness may be related to unobserved data. The reader is directed to other chapters in this book that discuss missing data in more detail for the theory, and in this section we will discuss only aspects that are specifically relevant to non-inferiority trials.

The prespecified primary analysis should be robust to some amount of missing data. A common strategy is to specify an analysis that is robust to an expected amount of data that are MAR. Such an analysis might include covariates that are predictive of missing data, longitudinal models that incorporate outcomes at timepoints other than the primary timepoint, and other features that use observed data to inform on missing data. A likelihood-based method to analyze longitudinal data is commonly used for superiority trials and can easily be adapted to non-inferiority trials (Wiens and Rosenkranz, 2013). Multiple imputations for non-inferiority trials have been studied when the primary endpoint is collected only at the conclusion of treatment, and may have some utility as a supportive analysis when the primary endpoint is collected longitudinally (Lipkovich and Wiens, 2018). Single imputation procedures such as last observation carried forward are not recommended for either non-inferiority or superiority trials. Because methods to handle missing data have not been studied as extensively for non-inferiority trials as for superiority trials, it is recommended that any method be studied extensively through simulations before finalizing as a primary analysis, to ensure control of the type I error rate and power under various scenarios.

A caution is required about using methods robust to missing data in non-inferiority trials. A method is most supportable if the same method was used as a primary analysis in the historical comparisons of the active control to placebo. Because the non-inferiority margin is obtained from these historical comparisons, the statistical methods (including robustness against missing data) should be identical. When the synthesis method is used for analysis, the need for identical analyses may be even greater. Unfortunately, the historical comparisons may have used various methods for primary analyses, may have used methods that are not up to current standards, or may have used methods that are not clear from published references. This makes it difficult to choose an analysis that supports the chosen non-inferiority margin. A tempting ad hoc adjustment is to make the margin smaller (or retain a larger fraction of the historical effect of the active control versus placebo, if using the synthesis method). However, the amount of adjustment is difficult to gauge and the implications are difficult to predict. It is better to find more information on the historical comparisons, even assess current methods on old datasets if possible, to better justify the proposed margin and analysis of the non-inferiority trial.

In addition to a prespecified primary analysis that is robust to missing data, sensitivity analyses should be considered. It is likely that more, and more complex, sensitivity analyses will be required for non-inferiority trials than for superiority trials, due to lingering concerns over any unknown aspects of how the active control would perform compared to placebo in the current trial. Obvious candidates include other models (different covariates or different covariance matrices for longitudinal data, as examples) or different estimates of variance, as discussed earlier. If some subjects are excluded from the primary analysis due to missing data, various ways of including them in the analysis should be investigated, such as tipping point graphs (Yan et al., 2009). Such subjects can also be investigated in an inflation analysis, in which the treatment difference that must be observed in such subjects (if outcome data was available) in order to change the primary conclusion of the study is calculated. If the treatment effect in such subjects must be markedly different from the observed treatment effect in subjects with available data in order to change a conclusion of non-inferiority, there is confidence that the missing data did not change the conclusion. Such assessments will be subjective, and a "markedly different" effect for one reviewer may not match a "markedly different" effect for another reviewer.

Finally, we come to a most important (and sometimes controversial) recommendation about missing data, which is to avoid it as much as possible. (Panel on Handling Missing Data in Clinical Trials, 2010) For a study subject who is generally compliant with the protocol, it is obvious that safety and outcome data should be collected. The desire to collect data is more misunderstood for the study subjects who did not meet inclusion and exclusion criteria for the study, who did not take study drug, or who otherwise deviated from the protocol guideline–this was especially understandable when the primary analysis set was per protocol, and other subjects did not contribute to evaluation. However, conclusions can be affected by missing data, even from subjects who are noncompliant, and perhaps even more from subjects who are noncompliant since lack of compliance can signal issues with safety and efficacy. Therefore, it data should be collected at all planned timepoints from all subjects, even those who did not follow the protocol perfectly. If such patients are uncooperative, guidance should be given to the investigators on the data that is most needed—typically, primary efficacy and safety data, and data that can give insight into reasons for missingness. Depending on the estimand chosen, some data collected after discontinuation of treatment or other intercurrent events (such as switching to alternative treatments) may not be used in the primary analysis, but it is beneficial to have it for supportive analyses.

24.3.3 Adaptive designs

Adaptive trials change some aspect of the trial design during the trial, based on data accumulated within the trial. Adaptive trials can increase efficiency and reduce development time without biasing the trial results, if designed and conducted rigorously.

Many adaptations of clinical trials have been proposed, most implicitly for superiority trials. Such studies can add or drop treatments, change endpoints, change inclusion and exclusion criteria, or stop the study early for a conclusion of success or failure, among other possibilities. Similar adaptations can be used in non-inferiority trials, but some details may differ. Additionally, some adaptations have been studied for superiority trials, not for non-inferiority trials, so details on the performance may not be well understood. (FDA, 2010)

A commonly used adaptive design is the group sequential design. At predetermined timepoints, data are summarized by randomized treatment groups. If the accumulated data support a conclusion, the study is stopped. With appropriate decision rules, the probability of a false positive conclusion from the study is controlled at the advertised α level. For superiority trials, the design has decades of experience going back to seminal articles by O'Brien and Fleming (1979) and Pocock (1978). A conclusion of superiority of an investigational treatment versus a control can be made based on less than the full amount of planned data if the nominal p-value at a planned interim timepoint is sufficiently small. For non-inferiority trials, there is less experience, but the mathematics work out directly. A practical consideration is that instead of comparing a p-value to an adjusted alpha, a confidence interval, with higher than $(1 - \alpha) \times 100\%$ nominal coverage, is calculated, with the upper bound compared to the non-inferiority margin δ.

An important difference between use of group sequential designs in superiority trials and in non-inferiority trials is the ethical imperative. If one treatment is superior to another, a group sequential trial can reduce the number of subjects who receive the inferior treatment, and reduce the time until the superior treatment is available commercially. A group sequential trial is especially efficient if the investigational treatment has a larger magnitude of benefit than minimally necessary, as such a treatment will have a better chance of showing benefit with fewer subjects and a smaller alpha, and when a decision can be made quickly enough to save time and effort of enrolling and following additional subjects. If the two treatments are identical (a positive outcome in non-inferiority trials), a group sequential trial will not change the number of subjects who receive the inferior treatment, because neither treatment is inferior. Alternatively, if the investigational treatment has safety advantages over the active control but identical efficacy, fewer subjects will be exposed to the less safe treatment. However, if the investigational is actually superior to the active control rather than identical, a group sequential trial may not demonstrate that if the trial is halted as soon as non-inferiority is demonstrated. Rather than stopping early for a conclusion of non-inferiority, a futility analysis (stopping the study because it has little chance of demonstrating non-inferiority) will reduce the number of subjects who receive the inferior treatment. Futility analyses will not reduce time to market, but will reduce cost and effort required in trials that are destined for negative results. Thus, futility analyses should be considered for non-inferiority trials when a futility decision can be made on accumulating data in time to preclude significant enrollment or follow-up of the entire planned sample size.

Sample size re-estimation exercises can change the number of subjects entered into the study based on results from the earliest enrolled subjects. In general, there are two kinds of sample size re-estimation: blinded (noncomparative) and unblinded (comparative). Blinded methods are generally preferred because they may result in less bias.

Blinded sample size re-estimation methods assess the observed data in amalgamation, and change the final sample size based on the observed data and (often) assumptions used to plan the study. These can be instead called noncomparative analyses, because comparison between groups is not required. A common application is to estimate the standard deviation (or background response rate for binary data), assuming a given difference in means (or proportions). Because data are not needed by treatment group, the opportunity to introduce bias is lower than for unblinded methods in which data are summarized by treatment group.

Unblinded sample size re-estimation methods assess the observed data by treatment group using partial enrollment or follow-up, and change the final sample size to ensure that the study is adequately powered. These may be instead called comparative methods, since comparison between groups is required. Such methods consider both the observed treatment difference and the observed variability in re-estimating sample size. Additionally, for a non-inferiority trial, reducing sample size may preclude the opportunity to demonstrate superiority (as mentioned above for group sequential trials). Thus, reducing sample size should be an option only if the variance is much smaller than expected, not if the treatment differences favor the investigational treatment, leading to another reason to prefer blinded sample size re-estimation.

The most important difference between blinded and unblinded methods is the increased potential to introduce bias with an unblinded sample size re-estimation. When unblinded summaries are generated, an independent committee is often required to review the data and prevent those involved in the daily activities of the study from knowing preliminary results. However, even with an independent committee, those involved in the study can infer preliminary results by back-calculating from the revised sample size. Thus, bias is always a concern. Admittedly, bias is also a concern for blinded sample size re-estimation, because some aspects of the data can be inferred by blinded results. (As an example, knowing the combined event rates for binary data can lead to a conclusion of the relative event rates, if the control treatment has a well-known and highly reproducible event rate.) As in superiority studies, dramatic changes in sample size requirements should not be made using these methods, and the sponsor should ensure that the resulting trial will provide sufficient evidence needed for registration. (Herson, 2009) Instead, modifications of 10–15% to ensure adequate power are generally the largest recommended change. A sponsor may add additional restraints to control timelines and budgets, or to ensure an adequate safety database is available at the conclusion of the study. Finally, the impact of sample size changes on secondary endpoints should also be considered.

Many other adaptations are possible based on accumulating data from the trial or from external data. Without considerable experience with any given adaptation in non-inferiority trials, a regulatory reviewer (or an informed clinical reviewer) will be necessarily skeptical of an adaptation. The trialist who implements a new adaptation will therefore be responsible for convincing the reviewer that the adaptation results in an unbiased conclusion on the relative efficacy of the test and investigational treatments.

24.4   Conclusions

Non-inferiority clinical trials have become common in the past few decades. With more exposure, the basic principles around design, analysis, and interpretation are widely known, and more complex enhancements have been investigated. We expect further development of issues in the coming years, so this chapter is not definitive. The reader is encouraged to learn the basics and maintain an understanding of new developments.

Bibliography

Akacha, M., Bretz, F., Ruberg, S. 2017. Estimands in clinical trials – broadening the perspective. *Statistics in Medicine* 36: 5–19.

Blackwelder, W. C. 1982. 'Proving the Null Hypothesis' in Clinical Trials. Controlled Clinical Trials 3: 345–353.

Choi, J.-H., Ahn, M.-J. Rhim, H.-C., et al. 2005. Comparison of WHO and RECIST Criteria for Response in Metastatic Colorectal Carcinoma. *Cancer Research and Treatment* 37:290–293.

Farrington, C.P, and Manning, G. 1990. Test statistics and sample size formulae for comparative binomial trials with null hypothesis of non-zero risk difference or non-unity relative risk. *Statistics in Medicine* 9:1447–1454.

Herson, J. 2009. Data and Safety Monitoring Committees in Clinical Trials, Boca Raton: Chapman & Hall / CRC.

Holmgren, E. B. 1999. Establishing equivalence by showing that a specified percentage of the effect of the active control over placebo is maintained. *Journal of Biopharmaceutical Statistics* 9:651–659.

Huitfeldt, B. and Hummel, J. 2011. The draft FDA Guideline on Non-inferiority clinical trials: a critical review from European pharmaceutical industry statisticians. *Pharmaceutical Statistics* 10:414–419.

International Conference on Harmonisation. 2017. Estimands and Sensitivity Analysis in Clinical Trials E9(R1). http://www.ich.org/fileadmin/Public_Web_Site/ICH_Products/Guidelines/Efficacy/E9/E9-R1EWG_Step2_Guideline_2017_0616.pdf, accessed August 15, 2019.

Le Hananff, A. Giraudeau, B., Baron, G., and Rivaud, P. 2006. Quality of reporting of noninferiority and equivalence randomized trials. *Journal of the American Medical Association* 295(10): 1147–1151.

Lipkovich, I., and Wiens, B.L. 2018. The role of multiple imputation in noninferiority trials for binary outcomes. *Statistics in Biopharmaceutical Research* 10:57–69.

Mühlbacher, A.C., Juhnke, C., Beyer, A.R. and Garner, S. 2016. Patient-Focused Benefit-Risk Analysis to Inform Regulatory Decisions: The European Union Perspective. *Value in Health* 19:734–740.

O'Brien, P. C. and Fleming, T. R. 1979. A multiple testing procedure for clinical trials. *Biometrics* 35:549–556.

Panel on Handling Missing Data in Clinical Trials. 2010. The Prevention and Treatment of Missing Data in Clinical Trials. Washington, DC: National Academies Press.

Piaggio, G., Elbourne, D. R., Altman, D. G., Pocock, S. J. and Evans S. J. W. 2006. Reporting of non-inferiority and equivalence randomized trials: An extension of the CONSORT statement. *Journal of the American Medical Association* 295:1152–1160.

Pocock, S. J. 1978. Size of cancer clinical trials and stopping rules. British *Journal of Cancer* 38:757–766.

Rothmann, M. D., Wiens, B. L. and Chan, I. S. F. 2012. Design and Analysis of Non-inferiority Trials. Boca Raton: Chapman & Hall/CRC.

Schmidt-Erfurth, U., Eldem, B., Guymer, R., Korobelnik, J.-F., Schlingemann, R.O., Axer-Siegel, R. Wiedemann, P., Simader, C., Gekkieva, M., and Weichselberger, A., EXCITE Study Group. 2011. Efficacy and Safety of Monthly versus Quarterly Ranibizumab Treatment in Neovascular Age-related Macular Degeneration: The EXCITE Study. *Ophthalmology*, 118 (5), 831–839.

Schuirmann, D. J. 1987. A comparison of the two one-sided tests procedure and the power approach for assessing the equivalence of average bioavailability. *Journal of Pharmacokinetics and Biopharmaceutics* 15:657–680.

Snapinn, S. and Jiang, Q. 2007. Controlling the type 1 error rate in non-inferiority trials. *Statistics in Medicine*, 27:371–381.

United States Food and Drug Administration. 2016. Guidance for industry: Non-inferiority Clinical Trials. http://www.fda.gov/downloads/Drugs/Guidances/UCM202140.pdf, accessed August 15, 2019.

United States Food and Drug Administration. 2018. Guidance for industry: Adaptive Design Clinical Trials for Drugs and Biologics. https://www.fda.gov/media/78495/download, accessed August 15, 2019.

Wellek, S. 2010. Testing Statistical Hypotheses of Equivalence and Non-inferiority. Boca Raton: Chapman & Hall/CRC.

Wiens, B.L. and Iglewicz, B. 1999. On Testing Equivalence of Three Populations. Journal of Biopharmaceutical Statistics 9:465–483.

Wiens, B.L. and Iglewicz, B. 2000. Design and Analysis of Three Treatment Equivalence Trials. *Controlled Clinical Trials* 21:127–137.

Wiens, B.L. and Rosenkranz, G.K. 2013. Missing data in non-inferiority trials. *Statistics in Biopharmaceutical Research* 5:383–393.

Wiens, B.L. and Zhao, W. 2007. The Role of Intention to Treat in Analysis of Noninferiority Studies. Clinical Trials: *Journal of the Society for Clinical Trials* 4: 286–291.

Yan, X., Lee, S. and Li, N. 2009. Missing data handling methods in medical device clinical trials. *Journal of Biopharmaceutical Statistics* 19:1085–1098.

25

Incorporating Historical Data into Randomized Controlled Trials

Heinz Schmidli, Sandro Gsteiger and Beat Neuenschwander

CONTENTS

The Bayesian framework facilitates use of historical control data in the analysis of a clinical trial, and hence to reduce the number of subjects randomized to control. This decreases costs, makes the trial shorter, facilitates recruitment, and may be more ethical. We focus on the *meta-analytic-predictive* (*MAP*) approach to historical data, but also describe alternatives such as bias models, commensurate and power priors. The prospective *MAP* approach derives a prior distribution from historical data based on a hierarchical model. We discuss mixture approximations of *MAP* priors, robustness to prior-data conflict, and prior effective sample sizes. A proof-of-concept study with placebo data from eight historical studies is used to illustrate the methodology. The *MAP* approach can be extended in various directions. We discuss two such extensions: over-dispersed count data in multiple sclerosis, where the historical placebo data consist of individual patient data and aggregate data; and non-inferiority trials, which depend critically on a proper integration of historical data.

25.1   Introduction

The Critical Path Initiative [15] by the United States Food and Drug Administration (FDA) and the Innovative Medicines Initiative [12] by the European Medicines Agency (EMA) highlight the continued need for innovation and increased productivity in drug development. Taking advantage of all relevant data sources should improve decision making, increase efficiency of drug development, and ultimately lead to better medicines. This is increasingly recognized as shown by various initiatives and programs that encourage fuller use of available information. For example, the TransCelerate initiative considers the use of historical subject-level data to minimize patient burden [30]. The incorporation of trial-external information with Bayesian approaches is also explicitly mentioned in FDA's Complex Innovative Designs pilot program [18]. A pediatric clinical trial participating in this pilot program clearly shows the advantages of borrowing strength from several trial-external sources (historical pediatric and adults trials), i.e. considerably reduced trial duration and enrolled less patients compared to a traditional design [48].

   We discuss here the use of historical control information for the design and analysis of clinical trials. The case study in Section 25.2 motivates the use of such data in a randomized trial, and shows that it saves costs, reduces study duration, and facilitates recruitment. Section 25.3 lays out the *MAP* approach to formally make use of historical data in a clinical trial. Section 25.4 reviews other approaches and explains similarities to the meta-analytic-predictive approach. Two extensions with applications are discussed in Section 25.5: the joint use of individual and aggregate over-dispersed count data, and the use of historical data in non-inferiority trials. We conclude the chapter with a discussion.

25.2   Case Study

In proof-of-concept (PoC) studies, test treatments are evaluated for the first time in patients. Baeten et al. [1] describe a PoC study, where patients with *ankylosing spondylitis* were randomized to the test treatment *secukinumab* or to placebo. The binary primary efficacy endpoint was response at week six. Traditionally, in such a PoC study, one would randomize patients equally to the two arms, and then use a significance test (e.g. Fisher's exact test).

   However, eight placebo-controlled studies in the same patient population (but with different test treatments) were available [34]. Table 25.1 shows similar placebo response rates in these studies. Therefore, Baeten and colleagues decided to make use of these data in their study. They used the Bayesian *MAP* approach (Section 25.3) to derive a $Beta(11, 32)$ prior for the placebo response rate; the prior has mean 26% and is worth 43 patients. Since the total number of historical placebo patients was 513, the *Beta* prior shows considerable discounting of the historical data, which results from between-trial heterogeneity. Because no information was available for the test treatment, its prior was set to a $Beta(0.5, 1)$, which is weakly informative (worth 1.5 patients) and has the same median as the placebo prior. In a traditional design without use of historical placebo information, 24 patients per arm would have achieved acceptable frequentist operating characteristics. The considerable prior information for placebo led to a substantial reduction in the number of placebo patients in the actual trial using a Bayesian primary analysis: 24 patients were randomized to the test treatment, but only six patients to placebo.

TABLE 25.1
Placebo data from eight historical placebo-controlled studies in patients with *ankylosing spondylitis.*

Study	1	2	3	4	5	6	7	8
Responders	23/107	12/44	19/51	9/39	39/139	6/20	9/78	10/35
	21%	27%	37%	23%	28%	30%	12%	29%

The results of the PoC study were as follows: one of the six patients on placebo (17%) and 14 of the 23 patients on *secukimumab* (61%) responded (one patient randomized to *secukinumab* was excluded due to a dosing error). Hence, the posterior distribution of the response rate was $Beta(12, 37)$ for placebo, and $Beta(14.5, 10)$ for *secukinumab*. From this, the probability that *secukinumab* has a higher response rate than placebo is 99.8% (obtained by numerical integration or simulation). The overwhelmingly positive result was confirmed in subsequent phase III trials.

In this PoC study, the use of historical data allowed the investigators to reduce the total number of patients by 38% (30 vs. 48), which greatly reduced costs and trial duration. In addition, for patients, the historical data study design was much more attractive compared to a traditional design, as fewer patients were randomized to placebo. This not only facilitated recruitment, but was also more ethical.

25.3 Meta-Analytic-Predictive Approach

We consider here the design and analysis of a new clinical study comparing a test treatment to control, and where control information from several historical trials is available. We describe the *MAP* approach to deriving a prior for the control in the new study based on the historical data, and show how this affects various aspects of the new study. To simplify the presentation, we focus on binary endpoints, although similar considerations apply to other endpoints [38, 47, 52]. The motivating example from Section 25.2, slightly simplified, is used to illustrate the methodology. The *MAP* approach can be implemented with standard software (Section 25.7). The R package RBesT (https://CRAN.R-project.org/package=RBesT) provides a comprehensive tool-set to support Bayesian evidence synthesis, including the derivation of (robust) *MAP* priors, the evaluation of operating characteristics, and the Bayesian analysis [60].

25.3.1 Hierarchical model

To borrow strength from historical data for a new study, one has to link the model parameters in the new and historical studies. Suppose that J historical studies are available, with the same population, control group and binary endpoint as in a planned new study. Let the number of responders in the control group be y_j, $j = 1, \ldots, J, \star$, for the historical and new study, respectively, noting that $y_\star$ is not yet observed at the design stage of the new study.

We assume a standard binomial model for the data,

$$y_j \mid \pi_j \sim Binomial(\pi_j, n_j), \quad j = 1, \ldots, J, \star, \tag{25.1}$$

where π_j and n_j denote response rate and sample size, respectively.

The similarity of control data across trials is described by a hierarchical model. The simplest model assumes exchangeability of the study parameters, more specifically that

$$\theta_1, \ldots, \theta_J, \theta_\star \mid \mu, \tau \sim N(\mu, \tau^2), \tag{25.2}$$

with log-odds $\theta_j = \log\{\pi_j/(1 - \pi_j)\}$. The *MAP* approach uses (25.1) and (25.2) to predict the parameter $\theta_\star$ in the new trial. This hierarchical model acknowledges that clinical trials are heterogeneous, but still allows us to borrow strength from historical information. The between-trial standard deviation τ characterizes the extent to which an individual trial deviates from the mean μ. One can also consider an extension of model (25.2) to partial exchangeability

$$\theta_j \mid \mu, \boldsymbol{\beta}, \tau \sim N(\mu + \boldsymbol{\beta}^T(\boldsymbol{x_j} - \bar{\boldsymbol{x}}), \tau^2) \,, \quad j = 1, \ldots, J, \star \,, \tag{25.3}$$

where $\boldsymbol{x_j}$ is a vector of predictors for study j and $\bar{\boldsymbol{x}}$ the mean.

In the Bayesian framework, prior distributions for μ and τ must be provided. The parameter μ is well informed by the data, and hence a weakly informative prior is typically used such as $N(0, 10^2)$. More care is needed for the between-trial standard deviation τ, especially if J is small. Weakly informative Half-Normal, Half-Cauchy or Half-t distributions that put most of the probability mass on realistic values of τ are recommended [23, 44, 52]. We will use a Half-Normal distribution with scale parameter 1, $HN(1)$. On the logit-scale, this prior covers the range from very small to very large ($\tau = 2$) between-trial heterogeneity [38].

For the models (25.1) to (25.2), the posterior probability density function $p(\theta_1, \ldots, \theta_J, \theta_\star, \mu, \tau \mid y_1, \ldots, y_J)$ is not tractable. However a sample from the posterior distribution using Markov chain Monte Carlo (MCMC) can be easily generated with standard software (Section 25.7). The *MAP* prior is the marginal posterior distribution for $\theta_\star$, which is used for the control in the new study:

$$p_{MAP}(\theta_\star) = p(\theta_\star \mid y_1, \ldots, y_J) \,. \tag{25.4}$$

Figure 25.1 shows a summary of the historical control parameters and the *MAP* prior (25.4) for the case study discussed in Section 25.2. The posterior median for both $\pi_\star$ and the transformed population mean $exp(\mu)/\{1 + exp(\mu)\}$ is 0.25. However, the 95% interval is wider for $\pi_\star$ (0.11, 0.47) than for the transformed population mean (0.18, 0.33), due to the between-trial standard deviation τ with a posterior median (95% interval) of 0.35 (0.04, 0.88).

25.3.2   Mixture approximation for priors

MCMC provides a sample from the *MAP* prior $p_{MAP}(\theta_\star)$ defined in (25.4), but no analytical form. This *MAP* prior sample is cumbersome to describe precisely in clinical study protocols and medical journals, and inconvenient for reviewers in ethical comittees or health authorities. However, it can be shown that every prior can be well approximated by a mixture of conjugate priors [6, 8]. For binary endpoints, this means that the *MAP* prior can be approximated by a mixture of *Beta* distributions, with density function $\sum_{k=1}^{K} w_k Beta(\theta_\star | a_k, b_k)$, where $Beta(\theta_\star | a, b)$ is the probability density function of a $Beta(a, b)$ random variate. For a given number of mixture components K, one can then find the parameters w_k, a_k, b_k such

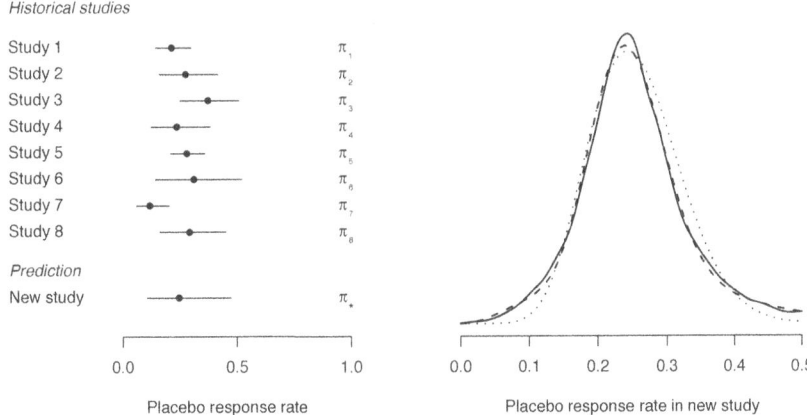

FIGURE 25.1

The left panel shows placebo response rates from eight historical studies in *ankylosing spondylitis*, and the derived MAP prior for the response rate in the new study (median and 95%-interval). The right panel shows the MAP prior density (solid line), an approximation by a conjugate $Beta(11, 32)$ density (dotted line), and an approximation by a mixture of conjugate densities $0.66Beta(17.2, 52.4) + 0.34Beta(3.1, 8.2)$ (dashed line).

that the mixture is closest to the MAP prior with respect to the Kullback-Leibler divergence. Technically, this can be achieved by calculating the maximum likelihood estimates for the parameters given the MCMC sample from the MAP prior [47]. The appropriate number of mixture components can be determined by numerical criteria or graphical tools.

For the example of Section 25.2, the MAP prior can be well approximated by a mixture of two $Beta$ distributions with probability density function:

$$0.66 \ Beta(\theta_\star | 17.2, 52.4) + 0.34 \ Beta(\theta_\star | 3.1, 8.2) \ . \tag{25.5}$$

Figure 25.1 shows the MAP prior and the approximating mixture of $Beta$ distributions. Also shown is the approximation by the conjugate $Beta(11, 32)$ distribution used in Baeten et al. [1].

The approximation of the MAP prior by a mixture of conjugate priors has two practical advantages. First, the prior can be easily communicated, and second the posterior distribution can be evaluated analytically [47]. For example, if the prior is a mixture of Beta distributions, then the posterior distribution is a also a mixture of Beta distributions.

25.3.3 Robustness to a prior-data conflict

Historical data will only be used as prior information in a clinical trial if they are considered relevant. Hence the selection of historical trials requires care and medical judgement [43]. Nevertheless, it is possible that, for some unforeseen reason, the selected historical data are not relevant for the current trial. In a clinical trial context, the most appropriate reaction to such a prior-data conflict is to discount the prior information, which we call robustness to a prior-data conflict.

The possibility that historical data are not relevant can be acknowledged within a Bayesian framework by placing some prior probability on this scenario. More formally, instead of the *MAP* prior $p_{MAP}(\theta_\star)$, a robust *MAP* prior is used as

$$p^R_{MAP}(\theta_\star) = p_R \, p_{MAP}(\theta_\star) + (1 - p_R) \, p_V(\theta_\star) \,, \tag{25.6}$$

where p_R is the probability that the historical data are relevant, and $p_V(\theta_\star)$ is a weakly informative prior that would be used when no historical data are available [47]. For example, if the approximate *MAP* prior is (25.5) as in Figure 25.1, and the probability that this prior is relevant is judged as $p_R = 0.5$, then a robust prior is a mixture of *Beta* distributions with probability density function:

$$0.5\{0.66 \; Beta(\theta_\star | 17.2, 52.4) + 0.34 \; Beta(\theta_\star | 3.1, 8.2)\} + 0.5 Beta(\theta_\star | 1, 1) \,. \tag{25.7}$$

Heavy-tailed prior distributions are robust to prior-data conflicts, while conjugate priors do not have this property [41]. To illustrate this, we compare results for three approximations to the *MAP* prior: the conjugate prior approximation $Beta(11, 32)$; the slightly heavy-tailed two component mixture approximation (25.5); the heavy-tailed robust *MAP* prior (25.6). If in the new trial all six patients randomized to placebo respond, then this can be considered a clear prior-data conflict, based on the prior predictive distribution [4]. Table 25.2 shows that for the conjugate prior, the posterior is then a compromise of prior and data, while for the robust prior, the posterior is similar to the one obtained with the weakly informative prior $Beta(1, 1)$ that would be used if no historical data were available. Hence the robust prior essentially discards the historical data in case of conflict. If data and prior are consistent (e.g. one of the six placebo patients responds), then results are very similar for all priors.

TABLE 25.2
Summary of posterior placebo response rate $\pi_\star$ (median, 95% interval) for different priors of the placebo response rate in case of prior-data conflict and prior-data consistency for *ankylosing spondylitis*.

Prior	Posterior $\pi_\star$
Prior-data conflict (6/6 respond)	
$Beta(11, 32)$	0.34 (0.22,0.48)
$0.66 Beta(17.2, 52.4) + 0.34 Beta(3.1, 8.2)$	0.49 (0.24,0.74)
$0.33 Beta(17.2, 52.4) + 0.17 Beta(3.1, 8.2) + 0.5 Beta(1, 1)$	0.90 (0.56,0.99)
$Beta(1, 1)$	0.91 (0.59,0.99)
Prior-data consistency (1/6 respond)	
$Beta(11, 32)$	0.24 (0.14,0.37)
$0.66 Beta(17.2, 52.4) + 0.34 Beta(3.1, 8.2)$	0.24 (0.11,0.39)
$0.33 Beta(17.2, 52.4) + 0.17 Beta(3.1, 8.2) + 0.5 Beta(1, 1)$	0.24 (0.07,0.49)
$Beta(1, 1)$	0.23 (0.04,0.58)

25.3.4 Prior effective sample size

The main motivation for the use of historical control information is the possibility to reduce the number of patients we need to randomize to control. To do this appropriately, the *effective sample size* (*ESS*) of the prior information has to be assessed. For conjugate priors, the *ESS* is well defined [2]. For example the conjugate prior $Beta(11, 32)$ used by Baeten and colleagues [1] has an *ESS* of 43 ($= 11 + 32$).

For non-conjugate priors, no exactly equivalent sample size can be derived, and several approaches have been proposed. Malec [33], Pennello and Thompson [42], and Neuenschwander et al. [38] note that variances are approximately inversely proportional to sample sizes. For *MAP* priors, the *ESS* can then be obtained by discounting the total sample size of the historical control data by the ratio of variances of the *MAP* prior and of the *MAP* prior assuming complete homogeity ($\tau = 0$) [38]. Alternatively, Morita et al.[35] define the *ESS* as the sample size such that the expected information of the sample using a weakly informative prior equals the information of the actual prior; the expected information is evaluated at the prior mean of the parameter.

Neuenschwander and colleagues [40] review these approaches, and point out that these may provide very different *ESS* values for nonconjugate settings. A desirable property of *ESS* is predictive consistency, i.e. the expected posterior-predictive *ESS* for a sample of size N should be the sum of the prior *ESS* and N. While the above mentioned methods are not predictively consistent, the *expected local-information-ratio ESS* (ELIR) has this property, and provides an appropriate *ESS* for one-parameter models [40].

For the *MAP* prior shown in Figure 25.1, the *ESS* is 25 based on Neuenschwander et al. [38], and 58 based on Morita et al.[35], while the predictively consistent ESS_{ELIR} is 36. For the robust *MAP* prior (25.7), the *ESS* is 3, 38 and 12 for the three approaches [38], [35] and [40], respectively. It should be noted that for the conjugate prior $Beta(11, 32)$ all three approaches give an *ESS* of 43.

For the case study (Section 25.2), 24 patients were randomized to the test treatment *secukinumab*. As the *ESS* of the placebo *MAP* prior is higher, one could consider a trial where no patients receive placebo. However this has two disadvantages. First, the lack of blinding in single arm trials may introduce biases in the conduct and evaluation of the study. Second, without placebo patients in the actual trial, the possibility of a prior-data conflict cannot be assessed. In the study by Baeten et al. [1], six patients were randomized to placebo. Hence the study has all the advantages of a randomized clinical trial, and allows one to evaluate whether the historical placebo information is consistent with the placebo data of the trial. In cases where the clinical team is less confident on the relevance of the historical control information, a less agressive reduction of the number of patients randomized to control may be preferable.

25.3.5 Operating characteristics

The evaluation of frequentist properties of Bayesian clinical trial designs is important, especially in a regulatory context. Frequentist metrics of interest include bias and mean squared error of point estimates (e.g. the posterior median), and type I error and power to meet the success criteria under various configurations of the parameters (e.g. that the posterior probability of a treatment effect has to be at least 95%). The evaluation of operating characteristics usually requires clinical trial simulation, where for given values of the parameters a large number of datasets are generated, and for which Bayesian analyses are performed [3]. Robust *MAP* priors have better frequentist properties compared to conjugate priors, as the historical information is discounted in case of a prior-data conflict [47].

25.3.6 Analysis

The Bayesian analysis of a randomized controlled clinical trial starts with the calculation of the posterior distribution of the model parameters. Often, as in the case study of Section 25.2, a weakly informative prior is used for the test treatment response rate, and an independent informative prior based on historical information for the control response rate. The posterior distribution can then be evaluated separately for test treatment and control.

For priors that are mixtures of conjugate distributions, the posterior is also a mixture of conjugate distributions and can be calculated analytically [2, 47]. For the case study, the robust MAP prior for the placebo response rate is given in (25.7). In the actual study [1], one of the six patients on placebo responded. The posterior for the placebo response rate has then probability density function

$$0.49 Beta(\theta_\star|18.2, 57.4) + 0.21 Beta(\theta_\star|4.1, 13.2) + 0.30 Beta(\theta_\star|2, 6) . \qquad (25.8)$$

The weights of the three mixture components depend on the prior mixture weights and the marginal probabilities of the data; for example, the first weight is proportional to $0.5 \cdot 0.66 B(18.2, 57.4)/B(17.2, 52.4)$, where $B(a, b)$ is the $Beta$ function. In the motivating example, the prior for the response rate of the test treatment was set as $Beta(0.5, 1)$, 14 of 23 patients responded, and hence the posterior is $Beta(14.5, 10)$. The posterior for the difference δ of the response rate on test treatment and control can then be obtained by numerical integration or simulation.

Table 25.3 shows the Bayesian analysis for the case study of Section 25.2, using different priors for the placebo response rate. The MAP prior (25.4), the approximating conjugate prior $Beta(11, 32)$, and the approximating mixture of conjugate priors (25.5) give almost identical results. (The results for the MAP prior were obtained by the meta-analytic-combined approach; see Section 25.4.1). The robust MAP prior (25.7) also gives very similar results, with slightly wider posterior probability intervals.

TABLE 25.3
Bayesian analysis of the clinical trial in *ankylosing spondylitis* for different priors of the placebo response rate $\pi_\star$. The posterior of the treatment effect δ is summarized by median and 95% interval.

Prior for $\pi_\star$	Posterior δ	$P(\delta > 0)$
MAP	0.36 (0.11,0.58)	0.997
$Beta(11, 32)$	0.35 (0.12,0.56)	0.998
$0.66 Beta(17.2, 52.4) + 0.34 Beta(3.1, 8.2)$	0.36 (0.11,0.58)	0.997
$Robust\ MAP^a$	0.36 (0.05,0.60)	0.987
$Beta(1, 1)$	0.35 (-0.03,0.64)	0.965

^{a} $0.33 Beta(17.2, 52.4) + 0.17 Beta(3.1, 8.2) + 0.5 Beta(1, 1)$

25.4　Other Approaches

In this section, we briefly review alternative ways to incorporate historical data in clinical trials. These are: the MAC approach, bias models, commensurate priors, and power priors. Like the MAP approach of Section 25.3, they are meta-analytic in nature and discount the historical data.

25.4.1　Meta-analytic-combined approach

The MAC approach is an analogue to the MAP approach of Section 25.3. However, instead of the two-step MAP approach (derivation of MAP prior, which is later combined with the trial data), historical and actual trial data are analyzed jointly. In other words, the inference for the parameter of interest $\theta_\star$ is based on the conditional distribution of $\theta_\star$ given historical *and* actual trial data,

$$\theta_\star | y_1, \dots, y_J, y_\star \ . \tag{25.9}$$

For any hierarchical model (like, for example, (25.1) and (25.2)), it can be shown that inference for $\theta_\star$ is the same for *MAC* and *MAP* [47]. In most clinical trials, specifications of priors at the design stage of a new trial will usually be required. On the other hand, for retrospective analyses with historical data, the *MAC* approach seems more natural. For a *MAC* analysis, non-Bayesian inference for $\theta_\star$ could be considered as well.

25.4.2 Bias models

Bias models [43] are similar to the standard hierarchical model used for *MAP* or *MAC*. However, they directly express historical parameters as biased versions of the parameter in the new trial:

$$\theta_j = \theta_\star + \delta_j, \quad j = 1, \dots, J \ . \tag{25.10}$$

The model is more general than the standard hierarchical model, since study-specific biases δ_j can be modelled flexibly. However, if no systematic biases are expected for all studies,

$$\delta_j | \tau_\delta \sim N(0, \tau_\delta^2) \tag{25.11}$$

should be considered, which is equivalent to the standard hierarchical model with $\tau_\delta^2 = 2\tau^2$. The prior for τ and hence also for τ_δ could be weakly informative as described in Section 25.3.1, but may also be based on expert judgement [56].

25.4.3 Commensurate priors

Commensurate priors [27] offer yet another way to account for historical data in a model-based framework. For the case of one historical trial with parameter θ_1, the commensurate prior for $\theta_\star$ is centered at θ_1,

$$\theta_\star | \theta_1, \tau_c \sim N(\theta_1, \tau_c^2) \ . \tag{25.12}$$

The parameter τ_c quantifies the commensurability of the historical and actual trial. For the case of a single historical trial, the commensurate prior approach is equivalent to the approaches of Sections 25.4.1 and 25.4.2; in particular, $\tau_c^2 = 2\tau^2$. For more than one historical trial, the suggested commensurate prior [27] is

$$\theta_\star | \mu, \tau_c \sim N(\mu, \tau_c^2), \tag{25.13}$$

which assumes equality of historical parameters,

$$\theta_1 = \dots = \theta_J = \mu \tag{25.14}$$

25.4.4 Power priors

The methods discussed so far build on models for the parameters in the actual and historical trials. *Power priors* do not explicitly consider parameters in historical trials [29]. Assuming one historical trial, the amount of borrowing is directly given by a power parameter α

$$\pi(\theta_\star | y_1) \propto L(y_1 | \theta_\star)^\alpha \pi_0(\theta_\star), \quad \alpha \in (0, 1) \ . \tag{25.15}$$

Here, L is the likelihood of the historical data, and π_0 is a weakly-informative prior. Values for α of 0 and 1 correspond to no borrowing (i.e. disregarding historical control data) and full borrowing (i.e. pooling historical and concurrent control data), respectively.

Two versions of power priors have been proposed. The first version assumes a fixed α, which determines a fixed degree of borrowing as a fraction of the historical data; similarly, fixed borrowing for the other approaches is obtained if values for τ, τ_δ, or τ_c are fixed. The relationship between power priors and hierarchical models has been discussed in [5]. For a hierarchical model with normally distributed data and trial parameters, the relation among the power parameter α, sample size n of a single historical trial, outcome standard deviation σ, and between-trial standard deviation τ is

$$\alpha = (1 + 2n\tau^2/\sigma^2)^{-1} . \tag{25.16}$$

The second version with unknown α poses severe problems for two reasons. First, in the original proposal [29]

$$\pi(\theta_\star, \alpha | y_1) \propto L(y_1 | \theta_\star)^\alpha \pi(\alpha) \pi_0(\theta_\star) \tag{25.17}$$

the normalizing constant

$$C(\alpha) = 1/ \int L(y_1 | \theta_\star)^\alpha \pi_0(\theta_\star) d\theta_\star \tag{25.18}$$

is missing [9, 37]. Therefore, (25.17) should be replaced by

$$\pi(\theta_\star, \alpha | y_1) = C(\alpha) L(y_1 | \theta_\star)^\alpha \pi(\alpha) \pi_0(\theta_\star) . \tag{25.19}$$

Since $C(\alpha)$ can usually not be calculated analytically, power priors with unknown α can be computationally challenging.

The second problem is more concerning: power priors with unknown α violate the likelihood principle [9]. For example, for the simple case of binary data with y responders in n subjects, it matters whether one uses individual data y_i or summary data $y = \sum_{i=1}^n y_i$. In fact, raising the individual data distribution $\pi^y(1 - \pi)^{n-y}$ or the summary data (sufficient statistic) distribution $\pi^y(1 - \pi)^{n-y} n!/(y!(n - y)!)$ to the power α and normalizing with (25.18) will lead to different conclusions.

Finally, for more than one historical trial, trial-specific power parameters α_j should be considered. From (25.16) it follows that assuming equal power parameters across studies would imply different assumptions for between-trial heterogeneity (τ). For example, $\alpha_j = 0.2$ can imply large (for small n_j) to small (for large n_j) between-trial heterogeneity.

In conclusion, given the complication arising from the need of the normalizing constant and the violation of the likelihood principle, the use of power priors with unknown power parameters should be avoided.

25.4.5 Test-then-pool

The methods discussed so far are similar meta-analytic dialects, which help to borrow information from historical data in a principled, model-based way. They strike a balance between two extremes: complete pooling (no discounting of historical data) and ignoring the historical data all together.

A fundamentally different approach to incorporating historical data is *test-then-pool*. First, equality of parameters in the historical and new trial is tested at a given significance level. If the test is not significant, historical and actual trial data will be pooled; otherwise, historical data are ignored. Some fine-tuning for the significance level is required to obtain reasonable frequentist operating characteristics; for a comparison of *test-then-pool* with some of the other approaches, see [57].

We discourage the use of *test-then-pool*. Even if operating characteristics are reasonable, they may not be sufficient for a good design. For example, consider the (not uncommon) situation where the amount of historical data is large (e.g. for placebo), and the new trial is small. Under *test-then-pool*, if the test fails, complete pooling would imply a very precise estimate, which is completely dominated by historical data. This seems unwarranted because it ignores the hierarchical structure of the data.

25.4.6 How much borrowing?

The approaches differ in how the data inform the heterogeneity parameters $\tau, \tau_\delta, \tau_c, \alpha$, which eventually determine the amount of borrowing from historical data. Irrespective of the chosen method, these parameters are notoriously difficult to estimate if the number of trials is small. Therefore, care is needed when specifying their prior distributions.

Under the hierarchical models for MAP (or MAC) and bias models, the heterogeneity parameters are estimated by taking into account the variability across all trials; for example, large variability in historical trials will necessarily lead to very little borrowing. For commensurate priors, learning about τ_c is more difficult, since information about heterogeneity stems from the difference between the actual and the pooled historical trials.

Priors for τ or τ_δ should be specified based on plausible assumptions for between-trial heterogeneity. For example, if little is known about τ, the prior should have most of its probability mass in a range representing small to large heterogeneity; for log-odds parameters, the corresponding range for τ is usually taken as 0 to 1 or 0 to 2, which are covered well by Half-Normal priors with scale parameter 0.5 or 1. It should be noted that the still popular $Gamma(\epsilon, \epsilon)$ priors (with small ϵ) for precisions ($= 1/\tau^2$) are not well-suited for a small number of trials, because they do not represent a-priori likely heterogeneity assumptions [23, 52]. In some cases, an informative prior on τ may be derived by leveraging external information on the expected magnitude of between-study heterogeneity [55]. For example, for studies in children, it may be possible to base the prior for between-trail heterogeneity on past studies in adults [48].

For the commensurability parameter τ_c, spike-and-slab priors with four tuning parameters have been recommended [27]. These parameters can be obtained through evaluation of frequentist design metrics (bias, mse, type-I error, power).

When specifying the prior for α in the second version of power priors, it may help to consider the respective heterogeneity among trial parameters. Normal approximations using (25.16) facilitate the understanding of prior assumptions about α in terms of between-trial heterogeneity. For example, for binary data and exchangeable log-odds parameters, between-trial standard deviations $\tau = 0.25$ and $\tau = 0.5$ are typical values for moderate and substantial between-trial heterogeneity [38]. For a binary outcome, the power parameters for historical trials of size 25, 50, and 100, are 0.56, 0.39, and 0.24 ($\tau = 0.25$), and 0.24, 0.14, and 0.07 ($\tau = 0.5$). This shows that under moderate heterogeneity, the absolute amount of data borrowed from the historical trial is 14, 19, and 24 subjects, respectively. Under substantial heterogeneity, these numbers are much lower (6–7 subjects).

25.5 Extensions

25.5.1 Individual patient data and aggregate data

Historical data arise in various forms. So far we have only considered the use of aggregate data, which is typical if one collects data from the literature, as for example in a systematic review. In some cases one may also have access to individual patient data for at least some trials. The additional information may allow for more accurate inference, such as individual level regression models or the estimation of nuisance parameters. Adjusting for population differences with individual patient covariates may not only reduce between-trial heterogeneity but also increase inferential precision for primary endpoints. In the context of the *MAP* approach of Section 25.3, individual patient data may ultimately lead to a more informative prior and larger effective sample size. The meta-analysis of individual patient and aggregate data is an active area of research, which gained increasing interest about twenty years ago with early papers from Stewart et al. [53] and Higgins et al. [26]; for a recent review see Dubrey et al. [7].

To formally combine individual patient and aggregate data in the *MAP* approach, we extend our notation. Let y_{jk} be the outcome for patient k in trial j, $k = 1, \ldots, n_j$. Let J_1 and J_2 denote the numbers of historical trials with individual patient and aggregate data, respectively. This means that for trials $j = J_1 + 1, \ldots, J_1 + J_2$ we only have aggregate data, typically the mean,

$$\bar{y}_j = \frac{1}{n_j} \sum_{k=1}^{n_j} y_{jk}, \tag{25.20}$$

so that the historical data are $y_{11}, \ldots, y_{1n_1}, y_{21}, \ldots, y_{2n_2}, \ldots, y_{J_1 1}, \ldots, y_{J_1 n_{J_1}}$ and $\bar{y}_{J_1+1}, \ldots, \bar{y}_{J_1+J_2}$.

To implement the *MAP* approach, we need a hierarchical model with three parts: sampling models for the individual and aggregate data, a parameter model for all trial parameters, and prior distributions. We briefly discuss each element in turn. Let

$$y_{jk} \sim f(\gamma_j, \kappa_j) \, , \, j = 1, \ldots, J_1, \star \tag{25.21}$$

denote a suitable sampling model, where γ_j are the (univariate) parameters of primary interest, and κ_j are q-dimensional nuisance parameters. Typically, the trial means $\gamma_j = \mathrm{E}(y_{jk})$ are of interest. The sampling model $f(\gamma_j, \kappa_j)$ applies to the individual patient data of the first J_1 trials. It also induces the distribution of the trial means

$$\bar{y}_j \sim \tilde{f}(\gamma_j, \kappa_j) \, , \, j = J_1 + 1, \ldots, J_1 + J_2 \, . \tag{25.22}$$

If $\tilde{f}$ is not available in closed form, normal approximations can be used instead. This will work well in practical applications unless study sizes n_j are small.

For simplicity, we treat trial means γ_j and nuisance parameters κ_j separately. We assume exchangeability of trial means. After transformation to a suitable scale $\theta_j = g(\gamma_j)$, we assume

$$\theta_1, \ldots, \theta_{J_1+J_2}, \theta_\star | \mu, \tau \sim N(\mu, \tau^2). \tag{25.23}$$

Often only the trials with individual data allow estimation of the nuisance parameters κ_j. We therefore need additional assumptions. The simplest situation arises if we can assume all κ_j be the same,

$$\kappa_1 = \ldots = \kappa_{J_1+J_2} = \kappa_\star = \kappa \, . \tag{25.24}$$

If needed, this assumption could be relaxed, and a more general exchangeability model including nuisance parameters could be considered.

The Bayesian set-up is completed by specifying priors for μ, τ, and κ. As before, μ and κ will typically be well informed by the data. Prior specification for the between-trial heterogeneity τ needs special care (Sections 25.3 and 25.4.6).

We now present an application to the case of historical trials with over-dispersed count data, originally discussed in Gsteiger et al. [25]. In this example, the objective is to derive a *MAP* prior for the placebo group of a new study in *Multiple Sclerosis*. The endpoint is the number of brain lesions counted on magnetic resonance imaging scans. For $J_1 = 3$ trials the individual patient data are available, while for $J_2 = 6$ trials the reported mean lesion counts were collected from the published literature; Table 25.4 shows the summary data for the nine historical trials from a total of $N = 1936$ patients.

TABLE 25.4
Summary placebo data from nine historical *Multiple Sclerosis* trials with over-dispersed counts endpoint.

	1	2	3	4	5	6	7	8	9
$\bar{y}_j$	2.2	1.3	1.4	1.6	1.7	1.3	1.4	1.1	0.9
n_j	81	373	50	32	548	315	35	65	437

A standard sampling model for over-dispersed lesion count data in *Multiple Sclerosis* is the negative-binomial (NB) distribution,

$$y_{jk} \sim NB(\gamma_j, \kappa) \ , \ j = 1, 2, 3, \star. \tag{25.25}$$

We parameterize the negative-binomial model with the mean γ_j and the dispersion κ, defined through

$$\mathrm{E}(y_{jk}) = \gamma_j, \quad \mathrm{Var}(y_{jk}) = \gamma_j + \kappa \gamma_j^2. \tag{25.26}$$

For the sampling model of the aggregate data $\bar{y}_j, j = 4, \ldots, 9$, it turns out that the convolution of NB distributions is again a NB,

$$n_j \bar{y}_j \sim NB(n_j \gamma_j, \kappa/n_j) \ , \ j = 4, \ldots, 9. \tag{25.27}$$

We assume exchangeability of trial means on the log-scale $\theta_j = \log(\gamma_j)$

$$\theta_1, \ldots, \theta_9, \theta_\star | \mu, \tau \sim N(\mu, \tau^2). \tag{25.28}$$

Finally, the priors were specified as follows:

$$\mu \sim N(0, 100^2), \ \tau \sim \mathrm{HN}(1), \kappa \sim \mathrm{Exp}(rate = 1/3) \tag{25.29}$$

as in [25]; see the discussion therein for more details.

The Bayesian analysis of the above negative-binomial model can be readily implemented in WinBUGS [32] or SAS (SAS Institute, Inc.; Cary NC) by noting its equivalent representation as Gamma-Poisson mixture. Indeed, if

$$Y | \lambda \sim \mathrm{Poisson}(\lambda), \quad \lambda | \alpha, \beta \sim \mathrm{Gamma}(\alpha, \beta), \tag{25.30}$$

then the marginal distribution $Y | \alpha, \beta$ is a negative-binomial distribution with mean= α/β and dispersion $1/\alpha$.

Fitting this model to the historical data leads to the posterior summaries in Table 25.5. The resulting *MAP* prior $p_{MAP}(\theta_\star)$ has median 0.32 and 95% interval $(-0.32, 0.97)$. The effective sample size (ESS) of this prior is about 45 patients, using the approach by Neuenschwander et al.[38] as described in Section 25.3. Since $\theta_\star$ corresponds to the log-transformed parameter, back-transformation to $\gamma_\star = \exp(\theta_\star)$ provides the *MAP* prior for the mean count in the new trial, which is easier to interpret for a clinical team. The corresponding $p_{MAP}(\gamma_\star)$ has median 1.4 and 95% interval $(0.7, 2.7)$. Between-trial heterogeneity τ was well informed by the nine trials, which can be seen from the prior and posterior summaries in Table 25.5. Sensitivity analyses show that the prior for τ has little impact on the *MAP* prior for $\theta_\star$ and on the effective sample size [25].

TABLE 25.5
Prior and posterior summaries from fitting the over-dispersed count model to combined individual patient and aggregate data in *Multiple Sclerosis*.

Parameter	Prior			Posterior		
	2.5%	50%	97.5%	2.5%	50%	97.5%
$\theta_\star$	-200	0	200	-0.32	0.32	0.97
$exp(\theta_\star)$	0	1	10^{85}	0.7	1.4	2.7
μ	-196	0	196	0.10	0.32	0.57
τ	0.03	0.67	2.24	0.12	0.25	0.55
κ	0.08	2.08	11.07	2.85	3.48	4.26

25.5.2   Non-inferiority trials

In the regulatory setting, the minimal efficacy requirement for a test treatment is its superiority to placebo. Ideally, the requirement is assessed in a randomized clinical trial with test treatment and placebo. In some cases, however, use of placebo may not be ethical (for example in a serious disease with an approved drug [13, 54]). The only option may then be a non-inferiority trial with the test treatment and an active-control. For such a trial, the minimal efficacy requirement can only be assessed indirectly by comparing the test treatment to historical placebo. In this setting, Bayesian approaches seem particularly attractive [22].

For illustration, we consider here the non-inferiority trial *SPORTIF V* discussed in the FDA guidance on non-inferiority [17]. In this trial for the prevention of strokes in patients with *nonvalvular atrial fibrillation*, $n_\star^T = 1960$ patients were randomized to the test treatment *ximelagatran* (T), and $n_\star^C = 1962$ patients to the active-control *warfarin* (C). The number of patients with undesirable events was $y_\star^T = 51$ for the test treatment and $y_\star^C = 37$ for the active-control.

Relevant historical information from six randomized trials comparing the active-control (C) with placebo (P) were available at the design stage. We denote the number of events and patients by y_j^C and n_j^C for active-control, and y_j^P and n_j^P for placebo, respectively. The statistical model for the data is

$$y_\star^T \mid \pi_\star^T \sim Binomial(\pi_\star^T, n_\star^T), \tag{25.31}$$
$$y_j^C \mid \pi_j^C \sim Binomial(\pi_j^C, n_j^C), \quad j = 1, \ldots, J, \star,$$
$$y_j^P \mid \pi_j^P \sim Binomial(\pi_j^P, n_j^P), \quad j = 1, \ldots, J.$$

In the following, it will be more convenient to work with the log-odds $\theta = \log\{\pi/(1-\pi)\}$ rather than the response rates π.

The minimal efficacy requirement for the test treatment *ximelagatran* is that it would have been superior to placebo, had placebo been included in the *SPORTIF V* trial, i.e. that $\theta_\star^T - \theta_\star^P > 0$ [51]. To adress this, we use the mathematical identity

$$\theta_\star^T - \theta_\star^P = (\theta_\star^T - \theta_\star^C) + (\theta_\star^C - \theta_\star^P). \tag{25.32}$$

Information on the first term ($\theta_\star^T - \theta_\star^C$) is directly obtained from the non-inferiority trial, while information on ($\theta_\star^C - \theta_\star^P$) can be derived from the historical trials, e.g. by using a meta-analytic-predictive (*MAP*) approach [50]. More formally, if $\delta_j^{CP} = (\theta_j^C - \theta_j^P)$ for $j = 1, \ldots, J, \star$, then the *MAP* approach links these parameters for the non-inferiority and historical trials as follows:

$$\delta_1^{CP}, \ldots, \delta_J^{CP}, \delta_\star^{CP} \mid \mu^{CP}, \tau \sim N(\mu^{CP}, \tau^2). \tag{25.33}$$

The hierarchical model allows for between-trial heterogeneity in the treatment effect (C vs. P) expressed by the between-trial standard deviation τ. With a weakly informative prior for the parameters, as in Section 25.3, a sample from the posterior distribution can be obtained with MCMC using standard software [31]. The posterior (median, 95% interval) for $\delta_\star^{CP}$ (odds ratio: 0.34, 0.15 to 0.79) is wider than for μ^{CP} (odds-ratio: 0.34, 0.22 to 0.52) due to between-trial heterogeneity τ (0.20, 0.01 to 0.84).

Figure 25.2 shows in the upper part the information obtained from the historical trials: the individual results for the six historical trials, the population mean μ^{CP} and the MAP prediction for $\delta_\star^{CP}$ in the non-inferiority trial. This historical information suggests that the active-control *warfarin* (C) would have been clearly superior to placebo (P) in the non-inferiority trial $SPORTIF\ V$, had placebo been included. Figure 25.2 also shows the results from the $SPORTIF\ V$ non-inferiority trial, suggesting that the test treatment *ximelagatran* (T) tends to be worse than the active-control *warfarin* (C). Finally, the information from the historical trials and the non-inferiority trial is combined to obtain the comparison of the test treatment with putative placebo in the non-inferiority trial. Although the test treatment appears to be better than placebo, $P(\theta_\star^T - \theta_\star^P > 0) = 0.96$, the evidence for the minimal efficacy requirement is not sufficient according to standard regulatory rules, which would require a probability of superiority of at least 0.975.

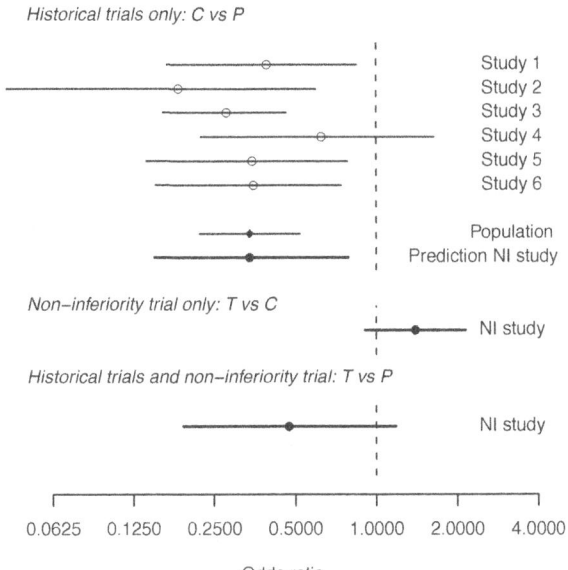

FIGURE 25.2
Summary results (posterior median and 95%-interval) from a Bayesian meta-analytic-predictive approach for a non-inferiority trial with the test treatment *ximelagatran* (T) and the active-control *warfarin* (C). Historical information from six trials comparing *warfarin* (C) with placebo (P) is included.

We considered here the simple case where historical information consists of summary data from several randomized trials comparing the active-control with placebo. To link the parameters of historical and NI trial, we used the hierarchical model (25.33). Additionally, one could also link historical data on the active control using the model (25.2) as described in [50]. Often several relevant historical clinical trials with multiple treatments are available which directly or indirectly link test treatment, active-control or placebo, and then a *network meta-analytic* approach can be used [50]. In some cases summary covariate information for each historical trial is known, which can be taken into account through *meta-regression* [61]. Although we focus here on non-inferiority trials, very similar approaches can also be applied for biosimilar trials [59].

25.6 Discussion

Incorporating historical data in clinical studies has economic and ethical advantages. It allows one to either reduce the number of patients, or to obtain more precise information for making decisions. This is particulary attractive in rare diseases [20, 21] or studies in children [45, 48, 58] where recruitment of patients is very challenging. These approaches are also commonly used in early phases of drug development [14, 62], where studies are typically small and fast decisions are needed. Using historical data in combination with adaptive designs further increases efficiency of clinical trials [24, 36].

Evidence synthesis methodology is needed for extracting information from historical data. We considered here mainly meta-analytic methods, as these are the standard tools in health technology assessment and evidence based medicine [11]. These methods are also very flexible and general, can incorporate data from multiple treatments (network meta-analysis) and include covariates (meta-regression), both for individual and aggregate level data. They also can be used with various endpoints, including time-to-event [46] or recurrent event [28] endpoints. Historical information on nuisance parameters such as variances may also be used [49]. Often several clinical trials on the same treatment run in parallel, which provides the opportunity to borrow strength from such co-data [39].

The use of historical data to borrow strength can be considered in all phases of drug development. In some areas, randomizing large numbers of patients to placebo is not feasible or not ethical, which makes historical information essential for study design and interpretation of results. At the time being, historical information is mostly used in phase I and II, where patient numbers are limited and external information can be of great value. For phase III trials, incorporation of historical information is essential for non-inferiority and biosimilar trials. Historical controls are occasionally accepted in phase III, for example for monotherapy treatment of epilepsy [19], in serious conditions with a high unmet medical need, or for medical devices [16]. Historical information is also valuable in the design and analysis of pediatric trials, which are difficult to conduct, but where strength can be gained through a considerable amount of adult data [10]. Historical information is also important for the design and analysis of studies after approval (phase IV), as much information is available at this stage.

Since the seminal paper of Pocock in 1976 [43], the scope for the use of historical data has been greatly expanded. Advances in methodology allow now for very general evidence synthesis models and complex data structures. Additionally, MCMC has revolutionized Bayesian computation and facilitates the routine implementation of these models. And finally, clinicians and health authorities increasingly accept Bayesian approaches and historical information in clinical trials.

25.7 Appendix

25.7.1 WinBUGS code

Code for the *MAP* analysis, using WinBUGS [32]; see [47] for more comprehensive code.

```
# Meta-analytic-predictive (MAP) approach;
model {
   for (j in 1:J) {
      y[j] ~ dbin(pi[j],n[j])
      pi[j] <- exp(theta[j])/(1+exp(theta[j]))
      theta[j] ~ dnorm(mu,tau_prec)
      }
  mu ~ dnorm(0,0.01)
  tau ~ dnorm(0,1)I(0,)
  tau_prec <- 1/(tau*tau)
  pi_star <- exp(theta_star)/(1+exp(theta_star))
  theta_star ~ dnorm(mu,tau_prec)
}


# Placebo data from 8 historical studies, y responders, n patients;
list(y=c(23,12,19,9,39,6,9,10),n=c(107,44,51,39,139,20,78,35),J=8)


# initial values
list(mu=0,tau=0.2,theta_star=0)
```

25.7.2 SAS code

Code for the *MAP* analysis, using SAS (SAS Institute, Inc.; Cary NC).

```
* Placebo data from 8 historical studies, y responders, n patients;
data hist; infile cards; input study y n @@; cards;
1 23 107   2 12 44   3 19 51   4 9 39   5 39 139   6 6 20   7 9 78   8 10 35
;run;


* Meta-analytic-predictive (MAP) approach;
proc mcmc data=hist outpost=map nbi=100000 nmc=1000000 thin=100
                  monitor=(mu tau theta_star pi_star);
 parms mu 0 tau 0.2 theta_star 0;
 prior mu ~ normal(0,sd=10);
 prior tau ~ normal(0,sd=1,lower=0.0001);
 prior theta_star ~ normal(mu, sd=tau);
 random eta ~ normal(0, sd=tau) subject=study;
 theta = mu + eta;
 pi = exp(theta)/(1+exp(theta));
 pi_star = exp(theta_star)/(1+exp(theta_star));
 model y ~ binomial(n,pi); run;


* Approximating the MAP prior by a mixture of conjugate priors;
proc fmm data=map plots=density(width=0.01);
 model pi_star = / k=2 dist=beta; run;
```

Bibliography

[1] D. Baeten, X. Baraliakos, J. Braun, J. Sieper, P. Emery, D. van der Heijde, I. McInnes, JM. van Laar, R. Landewé, P. Wordsworth, J. Wollenhaupt, H. Kellner, J. Paramarta, J. Wei, A. Brachat, S. Bek, D. Laurent, Y. Li, YA. Wang, AP. Bertolino, S. Gsteiger, AM. Wright, and W. Hueber. Anti-interleukin-17A monoclonal antibody secukinumab in treatment of ankylosing spondylitis: a randomised, double-blind, placebo-controlled trial. *Lancet*, 382(9906):1705–1713, 2013.

[2] JM. Bernardo and AFM. Smith. *Bayesian Theory*. Chichester: Wiley, 1994.

[3] SM. Berry, BP. Carlin, JJ. Lee, and P. Müller. *Bayesian Adaptive Methods for Clinical Trials*. Boca Raton: CRC Press, 2001.

[4] GEP. Box. Sampling and Bayes inference in scientific modeling and robustness. *Royal Statistical Society Series A*, 143:383–430, 1980.

[5] MH. Chen and JG. Ibrahim. The relationship between the power prior and hierarchical models. *Bayesian Analysis*, 1(3):551 – 574, 2006.

[6] S. Dallal and W. Hall. Approximating priors by mixtures of natural conjugate priors. *Journal of the Royal Statistical Society Series B*, 45:278–286, 1983.

[7] Thomas PA Debray, Karel GM Moons, Gert van Valkenhoef, Orestis Efthimiou, Noemi Hummel, Rolf HH Groenwold, Johannes B Reitsma, and GetReal Methods Review Group. Get real in individual participant data (IPD) meta-analysis: a review of the methodology. *Research synthesis methods*, 6(4):293–309, 2015.

[8] P. Diaconis and D. Ylvisaker. Quantifying prior opinion. *Bayesian Statistics (Proceedings of the Second Valencia International Meeting)*, 2:133–148, 1984.

[9] YY. Duan, K. Ye, and EP. Smith. Evaluating water quality using power priors to incorporate historical information. *Environmetrics*, 17(1):95 – 106, 2006.

[10] J. Dunne, WJ. Rodriguez, MD. Murphy, BN. Beasley, GJ. Burckart, JD. Filie, LL. Lewis, HC. Sachs, PH. Sheridan, P. Starke, and LP. Yao. Extrapolation of adult data and other data in pediatric drug-development programs. *Pediatrics*, 128(5):1242–1249, 2011.

[11] M. Egger, GD. Smith, and DG. Altman. *Systematic Reviews in Health Care: Meta-Analysis in Context*. Chichester: Wiley, 2001.

[12] European Comission. Innovative Medicines initiative 2: Europe's fast track to better medicines, 2014. [Online; accessed 10-June-2020].

[13] European Medicines Agency's Committee for Medicinal Products for Human Use (CHMP). Guideline on the choice of the non-inferiority margin, 2006. [Online; accessed 10-June-2020].

[14] R. Fisch, I. Jones, J. Jones, J. Kerman, GK. Rosenkranz, and H. Schmidli. Bayesian design of Proof-of-Concept trials. *Therapeutic Innovation & Regulatory Science*, 49(1):155–162, JAN 2015.

[15] Food and Drug Administration (FDA). Innovation/Stagnation - Challenge and Opportunity on the Critical Path to New Medical Products, 2004. [Online; accessed 10-June-2020].

[16] Food and Drug Administration (FDA). Guidance for the Use of Bayesian Statistics in Medical Device Clinical Trials, 2010. [Online; accessed 10-June-2020].

[17] Food and Drug Administration (FDA). Guidance for industry: non-inferiority clinical trials to establish effectiveness, 2016. [Online; accessed 10-June-2020].

[18] Food and Drug Administration (FDA). Complex Innovative Trial Designs Pilot Program, 2019. [Online; accessed 10-June-2020].

[19] JA. French, S. Wang, B. Warnock, and N. Temkin. Historical control monotherapy design in the treatment of epilepsy. *Epilepsia*, 51(10):1936–1943, 2010.

[20] T. Friede, C. Röver, S. Wandel, and B. Neuenschwander. Meta-analysis of few small studies in orphan diseases. *Research Synthesis Methods*, 8(1):79–91, 2017.

[21] T. Friede, C. Röver, S. Wandel, and B. Neuenschwander. Meta-analysis of two studies in the presence of heterogeneity with applications in rare diseases. *Biometrical Journal*, 59(4):658–671, 2017.

[22] M. Gamalo-Siebers, A. Gao, M. Lakshminarayanan, G. Liu, F. Natanegara, R. Railkar, H. Schmidli, and G. Song. Bayesian methods for the design and analysis of noninferiority trials. *Journal of Biopharmaceutical Statistics*, 26(5):823–841, 2016.

[23] A. Gelman. Prior distributions for variance parameters in hierarchical models. *Bayesian Analysis*, 1(3):515–534, 2006.

[24] T. Gsponer, F. Gerber, B. Bornkamp, D. Ohlssen, M. Vandemeulebroecke, and H. Schmidli. A practical guide to Bayesian group sequential designs. *Pharmaceutical Statistics*, 13(1, SI):71–80, JAN 2014.

[25] S. Gsteiger, B. Neuenschwander, F. Mercier, and H. Schmidli. Using historical control information for the design and analysis of clinical trials with overdispersed count data. *Statistics in Medicine*, 32(21):3609–3622, 2013.

[26] JPT. Higgins, A. Whitehead, RM. Turner, RZ. Omar, and SG. Thompson. Meta-analysis of continuous outcome data from individual patients. *Statistics in Medicine*, 20(15):2219–2241, 2001.

[27] BP. Hobbs, DJ. Sargent, and BP. Carlin. Commensurate priors for incorporating historical information in clinical trials using general and generalized linear models. *Bayesian Analysis*, 7(3):639–674, 2012.

[28] B. Holzhauer, C. Wang, and H. Schmidli. Evidence synthesis from aggregate recurrent event data for clinical trial design and analysis. *Statistics in Medicine*, 37(6):867–882, MAR 15 2018.

[29] JG. Ibrahim and MH. Chen. Power prior distributions for regression models. *Statistical Science*, 15(1):46 – 60, 2000.

[30] J. Lim, R. Walley, J. Yuan, J. Liu, A. Dabral, N. Best, A. Grieve, L. Hampson, J. Wolfram, Ph. Woodward, F. Yong, X. Zhang, and E. Bowen. Minimizing patient burden through the use of historical subject-level data in innovative confirmatory clinical trials: Review of methods and opportunities. *Therapeutic Innovation & Regulatory Science*, 52(5):546–559, 2018.

[31] DJ. Lunn, C. Jackson, N. Best, A. Thomas, and D. Spiegelhalter. *The BUGS Book: A Practical Introduction to Bayesian Analysis*. Boca Raton: CRC Press, 2012.

[32] DJ. Lunn, A. Thomas, N. Best, and D. Spiegelhalter. WinBUGS - a Bayesian modelling framework: Concepts, structure, and extensibility. *Statistics and Computing*, 10:325–337, 2000.

[33] D. Malec. A closer look at combining data among a small number of binomial experiments. *Statistics in Medicine*, 20(12):1811–1824, 2001.

[34] C. McLeod, A. Bagust, A. Boland, P. Dagenais, R. Dickson, Y. Dundar, RA. Hill, A. Jones, R. Mujica Mota, and T. Walley. Adalimumab, etanercept and infliximab for the treatment of ankylosing spondylitis: a systematic review and economic evaluation. *Health Technology Assessment*, 11(28):1–158, 2007.

[35] S. Morita, PF. Thall, and P. Müller. Determining the effective sample size of a parametric prior. *Biometrics*, 64(2):595–602, 2008.

[36] T. Muetze, H. Schmidli, and T. Friede. Sample size re-estimation incorporating prior information on a nuisance parameter. *Pharmaceutical Statistics*, 17(2):126–143, MAR-APR 2018.

[37] B. Neuenschwander, M. Branson, and DJ. Spiegelhalter. A note on the power prior. *Statistics in Medicine*, 28(28):3562 – 3566, 2009.

[38] B. Neuenschwander, G. Capkun-Niggli, M. Branson, and DJ. Spiegehalter. Summarizing historical information on controls in clinical trials. *Clinical Trials*, 7(1):5–18, 2010.

[39] B. Neuenschwander, S. Roychoudhury, and H. Schmidli. On the use of co-data in clinical trials. *Statistics in Biopharmaceutical Research*, 8(3):345–354, 2016.

[40] B. Neuenschwander, S. Weber, H. Schmidli, and A. O'Hagan. Predictively consistent prior effective sample sizes (with discussion). *Biometrics*, 76(2):578–587, 2020.

[41] A. O'Hagan and L. Pericchi. Bayesian heavy-tailed models and conflict resolution: A review. *Brazilian Journal of Probability and Statistics*, 26(4):372–401, 2012.

[42] G. Pennello and L. Thompson. Experience with reviewing Bayesian medical device trials. *Journal of Biopharmaceutical Statistics*, 18(1):81–115, 2008.

[43] SJ. Pocock. The combination of randomized and historical controls in clinical trials. *Journal of Chronic Diseases*, 29(3):175–188, 1976.

[44] NG. Polson and JG. Scott. On the Half-Cauchy prior for a global scale parameter. *Bayesian Analysis*, 7(4):887–902, 2012.

[45] C. Röver, S. Wandel, and T. Friede. Model averaging for robust extrapolation in evidence synthesis. *Statistics in Medicine*, 38(4):674–694, 2019.

[46] S. Roychoudhury and B. Neuenschwander. Bayesian leveraging of historical control data for a clinical trial with time-to-event endpoint. *Statistics in Medicine*, 39(7):984–995, 2020.

[47] H. Schmidli, S. Gsteiger, S. Roychoudhury, A. O'Hagan, D. Spiegelhalter, and B. Neuenschwander. Robust meta-analytic-predictive priors in clinical trials with historical control information. *Biometrics*, 70(4):1023–32, 2014.

[48] H. Schmidli, DA. Haering, M. Thomas, A. Cassidy, S. Weber, and F. Bretz. Beyond randomized clinical trials: Use of external controls. *Clinical Pharmacology & Therapeutics*, 107(4):806–816, 2020.

[49] H. Schmidli, B. Neuenschwander, and T. Friede. Meta-analytic-predictive use of historical variance data for the design and analysis of clinical trials. *Computational Statistics & Data Analysis*, 113:100–110, SEP 2017.

[50] H. Schmidli, S. Wandel, and B. Neuenschwander. The network meta-analytic-predictive approach to non-inferiority trials. *Statistical Methods in Medical Research*, 22(2):219–240, 2013.

[51] R. Simon. Bayesian design and analysis of active control clinical trials. *Biometrics*, 55(2):484–487, 1999.

[52] DJ. Spiegelhalter, KR. Abrams, and JP. Myles. *Bayesian Approaches to Clinical Trials and Health-Care Evaluation*. Chichester: Wiley, 2004.

[53] LA. Stewart and MJ. Clarke. Practical methodology of meta-analyses (overviews) using updated individual patient data. *Statistics in Medicine*, 14(19):2057–2079, 1995.

[54] R. Temple and SS. Ellenberg. Placebo-controlled trials and active-control trials in the evaluation of new treatments. Part 1: Ethical and scientific issues. *Annals of Internal Medicine*, 133(6):455–463, 2000.

[55] RM. Turner, D. Jackson, Y. Wei, SG. Thompson, and JPT. Higgins. Predictive distributions for between-study heterogeneity and simple methods for their application in bayesian meta-analysis. *Statistics in Medicine*, 34(6):984–998, 2015.

[56] RM. Turner, DJ. Spiegelhalter, GCS. Smith, and SG. Thompson. Bias modelling in evidence synthesis. *Journal of the Royal Statistical Society: Series A (Statistics in Society)*, 172(1):21–47, 2009.

[57] K. Viele, S. Berry, B. Neuenschwander, B. Amzal, F. Chen, N. Enas, B. Hobbs, JG. Ibrahim, N. Kinnersley, S. Lindborg, S. Micallef, S. Roychoudhury, and L. Thompson. Use of historical control data for assessing treatment effects in clinical trials. *Pharmaceutical Statistics*, 13(1):41–54, 2014.

[58] I. Wadsworth, LV. Hampson, T. Jaki, GJ. Sills, AG. Marson, and R. Appleton. A quantitative framework to inform extrapolation decisions in children. *Journal of the Royal Statistical Society: Series A (Statistics in Society)*, 2020.

[59] S. Wandel, H. Schmidli, and B. Neuenschwander. Use of historical data. In *Cancer Clinical Trials: Current and Controversial Issues in Design and Analysis (Ed. S.George)*. Boca Raton: CRC Press, 2015.

[60] S. Weber, Y. Li, J. Seaman, T. Kakizume, and H. Schmidli. Applying meta-analytic predictive priors with the R Bayesian evidence synthesis tools. *Journal of Statistical Software*, in press, 2021.

[61] S. Witte, H. Schmidli, A. O'Hagan, and A. Racine. Designing a non-inferiority study in kidney transplantation: a case study. *Pharmaceutical Statistics*, 10(5):427–432, 2011.

[62] H. Zheng, LV. Hampson, and S. Wandel. A robust Bayesian meta-analytic approach to incorporate animal data into phase I oncology trials. *Statistical Methods in Medical Research*, 29(1):94–110, 2020.

26

Evaluation of Surrogate Endpoints

Geert Molenberghs, Ziv Shkedy, Tomasz Burzykowski, Marc Buyse,
Ariel Alonso Abad and Wim Van der Elst

CONTENTS

26.1 Introduction

The rising costs of drug development and the challenges of new and re-emerging diseases are putting considerable demands on efficiency in the drug candidates selection process. A very important factor influencing duration and complexity of this process is the choice of endpoint used to assess drug efficacy. Often, the most sensitive and relevant clinical endpoint might be difficult to use in a trial. This happens if measurement of the clinical endpoint (1) is costly; (2) is difficult; (3) requires a long follow-up time; or (4) requires a large sample size because of low event incidence. An effective strategy is then selection and application of biomarkers for efficacy, replacing the clinical endpoint by a biomarker that is measured more cheaply, more conveniently, more frequently, or earlier. Of course, it needs to be valid and hence properly evaluated [13].

An important step came from the Biomarker Definitions Working Group [9, 38], their definitions nowadays being widely accepted and adopted. A clinical endpoint is considered the most credible indicator of drug response and defined as a characteristic or variable that reflects how a patient feels, functions, or survives. A surrogate endpoint is a biomarker, intended for substituting a clinical endpoint. A surrogate endpoint is expected to predict clinical benefit, harm, or lack of these.

Surrogate endpoints have been used in medical research for a long time [23, 40]. Owing to unfortunate historical events and in spite of potential advantages, their use has been surrounded by controversy [34, 35, 58].

The main reason behind failures was the incorrect perception that surrogacy simply follows from the association between a potential surrogate endpoint and the corresponding clinical endpoint, the mere existence of which is insufficient for surrogacy [23]. Even though the existence of an association between the potential surrogate and the clinical endpoint is undoubtedly a desirable property, what is required to replace the clinical endpoint by the surrogate is that the effect of the treatment on the surrogate endpoint reliably predicts the effect on the clinical endpoint. Owing to a large extent to the lack of appropriate methodology, this condition was not checked in the early attempts and, consequently, negative opinions about the use of surrogates in the evaluation of treatment efficacy emerged [23, 39, 41].

Currently, the steady advance in many medical and biological fields is dramatically increasing the number of biomarkers and hence potential surrogate endpoints. The genetics and 'omics revolutions have largely contributed to this [60]. There is also increasing public pressure for fast approval of promising drugs, so it is naturally to then base the approval process, at least in part, on biomarkers rather than on long-term, costly clinical endpoints [37]. Obviously, the pressure will be especially high when a rapidly increasing incidence of the targeted disease could become a serious threat to public health or the patient's (quality of) life. Shortening the duration of clinical trials not only can decrease the cost of the evaluation process but also limit potential problems with noncompliance and missing data, which are more likely in longer studies [13, 91].

Surrogate endpoints can play a role in the earlier detection of safety signals that could point to toxicity problems with new drugs. The duration and sample size of clinical trials aimed at evaluating the therapeutic efficacy of new drugs are often insufficient to detect rare or late adverse effects [50, 54]; using surrogate endpoints in this context might allow one to obtain information about such effects even during the clinical testing phase.

Evidently, surrogates should be used only when they have been properly evaluated. Sometimes, the term 'validation' is used, but this requires careful qualification [80]. Like in many clinical decisions, statistical arguments will play a major role, but ought to be considered in conjunction with clinical and biological evidence. At the same time, surrogate endpoints can play different roles in different phases of drug development. While it may be more acceptable to use surrogates in early phases of research, there should be much more restraint in using them as substitutes for the true endpoint in pivotal phase III trials, since the latter might imply replacing the true endpoint by a surrogate for all future studies as well, a far-reaching decision. For a biomarker to be used as a "valid" surrogate, a number of conditions must be fulfilled. The ICH Guidelines on Statistical Principles for Clinical Trials state that "In practice, the strength of the evidence for surrogacy depends upon (i) the biological plausibility of the relationship, (ii) the demonstration in epidemiological studies of the prognostic value of the surrogate for the clinical outcome, and (iii) evidence from clinical trials that treatment effects on the surrogate correspond to effects on the clinical outcome" [52].

A perspective on data from a single trial is given in Section 26.2. The meta-analytic evaluation framework is presented in Section 26.3, in the context of normally distributed outcomes. Extensions to a variety of non-Gaussian settings are discussed in Section 26.4. Some alternative computational techniques and validation paradigms is presented in Section 26.5. Implications for prediction of the effect in a new trial and for designing studies based on surrogates are the topics of Section 26.6. Three case studies are presented, in schizophrenia (Section 26.7.1), prostate cancer (Section 26.7.2), and gastric cancer (Section 26.7.3), respectively. The developments presented here are based to a large extent on Alonso *et al* [13] and Molenberghs *et al* [67], Buyse, Molenberghs, and Alonso Abad [16], Molenberghs et al. [66], and Alonso Abad, Molenberghs, and van Breukelen [4].

26.2   Data from a Single Trial

In this section, we will discuss the single trial setting (e.g., occasionally, it could be a different unit, such as a center, but for clarity we will use "trial"). The notation and modeling concepts introduced are useful to present and critically discuss the key ingredients of the Prentice–Freedman framework. Therefore, this section should not be seen as setting the scene for the rest of the chapter. For that, we refer to the multi-trial case (Section 26.3).

Throughout the chapter, we will adopt the following notation: T and S are random variables that denote the true and surrogate endpoints, respectively, and Z is an indicator variable for treatment. For ease of exposition, we will assume that S and T are normally distributed. The effect of treatment on S and T can be modeled as follows:

$$S_j = \mu_S + \alpha Z_j + \varepsilon_{Sj}, \qquad (26.1)$$
$$T_j = \mu_T + \beta Z_j + \varepsilon_{Tj}, \qquad (26.2)$$

where $j = 1, \ldots, n$ indicates patients, and the error terms have a joint zero-mean normal distribution with covariance matrix

$$\Sigma = \begin{pmatrix} \sigma_{SS} & \sigma_{ST} \\ & \sigma_{TT} \end{pmatrix}. \qquad (26.3)$$

In addition, the relationship between S and T can be described by a regression of the form

$$T_j = \mu + \gamma S_j + \varepsilon_j. \qquad (26.4)$$

Note that this model is introduced because it is a component of the Prentice–Freedman framework.

We will assume later (Section 26.3) that the n patients come from N different experimental trials or centers, but for now the simple situation of a single experiment will suffice to explore some fundamental difficulties with the validation of surrogate endpoints.

26.2.1 Definition and criteria

Prentice [74] proposed to define a surrogate endpoint as "a response variable for which a test of the null hypothesis of no relationship to the treatment groups under comparison is also a valid test of the corresponding null hypothesis based on the true endpoint" ([74] p. 432). In terms of our simple model (26.1)–(26.2), the definition states that for S to be a valid surrogate for T, parameters α and β must simultaneously be equal to, or different from, zero. This definition is not consistent with the availability of a single experiment only, since it requires a large number of experiments to be available, each with tests of hypothesis on both the surrogate and true endpoints. An important drawback is also that evidence from trials with non-significant treatment effects cannot be used, even though such trials may be consistent with a desirable relationship between both endpoints. Prentice derived operational criteria that are equivalent to his definition. These criteria require that

- (criterion 1) treatment has a significant impact on the surrogate endpoint (parameter α differs significantly from zero in (26.1)),

- (criterion 2) treatment has a significant impact on the true endpoint (parameter β differs significantly from zero in (26.2)),

- (criterion 3) the surrogate endpoint has a significant impact on the true endpoint (parameter γ differs significantly form zero in (26.4)), and

- (criterion 4) the full effect of treatment upon the true endpoint is captured by the surrogate.

This last criterion, referred to as "the fourth criterion," is verified through the conditional distribution of the true endpoint, given treatment *and* surrogate endpoint, derived from (26.1)–(26.2):

$$T_j = \tilde{\mu}_T + \beta_S Z_j + \gamma_Z S_j + \tilde{\varepsilon}_{Tj}, \tag{26.5}$$

where the treatment effect (corrected for the surrogate S), β_S, and the surrogate effect (corrected for treatment Z), γ_Z, are

$$\beta_S = \beta - \sigma_{TS}\sigma_{SS}^{-1}\alpha, \tag{26.6}$$

$$\gamma_Z = \sigma_{TS}\sigma_{SS}^{-1}, \tag{26.7}$$

and the variance of $\tilde{\varepsilon}_{Tj}$ is given by $\sigma_{TT} - \sigma_{TS}^2\sigma_{SS}^{-1}$.

It is usually stated that the fourth criterion requires that the parameter β_S be equal to zero (we return to this notion in Section 26.2.3). Essentially, this fourth criterion states that the true endpoint T is completely determined by knowledge of the surrogate endpoint S. Buyse and Molenberghs [18] showed that the criteria are necessary and sufficient for binary responses, but not in general. Several authors, including Prentice, pointed out that the criteria are too stringent to be fulfilled in real situations [74].

In spite of these criticisms, the spirit of the fourth criterion is very appealing. This is especially true if it can be considered in the light of an underlying biological mechanism. For example, it is interesting to explore whether the surrogate is part of the causal chain leading from treatment exposure to the true endpoint.

26.2.2 The proportion explained

Freedman, Graubard, and Schatzkin [44] argued that the fourth Prentice criterion raises a conceptual difficulty since it requires the statistical test for treatment effect on the true endpoint to be *non*-significant after adjustment for the surrogate. The non-significance of this test does not prove that the effect of treatment upon the true endpoint is *fully* captured by the surrogate, and therefore Freedman, Graubard, and Schatzkin [44] proposed the proportion of the treatment effect mediated by the surrogate:

$$PE = \frac{\beta - \beta_s}{\beta},$$

with β_s and β obtained, respectively, from (26.5) and (26.2). In this paradigm, a valid surrogate would be one for which the proportion explained (PE) is equal to one. In practice, a surrogate would be deemed acceptable if the lower limit of its confidence interval of PE was "sufficiently" large.

Important difficulties surrounding the PE have been described in the literature [18, 24, 33, 42, 63, 92]: PE is unstable when β lies near 0; its confidence limits tend to be wide and are sometimes even unbounded. Most importantly though, Buyse et al [21] showed that the PE suffers from ill-conditioning, in the sense that a variance ratio (details in the next section), while irrelevant for the relationship between S and T, can move the PE along the real line. As a consequence, it is frequently seen to fall outside the unit interval.

26.2.3 The relative effect

Buyse and Molenberghs [18] proposed another quantity for the validation of a surrogate endpoint: the relative effect (RE), which is the ratio of the effects of treatment upon the true and the surrogate endpoint. Formally:

$$RE = \frac{\beta}{\alpha}, \tag{26.8}$$

They also suggested the treatment-adjusted association between the surrogate and the true endpoint, ρ_Z:

$$\rho_Z = \frac{\sigma_{ST}}{\sqrt{\sigma_{SS}\sigma_{TT}}}. \tag{26.9}$$

Now, a simple relationship can be derived between PE, RE, and ρ_Z. Let us define $\lambda^2 = \sigma_{TT}\sigma_{SS}^{-1}$. It follows that $\lambda\rho_Z = \sigma_{ST}\sigma_{SS}^{-1}$ and, from (26.6), $\beta_s = \beta - \rho_Z\lambda\alpha$. As a result, we obtain

$$PE = \lambda\rho_Z\frac{\alpha}{\beta} = \lambda\rho_Z\frac{1}{RE}. \tag{26.10}$$

A similar relationship was derived by Buyse and Molenberghs [18] and by Begg and Leung [8] for standardized surrogate and true endpoints. Let us now turn to the more promising meta-analytic framework.

26.3 A Meta-analytic Framework for Normally Distributed Outcomes

26.3.1 A meta-analytic approach

Although the single trial based methods are relatively easy in terms of implementation, they are surrounded with the difficulties alluded to at the end of the previous section. Therefore, several authors, such as Daniels and Hughes [33], Buyse et al [19], and Gail et al [45] have introduced the meta-analytic approach. This section briefly outlines the methodology, followed by simplified modeling approaches as suggested by Tibaldi et al [88].

The meta-analytic approach was formulated originally for two continuous, normally distributed outcomes, and extended in the meantime to various outcome types, ranging from continuous, binary, ordinal, time-to-event, and longitudinally measured outcomes [13]. First, we focus on the continuous case, where the surrogate and true endpoints are jointly normally distributed.

The method is based on a hierarchical two-level model. Both a fixed-effects and a random-effects view can be taken. Let T_{ij} and S_{ij} be the random variables denoting the true and surrogate endpoints for the jth subject in the ith trial, respectively, and let Z_{ij} be the indicator variable for treatment. First, consider the following fixed-effects models:

$$S_{ij} = \mu_{si} + \alpha_i Z_{ij} + \varepsilon_{sij}, \tag{26.11}$$
$$T_{ij} = \mu_{Ti} + \beta_i Z_{ij} + \varepsilon_{Tij}, \tag{26.12}$$

where μ_{si} and μ_{Ti} are trial-specific intercepts, α_i and β_i are trial-specific effects of treatment Z_{ij} on the endpoints in trial i, and ε_{si} and ε_{Ti} are correlated error terms, assumed to be zero-mean normally distributed with covariance matrix

$$\mathbf{\Sigma} = \begin{pmatrix} \sigma_{SS} & \sigma_{ST} \\ & \sigma_{TT} \end{pmatrix}. \tag{26.13}$$

In addition, we can decompose

$$\begin{pmatrix} \mu_{si} \\ \mu_{Ti} \\ \alpha_i \\ \beta_i \end{pmatrix} = \begin{pmatrix} \mu_s \\ \mu_T \\ \alpha \\ \beta \end{pmatrix} + \begin{pmatrix} m_{si} \\ m_{Ti} \\ a_i \\ b_i \end{pmatrix}, \tag{26.14}$$

where the second term on the right hand side of (26.14) is assumed to follow a zero-mean normal distribution with covariance matrix

$$D = \begin{pmatrix} d_{SS} & d_{ST} & d_{sa} & d_{sb} \\ & d_{TT} & d_{Ta} & d_{Tb} \\ & & d_{aa} & d_{ab} \\ & & & d_{bb} \end{pmatrix}. \tag{26.15}$$

A classical hierarchical, random-effects modeling strategy results from the combination of the above two steps into a single one:

$$S_{ij} = \mu_s + m_{si} + \alpha Z_{ij} + a_i Z_{ij} + \varepsilon_{sij}, \tag{26.16}$$
$$T_{ij} = \mu_T + m_{Ti} + \beta Z_{ij} + b_i Z_{ij} + \varepsilon_{Tij}. \tag{26.17}$$

Here, μ_s and μ_T are fixed intercepts, α and β are fixed treatment effects, m_{si} and m_{Ti} are random intercepts, and a_i and b_i are random treatment effects in trial i for the surrogate and true endpoints, respectively. The vector of random effects $(m_{si}, m_{Ti}, a_i, b_i)$ are assumed to be mean-zero normally distributed with covariance matrix (26.15). The error terms ε_{sij} and ε_{Tij} follow the same assumptions as in the fixed effects models.

After fitting the above models, surrogacy is captured by means of two quantities: trial-level and individual-level coefficients of determination. The former quantifies the association between the treatment effects on the true and surrogate endpoints at the trial level, while the latter quantifies the association at the level of the individual patient, after adjustment for the treatment effect. The former is given by:

$$R_{\text{trial}}^2 = R_{b_i|m_{Si},a_i}^2 = \frac{\begin{pmatrix} d_{sb} \\ d_{ab} \end{pmatrix}^T \begin{pmatrix} d_{ss} & d_{sa} \\ d_{sa} & d_{aa} \end{pmatrix}^{-1} \begin{pmatrix} d_{sb} \\ d_{ab} \end{pmatrix}}{d_{bb}}. \tag{26.18}$$

The above quantity is unitless and, under the condition that the corresponding variance-covariance matrix is positive definite, lies within the unit interval.

Apart from estimating the strength of surrogacy, the above model can also be used for prediction purposes. To this end, observe that $(\beta+b_0|m_{s0},a_0)$ (the quantities with subscript 0 are the trial-specific effects for a new trial) follows a normal distribution with mean and variance:

$$E(\beta + b_0|m_{s0}, a_0)$$
$$= \beta + \begin{pmatrix} d_{sb} \\ d_{ab} \end{pmatrix}^T \begin{pmatrix} d_{ss} & d_{sa} \\ d_{sa} & d_{aa} \end{pmatrix}^{-1} \begin{pmatrix} \mu_{s0} - \mu_s \\ \alpha_0 - \alpha \end{pmatrix}, \tag{26.19}$$

$$\text{Var}(\beta + b_0|m_{s0}, a_0)$$
$$= d_{bb} \begin{pmatrix} d_{sb} \\ d_{ab} \end{pmatrix}^T \begin{pmatrix} d_{ss} & d_{sa} \\ d_{sa} & d_{aa} \end{pmatrix}^{-1} \begin{pmatrix} d_{sb} \\ d_{ab} \end{pmatrix}. \tag{26.20}$$

A prediction can be made using (26.19), with prediction variance (26.20). Of course, one has to properly acknowledge the uncertainty resulting from the fact that parameters are not known but merely estimated. We return to this issue in Section 26.6.

Models (26.16) and (26.17) are referred to as the full mixed-effects models. It is sometimes necessary, for computational reasons, to consider a simplified version. A simplified version of these models is obtained by replacing the fixed trial-specific intercepts by a common one. Thus, the reduced mixed effect models result from removing the random trial-specific intercepts m_{Si} and m_{Ti} from models (26.16) and (26.17). The R^2 for the reduced models then is:

$$R_{\text{trial(r)}}^2 = R_{b_i|a_i}^2 = \frac{d_{ab}^2}{d_{aa}d_{bb}}.$$

A surrogate could be adopted when R_{trial}^2 is sufficiently large. Arguably, rather than using a fixed cutoff above which a surrogate would be adopted, there always will be clinical and other judgment involved in the decision process. The individual-level coefficient of determination, R_{indiv}^2, is based on (26.13) and takes the following form:

$$R_{\text{indiv}}^2 = R_{\varepsilon_{Ti}|\varepsilon_{Si}}^2 = \frac{\sigma_{ST}^2}{\sigma_{SS}\sigma_{TT}}. \tag{26.21}$$

Though the above hierarchical modeling is elegant, it often poses computational challenges [13]. To address this problem, Tibaldi et al [88] suggested several simplifications.

26.4 Non-Gaussian Endpoints

As is clear from the formalism in Section 26.3, one needs the joint distribution of the random variables governing the surrogate and true endpoints. The easiest, though not the

TABLE 26.1
Examples of possible surrogate endpoints in various diseases (Abbreviations: AIDS = acquired immune deficiency syndrome; ARMD = age-related macular degeneration; HIV = human immunodeficiency virus).

Disease	Surrogate Endpoint	Type	Final Endpoint	Type
Resectable solid tumor	Time to recurrence	Censored	Survival	Censored
Advanced cancer	Tumor response	Binary	Time to progression	Censored
Osteoporosis	Bone mineral density	Longitudinal	Fracture	Binary
Cardiovasc. disease	Ejection fraction	Continuous	Myocardial infraction	Binary
Hypertension	Blood pressure	Longitudinal	Coronary heart disease	Binary
Arrhythmia	Arrhythmic episodes	Longitudinal	Survival	Censored
ARMD	6-month visual acuity	Continuous	24-month visual acuity	Continuous
Glaucoma	Intraoccular pressure	Continuous	Vision loss	Censored
Depression	Biomarkers	Multivariate	Depression scale	Continuous
HIV infection	CD4 counts + viral load	Multivariate	Progression to AIDS	Censored

only, situation is where both are Gaussian random variables, but one also encounters binary (e.g., CD4+ counts over 500/mm3, tumor shrinkage), categorical (e.g., cholesterol levels <200 mg/dl, 200–299 mg/dl, 300+ mg/dl; tumor response as complete response, partial response, stable disease, progressive disease), censored continuous (e.g., time to undetectable viral load, time to cardiovascular death), longitudinal (e.g., CD4+ counts over time, blood pressure over time), and multivariate longitudinal (e.g., CD4+ and viral load over time jointly, various measures of quality of life over time) endpoints. The models used to validate a surrogate for a clinical endpoint will depend on the type of variables observed in the problem at hand. Table 26.1 shows some examples of potential surrogate endpoints in various diseases. In what follows, we will briefly discuss the settings of binary endpoints, failure-time endpoints, the combination of an ordinal and a survival endpoint, and longitudinal endpoints.

26.4.1 Two binary endpoints

Renard et al [75] have shown that extension to this situation is easily done using a latent variable formulation. That is, one posits the existence of a pair of continuously distributed latent variable responses $(\widetilde{S}_{ij}, \widetilde{T}_{ij})$ that produce the actual values of (S_{ij}, T_{ij}). These unobserved variables are assumed to have a joint normal distribution and the realized values follow by double dichotomization. On the latent-variable scale, we obtain a model similar to (26.11)–(26.12) and in the matrix (26.13) the variances are set equal to unity in order to ensure identifiability. This leads to the following model:

$$\begin{cases} \Phi^{-1}(P[S_{ij} = 1 | Z_{ij}, m_{S_i}, a_i, m_{T_i}, b_i]) & = \mu_S + m_{S_i} + (\alpha + a_i)Z_{ij}, \\ \Phi^{-1}(P[T_{ij} = 1 | Z_{ij}, m_{S_i}, a_i, m_{T_i}, b_i]) & = \mu_T + m_{T_i} + (\beta + b_i)Z_{ij}, \end{cases}$$

where Φ denotes the standard normal cumulative distribution function. Renard et al [75] used pseudo-likelihood methods to estimate the model parameters. Similar ideas have been used in the case one of the endpoints is continuous, with the other one binary or categorical[13, Ch. 6]. The case of two binary outcomes has received further attention, encompassing flexible software implementation [89].

26.4.2   Two failure-time endpoints

Assume now that S_{ij} and T_{ij} are failure-time endpoints. Model (26.11)–(26.12) is replaced by a model for two correlated failure-time random variables. Burzykowski et al [12] used copulas to this end [25, 51]. One then assumes that the joint survivor function of (S_{ij}, T_{ij}) can be written as:

$$F(s, t) = P(S_{ij} \geq s, T_{ij} \geq t) = C_\delta\{F_{sij}(s), F_{Tij}(t)\}, \quad s, t \geq 0, \qquad (26.22)$$

where (F_{sij}, F_{Tij}) denote marginal survivor functions and C_δ is a copula, i.e., a distribution function on $[0, 1]^2$ with $\delta \in R^1$.

When the hazard functions are specified, estimates of the parameters for the joint model can be obtained using maximum likelihood. Shih and Louis [85] discuss alternative estimation methods. The association parameter is generally hard to interpret. However, it can be shown [49] that there is a link with Kendall's τ:

$$\tau = 4 \int_0^1 \int_0^1 C_\delta(u, v) C_\delta(du, dv) - 1,$$

providing an easy measure of surrogacy at the individual level. At the second stage R^2_{trial} can be computed based on the pairs of treatment effects estimated at the first stage.

26.4.3   An ordinal surrogate and a survival endpoint

Assume that T is a failure-time random variable and S is a categorical variable with K ordered categories. To propose validation measures, similar to those introduced in the previous section, Burzykowski, Molenberghs, and Buyse [12] also used bivariate copulas, combining ideas of Molenberghs, Geys, and Buyse [68] and Burzykowski et al [12]. One marginal distribution is a proportional odds logistic regression, while the other is a proportional hazards model. The Plackett copula [31] was chosen to capture the association between both endpoints. The ensuing global odds ratio is relatively easy to interpret.

26.4.4   Binary and normally distributed endpoints

Alonso *et al* [13] review extensions of the meta-analytic approach, ranging over continuous, binary, ordinal, time-to-event, and longitudinally measured outcomes. Here, we focus on the combination of continuous and binary outcomes. In view of quantifying surrogacy, it is important to choose a modeling framework that allows for this.

Generalized linear mixed models for endpoints of different data types are challenging [69]. Hence, we concentrate on two-stage fixed-effects models. In the first stage, let $\widetilde{S}_{ij}$ be a latent variable of which S_{ij} is the dichotomized version. A bivariate normal model for $\widetilde{S}_{ij}$ and T_{ij} is given by [68]:

$$\widetilde{S}_{ij} = \mu_{si} + \alpha_i Z_{ij} + \varepsilon_{sij}, \tag{26.23}$$

$$T_{ij} = \mu_{Ti} + \beta_i Z_{ij} + \varepsilon_{Tij}, \tag{26.24}$$

where μ_{si} and μ_{Ti} are trial-specific intercepts, α_i and β_i are trial-specific effects of treatment Z_{ij} on the endpoints in trial i, and ε_{si} and ε_{Ti} are correlated error terms, assumed to be zero-mean normally distributed with covariance matrix

$$\Sigma = \begin{pmatrix} \frac{1}{(1-\rho^2)} & \frac{\rho\sigma}{\sqrt{(1-\rho^2)}} \\ & \sigma \end{pmatrix}, \tag{26.25}$$

where σ is the variance of the continuous outcome and ρ is the correlation between both outcomes. The variance of $\widetilde{S}_{ij}$ is chosen for computational reasons. Using a probit formulation like Molenberghs Geys, and Buyse [68] and owing to the replication at the trial level, we can impose a distribution on the trial-specific parameters. At the second stage, we assume

$$\begin{pmatrix} \mu_{si} \\ \mu_{Ti} \\ \alpha_i \\ \beta_i \end{pmatrix} = \begin{pmatrix} \mu_s \\ \mu_T \\ \alpha \\ \beta \end{pmatrix} + \begin{pmatrix} m_{si} \\ m_{Ti} \\ a_i \\ b_i \end{pmatrix}, \tag{26.26}$$

where the second term on the right hand of (26.26) is assumed to follow a zero-mean normal distribution with dispersion matrix (26.15). Measures to assess the quality of the surrogate both at the trial and individual level are then obtained. This case has received full attention in Assam et al [7].

26.4.5   Longitudinal endpoints

Alonso et al [1] showed that going from a univariate setting to a multivariate framework represents new challenges. The R^2 measures proposed by Buyse et al [19], are no longer applicable. Alonso et al [1] based their calculations of surrogacy measures on a two-stage approach. They assume that information from $i = 1, \ldots, N$ trials is available, in the ith of which, $j = 1, \ldots, n_i$ subjects are enrolled and they denote further the time at which subject j in trial i is measured as t_{ijk}. If T_{ijk} and S_{ijk} represent the associated true and surrogate endpoints, respectively, and Z_{ij} is a binary indicator variable for treatment, then along the ideas of Galecki [46], they proposed the following joint model, at the first stage, for both responses

$$\begin{aligned} T_{ijk} &= \mu_{Ti} + \beta_i Z_{ij} + g_{Tij}(t_{ijk}) + \varepsilon_{Tijk}, \\ S_{ijk} &= \mu_{si} + \alpha_i Z_{ij} + g_{sij}(t_{ijk}) + \varepsilon_{sijk}, \end{aligned} \tag{26.27}$$

where μ_{Ti} and μ_{si} are trial-specific intercepts, β_i and α_i are trial-specific effects of treatment Z_{ij} on the two endpoints and g_{Tij} and g_{sij} are trial-subject-specific time functions that can include treatment-by-time interactions. They also assume that the vectors, collecting all information over time for patient j in trial i, $\widetilde{\varepsilon}_{T_{ij}}$, and $\widetilde{\varepsilon}_{S_{ij}}$ are correlated error terms, following a mean-zero multivariate normal distribution with covariance matrix

$$\Sigma_i = \begin{pmatrix} \Sigma_{TTi} & \Sigma_{TSi} \\ \Sigma'_{TSi} & \Sigma_{SSi} \end{pmatrix} = \begin{pmatrix} \sigma_{TTi} & \sigma_{TSi} \\ \sigma_{TSi} & \sigma_{SSi} \end{pmatrix} \otimes R_i. \tag{26.28}$$

Here, R_i is a correlation matrix for the repeated measurements.

If treatment effect can be assumed constant over time, then (26.18) can still be useful to evaluate surrogacy at the trial level. However, at the individual level, the situation is totally different, the R_{ind}^2 is no longer applicable, and new concepts are needed.

Using multivariate ideas, Alonso et al [1] proposed the *variance reduction factor* (*VRF*) to capture individual-level surrogacy in this more elaborate setting. They quantified the relative reduction in the true endpoint variance after adjustment by the surrogate as

$$VRF_{\text{ind}} = \frac{\sum_i \{\text{tr}(\Sigma_{TTi}) - \text{tr}(\Sigma_{(T|S)i})\}}{\sum_i \text{tr}(\Sigma_{TTi})}, \tag{26.29}$$

with notation as in (26.28) and $\Sigma_{(T|S)i} = \Sigma_{TTi} - \Sigma_{TSi}\Sigma_{SSi}^{-1}\Sigma'_{TSi}$. Alonso et al [1] showed that the VRF_{ind} ranges between zero and one, and that $VRF_{\text{ind}} = R_{\text{ind}}^2$ when the endpoints are measured only once.

An alternative proposal is

$$\theta_p = \sum_i \frac{1}{Np_i} \text{tr}\left\{\left(\Sigma_{TTi} - \Sigma_{(T|S)i}\right)\Sigma_{TTi}^{-1}\right\}. \tag{26.30}$$

While structurally similar, the VRF is not symmetric in S and T and it is only invariant with respect to linear orthogonal transformations, whereas θ_p is both symmetric and invariant with respect to the broader class of linear bijective transformations.

An extension that allows for non-normal outcomes [3] is

$$R_{\Lambda}^2 = \frac{1}{N}\sum_i (1 - \Lambda_i), \tag{26.31}$$

where: $\Lambda_i = |\Sigma_i| / \{|\Sigma_{TTi}| |\Sigma_{SSi}|\}$. This parameter not only allows the detection of more general patterns of association but can also be extended to more general settings than those defined by the normal distribution. They proved that R_{Λ}^2 ranges between zero and one, and that in the cross-sectional case $R_{\Lambda}^2 = R_{\text{ind}}^2$. These authors have shown that $R_{\Lambda}^2 = 1$ whenever there is a deterministic relationship between two linear combinations of both endpoints, allowing the detection of strong associations in cases where the VRF or θ_p would fail in doing so.

Several extensions of this approach have been considered [2, 3], to allow for Gaussian as well as non-Gaussian outcomes, to improve computational performance [88], and to offer a unified framework based on information theory and causal inference [2, 5, 83].

26.5   Alternatives and Extensions

As a result of the aforementioned computational problems, several alternative strategies have been considered. For example, Shkedy and Torres Barbosa [86] and Renfro [76] study in detail the use of Bayesian methodology and conclude that even relatively non-informative prior have a strongly beneficial impact on the algorithms' performance.

Cortiñas, Shkedy, and Molenberghs [27] start from the information-theoretic approach, in the contexts of: (1) normally distributed endpoints; (2) a copula model for a categorical surrogate and a survival true endpoint; and (3) a joint modeling approach for longitudinal surrogate and true endpoints. Rather than fully relying on the methods described in Section 26.4, they use cross-validation to obtain adequate estimates of the trial-level surrogacy measure. Also, they explore the use of regression tree analysis, bagging regression analysis, random forests, and support vector machine methodology. They concluded that performance of such methods, in simulations and case studies, in terms of point and interval estimation, ranges from very good to excellent.

The above are variations to the meta-analytic theme, as described here, in Alonso *et al* [13], and of which Daniels and Hughes [33] is an early instance. There are a number of alternative paradigms. Frangakis and Rubin [43] employ so-called principal stratification, still using the data from a single trial only. Drawing from the causality literature, Robins and Greenland [77], Pearl [72], and Taylor, Wang, and Thiébaut [87] use the direct/indirect-effect machinery.

It took two decades after the publication of Prentices seminal paper until an attempt was made to review, classify, and study similarities and differences between the various paradigms [53]. Joffe and Greene saw two important dimensions. First, some methods are based on a single trial while others use several trials, i.e., meta-analysis. Second, some approaches are based on association, while others are based on causation. Because the meta-analytic framework described earlier is based on association and uses multiple trials, on the one hand, and because the causal framework initially used a single trial, on the other, the above dimensions got convoluted and it appeared that correlation/meta-analysis had to be a pair, just like causal/single trial. However, it is useful to disentangle the two dimensions and to keep in mind that proper evaluation of the relationship between the treatment effect on the surrogate and true endpoints is ideally based on meta-analysis. Joffe and Green state that the meta-analytic approach is essentially causal in so far as the treatment effects observed in all trials are in fact average causal effects. If a meta-analysis of several trials is not possible, then causal effects must be estimated for individual patients, which requires strong and unverifiable assumptions to be made. Recently, progress has been made regarding the relationship between the association and causal frameworks [5]. Alonso et al. [5] consider a quadruple $Y_{ij} = [T_{ij}(Z_{ij} = 0), T_{ij}(Z_{ij} = 1), S_{ij}(Z_{ij} = 0), S_{ij}(Z_{ij} = 1)]'$, which is observable only if patient j in trial i would be assessed under both control and experimental treatment. Clearly, this is not possible and hence some of the outcomes in the quadruple are "counterfactual." Counterfactuals are essential to the causal-inference framework, while the above equation also carries a meta-analytic structure. Alonso et al. [5] assume a multivariate normal for Y_{ij}, to derive insightful expressions. It is clear that both paradigms base their validation approach upon causal effects of treatment. However, there is an important difference. While the causal inference line of thinking places emphasis on individual causal effects, in a meta-analytic approach the focus is on the expected causal treatment effect. These authors show that, under broad circumstances, when a surrogate is considered acceptable from a meta-analytic perspective, at both the trial and individual level, then it would be good as well from a causal-inference angle. These authors also carefully show, in line with comments made earlier, that a surrogate, valid from a single-trial framework perspective, using individual causal effects, may not pass the test from a meta-analytic view-point, when heterogeneity from one trial to another is large and the causal association is low. Evidently, more work is needed, especially for endpoints of a different type, but at the same time it is comforting that, when based on multiple trials, the frameworks appear to show a good amount of agreement.

26.6   Prediction and Design Aspects

Until now, we have focused on quantifying surrogacy through a slate of measures, culminating in the information-theoretic ones. In practice, one may want to go beyond merely quantifying the strength of surrogacy, and further use a surrogate endpoint to predict the treatment effect on the true endpoint *without measuring the latter*. Put simply, the issue then is to obtain point and interval predictions for the treatment effect on the true endpoint based on the surrogate. This issue has been studied by Burzykowski and Buyse [10] for the original meta-analytic approach for continuous endpoints and will be reviewed here.

The key motivation for validating a surrogate endpoint is the ability to predict the effect of treatment on the true endpoint based on the observed effect of treatment on the surrogate endpoint. It is essential, therefore, to explore the quality of prediction by (a) information obtained in the validation process based on trials $i = 1, \ldots, N$ and (b) the estimate of the effect of Z on S in a new trial $i = 0$. Fitting the mixed-effects model (26.16)–(26.17) to data from a meta-analysis provides estimates for the parameters and the variance components. Suppose then that a new trial $i = 0$ is considered for which data are available on the surrogate endpoint but not on the true endpoint. We can then fit the following linear model to the surrogate outcomes S_{0j}:

$$S_{0j} = \mu_{s0} + \alpha_0 Z_{0j} + \varepsilon_{s0j}. \tag{26.32}$$

We are interested in an estimate of the effect $\beta + b_0$ of Z on T, given the effect of Z on S. To this end, one can observe that $(\beta + b_0 | m_{s0}, a_0)$, where m_{s0} and a_0 are, respectively, the surrogate-specific random intercept and treatment effect in the new trial follows a normal distribution with mean linear in μ_{s0}, μ_s, α_0, and α, and variance

$$\text{Var}(\beta + b_0 | m_{s0}, a_0) \quad = \quad (1 - R^2_{\text{trial}}) \text{Var}(b_0). \tag{26.33}$$

Here, $\text{Var}(b_0)$ denotes the unconditional variance of the trial-specific random effect, related to the effect of Z on T (in the past or the new trials). The smaller the conditional variance (26.33), the higher the precision of the prediction, as captured by R^2_{trial}. Let us use ϑ to group the fixed-effects parameters and variance components related to the mixed-effects model (26.11)–(26.12), with $\widehat{\vartheta}$ denoting the corresponding estimates. Fitting the linear model (26.32) to data on the surrogate endpoint from the new trial provides estimates for m_{s0} and a_0. The prediction variance can be written as:

$$
\begin{aligned}
&\text{Var}(\beta + b_0 | \mu_{s0}, \alpha_0, \vartheta) \\
&\approx \quad f\{\text{Var}(\widehat{\mu}_{s0}, \widehat{\alpha}_0)\} + f\{\text{Var}(\widehat{\vartheta})\} + (1 - R^2_{\text{trial}}) \text{Var}(b_0),
\end{aligned} \tag{26.34}
$$

where $f\{\text{Var}(\widehat{\mu}_{s0}, \widehat{\alpha}_0)\}$ and $f\{\text{Var}(\widehat{\vartheta})\}$ are functions of the asymptotic variance-covariance matrices of $(\widehat{\mu}_{s0}, \widehat{\alpha}_0)^T$ and $\widehat{\vartheta}$, respectively. The third term on the right hand side of (26.34), i.e., (26.33), describes the prediction's variability if μ_{s0}, α_0, and ϑ were known. The first two terms describe the contribution to the variability due to the use of the estimates of these parameters. It is useful to consider three scenarios.

Scenario 1. Estimation error in both the meta-analysis and the new trial. If the parameters of models (26.11)–(26.12) and (26.32) have to be estimated, as is the case in reality, the prediction variance is given by (26.34). From the equation it is clear that in practice, the reduction of the variability of the estimation of $\beta + b_0$, related to the use of the information on m_{s0} and a_0, will always be smaller than that indicated by R^2_{trial}. The latter coefficient can thus be thought of as measuring the "potential" validity of a surrogate endpoint at the trial-level, assuming precise knowledge (or infinite numbers of trials and sample sizes per trial available for the estimation) of the parameters of models (26.11)–(26.12) and (26.32). See also Scenario 3.

Scenario 2. Estimation error only in the meta-analysis. This scenario is possible only in theory, as it would require an infinite sample size in the new trial. But it can provide information of practical interest since, with an infinite sample size, the parameters of the single-trial regression model (26.32) would be known. Consequently, the first term on the right hand side of (26.34), $f\{\mathrm{Var}(\widehat{\mu}_{s0}, \widehat{\alpha}_0)\}$, would vanish and (26.34) would reduce to

$$\mathrm{Var}(\beta + b_0 | \mu_{s0}, \alpha_0, \vartheta) \approx f\{\mathrm{Var}(\widehat{\vartheta})\} + (1 - R^2_{\mathrm{trial}})\mathrm{Var}(b_0). \tag{26.35}$$

Expression (26.35) can thus be interpreted as indicating the minimum variance of the prediction of $\beta + b_0$, achievable in the actual application of the surrogate endpoint. In practice, the size of the meta-analytic data providing an estimate of ϑ will necessarily be finite and fixed. Consequently, the first term on the right hand side of (26.35) will always be present. Based on this observation, Gail et al [45] conclude that the use of surrogates validated through the meta-analytic approach will always be less efficient than the direct use of the true endpoint. Of course, even so, a surrogate can be of great use in terms of reduced sample size, reduce trial length, gain in number of life years, etc.

Scenario 3. No estimation error. If the parameters of the mixed-effects model (26.11)–(26.12) and the single-trial regression model (26.32) were known, the prediction variance for $\beta + b_0$ would only contain the last term on the right hand side of (26.34). Thus, the variance would be reduced to (26.33). While this situation is, strictly speaking, of theoretical relevance only, as it would require infinite numbers of trials and sample sizes per trial available for the estimation in the meta-analysis and in the new trial, it provides important insight.

Based on the above scenarios one can argue that in a particular application the size of the minimum variance (26.35) is of importance. The reason is that (26.35) is associated with the minimum width of the prediction interval for $\beta + b_0$ that might be approached in a particular application by letting the sample size for the new trial increase toward infinity. This minimum width will be responsible for the loss of efficiency related to the use of the surrogate, pointed out in Gail et al [45]. It would thus be important to quantify the loss of efficiency, since it may be counter-balanced by a shortening of trial duration. One might consider using the ratio of (26.35) to $\mathrm{Var}(b_0)$, the unconditional variance of $\beta + b_0$. However, Burzykowski and Buyse [10] considered another way of expressing this information, which should be more meaningful clinically.

26.7   Case Studies

26.7.1   A meta-analysis of five clinical trials in schizophrenia

The data come from a meta-analysis of five double-blind randomized clinical trials, comparing the effects of risperidone to conventional anti psychotic agents for the treatment of chronic schizophrenia. The treatment indicator for risperidone versus conventional treatment will be denoted by Z (denoted as $0/1$ or $-1/1$). So-called negative symptoms are characterized by deficits in cognitive, affective and social functions, for example poverty of speech, apathy and emotional withdrawal. Positive symptoms entail more florid symptoms such as delusions, hallucinations and disorganized thinking, which are superimposed on mental status [55]. Several measures can be considered to asses a patient's global condition. Clinician's Global Impression (CGI) is generally accepted as a clinical measure of change, even though it is somewhat subjective. Here, the change of CGI versus baseline will be considered as the true endpoint T. It is scored on a 7-grade scale used by the treating physician to characterize how well a subject has improved since baseline. Another useful and sufficiently sensitive assessment scales is the Positive and Negative Syndrome Scale (PANSS) [56]. The PANSS consists of 30 items that provide an operationalized, drug-sensitive instrument, which is highly useful for both typological and dimensional assessment of schizophrenia. We will use the change versus baseline in PANSS as our surrogate S. The data contain five trials and in all trials, information is available on the investigators that treated the patients. This information is helpful to define group of patients that will become units of analysis.

26.7.1.1 Analysis of continuous endpoints

Due to the small number of trials we choose as unit of analysis, "investiagor" is used instead. There were 176 investigators, each treating between 2 and 60 patients. The use of investigator as unit of analysis is also surrounded with problems. Although a large number of investigators is convenient to explain the between investigator variability, because some investigators treated few patients, the resulting within-unit variability might not be estimated correctly.

TABLE 26.2
Schizophrenia study. Results of the trial-level (R^2_{trial}) surrogacy analysis.

Unit	Fixed effects		Random effects	
	Unwt.	Wt.	Unwt.	Wt.
Full Model				
Univariate approach				
Investigator	0.5887	0.5608	0.5488	0.5447
Trial	0.9641	0.9636	0.9849	0.9909
Bivariate approach				
Investigator	0.5887	0.5608	0.9898*	
Trial	0.9641	0.9636	—	
Reduced Model				
Univariate approach				
Investigator	0.6707	0.5927	0.5392	0.5354
Trial	0.8910	0.8519	0.7778	0.8487
Bivariate approach				
Investigator	0.6707	0.5927	0.9999*	
Trial	0.7418	0.8367	0.9999*	

*: The variance-covariance matrix is ill-conditioned; in particular, at least one eigenvalue is very close to zero. The condition numbers for the three models with ill-conditioned matrices, from top to bottom are 3.415E+18, 2.384E+18, and 1.563E+18, respectively.

The basic meta-analytic approach and the corresponding simplified strategies have been applied, with results displayed in Table 26.2. Investigator and trial were both used as units of analysis. However, as there were only five trials, it became difficult to base the analysis on trial as unit of analysis in the case of the full bivariate random-effects approach. The results have shown a remarkable difference in the two cases. Consistently, in all of the different simplifications, the R^2_{trial} values were found to be higher when trial was used as unit of analysis. The bivariate full random-effects model does not converge when trial is used as the unit of analysis. This might be due to lack of sufficient information to compute all sources of variability, or to the fact that sample sizes tend to vary across trials. The reduced bivariate random effects model converged for both cases, but the resulting variance-covariance matrices were not positive-definite and were ill-conditioned, as can be seen from the very large value of the condition number. Consequently, the results of the bivariate random effects model should be treated with caution. Such issues are the topic of ongoing research. If we concentrate on the results based on investigator as unit of analysis, we observe a low level of surrogacy of PANSS for CGI, with R^2_{trial} ranging roughly between 0.5 and 0.68 for the different simplified models. This result, however, has to be coupled with other findings based on expert opinion to fully guarantee the validation of PANSS as possible surrogate for CGI. Turning to R^2_{indiv}, it ranges between 0.4904 and 0.5230, depending on the method of analysis, which is relatively low. To conclude, based on the investigators as unit of analysis, PANSS does not seem a promising surrogate for CGI.

26.7.1.2 Analysis of the categorical endpoints

We will now treat CGI as the true endpoint and PANSS as surrogate, although the reverse would be sensible, too. In practice, these endpoints are frequently dichotomized in a clinically meaningful way. Our binary true endpoint $T = \text{CGId} = 1$ for patients classified from "Very much improved" to "Improved", and 0 otherwise. The binary surrogate $S = \text{PANSSd} = 1$ for patients with at least 20 points reduction versus baseline, and 0 otherwise. We will start from probit and Plackett-Dale models and compare results with the ones from the information-theoretic approach.

In line with Section 26.4.1, we formulate two continuous latent variables $(\widetilde{\text{CGI}}_{ij}, \widetilde{\text{PANSS}}_{ij})$ assumed to follow a bivariate normal distribution. The following probit model can be fitted

$$
\begin{pmatrix}
\widetilde{\mu}_{ij}^T \\
\widetilde{\mu}_{ij}^S \\
\ln(\sigma^2) \\
\ln\left(\dfrac{1+\widetilde{\rho}}{1+\widetilde{\rho}}\right)
\end{pmatrix}
=
\begin{pmatrix}
\widetilde{\mu}_{T_i} + \widetilde{\beta}_i Z_{ij} \\
\widetilde{\mu}_{S_i} + \widetilde{\alpha}_i Z_{ij} \\
c_{\sigma^2} \\
c_{\widetilde{\rho}}
\end{pmatrix},
\tag{26.36}
$$

where $\widetilde{\mu}_{ij}^T = E(\widetilde{\text{CGI}}_{ij})$, $\widetilde{\mu}_{ij}^S = E(\widetilde{\text{PANSS}}_{ij})$, $\text{Var}(\widetilde{\text{CGI}}_{ij}) = 1$, $\sigma^2 = \text{Var}(\widetilde{\text{PANSS}}_{ij})$ and $\widetilde{\rho} = \text{corr}(\widetilde{\text{CGI}}_{ij}, \widetilde{\text{PANSS}}_{ij})$ denotes the correlation between the true and surrogate endpoint latent variables. We can then use the estimated values of $(\widetilde{\mu}_{S_i}, \widetilde{\alpha}_i, \widetilde{\beta}_i)$ to evaluate trial level surrogacy through the R^2_{trial}. At the individual level, $\widetilde{\rho}^2$ is used to capture surrogacy.

Alternatively, the Dale [31] formulation can be used, based on

$$
\begin{pmatrix}
\text{logit}(\pi_{ij}^T) \\
\text{logit}(\pi_{ij}^S) \\
\ln(\psi)
\end{pmatrix}
=
\begin{pmatrix}
\mu_{T_i} + \beta_i Z_{ij} \\
\mu_{S_i} + \alpha_i Z_{ij} \\
c_{\psi}
\end{pmatrix}
\tag{26.37}
$$

where $\pi_{ij}^T = \text{E}(\text{CGId}_{ij})$, $\pi_{ij}^S = \text{E}(\text{PANSSd}_{ij})$ and ψ is the global odds ratio associated to both endpoint. As before, the estimated values of $(\mu_{S_i}, \alpha_i, \beta_i)$ can be used to evaluate surrogacy at the trial level and the individual level surrogacy is quantified using the global odds ratio.

In the information-theoretic approach the following three models are fitted independently

$$
\Phi(\pi_{ij}^T) = \mu_{T_i} + \beta_i Z_{ij}, \tag{26.38}
$$

$$
\Phi(\pi_{ij}^{T|S}) = \mu_{T_i}^S + \beta_i^S Z_{ij} + \gamma_{ij} S_{ij}, \tag{26.39}
$$

$$
\Phi(\pi_{ij}^S) = \mu_{S_i} + \alpha_i Z_{ij}, \tag{26.40}
$$

where $\pi_{ij}^T = \text{E}(\text{CGId}_{ij})$, $\pi_{ij}^{T|S} = \text{E}(\text{CGId}_{ij}|\text{PANSSd}_{ij})$, $\pi_{ij}^S = \text{E}(\text{PANSSd}_{ij})$ and Φ denotes

TABLE 26.3
Schizophrenia study. Trial-level and individual-level validation measures (95% confidence intervals). Binary-binary case.

Parameter	Estimate	95% C.I.
Trial-level R^2_{trial} measures		
1.1 Information-theoretic	0.49	(0.21,0.81)
1.2 Probit	0.51	(0.18,0.78)
1.3 Plackett-Dale	0.51	(0.21,0.81)
Individual-level measures		
R^2_h	0.27	(0.24,0.33)
$R^2_{h\text{max}}$	0.39	(0.35,0.48)
Probit	0.67	(0.55,0.76)
Plackett-Dale ψ	25.12	(14.66;43.02)
Fano's lower-bound	0.08	

the cumulative standard normal distribution. At the trial level, the estimated values of $(\mu_{S_i}, \alpha_i, \beta_i)$ obtained from (26.38) and (26.40) can be used to calculate the R^2_{trial}, whereas at the individual level we can quantify surrogacy using R^2_h. As it was stated before, the likelihood reduction factor (LRF) is a consistent estimator of R^2_h, however, in principle other estimators could be used as well. We will then quantify surrogacy at the individual level by $\hat{R}^2_h = 1 - \exp\left(-G^2/n\right)$, where G^2 is the loglikelihood ratio test to compare (26.38) with (26.39) and n denotes total number of patients. Furthermore, when applied to the binary-binary setting, Fanos's inequality takes the form

$$P(T \neq S) \geq \frac{1}{\log|\Psi|}\left[H(T) - 1 + \frac{1}{2}\ln(1 - R^2_h)\right],$$

where $\Psi = \{0, 1\}$ and $|\Psi|$ denotes the cardinality of Ψ. Here, again, Fano's inequality gives a lower bound for the probability of incorrect prediction.

Table 26.3 shows the results at the trial and individual level obtained with the different approaches described above, where the dichotomization used is the one introduced at the start of this section. At the trial level, all the methods produced very similar values for the validation measure. In all cases, $R^2_{\text{trial}} \simeq 0.50$. It is also remarkable that the probit approach, in spite of being based on treatment effects defined at a latent level, produced a R^2_{trial} value similar to the ones obtained with the information–theoretic and Plackett-Dale approaches. However, as Alonso et al (2003) showed, there is a linear relationship between the mean parameters defined at the latent level and the mean parameters of the model based on the observable endpoints and that could explain the agreement between the probit and the other two procedures. Therefore, at the trial level, we could conclude that knowing the treatment effect on the surrogate will reduce our uncertainty about the treatment effect on the true endpoint by 50%.

At the individual level, the probit approach gives the strongest association between the surrogate and the true endpoint. Nevertheless, this value describes the association at an unobservable latent level, rendering its interpretation more awkward than with information theory, since it is not clear how this latent association could be relevant from a clinical point of view or how it could be translated into an association for the observable endpoints. The Plackett-Dale procedure quantifies surrogacy using a global odds ratio, making the comparison between this method and the others more difficult. Note that even though odds

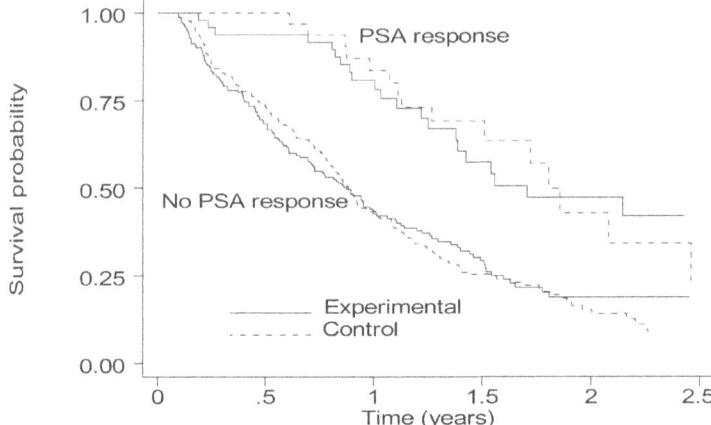

FIGURE 26.1
Survival by PSA response for patients with androgen sensitive metastatic prostate cancer randomized to a control or an experimental hormonal therapy.

ratios are widely used in biomedical fields the lack of an upper bound makes difficult their interpretation in this setting.

On the other hand, the value of the $R^2_{h\text{max}}$ illustrates that the surrogate can merely explain 39% of our uncertainty about the true endpoint, a relatively low value. Additionally, the lower bound for Fano's inequality clearly shows that using the value of PANSS to predict the outcome on CGI would be misleading in at least 8% of the cases. Even though this value is relatively low, it is only a lower bound and the real probability of error could be much larger.

At the trial level, the information-theoretic approach produces results similar to the ones from the conventional methods, but does so by means of models that are generally much easier to fit. At the individual level, the information-theoretic approach avoids the problem common with the probit model in that the correlation of the latter is formulated at the latent scale and therefore less relevant for practice. In addition, the information-theoretic measure ranges between 0 and 1, circumventing interpretational problems arising from using the unbounded Plackett-Dale based odds ratio.

26.7.2   Prostate-specific antigen (PSA)

PSA is an example of a longitudinal biomarker [79]. PSA is one of the most extensively studied cancer biomarker, in part because it is useful for the management of the disease. Changes in PSA can be used as a sensitive pharmacodynamic indicator of a treatment activity. Many attempts have been made to validate PSA-based endpoints as surrogates for cancer-specific survival (in early disease) or overall survival (in metastatic disease), but so far solid evidence for surrogacy is lacking. We use data from two trials in patients with androgen sensitive metastatic prostate cancer who were randomized to receive an experimental or a control hormonal therapy [22, 75]. Figure 26.1 shows the survival curves by randomized treatment and by PSA response, defined as a decrease of 50% or more in PSA compared with the baseline level, or a decrease in PSA to < 4 ng/mL confirmed by a second PSA measurement at least four weeks later. Clearly, PSA response had a major prognostic impact on survival. In contrast, in these two trials, there was no significant effect of treatment on survival overall, neither among patients with PSA, nor among patients without such response. We will perform further analyses of these trials in Section 26.7.2.1.

TABLE 26.4
Potential PSA-based surrogate endpoints in androgen sensitive metastatic prostate cancer (Abbreviations: surrogate type, B = Binary, L = Longitudinal, T = Time to failure).

Surrogate Endpoint	Surrogate Type	Individual-level Surrogacy Measure	Reference
PSA Response	B	Odds Ratio	[22]
Time to PSA Progression	T	Kendall's τ	
PSA Repeated Measures	L	$R^2_{indiv}(t)$	
PSA Response	B	Odds Ratio	[29]
PSA Normalization	B	Odds Ratio	
Time to PSA Progression	T	Kendall's τ	
PSA Repeated Measures	L	$R^2_{indiv}(t)$	

26.7.2.1 PSA as a surrogate in multiple trials

Table 26.4 shows several PSA-based endpoints that have been evaluated as surrogates for survival in patients with androgen sensitive metastatic prostate cancer. The analyses were based on two meta-analyses of patient-level data from five randomized trials testing experimental *versus* standard hormonal therapies [22, 29, 75]. Since there were only two trials in one meta-analysis and three in the other, each trial was split by country into smaller "units" (*i.e.*, groups of patients), to which the meta-analytic approach was applied. Note that the choice of appropriate units of analysis is a matter of debate. Technically, the optimal conditions are to have both a large number of units and a large number of patients per unit [26]. Splitting trials into smaller units may achieve the former condition at the expense of the latter. Some studies evaluating surrogates in a small number of trials have used clinical site as the unit of analysis [14, 17, 59].

As indicated in Table 26.4, the PSA-based endpoints were of different types, calling for the use of methods specific to each variable type, as described in Sections 26.4.1 to 26.4.5. We discuss the results of one meta-analysis [22], but the other meta-analysis showed very similar results [29]. There was no significant difference in overall survival between the randomized treatment groups, so the Prentice criteria could not be used. In contrast, the meta-analytic approach could be used and was informative.

One of the potential surrogates was PSA response, defined in Section 26.7.2. At the individual level, there was a strong impact of PSA response on survival, as shown in Figure 26.1. Several measures can be used to quantify the individual level association. The survival odds ratio is one such measure: it is the ratio, assumed constant, of the odds of surviving beyond time t for PSA responders as compared with nonresponders. The survival odds ratio for PSA response was equal to 5.5 (95% C.I. $2.7 - -8.2$). At the country level, there was no association between the treatment effects on PSA response and on survival ($R^2_{country} = 0.05$, Figure 26.2).

The treatment effects on other potential PSA-based surrogates such as PSA normalization or time to PSA progression did not reach much stronger associations with the treatment effects on survival at the country level. Even when the full vector of longitudinal PSA measurements was used for each patient as suggested in Section 26.4.5, the association between treatment effects remained quite weak (Figure 26.3), suggesting that PSA alone cannot be used to define a valid surrogate endpoint for survival in prostate cancer.

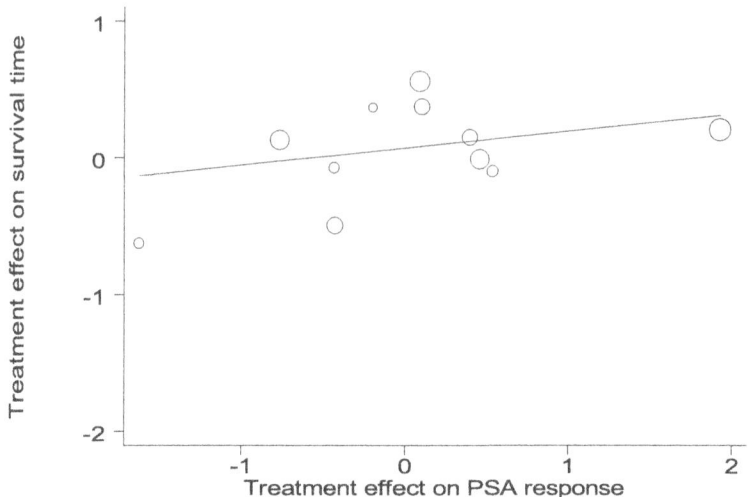

FIGURE 26.2
Association between treatment effects on PSA response and survival in patients with androgen sensitive prostate cancer. Each bubble represents a country, with bubble size proportional to the sample size in the corresponding country.

26.7.3 Surrogate endpoints in gastric cancer

To illustrate the concepts and difficulties inherent in the evaluation of surrogate endpoints, we will use two meta-analyses of randomized clinical trials conducted by the GASTRIC (Global Advanced/Adjuvant Stomach Tumor Research International Collaboration) Group. The first meta-analysis was carried out using individual data from patients with curatively resected gastric cancer. It confirmed the benefit of adjuvant chemotherapy as compared with no adjuvant treatment in terms of both disease-free survival (DFS) and overall survival (OS) [47]. The second meta-analysis was carried out using individual data from patients with advanced or recurrent gastric cancer. It confirmed the benefit of adding experimental agents to standard chemotherapy regimens in terms of both progression-free survival (PFS) and OS [48]. We will apply the methods described in the previous sections to investigate whether DFS and PFS are acceptable surrogates for OS. Even though both of these early endpoints are *a priori* plausible surrogates for OS, we will show that DFS can be used as a reasonable surrogate for OS in localized disease, while PFS cannot be used reliably as a surrogate in advanced disease.

26.7.3.1 Resectable gastric cancer: can DFS be used a surrogate for OS?

The meta-analysis of trials for patients with resected gastric cancer was used to evaluate DFS as a surrogate for OS. Data were available on 3,288 patients from 14 trials with documented overall survival and disease free survival [70].

At the individual level, a Plackett copula was fitted on the joint distribution of DFS and OS. The individual level association, quantified by Spearman's rank correlation coefficient (i.e., the Pearson correlation coefficient between the corresponding ranks), was equal to 0.974 (95% CI [0.971, 0.976]), indicating a very tight correlation between DFS and OS for a given patient. The tight correlation between endpoints was confirmed using the information-theoretic approach (Section 2.65), with R_h^2 equal to 0.935 (95% CI [0.924, 0.938]).

At the trial level, there was also a tight association between the treatment effects on DFS and on OS (Figure 26.4). Without adjustment for the estimation error in treatment

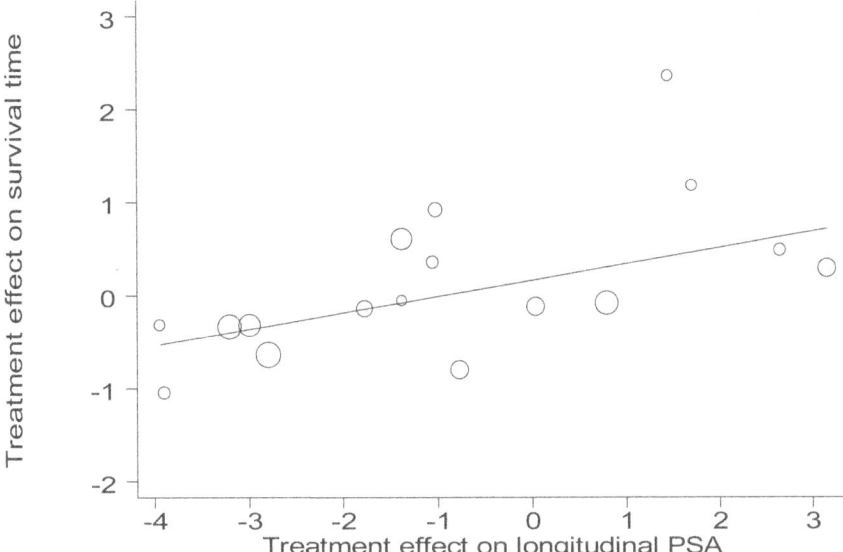

FIGURE 26.3
Association between treatment effects on longitudinal measurements of PSA and survival in patients with androgen sensitive prostate cancer. Each bubble represents a country, with bubble size proportional to the sample size in the corresponding country.

effects, R^2_{trial} was equal to 0.964 (95% CI [0.926, 1.000]). After adjusting for the estimation error [14], R^2_{trial} was estimated to be almost equal to 1 (95% CI [0.999, 1.000]). It is worth noting that because the estimated R^2_{trial} value was very close to the upper limit of 1, the numerical results need to be treated with caution. The tight correlation between treatment effects on the two endpoints was also confirmed using the information-theoretic approach, with $R^2_{\text{h,trial}}$ equal to 0.969 (95% CI [0.879, 0.991]).

The linear regression model adjusted for estimation errors was

$$\ln(HR_{OS}) = 0.047(0.023) + 1.239(0.151) \times \ln(HR_{DFS}) \qquad (26.41)$$

where the numbers in parentheses are the standard errors of the intercept and slope, respectively. Figure 26.4 shows regression line (26.41) and the treatment effects in the 14 trials included in the analysis. Each trial is represented by a bubble whose size is proportional to the trial sample size. The 95% prediction limits indicate the range of effects on OS that can be expected for a given effect on DFS.

The STE was equal to 0.92, hence in a future trial using similar treatment modalities as in the set of trials in the meta-analysis, a HR_{DFS} smaller than 0.92 would predict a HR_{OS} smaller than 1. This fact can be used to design a trial based on the surrogate, rather than the true endpoint. The size of this new trial can be calculated so that the 95% confidence interval of the estimated HR_{DFS} lies entirely under 0.92. In other words, the null and alternative hypotheses of interest for DFS are $H_0 : HR_{DFS} \geq 0.92$ *versus* $H_A : HR_{DFS} < 0.92$. The test of hypothesis for DFS is more stringent than the test of hypothesis for OS, which is based on the conventional null and alternative hypotheses $H_0 : HR_{OS} \geq 1.0$ *versus* $H_A : HR_{OS} < 1.0$. Even so, the test of hypothesis for DFS may require less patients and less follow-up time to reach the same statistical power than the test of hypothesis for OS, since the treatment effect may be larger on DFS than on OS, and the events are observed earlier.

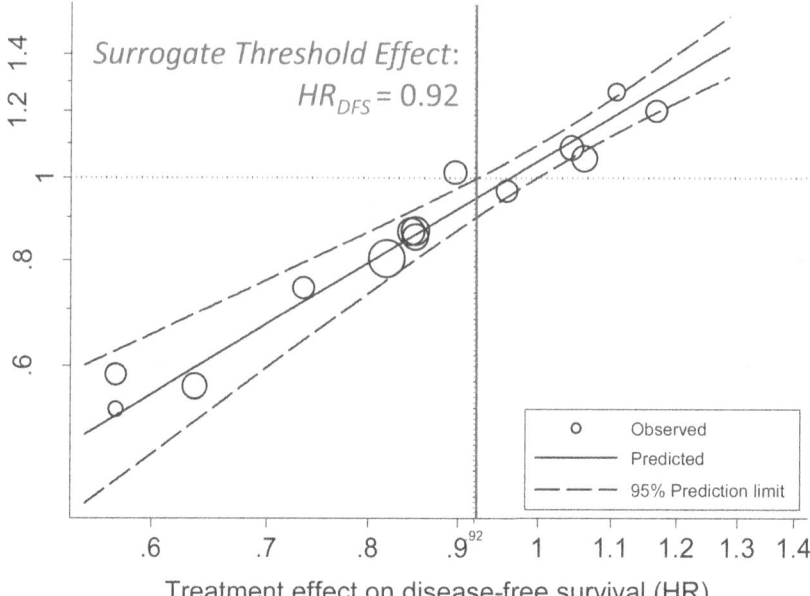

FIGURE 26.4
Trial level association between treatment effects on DFS and OS in resectable gastric cancer (both axes on a log scale).

The results of our surrogate evaluation could be externally validated using five trials not included in the meta-analysis, three for which the treatment effects were extracted from reports published in the literature, and two for which individual patient data became available after the meta-analysis was completed (Table 26.5). Figure 26.5 shows regression line (26.41), the observed treatment effects on survival (HR_{OS}) in these five trials, and the treatment effects on survival predicted from the treatment effects on the surrogate (HR_{DFS}), along with their 95% prediction intervals. There is excellent agreement between the observed and the predicted treatment effects, and in the three trials for which the prediction limits of HR_{OS} excluded one, the observed effects on survival actually reached statistical significance [70].

26.7.3.2 Advanced gastric cancer: can PFS be used as a surrogate for OS?

The meta-analysis of trials in advanced disease was used for the purposes of evaluating PFS as a surrogate for OS. Data were available on 4,069 patients from 20 eligible randomized trials with documented overall survival and progression-free survival [71].

The individual level association, quantified by Spearman's rank correlation coefficient, was equal to 0.853 (95% CI [0.852, 0.854]), indicating substantial correlation between PFS and OS for a given patient. The correlation between endpoints was confirmed using the information-theoretic approach (Section 2.65), with R_{h}^2 equal to 0.859 (95% CI [0.842, 0.862]).

At the trial level, the association between the treatment effects on PFS and on OS was only moderate (Figure 26.6). R_{trial}^2, adjusted for the estimation errors, was equal to 0.61 (95% CI [0.04, 1.00]). The large confidence interval reflects the uncertainty around this estimate, due in part to the small sample sizes of some of the trials included in the

TABLE 26.5
Observed treatment effects [95% confidence intervals] on DFS and OS, and predicted treatment effect on OS [95% prediction intervals] in five validation trials in resectable gastric cancer. Asterisks mark intervals that exclude no effect.

Trial [70]	Observed HR_{DFS}	Observed HR_{OS}	Predicted HR_{OS}
Cirera	0.55 [0.36, 0.85]∗	0.60 [0.39, 0.93]∗	0.50 [0.28, 0.87]∗
ACTS-GC	0.65 [0.54, 0.79]∗	0.67 [0.54, 0.83]∗	0.61 [0.47, 0.81]∗
INT-1018	0.66 [0.53, 0.82]∗	0.75 [0.61, 0.92]∗	0.63 [0.46, 0.84]∗
GOIM-9602	0.88 [0.66, 1.17]	0.91 [0.69, 1.21]	0.89 [0.62, 1.28]
GOIRC	0.92 [0.66, 1.27]	0.90 [0.64, 1.26]	0.94 [0.63, 1.42]

meta-analysis. The poor correlation between treatment effects on the two endpoints was also confirmed using the information-theoretic approach, with $R^2_{\text{h,trial}}$ equal to 0.45 (95% CI [0.16, 0.70]). Of note, the estimate $R^2_{\text{h,trial}} = 0.45$ was not adjusted for measurement errors, while the estimate $R^2_{\text{trial}} = 0.61$ was. The issue of measurement error is important. Not taking into account can have a profound impact on the R^2-based estimates obtained. Precisely, failure to adjust for measurement error can lead to overly optimistic R^2 values. Models that adjust for measurement error are more complicated in terms of convergence, but this does not mean that the issue should be overlooked.

The linear regression model adjusted for estimation errors was

$$\ln(HR_{OS}) = 0.042(0.079) + 0.779(0.295) \times \ln(HR_{PFS}) \tag{26.42}$$

where the numbers in parentheses are the standard errors of the intercept and slope, respectively. Figure 26.6 shows regression line (26.42) and the treatment effects in the 20 trials included in the analysis. Each trial is represented by a bubble whose size is proportional to the trial sample size. The 95% prediction limits indicate the range of effects on OS that can be expected for a given effect on PFS.

The moderate correlation at the trial level is reflected by a surrogate threshold effect (STE) equal to 0.56: hence, in a future trial using similar treatment modalities as in the set of trials in the meta-analysis, a HR_{DFS} smaller than 0.56 would predict a HR_{OS} smaller than 1.

The results of our surrogate evaluation could be externally validated using twelve trials not included in the meta-analysis, using treatment effects extracted from reports published in the literature after the meta-analysis was completed (Table 26.6). Figure 26.7 shows regression line (26.42), the observed treatment effects on survival (HR_{OS}), and the treatment effects on survival predicted from equation the treatment effects on the surrogate (HR_{PFS}) in these trials, along with their 95% prediction intervals. The prediction intervals are wide and include one (no treatment effect on OS) in all trials, which means that the observed effects on PFS would not have allowed predicting an effect on OS in any of these twelve trials. Yet, three of the twelve trials showed a statistically significant effect of treatment on survival [71].

26.7.3.3 Contrasting conclusions about DFS and PFS

The analyses shown in Section 26.7.3.1 suggest that DFS is a good surrogate for OS in patients with gastric tumors that can be surgically resected or locally treated. These findings parallel those in resectable colon cancer, as well as operable or locally advanced head and neck and lung cancers [64, 65, 78]. Taken together, these findings suggest that in early forms of cancer that are amenable to local treatment, disease-free survival can be used as a reliable surrogate for overall survival.

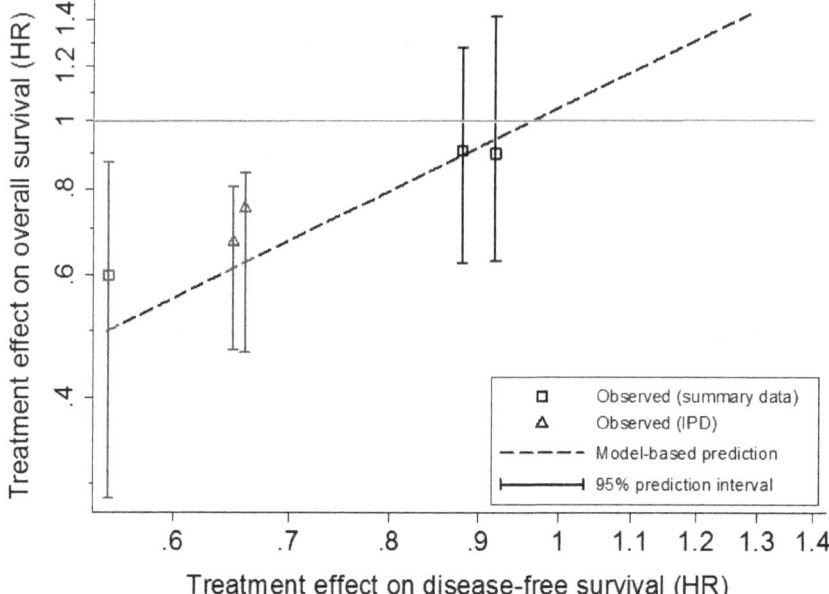

FIGURE 26.5

Observed treatment effects on OS (squares, published summary data; triangles, individual patient data) and 95% prediction intervals in five validation trials in resectable gastric cancer (both axes on a log scale). Prediction intervals are depicted in red for the three trials that showed a significant treatment effect on OS (see Table 26.5).

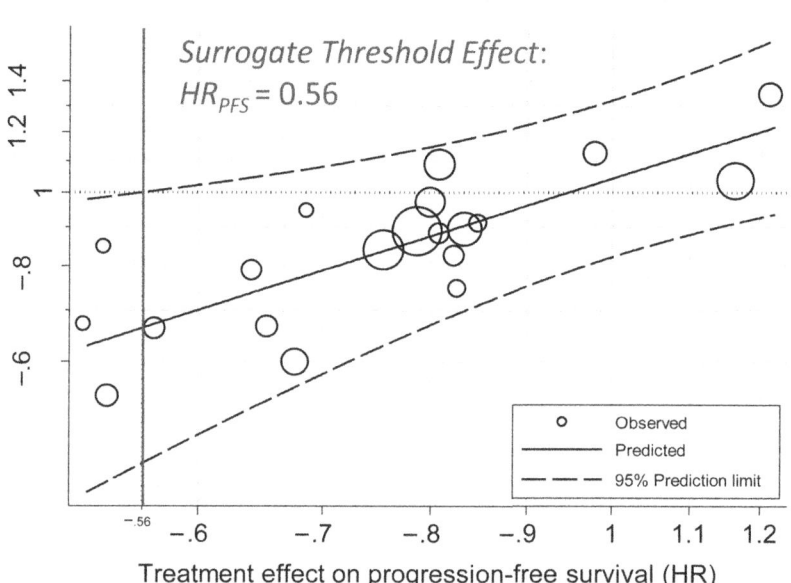

FIGURE 26.6

Trial level association between treatment effects on PFS and OS in advanced gastric cancer (both axes on a log scale).

TABLE 26.6
Observed treatment effects [95% confidence intervals] on PFS and OS, and predicted treatment effect on OS [95% prediction intervals] in twelve validation trials in advanced gastric cancer. Asterisks mark intervals that exclude no effect.

Trial [71]	Observed HR_{PFS}	Observed HR_{OS}	Predicted HR_{OS}
Jeung	0.63 [0.38, 1.05]	0.56 [0.35, 0.88]*	0.73 [0.46, 1.04]
AIO	0.67 [0.43, 1.04]	0.82 [0.47, 1.45]	0.76 [0.53, 1.07]
ToGA	0.71 [0.59, 0.85]*	0.74 [0.60, 0.91]*	0.80 [0.58, 1.09]
AVAGAST	0.80 [0.68, 0.93]*	0.87 [0.73, 1.03]	0.88 [0.76, 1.14]
Kang	0.80 [0.63, 1.03]	0.85 [0.64, 1.13]	0.88 [0.76, 1.14]
Park	0.86 [0.54, 1.37]	0.96 [0.60, 1.52]	0.93 [0.71, 1.18]
REAL(a)	0.92 [0.80, 1.04]	0.92 [0.80, 1.10]	0.98 [0.77, 1.22]
REAL(b)	0.92 [0.81, 1.05]	0.86 [0.80, 0.99]*	0.98 [0.77, 1.22]
Ross	0.95 [0.80, 1.08]	0.91 [0.76, 1.04]	1.00 [0.79, 1.29]
FLAGS	0.99 [0.86, 1.14]	0.92 [0.80, 1.05]	1.03 [0.81, 1.31]
Rao	1.13 [0.63, 2.01]	1.02 [0.61, 1.70]	1.14 [0.89, 1.46]
Moehler	1.14 [0.59, 2.21]	0.77 [0.51, 1.17]	1.15 [0.90, 1.48]

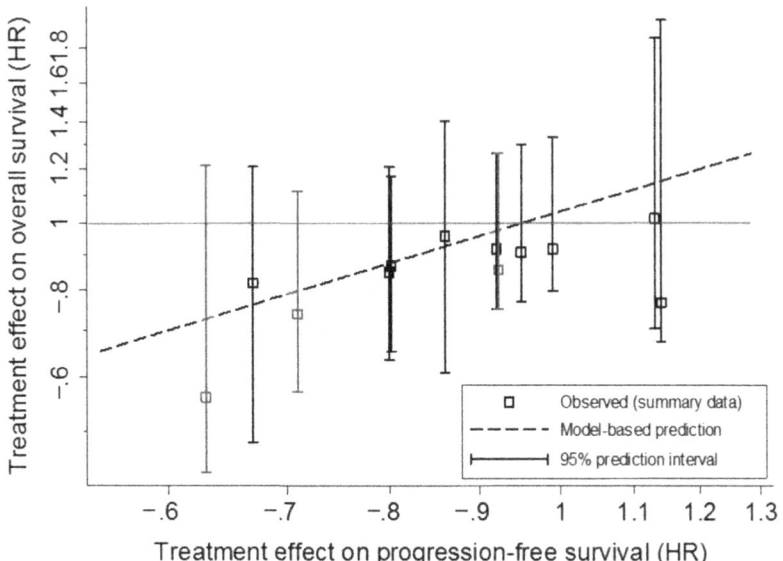

FIGURE 26.7
Observed treatment effects on OS (squares) and 95% prediction intervals in twelve validation trials in advanced gastric cancer (both axes on a log scale). Prediction intervals are depicted in red for the three trials that showed a significant treatment effect on OS (see Table 26.6).

In contrast, the analyses shown in Section 26.7.3.2 suggest that PFS is not a useful surrogate for OS in advanced gastric cancer. PFS was also found to be a poor surrogate for OS in advanced breast cancer [11]. In contrast, PFS was a good surrogate in advanced ovarian cancer [14]. PFS appeared to be a good surrogate for OS in advanced colorectal cancer treated with fluoropyrimidines [15] but not with more recent therapies [84]. In advanced lung cancer, the value of PFS as a surrogate for OS is questionable [59]. All in all, progression-free survival tends to be a poor surrogate for overall survival in advanced solid tumors.

These contrasting findings for DFS and PFS may seem counterintuitive. If anything, one would expect PFS to be a better surrogate for OS than DFS, because the time between tumor progression and death is much shorter in advanced disease than after surgical resection of the tumor. However, from a biological standpoint, the reappearance of a tumor after a long disease-free period may be a far more consequential event than an increase in size of a measurable tumor mass.

26.8 Concluding Remarks

Over the years, a variety of surrogate marker evaluation strategies have been proposed, cast within a meta-analytic framework. With an increasing range of endpoint types considered, such as continuous, binary, time-to-event, and longitudinal endpoints, also the scatter of types of measures proposed has increased. Some of these measures are difficult to calculate from fully specified hierarchical models, which has sparked off the formulation of simplified strategies. We reviewed the ensuing divergence of proposals, which then has triggered efforts of convergence, eventually leading to the information-theoretic approach, which is both general and simple to implement. These developments have been illustrated using data from clinical trials in schizophrenia.

While quantifying surrogacy is important, so is prediction of the treatment effect in a new trial based on the surrogate. Work done in this area has been reviewed, with emphasis on the so-called surrogate threshold effect and the sources of variability involved in the prediction process. A connection with the information-theoretic approach is pointed out.

Even though more work is called for, we believe the information-theoretic approach and the surrogate threshold effect are promising paths toward effective assessment and use of surrogate endpoints in practice. Software implementations for methodology described here and beyond are available from www.ibiostat.be. See also [reference to book Alonso et al.].

A key issue is whether a surrogate is still valid if, in a new trial, the same surrogate and true endpoints, but a different drug is envisaged. This is the so-called "class" question. It is usually argued that a surrogate could still be used if the new drug belongs to the same class of drugs as the ones in the evaluation exercise. Of course, the most important point is that a surrogate is sensible, e.g., physiologically plausible, toward the true endpoint.

Acknowledgment

The authors gratefully acknowledge support from IAP research Network P7/06 of the Belgian Government (Belgian Science Policy). They thank the GASTRIC (Global Advanced/Adjuvant Stomach Tumor Research International Collaboration) Group for permission to use their data. The investigators who contributed to GASTRIC are listed in references [47, 48, 70, 71]. The GASTRIC Group data can be downloaded for further analyses as specified in Buyse et al [20]. Software implementations for the methods described in this chapter are available from www.ibiostat.be/software.

Bibliography

[1] Alonso, A., Geys, H., Molenberghs, G., and Vangeneugden, T. 2003. Validation of surrogate markers in multiple randomized clinical trials with repeated measurements. *Biometrical Journal*, 45:931–945.

[2] Alonso, A., and Molenberghs, G. 2007. Surrogate marker evaluation from an information theoretic perspective. *Biometrics*, 63:180–186.

[3] Alonso, A., Molenberghs, G., Geys, H., and Buyse, M. 2005. A unifying approach for surrogate marker validation based on Prentice's criteria. *Statistics in Medicine*, 25:205–211.

[4] Alonso, A., Molenberghs, G., and van Breukelen, G. 2013. Statistical validation of surrogate markers in clinical trials. In K. van Montfort, J.H.L. Oud, and W. Ghidey (eds.), *Developments in Statistical Evaluation of Clinical Trials*, pp. 227–246. Springer, New York.

[5] Alonso, A., Van der Elst, W., Molenberghs, G., Buyse, M. and Burzykowski, T. 2015. On the relationship between the causal-inference and meta-analytic paradigms for the validation of surrogate endpoints. *Biometrics*, 71:15–24.

[6] Armstrong, A.J., Garrett–Mayer, E., Yang, Y.C.O., Carducci, M.A., Tannock, I., de Wit, R., and Eisenberger, M. 2007. Prostate-specific antigen and pain surrogacy analysis in metastatic hormone-refractory prostate cancer. *Journal of Clinical Oncology*, 25:3965–2970.

[7] Assam, P., Tilahun, A., Alonso, A., and Molenberghs, G. 2007. Information-theory based surrogate marker evaluation from several randomized clinical trials with continuous true and binary surrogate endpoints. *Clinical Trials*, 4:587–597.

[8] Begg, C. and Leung, D. 2000. On the use of surrogate endpoints in randomized trials. *Journal of the Royal Statistical Society , Series A*, 163:26–27.

[9] Biomarkers Definition Working Group 2001. Biomarkers and surrogate endpoints: preferred definitions and conceptual framework. *Clinical Pharmacological Therapy*, 69:89–95.

[10] Burzykowski, T. and Buyse, M. 2006. Surrogate threshold effect: an alternative measure for meta-analytic surrogate endpoint validation. *Pharmaceutical Statistics*, 5:173—186.

[11] Burzykowski, T., Buyse, M., Piccart-Gebhart, M.J., Sledge, G., Carmichael, J., Lück, H.J., Mackey, J.R., Nabholtz, J.M., Paridaens, R., Biganzoli, L., Jassem, J., Bontenbal, M., Bonneterre, J., Chan, S., Basaran, G.A., and Therasse, P. 2008. Evaluation of tumor response, disease control, progression-free survival, and time to progression as potential surrogate endpoints in metastatic breast cancer. *Journal of Clinical Oncology*, 26:1987–1992.

[12] Burzykowski, T., Molenberghs, G., and Buyse, M. 2004. The validation of surrogate endpoints using data from randomized clinical trials:a case-study in advanced colorectal cancer. *Journal of the Royal Statistical Society, Series A*, 167:103–124.

[13] Alonso, A., Bigirumurame, T., Burzykowski, T., Buyse, M., Molenberghs, G., Muchene, L., Perualila, N.J., Shkedy, Z., and Van der Elst, W. (2017). Applied Surrogate Endpoint Evaluation with SAS and R. Boca Raton: ChapmanHall/CRC.

[14] Burzykowski, T., Molenberghs, G., Buyse, M., Geys, H. and Renard, D. 2001. Validation of surrogate endpoints in multiple randomized clinical trials with failure-time endpoints. *Journal of the Royal Statistical Society, Series C*, 50:405–422.

[15] Buyse, M., Burzykowski, T., Carroll, K., Michiels, S., Sargent, D.J., Miller, L.L., Elfring, G.L., Pignon, J.P., and Piedbois, P. 2007. Progression-free survival is a surrogate for survival in advanced colorectal cancer. *Journal of Clinical Oncology*, 25: 5218–5224.

[16] Buyse, M., Burzykowski, T., Molenberghs, G., and Alonso Abad, A. 2015. Biomarker-based surrogate endpoints. In S. Matsui, M. Buyse, and R. Simon (eds.), *Design and Analysis of Clinical Trials for Predictive Medicine*, pp. 497–535. Chapman % Hall/CRC Press, New York.

[17] Buyse, M., Michiels, S., Squifflet, P., Lucchesi, K.J., Hellstrand, K., Brune, M.L., Castaigne, S., and Rowe, J.M. 2011. Leukemia-free survival as a surrogate endpoint for overall survival in the evaluation of maintenance therapy for patients with acute myeloid leukemia in complete remission. *Haematologica*, 96:1106–12.

[18] Buyse, M. and Molenberghs, G. 1998. Criteria for the validation of surrogate end-points in randomized experiments. *Biometrics*, 54:1014–1029.

[19] Buyse, M., Molenberghs, G., Burzykowski, T., Renard, D., and Geys, H. 2000. The validation of surrogate endpoints in meta-analyses of randomized experiments. *Biostatistics*, 1:49–68.

[20] Buyse, M., Molenberghs, G., Paoletti, X., Oba, K., Alonso, A., Van der Elst, W., and Burzykowski, T. 2016. Statistical evaluation of surrogate endpoints with examples from cancer clinical trials. *Biometrical Journal*, 58:104–132.

[21] Buyse, M., Molenberghs, G., Burzykowski, T., Renard, D., and Geys, H. 2000. Statistical validation of surrogate endpoints: problems and proposals. *Drug Information Journal*, 34, 447–454.

[22] Buyse, M., Vangeneugden, T., Bijnens, L., Renard, D., Burzykowski, T., Geys, H., and Molenberghs, G. 2003. Validation of biomarkers as surrogates for clinical endpoints. In: J.C. Bloom, R.A. Dean, (eds.), *Biomarkers in Clinical Drug Development*, pp. 149–168. Marcel Dekker, New York.

[23] Cardiac Arrhythmia Suppression Trial (CAST) Investigators 1989. Preliminary Report: effect of encainide and flecainide on mortality in a randomized trial of arrhythmia suppression after myocardial infraction. *New England Journal of Medicine*, 321:406–412.

[24] Choi, S., Lagakos, S., Schooley, R.T., and Volberding, P.A. 1993. CD4+ lymphocytes are an incomplete surrogate marker for clinical progression in persons with asymptomatic HIV infection taking zidovudine. *Annals of Internal Medicine*, 118:674–680.

[25] Clayton, D.G. 1978. A model for association in bivariate life tables and its application in epidemiological studies of familial tendency in chronic disease incidence. *Biometrika*, 65:141–151.

[26] Cortiñas Abrahantes, J., Molenberghs, G., Burzykowski, T., Shkedy, Z., and Renard, D. 2004. Choice of units of analysis and modeling strategies in multilevel hierarchical models. *Computational Statistics and Data Analysis*, 47:537–563.

[27] Cortiñas, J., Shkedy, Z., and Molenberghs, G. 2008. Alternative methods to evaluate trial level surrogacy. *Clinical Trials*, 5:194–208.

[28] Collette, L., Burzykowski, T. and Buyse, M. 2007. Are prostate-specific antigen changes valid surrogates for survival in hormone-refractory cancer? A meta-analysis is needed! [Letter to the Editor]. *Journal of Clinical Oncology*, 25:5673–5674.

[29] Collette, L., Burzykowski, T., Carroll, K., Newling, D., Morris, T., and Schröder, F.H. 2005. Is prostate-specific antigen a valid surrogate end point for survival in hormonally treated patients with metastatic prostate cancer? *Journal of Clinical Oncology*, 23:6139–6148.

[30] Cover, T. and Tomas, J. 1991. *Elements of Information Theory*. John Wiley & Sons, New York.

[31] Dale, J.R. 1986. Global cross ratio models for bivariate, discrete, ordered responses. *Biometrics*, 42:909–917.

[32] D'Amico, A.V., Chen, M.H., de Castro, M., Loffredo, M., Lamb, D.S., Steigler, A., Kantoff, P.W., and Denham, J.W. 2012. Surrogate endpoints for prostate cancer-specific mortality after radiotherapy and androgen suppression therapy in men with localized or locally advanced prostate cancer: an analysis of 2 randomized trials. *Lancet Oncology*, 13:189–195.

[33] Daniels, M.J. and Hughes, M.D. 1997. Meta-analysis for the evaluation of potential surrogate markers. *Statistics in Medicine*, 16:1515–1527.

[34] DeGruttola, V. and Tu, X.M. 1994. Modelling progression of CD-4 lymphocyte count and its relationship to survival time. *Biometrics*, 50:1003–1014.

[35] DeGruttola, V., Fleming, T.R., Lin, D.Y., and Coombs, R. 1997. Validating surrogate markers - are we being naive? *Journal of Infectious Diseases*, 175:237–246.

[36] Denham, J.W., Steigler, A., Wilcox, C., Lamb, D.S., Joseph, D., Atkinson, C., Matthews, J., Tai, K.-H., Spry, N.A., Christie, D., Gleesona, P.S., Greer, P.B., D'Este, C., on behalf of the Trans-Tasman Radiation Oncology Group 96.01 Trialists 2008. Time to biochemical failure and prostate-specific antigen doubling time as surrogates for prostate cancer-specific mortality: evidence from the TROG 96.01 randomised controlled trial. *Lancet Oncology*, 9:1058–1068.

[37] Dunn, N. and Mann, R.D. 1999. Prescription-event and other forms of epidemiological monitoring of side-effects in the UK. *Clinical and Experimental Allergy*, 29:217–239.

[38] Ellenberg, S.S. and Hamilton, J.M. 1989. Surrogate endpoints in clinical trials: cancer. *Statistics in Medicine*, 8:405–413.

[39] Ferentz, A.E. 2002 Integrating pharmacogenomics into drug development. *Pharmacogenomics*, 3:453–467.

[40] Fleming, T.R., DeMets, D.L. 1996. Surrogate end points in clinical trials: are we being misled? *Annals of Internal Medicine*, 125:605–613.

[41] Fleming, T.R. 1994. Surrogate markers in AIDS and cancer trials. *Statistics in Medicine*, 13:1423–1435.

[42] Flandre, P. and Saidi, Y. 1999. Letter to the editor:estimating the proportion of treatment effect explained by a surrogate marker. *Statistics in Medicine*, 18:107–115.

[43] Frangakis, C.E. and Rubin, D.B. 2004. Principal stratification in causal inference. *Biometrics*, 58:21–29.

[44] Freedman, L.S., Graubard, B.I., Schatzkin, A. 1992. Statistical validation of intermediate endpoints for chronic diseases. *Statistics in Medicine*, 11:167–178.

[45] Gail, M.H., Pfeiffer, R., van Houwelingen, H.C., and Carroll, R.J. 2000. On meta-analytic assessment of surrogate outcomes. *Biostatistics*, 1:231–246.

[46] Galecki, A. 1994. General class of covariance structures for two or more repeated factors in longitudinal data analysis. *Communications in Statistics:theory and methods*, 23:3105–3119.

[47] The GASTRIC (Global Advanced/Adjuvant Stomach Tumor Research International Collaboration) Group. 2010. Benefit of adjuvant chemotherapy for resectable gastric cancer: a meta-analysis. *Journal of the American Medical Association*, 303:1729–1737.

[48] The GASTRIC (Global Advanced/Adjuvant Stomach Tumor Research International Collaboration) Group. 2013. Role of chemotherapy for advanced/recurrent gastric cancer: an individual-patient-data meta-analysis. *European Journal of Cancer*, 49: 1565–1577.

[49] Genest, C. and McKay, J. 1986. The joy of copulas: bivariate distributions with uniform marginals. *American Statistician*, 40:280–283.

[50] Heise, C., Sampson-Johannes, A., Williams, A., McCormick, F., Von Hoff, D.D., and Kirn, D.H. 1997. ONYX-015, an E1B gene-attenuated adenovirus, causes tumor-specific cytolysis and antitumoral efficacy that can be augmented by standard chemotherapeutic agents. *Nature Medicine*, 3:639–645.

[51] Hougaard, P. 1986. Survival models for heterogeneous populations derived from stable distributions. *Biometrika*, 73:387–396.

[52] International Conference on harmonisation of technical requirements for registration of pharmaceuticals for human use. 1998. ICH Harmonised Tripartite Guideline. Statistical principles for clinical trials. (http://www.ich.org/pdfICH/e9.pdf), *Federal Register 63*, No. 179, 49583.

[53] Joffe, M.M. and Greene, T. 2008. Related causal frameworks for surrogate outcomes. *Biometrics*, 64; 1–10.

[54] Jones, T.C. 2001. Call for a new approach to the process of clinical trials and drug registration. *British Medical Journal*, 322:920–923.

[55] Kay, S.R., Fiszbein, A., and Opler, L.A. 1987. The positive and negative syndrome scale (PANSS) for schizophrenia. *Schizophrenia Bulletin*, 13:261–276.

[56] Kay, S.R., Opler, L.A., and Lindenmayer, J.P. 1988. Reliability and validity of the Positive and Negative Syndrome Scale for schizophrenics. *Psychiatric Research*, 23:99–110.

[57] Kent, J. 1983. Information gain and a general measure of correlation. *Biometrika*, 70:163–173.

[58] Lagakos, S.W. and Hoth, D.F. 1992. Surrogate markers in AIDS: Where are we? Where are we going? *Annals of Internal Medicine*, 116:599–601.

[59] Laporte, S., Squifflet, P., Baroux, N., et al. 2013. Prediction of survival benefits from progression-free survival benefits in advanced non small cell lung cancer: evidence from a pooled analysis of 2,334 patients randomized in 5 trials. *BMJ Open*, 3:3.03.

[60] Lesko, L.J. and Atkinson, A.J. 2001. Use of biomarkers and surrogate endpoints in drug development and regulatory decision making:criteria, validation, strategies. *Annual Review of Pharmacological Toxicology*, 41:347–66.

[61] Li, Y., Taylor, J.M.G. and Elliott, M.R. 2010. A Bayesian approach to surrogacy assessment using principal stratification in clinical trials. *Biometrics*, 58: 21–9.

[62] Li, Y., Taylor, J.M.G., Elliott, M.R. and Sargent, D.R. 2011. Causal assessment of surrogacy in a meta-analysis of colorectal clinical trials. *Biostatistics*, 12: 478–92.

[63] Lin, D.Y., Fleming, T.R., and DeGruttola, V. 1997. Estimating the proportion of treatment effect explained by a surrogate marker. *Statistics in Medicine*, 16:1515–1527.

[64] Mauguen, A., Pignon, J.P., Burdett, S., Domerg, C., Fisher, D., Paulus, R., Mandrekar, S.J., Belani, C.P., Shepherd, F.A., Eisen, T., Pang, H., Collette, L., Sause, W.T., Dahlberg, S.E., Crawford, J., O'Brien, M., Schild, S.E., Parmar, M., Tierney, J.F., Le Pechoux, C., and Michiels, S., on behalf of the Surrogate Lung Project Collaborative Group 2013. Surrogate endpoints for overall survival in chemotherapy and radiotherapy trials in operable and locally advanced lung cancer: a re-analysis of meta-analyses of individual patients' data. *Lancet Oncology*, 14:619–626.

[65] Michiels, S., Le Maître, A., Buyse, M., Burzykowski, T., Maillard, E., Bogaerts, J., Vermorken, J.B., Budach, W., Pajak, T.F., Ang, K.K., Bourhis, J., Pignon, J.P., on behalf of the MARCH and MACH-NC Collaborative Groups. 2009. Surrogate endpoints for overall survival in locally advanced head and neck cancer: meta-analyses of individual patient data. *Lancet Oncology*, 10: 341–350.

[66] Molenberghs, G., Alonso Abad, A., Van der Elst, W., Burzykowski, T., and Buyse, M. 2014. Statistical evaluation of surrogate endpoints in clinical studies. In: W. Young and D.-G. Chen (eds.), *Clinical Trial Biostatistics and Biopharmaceutical Applications*, pp. 497–536. Chapman & Hall/CRC, New York.

[67] Molenberghs, G., Burzykowski, T., Alonso, A., Assam, P., Tilahun, A., and Buyse, M. 2010. A unified framework for the evaluation of surrogate endpoints in clinical trials. *Statistical Methods in Medical Research*, 19:205–236.

[68] Molenberghs, G., Geys, H., and Buyse, M. 2001. Evaluation of surrogate end-points in randomized experiments with mixed discrete and continuous outcomes. *Statistics in Medicine*, 20:3023–3038.

[69] Molenberghs, G. and Verbeke, G. 2005. *Models for Discrete Longitudinal Data*. Springer, New York.

[70] Oba, K., Paoletti, X., Alberts, S., Bang, Y.J., Benedetti, J., Bleiberg, H., Catalano, P., Lordick, F., Michiels, S., Morita, S., Ohashi, Y., Pignon, J.P., Rougier, P., Sasako, M., Sakamoto, J., Sargent, D., Shitara, K., Cutsem, E.V., Buyse, M., Burzykowski, T., on behalf of the GASTRIC group. 2013. Disease-free survival as a surrogate for overall survival in adjuvant trials of gastric cancer: a meta-analysis. *Journal of the National Cancer Instite*, 5:1600–1607.

[71] Paoletti, X., Oba, K., Bang, Y.J., Bleiberg, H., Boku, N., Bouché, O., Catalano, P., Fuse, N., Michiels, S., Moehler, M., Morita, S., Ohashi, Y., Ohtsu, A., Roth, A., Rougier, P., Sakamoto, J., Sargent, D., Sasako, M., Shitara, K., Thuss-Patience, P., Van Cutsem, E., Burzykowski, T., Buyse, M., on behalf of the GASTRIC group. 2013. Progression-free survival as a surrogate for overall survival in patients with advanced/recurrent gastric cancer: a meta-analysis. *Journal of the National Cancer Institute*, 5:1608–1612.

[72] Pearl, J. 2001. *Causality: Models, Reasoning, and Inference.* Cambridge University Press, Cambridge.

[73] Petrylak, D.P., Ankerst, D.P., Jiang, C.S., Tangen, C.M., Hussain, M.H.A., Lara, P.N. Jr., Jones, J.A., Taplin, M.E., Burch, P.A., Kohli, M., Benson, M.C., Small, E.J., Raghavan, D., and Crawford, E.D. 2006. Evaluation of prostate-specific antigen declines for surrogacy in patients treated on SWOG 99–16. *Journal of the National Cancer Institute*, 98:516–521.

[74] Prentice, R.L. 1989. Surrogate endpoints in clinical trials: definitions and operational criteria. *Statistics in Medicine*, 8:431–440.

[75] Renard, D., Geys, H., Molenberghs, G., Burzykowski, T., and Buyse, M. 2002. Validation of surrogate endpoints in multiple randomized clinical trials with discrete outcomes. *Biometrical Journal*, 44:1–15.

[76] Renfro, L.A., Shi, Q., Sargent, D.J., and Carlin, B.P. 2012. Bayesian adjusted R^2 for the meta-analytic evaluation of surrogate time-to-event endpoints in clinical trials. *Statistics in Medicine*, 31:743–761.

[77] Robins, J.M. and Greenland, S. 1992. Identifiability and exchangeability for direct and indirect effects. *Epidemiology* 3:143–155.

[78] Sargent, D.J., Wieand, H.S., Haller, D.G., Gray, R., Benedetti, J.K., Buyse, M., Labianca, R., Seitz, J.F., O'Callaghan, C.J., Francini, G., Grothey, A., O'Connell, M., Catalano, P.J., Blanke, C.D., Kerr, D., Green, E., Wolmark, N., Andre, T., Goldberg, R.M., De Gramont, A. 2005. Disease-free survival versus overall survival as a primary end point for adjuvant colon cancer studies: individual patient data from 20,898 patients on 18 randomized trials. *Journal of Clinical Oncology*, 23:8664–8670.

[79] Scher, H.I., Eisenberger, M., D'Amico, A.V., et al. 2004. Eligibility and outcomes reporting guidelines for clinical trials for patients in the state of a rising prostate-specific antigen: recommendations from the Prostate-Specific Antigen Working Group. *Journal of Clinica Oncology*, 22:537–56.

[80] Schatzkin, A and Gail, M. 2002. The promise and peril of surrogate end points in cancer research. *Nature Reviews Cancer*, 2:19–27.

[81] Schemper, M. 1993. The relative importance of prognostic factors in studies of survival. *Statistics in Medicine*, 12:2377–2382.

[82] Schemper, M. and Stare, J. 1996. Explained variation in survival analysis. *Statistics in Medicine*, 15:1999–2012.

[83] Shannon, C. 1948. A mathematical theory of communication, *Bell System Technical Journal*, 27:379–423 and 623–656.

[84] Shi, Q., de Gramont, A., Grothey, A. *et al.*, for the Analysis and Research in CAncers of the Digestive System (ARCAD) Group. 2015. Individual patient data analysis of progression-free versus overall survival as a first-line endpoint for metastatic colorectal cancer in modern randomized trials: Findings from 16,700 patients from the ARCAD database. *Journal of Clinical Oncology*, 33:22–28.

[85] Shih, J.H. and Louis, T.A. 1995. Inferences on association parameter in copula models for bivariate survival data. *Biometrics*, 51:1384–1399.

[86] Shkedy, Z. and Torres Barbosa, F. 2005. Bayesian evaluation of surrogate endpoints. In: T. Burzykowski, G. Molenberghs, and M. Buyse (eds.), *The Evaluation of Surrogate Endpoints*, pp. 253–270. Springer, New York.

[87] Taylor, J.M.G., Wang, Y., and Thiébaut, R. 2005. Counterfactual links to the proportion of treatment effect explained by a surrogate marker. *Biometrics*, 61:1102–1111.

[88] Tibaldi, F.S., Cortiñas Abrahantes, J., Molenberghs, G., Renard, D., Burzykowski, T., Buyse, M., Parmar, M., Stijnen, T., Wolfinger, R. 2003. Simplified hierarchical linear models for the evaluation of surrogate endpoints. *Journal of Statistical Computation and Simulation* 73:643–658.

[89] Tilahun, A., Assam, P., Alonso, A., and Molenberghs, G. 2008. Information theory-based surrogate marker evaluation from several randomized clinical trials with binary endpoints, using SAS. *Journal of Biopharmaceutical Statistics*, 18:326–341.

[90] Valicenti R.K., DeSilvio M., Hanks G.E., Porter, A., Brereton, H., Rosenthal, S.A., Shipley, W.U., and Sandler, H.M. 2006. Posttreatment prostatic-specific antigen doubling time as a surrogate endpoint for prostate cancer-specific survival: an analysis of Radiation Therapy Oncology Group Protocol 92-02. *International Journal of Radiation Oncology and Biological Physiology*, 66:1064–1071.

[91] Verbeke, G. and Molenberghs, G. 2000. *Linear Mixed Models for Longitudinal Data*. Springer, New York.

[92] Volberding, P.A., Lagakos, S.W., Koch, M.A., et al. 1990. Zidovudine in asymptomtic human immunodeficiency virus infection:a controlled trial in persons with fewer than 500 CD4-positive cells per cubic millimeter. *New England Journal of Medicine*, 322:941–949.

Index

Note: Locators in *italics* represent figures and **bold** indicate tables in the text.

Printed in the United States
by Baker & Taylor Publisher Services